PROGRESS IN

Molecular Biology and Translational Science

Volume 104

PROGRESS IN

Molecular Biology and Translational Science

Nanoparticles in Translational Science and Medicine

edited by

Antonio Villaverde

Professor of Microbiology, Institute of Biotechnology and Biomedicine
Department of Genetics and Microbiology
Universitat Autònoma de Barcelona, Spain
Networking Biomedical Research Center in Bioengineering
Biomaterials and Nanomedicine (CIBER-BBN), Spain

Volume 104

ELSEVIER

AMSTERDAM • BOSTON • HEIDELBERG • LONDON
NEW YORK • OXFORD • PARIS • SAN DIEGO
SAN FRANCISCO • SINGAPORE • SYDNEY • TOKYO
Academic Press is an imprint of Elsevier

Academic Press is an imprint of Elsevier
32 Jamestown Road, London, NW1 7BY, UK
Radarweg 29, PO Box 211, 1000 AE Amsterdam, The Netherlands
225 Wyman Street, Waltham, MA 02451, USA
525 B Street, Suite 1900, San Diego, CA 92101-4495, USA

This book is printed on acid-free paper. ∞

Library of Congress Cataloging-in-Publication Data
A catalog record for this book is available from the Library of Congress

British Library Cataloguing in Publication Data
A catalogue record for this book is available from the British Library

ISBN: 978-0-12-416020-0
ISSN: 1877-1173

Transferred to digital print 2012

Contents

Contributors xi
Preface xv

Liposomes and Other Vesicular Systems: Structural Characteristics, Methods of Preparation, and Use in Nanomedicine 1

Elisa Elizondo, Evelyn Moreno, Ingrid Cabrera, Alba Córdoba, Santiago Sala, Jaume Veciana, and Nora Ventosa

I. Introduction 2
II. Description and Classification of Vesicles 6
III. Techniques for Vesicle Structural Characterization 9
IV. Vesicles as Carriers for the Delivery of Bioactive Molecules 27
V. Methods of Preparation 34
VI. Conclusions and Future Challenges 43
References 43

Luminescent Quantum Dots, Making Invisibles Visible in Bioimaging 53

Mitsuru Ishikawa and Vasudevanpillai Biju

I. Introduction 54
II. Preparation of QDs for Bioconjugation 57
III. Labeling Cell Surfaces with QDs 60
IV. Delivery of QDs in Living Cells 67
V. Photodynamic Therapy by Quantum Dots 82
VI. Concluding Remarks 92
References 92

Fabrication of Dendrimers Toward Biological Application 101

Hsieh-Chih Tsai and Toyoko Imae

I. Introduction 102
II. Are Dendrimers Attractive in Biomedical Sciences? 103
III. Drug Delivery Systems of Dendrimers 110
IV. Targeting Delivery of Dendrimers 116
V. Dendrimers for Therapy 119
VI. Dendrimers as Imaging Agents for Inspection 126
VII. Conclusions 132
References 132

Carbohydrate-Based Nanoparticles for Potential Applications in Medicine 141

Marco Marradi, Isabel García, and Soledad Penadés

I. Introduction 142
II. Metallic Carbohydrate-Based Nanoparticles for *In Vitro* Detection of Cells and Pathogens 144
III. Metallic Carbohydrate-Based Nanoparticles for *In Vivo* Diagnosis by Magnetic Resonance Imaging 151
IV. Metallic Carbohydrate-Based Nanoparticles for *In Vivo* Diagnosis by Optical Imaging 156
V. Metallic Carbohydrate-Based Nanoparticles in Therapy 158
VI. NonMetallic Polysaccharide-Based Nanoparticles in Therapy 164
VII. Conclusions 165
References 166

Carbon Nanotubes: Engineering Biomedical Applications 175

Diego A. Gomez-Gualdrón, Juan C. Burgos, Jiamei Yu, and Perla B. Balbuena

I. Introduction 176
II. Characterization 176
III. Synthesis 194
IV. Post-Synthesis 207
V. Biomedical Applications 223
VI. Summary and Outlook 233
References 234

Engineered Biological Entities for Drug Delivery and Gene Therapy: Protein Nanoparticles. 247

Joan Domingo-Espín, Ugutz Unzueta, Paolo Saccardo, Escarlata Rodríguez-Carmona, José Luís Corchero, Esther Vázquez, and Neus Ferrer-Miralles

I. Introduction.. 248
II. Protein Nanoparticle Formulation and Biological Barriers 249
III. Multifunctional Proteins.. 257
IV. Natural Self-Assembling Protein Nanoparticles: VLPs......................... 264
V. Nonviral Self-Assembling Proteins.. 270
VI. Medical Applications of Protein Nanoparticles.................................. 274
VII. Conclusions 281
References.. 282

Polymer Nanoparticles 299

Xiao-Yun Lu, Dao-Cheng Wu, Zheng-Jun Li, and Guo-Qiang Chen

I. Definition and General Ingredients of Polymer Nanoparticles.............. 300
II. Production Procedures 302
III. Characterization, Quality, and Stability of Polymer Nanoparticles.......... 305
IV. Applications of Microbial PHA-Based Nanoparticles 308
V. Summary and Outlook.. 318
References 320

Nanobiotechnology and Personalized Medicine 325

K.K. Jain

I. Introduction 326
II. Role of Nanodiagnostics in Personalized Medicine............................ 326
III. Nanobiotechnology, Biomarkers, and Personalized Medicine 332
IV. Nanoparticle-Based Combination of Diagnostics and Therapeutics 333
V. Discovery, Development, and Delivery of Personalized Nanomedicines... 334
VI. Personalized Nanooncology 341
VII. Personalized Nanoneurology.. 347
VIII. Personalized Nanocardiology 349
IX. Future Prospects of Personalized Nanomedicine 351
References 352

Tissue Engineering Using Magnetite Nanoparticles 355

Akira Ito and Masamichi Kamihira

I. Introduction 356
II. Magnetofection 359
III. Magnetic Patterning of Cell 362
IV. Construction of 3D Tissue-Like Structures 368
V. Conclusion 388
References 389

Peptide Nanoparticles for Oligonucleotide Delivery. . . . 397

Taavi Lehto, Kariem Ezzat, and Ülo Langel

I. Introduction 397
II. Gene Therapy 399
III. Uptake Mechanisms of CPPs 401
IV. CPPs for Noncovalent Nanoparticle Formation 404
V. CPP-Based Nanoparticles and Their Applications 407
VI. Future Directions and Perspective with CPP-Based Nanoparticles 420
References 422

Nanoparticles in Molecular Diagnostics 427

Pedro V. Baptista, Gonçalo Doria, Pedro Quaresma, Miguel Cavadas, Cristina S. Neves, Inês Gomes, Peter Eaton, Eulália Pereira, and Ricardo Franco

I. Unique Properties of Nanoparticles for Molecular Diagnostics 428
II. Molecular Diagnostics by Nucleic Acid Detection 445
III. Molecular Diagnostics by Immunoassays 461
IV. Conclusions and Future Perspectives 471
References 473

Nanoparticles in Cancer Chemotherapy 489

Deboshri Banerjee and Shiladitya Sengupta

I. Introduction 490
II. Physicochemical Properties of Nanoparticles and Impact on Tumor Targeting 490
III. Engineering Nanoparticles for Therapeutic and Imaging Purposes 496
IV. Shape–Size Effect of Nanoparticles 502

V. Current Status of Nanoparticles in Clinics 504
VI. Conclusion 504
References 504

Nanoparticles in Gene Therapy: Principles, Prospects, and Challenges 509

Chunxi Liu and Na Zhang

I. Introduction 510
II. Barriers and Strategies to Gene Delivery 511
III. NPs Used in Gene Therapy 521
IV. Challenges and Prospects 551
References 552

Cell Delivery of Therapeutic Nanoparticles 563

JoEllyn McMillan, Elena Batrakova, and Howard E. Gendelman

I. Translational Pathways for Cell-Based Nanomedicines 564
II. Nanomedicines for the Diagnosis and Treatment of Cancer, Infectious and Degenerative Diseases, and Tissue Repair 567
III. Chemical Composition, Structure, Function, and Manufacture 572
IV. Targeted Drug Delivery 576
V. Nanomedicine and Vaccines 577
VI. Nanodevices and Cell Reprogramming 577
VII. Nanomedicine and Stem Cells 578
VIII. Potential of Nanotherapeutics 579
IX. Nanotoxicology: Immunogenicity, Cytotoxicity, and Generation of ROS 581
X. Unique Challenges for Translational Nanomedicine 583
XI. Development of Cell-Based Nanomedicine: A Perspective 585
References 588

Index 603

Contributors

Numbers in parentheses indicate the pages on which the authors' contributions begin.

Perla B. Balbuena, Department of Chemical Engineering and Materials Science and Engineering Program, Texas A&M University, College Station, Texas, USA (175)

Deboshri Banerjee, BWH-HST Center for Biomedical Engineering, Department of Medicine, Brigham and Women's Hospital, Harvard Medical School, Harvard-MIT Division of Health Science and Technology, Cambridge, Massachusetts, USA (489)

Pedro V. Baptista, Departamento de Ciências da Vida, Faculdade de Ciências e Tecnologia, Centro de Investigação em Genética Molecular Humana (CIGMH), Universidade Nova de Lisboa, Caparica, Portugal (427)

Elena Batrakova, Department of Pharmaceutical Sciences, Nebraska Medical Center, Omaha, Nebraska, USA (563)

Vasudevanpillai Biju, National Institute of Advanced Industrial Science and Technology (AIST), Hayashi-cho, Takamatsu, Kagawa, Japan (53)

Juan C. Burgos, Department of Chemical Engineering and Materials Science and Engineering Program, Texas A&M University, College Station, Texas, USA (175)

Alba Córdoba, Department of Molecular Nanoscience and Organic Materials, Institut de Ciència de Materials de Barcelona (ICMAB-CSIC), Campus de la Universitat Autònoma de Barcelona (UAB), Bellaterra; and Networking Research Center on Bioengineering, Biomaterials and Nanomedicine, CIBER-BBN, Barcelona, Spain (1)

Ingrid Cabrera, Department of Molecular Nanoscience and Organic Materials, Institut de Ciència de Materials de Barcelona (ICMAB-CSIC), Campus de la Universitat Autònoma de Barcelona (UAB), Bellaterra; and Networking Research Center on Bioengineering, Biomaterials and Nanomedicine, CIBER-BBN, Barcelona, Spain (1)

Miguel Cavadas, REQUIMTE, Departamento de Química, Faculdade de Ciências e Tecnologia, Universidade Nova de Lisboa, Caparica, Portugal (427)

Guo-Qiang Chen, School of Life Sciences, Tsinghua University, Beijing, China (299)

José Luís Corchero, Institute for Biotechnology and Biomedicine; Department of Genetics and Microbiology, Universitat Autònoma de Barcelona; and CIBER de Bioingeniería, Biomateriales y Nanomedicina (CIBER-BBN), Bellaterra, Barcelona, Spain (247)

Joan Domingo-Espín, Institute for Biotechnology and Biomedicine; Department of Genetics and Microbiology, Universitat Autònoma de Barcelona; and CIBER de Bioingeniería, Biomateriales y Nanomedicina (CIBER-BBN), Bellaterra, Barcelona, Spain (247)

Gonçalo Doria, Departamento de Ciências da Vida, Faculdade de Ciências e Tecnologia, Centro de Investigação em Genética Molecular Humana (CIGMH), Universidade Nova de Lisboa, Caparica, Portugal (427)

Peter Eaton, REQUIMTE, Departamento de Química e Bioquímica, Faculdade de Ciências da Universidade do Porto, R. Campo Alegre, Porto, Portugal (427)

Elisa Elizondo, Department of Molecular Nanoscience and Organic Materials, Institut de Ciència de Materials de Barcelona (ICMAB-CSIC), Campus de la Universitat Autònoma de Barcelona (UAB), Bellaterra; and Networking Research Center on Bioengineering, Biomaterials and Nanomedicine, CIBER-BBN, Barcelona, Spain (1)

Kariem Ezzat, Department of Neurochemistry, The Arrhenius Laboratories for Natural Sciences, Stockholm University, Stockholm, Sweden (397)

Neus Ferrer-Miralles, Institute for Biotechnology and Biomedicine; Department of Genetics and Microbiology, Universitat Autònoma de Barcelona; and CIBER de Bioingeniería, Biomateriales y Nanomedicina (CIBER-BBN), Bellaterra, Barcelona, Spain (247)

Ricardo Franco, REQUIMTE, Departamento de Química, Faculdade de Ciências e Tecnologia, Universidade Nova de Lisboa, Caparica, Portugal (427)

Isabel García, Laboratory of GlycoNanotechnology, Biofunctional Nanomaterials Unit, CIC biomaGUNE, Parque Tecnológico de San Sebastián; and Biomedical Research Networking Center in Bioengineering, Biomaterials and Nanomedicine (CIBER-BBN), Paseo Miramón 182, San Sebastián, Spain (141)

Howard E. Gendelman, Department of Pharmacology and Experimental Neuroscience, Nebraska Medical Center, Omaha, Nebraska, USA (563)

Inês Gomes, REQUIMTE, Departamento de Química, Faculdade de Ciências e Tecnologia, Universidade Nova de Lisboa, Caparica, Portugal (427)

Diego A. Gomez-Gualdrón, Department of Chemical Engineering and Materials Science and Engineering Program, Texas A&M University, College Station, Texas, USA (175)

Toyoko Imae, Graduate Institute of Applied Science and Technology; and Department of Chemical Engineering, National Taiwan University of Science and Technology, Taipei, Taiwan, ROC (101)

Mitsuru Ishikawa, National Institute of Advanced Industrial Science and Technology (AIST), Hayashi-cho, Takamatsu, Kagawa, Japan (53)

Akira Ito, Department of Chemical Engineering, Faculty of Engineering, Kyushu University, Nishi-ku, Fukuoka, Japan (355)

K.K. Jain, Jain PharmaBiotech, Basel, Switzerland (325)

Masamichi Kamihira, Department of Chemical Engineering, Faculty of Engineering, Kyushu University, Nishi-ku, Fukuoka, Japan (355)

Ülo Langel, Laboratory of Molecular Biotechnology, Institute of Technology, University of Tartu, Tartu, Estonia; and Department of Neurochemistry, The Arrhenius Laboratories for Natural Sciences, Stockholm University, Stockholm, Sweden (397)

Taavi Lehto, Laboratory of Molecular Biotechnology, Institute of Technology, University of Tartu, Tartu, Estonia (397)

Zheng-Jun Li, College of Life Science and Technology, Beijing University of Chemical Technology, Beijing, China (299)

Chunxi Liu, Department of Pharmaceutics, School of Pharmaceutical Science, Shandong University, Shandong Province, People's Republic of China (509)

Xiao-Yun Lu, School of Life Science and Technology, Xi'an Jiaotong University, Xi'an, China (299)

Marco Marradi, Laboratory of GlycoNanotechnology, Biofunctional Nanomaterials Unit, CIC biomaGUNE, Parque Tecnológico de San Sebastián; and Biomedical Research Networking Center in Bioengineering, Biomaterials and Nanomedicine (CIBER-BBN), Paseo Miramón 182, San Sebastián, Spain (141)

JoEllyn McMillan, Department of Pharmacology and Experimental Neuroscience, Nebraska Medical Center, Omaha, Nebraska, USA (563)

Evelyn Moreno, Department of Molecular Nanoscience and Organic Materials, Institut de Ciència de Materials de Barcelona (ICMAB-CSIC), Campus de la Universitat Autònoma de Barcelona (UAB), Bellaterra; and Networking Research Center on Bioengineering, Biomaterials and Nanomedicine, CIBER-BBN, Barcelona, Spain (1)

Cristina S. Neves, REQUIMTE, Departamento de Química e Bioquímica, Faculdade de Ciências da Universidade do Porto, R. Campo Alegre, Porto, Portugal (427)

Soledad Penadés, Laboratory of GlycoNanotechnology, Biofunctional Nanomaterials Unit, CIC biomaGUNE, Parque Tecnológico de San Sebastián; and Biomedical Research Networking Center in Bioengineering, Biomaterials and Nanomedicine (CIBER-BBN), Paseo Miramón 182, San Sebastián, Spain (141)

Eulália Pereira, REQUIMTE, Departamento de Química e Bioquímica, Faculdade de Ciências da Universidade do Porto, R. Campo Alegre, Porto, Portugal (427)

Pedro Quaresma, Departamento de Ciências da Vida, Faculdade de Ciências e Tecnologia, Centro de Investigação em Genética Molecular Humana (CIGMH), Universidade Nova de Lisboa, Caparica; and REQUIMTE, Departamento de Química e Bioquímica, Faculdade de Ciências da Universidade do Porto, R. Campo Alegre, Porto, Portugal (427)

Escarlata Rodríguez-Carmona, Institute for Biotechnology and Biomedicine; Department of Genetics and Microbiology, Universitat Autònoma de

Barcelona; and CIBER de Bioingeniería, Biomateriales y Nanomedicina (CIBER-BBN), Bellaterra, Barcelona, Spain (247)

Paolo Saccardo, Institute for Biotechnology and Biomedicine; Department of Genetics and Microbiology, Universitat Autònoma de Barcelona; and CIBER de Bioingeniería, Biomateriales y Nanomedicina (CIBER-BBN), Bellaterra, Barcelona, Spain (247)

Santiago Sala, Department of Molecular Nanoscience and Organic Materials, Institut de Ciència de Materials de Barcelona (ICMAB-CSIC), Campus de la Universitat Autònoma de Barcelona (UAB), Bellaterra; and Networking Research Center on Bioengineering, Biomaterials and Nanomedicine, CIBER-BBN, Barcelona, Spain (1)

Shiladitya Sengupta, BWH-HST Center for Biomedical Engineering, Department of Medicine, Brigham and Women's Hospital, Harvard Medical School, Harvard-MIT Division of Health Science and Technology, Cambridge, Massachusetts, USA (489)

Hsieh-Chih Tsai, Graduate Institute of Applied Science and Technology, National Taiwan University of Science and Technology, Taipei, Taiwan, ROC (101)

Ugutz Unzueta, Institute for Biotechnology and Biomedicine; Department of Genetics and Microbiology, Universitat Autònoma de Barcelona; and CIBER de Bioingeniería, Biomateriales y Nanomedicina (CIBER-BBN), Bellaterra, Barcelona, Spain (247)

Esther Vázquez, Institute for Biotechnology and Biomedicine; Department of Genetics and Microbiology, Universitat Autònoma de Barcelona; and CIBER de Bioingeniería, Biomateriales y Nanomedicina (CIBER-BBN), Bellaterra, Barcelona, Spain (247)

Jaume Veciana, Department of Molecular Nanoscience and Organic Materials, Institut de Ciència de Materials de Barcelona (ICMAB-CSIC), Campus de la Universitat Autònoma de Barcelona (UAB), Bellaterra; and Networking Research Center on Bioengineering, Biomaterials and Nanomedicine, CIBER-BBN, Barcelona, Spain (1)

Nora Ventosa, Department of Molecular Nanoscience and Organic Materials, Institut de Ciència de Materials de Barcelona (ICMAB-CSIC), Campus de la Universitat Autònoma de Barcelona (UAB), Bellaterra; and Networking Research Center on Bioengineering, Biomaterials and Nanomedicine, CIBER-BBN, Barcelona, Spain (1)

Dao-Cheng Wu, School of Life Science and Technology, Xi'an Jiaotong University, Xi'an, China (299)

Jiamei Yu, Department of Chemical Engineering and Materials Science and Engineering Program, Texas A&M University, College Station, Texas, USA (175)

Na Zhang, Department of Pharmaceutics, School of Pharmaceutical Science, Shandong University, Shandong Province, People's Republic of China (509)

Preface

In contrast to conventional therapies, mostly based in the generic administration of chemicals, radiation, or surgery, innovative medicines propose the cell- or tissue-targeted delivery of advanced therapeutics with improved stability, pharmacokinetics, and bioavailability, as well as diminished side effects. Desirably, the therapeutic strategy should be designed upon a precise diagnostic at the molecular level to reveal specific traits of disease, thus making possible a patient-oriented (and therefore highly efficient) treatment. Both fine diagnosis and personalized molecular therapies require not only integrative biological approaches but also appropriate instruments and vehicles with high stability, organic diffusion, tissue and cell penetrability, and sufficient manipulability for application-focused tailoring.

In this context, the design, fabrication, and application of nano-objects (ranging from 1 to 100 nm in diameter) are of bubbling interest in a wide spectrum of fields, including material sciences, electronics, but especially biomedicine. Nanoparticles possess unusual physical properties that make them highly convenient for innovative and personalized medicines. With a size in the nanoscale as a common attribute, diverse categories of nanoparticles have been produced by either chemical or biological synthesis, including liposomes, quantum dots, carbon nanotubes, and those formed by dendrimers, carbohydrates, polymers, or proteins. The fabrication, tailoring, and specific properties and biomedical potentials of those types of nanoparticles are dissected in detail in the first part of this volume (Chapters 1-7), by top scientists leading the supporting technological platforms in their respective fields.

Further, how nanoparticles are used in novel therapeutic approaches, the properties of different types of nanoparticles (enabling them for specific applications), and specially, the envisaged, not yet fully exploited potential of these emerging instruments in drug delivery, diagnosis, and personalized medicines is summarized in the second part of this volume (Chapters 8-14), exemplified by relevant medical applications. Again, leading scientists in emerging nanomedical fields describe how nanoparticles can be exploited to address specific needs in tissue engineering, gene therapy, drug delivery, molecular diagnosis and imaging for a diversity of conditions including infectious diseases, genetic and metabolic disorders, inflammation, neurological disorders, cancer and cardiovascular diseases.

Despite still unsolved controversial issues such as the potential toxicity of some types of nanoparticles and the environmental impact eventually derived from their large-scale implementation as therapeutic agents, fast advances in the chemistry, physics, and biology of nanoparticle fabrication should generically enhance their biotolerance, expand their functionalities, and stress their versatility and ease of operation for biomedical applications. The progressive processing of the huge amount of genetic data that have been accumulated in the past decade, together with the full exploitation of "omics" and the implementation of systems biology approaches, offers an unusually favorable landscape for the development of personalized medicines in which improved nanoparticles, resulting from the integrated application of nanotechnology and biotechnology concepts, are already exhibiting a prominent position as pivotal tools.

ANTONIO VILLAVERDE

Liposomes and Other Vesicular Systems: Structural Characteristics, Methods of Preparation, and Use in Nanomedicine

Elisa Elizondo,[*,†] Evelyn Moreno,[*,†] Ingrid Cabrera,[*,†] Alba Córdoba,[*,†] Santiago Sala,[*,†] Jaume Veciana,[*,†] and Nora Ventosa[*,†]

[*]*Department of Molecular Nanoscience and Organic Materials, Institut de Ciència de Materials de Barcelona (ICMAB-CSIC), Campus de la Universitat Autònoma de Barcelona (UAB), Bellaterra, Barcelona, Spain*

[†]*Networking Research Center on Bioengineering, Biomaterials and Nanomedicine, CIBER-BBN, Barcelona, Spain*

I. Introduction 2
  A. Molecular Self-assembly: Principles 3
  B. Vesicles: Supramolecular Structures from Self-assembled Amphiphilic Molecules 5
II. Description and Classification of Vesicles 6
III. Techniques for Vesicle Structural Characterization 9
  A. Size, Size Distribution, and Concentration 10
  B. Zeta Potential 13
  C. Morphology 15
  D. Membrane Supramolecular Organization 18
  E. Interaction Strength of Biomolecules with Vesicles 20
  F. Drug Content Analysis 25
IV. Vesicles as Carriers for the Delivery of Bioactive Molecules 27
V. Methods of Preparation 34
  A. Conventional Methods for the Preparation of Drug-Loaded Vesicles 35
  B. Compressed Fluid-Based Methods for the Preparation of Drug-Loaded Vesicles 37
VI. Conclusions and Future Challenges 43
  References 43

Vesicular systems, especially liposomes, have generated a great deal of interest as intelligent materials for the delivery of bioactive molecules since they can be used as sensitive containers that respond to external stimuli,

DOI: 10.1016/B978-0-12-416020-0.00001-2

such as pressure, pH, temperature, or concentration changes in the medium, triggering modifications in their supramolecular structure. The control of the nanostructure—particle size and size distribution, membrane morphology, and supramolecular organization—of these self-assembled systems is of profound importance for their application in drug delivery and the discovery of new nanomedicines. This chapter will describe the chemical structure of vesicles and their pharmacological properties, conventional and new vesicle preparation methods and structural characterization, as well as their use in the rational design and fabrication of nanomedicines.

I. Introduction

Vesicles are nano/microparticulate colloidal carriers, usually 0.05–5.0 μm in diameter, which form spontaneously when certain lipids are hydrated in aqueous media.[1] They consist of a small enclosed liquid compartment separated from its surroundings by one or more lipid bilayers (Fig. 1).

Vesicles constitute one of the most developed and already commercially used systems for drug delivery. A better comprehension and, therefore, a more effective design of these bilayered vehicles as pharmaceutical carriers entail a look at the basics, that is, understanding the way in which these supramolecular entities are formed. In order to do that, concepts such as supramolecular chemistry and, particularly, self-assembly become crucial.

Supramolecular chemistry, which dates back to the late 1960s, has been described as "chemistry beyond the molecule," where a "supermolecule" is a species that is held together by noncovalent interactions between two or more covalent molecules or ions. It can also be described as "lego™ chemistry" in which each lego™ brick represents a molecular building block and these

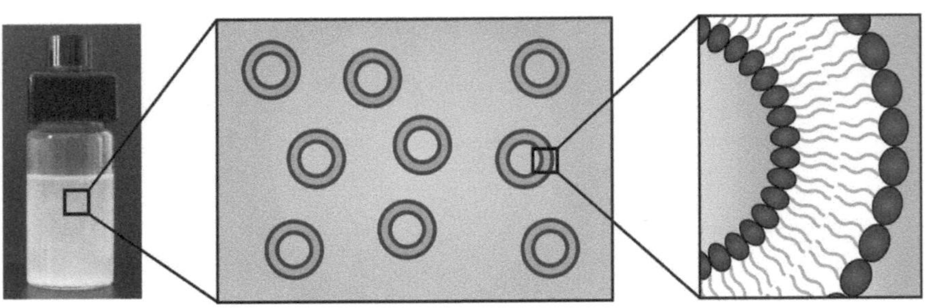

FIG. 1. Left: photograph showing the macroscopic aspect of a vesicular system. Zooms: schematic representation of vesicles and their lipid bilayer. (For color version of this figure, the reader is referred to the Web version of this chapter.)

blocks are held together by intermolecular interactions (bonds) of a reversible nature, to form a supramolecular aggregate.[2] Supramolecular chemistry can be split into two broad categories: *host–guest chemistry* and *self-assembly*, the latter of which is responsible for vesicle formation and will be therefore described in the following section.

A. Molecular Self-assembly: Principles

Molecular self-assembly can be defined as the spontaneous and reversible association of molecular species to form larger, more complex supramolecular entities according to the intrinsic information contained in the components.[2–5] The formations of molecular crystals,[6] colloids,[7] lipid bilayers,[8] phase-separated polymers,[9] and self-assembled monolayers[10] are all examples of molecular self-assembly, such as the folding of polypeptide chains into proteins[11] or the folding of nucleic acids into their functional forms[12] in Nature. The concepts of self-assembly historically have come from studying molecular processes and, as it occurs in the majority of fields, the inspiration for many supramolecular species designed and developed by scientists has come from biological systems (Fig. 2).

The success of self-assembly in a molecular system is determined by five characteristics of such systems[4]:

- *Components*: A self-assembling system consists of a group of molecules or segments of a macromolecule that interact with one another. Their interaction leads from some less ordered state (a solution, disordered aggregate, or random coil) to a final state (a crystal or folded macromolecule) that is more ordered.
- *Interactions*: Self-assembly occurs when molecules interact with one another through a balance of attractive and repulsive interactions. These interactions are generally weak (i.e., comparable to thermal energies) and noncovalent (van der Waals and Coulomb interactions, hydrophobic interactions, and hydrogen bonds).
- *Reversibility (or adjustability)*: For self-assembly to generate ordered structures, the association either must be reversible or must allow the components to adjust their positions within an aggregate once it has formed. The strength of the bonds between the components, therefore, must be comparable to the forces tending to disrupt them. For molecules, the forces are generated by thermal motion.
- *Environment*: The self-assembly of molecules is carried out normally in solution or at an interface to allow the required motion of the components. The interaction of the components with their environment can strongly influence the course of the process.

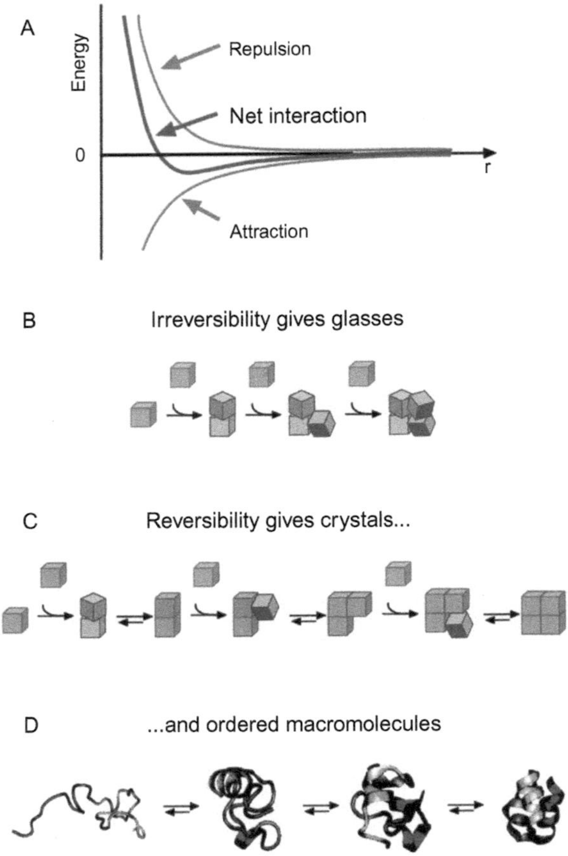

FIG. 2. (A) Aggregation occurs when there is a net attraction and an equilibrium separation between the components. The equilibrium separation normally represents a balance between attraction and repulsion. (B and C) Schematic illustration of the essential differences between irreversible aggregation and ordered self-assembly. (B) Components (shown in blue) that interact with one another irreversibly form disordered glasses (shown in green). (C) Components that can equilibrate, or adjust their positions once in contact, can form ordered crystals if the ordered form is the lowest-energy form (shown in red). (D) Biology provides many examples of self-assembly (here, the formation of a protein, an asymmetric, catalytically active nanostructure); these examples will stimulate the design of biomimetic processes. Reproduced with permission from Ref. 4. (See Color Insert.)

- *Mass transport and agitation*: For self-assembly to occur, the molecules must be mobile. In solution, thermal motion provides the major part of the energy required to bring the molecules into contact.

B. Vesicles: Supramolecular Structures from Self-assembled Amphiphilic Molecules

Most of the self-assembling molecules possess an amphiphilic character, that is, they contain hydrophilic and hydrophobic domains in their structure. Phospholipids, surfactants, and block copolymers, which are the normally used self-assembling monomers for the production of vesicles, are usually constituted by a long hydrophobic tail and a polar hydrophilic head group. Under aqueous conditions, this hydrophobic–hydrophilic dual character promotes their association through weak, noncovalent interactions to form ordered assemblies with different morphologies and sizes that range from nanometers to microns[13–16] (Fig. 3). The thermodynamic driving force of this molecular self-assembly is provided by the desolvation, collapse, and intermolecular association of the hydrophobic part of monomers. This tendency of the hydrophobic tail to minimize contact with water, called the hydrophobic effect, is mainly due to the entropic gain of the water structure by not being in contact with the hydrophobic part.[17] Intermolecular polar interactions, such as electrostatics and hydrogen bonding, can also occur and help define the structural specificity. A variety of different aggregate shapes are observed that range from spherical and rodlike micelles to amphiphilic bilayers (vesicles).[18]

The actual form assumed by an aggregate depends on the molecular constitution of the amphiphile and is highly dependent on its geometry (see Fig. 3). In a first-order approximation, the geometry of an amphiphile is

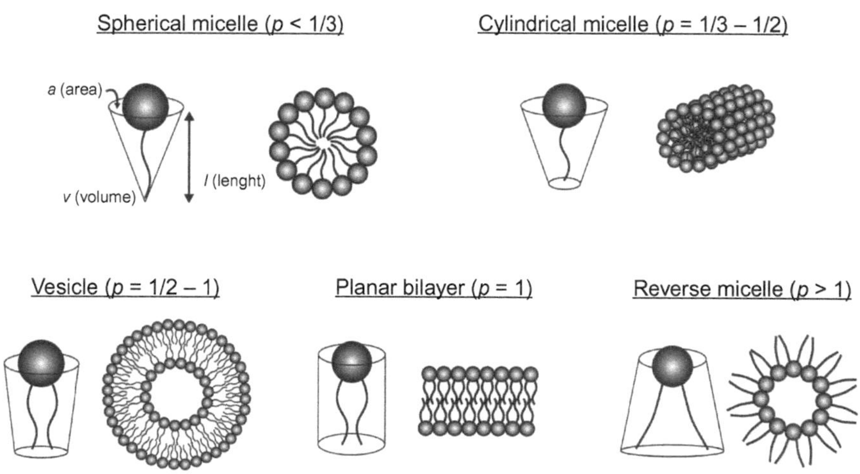

FIG. 3. Relationship between the packing parameter (p) and the self-assembled morphology of amphiphilic molecules under aqueous conditions.

described by its packing parameter (p), which is defined as the ratio of the hydrophobic volume (v) to the product of the head group area (a) and chain length (l)[19]:

$$p = \frac{v}{al} \tag{1}$$

The packing parameter determines the preferred curvature of the aggregates formed: $p < 1/3$ spherical micelles; $1/3 < p < 1/2$ cylindrical micelles; $1/2 < p < 1$ vesicles; $p = 1$ planar bilayers; $p > 1$ reverse micelles.

II. Description and Classification of Vesicles

Among the different types of vesicular systems used for drug delivery, liposomes (constituted mainly by phospholipids) and vesicles containing either nonionic (niosomes), cationic (cationic vesicles), or both, cationic and anionic surfactants (catanionic vesicles) are the most investigated for this purpose. Some examples of the most used phospholipids and surfactants are depicted in Fig. 4.

Liposomes are, without any doubt, the most studied vesicular systems for drug delivery. Since the pioneering observation of Alec Bangham roughly 45 years ago that phospholipids in aqueous systems can form closed bilayered

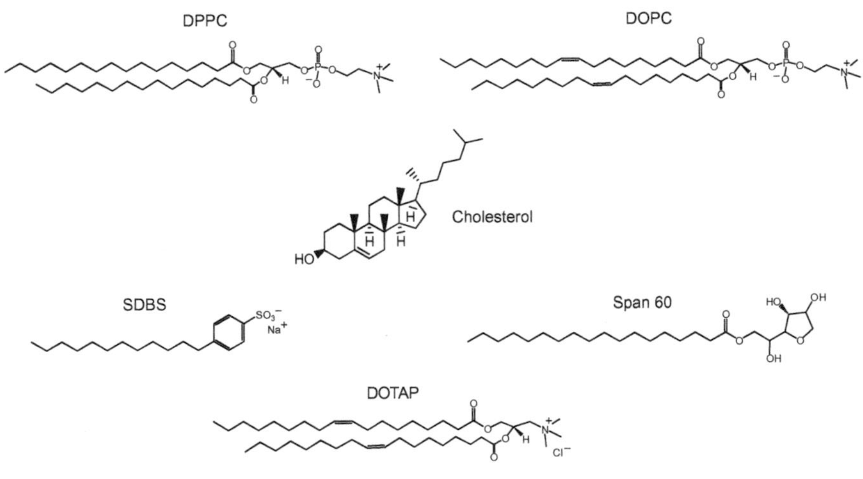

FIG. 4. Molecular structures of some common lipids used for vesicle preparation. Phospholipids, 1,2-dipalmitoyl-*sn*-glycero-3-phosphocholine (DPPC) and 1,2-dioleoyl-*sn*-glycero-3-phosphocholine (DOPC); natural lipid, cholesterol; anionic surfactant, sodium dodecylbenzene sulfonate (SDBS); nonionic surfactant, sorbitane monostearate (Span 60); cationic surfactant, 1,2-dioleoyl-3-trimethylammonium-propane (chloride salt) (DOTAP).

structures,[1] liposomes have moved a long way from being just another exotic object of biophysical research to becoming a pharmaceutical carrier of choice for numerous practical applications.[20] They are the most evolved vesicular delivery systems, with the liposomes combining external stimuli sensitivity, long circulation, and targeting characteristics being already a reality.[21,22] Despite their versatility and biocompatibility, liposomes suffer chemical and physical alterations, which shorten their shelf-life and limit, as a consequence, drug stability and conditions of storage.[23] Liposomes' physical instability leads to problems of high permeability which induce undesired or too fast leakage rates. Indeed, the addition of cholesterol is often required for modulating membrane rigidity and stability.[24,25] Their chemical instability is related to hydrolysis, oxidation, or peroxidation of the constitutive lipid molecules and is normally prevented by adding antioxidants and low-temperature storage. These special conditions of preparation and storage together with the elevated price of lipids have promoted the research in new vesicular formulations comprising cheaper and more stable membrane components.

In this sense, the use of surfactants as cheaper and more stable substitutive molecules than phospholipids has emerged as a really interesting choice for the industrial production of vesicular systems in pharmaceutical and cosmetic applications. Niosomes, nonionic surfactant vesicles, are widely studied as an inexpensive alternative of nonbiological origin.[26–28] As their name indicates, niosomes are constituted by nonionic surfactants, which, depending on their nature, require the addition of a certain percentage of cholesterol to form vesicle-like supramolecular assemblies.[26,29] Apart from this reason, cholesterol is added to many formulations, as is the case with liposomes, to modulate the bilayer fluidity, reduce the leakage, and enhance the encapsulation efficiency (EE) of drugs.[26] Cationic surfactants, such as 1,2-dioleoyl-3-trimethylammonium-propane, can also be combined with cholesterol (with or without other lipids), leading to cationic vesicles[30] which are preferentially used for gene delivery via cationic vesicles/DNA complexes[31] or through the encapsulation of genes inside the lumen of the vesicles.[32] The combination of cationic and anionic surfactants generates the so-called catanionic vesicles, which are also used for drug and gene delivery.[33–35] More recently, polymersomes,[36] which are constituted by block copolymers, are also gaining interest as an alternative to liposomes in drug delivery.[37]

Independent of the monomer nature (phospholipid, surfactant, or polymer), different morphologies and sizes can result from the hydration of such amphiphiles in aqueous media. The morphology of the final assembled structures is dependent not only on the geometry of the monomer but also on the external environment in which self-assembly occurs.[4,38] Temperature, pH, and ionic strength of the solution, as well as the concentration of the monomer and the preparation method, can dictate the formation of a variety of structures

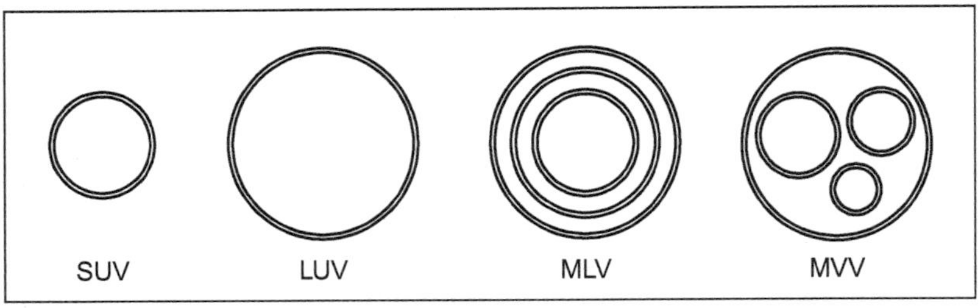

FIG. 5. Types of vesicles depending on size and lamellarity. SUV, small unilamellar vesicle; LUV, large unilamellar vesicle; MLV, multilamellar vesicle; MVV, multivesicular vesicle.

formed by a single, distinct amphiphile. For instance, regarding size and lamellarity, different vesicular structures can be achieved from the self-assembly of a certain amphiphilic molecule (Fig. 5).

Thus, vesicles can be classified into small unilamellar vesicles (SUVs, $d \leq 200$ nm),[1] large unilamellar vesicles (LUVs, $d \geq 200$ nm), multilamellar vesicles (MLVs), and multivesicular vesicles. Size and lamellarity are important structural parameters that need to be controlled, since they are crucial factors affecting the performance of vesicles, for instance, as pharmaceutical carriers.[41] Apart from this application of vesicles, which probably is the most studied one since their discovery in the mid-1960s,[1] vesicles, as other self-assembled structures, offer an exceptional versatility since, by an appropriate monomer design, the properties of these materials can be specifically tailored for their intended application.[3] For instance, vesicles have also been used as cell membrane models,[42] reaction vessels,[43,44] or dye dispersants.[45]

Despite their versatility, in order to achieve optimal performance of these self-assembled structures as functional materials, a high degree of structural homogeneity is required. Since vesicle membrane plays an important role in terms of its stability, rigidity, permeability, functionalization, and response to external stimuli,[46–48] the behavior of vesicles as drug delivery systems (DDS) is highly affected by its homogeneity, not only in size or morphology, but also in the membrane composition and supramolecular organization. For instance, structural homogeneity between the different vesicles forming a certain system is a crucial issue in order to have sharp responses that allow the homogeneous release of the drug at the site of action.

Figure 6 shows a schematic illustration of the response to an external stimulus presented by two hypothetical vesicular systems with homogeneous and heterogeneous vesicle-to-vesicle structural characteristics in terms of size

[1]The margin of size between SUVs and LUVs is not clear. Some authors claim that SUV diameter should be less than 50 nm,[39] whereas others refer to SUVs as large as 200 nm.[40] In this chapter, the latter definition is adopted.

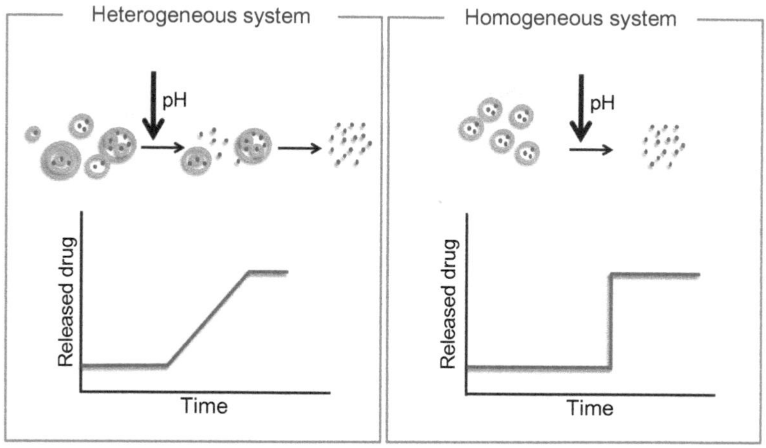

FIG. 6. Schematic illustration of the response to an external stimulus presented by a vesicular DDS with homogeneous (right) and heterogeneous (left) vesicle-to-vesicle structural characteristics in terms of size and lamellarity. (For color version of this figure, the reader is referred to the Web version of this chapter.)

and lamellarity. While the homogeneous sample would have a sharp response to such a stimulus and the entire encapsulated drug would be released at the same time, in the case of the heterogeneous sample, the release would be prolonged in time and probably part of the drug would not be liberated when required.

Because of this, methods for the preparation of structurally homogeneous vesicular systems, not only in terms of size and morphology but also in the supramolecular arrangement of the membrane constituents, are required for fully exploiting the potential of these self-assembled structures as functional materials.

III. Techniques for Vesicle Structural Characterization

As mentioned in previous sections, the good performance of vesicular systems as drug carriers is tightly related to their structure at supramolecular, microscopic, and nanoscopic level. For instance, stability, membrane leakage, targeting efficiency, or drug release precision are properties closely relayed to vesicle structural characteristics at these scales. The deep characterization of these structural features is of profound importance in order to well understand the performance of these devices and well establish structure–activity correlations. The following section attempts to highlight some of the most relevant structural parameters of these lipid structures and show some analytical techniques for their measurement.

A. Size, Size Distribution, and Concentration

Vesicle size and size distribution, whose range gives an idea of the size homogeneity between vesicles forming the same system, are essential parameters for a good characterization of a vesicular formulation. Specifically, concerning the use of vesicles as DDS, size plays an important role in the encapsulation of hydrophilic drugs, whose loading depends on the entrapped aqueous volume, which is directly related to the size of the vesicle. Moreover, vesicle size determines to a great extent its *in vivo* performance. Indeed, the significant restriction of vesicles to the circulation and their rate and extent of distribution into mononuclear phagocyte system (MPS) cells is, for most tissues, dependent primarily upon vesicle size. There is an inverse relationship between the size and the rate at which vesicles are distributed to MPS cells such as Kupffer cells.[49] For instance, small vesicles can pass through the fenestrae of the liver sinusoids, intermediate-sized ones are retained within the blood compartment circulating for long periods, and larger vesicles are rapidly taken up by Kupffer cells.

Several techniques are available for assessing submicrometer vesicle size and size distribution.[50] These include static and dynamic light scattering (DLS)[51,52] several types of microscopy techniques,[53–55] size-exclusion chromatography (SEC),[56] field-flow fractionation,[57–59] and analytical centrifugation.[53] Among them, DLS, sometimes referred to as photon correlation spectroscopy or quasi-elastic light scattering, constitutes an appropriate and particularly quick method for size measurements and covers a size range from a few nanometers to several microns. DLS is a noninvasive, well-established technique that allows measuring the distribution of tens of thousands of particles at a time, being easier and quicker than other methods like microscopic techniques, that are time-consuming to obtain a representative size distribution of the population and not amenable to being a routine size measurement.

DLS measures the time-dependent fluctuations of light scattered from particles experiencing Brownian motion, which results from collisions between suspended particles and solvent molecules. Since the Brownian motion depends on the size of the particles, the rate at which the intensity fluctuations occur also depends on this parameter. Thus, small particles, which are "kicked" further by the solvent molecules and have faster Brownian motions, cause the intensity to fluctuate more rapidly than the large ones.[60–62] Analysis of these intensity fluctuations enables the determination of the diffusion coefficients (D) of the particles which are converted into a size distribution through the Stokes–Einstein equation:

$$r_{\mathrm{h}} = \frac{kT}{6\pi\eta D} \tag{2}$$

where r_h is the hydrodynamic radius of the particles under study, k is the Boltzmann constant, T is the temperature, and η is the solvent viscosity.

The hydrodynamic radius (Stokes radius) of the particle is defined as the apparent size of the hydrated sphere and is calculated from the radius of a sphere that diffuses at the same rate, that is, that has the same translational diffusion coefficient as the particle being measured. The translational diffusion coefficient will depend not only on the size of the particle "core" but also on any surface structure, as well as the concentration and type of ions in the medium. This means that the size can be larger than measured by, for example, electron microscopy (EM) where the particle is removed from its native environment. Figure 7 shows a size distribution obtained by DLS corresponding to a 140-nm-sized vesicular formulation.

As explained above, vesicle size plays an important role in its *in vivo* performance when used as a pharmaceutical carrier. Some interesting studies have recently been performed in this area, where liposome size has been shown to be an important factor in the efficient delivery of an antitumor agent. The review of Nagayasu *et al.*[63] reports that size is a key property in the optimization of liposome pharmacokinetic factors like blood circulation, tumor accumulation, and *in vivo* drug release and concludes that 100 nm appears to be the most suitable size for obtaining an effective liposomal antitumor preparation.

Size analysis can also be used for studying vesicle stability. As an example, Ruozi *et al.*[52] used DLS together with atomic force microscopy for evaluating the stability of liposomal formulations with time. They found out that, in that particular case, the aggregation of liposomes depended more on the method used for their preparation than on the liposomal membrane composition.

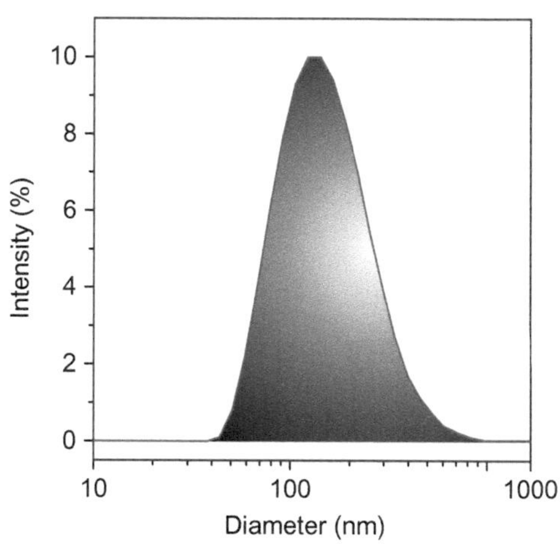

FIG. 7. Particle size distribution obtained by dynamic light scattering corresponding to a 140-nm-sized vesicular formulation.

Recently, a new methodology called nanoparticle tracking analysis (NTA) was developed for the direct and real-time visualization and analysis of nanoparticles in liquids,[64,65] being suitable for the characterization of vesicular systems. This method allows for determining size and size distribution and counting the number of particles to determine the concentration of particles in terms of number of particles per volume. Additionally, the methodology allows for the distinguishing of particles as a function of their light-scattering power (*Ri*), and by using different illumination laser wavelengths, suitably labeled particles can be discriminated and analyzed from the nonlabeled background.[66]

NTA currently operates for particles from 10 nm (high refractive index particles, such as gold and silver) or 30 nm (organic materials) to 1 μm, depending on particle type. The NTA-based method is a laser-illuminated microscopic technique in which the sample is injected into a viewing chamber ($120 \times 80 \times 20$ μm of 0.3 ml in volume) and particles contained in the sample are visualized by virtue of the light that they scatter when illuminated by a laser light. Brownian motion of nanoparticles is analyzed in real-time by a charge coupled device (CCD) camera, and each particle is simultaneously but separately visualized and tracked by a particle-tracking image analysis program. The rate of movement is related only to the viscosity of the liquid, the temperature, and the size of the particle and is not influenced by particle density or refractive index. Also, since video clips form the basis of the analysis, accurate characterization of real-time events such as aggregation and dissolution is possible. Because each and every particle is visualized and analyzed separately, the resulting estimate of particle size and particle size distribution does not suffer from the limitation of being an intensity-weighted distribution, which is normal in conventional methods of particle sizing in this size regime.

The ability of NTA to simultaneously measure particle size and particle-scattering intensity allows distinguishing particles of different refractive index within the same sample solution and, importantly, it allows the direct estimation of particle concentration. These features—counting and separating particles by virtue of their scattering intensity—can be used, for example, for probing whether nanoscale drug delivery structures such as liposomes vary in their contents, that is, empty liposomes may have a lower *Ri* (light-scattering power) than those loaded with a higher *Ri* material. This would allow them to be discriminated even though they may be of very similar size.

In the example displayed in Fig. 8, a sample of clinically significant cellular vesicles was specifically labeled with an appropriate fluorescent marker. In this case, the methodology allowed a particular subpopulation of labeled vesicles to be discriminated from the total population of vesicles. The red plot in Fig. 8 corresponds to the labeled vesicle population as viewed under fluorescence analysis; the green plot is of the entire population of labeled and unlabeled structures. The similar-sized peaks shown below indicate that nearly

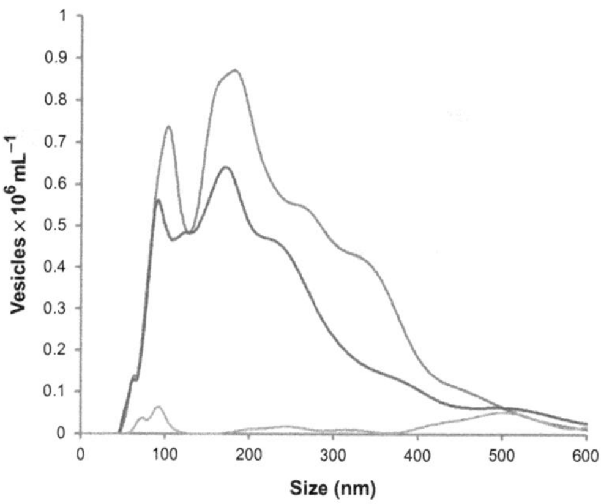

FIG. 8. Results from NTA in light-scattering mode (blue line) and fluorescent mode (red line). (The green line corresponds to a control sample). Reproduced with permission from Ref. 67. (See Color Insert.)

all the vesicles seen under the light-scattering mode (blue line) have been labeled with fluorophores (red line, as visualized under a suitable excitation wavelength).[67]

In a recent publication, the group of Cindy Gutzeit used this methodology to size and count exosomes of different origins in order to establish whether exosomes of different origins and functions could selectively target different immune cells in human peripheral blood.[68]

B. Zeta Potential

As well as size, surface properties of vesicles are critical in determining their drug carrier potential since they are also related to vesicle fate *in vivo*. Zeta potential measurements help to characterize the surface of colloidal drug carrier systems, such as vesicles, and predict their stability and behavior in different environments.

Zeta potential is a physical property that is exhibited by any particle in suspension. Most particles dispersed in an aqueous system will acquire a surface charge, principally either by ionization of surface groups or by adsorption of charged species. These surface charges modify the distribution of the surrounding ions, resulting in an inner region (Stern layer) where the ions are strongly bound to the particle and an outer (diffuse) region where they are less firmly associated. Within the diffuse layer, there is a notional boundary inside which the ions and particles form a stable entity. Thus, when a particle moves (i.e., due to Brownian motion), ions within the boundary move as part of the particle. The zeta potential is the potential at the boundary (surface of hydrodynamic shear) beyond which ions stay with the bulk dispersant (see Fig. 9).

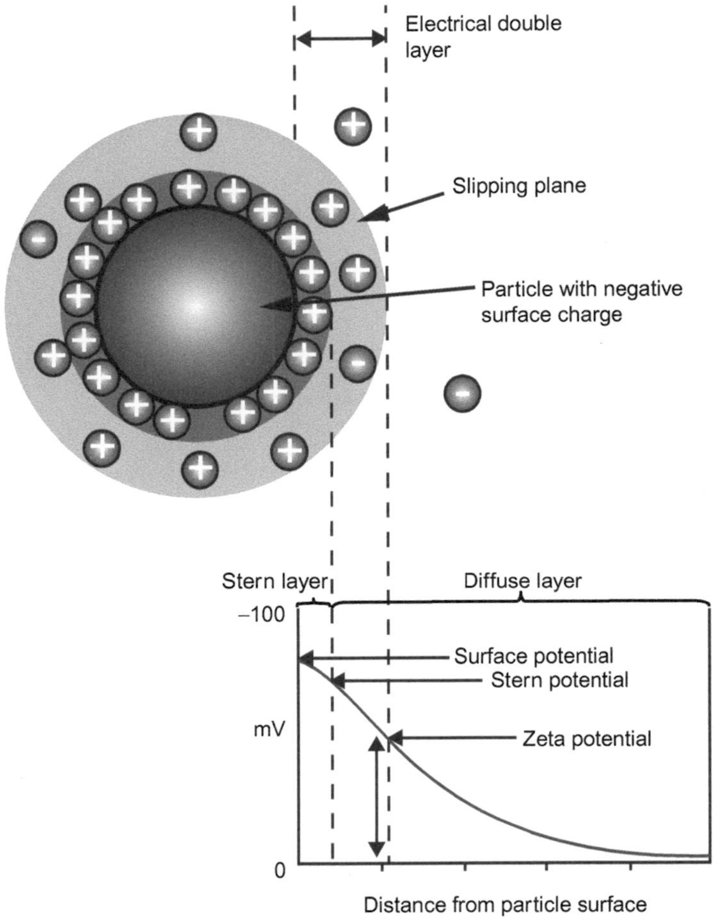

FIG. 9. Schematic representation of zeta potential. Courtesy of Malvern Instruments Ltd.

Zeta potential is measured by applying a voltage across a pair of electrodes at either end of a cell containing the particle dispersion. Particles will migrate toward the oppositely charged electrode with a velocity proportional to the magnitude of their zeta potential. This velocity is measured and expressed as particle velocity in a unit electric field (electrophoretic mobility) and can be conveniently investigated by microelectrophoresis, in which the movement of the vesicles in an electric field is observed with a light microscope[69] or by electrophoretic light scattering,[70,71] where the frequency shift or phase shift of an incident laser beam caused by particles moving under an electric field is measured as the particle electrophoretic mobility. This mobility (U_E) is converted to the zeta potential (ζ) using the Henry equation (see Eq. (3)) by inputting the dielectric constant (ε) and viscosity (η) of the dispersant and Henry's function ($F(\kappa a)$), which according to the Smoluchowski approximation is 1.5 for aqueous media.

$$U_E = \frac{2\varepsilon\zeta F(\kappa a)}{3\eta} \tag{3}$$

Zeta potential, which gives information about the overall charge that a particle acquires in a certain medium, indicates the degree of repulsion between adjacent, similarly charged particles in dispersion, providing, as a result, important information about the stability of vesicular and other colloidal systems. Colloids with high absolute zeta potential values (normally above 30 mV) are electrically stabilized, while those with low zeta potentials are not stable and tend to coagulate or flocculate. Since zeta potential depends both on the characteristics of the dispersing medium and the nature of the vesicle membrane constituents, it is a potent tool to monitor the effect of both in the stability of a vesicular formulation. For instance, Carrion *et al.*[72] studied the influence of ionic strength and lipid bilayer charge on the stability of various liposomal formulations.

Zeta potential measurements can also be used for studying interactions of vesicles with different molecules present in the dispersing medium, such as counter ions, surfactants, plasma proteins, charged drugs, or DNA. Indeed, it has become a popular characterization tool in liposome-mediated gene delivery, since it allows evaluation of the extent of interaction of cationic liposomes with the anionic charges of DNA.[73] Changes in the surface of vesicles can also be monitored by zeta potential measurements. For instance, zeta potential data have been used for evaluating charge shielding by polyethylene glycol (PEG), providing PEG-coating thickness and giving information about the effectiveness of this barrier against opsonization *in vivo*.[74]

C. Morphology

Beside size and surface properties, the analysis of vesicle morphological features, such as shape and lamellarity, is essential for a proper characterization of vesicular formulations and the establishment of structure–activity correlations. Microscopy is the preferred tool for this purpose, since it allows the direct observation of the sample and, therefore, its morphological characteristics.

There are three well-known branches of microscopy: optical, electron, and scanning probe microscopy, but perhaps the widest used to visualize vesicle morphology is electron microscopy (EM). In this technique, a particle beam of electrons is used to illuminate the specimen and produce a magnified image. Materials to be viewed under an electron microscope may require processing to produce a suitable sample and the technique varies depending on the specimen and the analysis required.

1. Freeze-Fracture EM

This type of EM consists of physically breaking apart (fracturing) a frozen sample. Structural details exposed by the fracture plane are then visualized by vacuum deposition of platinum/carbon on the fractured sample in order to make a replica for examination in the transmission electron microscope. The key steps in making a freeze-fracture replica are (i) rapid freezing, (ii) fracturing, (iii) replication, and (iv) replica cleaning. In routine protocols, a pretreatment step is carried out before freezing, typically comprising fixation in glutaraldehyde followed by cryoprotection with glycerol. Resorting to these pretreatment steps falls short of the ideal method of directly freezing samples, a limitation that has been the driving force behind the development of more specialized techniques. Freeze-fracture is unique among EM techniques in providing planar views of the internal organization of membranes.[75] For instance, this microscopy has been used to examine the morphology of vesicles made of PEO-*b*-PMA/dC_{16}SU complexes (Fig. 10) that can be used as pharmaceutical drug carriers.[76]

2. Negative Stain EM

In this type of EM, vesicular suspensions are briefly mixed with a dilute solution of an electron-opaque material, such as ammonium molybdate, uranyl acetate (or formate), or phosphotungstic acid. This mixture is applied to a

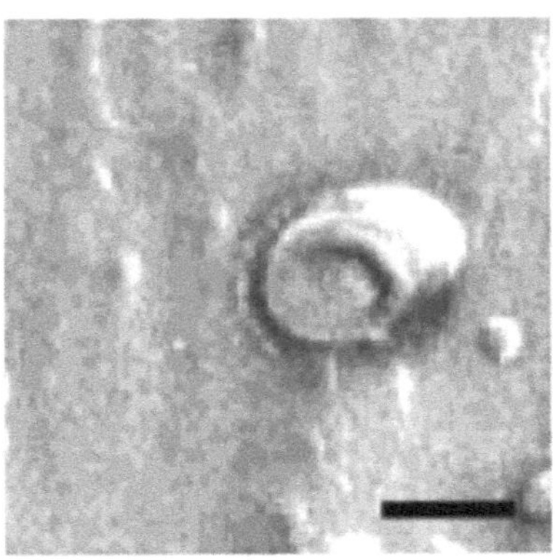

Fig. 10. Microphotograph of PEO-*b*-PMA/dC_{16}SU complexes obtained using freeze-fracture techniques (scale bar corresponds to 100 nm). Reproduced with permission from Ref. 76.

suitably coated EM grid, blotted, and then allowed to dry. Viewing of this preparation in the transmission electron microscope should be carried out without delay for best results. Negative staining is used for fast but crude morphological identification, but can also be used as the basis for high-resolution 3D reconstruction using EM tomography methodology when carbon films are used for support. Nanovesicles formed from self-assembled porphyrin bilayers (porphysomes), which have application as multimodal biophotonic contrast agents in the sensitive visualization of lymphatic systems using photo-acoustic tomography, were negatively stained and imaged by EM as is shown in Fig. 11.[77]

3. Cryo-transmission EM

The main difference of this variety of EM is that the sample is studied at cryogenic temperatures (generally liquid nitrogen temperatures). For sample preparation, the material is spread on an EM grid and is preserved in a frozen-hydrated state by rapid freezing, usually in liquid ethane near liquid nitrogen temperature. By maintaining specimens at liquid nitrogen temperature or colder, they can be introduced into the high vacuum of the electron microscope column and, thus, images can be obtained. The main advantage of this type of EM is the possibility to straightforwardly observe vesicles that have not been stained or fixed in any way, showing them in their native environment and achieving a resolution below 0.5 Å. Among other examples, cryo-transmission electron microscopy (cryo-TEM) has been used to image a dispersed system processed with CO_2 demonstrating the extremely uniform spherical and

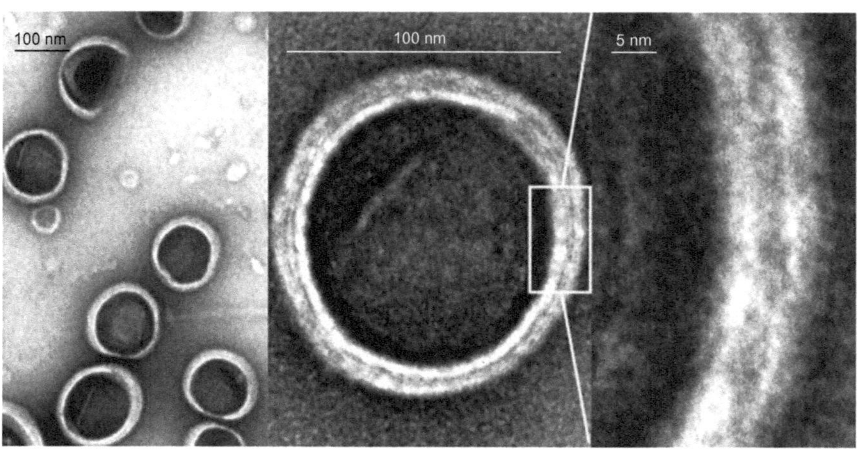

FIG. 11. Electron micrographs of negatively stained porphysomes. Reproduced with permission from Ref. 77.

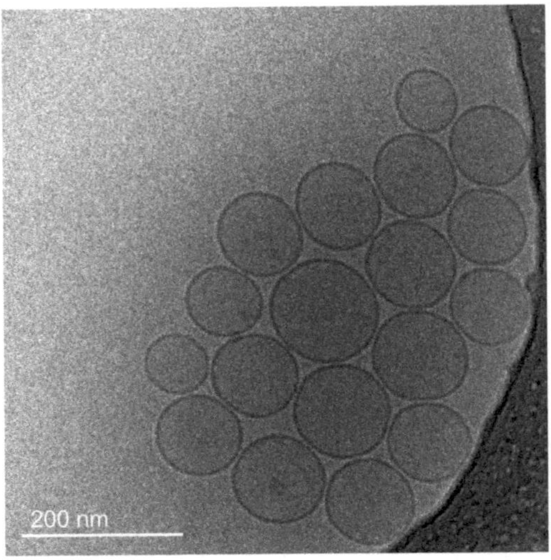

FIG. 12. Cryo-TEM image of cholesterol-rich vesicular system obtained using CO_2-expanded solvents.

unilamellar vesicles obtained with this method (Fig. 12) and confirming that the mean size was centered around 200 nm, in good agreement with DLS measurements.[78] A version of electron cryomicroscopy is cryo-electron tomography where a 3D reconstruction of a sample is created from tilted 2D images.

There are other microscopies, such as confocal laser scanning microscopy, immunomicroscopy, or total internal reflection microscopy, that allow not only imaging of the morphology of the vesicles but also the performance of different studies like cellular uptake, vesicle stability, or protein localization inside the vesicles.[79]

D. Membrane Supramolecular Organization

The fluid–lipid bilayer component of vesicles and cell membranes is a ubiquitous structural element of drug nanocarriers and living matter, respectively. A structural characterization of this bilayer and a determination of its physical properties are essential for providing a basis for understanding structure–function relationships in biological systems as well as providing insight into the interactions of peptides, proteins, and drugs with lipid bilayers. Because vesicles and cellular membranes must be in a fluid state for their normal function, it is the structure of fluid bilayers that is relevant. Unfortunately, the high thermal disorder of fluid bilayers hinders the application of atomic-resolution three-dimensional crystallography. Nevertheless, useful structural information can be obtained by diffraction methods because vesicle membranes can adopt multilamellar configurations in which a large number of layers stack in the form of approximately concentric bilayers that have an aqueous space in between and are highly periodic along the normal bilayer (Fig. 13). This one-dimensional crystallinity allows the distribution of matter along the normal bilayer to be determined from combined X-ray and neutron diffraction measurements using small and wide-angle X-ray scattering (SAXS/WAXS) and small-angle neutron scattering (SANS) techniques, either from synchrotron or laboratory sources.[80,81]

X-ray scattering methods (SAXS/WAXS, SANS) are accurate, nondestructive, and usually require only a minimum of sample preparation, the major problem being the poor contrast between the vesicle components and the dispersing media. Low-angle diffraction from the stacking of bilayers in multillamellar vesicles provides several kinds of average structural information about the bilayer. The most direct result is obtained from the position of the reflection (q), from which the distance D from the center of one bilayer to the next in the stack as represented in Fig. 13 can be derived by the following equation:

$$D = \frac{2\pi}{q} \tag{4}$$

This distance D increases with the amount of water added until it becomes constant when water is added in excess. Determining the amount of water at the point of D becoming constant yields the number of water per lipid, n_W, for the fully hydrated system. This quantity n_W can be related to the area of each lipid in the plane of the membrane (A) by

$$A = \frac{2(V_L + n_W V_W)}{D} \tag{5}$$

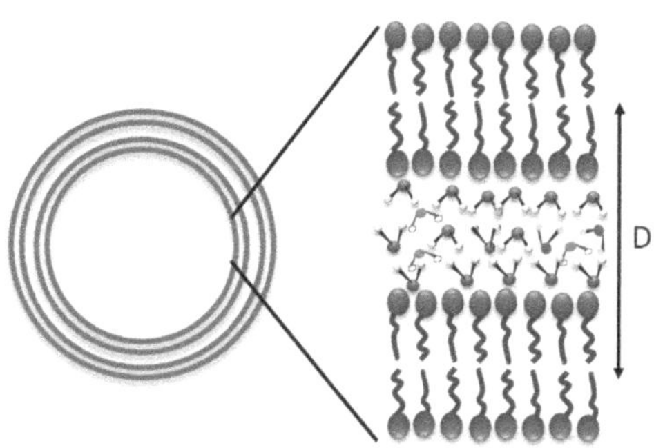

FIG. 13. Schematic illustration of a multilamellar array of lipid bilayers. D, lamellar repeat distance.

where V_W is the molecular volume of water and V_L is the volume of the lipid molecule in a fully hydrated system. Weiner *et al.* determined all these parameters for a model lipid bilayer of dipalmitoylphosphatidylcholine (DPPC).[82] In addition to determining *D* and *A*, an accurate fitting of the low-angle diffraction data can provide information on the average number of bilayers forming the liposome and an estimate of the boundary between the head groups and the chain regions (i.e., the thickness of the hydrocarbon chain region).[83] Additional information regarding the fluidity and orientation of the acyl chains in the lipid membrane can be inferred from the wide-angle diffraction (WAXS) region.

Talsma *et al.* described the possibilities and limitations of studying vesicles from SAXS data using DPPC and cholesterol hemisuccinate liposomes as model systems.[84] In a recent publication,[85] high-resolution SAXS, complemented by SANS and DLS experiments, was used to study the effect of curvature on the bilayer structure of dioleoyl-phosphatidylcholine and dioleoyl-phosphatidylserine unilamellar vesicles. Bilayer curvature, as a result of finite vesicle size, was varied as a function of vesicle radius and determined by DLS and SANS measurements.

Although scattering techniques provide valuable information on packing parameters of the membrane components, there is still a long way to fully determine the structure of soft matter in the same way as the structure of condensed matter is done routinely.

E. Interaction Strength of Biomolecules with Vesicles

Impressive progress has been made in detecting and imaging structural properties of biological systems. However, structure data are only a first step toward an understanding of physiological processes. Insight into the functions of biological macromolecules requires additional information on dynamics and on the interactions governing their behavior. Such issues are tackled by biothermodynamics and calorimetry, and an increasing number of researchers recognize the great potential of these techniques. Excellent microcalorimeters and a broad variety of calorimetric techniques and assays have been developed over the past decade and are now available to a broad spectrum of users.

The crucial challenge is to combine insights from biochemistry and physiology with those from structural biology and from biothermodynamics to derive an integral picture of membranes and their functions. Unfortunately, the recent success in tackling this problem is rather limited. Molecular dynamics simulations may be of great value to overcome the gap between the disciplines. Further, substantial progress of both structural and thermodynamic insight will be required for deriving a general understanding of membranes and other systems.

Calorimeters measure the heat consumed or released by a sample upon reequilibration after a perturbation. Such perturbations can be caused by a change in temperature (differential scanning calorimetry, DSC), addition of material (isothermal titration calorimetry, ITC), a change in pressure (pressure perturbation calorimetry, PPC), or a change in the mass of the system (quartz crystal microbalance, QCM; QCM with dissipation, QCM-D).

1. Differential Scanning Calorimetry

A detailed introduction to DSC is given by Leharne and Chowdhry[86] and an instrument is described by Plotnikov *et al.* and Privalov *et al.*[87,88] Briefly, DSC directly records the temperature-dependent isobaric heat capacity, Cp (T), of a sample:

$$Cp = \partial \mathrm{H}/\partial \mathrm{T}|_{\mathrm{p}} \tag{6}$$

Equation (6) shows that the integral of a DSC curve from an appropriate base line directly yields the enthalpy change of the underlying phenomena (transition, melting, folding, unfolding, interaction, etc.), ΔH, at a temperature T of thermally induced structural transitions in solution. This information gives valuable insights into the factors that stabilize or destabilize proteins, nucleic acids, micellar complexes, and many other macromolecular systems, alone or as complexes. Since the pioneering studies of Chapman,[89,90] DSC is the standard technique to monitor phase transition in lipids.[91,92] In addition, ligand affinities to a target system (protein, liposome) can be ranked rapidly.[93–96] For example, the interaction of biomolecules such as cell-penetrating peptides with the liposome lipid bilayer can be investigated by means of DSC as Alves *et al.* reported.[97] In this study, the peptide effects on the thermotropic phase behavior of large MLVs of dimyristoylphosphatidylcholine (DMPC) and dimyristoylphosphatidylglycerol (DMPG) were investigated using two basic peptides—penetratin (nonamphipathic) and RL16, a secondary amphipathic peptide—with similar length but distinct charge distribution.

The results in Fig. 14 show that even though both peptides are cationic, their interaction with zwitterionic versus anionic lipids is markedly distinct. Penetratin greatly affects the temperature, enthalpy, and cooperativity of DMPG main phase transition but does not affect those of DMPC while RL16 presents opposite effects.

2. Isothermal Titration Calorimetry

ITC measures the heat absorbed or generated when molecules interact. It is based on a series of consecutive injections (i.e., 20 of 10 μl each) of a liquid sample (typically a ligand molecule) into the calorimeter cell under isothermal

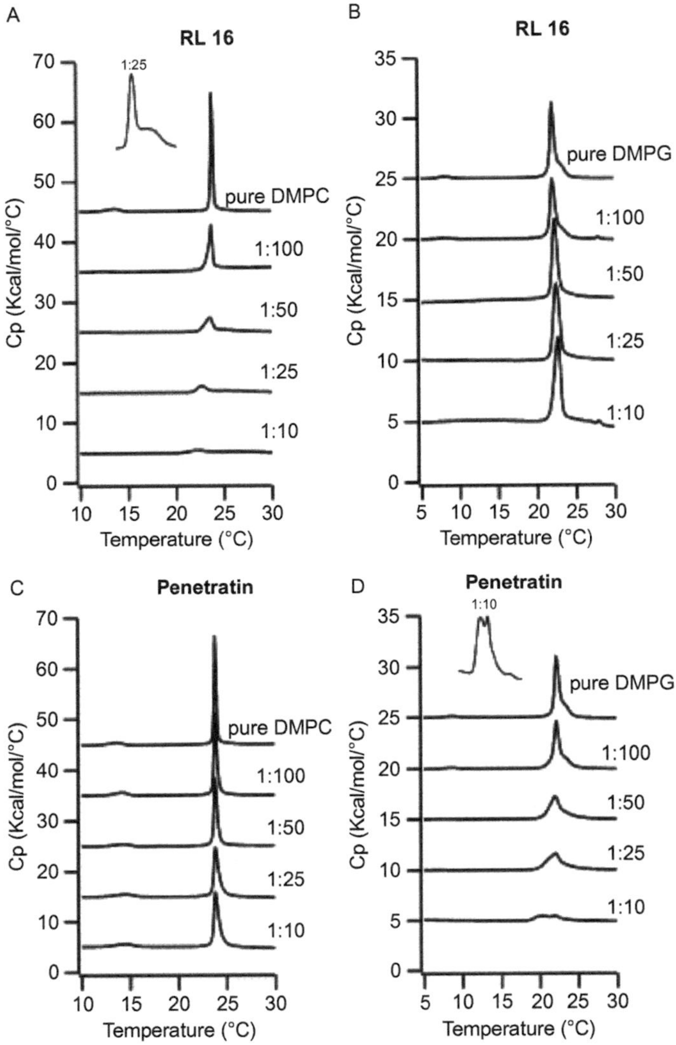

FIG. 14. High-sensitivity DSC heating scans illustrating the effect of the addition of increasing quantities of RL16 (A and B) and penetratin (C and D) on the thermotropic phase behavior of DMPC and DMPG vesicles. The top scan corresponds to the lipid alone and the peptide/lipid molar ratios of the lower scans are indicated in the figure. Reproduced with permission from Ref. 97.

conditions. Upon injection, the heat that is released or absorbed as a consequence of the interaction is measured as a function of the injection number. The reaction of the system may be transitions or transfers of molecules between different chemical or physical states. Considering that the injection causes ΔN^{trans} moles of a compound to undergo a transition that is accompanied by a molar enthalpy change of ΔH^{trans}, the measured heat $q - q_{\text{dil}}$, where q_{dil} corresponds to the dilution heat, may be written as a sum of the enthalpy changes of all i processes induced by the injection:

$$q - q_{\text{dil}} = \sum_i \Delta N_i^{\text{trans}} \Delta H_i^{\text{trans}} \tag{7}$$

Once titration is completed, an appropriate binding model is chosen and the isotherm is fitted to yield information about relative binding affinity (KD), stoichiometry (n), enthalpy (ΔH), entropy (ΔS), and consequently the Gibb's free energy, ΔG. This information related to the enthalpy–entropy compensation phenomena provides valuable information on the nature of the mechanism of binding, including hydrogen bonding, van der Waals interactions, and hydrophobic interactions.

Different types of assays can be performed:

(i) *Reaction assays* study interactions between two compounds (one originally loaded into the measurement cell and the other into the syringe) by bringing them into contact with each other in the cell. A binding assay can be used to determine the equilibrium constant (as described by the mass action law) and ΔH. The uptake protocol for membrane partitioning serves to determine the membrane–water partition coefficient, K, and ΔH.

(ii) *Dilution assays* are based on injections of a sample into a large excess volume of water or buffer. Bimolecular interactions between molecules (or particles such as vesicles) in solution give rise to heats of dilution that depend only slightly on concentration, that is, ITC curves are constant or decrease gradually during the titration. Such experiments are often performed as "blank" measurements to correct other data sets for dilution effects. If oligomers or micelles (demicellization assay) are diluted in water, more cooperative dissociation curves are obtained that allow determining the equilibrium constant K or critical micelle concentration (CMC), and the enthalpy change of demicellization, $\Delta H^{\text{m}\rightarrow\text{w}}$ (entalphy change from micelle to water). The release assay serves to characterize membrane partitioning of solutes by injecting solute-containing vesicles into water/buffer.

(iii) *Partial enthalpy assays* are performed under conditions ensuring that all injected material undergoes a certain transition. For membrane partitioning, this condition is realized at lipid concentrations that are large compared to the dissociation constant. Such experiments are performed in order to obtain independent information on the molar enthalpy change of solute transfer from water into lipid bilayers. Concentration-dependent enthalpies of membrane insertion can also be interpreted in terms of nonideal lipid–solute mixing thermodynamics and show sudden changes at solute-induced phase transitions.

Researchers at the Max Planck Institute of Colloids and Interfaces studied the interaction of chitosan with small liposomes and the binding and organization of the polysaccharide on the membrane of the vesicles by means of ITC, combined with DLS and zeta potential techniques.[98] From the experiments, they obtained the enthalpy variations arising from binding of the positively ionized chitosan to neutral and negatively charged liposomes (Fig. 15). They concluded that the strong exothermic signal indicated that the electrostatically driven binding of chitosan to the membrane was energetically favored, leading to further stabilization of the vesicle suspension and that the higher the net negative charge of the vesicles, the more pronounced was the adsorption of chitosan, leading to weaker chain organization of the adsorbed chitosan at the membrane.

3. Pressure Perturbation Calorimetry

Different calorimeters have been designed for the measurement of the heat accompanying an isothermal pressure change, $dQ/dp|_T$, to determine the temperature-dependent, isobaric volume expansion of a sample.

Such techniques have been referred to as piezothermal analysis,[99] scanning transitiometry,[100,101] pressure jump calorimetry,[102,103] or pressure perturbation calorimetry (PPC).[104] A related adiabatic technique has been termed volume perturbation calorimetry.[105–107]

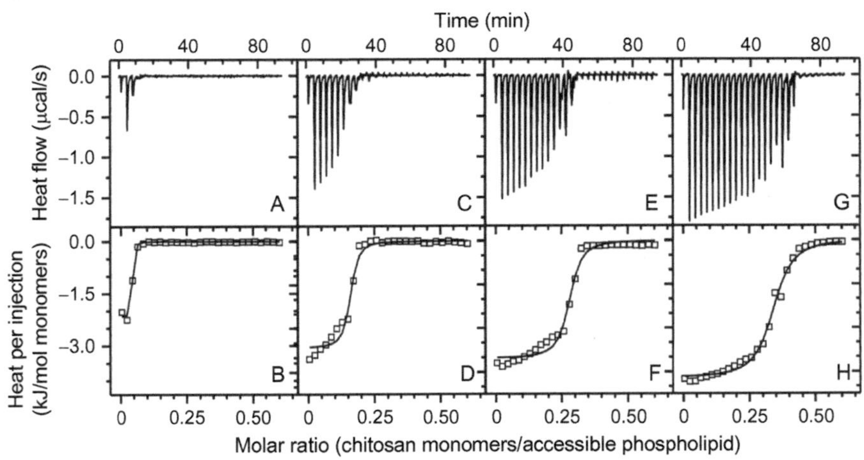

Fig. 15. ITC data for the titration of chitosan into DOPC/DOPG liposomes in an 80 nM acetate buffer at pH 4.48 (25 °C). The heat flow (A, C, E, G) and the respective integrated binding heat corrected for chitosan dilution (B, D, F, H) are plotted for vesicles with various molar fractions of DOPG: 10% (A, B), 20% (C, D), 30% (E, F), and 40% (G, H). Reproduced with permission from Ref. 98.

Over many years, mainly bulk liquids or solutions were studied on home-built, heat flow calorimeters, mostly using high pressures. Recently, a new generation of PPC instruments has become commercially available as an accessory to highly sensitive scanning calorimeters of the power compensation type.[104] The extremely high sensitivity of the calorimeter makes it possible to study changes in the partial volume of as little as ≈ 1 mg of a protein using only very small pressure jumps of 5 bars. The first applications of the technique to lipids were studies of the kinetics of phase transitions on the basis of the relaxation of the temperature or heat changes following a pressure variation. Volumetric investigations were performed characterizing lipid melting[108,109] and domain formation in membranes.[110]

4. Quartz Crystal Microbalance

This technique is used to determine the nature of binding interactions in real time occurring in surfaces within a label-free environment.[111] The sensing heart of the QCM is based on a quartz crystal sensor that oscillates at a specific frequency when voltage is applied. The frequency of the oscillation changes as the mass on the sensor changes affording high sensitivity to mass addition (ng/cm^3) and loss. A modification to QCM is the *quartz crystal microbalance* with dissipation (QCM-D) that allows turning off the voltage in continuous measuring. Turning off the voltage causes the oscillation to decay. The decay rate or dissipation factor is related to the elasticity and viscosity of the molecular layer on the sensor. By measuring the frequency and dissipation, it is possible to analyze the state of molecular layers bound to the sensor surface, their mass, thickness, and structural (viscoelastic) properties.

QCM and QCM-D are ideally suited to the study of nanomedicine where investigation of nanovehicle performance in the biological field has gained increased clinical significance. Primary issues include activation of the complement system and induction of apoptosis.[112]

Among other interesting applications,[113,114] QCM has also been applied to study the interaction of liposomes with chemically modified surfaces.[115] These studies are completely label-free in contrast to many other analytical techniques (fluorescent, colorimetric, etc.) and have the capacity to determine molecular interaction in real time in the form of mass addition or loss from the crystal surface.

F. Drug Content Analysis

Apart from the exhaustive structural characterization described above, the ability of a vesicular formulation to encapsulate the desired bioactive molecule is basic for its evaluation as a DDS. The most used terminology for defining such an ability is expressed in percentage through terms like encapsulation efficiency (EE), incorporation efficiency, or trapping efficiency.[116–118] Although some ambiguity exists around this terminology, and these terms are sometimes indistinctly used, EE would refer to the percentage of drug encapsulated inside the vesicle aqueous core with respect to the initial amount of drug used for the encapsulation procedure (see Eq. (8)), whereas trapping or incorporation efficiencies would be more recommendable for cases in which it is not sure whether the drug is encapsulated or adsorbed/integrated in the vesicular membrane.

$$\text{Encapsulation efficiency}(\%) = \frac{\text{mass of encapsulated drug}}{\text{initial mass of drug}} \times 100 \qquad (8)$$

Other authors define the encapsulation/trapping efficiency as the molar amount of encapsulant per mole of lipid,[119,120] which is obtained by dividing the concentration of the encapsulant by the concentration of the lipid. Another commonly used parameter, especially for those vesicular formulations in which the payload does not interact with the vesicular membrane, is the captured (or encapsulation) volume, defined as μL of entrapped volume/μmol of lipid.[121]

Methods for determining the amount of payload encapsulated/entrapped within vesicles typically rely on the subsequent steps: separation of the nonencapsulated payload from the loaded vesicles, disruption of the vesicular membrane for the release of the entrapped payload, and quantification of the released material. Different methods can be used for the separation of loaded vesicles from the nonencapsulated payload, comprising centrifugation, size exclusion chromatography (SEC),[116] dialysis,[117] and other more specific methods that depend on the characteristics of the active molecule. An example of that includes ion exchange membranes in which the nonencapsulated payload, such as free DNA, is retained in the membrane while the loaded vesicles pass through it.[122] Once separated, the disruption of the loaded vesicles is normally performed by diluting the aqueous vesicular suspension with alcohols such as 2-propanol, ethanol, or methanol[123,124]; extraction of lipids into chloroform solvent mixtures[124–126]; or surfactant-induced bilayer disruption.[127] After release of the cargo, the techniques used for payload quantification depend on its nature and include spectrophotometry,[128,129] fluorescence spectroscopy,[130] enzyme-based methods,[131] and electrochemical techniques.

Since techniques used to separate free from vesicle-encapsulated contents can potentially cause their leakage and, in some cases, ambiguity in the extent of separation, methods that do not rely on separation are of interest. Reported methods include ^{1}H NMR where free markers exhibited pH-sensitive resonance shifts in the external medium versus encapsulated markers,[132] fluorescence methods where the signal from nonencapsulated fluorophores was quenched by substances present in the external solution,[133] and electron spin resonance methods which rely on the signal broadening of nonencapsulated markers by the addition of a membrane-impermeable agent.[134,135]

IV. Vesicles as Carriers for the Delivery of Bioactive Molecules

Among the different approaches that have been attempted in the field of nanomedicine to overcome the inability of most conventional therapies to deliver therapeutic drug concentrations without provoking severe toxic effects on normal organs and tissues, vesicular systems have received the maximum attention during the past decades. Vesicles, especially liposomes, are the most clinically established nanometre-scale systems used for drug delivery and in the development of new nanomedicines.[136] Liposomes are responsible for a very significant part of the market already initiated into nanomedicine and it is expected that this trend will be easily prolonged in the future. Quantification of companies that are developing new liposome-based nanomedicines can be easily underestimated because the number of companies that have filed patents relating to health-care applications of liposomes is twice as high as the number of companies that openly communicate their involvement in this technology.[137]

In the specific context of cancer treatment, liposome nanotherapeutics have been on the market from the 1990s, whereas a variety of liposomal drugs for the treatment of other diseases are in various stages of clinical development.[138] Some examples of available cancer drugs that have higher efficacy and lower toxicity than nonliposomal preparations are stealth liposomal doxorubicin (brand names: Doxil®/Caelyx®), liposomal daunorubicine (brand name: DaunoXome®), and liposomal cytosine β-arabinoside (brand name: DepoCyte®).

Of these examples, Doxil®, a doxorubicin-based liposome injection, is perhaps one of the best known cases where nanotechnology has acted as driving innovation in targeted oncology. Doxorubicin is an anticancer drug that is widely used for the treatment of different types of tumors such as breast cancer, Kaposi sarcoma, and ovarian cancer. This active agent is a highly toxic compound affecting not only tumor tissue but also heart and kidney, a fact that limits its therapeutic applications. The development of doxorubicin enclosed in liposomes culminated in an approved nanomedical DDS, Doxil®, that resulted in a reduced delivery of doxorubicin to the heart and renal system, with an elevated accumulation of the drug in tumor tissue. Doxil® was approved by the Food and Drug Administration (FDA) and the European Medicines Agency (EMA) in 1995, being marketed in the United States by its originator, ALZA, and in 15 European countries by Schering-Plough under the name Caelyx®. ALZA has developed its own branded generic version of liposomal DDS using the company's trademark STEALTH. It is estimated that Doxil® has generated revenues of $550M.[139]

Another very significant example of liposomal formulation in the market, not related with cancer, is liposomal amphotericin B (brand names: AmBisome®, Amphotec®, Abelcet®). Amphotericin B is an antifungal agent which is indicated for different fungal infections and as an empirical therapy for presumed fungal infection in febrile neutropenic patients. It can also be used for the treatment of visceral leishmaniasis. AmBisome® is composed of very rigid, small unilamellar liposomes with less than 100 nm diameter that contain amphotericin B intercalated within the membrane. AmBisome® shows less toxicity and fewer side effects than free amphotericin B due to a longer circulation time, which allows accumulation in the required areas, and the retention, at the same time, of the full spectrum of antifungal activity. Liposomal amphotericin B also shows higher stability. AmBisome® was approved in 1990 by the FDA and marketed by Gilead in the United States and by Fujisawa in Japan. A figure of this commercialization is given by $400M in revenues.[139]

Visudyne® is also a marketed liposomal formulation of verteporfin for the treatment of one type of wet age-related macular degeneration called predominantly classic subfoveal choroidal neovascularization, which is the main cause of blindness in people older than 50. Visudyne® therapy is the first treatment ever approved (2000) by the FDA for this disease. The estimated revenues obtained by this product are $90M.[139]

The list of approved liposome-based drug products include also Definity® (octafluoropropane) approved in 2001 as a diagnostic drug for contrast enhancement during echocardiographic procedures, DepoDur® (morphine) approved in 2004 for extended postoperative pain relief, and Octocog alfa® (factor VIII) approved in 2009 for the treatment of hemophilia A.

Concerning the pipeline of liposomal formulations in clinical trials, there is a long list of products under development.[139] Apart from those related to cancer treatment, which continue to be predominant, some of them are related to treatment of Alzheimer, wound healing, or in use as vaccine adjuvants.

The main features that promote the use of vesicles and especially liposomes as drug carriers can be summarized as follows:[140]

- *Protection*: Drugs incorporated in vesicles are protected against the action of detrimental factors (i.e., degradative enzymes) present in the host. Conversely, the patient can be protected against detrimental toxic effects of drugs.
- *Duration*: Vesicular systems can act as a depot from which the entrapped compound is slowly released over time. Thus, an increased duration of action and a decreased frequency of administration are beneficial consequences.
- *Internalization*: Vesicles can interact with target cells in various ways and are therefore able to promote the intracellular delivery of drug molecules that in their "free" form (i.e., nonencapsulated) would not be able to enter the cell interior due to unfavorable physicochemical characteristics (i.e., DNA molecules).

- *Direction*: Vesicles can target a drug to the intended site of action in the body, enhancing, as a result, its therapeutic efficacy (drug targeting, site-specific delivery).

The achievement of all these features can be accomplished through appropriate changes in the vesicular membrane which convert these vesicles into very versatile pharmaceutical carriers. These modifications allow the alteration of their physicochemical properties and, as a consequence, their performance as DDS. Figure 16 summarizes most of the modifications or improvements that a vesicle can experience for different applications in drug delivery and diagnosis.

These changes consist of the incorporation of a small percentage of "new ingredients" into the vesicular membrane with the following objectives:

- *Increasing stability and circulation time*: Different methods have been suggested to achieve long circulation times for vesicles *in vivo*.[142] The main approach includes coating their surface with inert, biocompatible polymers, such as PEG. An important feature of protective polymers is their flexibility, which allows a relatively small number of surface-grafted polymer molecules to create an impermeable layer over the vesicle surface.[143] This layer slows down their recognition by opsonins and the subsequent clearance of vesicles, increasing, as a result, their circulation half-life.[144]

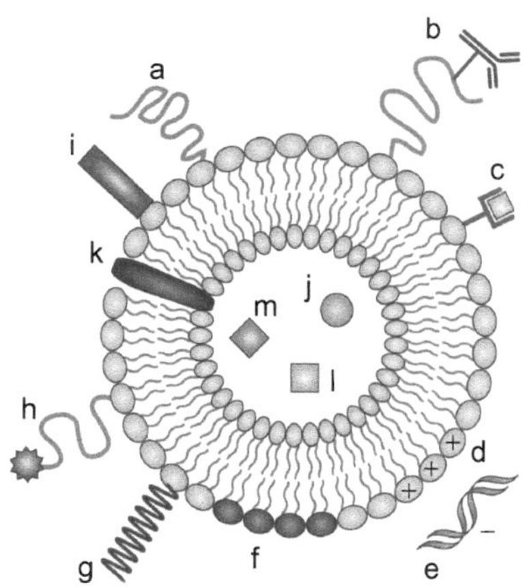

Fig. 16. Schematic representation of different modifications that can be performed on a vesicle membrane for drug delivery and diagnostic purposes. Among these modifications are the attachment of protective polymer (a) or protective polymer and targeting ligand, such as antibody (b); the attachment/incorporation of the diagnostic label (c); the incorporation of positively charged lipids (d) allowing for the complexation with DNA (e); the incorporation of stimuli-sensitive lipids (f); the attachment of stimuli-sensitive polymer (g); the attachment of cell-penetrating peptide (h); and the incorporation of viral components (i). In addition to a drug, (j) and (k), vesicles can be loaded with magnetic particles (l) for magnetic targeting and/or with colloidal gold or silver particles (m) for electron microscopy. Adapted from Ref. 141. (For color version of this figure, the reader is referred to the Web version of this chapter.)

- *Specific targeting to the affected organs and tissues*: The use of targeted vesicles with surface-attached ligands capable of recognizing and binding to the cells of interest has been suggested to increase drug accumulation in the desired tissues and organs. The search for ligands used in vesicle targeting is focused on specific molecules capable of recognizing receptors that are overexpressed on target cells (particularly in tumors) and certain specific components of such pathological cells. Antibodies are the most commonly used targeting moieties. They can be directly anchored to the vesicles' membrane or bound via a spacer arm like PEG molecules in order to achieve long circulating immunovesicles.[145,146] Folate- and tranferrin-mediated targeting represent other exploited ways for locally addressing diseases, especially tumors.[147–149]
- *Targeted release upon response to external stimuli*: The effective combination of functional head groups, lipid chains, and linker groups in membrane components can achieve controlled stability of the vesicular membrane and selective release of encapsulated material under specific environmental conditions.[48,150] One of the most popular responsive vesicles are the pH-sensitive ones (Fig. 17), which after being endocytosed in the "intact form" fuse with the endovacuolar membrane due to the lower pH inside the endosome, and release their contents into the cytoplasm.[151]
- *DNA delivery*: Incorporation of positively charged lipid derivatives or positively charged polymers allows for DNA binding to vesicles through electrostatic interactions and can enable cell transfection.[34,152]

Although targeting and response to external stimuli are desired characteristics for the optimum performance of vesicles as DDS, the main requisite for the successful use of these entities as pharmaceutical carriers is their efficient and rapid loading with sufficient quantities of payload.

The EE of a certain drug in a vesicular system is influenced by the nature of both drug and vesicle and the drug-trapping procedure chosen for the encapsulation.[41] Thus, the appropriate combination of a vesicular formulation and a trapping method is needed for accomplishing high encapsulation efficiencies of a drug, which is desired for achieving cost-effective DDS. According to this, factors affecting the encapsulation of drugs into vesicles can be summarized in three aspects: characteristics of the drug to be encapsulated, characteristics of the vesicular system, and the method of encapsulation. These three aspects affect the encapsulation of drugs as follows:

- *Characteristics of the drug to be encapsulated.* The chemical nature of the drug to be encapsulated directly restricts the use of a certain type of vesicle for its successful trapping. Different types of drugs vary in the manner they interact with vesicles, and depending on this interaction, Talsma and Crommelin[153] divided drugs into four classes:

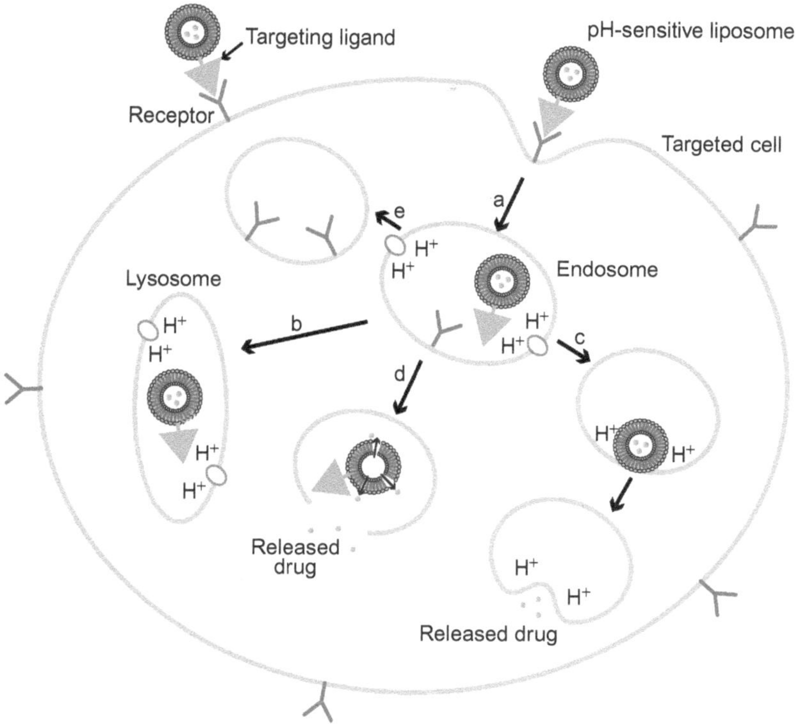

FIG. 17. Potential fates of pH-sensitive liposomes following binding to a target cell. Upon binding to cell surface receptors, liposomes can remain bound at the cell surface, disassociate from the receptor, or accumulate in coated or noncoated invaginations. Following endocytosis (a), liposomes can be delivered to lysosomes (b) where they and their contents may be degraded by lysosomal peptidases and hydrolases. Following acidification of the endosomal lumen, pH-sensitive liposomes are designed to either fuse with the endosomal membrane (c), releasing their content directly into the cytoplasm, or become destabilized and subsequently destabilize the endosomal membrane (d), resulting in leakage of the endosomal contents into the cytosol. Receptors may be recycled back to the cell surface (e) or targeted for degradation in the lysosome (b). Adapted from Ref. 151. (For color version of this figure, the reader is referred to the Web version of this chapter.)

(1) Water-soluble, nonbilayer-interacting drugs
(2) Hydrophobic, bilayer-interacting drugs that are bound inside the hydrophobic region of the bilayers
(3) Drugs in the aqueous domain that can be associated with the bilayer via electrostatic interaction
(4) Drugs that are neither water-soluble nor bilayer-bound or bilayer-associated.

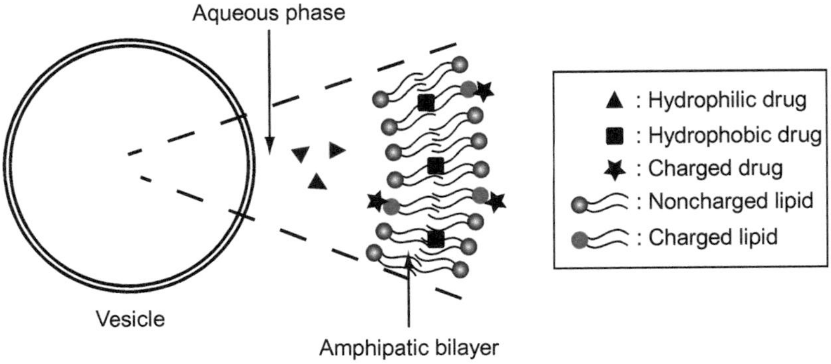

FIG. 18. Schematic representation of different possibilities of drug encapsulation into vesicles depending on the chemical nature of the drug. (For color version of this figure, the reader is referred to the Web version of this chapter.)

Based on this classification, the encapsulation of water-soluble drugs, which remain in the aqueous phase (see Fig. 18), depends on the entrapped aqueous volume, and, therefore, on the size and lamellarity of vesicles and lipid concentration. Although the encapsulation of hydrophilic drugs does not directly depend on the nature of the lipids, an appropriate permeability of the vesicle membrane is encouraged and certain quantities of cholesterol are recommended to avoid leakage of the encapsulated active molecules. Conversely, the encapsulation of bilayer-interacting hydrophobic drugs (also depicted in Fig. 18) largely depends on the amount of lipids, their chain length, and the properties (gel or fluid state) of the bilayers. Generally, these drugs tend to incorporate more efficiently in fluid membranes where the fatty acyl chains have considerable freedom of movement.

Regarding ionic drugs, whose interaction with the membrane is based on electrostatic forces, the encapsulation is affected by the density of charged or charge-inducing bilayer constituents and the ionic strength of the aqueous medium. Finally, those drugs that are neither water-soluble nor bilayer-bound or bilayer-associated exhibit very low encapsulation efficiencies and more sophisticated trapping methods are demanded.

- *Characteristics of the vesicular system*: Apart from the chemical nature of the membrane constituents, which as described above influences the encapsulation of some drugs, both vesicle size and lamellarity also play an important role in their loading with active ingredients. In the case of hydrophilic drugs, which remain dissolved in the aqueous phase, unilamellar vesicles are preferred to multilamellar ones due to their larger entrapped aqueous volume. Correspondingly, LUVs ($d \geq 200$ nm) are the most useful type of vesicles for the entrapment of hydrophilic drugs, since their volume of encapsulated water is relatively high compared to SUVs ($d \leq 200$ nm).[154,155] However, vesicle size is a critical parameter in determining their circulation half-life, being one of the most important factors that define their clearance by the reticuloendothelial system (RES).[156,157] The rate of vesicle uptake by RES increases with size, giving small vesicles longer circulation times than larger ones and therefore being preferred for higher accumulation in the affected tissues. Hence, a compromise between encapsulation and circulation half-life is recommended, promoting in most cases the use of SUVs for this purpose. For efficient entrapping of hydrophobic compounds, which are located in the lipid bilayer, small MLVs or SUVs may be applied due to their higher surface to volume ratio.

- *Method of encapsulation*: Two approaches of drug encapsulation can be taken: passive or active.[158] Passive trapping of drugs occurs during vesicle formation, in such a manner that the EE depends on the type of vesicle (MLV, LUV, or SUV) obtained.
- Conversely, active trapping techniques rely on the ability of certain drugs, normally amphipatic weak acids or bases, to redistribute across the lipid bilayer in response to a transmembrane ion gradient (Fig. 19).

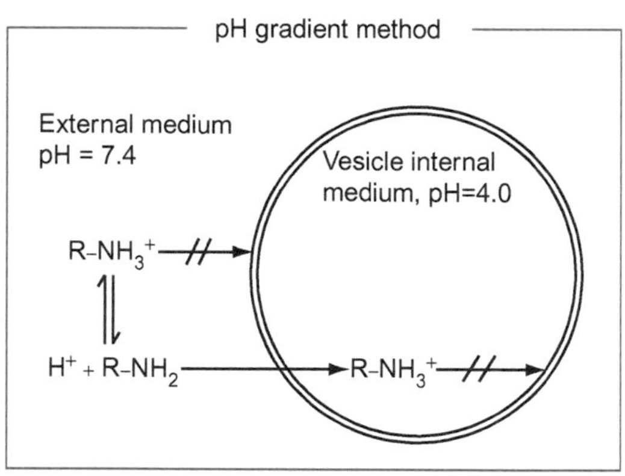

FIG. 19. Example of an active trapping technique for loading drugs into vesicles. In the pH gradient method, a weak acid (not depicted) or a weak base in the neutral form (R-NH_2), but not the charged form (R-NH_3^+), will penetrate the vesicle bilayer and reestablish an equilibrium in favor of the charged form in the acid environment of the vesicle interior. The charged form of the drug is then trapped in the vesicle interior and will be released as the pH gradient dissipates. Adapted from Ref. 159.

These methods,[160–162] which use preformed vesicles for their "remote" loading, provide higher encapsulation efficiencies than passive trapping techniques. Indeed, 100% of doxorubicin could be encapsulated in the case of Doxil®. Despite their important benefits, active techniques are only applicable to certain drugs and require more complicated protocols than passive methods, making the latter more commonly used for the encapsulation of active ingredients into vesicles.

V. Methods of Preparation

In order to achieve optimal performance of vesicular systems as functional materials, a tight control over the structural characteristics of the resulting self-assembled system is desired. Beside environmental conditions, such as temperature, pH, and ionic strength of the dispersing medium or type and concentration of lipid, the path followed for the self-assembly of the monomer units exerts an important influence on the structural properties of the obtained vesicular system and its homogeneity. Methods for the preparation of homogeneous vesicular formulations that present controlled size, morphology, and supramolecular organization are required for fully exploiting the potential of these self-assembled structures. Regarding their use as DDS, although maximum trapping efficiencies are obviously desirable, this property is often limited by the type of formulation (SUVs, LUVs, or MLVs) required for *in vivo* applications, which normally corresponds to unilamellar vesicles in the range of 100 nm. In this area, techniques for the rapid and easy generation of drug-loaded unilamellar vesicles possessing an optimal size and narrow size distribution are required to take full advantage of these entities as DDS.

Apart from controlling the structural characteristics of the final material, an adequate manufacturing process for pharmaceutical products (i.e., vesicles) should be easily compliant with the regulations imposed by the FDA in the United States or the EMA in Europe. These regulations involve the development and implementation of good manufacturing practices (GMPs), which are rules that ensure the quality, safety, and traceability of the obtained product.[163] Regarding this, it is important to note that the majority of vesicle preparation methods use organic solvents and that, depending on the application, they should be reduced below the allowed limits.[164] Simple methods for organic solvent reduction would comprise the change of the dispersing medium by centrifugation, lyophilization, or SEC.

In the following section, methods for vesicle production will be described, paying special attention to important characteristics such as size, lamellarity, and encapsulation efficiencies that they can provide, as well as their industrial application potential.

A. Conventional Methods for the Preparation of Drug-Loaded Vesicles

Conventional methods for vesicle production primarily involve the dispersion of lipids in an aqueous phase. Figure 20 shows a schematic summary of the most commonly used conventional techniques for the production and postformation processing of vesicles in order to achieve SUVs or LUVs.

One of the most widely used techniques for vesicle formation is the *thin film hydration or Bangham method*,[1,165,166] in which a dried lipid film is hydrated with an aqueous phase. This technique provides large and nonhomogeneous MLVs with low encapsulation efficiencies. *Freeze/thaw cycles* are often used for improving the loadings after formation of the vesicles.[167] In this method, there is no control over the vesicle formation step and therefore, the resulting MLVs need to be sonicated or extruded for their conversion to more homogeneous SUVs. Although it is a simple technique at laboratory scale, its use in bulk industrial production is not easily viable since its reproducibility and yield are low, requiring multiple time- and energy-consuming steps to reach the final product.

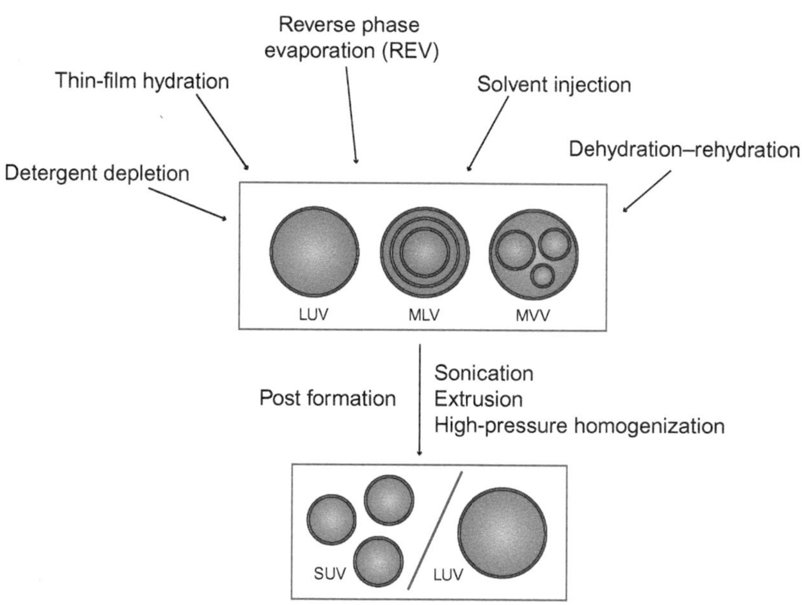

FIG. 20. Schematic representation of the most commonly used conventional methods for the production and postformation processing of vesicles. (For color version of this figure, the reader is referred to the Web version of this chapter.)

In the so-called *reverse-phase methods*, which include *reverse-phase evaporation (REV)* or *solvent injection* methods, the hydration of the lipids is performed directly from an organic solvent. The *REV method*[168] is based on the formation of inverted micelles upon the addition of a small volume of aqueous phase to an organic solution of the lipids. Subsequent removal of the organic solvent generates a viscous gel that collapses, yielding an aqueous suspension of MLVs. Using *solvent injection methods*,[169,170] unilamellar vesicles can be achieved by injecting a solution of lipids in ethanol or ether into an aqueous phase. The formation of LUVs occurs upon solvent dilution in the case of ethanol, whereas in the ether procedure, hydration takes place as solvent evaporates. Although reverse-phase techniques present higher encapsulation efficiencies than thin film hydration, they are limited by the solubility of the lipids in the organic phase and the subsequent removal of the solvent from the final preparation.

LUVs can also be produced by the *detergent depletion method,*[165,171,172] which involves the formation of detergent–lipid micelles by the hydration of a lipid film with a detergent solution. As detergent is removed, the micelles coalesce and the phospholipids adopt the bilayer configuration, resulting in sealed vesicles. Generally, problems associated with detergent removal, which include low trapping efficiency and length of preparation, limit the use of detergent-based techniques in producing LUVs.

A method for producing MLVs with high trapping efficiency specifically designed for pharmacological applications is the *dehydration–rehydration procedure* described by Kirby and Gregoriadis.[173] The technique involves the drying or the lyophilization of preformed vesicles in an aqueous medium containing the solute- and the subsequent-controlled rehydration. During the drying process, lipid vesicles are concentrated concomitantly with solute, and at some point, fuse into large aggregates containing the material to be trapped. Since the solute is already present in many of the interbilayer spaces when rehydrating, efficient trapping occurs. Beside the higher entrapment efficiencies, the dehydration–rehydration method usually gives heterogeneous systems with low sample reproducibility, which would become inconvenient for its production under GMPs.

With the majority of the techniques explained above, big MLVs are usually obtained; therefore, postformation processing is required for achieving the desired size, lamellarity, and homogeneity properties that permit the best performance of the vesicular systems prepared. The most common methods for postformation processing are *sonication*,[174] *extrusion*,[175–177], and *high-pressure homogenization.*[178,179] In these methods, mechanical force is used to disrupt large MLVs into small membrane patches that ultimately form SUVs. The need for using additional mechanical processing limits to a great extent the application of conventional methods for the encapsulation of fragile molecules due to the additional exposure to sonication or high shear homogenization.

The main drawbacks of most conventional preparation methods include being generally complex and time-consuming, comprising multiple steps, having equipment limitations, and being unreliable for the reproducible bulk production of homogeneous vesicles on a large scale.[180]

In order to expand the possibilities of vesicles as DDS, great effort is being invested for the development of new methods that minimize the issues inherent in conventional vesicle production techniques.[180]

Recently, Howse *et al.* reported a new methodology for the preparation of micron-sized giant unilamellar vesicles with controlled narrow size distributions by *photolithography/dewetting and posterior hydration.*[181] Due to their bigger size, these giant vesicles could allow picoliter dosage DDS.

Another recent technique for vesicle production is based on *microfluidics*. A measured flow of lipids dissolved in alcohol is hydrodynamically focused between two sheathed aqueous streams in a microfluidic channel with an accurate design, enabling a controlled diffusive mixing at the two liquid interfaces where the lipids self-assemble into vesicles.[182] By changing the flow conditions, it is possible to obtain vesicles of a particular size and size distribution, without the need of postprocessing steps. Large biological samples, like cancer cells and yeast cells, micron-sized beads, and nanoscopic proteins have been encapsulated using this method.[183] Microfluidization can be scaled up by increasing the number of channels in a parallel configuration. Nevertheless, it can be problematic for the processing of thermolabile compounds due to the exposure to high shears, and, also, processing problems, such as channel blocking, might take place. Another aspect to take into account for this kind of sophisticated equipment is how to successfully perform the required cleaning step after each batch production. A GMP-compliant procedure should include cleaning and cleaning validation procedures to avoid batch-to-batch or product-to-product cross-contamination. It is not clear whether at industrial scale it would be possible to ensure the cleanliness of multiple nanometer-sized channels.

B. Compressed Fluid-Based Methods for the Preparation of Drug-Loaded Vesicles

In the early 1990s, compressed fluid (CF)-based technologies emerged as an alternative to conventional processes using liquid solvents, attracting enormous interest for the production of micro- and nanoparticulate materials. Up to now, the main effort has been invested in the production of nanoparticulate solids; however, although still in the early stages of its research, CFs have already proved to be a good alternative for vesicle production. A CF—or *dense gas*—is defined as a substance that at normal conditions of pressure and temperature exists as a gas but that presents properties between those of liquid and gas when it is near or above its critical temperature and pressure. The most used CF is carbon dioxide (CO_2), which is cheap, available worldwide, and considered a green solvent, since it is nontoxic, nonflammable, and easily recyclable. The critical pressure and temperature of CO_2 are 73 bar and 31 °C, respectively. At atmospheric pressure and ambient temperature, CO_2 exists as a gas, but after compression above 73 bar, it becomes a liquid, allowing some solvating phenomena to occur. If kept at that pressure while increasing temperature above 31 °C, CO_2 turns supercritical. At the supercritical state and its surroundings, the CF has liquid-like density and, hence, solvating characteristics that are similar to those of liquids, but gas-like diffusivity, presenting mass transfer properties more similar to those of gases.[184,185] In contrast to conventional liquid solvents, solvating properties of CFs might be tuned by pressure changes, which transmit much faster than temperature or composition modifications, allowing a better control on material production processes. Due to their special characteristics, CFs enable the achievement of highly homogeneous formulations without the necessity of postforming stages, and it is often possible to synthesize materials with unique physicochemical characteristics (size, porosity, polymorphic nature, morphology, etc.) unachievable with classical liquid media.[186–189]

The use of CFs enables complete or partial reduction of organic solvents required in conventional production methods and can be easily separated from products just by decreasing the pressure of the system until reaching the gas phase of the CF. Moreover, these methods allow one-step achievement of the desired final products and can be easily transferred to large-scale operation.[190] These features, together with the sterile operating conditions provided by CF processing, are particularly important and attractive for food and pharmaceutical applications.

Regarding vesicle production, the most significant advantages of using CF-based technologies are the dramatic reduction in processing time, due to the reduced number of operating steps and the improved mass transfer properties intrinsic to the use of CFs, and the simple achievement of large quantities of vesicular systems with controlled characteristics. In addition to this, in comparison with conventional processes, with some of the CF methods developed nowadays, higher production yields and reduced costs can be achieved.

Most of the methods using compressed CO_2 involve a mixture between this dense gas, the membrane constituents, and an organic solvent at a certain pressure for producing vesicles upon contact with an aqueous phase. However, changes in the experimental procedure, parameters, and equipment result in vesicular systems with differentiated characteristics. Regarding encapsulation, hydrophobic drugs can be dissolved in the compressed mixture, whereas hydrophilic compounds would be incorporated in the aqueous phase. The major techniques described using CF technology for the formation of vesicles include the *injection method*,[191] the *decompression method*,[191] the *supercritical liposome method*[192,193], the *supercritical reverse phase evaporation (scRPE) method*,[190] and more recently the *DELOS-susp*[78] and *DESAM*[194] methods.

The *injection* and *decompression methods*,[191] described by Castor and Chu in 1994, were the first dense gas techniques for the formation of liposomes. The major distinction between the processes is the incorporation of the aqueous solution. While in the injection method, a mixture of lipid, organic cosolvent, and compressed gas is injected through a nozzle into an aqueous solution, the decompression method involves a mixture of lipid, organic cosolvent, compressed gas, and aqueous solution being decompressed into air through a nozzle to form the vesicles. Castor and Chu claim that the injection and decompression processes are capable of producing sterile, pharmaceutical-grade vesicles of a predetermined size that have a narrow particle size distribution and are substantially solvent-free.[195]

Based on the injection and decompression methods, Castor developed in 2005 the *super fluids phospholipid nanosome (SFS-CFN) method.*[196] The high temperatures (333 K), pressures (up to 35 MPa), and time needed for solubilizing the raw materials in the SFS-CFN method are still challenges for large-scale production of vesicles using this technique.[180]

The *supercritical liposome method*[192,193] is similar to the injection method. The low encapsulation efficiencies achieved, together with the high pressures (up to 25 MPa) and low yield, are disadvantages that compete with the low organic solvent consumption and the possible scalability of this method.

Otake *et al.*[190] developed the *scRPE method*, in which the lipid, organic cosolvent, and compressed gas are combined in a stirred, variable volume cell at 20 MPa and 333 K. An aqueous solution is then slowly introduced to the cell. The pressure is reduced by the release of the compressed gas, and vesicles are formed. Recently, the same authors developed an organic solvent-free version of the scRPE process and referred to it as the *improved supercritical reverse phase evaporation method.*[197,198] In this method, an inhomogeneous mixture of phospholipid and aqueous solution is pressurized by CO_2 to 20 MPa at 333 K and then depressurized to generate liposomes. The drug-loading efficiency and stability are improved compared with the scRPE method because of the absence of ethanol in the processing and final formulation.

The introduction of CF processing has enabled the rapid production of sterile liposomes of similar or improved quality compared to those produced by conventional methods, reducing or avoiding the use of toxic organic solvents. Despite the clear advantages of these methods, the CF-based processes described above generally require elevated pressures, commonly around 20–25 MPa, and temperatures of 333 K because of the low solubility of lipids in CFs. In the past years, efforts have been made in order to take full advantage of CF processing without using severe working conditions, and new CF-based methods have been developed for vesicle production.[78,194]

One of these methods, *DELOS-susp*,[78] involves the depressurization of a CO_2-expanded liquid solution of lipids into an aqueous phase using milder conditions of pressure (10 MPa) and temperature (308 K) than the majority of other methodologies using CFs. The lower temperatures would permit the processing of heat-labile compounds, while the milder pressures would reduce the capital cost of a high-pressure plant when scaling up the process.

As depicted in Fig. 21, in a DELOS-susp procedure, the lipid or lipids are dissolved in a conventional organic solvent, at atmospheric pressure and at a working temperature, T_w. A CF (normally CO_2) is then added in order to obtain a volumetric expanded solution of the lipids, at T_w and at high pressure (P_w), with a given molar fraction of CO_2, X_w. It is important that the lipids be

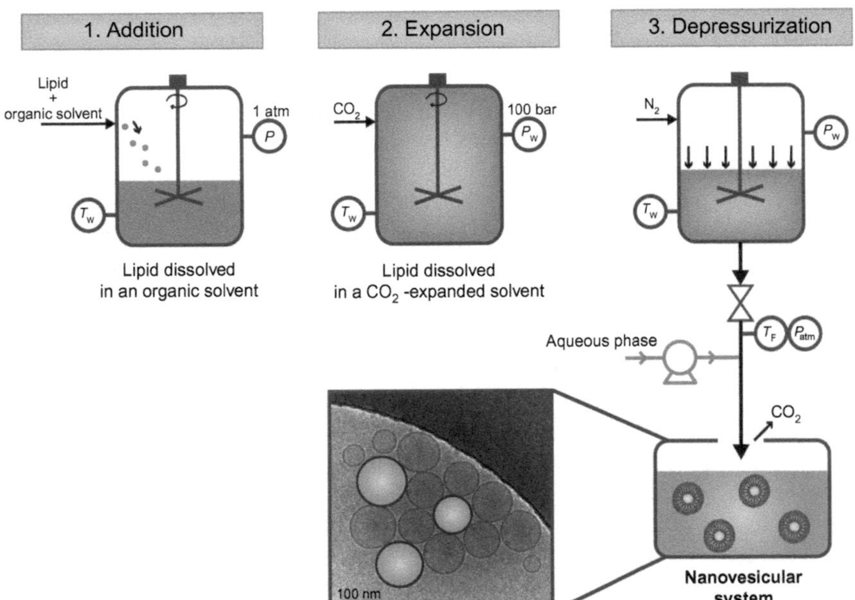

FIG. 21. Schematic representation of the DELOS-susp procedure for vesicles preparation. (1) Addition of an organic solution of the lipid to the vessel at T_w and atmospheric pressure (P_{atm}). (2) Expansion of the organic solution of the lipid by adding CO_2 until it reaches X_w at P_w and T_w. (3) Depressurization of the CO_2-expanded solution over a flow of aqueous phase from P_w to P_{atm}. Depending on the hydrophilic or hydrophobic nature of the drug or active substance to be entrapped, it will be dissolved in the initial organic solution (step 1) or in the aqueous phase (step 3), allowing a single-step preparation of loaded vesicles.

completely soluble in the CO_2-expanded solvent in order to assure the presence of only one phase inside the high-pressure chamber. Finally, this CO_2-expanded solution is depressurized, from P_w to atmospheric pressure, over a continuous aqueous flow, containing a surfactant if necessary, to give uniform unilamellar lipid nanovesicles (see cryo-TEM image in Fig. 21).

By this methodology, the lipidic solution experiences a large, abrupt, and homogeneous temperature decrease, from T_w to T_F, during depressurization. This ideal homogeneous temperature decrease is probably the reason why this procedure provides more homogeneous vesicles in terms of size, morphology, and supramolecular organization compared to conventional procedures.[78]

Incorporation of cholesterol in the lipid membrane of vesicles is sometimes required in order to reduce their permeability and avoid leakage of the encapsulated active molecule. Due to the low aqueous solubility of cholesterol, the preparation of cholesterol-rich vesicles by conventional hydration procedures is problematic. By DELOS-susp, nanoscopic cholesterol/CTAB and cholesterol/DPPC vesicles with a high content in cholesterol have been prepared (1:1 mol/mol in both cases). These formulations are very homogeneous, stable in time, and with medium vesicle diameters between 100 nm and 200 nm.

Regarding the entrapment of active substances within vesicles by DELOS-susp, the drug can be dissolved either in the expanded mixture or in the aqueous phase depending on its hydrophobic or hydrophilic nature, respectively. For instance, an hydrophilic antibiotic, gentamicin sulphate, has been incorporated into cholesterol-rich vesicles by this method, yielding in one step structurally homogeneous antibiotic-loaded vesicular systems with comparable entrapment efficiencies to those prepared by a multistep conventional passive method, which provided a less homogeneous vesicular formulation (see Fig. 22).[199]

One of the major problems that prevent the broad use of vesicles as DDS is the lack of reproducibility and high cost related to the scaling up of vesicle preparation methods. Since they are subtended by noncovalent interactions, the maintenance of their structural properties from lab to industrial scale becomes a difficult issue to overcome. CF-based technologies have arisen as a promising alternative mainly due to an easier and more reproducible scalability than other material processing methods. An example of this is the scalability of the DELOS-susp process for the encapsulation of gentamicin sulphate in cholesterol/CTAB vesicles.[199] Figure 23 shows cryo-TEM images of unilamellar antibiotic-loaded liposomes obtained by DELOS-susp in different-sized equipments ranging from the milliliter scale to the liter scale by a 50-fold scale-up factor. In both scales, reproducible results were obtained, showing that the DELOS-susp method can be easily scaled up, yielding similar structural homogeneity—vesicle size, size distribution, morphology—and entrapment efficiencies with equipments of increased working volumes.

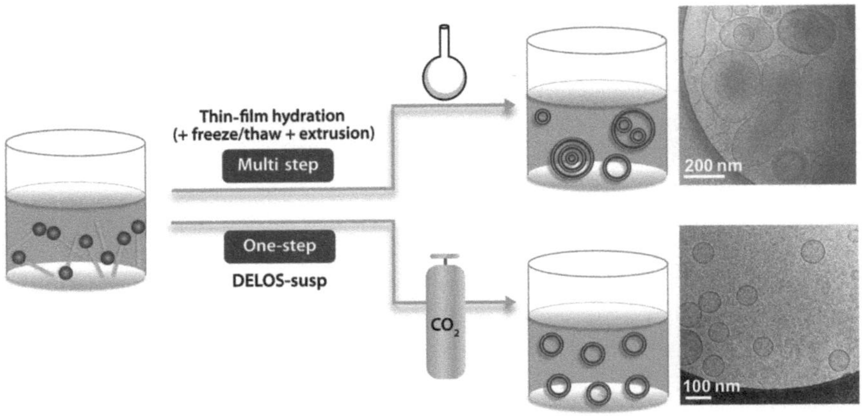

FIG. 22. Impact of the preparation procedure on the homogeneity and structural organization of vesicles with the same chemical composition. Cryo-TEM images correspond to antibiotic-loaded vesicular systems obtained by thin film hydration (top image) and by DELOS-susp (bottom image). (For color version of this figure, the reader is referred to the Web version of this chapter.)

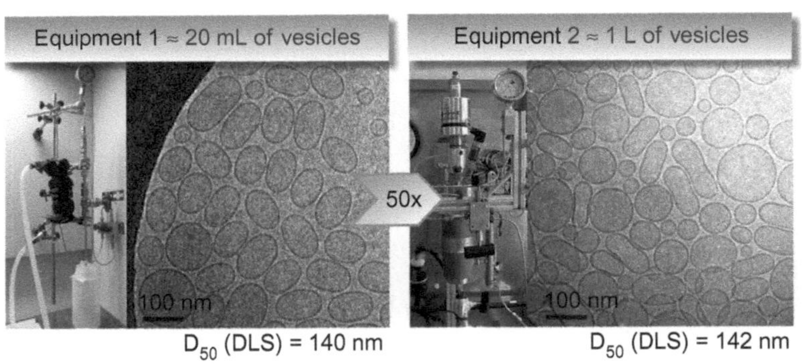

FIG. 23. DELOS-susp scale-up: preparation of antibiotic-loaded cholesterol/CTAB unilamellar vesicles with two different equipments of different capacity. Equipment 2 produces 50 times more vesicles than equipment 1. Cryo-TEM images and average DLS particle size of vesicles prepared in equipments 1 and 2 show similar morphology and particle size. (For color version of this figure, the reader is referred to the Web version of this chapter.)

As mentioned earlier, from the large-scale application point of view, a manufacturing process for obtaining nanomedicines should be compliant with the GMPs, which are strict rules that govern the manufacturing of pharmaceutical products, ensuring their quality and traceability.[163] The DELOS-susp method can accomplish these quality requirements, since it is simple, rapid, easily controlled, and provides batch-to-batch reproducibility. Moreover, it is easily scalable, does not require complex equipment, provides elevated production yields, and cleaning issues are easy to validate.

VI. Conclusions and Future Challenges

Liposomes and other vesicular systems are widely accepted as versatile nanocarriers of bioactive molecules with poor pharmacological properties. Their use in the field of drug delivery has been continuously increasing over the past years, and a large number of research groups all over the world are using these lipid nanoparticles to ameliorate the pharmacokinetics and the biodistribution of biologically active ingredients with low therapeutic indexes. Recent advances in microscopic, spectroscopic, and diffraction techniques have allowed a significant improvement in the characterization of these lipid structures at nanoscopic and supramolecular scales. However, further improvements in the characterization instruments and methodologies are still required in order to gain a complete insight into the organization, at molecular level, of these noncrystalline ordered submicroscopic and nanoscopic colloidal structures. This deep structural characterization is absolutely necessary, for instance, to control and fine-tune the functionalization of vesicle membranes with ligands, such as peptides or antibodies, which specifically will recognize receptors overexpressed in damaged tissues or cells, contributing to the development of nanomedicines with high therapeutic indexes and very few or no side effects. A better comprehension of the supramolecular interactions responsible for the self-assembling of the lipid-like molecules is necessary for the development of vesicle-based nanomedicines with specific delivery of active molecules in targeted diseased cells.

Last but not least, further efforts should be dedicated to establish adequate production platforms for large-scale green and economical precise synthesis of vesicles with reproducible structural characteristics (size, shape, chemical composition, supramolecular self-assembling). These processes should also allow the production of these colloidal nanomedicines under the normative of GMPs.

References

1. Bangham AD, Standish MM, Watkins JC. Diffusion of univalent ions across the lamellae of swollen phospholipids. *J Mol Biol* 1965;**13**:238–52.
2. Steed JW, Turner DR, Wallace KJ. *Core concepts in supramolecular chemistry and nanochemistry*. Chichester (England): John Wiley & Sons, Ltd.; 2007.

3. Rajagopal K, Schneider JP. Self-assembling peptides and proteins for nanotechnological applications. *Curr Opin Struct Biol* 2004;**14**:480–6.
4. Whitesides GM, Boncheva M. Beyond molecules: self-assembly of mesoscopic and macroscopic components. *Proc Natl Acad Sci USA* 2002;**99**:4769–74.
5. Whitesides GM, Grzybowski B. Self-assembly at all scales. *Science* 2002;**295**:2418–21.
6. Desiraju GR. *Crystal engineering: the design of organic solids.* New York: Elsevier; 1989.
7. Evans DF, Wennerstrom H. *The colloidal domain: where physics, chemistry, biology, and technology meet.* New York: Wiley; 1999.
8. Jones MN, Chapman D. *Micelles*, monolayers and biomembranes. New York: Wiley-Liss; 1995.
9. Thomas EL. Materials Science:Enhanced: The ABCs of Self-Assembly. *Science* 1999;**286**:1307.
10. Kumar A, Abbott NL, Kim E, Biebuyck HA, Whitesides GM. Patterned self-assembled monolayers and mesoscale phenomena. *Acc Chem Res* 1995;**28**:219–26.
11. Grantcharova V, Alm EJ, Baker D, Horwich AL. Mechanisms of protein folding. *Curr Opin Struct Biol* 2001;**11**:70–82.
12. Neidle S. *Oxford handbook of nucleic acid structure.* Oxford, UK: Oxford University Press; 1999.
13. Tu RS, Tirrell M. Bottom-up design of biomimetic assemblies. *Adv Drug Deliv Rev* 2004;**56**:1537–63.
14. Ekwall P, Mandell L, Fontell K. Solubilization in micelles and mesophases and transition from normal to reserved structures. *Mol Crystals Liquid Crystals* 1969;**8**:157–213.
15. Wennerström H, Lindman B. Micelles. Physical chemistry of surfactant association. *Phys Reps* 1979;**52**:1–86.
16. Chevalier Y, Zemb T. The structure of micelles and microemulsions. *Rep Prog Phys* 1990;**53**:279–371.
17. Tanford C. *The hydrophobic effect: formation of micelles and biological membranes.* New York: Wiley; 1980.
18. Lee JH, Choi YJ, Lim YB. Self-assembled filamentous nanostructures for drug/gene delivery applications. *Expert Opin Drug Deliv* 2010;**7**:341–51.
19. Israelachvili JN, Mitchell DJ, Ninham BW. Theory of self-assembly of hydrocarbon amphiphiles into micelles and bilayers. *J Chem Soc Faraday Trans II* 1976;**72**:1525–68.
20. ElBayoumi TA, Torchilin VP. *Current trends in liposome research. Liposomes: methods and protocols, vol. 1. Pharmaceutical nanocarriers*. 2010, pp. 1–27.
21. Roux E, Passirani C, Scheffold S, Benoit J-P, Leroux J-C. Serum-stable and long-circulating, pegylated, PH-sensitive liposomes. *J Control Release* 2004;**94**:447–51.
22. Shi G, Guo W, Stephenson SM, Lee RJ. Efficient intracellular drug and gene delivery using folate receptor-targeted PH-sensitive liposomes composed of cationic/anionic lipid combinations. *J Control Release* 2002;**80**:309–19.
23. Drulis-Kawa Z, Dorotkiewicz-Jach A. Liposomes as delivery systems for antibiotics. *Int J Pharm* 2010;**387**:187–98.
24. Gregoriadis G, Davis C. Stability of liposomes in vivo and in vitro is promoted by their cholesterol content and the presence of blood-cells. *Biochem Biophys Res Commun* 1979;**89**:1287–93.
25. Ulrich AS. Biophysical aspects of using liposomes as delivery vehicles. *Biosci Rep* 2002;**22**:129–50.
26. Uchegbu IF, Florence AT. Non-ionic surfactant vesicles (niosomes): physical and pharmaceutical chemistry. *Adv Colloid Interface Sci* 1995;**58**:1–55.
27. Kaur IP, Garg A, Singla AK, Aggarwal D. Vesicular systems in ocular drug delivery: an overview. *Int J Pharm* 2004;**269**:1–14.

28. Uchegbu IF, Vyas SP. Non-ionic surfactant based vesicles (niosomes) in drug delivery. *Int J Pharm* 1998;**172**:33–70.
29. Manosroi A, Wongtrakul P, Manosroi J, Sakai H, Sugawara F, Yuasa M, et al. Characterization of vesicles prepared with various non-ionic surfactants mixed with cholesterol. *Colloids Surf B Biointerfaces* 2003;**30**:129–38.
30. Cano-Sarabia M, Angelova A, Ventosa N, Lesieur S, Veciana J. Cholesterol induced CTAB micelle-to-vesicle phase transitions. *J Colloid Interface Sci* 2010;**350**:10–5.
31. Templeton NS, Lasic DD, Frederik PM, Strey HH, Roberts DD, Pavlakis GN. Improved DNA: liposome complexes for increased systemic delivery and gene expression. *Nat Biotechnol* 1997;**15**:647–52.
32. Perrie Y, Frederik PM, Gregoriadis G. Liposome-mediated DNA vaccination: the effect of vesicle composition. *Vaccine* 2001;**19**:3301–10.
33. Bramer T, Dew N, Edsman K. Pharmaceutical applications for catanionic mixtures. *J Pharm Pharmacol* 2007;**59**:1319–34.
34. Liu Y-C, Le Ny A-LM, Schmidt J, Talmon Y, Chmelka BF, Lee CT. Photo-assisted gene delivery using light-responsive catanionic vesicles. *Langmuir* 2009;**25**:5713–24.
35. Danoff EJ, Wang X, Tung S-H, Sinkov NA, Kemme AM, Raghavan SR, et al. Surfactant vesicles for high-efficiency capture and separation of charged organic solutes. *Langmuir* 2007;**23**:8965–71.
36. Discher BM, Won Y-Y, Ege DS, Lee JC-M, Bates FS, Discher DE, et al. Polymersomes: tough vesicles made from diblock copolymers. *Science* 1999;**284**:1143–6.
37. Ahmed F, Discher DE. Self-porating polymersomes of PEG-PLA and PEG-PCL: hydrolysis-triggered controlled release vesicles. *J Control Release* 2004;**96**:37–53.
38. Kita-Tokarczyk K, Grumelard J, Haefele T, Meier W. Block copolymer vesicles-using concepts from polymer chemistry to mimic biomembranes. *Polymer* 2005;**46**:3540–63.
39. Gradzielski M. Vesicles and vesicle gels—structure and dynamics of formation. *J Phys Condens Matter* 2003;**15**:R655–97.
40. Lorin A, Flore C, Thomas A, Brasseur R. Les liposomes: description, fabrication et applications. *Biotechnol Agron Soc Environ* 2004;**8**:163–76.
41. Kulkarni SB, Betageri GV, Singh M. Factors affecting microencapsulation of drugs in liposomes. *J Microencapsul* 1995;**12**:229–46.
42. Veatch SL, Keller SL. Separation of liquid phases in giant vesicles of ternary mixtures of phospholipids and cholesterol. *Biophys J* 2003;**85**:3074–83.
43. Kulin S, Kishore R, Helmerson K, Locascio L. Optical manipulation and fusion of liposomes as microreactors. *Langmuir* 2003;**19**:8206–10.
44. Bolinger PY, Stamou D, Vogel H. Integrated nanoreactor systems: triggering the release and mixing of compounds inside single vesicles. *J Am Chem Soc* 2004;**126**:8594–5.
45. Delamaza A, Manich AM, Coderch L, Parra JL. Multilamellar liposomes including cholesterol as carriers of azobenzene disperse dyes in wool dyeing. *Textile Res J* 1995;**65**:163–70.
46. Rawicz W, Olbrich KC, McIntosh T, Needham D, Evans E. Effect of chain length and unsaturation on elasticity of lipid bilayers. *Biophys J* 2000;**79**:328–39.
47. Sapra P, Allen TM. Ligand-targeted liposomal anticancer drugs. *Prog Lipid Res* 2003;**42**:439–62.
48. Guo X, Szoka FC. Chemical approaches to triggerable lipid vesicles for drug and gene delivery. *Acc Chem Res* 2003;**36**:335–41.
49. Allen TM, Hansen CB, de Menezes DEL. Pharmacokinetics of long-circulating liposomes. *Adv Drug Deliv Rev* 1995;**16**:267–84.
50. Edwards KA, Baeumner AJ. Analysis of liposomes. *Talanta* 2006;**68**:1432–41.
51. Ruf H, Georgalis Y, Grell E. Dynamic laser-light scattering to determine size distributions of vesicles. *Methods Enzymol* 1989;**172**:364–90.

52. Ruozi B, Tosi G, Forni F, Fresta M, Vandelli MA. Atomic force microscopy and photon correlation spectroscopy: two techniques for rapid characterization of liposomes. *Eur J Pharm Sci* 2005;**25**:81–9.
53. McCracken MS, Sammons MC. Sizing of a vesicle drug formulation by quasi-elastic light-scattering and comparison with electron-microscopy and ultracentrifugation. *J Pharm Sci* 1987;**76**:56–9.
54. Frederik PM, Hubert DHW. Cryoelectron microscopy of liposomes. *Methods Enzymol* 2005;**391**:431–48.
55. Jass J, Tjarnhage T, Puu G. Atomic force microscopy imaging of liposomes. *Methods Enzymol* 2003;**367**:199–213.
56. Grabielle-Madelmont C, Lesieur S, Ollivon M. Characterization of loaded liposomes by size exclusion chromatography. *J Biochem Biophys Methods* 2003;**56**:189–217.
57. Moon MH, Giddings JC. Size distribution of liposomes by flow field-flow fractionation. *J Pharm Biomed Anal* 1993;**11**:911–20.
58. Arifin DR, Palmer AF. Determination of size distribution and encapsulation efficiency of liposome-encapsulated hemoglobin blood substitutes using asymmetric flow field-flow fractionation coupled with multi-angle static light scattering. *Biotechnol Prog* 2003;**19**:1798–811.
59. Korgel BA, van Zanten JH, Monbouquette HG. Vesicle size distributions measured by flow field-flow fractionation coupled with multiangle light scattering. *Biophys J* 1998;**74**:3264–72.
60. Pecora R, editor. *Dynamic light scattering: applications of photon correlation spectroscopy.* New York: Plenum Press; 1985.
61. Dahneke BE, editor. *Measurements of suspended particles by quasi-elastic light scattering.* New York: Wiley; 1983.
62. Brown W, editor. *Dynamic light scattering: the method and some applications*. Oxford: Clarendon Press; 1993.
63. Nagayasu A, Uchiyama K, Kiwada H. The size of liposomes: a factor which affects their targeting efficiency to tumors and therapeutic activity of liposomal antitumor drugs. *Adv Drug Deliv Rev* 1999;**40**:75–87.
64. Filipe V, Hawe A, Jiskoot W. Critical evaluation of nanoparticle tracking analysis (NTA) by nanosight for the measurement of nanoparticles and protein aggregates. *Pharm Res* 2010;**27**:796–810.
65. Carr B, Hole P, Malloy A, Nelson P, Wright M, Smith J. Applications of nanoparticle tracking analysis in nanoparticle research—a mini review. *Eur J Parenter Pharm Sci* 2009;**14**:45–50.
66. Filipe V, Poole R, Kutscher M, Forier K, Braeckmans K, Jiskoot W. Fluorescence single particle tracking for the characterization of submicron protein aggregates in biological fluids and complex formulations. *Pharm Res* 2011;**28**:1112–20.
67. Dragovic RA, Gardiner C, Brooks AS, Tannetta DS, Ferguson DJP, Hole P, et al. Sizing and phenotyping of cellular vesicles using nanoparticle tracking analysis. *Nanomedicine* 2011; doi:10.1016/j.nano.2011.04.003.
68. Vallhov H, Gutzeit C, Johansson SM, Nagy N, Paul M, Li Q, et al. Exosomes containing glycoprotein 350 released by EBV-transformed B cells selectively target B cells through CD21 and block EBV infection in vitro. *J Immunol* 2011;**186**:73–82.
69. Bangham AD, Heard DH, Flemans R, Seaman GVF. Apparatus for microelectrophoresis of small particles. *Nature* 1958;**182**:642–4.
70. Earnshaw JC, Steer MW, editor. *The application of laser light scattering to the study of biological motion. NATO ASI Series Series A: Life Sciences*. New York: Plenum Press; 1983.
71. Jones MN. The surface properties of phospholipid liposome systems and their characterisation. *Adv Colloid Interface Sci* 1995;**54**:93–128.
72. Carrion FJ, Delamaza A, Parra JL. The influence of ionic-strength and lipid bilayer charge on the stability of liposomes. *J Colloid Interface Sci* 1994;**164**:78–87.

73. Perrie Y, Gregoriadis G. Liposome-entrapped plasmid DNA: characterisation studies. *Biochim Biophys Acta* 2000;**1475**:125–32.
74. Garbuzenko O, Zalipsky S, Qazen M, Barenholz Y. Electrostatics of Pegylated micelles and liposomes containing charged and neutral lipopolymers. *Langmuir* 2005;**21**:2560–8.
75. Severs NJ. Freeze-fracture electron microscopy. *Nat Protoc* 2007;**2**:547–76.
76. Bronich TK, Ouyang M, Kabanov VA, Eisenberg A, Szoka FC, Kabanov AV. Synthesis of vesicles on polymer template. *J Am Chem Soc* 2002;**124**:11872–3.
77. Lovell JF, Jin CS, Huynh E, Jin HL, Kim C, Rubinstein JL, et al. Porphysome nanovesicles generated by porphyrin bilayers for use as multimodal biophotonic contrast agents. *Nat Mater* 2011;**10**:324–32.
78. Cano-Sarabia M, Ventosa N, Sala S, Patino C, Arranz R, Veciana J. Preparation of uniform rich cholesterol unilamellar nanovesicles using CO_2-expanded solvents. *Langmuir* 2008;**24**:2433–7.
79. Zhao H, Wang JC, Sun QS, Luo CL, Zhang Q. Rgd-based strategies for improving antitumor activity of paclitaxel-loaded liposomes in nude mice xenografted with human ovarian cancer. *J Drug Target* 2009;**17**:10–8.
80. White SH, Wiener MC. Determination of the structure of fluid lipid bilayer membranes. In: Disalvo EA, Simon SA, editors. *Permeability and stability of lipid bilayers*. Boca Raton: CRC Press; 1995. pp. 1–19.
81. White SH, Wiener MC. The liquid-crystallographic structure of fluid lipid bilayer membranes. In: Merz KM, Roux B, editors. *Membrane structure and dynamics*. Boston: Birkhäuser; 1996. pp. 127–44.
82. Wiener MC, Suter RM, Nagle JF. Structure of the fully hydrated gel phase of dipalmitoylphosphatidylcholine. *Biophys J* 1989;**55**:315–25.
83. King GI, White SH. Determining bilayer hydrocarbon thickness from neutron-diffraction measurements using strip-function models. *Biophys J* 1986;**49**:1047–54.
84. Bouwstra JA, Gooris GS, Bras W, Talsma H. Small-angle X-ray-scattering—possibilities and limitations in characterization of vesicles. *Chem Phys Lipids* 1993;**64**:83–98.
85. Kucerka N, Pencer J, Sachs JN, Nagle JF, Katsaras J. Curvature effect on the structure of phospholipid bilayers. *Langmuir* 2007;**23**:1292–9.
86. Leharne SA, Chowdhry BZ. Thermodynamic background to differential scanning calorimetry. In: Ladbury J, Chowdhry BZ, editors. *Biocalorimetry: Applications of calorimetry in the biological sciences*. Chichester: Wiley; 1998 pp. 157–82.
87. Plotnikov VV, Brandts JM, Lin LN, Brandts JF. A new ultrasensitive scanning calorimeter. *Anal Biochem* 1997;**250**:237–44.
88. Privalov G, Kavina V, Freire E, Privalov PL. Precise scanning calorimeter for studying thermal-properties of biological macromolecules in dilute-solution. *Anal Biochem* 1995;**232**:79–85.
89. Ladbrook BD, Chapman D. Thermal analysis of lipids, proteins and biological membranes. A review and summary of some recent studies. *Chem Phys Lipids* 1969;**3**:304–56.
90. Chapman D, Urbina J, Keough KM. Biomembrane phase-transitions—studies of lipid-water systems using differential scanning calorimetry. *J Biol Chem* 1974;**249**:2512–21.
91. Taylor KMG, Morris RM. Thermal-analysis of phase-transition behavior in liposomes. *Thermochim Acta* 1995;**248**:289–301.
92. Griffin KL, Cheng CY, Smith EA, Dea PK. Effects of pentanol isomers on the phase behavior of phospholipid bilayer membranes. *Biophys Chem* 2010;**152**:178–83.
93. Lohner K, Latal A, Lehrer RI, Ganz T. Differential scanning microcalorimetry indicates that human defensin, HNP-2, interacts specifically with biomembrane mimetic systems. *Biochemistry* 1997;**36**:1525–31.

94. Lo YL, Rahman YE. Protein location in liposomes, a drug carrier—a prediction by differential scanning calorimetry. *J Pharm Sci* 1995;**84**:805–14.
95. Zhao LY, Feng SS. Effects of lipid chain length on molecular interactions between paclitaxel and phospholipid within model biomembranes. *J Colloid Interface Sci* 2004;**274**:55–68.
96. Filippov A, Munavirov B, Sparrman T, Ishmuhametova V, Rudakova M, Shriram P, et al. Interaction of a poly(acrylic acid) oligomer with dimyristoylphosphatidylcholine bilayers. *Langmuir* 2011;**27**:3754–61.
97. Alves ID, Goasdoue N, Correia I, Aubry S, Galanth C, Sagan S, et al. Membrane interaction and perturbation mechanisms induced by two cationic cell penetrating peptides with distinct charge distribution. *Biochim Biophys Acta* 2008;**1780**:948–59.
98. Mertins O, Dimova R. Binding of chitosan to phospholipid vesicles studied with isothermal titration calorimetry. *Langmuir* 2011;**27**:5506–15.
99. Pruzan P, Minassian LT, Figuiere P, Szwarc H. High-pressure calorimetry as applied to piezothermal analysis. *Rev Sci Instrum* 1976;**47**:66–71.
100. Randzio SL. Scanning transitiometry for science and industry. *J Thermal Anal* 1997;**48**:573–83.
101. Randzio SL, Grolier JPE, Quint JR. An isothermal scanning calorimeter controlled by linear pressure variations from 0.1-MPa to 400-MPa—calibration and comparison with the piezothermal technique. *Rev Sci Instrum* 1994;**65**:960–5.
102. Heimburg T, Grabitz P, Ivanova VP. The coupling of relaxation times to the heat capacity of lipid vesicles studied by pressure jump calorimetry. *Biophys J* 2001;**80**:2148.
103. Grabitz P, Ivanova VP, Heimburg T. Relaxation kinetics of lipid membranes and its relation to the heat capacity. *Biophys J* 2002;**82**:299–309.
104. Lin LN, Brandts JF, Brandts JM, Plotnikov V. Determination of the volumetric properties of proteins and other solutes using pressure perturbation calorimetry. *Anal Biochem* 2002;**302**:144–60.
105. Johnson ML, Vanosdol WW, Biltonen RL. The measurement of the kinetics of lipid phase-transitions—a volume-perturbation kinetic calorimeter. *Methods Enzymol* 1986;**130**:534–51.
106. Vanosdol WW, Johnson ML, Ye Q, Biltonen RL. Relaxation dynamics of the gel to liquid-crystalline transition of phosphatidylcholine bilayers—effects of chainlength and vesicle size. *Biophys J* 1991;**59**:775–85.
107. Vanosdol WW, Ye Q, Johnson ML, Biltonen RL. Effects of the anesthetic dibucaine on the kinetics of the gel-liquid crystalline transition of dipalmitoylphosphatidylcholine multilamellar vesicles. *Biophys J* 1992;**63**:1011–7.
108. Heerklotz H, Seelig J. Application of pressure perturbation calorimetry to lipid bilayers. *Biophys J* 2002;**82**:1445–52.
109. Ebel H, Grabitz P, Heimburg T. Enthalpy and volume changes in lipid membranes. I. The proportionality of heat and volume changes in the lipid melting transition and its implication for the elastic constants. *J Phys Chem B* 2001;**105**:7353–60.
110. Heerklotz H. Triton promotes domain formation in lipid raft mixtures. *Biophys J* 2002;**83**:2693–701.
111. Hunter AC. Application of the quartz crystal microbalance to nanomedicine. *J Biomed Nanotechnol* 2009;**5**:669–75.
112. Braunhut SJ, McIntosh D, Vorotnikova E, Zhou T, Marx KA. Detection of apoptosis and drug resistance of human breast cancer cells to taxane treatments using quartz crystal microbalance biosensor technology. *Assay Drug Dev Technol* 2005;**3**:77–88.
113. Pastorino L, Soumetz FC, Giacomini M, Ruggiero C. Development of a piezoelectric immunosensor for the measurement of paclitaxel. *J Immunol Methods* 2006;**313**:191–8.
114. Elsom J, Lethem MI, Rees GD, Hunter AC. Novel quartz crystal microbalance based biosensor for detection of oral epithelial cell-microparticle interaction in real-time. *Biosens Bioelectron* 2008;**23**:1259–65.

115. Reimhult E, Kasemo B, Hook F. Rupture pathway of phosphatidylcholine liposomes on silicon dioxide. *Int J Mol Sci* 2009;**10**:1683–96.
116. Kaiser N, Kimpfler A, Massing U, Burger AM, Fiebig HH, Brandl M, et al. 5-Fluorouracil in vesicular phospholipid gels for anticancer treatment: entrapment and release properties. *Int J Pharm* 2003;**256**:123–31.
117. Xia S, Xu S. Ferrous sulfate liposomes: preparation, stability and application in fluid milk. *Food Res Int* 2005;**38**:289–96.
118. Virto MR, Elorza B, Torrado S, Elorza MDA, Frutos G. Improvement of gentamicin poly(D, L-lactic-co-glycolic acid) microspheres for treatment of osteomyelitis induced by orthopedic procedures. *Biomaterials* 2007;**28**:877–85.
119. Kallinteri P, Fatouros D, Klepetsanis P, Antimisiaris SG. Arsenic trioxide liposomes: encapsulation efficiency and in vitro stability. *J Liposome Res* 2004;**14**:27–38.
120. Betageri GV, Parsons DL. Drug encapsulation and release from multilamellar and unilamellar liposomes. *Int J Pharm* 1992;**81**:235–41.
121. Perkins WR, Minchey SR, Ahl PL, Janoff AS. The determination of liposome captured volume. *Chem Phys Lipids* 1993;**64**:197–217.
122. Jeffs LB, Palmer LR, Ambegia EG, Giesbrecht C, Ewanick S, MacLachlan I. A scalable, extrusion-free method for efficient liposomal encapsulation of plasmid DNA. *Pharm Res* 2005;**22**:362–72.
123. Lang JK. Quantitative-determination of cholesterol in liposome drug products and raw-materials by high-performance liquid-chromatography. *J Chromatogr* 1990;**507**:157–63.
124. Samuni AM, Lipman A, Barenholz Y. Damage to liposomal lipids: protection by antioxidants and cholesterol-mediated dehydration. *Chem Phys Lipids* 2000;**105**:121–34.
125. Jaaskelainen I, Urtti A. Liquid-chromatography determination of liposome components using a light-scattering evaporative detector. *J Pharm Biomed Anal* 1994;**12**:977–82.
126. Bligh EG, Dyer WJ. A rapid method of total lipid extraction and purification. *Can J Biochem Physiol* 1959;**37**:911–7.
127. Grohganz H, Schlafli O, Rischer M, Brandl M. Development and validation of a HPLC method for routine quantification of the decapeptide Cetrorelix in liposome dispersions. *J Pharm Biomed Anal* 2004;**34**:963–9.
128. Manosroi A, Kongkaneramit L, Manosroi J. Characterization of amphotericin B liposome formulations. *Drug Dev Ind Pharm* 2004;**30**:535–43.
129. Sezer AD, Bas AL, Akbuga J. Encapsulation of enrofloxacin in liposomes I: preparation and in vitro characterization of LUV. *J Liposome Res* 2004;**14**:77–86.
130. Guo W, Ahmad A, Khan S, Dahhani F, Wang YF, Ahmad I. Determination by liquid chromatography with fluorescence detection of total 7-ethyl-10-hydroxy-camptothecin (SN-38) in beagle dog plasma after intravenous administration of liposome-based SN-38 (LE-SN38). *J Chromatogr B Analyt Technol Biomed Life Sci* 2003;**791**:85–92.
131. Ramaldes GA, Deverre JR, Grognet JM, Puisieux F, Fattal E. Use of an enzyme immunoassay for the evaluation of entrapment efficiency and in vitro stability in intestinal fluids of liposomal bovine serum albumin. *Int J Pharm* 1996;**143**:1–11.
132. Zhang XM, Patel AB, de Graaf RA, Behar KL. Determination of liposomal encapsulation efficiency using proton NMR spectroscopy. *Chem Phys Lipids* 2004;**127**:113–20.
133. Oku N, Kendall DA, Macdonald RC. A simple procedure for the determination of the trapped volume of liposomes. *Biochim Biophys Acta* 1982;**691**:332–40.
134. Mehlhorn RJ, Candau P, Packer L. Measurements of volumes and electrochemical gradients with spin probes in membrane-vesicles. *Methods Enzymol* 1982;**88**:751–62.
135. Vistnes AI, Puskin JS. A spin label method for measuring internal volumes in liposomes or cells, applied to Ca-dependent fusion of negatively charged vesicles. *Biochim Biophys Acta* 1981;**644**:244–50.

136. Riehemann K, Schneider SW, Luger TA, Godin B, Ferrari M, Fuchs H. Nanomedicine-challenge and perspectives. *Angew Chem Int Ed* 2009;**48**:872–97.
137. Wagner V, Dullaart A, Bock AK, Zweck A. The emerging nanomedicine landscape. *Nat Biotechnol* 2006;**24**:1211–7.
138. Chen Z. Small-molecule delivery by nanoparticles for anticancer therapy. *Trends Mol Med* 2010;**16**:594–602.
139. Crommelin D. Liposomes, In: *EMA 1st international workshop on nanomedicines*; 2010, London.
140. Storm G, Crommelin DJA. Liposomes: quo vadis? *Pharm Sci Technol Today* 1998;**1**:19–31.
141. Torchilin VP. Recent advances with liposomes as pharmaceutical carriers. *Nat Rev Drug Discov* 2005;**4**:145–60.
142. Lasic DD, Needham D. The "Stealth" liposome: a prototypical biomaterial. *Chem Rev* 1995;**95**:2601–28.
143. Torchilin VP, Omelyanenko VG, Papisov MI, Bogdanov Jr. AA, Trubetskoy VS, Herron JN, et al. Poly(ethylene glycol) on the liposome surface: on the mechanism of polymer-coated liposome longevity. *Biochim Biophys Acta* 1994;**1195**:11–20.
144. Klibanov AL, Maruyama K, Torchilin VP, Huang L. Amphipathic polyethyleneglycols effectively prolong the circulation time of liposomes. *FEBS Lett* 1990;**268**:235–7.
145. Torchilin VP. Liposomes as targetable drug carriers. *Crit Rev Ther Drug Carrier Syst* 1985;**2**:65–115.
146. Torchilin VP, Klibanov AL, Huang L, Odonnell S, Nossiff ND, Khaw BA. Targeted accumulation of polyethylene glycol-coated immunoliposomes in infarcted rabbit myocardium. *FASEB J* 1992;**6**:2716–9.
147. Pan XQ, Wang HQ, Lee RJ. Antitumor activity of folate receptor-targeted liposomal doxorubicin in a Kb oral carcinoma murine xenograft model. *Pharm Res* 2003;**20**:417–22.
148. Hatakeyama H, Akita H, Maruyama K, Suhara T, Harashima H. Factors governing the in vivo tissue uptake of transferrin-coupled polyethylene glycol liposomes in vivo. *Int J Pharm* 2004;**281**:25–33.
149. Dufes C, Schätzlein AG, Tetley L, Gray AI, Watson DG, Olivier J-C, et al. Niosomes and polymeric chitosan based vesicles bearing transferrin and glucose ligands for drug targeting. *Pharm Res* 2000;**17**:1250–8.
150. Sawant RR, Torchilin VP. Liposomes as 'smart' Pharmaceutical nanocarriers. *Soft Matter* 2010;**6**:4026–44.
151. Drummond DC, Zignani M, Leroux J-C. Current status of PH-sensitive liposomes in drug delivery. *Prog Lipid Res* 2000;**39**:409–60.
152. Matsuura M, Yamazaki Y, Sugiyama M, Kondo M, Ori H, Nango M, et al. Polycation liposome-mediated gene transfer in vivo. *Biochim Biophys Acta* 2003;**1612**:136–43.
153. Talsma H, Crommelin DJA. Liposomes as drug delivery systems. Part 1. Preparation.. *Pharm Technol* 1992;**16**:96–106.
154. Gregoriadis G. Engineering liposomes for drug delivery: progress and problems. *Trends Biotechnol* 1995;**13**:527–37.
155. Sharma A, Sharma US. Liposomes in drug delivery: progress and limitations. *Int J Pharm* 1997;**154**:123–40.
156. Hwang K. Liposome pharmacokinetics. In: Ostro MJ, editor. *Liposomes: from biophysics to therapeutics*. New York: Marcel Dekker; 1987. pp. 109–56.
157. Harashima H, Sakata K, Funato K, Kiwada H. Enhanced hepatic-uptake of liposomes through complement activation depending on the size of liposomes. *Pharm Res* 1994;**11**:402–6.
158. Mayer LD, Bally MB, Hope MJ, Cullis PR. Techniques for encapsulating bioactive agents into liposomes. *Chem Phys Lipids* 1986;**40**:333–45.

159. Allen TM. Liposomal drug formulations—rationale for development and what we can expect for the future. *Drugs* 1998;**56**:747–56.
160. Haran G, Cohen R, Bar LK, Barenholz Y. Transmembrane ammonium-sulphate gradients in liposomes produce efficient and stable entrapment of amphipathic weak bases. *Biochim Biophys Acta* 1993;**1151**:201–15.
161. Cullis PR, Hope MJ, Bally MB, Madden TD, Mayer LD, Fenske DB. Influence of PH gradients on the transbilayer transport of drugs, lipids, peptides and metal ions into large unilamellar vesicles. *Biochim Biophys Acta* 1997;**1331**:187–211.
162. Barenholz Y. Liposome application: problems and prospects. *Curr Opin Colloid Interface Sci* 2001;**6**:66–77.
163. The Rules Governing Medicinal Products in the European Union. Vol 4: EU Guidelines to Good Manufacturing Practice. Medicinal Products for Human and Veterinary Use, **2005**.
164. ICH Harmonished tripartite guideline. Impurities: Guideline for Residual Solvents Q3C(R4), International conference on harmonisation of technical requirements for registration of pharmaceuticals for human use, **2009**.
165. New RCC. Preparation of liposomes. In: New RCC, editor. *Liposomes: a practical approach*. New York: Oxford University Press; 1990.
166. Bangham AD, Standish MM, Weissmann G. The action of steroids and streptolysin S on the permeability of phospholipid structures to cations. *J Mol Biol* 1965;**13**:253–9, IN28.
167. Mayer LD, Hope MJ, Cullis PR, Janoff AS. Solute distributions and trapping efficiencies observed in freeze-thawed multilamellar vesicles. *Biochim Biophys Acta* 1985;**817**:193–6.
168. Szoka F, Papahadjopoulos D. Procedure for preparation of liposomes with large internal aqueous space and high capture by reverse-phase evaporation. *Proc Natl Acad Sci USA* 1978;**75**:4194–8.
169. Batzri S, Korn ED. Single bilayer liposomes prepared without sonication. *Biochim Biophys Acta* 1973;**298**:1015–9.
170. Deamer D, Bangham AD. Large volume liposomes by an ether vaporization method. *Biochim Biophys Acta* 1976;**443**:629–34.
171. Lasch J, Weissig V, Brandl M. Preparation of liposomes. In: Torchilin V, Weissig V, editors. *Liposomes: a practical approach*. New York: Oxford University Press; 2003.
172. Brunner J, Skrabal P, Hausser H. Single bilayer vesicles prepared without sonication physico-chemical properties. *Biochim Biophys Acta* 1976;**455**:322–31.
173. Kirby C, Gregoriadis G. Dehydration-rehydration vesicles—a simple method for high-yield drug entrapment in liposomes. *Nat Biotechnol* 1984;**2**:979–84.
174. Woodbury DJ, Richardson ES, Grigg AW, Welling RD, Knudson BH. Reducing liposome size with ultrasound: bimodal size distributions. *J Liposome Res* 2006;**16**:57–80.
175. Hope MJ, Bally MB, Webb G, Cullis PR. Production of large unilamellar vesicles by a rapid extrusion procedure—characterization of size distribution, trapped volume and ability to maintain a membrane-potential. *Biochim Biophys Acta* 1985;**812**:55–65.
176. Berger N, Sachse A, Bender J, Schubert R, Brandl M. Filter extrusion of liposomes using different devices: comparison of liposome size, encapsulation efficiency, and process characteristics. *Int J Pharm* 2001;**223**:55–68.
177. Macdonald RC, Macdonald RI, Menco BPM, Takeshita K, Subbarao NK, Hu LR. Small-volume extrusion apparatus for prepration of large, unilamellar vesicles. *Biochim Biophys Acta* 1991;**1061**:297–303.
178. Bachmann D, Brandl M, Gregoriadis G. Preparation of liposomes using a mini-lab 8.30 H high-pressure homogenizer. *Int J Pharm* 1993;**91**:69–74.
179. Pupo E, Padrón A, Santana E, Sotolongo J, Quintana D, Dueñas S, et al. Preparation of plasmid DNA-containing liposomes using a high-pressure homogenization-extrusion technique. *J Control Release* 2005;**104**:379–96.

180. Meure LA, Foster NR, Dehghani F. Conventional and dense gas techniques for the production of liposomes: a review. *AAPS PharmSciTech* 2008;**9**:798–809.
181. Howse JR, Jones RAL, Battaglia G, Ducker RE, Leggett GJ, Ryan AJ. Templated formation of giant polymer vesicles with controlled size distributions. *Nat Mater* 2009;**8**:507–11.
182. Jahn A, Vreeland WN, DeVoe DL, Locascio LE, Gaitan M. Microfluidic directed formation of liposomes of controlled size. *Langmuir* 2007;**23**:6289–93.
183. Tan YC, Hettiarachchi K, Siu M, Pan YP. Controlled microfluidic encapsulation of cells, proteins, and microbeads in lipid vesicles. *J Am Chem Soc* 2006;**128**:5656–8.
184. Jessop PG, Leitner W, editors. *Chemical synthesis using supercritical fluids*. Weinheim: Wiley-VCH; 1999.
185. Sun Y-P, editor. *Supercritical fluid technology in materials science and engineering: syntheses, properties, and applications*. Boca Raton, FL: CRC Press; 2002.
186. Wells SL, DeSimone J. Co_2 technology platform: an important tool for environmental problem solving. *Angew Chem Int Ed* 2001;**40**:519–27.
187. Woods HM, Silva M, Nouvel C, Shakesheff KM, Howdle SM. Materials processing in supercritical carbon dioxide: surfactants, polymers and biomaterials. *J Mater Chem* 2004;**14**:1663–78.
188. Johnston KP, Shah PS. Materials science—making nanoscale materials with supercritical fluids. *Science* 2004;**303**:482–3.
189. Sala S, Elizondo E, Moreno E, Calvet T, Cuevas-Diarte MA, Ventosa N, et al. Kinetically driven crystallization of a pure polymorphic phase of stearic acid from Co_2-expanded solutions. *Crystal Growth Des* 2010;**10**:1226–32.
190. Otake K, Imura T, Sakai H, Abe M. Development of a new preparation method of liposomes using supercritical carbon dioxide. *Langmuir* 2001;**17**:3898–901.
191. Castor TP. *Methods and Apparatus for Making Liposomes.* 1994, WO9427581.
192. Anton K, Van Hoogevest P, Frederiksen L. *Preparation of a liposome dispersion containing an active agent by compression-decompression.* 1994, EP616801.
193. Frederiksen L, Anton K, vanHoogevest P, Keller HR, Leuenberger H. Preparation of liposomes encapsulating water-soluble compounds using supercritical carbon dioxide. *J Pharm Sci* 1997;**86**:921–8.
194. Meure LA, Knott R, Foster NR, Dehghani F. The depressurization of an expanded solution into aqueous media for the bulk production of liposomes. *Langmuir* 2009;**25**:326–37.
195. Chu L, Castor TP. *Method of forming liposomes containing hydrophobic drugs—comprising forming solution or mixture of phospholipid, drugs and aqueous phase in fluid.* 1998, US5776486-A.
196. Castor TP. Phospholipid nanosomes. *Curr Drug Deliv* 2005;**2**:329–40.
197. Otake K, Shimomura T, Goto T, Imura T, Furuya T, Yoda S, et al. Preparation of liposomes using an improved supercritical reverse phase evaporation method. *Langmuir* 2006;**22**:2543–50.
198. Otake K, Shimomura T, Goto T, Imura T, Furuya T, Yoda S, et al. One-step preparation of chitosan-coated cationic liposomes by an improved supercritical reverse-phase evaporation method. *Langmuir* 2006;**22**:4054–9.
199. Elizondo E. *New molecule-based nanostructured drug carriers prepared using compressed fluids.* 2010, Ph.D Thesis, Autonomous University of Barcelona, Bellaterra.

Luminescent Quantum Dots, Making Invisibles Visible in Bioimaging

Mitsuru Ishikawa and Vasudevanpillai Biju

National Institute of Advanced Industrial Science and Technology (AIST), Hayashi-cho, Takamatsu, Kagawa, Japan

I. Introduction 54
A. QDs, Go-Between Nanotechnology, and Biotechnology 54
B. Advantages of QDs Over Fluorescent Dyes in Bioimaging 55
II. Preparation of QDs for Bioconjugation 57
A. Synthesis of Core QDs 57
B. Formation of Shells for Bioconjugation 59
III. Labeling Cell Surfaces with QDs 60
A. Nonspecific Labeling of Cell Surfaces 60
B. Specific Labeling of Cell Surfaces 62
IV. Delivery of QDs in Living Cells 67
A. Nonspecific Delivery 68
B. Specific Delivery 78
V. Photodynamic Therapy by Quantum Dots 82
A. ROS, Key to PDT 83
B. QDs for PDT of Cancer 84
C. QDs Versus Conventional PS Drugs for PDT 85
D. QD and Photosensitizer Hybrids as Drugs for PDT 87
VI. Concluding Remarks 92
References 92

Semiconductor quantum dots (QDs), their size-dependent tunable absorption and emission in the visible and near-infrared regions, narrow emission and broad absorption bands, large one- and multiphoton absorption cross sections, brightness, and photostability make them powerful light sources in the nanoscale region that will help to advance biotechnologies. This chapter overviews the present status and prospects of QDs by focusing on bioconjugation of QDs, and applications of bioconjugated QDs to surface-cell as well as inside-cell labeling for imaging and photodynamic therapy (PDT). In particular, it highlights the synthesis of biocompatible QDs, surface modification of QDs, conjugation of QDs with a variety of ligands and antibodies, potential targets of bioconjugated QDs in cells, nonspecific and specific surface-cell labeling, nonspecific and specific delivery of QDs in cells, and PDT based on QDs.

Progress in Molecular Biology
and Translational Science, Vol. 104
DOI: 10.1016/B978-0-12-416020-0.00002-4

Such information will be helpful for both applications of QDs for bioimaging and extension of QD technologies to other nanoparticles.

I. Introduction

Nanomaterials are fundamental building blocks for modern scientific and technological excellence. As the size of inorganic and organic materials becomes smaller and smaller down to the nanometer scale, their optical and electronic properties largely vary from those in the bulk and become size- and shape-dependent ones. Such properties of nanomaterials are the keys to the excellence of nanoscience and nanotechnology. The basic concepts and emerging applications of semiconductor nanoparticles, or quantum dots (QDs), make them excellent nanomaterials. Systematic advancements in the science and technology of QDs started in 1984. Brus derived a close relationship between size and band gap for semiconductor nanoparticles by applying a sphere model approximation to the wave function for bulk semiconductors.[1] Following this substantial achievement, the quantum confinement of charge carriers in nanoparticles was verified by applying top-down and bottom-up approaches to obtain nanoparticles. However, it took nearly a decade for a breakthrough in the nanoparticle research until Murray *et al.* accomplished the synthesis of colloidal CdX (X = S, Se, and Te) QDs having size-tunable band-edge absorption and emission.[2] Through the systematic refinement of various parameters in nanocrystal growth by many researchers, synthesis of QDs is now established. CdX has so far widely been investigated owing to the wide availability of precursors, well-defined technology of crystal growth, and size-tunable fluorescence in the ultraviolet (UV), visible, and near-infrared (NIR) regions. As a result of the widespread and rapidly growing applications of QDs, comprehensive reviewing from theories to technologies is difficult and not attempted here. Instead, the focus is on surfing QD applications in bioimaging and its related research subjects.

A. QDs, Go-Between Nanotechnology, and Biotechnology

Nanobiotechnology has rapidly advanced in recent years by the integration of nanomaterials with biotechnology. Now, bioconjugated QDs are indispensable for extended imaging of cells and the structures and functions of subcellular organelles. Thus, synthesis, optimization of optical properties, and bioconjugation of QDs have emerged into a great research area. A definite demand for new materials and methodologies is continuously asked to bridge the gap between nanomaterial technology and biotechnology. Figure 1 shows key steps in the integration of nanomaterials with biotechnology. Soon after colloidal synthesis of nanoparticles turns easily available, a variety of size- and

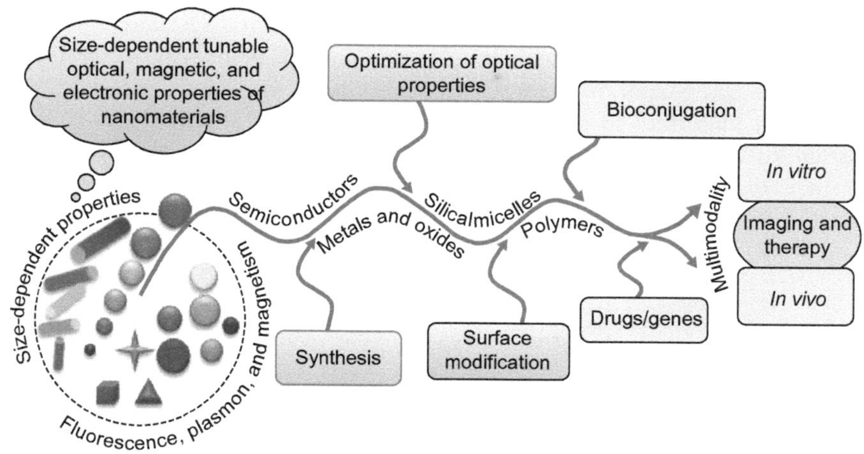

FIG. 1. Key steps in interfacing nanomaterials with biotechnology and biomedical applications. (For color version of this figure, the reader is referred to the Web version of this chapter.)

shape-controlled nanomaterials from various chemical precursors was developed. In parallel with QDs, inorganic, polymer, and hybrid nanomaterials for multimodal imaging and drug delivery were developed by utilizing the technology of shell growth and chemistry of bioconjugation in QDs. By the fine-tuning of all these steps, it is now possible to formulate bioconjugated nanomaterials of choice and test specific multimodal imaging and therapy of major diseases such as cancer. Among various nanomaterials, QDs attracted much attention in nanobiotechnology and its biomedical applications owing to the wide availability of precursors, straightforward synthesis, and unique optical properties. In particular, biosensing, drug delivery, and *in vitro* and *in vivo* imaging are the most benefited areas from bioconjugated QDs.

B. Advantages of QDs Over Fluorescent Dyes in Bioimaging

Fluorescent dye molecules are basic tools to make visible the structures and functions of biomolecules and cells. However, despite the small size and widespread availability in various colors, organic dyes suffer from limitations, such as narrow excitation bands, small Stokes shifts, broad fluorescence bands, and photobleaching. From these limitations, dye molecules are less attractive for durable, sensitive, and multiplexed imaging. Semiconductor QDs, on the other hand, are exceptionally bright and stable against photobleaching. Other favorable optical properties of QDs are size-dependent tunable absorption and emission in the visible and NIR regions, narrow emission and broad absorption bands, and large one- and multiphoton absorption cross sections.[2–5] These unique optical properties come from a combination of bulk semiconductor properties and quantum confinement of excitons. For most materials, the dimension important to quantum confinement is on the scale of 2–10 nm. Luminescence from near UV to NIR regions are obtained by simply varying the size of the quantum confinement. Figure 2 shows representative size-tunable absorption and luminescence spectra of CdSe/ZnS QDs. The advantage of size-tunable luminescence is that the methods for surface modification

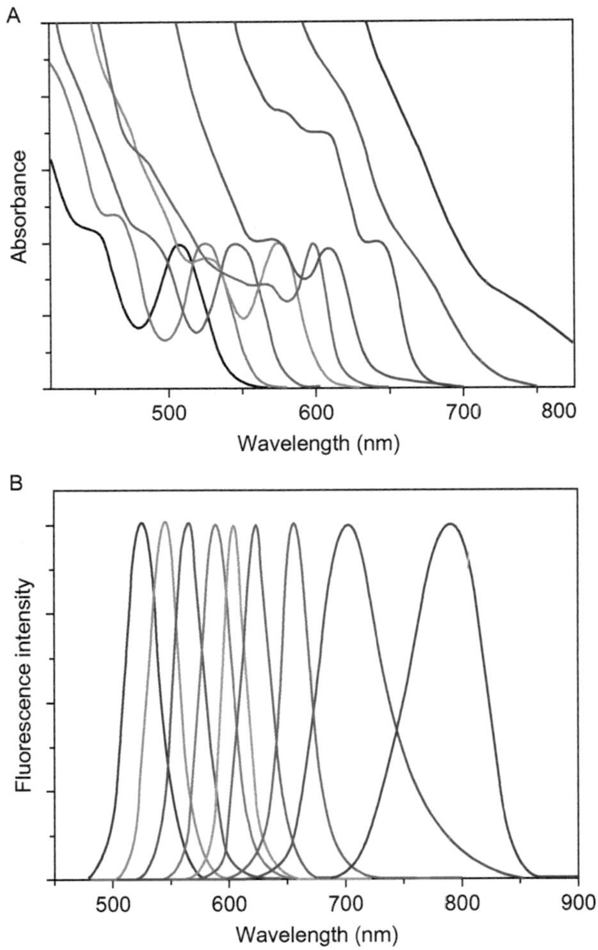

FIG. 2. (A) Absorption and (B) luminescence spectra of CdSe/ZnS core/shell QDs: from left to right. QD525, QD545, QD565, QD585, QD605, QD625, QD655, QD705, and QD800. Data were kindly provided by Invitrogen Corporation. (See Color Insert.)

or bioconjugation are commonly applied to obtain multicolor probes. Broad absorption bands of QDs provide two advantages: one is freedom to select any excitation wavelength below the band gap energy; the other is minimum background favored by large Stokes shifts. Narrow luminescence bands of QDs are advantageous for minimizing bleed during multiplexed imaging. Bright and stable luminescence of QDs permits extended and sensitive bioimaging even at the single-molecule level. However, blinking in luminescence[6–8] is the limitation on the advancement in QD technology for single-molecule imaging. Completely nonblinking ternary core/shell CdZnSe/ZnSe QDs were recently synthesized by alloying core CdZnSe into shell ZnSe.[9]

A large two-photon absorption cross section 10^3–10^4 in Goeppert–Mayer unit[10] of QDs is utilized for minimizing background by selecting NIR excitation of visible QDs. The advantage of NIR excitation over UV and visible excitation is that it provides deeper tissue penetration. This property is preferable for *in vivo* imaging[3,11–14] and photodynamic therapy (PDT) (Ref. 15 and Section V). In addition to these gifted optical properties, the large surface area of QDs is likely to be useful for multiple bioconjugation and the preparation of multifunctional and multimodal probes. The unique optical and structural properties formulate QDs to be ideal alternatives for organic dyes, in particular, for multimodal and multiplexed imaging of single molecules, cells, tissues, and animals. Additional information on the optical properties of QDs relevant to bioimaging will be obtained from recent review articles.[10–15]

II. Preparation of QDs for Bioconjugation

Preparation of bioconjugated QDs is a prerequisite for bioimaging. Selection of a biomolecule on the surface of QDs depends on a particular application, such as biosensing, surface-cell labeling, inside-cell delivery, inside-cell labeling, and *in vivo* imaging. Reactive functional groups or molecules, such as streptavidin, biotin, primary amine, thiol, maleimide, succinimide, or carboxylic acid, are essential to biolabeling on the surface of core or core/shell QDs. Without particular functional groups, bioconjugated QDs are prepared by noncovalent coating of peptides, proteins, or liposomes on the QD surface. Here, the preparation of core-and-shell QDs is outlined before a comprehensive description of bioconjugated QDs and their biological applications in Sections III–V.

A. Synthesis of Core QDs

In 1993, Murray *et al.* achieved a breakthrough in the basic research and applications of QDs by developing colloidal synthesis of cadmium chalcogenide (CdX, X = S, Se, and Te) QDs.[2] In its key reaction, CdSe QDs were synthesized by the pyrolysis of dimethyl cadmium ($CdMe_2$) and trioctylphosphine selenide (TOPSe) in a mixture of coordinating solvents, namely TOP and its oxide (TOPO). The synthesis was carried out in an inert atmosphere by injecting cadmium and selenium precursors dissolved in TOP into hot (300 °C) TOPO, followed by growing the nanocrystals at ~230–260 °C. The synthesis provided size-distributed QDs showing white luminescence. Narrow size-distributed QDs showing each luminescence color were isolated from size-distributed QDs by size-selective precipitation using a mixture of methanol and *n*-butanol. This approach to the synthesis of CdX QDs established the fundamentals of colloidal synthesis of various QDs. CdS or CdTe QDs also were prepared by simply replacing TOPSe with bis(trimethylsilyl) sulfide or TOPTe. Katari *et al.* refined the classic method for the synthesis of CdX QDs by selecting suitable injection temperature and nanocrystal growth temperature, and obtained size-selected QDs.[16] Although the use of pyrophoric and volatile $CdMe_2$ was an occupational limitation, these two methods are widely accepted for the synthesis of CdX QDs even today. Peng *et al.*[17] and Talapin *et al.*[18] significantly improved the synthesis of CdX QDs by supplementing the solvent mixture (TOPO and TOP) with alkylphosphonic acids or alkyl amines. The use of $CdMe_2$ was replaced by nonvolatile cadmium precursors in the synthesis of high-quality and size-selected CdX QDs. Vossmeyer *et al.* accomplished the first alternative synthesis of colloidal QDs by replacing $CdMe_2$ with cadmium perchlorate ($CdClO_4$). This achievement is the first report on colloidal synthesis of QDs in aqueous phase.[19] Peng and coworkers significantly contributed to the synthesis of CdX QDs by introducing alternative cadmium precursors, such as cadmium chloride ($CdCl_2$), cadmium oxide (CdO), cadmium acetate ($CdAc_2$), and cadmium carbonate ($CdCO_3$).[17,20,21] Synthesis of CdX QDs is now simplified and optimized by varying the precursors, temperatures, and chelating ligands.[3,21–24]

Among CdX QDs, CdSe and CdTe have attracted much attention in bioimaging owing to their tunable and stable luminescence in the visible to NIR region. Gaponik *et al.* accomplished the synthesis of biocompatible CdTe QDs by passing H_2Te gas through an aqueous solution of cadmium perchlorate hexahydrate ($CdClO_4{\cdot}6H_2O$) and thioglycolic acid (TGA) kept at 100 °C and pH ~11.5.[22] InP QDs are also highly suitable for bioimaging applications, although optical properties of CdSe and CdTe QDs are excellent compared with other fluorophores. Due to their nontoxic nature, InP QDs promise well for *in vivo* imaging and PDT. InP QDs are synthesized by modifying the high temperature synthesis for CdX QDs, which is described in the previous paragraph. Bharali *et al.* synthesized high-quality InP QDs by injecting tris (trimethylsilyl) phosphine into a dispersion of indium (III) myristate [$In(Mt)_3$] in 1-octadecene at ~280 °C.[23] Following the injection, the reaction mixture was diluted with 1-octadecene, and then QDs were grown at 180 °C. This synthesis provided hydrophobic-capped InP QDs in good yield. Detailed procedures for the synthesis of various QDs suitable for biological applications are provided in recent review articles[3,22] and the references therein.

B. Formation of Shells for Bioconjugation

Core-only QDs are less attractive materials for bioconjugation and bioimaging. Indeed, the core surface is poorly accessible to bioconjugate reactions. Also, the optical properties of core-only QDs are highly unstable against chemical reactions. More importantly, the surfaces of most as-synthesized QDs are hydrophobically capped. This fact makes them insoluble in aqueous solutions. Thus, surface modification of QDs is the fundamental procedure for biological applications and divided into two steps: preparing shells on core-only QDs; and modifying the surface of shells on QDs with biocompatible molecules.

Shells on core-only QDs are multipurpose. In general, shells protect the core against surface oxidation as well as leaching and dissolution in the form of toxic ions, and serve as platforms for ligand exchange and bioconjugate reactions. Indeed, shells from suitable semiconductor materials preserve and improve the optical properties of the core. Reiss *et al.* summarized the preparation of semiconductor shells on core-only QDs.[24] Their analysis indicates that in type-I and type-II core/shell QDs, such as CdS/ZnS, CdSe/ZnS, CdTe/ZnS, InP/ZnS, CdSe/CdS, CdTe/CdS, CdTe/CdSe, CdS/ZnSe, and CdSe/ZnTe, shells not only serve as protecting layers but also provide improved luminescence quantum efficiencies and red-shifted luminescence spectra. In particular, higher band-gap materials are preferred for shell preparation to minimize surface defects and improve optical properties of the core.[3,18,24] Likewise, thiols,[22,25] polymers,[12,26] or silica coatings on QDs serve as protecting layers and platforms for bioconjugation.

Bioconjugated core/shell QDs become fundamentals of imaging of molecules, cells, tissues, and animals.[3,11,12,15,27,28] Covalent and noncovalent conjugation of QDs with antibodies, proteins, peptides, aptamers, nucleic acids, small molecules, and liposomes are bioconjugated QDs. They are widely used for direct and indirect labeling of surface-cell proteins and subcellular organelles. Many examples of such applications are described in Sections III and IV of this chapter. Indeed, bioconjugated QDs are ideal substitutes for organic dyes when photostability or multiplexing is a requirement and excitation laser source is a limitation. Furthermore, QDs offer a spacious and flexible surface for the conjugation of multiple tags. Advantages of QDs over organic dyes for bioimaging are discussed in Section I.B.

III. Labeling Cell Surfaces with QDs

Labeling of cells with bioconjugated QDs can be classified as nonspecific or specific. While nonspecific labeling considerably contributes to an interface between biology and QDs, specific labeling offers specific imaging of the structures and functions of subcellular molecules and organelles. For example, specific imaging of cells and tissues using QDs conjugated with anticancer antibody and peptides has considerably improved the detection limit of cancer. Michalet *et al.*,[11] Walling *et al.*,[27] and Smith *et al.*[28] summarized *in vivo* applications of QDs. Conjugation of QDs with specified tags enables the imaging of surface-cell receptors, cargo transport, gene and drug delivery, membrane dynamics, cancer cells, and embryonic development. This section summarizes first nonspecific and then specific surface-cell labeling.

A. Nonspecific Labeling of Cell Surfaces

Nonspecific binding of QDs on cell membranes was initially exploited for surface-cell labeling followed by the delivery of QDs in cells. It is hostile to high-contrast imaging of cells and worsens the efficiencies in specific surface-cell labeling and delivery. Nonspecific binding of QDs on cell membranes occurs owing to hydrophobic and electrostatic interactions between capping molecules on the surface of QDs and biomolecules on the cell membrane. For this reason, surface chemistry of QDs and the type of cells play a crucial role in the nonspecific adsorption of QDs.

1. Electrostatic Nonspecific Adsorption

Jaiswal *et al.* identified considerable nonspecific adsorption of dihydrolipoic acid (DHLA)-capped CdSe/ZnS QDs on HeLa cells at 4 °C.[29] However, negative charge of DHLA does not support charge-based adsorption of QDs on a negatively charged cell membrane. Similarly, Gomez *et al.* detected strong adsorption of TGA-capped CdS QDs on neuroblastoma cells (SK-N-SH) under the physiological pH (6.8–7.0).[30] The extent of QD adsorption increased with particle size. A possible relationship between size of QDs and their nonspecific adsorption on a cell membrane is that larger QDs carry more ligands. They induce electrostatic interaction between them and a cell membrane, in particular, under acidic conditions. However, the potential binding of DHLA- and TGA-capped QDs on a cell membrane at the physiological pH remains an open question. Analogous to these observations is the nonspecific binding of zwitterionic molecules, such as D-penicillamine on human monocytic cells.[31] Gomez *et al.* found that compared with SK-N-SH cells, the degree of nonspecific adsorption of CdS QDs was low for rat adrenal pheochromocytoma cells (PC 12) and neonatal cortical cells.[30] The difference in the degree of nonspecific adsorption depends on cell culture medium, cell membrane composition, and excreted surface-cell matrix components. Benzen *et al.* found that CdSe/ZnS QDs coated with amphiphilic poly(acrylic acid) (AMP) nonspecifically are bound on human epithelial kidney (HEK) cells to a greater extent than to mouse fibroblast cells (NIH3T3).[26] Like polymer-capped QDs, QDs coated with cationic lipids result in their electrostatic binding on a cell membrane. Murcia *et al.* exploited nonspecific binding of phospholipid-conjugated CdSe/ZnS QDs for single-molecule imaging and tracking of QDs in phospholipid bilayers, or in the membranes of normal rat kidney fibroblasts, African Green Monkey SV40-transfected kidney fibroblasts (COS 7), HEK 293 cells, or embryonic mouse fibroblasts (Swiss 3T3).[32] In addition, QDs conjugated with other molecules, such as silanes, amines, cationic peptides, carboxylic acids, proteins, polymers, and surfactants, are bound on cell membranes, thereby indicating that multiple modes of interaction are possible in the nonspecific adsorption of QDs on cells. Figure 3 shows representative examples of nonspecific interactions between QDs and cell membranes.

2. Nonspecific Adsorption as Annoyance

Nonspecific surface-cell binding of QDs is advantageous in the sense that cells are labeled without labor-intensive and time-consuming bioconjugation. However, the overall strength of nonspecific binding is mostly unknown, leaving a possibility that QDs dissociate with changes in pH, temperature, as

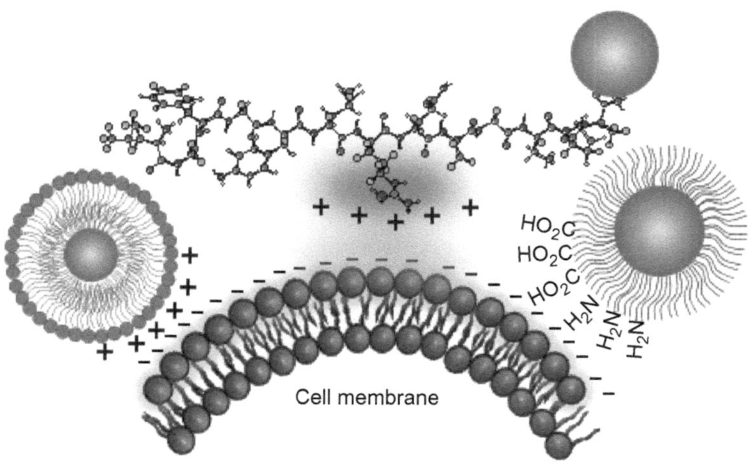

FIG. 3. Schematic of nonspecific interaction and adsorption of QDs coated with peptides, lipids, liposomes, amines, and carboxylic acids on a cell membrane. (For color version of this figure, the reader is referred to the Web version of this chapter.)

well as chemical and biochemical environment. Surface chemistry of QDs and bioconjugate chemistry have well advanced; thus, selective labeling of surface-cell molecules and subcellular organelles is not a concern anymore. In other words, boundaries among QDs, bioconjugate chemistry, information about target biomolecules in cells, and cell biology have rapidly advanced, and nonspecific adsorption of QDs on cells has turned into an undesirable property. For example, Ryman-Rasmussen *et al.*[33] and Mortensen *et al.*[34] showed that dermal contact and subsequent uptake of QDs, the extent of which depends on the size and surface chemistry of QDs, are crucial to occupational safety. Several researchers recognized that coating or conjugation of polyethylene glycol (PEG) molecules on the surface of QDs is effective in minimizing or preventing nonspecific interactions of QDs with biomolecules, cells, and tissues. For example, Bentzen *et al.* found that nonspecific binding of AMP-coated CdSe/ZnS QDs on HEK and NIH3T3 cells is suppressed by 18% and 90%, respectively, by overlaying QD-AMP with PEG.[26] They also found that nonspecific binding of QD-AMP on cells is suppressed by increasing the amount of PEG. However, if the construction of QD–PEG conjugates is not perfect, a PEG layer acts as a barrier against the delivery of the conjugates in cells. Other molecules, such as cationic lipids, hold QDs in the cell membrane. For example, Gopalakrishnan *et al.* found an interplay between membrane labeling and delivery of CdSe QDs coated with two lipid constructs.[35] In construct I, QDs were encompassed in vesicles composed of 1,2-dimyristoyl-*sn*-glycero-3-phosphatidylcholine (DMPC) and 1,2-dioleoyl-3-trimethylammonium propane chloride salt (DOTAP). In construct II, QDs were encompassed in vesicles composed of DMPC, DOTAP, and 1,2-distearoyl-*sn*-glycero-3-phosphatidyl ethanolamine-*N*-(biotinyl-PEG2000). Although both the constructs, through the presence of 25% cationic DOTAP, were expected to deliver QDs inside the cytosol, construct II was stuck in the cell membrane owing to a barrier created by PEG molecules for direct interaction between liposome and cell membrane. Nonetheless, systematic coating of QDs with PEG has firmly been established, and the potential of PEG to prevent nonspecific adsorption of QDs on proteins, cells, and tissues is widely appreciated.[12,26,35–41]

B. Specific Labeling of Cell Surfaces

Specific or targeted labeling of cell membrane using QDs will be accomplished by targeting membrane proteins and receptors with QD–antibody or QD–ligand conjugates. Section III.B summarizes specific surface-cell labeling following the type of antibodies and ligands conjugated with QDs and that of target proteins and receptors in the cell membrane. Note that specific surface-cell labeling is not completely independent of specific delivery of QDs in cells. Indeed, most membrane proteins and receptors, which just give the cells specificity, are taken and recycled by the cells. For this reason, surface-cell labeling should be considered as an early event in specific delivery of bioconjugated QDs.

1. Antibodies and Secondary Antibodies

Site-specific surface-cell labeling is essential in unraveling the function of cells and membrane proteins. For this purpose, antibodies are an ultimate tool for the highly specific detection of target proteins on a cell membrane. Membrane proteins in living cells are labeled either directly with QD–antibody conjugates or indirectly with QD–secondary antibody conjugates. Indeed, labeling of cells with QD–antibody and QD–secondary antibody conjugates means taking advantage of the unique optical properties of QDs by replacing fluorescent dyes in standard immune fluorescence methods. Chan and Nie tested the selectivity and effectiveness of QD–antibody conjugates for antigen detection by binding CdSe/ZnS QD–human immunoglobulin G (IgG) conjugate on a polyclonal serum, but without any cells.[25] Winter *et al.* first investigated applications of QD–antibody conjugates to cell labeling and imaging, targeting α_v portion of the $\alpha_v\beta_1$ and $\alpha_v\beta_3$ integrins in human neurons (SK-N-SH).[42] Antibody- and secondary antibody-based labeling of cells using QDs has widely been appreciated for cell imaging, such as simple imaging of cells,[29,43] selective detection followed by imaging of overexpressed receptors in cancer cells,[12,44–52] discriminating cancer cells from each other,[12,45,47] investigating the spatiotemporal distributions of receptors at the ensemble and single-molecule level,[51,53,54] and analyzing the growth and differentiation of cells.[29,55] Surface-cell labeling using antibodies can be classified into four categories (Fig. 4).

Jaiswal *et al.* showed the usefulness of primary antibody for surface-cell labeling using QDs by labeling the surface-cell epitope of the multidrug transporter P-glycoprotein in HeLa cells.[29] Jayagopal *et al.* showed that QDs conjugated with monoclonal antibodies for cell adhesion molecules (CAMs), such as platelet endothelial cell adhesion molecule-1, intercellular adhesion molecule-1, and vascular cell adhesion molecule-1, are ideal for surface-cell labeling as well as fluorescence imaging of cells.[39] The QD–antibody conjugates were selectively bound on CAMs on the rat endothelial cells stimulated by tumor necrosis factor (TNF)-α, leukocytes, or retinal endothelium. In contrast to these generalized approaches for cell labeling, QDs conjugated with antibodies against the Her2 receptor, overexpressed in many human breast cancer cell lines such as SK-BR-3,[44,45,52] MCF 7,[45,46,48] BT 474,[46] and MDA-MB-231,[46] have extensively been utilized for selective surface-cell labeling. Wu *et al.* accomplished selective labeling of Her2 receptor in SK-BR-3 cells by treating the cells with trastuzumab (herceptin), an anti-Her2 antibody, followed by QD–IgG conjugates.[44] Alternatively, SK-BR-3 cells were incubated sequentially

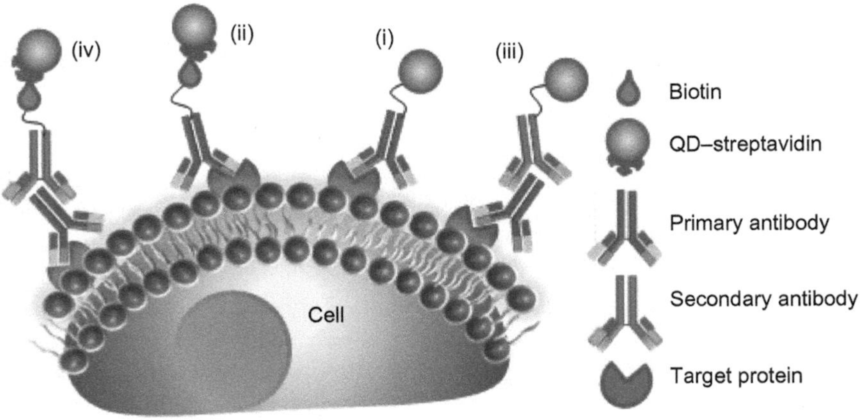

FIG. 4. Schematic of antibody-based labeling of cells with QDs: (i) direct labeling of a target protein with a QD–primary antibody conjugate, (ii) labeling of a target protein with a biotinylated primary antibody followed by a QD–streptavidin conjugate, (iii) labeling of a target protein with a primary antibody followed by a QD–secondary antibody conjugate, and (iv) sequential labeling of a target protein with a primary antibody, biotinylated secondary antibody, and QD–streptavidin conjugate. (For color version of this figure, the reader is referred to the Web version of this chapter.)

with humanized anti-Her2 antibody, biotinylated antihuman IgG, and QD–streptavidin conjugates. Conjugates of QDs and antibodies were also tested for labeling prostate-specific membrane antigens (PSMA), prostate stem cell antigens (PSCA), and Her2 receptors, all of which are overexpressed in human prostate cancer cells, such as C4-2 and LNCaP.[12,48] Gao *et al.* accomplished selective surface-cell labeling of PSMA-positive C4-2 cells using QDs conjugated with monoclonal anti-PSMA antibody. However, the conjugates did not recognize PSMA-negative human prostate adenocarcinoma cells (PC 3) and QD alone labeled C4-2 or PC 3 cells.[12] Barat *et al.* successfully labeled LNCaP cells by simultaneously targeting PSCA and Her2 receptors using NIR QDs conjugated with PSCA antibody and anti-Her2 antibody fragments (anti-Her2 cys-diabody).[48] This report suggests that both small fragments of antibodies and multiple targets can be exploited for effective detection of cancer cells. Similarly, Yezhelyev *et al.* and Kawashima *et al.* accomplished surface-cell labeling of cancer cells, such as MCF 7, BT 474, and MDA-MB-231,[46] and A431 using a combination of QDs and anti-EGFR antibody.[51] Other examples for surface-cell labeling of cancer cells using QD–antibody combinations targeted to the CD56 receptor in natural killer cells (NK92MI) using QD–anti-CD56 antibody conjugates[49] discriminate the level of epidermal growth factor receptor (EGFR) and E-cadherin (E-cad) in cancer cells derived from non-small cell lung carcinoma and non-squamous cell carcinoma of the head and neck by labeling the cells with anti-EGFR or anti-E-cad antibody followed by QD605–IgG conjugates.[47] Zhang *et al.* realized that CdTe QDs conjugated with an antibody (AVE-1642) to type 1 insulin-like growth factor receptor (IGF1R) is ideal for selective detection of overexpressed IGF1R in MCF 7 cells.[50]

Beside the reports in this section on labeling and imaging of cells, QD–antibody conjugates are widely utilized for investigating the functioning of cells and biomolecules as well. Dahan *et al.* showed that QD–antibody conjugates are utilized for understanding the diffusion dynamics of glycine receptors (GlyRs) in neurons at single-molecule resolution.[53] Blinking fluorescence of QDs was adopted as a marker for single QD–GlyR conjugates. Owing to the exceptional photostability of QDs, lateral diffusion of GlyRs between synaptic and extrasynaptic domains was realized by extended imaging at single-molecule level. Similarly, Chen *et al.* showed that photostability of QDs is favorable for extended imaging and analysis of integrin dynamics during osteogenic differentiation of human bone marrow-derived progenitor cells.[55] Bouzigues *et al.* utilized QD–antibody conjugates for real-time single-molecule imaging of gamma-aminobutyric acid (GABA) (A) receptors in rat spinal neuron membrane cells.[54] The γ_2 subunit of the GABA receptor was labeled with QDs by treating neurons sequentially with guinea pig anti-γ_2 antibody, biotinylated anti-guinea pig antibody, and QD605–streptavidin conjugates. Photostability of QDs facilitated the detection of transient interaction between receptors and microtubules as well as asymmetric receptor redistribution across growth cones. Although antibodies offer efficient labeling of surface-cell proteins and receptors, cells gradually uptake QD–antibody–receptor complexes by receptor-mediated endocytosis. Thus, many reports related to labeling of cells using QDs and antibodies are further discussed for specific delivery of QD–antibody conjugates in cells.

2. Ligands for Fusion Proteins

Despite the selectivity of QD–antibody conjugates for specific labeling of cells, antibodies are expensive. Furthermore, antibody-conjugated large nanoparticles, such as QDs, are not equally active as in their native state. Thus, alternative methods for specific surface-cell labeling were sought. A valuable alternative is the fusion of a 15-amino acid acceptor peptide (AP) to any target proteins followed by selective biotinylation of the lysine side chain in AP using the *Escherichia coli* enzyme biotin ligase (BirA). Biotinylated AP is labeled with QD–streptavidin conjugate. Thus, AP-BirA combination has been widely accepted for specific labeling of membrane proteins. Howarth *et al.* introduced this approach to the QD application by fusing AP with cyanine fluorescence protein (CFP) or EGFR in HeLa cells.[56] They also validated the versatility of this method by labeling the tetrameric a-amino-3-hydroxy-5-methyl-4-isoxasolepropionate (AMPA) glutamate receptor (GluR2) in hippocampal neurons at the ensemble and single-molecule levels. For this labeling, the neurons were first cotransfected with AP–GluR2, and the synaptic marker postsynaptic density-95 was fused with yellow fluorescent protein (YFP). The AP was then biotinylated using BirA and successively labeled with QD–streptavidin. Finally, the AP–BirA combination was utilized for labeling QDs to various membrane proteins, such as glutamate receptors, neuroligin, and EGFR in HeLa cells,[57] and type I interferon receptors in COS 7 cells.[58]

Like the AP–BirA pair, selective labeling of fusion proteins, such as cutinase, polyhistidine, and HaloTag, using corresponding ligands is an alternative for antibody-based labeling. Bonasio *et al.* accomplished the fusion of cutinase to integrin lymphocyte function-associated antigen 1 (LFA-1) in K562 cells, and then specific cutinase using QDs conjugated with *p*-nitrophenyl phosphonate (pNPP).[59] Here, selective binding of pNPP on LFA-1 was used for real-time imaging of LFA-1-assisted cell migration. Similarly, Kim *et al.* accomplished selective labeling of a hexahistidine sequence genetically fused to the surface-cell domain of 5HT2C serotonin receptor in HEK-293 cells.[60] In this approach, they used a QD–Ni–nitrilotriacetic acid (QD–Ni–NTA) conjugate as a ligand for hexahistidine. The key to the Ni–NTA-based targeting is that Ni ions form a complex with polyhistidine. Roullier *et al.* and Dif *et al.* applied the Ni–NTA-based targeting method for labeling a decahistidine unit fused with CFP in HeLa cells[58] or polyhistidine-tagged (histag) proteins inherent to HeLa cells.[61] In an analogous approach, So *et al.* accomplished surface-cell labeling of living COS 7 cells with QDs by expressing an engineered bacterial enzyme, haloalkane dehalogenase, also called HaloTag protein, followed by incubating the cells with QD–HaloTag conjugates.[62]

3. Peptides for Membrane Proteins

Peptides are cost-effective and stable ligands for selective and efficient labeling of cells. However, the availability of peptides for targeting inside-cell organelles is limited. Peptides containing arginine–glycine–aspartic acid (RGD) sequence offer selective labeling of $\alpha_v\beta_1$ and $\alpha_v\beta_3$ integrins. They are overexpressed in angiogenetic endothelial cells and many cancer cells including metastatic cancer cells. Winter *et al.*[42] and Mulder *et al.*[63] utilized QD–RGD conjugates for selective labeling of $\alpha_v\beta_3$ integrins in SK-N-SH cells and human umbilical vein endothelial cells (HUVEC). In this labeling scheme, RGD peptides were found to be as good as a combination of anti-CD51 antibody and QD–IgG conjugates.[42] Cai *et al.* selectively detected cancer cells using CdTe/ZnS QD–RGD conjugates by targeting and imaging overexpressed $\alpha_v\beta_5$ and $\alpha_v\beta_3$ integrins in human breast cancer cells (MDA-MB-435) and human glioblastoma cells (U87MG), not MCF 7 cells, simply because they do not express $\alpha_v\beta_5$ or $\alpha_v\beta_3$ integrins.[13] Akerman *et al.* accomplished surface-cell labeling using QDs conjugated with specific peptides, such as CGFECVRQCPERC (GFE) for targeting the membrane dipeptidase in endothelial cells, KDEPQRRSARLSAKPAPPKPEPKPKKAPAKK (F3) for targeting blood vessels and tumor cells, and CGNKRTRGC (LyP-1) for targeting lymphatic vessels and tumor cells.[14] With the QD–GFE conjugates, they selectively labeled lung endothelial cells. On the other hand, QD–GFE and QD–LyP-1 were not effective in labeling lung endothelial cells. Similarly, using QD–F3 and QD–LyP-1 conjugates, not QD–GFE, they labeled MDA-MB-435 cells. In another example, for peptide-based labeling, Pinaud *et al.* showed that QDs conjugated with a series of synthetic peptides related to phytochelatin target certain surface-cell proteins.[64] Thus, although these synthetic peptides bridge the gap between QDs and cell surfaces, this method is equivalent to biotin–avidin-based labeling. While the potentials of various peptides for delivery of QDs in cells and other nanoparticles are being investigated more and more, QD–RGD conjugates rank top in their efficiency in labeling.

IV. Delivery of QDs in Living Cells

Delivery of QDs in cells can be divided into two classes: one is applied by the hands; the other is adopted by the cells themselves. Applied methods imply forced delivery of QDs into the cytosol or the nucleus by transiently destabilizing the cell and nuclear membrane. Applied methods include micro- and nanoinjection, electroporation, and osmotic lysis. Adopted methods follow the general routes of endocytosis, such as phagocytosis, macropinocytosis, clathrin-dependent endocytosis, caveolae-dependent endocytosis, and routes other than clathrin—or caveolae-dependent endocytosis. Section IV is exclusively involved in the adopted methods according to nonspecific or specific delivery, which depends on the surface chemistry of QDs. Before moving on to the core subjects, the process of endocytosis is briefly described.

Phagocytosis is the process of swallowing large foreign particles, such as pathogens, minerals, or cell debris by specialized cells, such as phagocytes and protists. Thus, phagocytosis is less important in the cellular delivery of nanoparticles like QDs. In macropinocytosis, membrane protrusions fuse into macropinosomes by engulfing foreign materials along with surrounding fluid. Macropinocytosis is a mechanism of nonspecific uptake of small particles common to most cells. In clathrin-mediated endocytosis, foreign materials, such as biomolecules and nanoparticles, are first bound on the membrane receptors or proteins, and then get trapped along with clustered receptors into $\sim$ 100 nm-sized clathrin-coated pits. These pits are pinched off the membrane in the form of clathrin-coated vesicles (CCV). The regulatory enzyme phosphoinositide 3-kinase (PI3K) assists the formation of clathrin-coated pits, and a hydrolase enzyme dynamin supports the detachment of CCV. Clathrin-mediated endocytosis is common to all the type of cells. In caveolae-dependent endocytosis, nanoparticles or biomolecules are trapped into small (~50 nm) caves formed by lipid raft invaginations in the cell membrane. The transmembrane protein caveolin assists the formation of caveolae, which are distinguished by specific proteins, cholesterol, and sphingolipids. Caveolae-dependent endocytosis is common to smooth muscle cells, pneumocytes, fibroblasts, adipocytes, and endothelial cells. In short, phagocytosis and macropinocytosis are completely nonspecific endocytosis, whereas clathrin- and caveolae-dependent endocytosis are either nonspecific or receptor-mediated pathways. Nonspecific endocytosis is not mediated by any target molecule; thus, this fact will neither pose obstacles to the delivery of QDs in a particular type of cells nor facilitate targeting of a particular organelle/molecule in cells. Thus, nonspecific endocytosis of QDs is important in the extended imaging of cell samples and investigation of endocytosis mechanism. Nonspecific delivery of QDs is not completely independent of nonspecific surface-cell labeling. Indeed, in a cell sample, the fate of QDs coated with liposomes, certain peptides, surfactants, carboxylic acids, silanes, polymers, amines, and certain proteins depends mostly on temperature. At 4 °C, such QDs stick in the cell membrane, whereas at 37 °C, endocytosis will be operative.

A. Nonspecific Delivery

Surface charge and hydrophobicity are two important factors in the nonspecific endocytosis of QDs coated with cationic lipids and liposomes, amines and polyamines, peptides, proteins, and carbohydrates. Such QDs are taken by cells through one or multiple pathways outlined in the previous paragraph. Here, the nonspecific endocytosis of QDs in terms of the surface functions of QDs is considered.

1. Lipids and Liposomes

Cationic lipids and liposomes are extensively used for *in vitro* and *in vivo* delivery of a wide spectrum of drugs, genes, and nanomaterials (see the chapter by Moreno and coauthors in this volume). The advantages of cationic lipids and liposomes over peptides and other ligands are high efficiency in delivery of nanomaterials in cells, wide availability, cost effectiveness, chemical and physical stability, and biocompatibility. Although the mechanism underlying the delivery of cargos by cationic liposomes is controversial, the general concept is that lipid molecules electrostatically bind on the cell membrane, then fuse with the membrane, and finally release the contents into the cytoplasm. Indeed, the efficiency in liposome-mediated delivery of cargos depends on the type of cells.

Derfus *et al.* accomplished efficient delivery of PEG-conjugated CdSe/ZnS QDs in HeLa cells by mixing the QDs with transfection reagents, such as lipofectamine 2000, activated dendrimers, or translocation peptides.[65] Among these reagents, cationic liposomes derived from lipofectamine 2000 provided highest efficiency in the delivery. Endocytosis of QDs encapsulated in liposomes was evidenced by both the presence of large aggregates in the cytosol and the colocalization of QDs with fluorescent-labeled epidermal growth factor (EGF), which is known as an endosome marker in receptor (EGFR)-mediated endocytosis. Similarly, Hsieh *et al.* and Lagerholm *et al.* accomplished high efficiency in the delivery of QDs encapsulated in lipofectamine PLUS™ liposomes,[66] or QDs coated with cationic surfactants, such as didodecyl and dihexadecyl dimethylammonium bromide.[67] Despite the high efficiency in the delivery of QDs in cells by lipids and liposomes, endosomal arrest, endolysosomal degradation, and exocytosis of QDs are not yet completely addressed.

2. Polymers

Nonspecific endocytosis of QDs coated with polymers is a matter of great concern for both occupational safety[33,34,68] and specific delivery of QDs in cells. Skin and respiratory organs are two potential routes for occupational exposure. Thus, nonspecific delivery and cytotoxicity of QDs were investigated in human skin and lung fibroblast cells. Hsieh *et al.* investigated effects of silica-shelled and PEG-coated CdSe/ZnS QDs on gene expression and cytotoxicity as a function of nonspecific uptake in human skin (HSF 42) and lung epithelial (IMR 90) cells.[66] While IMR 90 cells were not detectably affected, genes related to cell cycle progression were slightly downregulated in HSF 42 cells. In another example among QDs conjugated with PEG, —NH_2, and —COOH, Ryman-Rasmussen *et al.* curiously found fast uptake of QD–COOH conjugates by normal human epidermal keratinocytes.[33] Nonspecific endocytosis was also detected for QDs coated with polymers, such as PEG-grafted–polyethyleneimine (PEG-g–PEI),[69] poly(L-lysine)–PEG–citraconic amide,[70] poly(maleic anhydride/acid–alt-1-tetradecane) (PMAD),[71] thiolated polyacrylamide,[72] and poly(D,L-lactide–co-glycolide) (QDNC).[52] These reports observe that QD–polymer conjugates are taken through endocytosis by various types of cells.

One of the hopeful aspects of nonspecific endocytosis of QDs conjugated with certain engineered polymers is the endosome-disrupting property of the polymer and uniform distribution of QDs in the cytosol. For example, Duan and Nie reconstructed an endosome-disrupting PEG-g–PEI copolymer and conjugated it to CdSe/CdS/ZnS QDs.[69] The PEI segment is known to provide a proton sponge effect, with which the cargos escape from the endolysosomal compartments. The QD–PEG-g–PEI conjugate was strongly bound to HeLa cells by electrostatic interactions and then delivered in the cytosol by nonspecific endocytosis or acropinocytosis. The fate of QDs in the cytosol depends on the ratio between PEG and PEI. For example, QDs conjugated with PEI-g–PEG4 were preferentially trapped in endosomes whereas QDs conjugated with PEI-g–PEG2 successfully escaped from endosomes and were uniformly distributed in the cytoplasm. Proton sponge effect facilitates enhanced delivery of the contents trapped in endosomes by buffering H^+ and subsequent accumulation of Cl^-. Molecules such as PEI having multiple amino groups efficiently buffer H^+. From an increase in the ionic concentration, the endosome swells by osmosis and then breaks. Figure 5 schematically presents proton sponge effect of PEI and its endosomal escape. Yezhelyev *et al.* extended the proton sponge effect to pH-dependent siRNA delivery using CdSe/ZnS QDs coated with PMAD.[71] Similarly, Qi and Gao accomplished

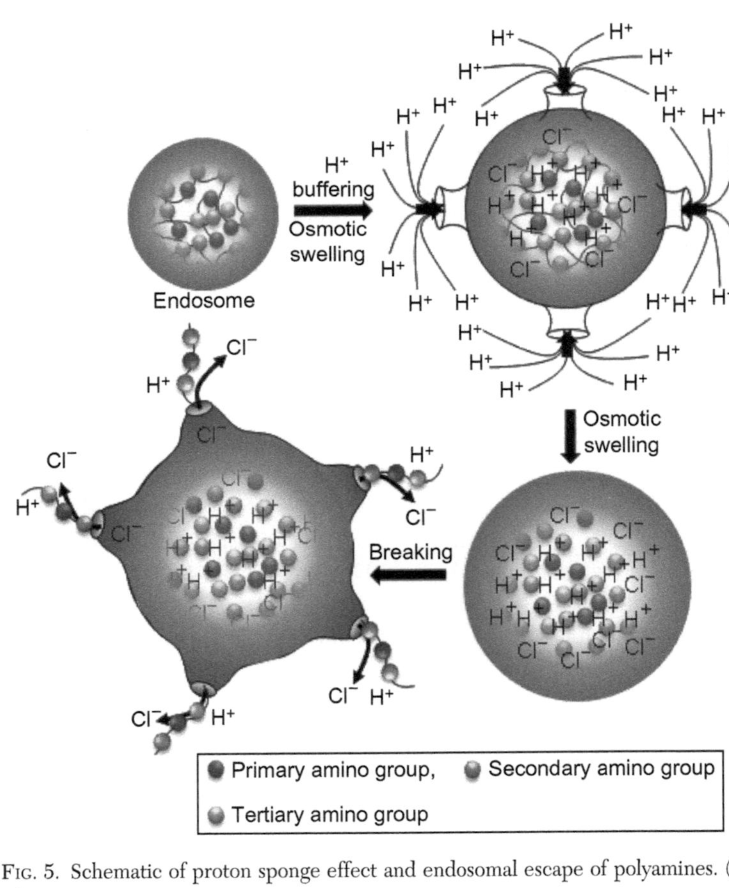

FIG. 5. Schematic of proton sponge effect and endosomal escape of polyamines. (See Color Insert.)

the delivery and endosomal escape of siRNA by utilizing the proton sponge effect from maleic acid groups in an amphiphol.[73] Kim *et al.* took a step ahead in the proton sponge effect by successfully targeting actin filaments in cells using QDs coated by an endosome-disrupting lactide–co-glycolide polymer (QDNC).[52]

3. Peptides

Peptides are ligands commonly used for the delivery of QDs in cells. The advantage of peptides, in particular cationic peptides, is that electrostatic interactions between peptides and cell membranes facilitate initial binding of QDs on cells. Following the initial step, QD–peptide conjugates are translocated into the cytosol by macropinocytosis, direct cell penetration, or clathrin- or caveolae-dependent endocytosis. The efficiency in and the mechanism of delivery of QD–peptide conjugates depend on various factors, such as charge, amino acid sequence of peptides, hydrodynamic size of the conjugates, and the type of cells. The following sections provide details in three representative peptides.

a. Arginine-Rich Peptides. To facilitate high-efficiency transfection, researchers have developed nonviral vectors based on peptides containing polyarginine sequences or multiple arginine units (see the chapter by Domingo-Espín and coauthors in this volume). The concept of arginine-rich peptides in the delivery of QDs and other cargos originated from the uptake of transactivating transcriptional activator (Tat) by many types of cells. Simply, peptides containing multiple arginine moieties carry net positive charge, are bound on the cell membrane through electrostatic interactions, and translocate into the cytosol. Thus, arginine-rich peptides[67,74–77] and Tat peptides[78–80] are extensively utilized for the delivery of QDs. Lagerholm *et al.* accomplished efficient delivery of poly-L-arginine (L-Arg9)-conjugated QD565, QD605, and QD655 in Swiss 3T3 cells, HeLa cells, and MG63 cells.[67] The efficiency in delivery of QD–L-Arg9 conjugate is equal to that of QDs coated with cationic surfactants. However, aggregation of QD–polyarginine conjugates in endolysosomal compartments is a limitation.[74] Formation of the aggregation is evidenced by colocalization experiments using fluorescent-labeled antibody to lysobisphosphatidic acid, or LysoTracker™. Delahanty *et al.* accomplished the delivery of QDs by coating a bifunctional octaarginine–co-octahistidine peptide on the QD surface.[75] In this approach, the octahistidine residue due to its affinity for metal atoms facilitated noncovalent conjugation to CdSe/ZnS QDs; the octaarginine residue due to its positive charge facilitated endocytosis in HEK 293T/17 and COS-1 cells. Medintz *et al.* extended this approach to the delivery of small and large protein cargos in cells, such as monomeric YFP and streptavidin-tagged B-phycoerythrin.[77] The key mechanism underlying the delivery of QD–polyarginine conjugates is endocytosis, which was evidenced by colocalization of QDs with AlexaFluor™–transferrin conjugate.[75] Transferrin is a widely known protein for the endocytosis mediated by the transferrin receptor. Later, Zhang *et al.* validated the role of arginine in the delivery of QDs by analyzing matrix metalloprotease-2 (MMP-2)- and MMP-7-modulated deprotection of arginine segments followed by the uptake of QDs by human fibrosarcoma cells (HT 1080).[76] The delivery of the QD–peptide conjugates was facilitated only after de-protecting the arginine residue by MMP-2/MMP-7-assisted cleavage of the substrates.

Ruan *et al.* evaluated the mode of delivery of QDs by Tat.[78] They found that QDs conjugated with multiple Tat peptides (QD/Tat = 1:20) efficiently are bound on the cell membrane, and then taken by HeLa cells. The mechanism of Tat-mediated delivery of QDs was investigated by treating the cells with QD–Tat conjugates at 4 °C or after disrupting the cytoskeleton. Delivery of QDs was blocked in both cases, indicating that the conjugates were taken by macropinocytosis. The QD–Tat conjugates were eventually transported along the microtubule tracks to the microtubule organizing center in the perinuclear region. Lei *et al.* sophisticated Tat-mediated delivery of QDs by first encompassing CdSe/ZnS QDs in lipid micelles followed by conjugating Tat peptide to the lipid coat.[79] The QD–micelle–Tat assembly was efficiently endocytosed by mesenchymal stem cells. Chen *et al.* investigated the delivery of QDs in human alveolar basal epithelial carcinoma cells (A549) as a function of QD/Tat ratio.[80] They prepared the QD–Tat conjugate by tethering biotinylated Tat to streptavidin-coated CdSe/ZnS QDs up to 1:40 as a ratio of QD/Tat. The efficiency in delivery was increased with increasing ratio of Tat. Endocytosis of the QD–Tat conjugate was evidenced by a considerable suppression in its delivery either at 4 °C or when ATP was inhibited by treating the cells with NaN3/2-deoxy-D-glucose. The contribution of clathrin-mediated endocytosis to the delivery of QD–Tat conjugates was investigated by either depleting K^+ in the cell culture medium or treating the cells with hypertonic sucrose. These treatments result in the inhibition of delivery of transferrin except QD–Tat. On the other hand, the delivery of the conjugate was arrested when lipid rafts associated with the membrane cholesterol were depleted, indicating that caveolae-dependent endocytosis is the most important endocytic pathway for QD–Tat. All the examples in Section IV.A.3.a show that efficient delivery of QDs in cells is accomplished by conjugating QDs with arginine-rich peptides. However, such QD–peptide conjugates are mostly trapped in endolysosomal vesicles. Thus, a combination of arginine-rich peptides and endosome-disrupting coating is necessary for further advancement of the applications of QDs for use in delivery.

b. Nuclear Localization Signals. Peptide ligand nuclear localization signal (NLS) deliver foreign materials into the cytosol and nucleus (see the chapter by Domingo-Espín and coauthors in this volume). From the name, the main function of NLS is inside-nucleus signaling through interactions with importin in the nuclear pore complex. The positive charge of NLS allows it alone or with other peptides to be utilized for the delivery of QDs in cells and their nuclei. Chen and Gerion accomplished the delivery of CdSe/ZnS QDs conjugated with NLS derived from SV40 T antigen in cells and their nuclei.[36] On the other hand, only inside-cell not inside-nucleus delivery was detected for QDs conjugated with a random peptide sequence. Similarly, Rozenzhak *et al.* utilized a carrier peptide (Pep-1) for the delivery of QD–streptavidin, QD–NLS, and QD–GH3 domain sequence (derived from the Grim protein) conjugates in cells.[81] They found that Pep-1 serves only as a transmembrane carrier of QDs. The complex between bioconjugated QDs and Pep-1 dissociated in the cytoplasm and resulted in the release of QD–streptavidin, QD–NLS, and QD–GH3. Thus, QD–NLS and QD–GH3 conjugates delivered in the cytoplasm further target nucleus and mitochondria, respectively. Hoshino *et al.* utilized NLS for the delivery of CdSe/ZnS QD–eGFP conjugates in HEK293T cells.[82] They successfully achieved delivery of QD–eGFP conjugates for the expression of the GFP gene in the cells.

c. Insect Neuropeptides. The potential of natural peptides other than Tat-based peptides has been investigated for delivery of QDs and other nanoparticles in cells. It was shown that allatostatin I (AST1, APSGAQRLYG FGL-NH2), a neuropeptide present in insects and crustaceans, is encouraging for delivery of QDs in living mammalian cells, such as human epidermoid ovarian carcinoma cells (A431) and NIH3T3 cells.[83,84] By selecting AST1 as a carrier peptide for QDs, multiple pathways in the delivery of QDs mediated by the peptide were identified. QD–AST1 conjugates were delivered efficiently in the cytosol and sparingly in the nucleus of NIH3T3 (Fig. 6A) and A431 cells (Fig. 6B). Apparently, the QD–AST1 conjugate was first bound on the cell membrane, and successively transported into the cytosol and nucleus. In the cytosol, QDs will mostly be present in aggregates, indicating endosomal arrest. Although details about the endosomal escape of QD–AST1 conjugates are poorly understood, the presence of QDs in the nucleus indicates that the aggregates of QDs partly dissociate and successively cross the nuclear pore complex.

To obtain a general picture of the delivery of QDs by peptides, flow cytometry and fluorescence microscopy have been combined, and expected pathways investigated, such as clathrin-mediated endocytosis, galanin receptor-mediated endocytosis, charge-based cell penetration, and temperature-dependent endocytosis involved in the delivery. Clathrin-mediated endocytosis might be either a receptor-mediated or a nonspecific pathway. It was found that the delivery of QD–AST1 in cells is considerably suppressed by inhibiting the regulatory enzyme PI3K with wortmannin and blocking the formation of CCV.

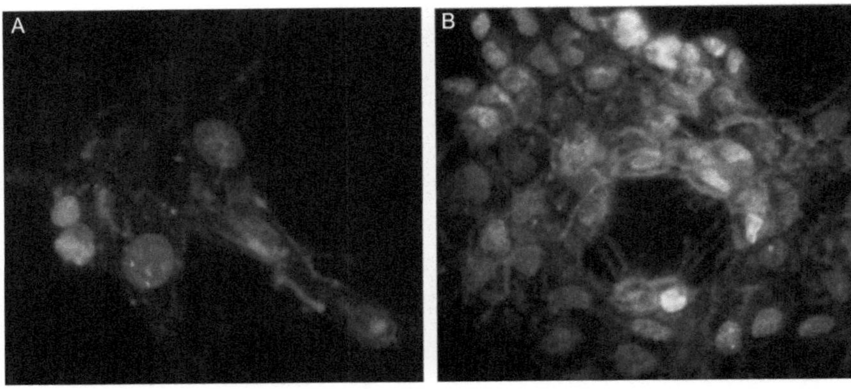

FIG. 6. Fluorescence images of (A) 3T3 and (B) A431 cells incubated with a 5 mM solution of Syto16 dye for 10 min followed by a 1-nM solution of QD605–AST1 for 30 min. The cell nucleus is preferentially stained green by Syto16 due to its cell permeability and ability for intercalation with DNA. The yellow-orange color indicates colocalized QD–AST1 and Syto16. Reprinted with permission from Ref. 84; copyright (2009), American Chemical Society. (See Color Insert.)

Figure 7A and B shows that clathrin-mediated endocytosis was supported by the colocalization of QD605–AST1 with anti-clathrin antibody labeled with QD565. Figure 7C shows a scheme of clathrin-mediated endocytosis and its inhibition by wortmannin. Furthermore, receptor-mediated endocytosis was investigated because the target receptors for AST1 in insects are analogous to galanin in mammals. However, the delivery of QD–AST1 was not substantially affected in the presence of a galanin antagonist. Indeed, the level of galanin receptor is low in A431 and 3T3 cells, indicating that galanin receptor-mediated endocytosis of QD–AST1 is negligible in these cell lines. As AST1 contains an arginine moiety, which carries a net positive charge, charge-based cell penetration of QD–AST1 was investigated. Replacement of the arginine unit with an alanine unit (MAST1) reduced the efficiency in the delivery of QD by $\sim 13\%$ for A431 and $\sim 27\%$ for 3T3 cells from the efficiency in the delivery of QD–AST1 at 37 °C (100%). Similarly, when the negative charge of cell membrane due to heparan sulfate was suppressed by preincubating the cells with heparinase enzyme, the efficiency in delivery of QD–AST1 was reduced by $\sim 18\%$ for A431 and $\sim 30\%$ for 3T3 cells. These observations indicate substantial contributions from charge-dependent delivery. Because both initial charge-driven binding of QD–AST1 on cell membranes and subsequent clathrin-mediated endocytosis are tightly combined, it is difficult to do a direct comparison of efficiency in inside-cell delivery with charge-driven delivery. Nonendocytic direct cell penetration is an energy-independent process unaffected at low temperatures. Thus, the difference between cell penetration and

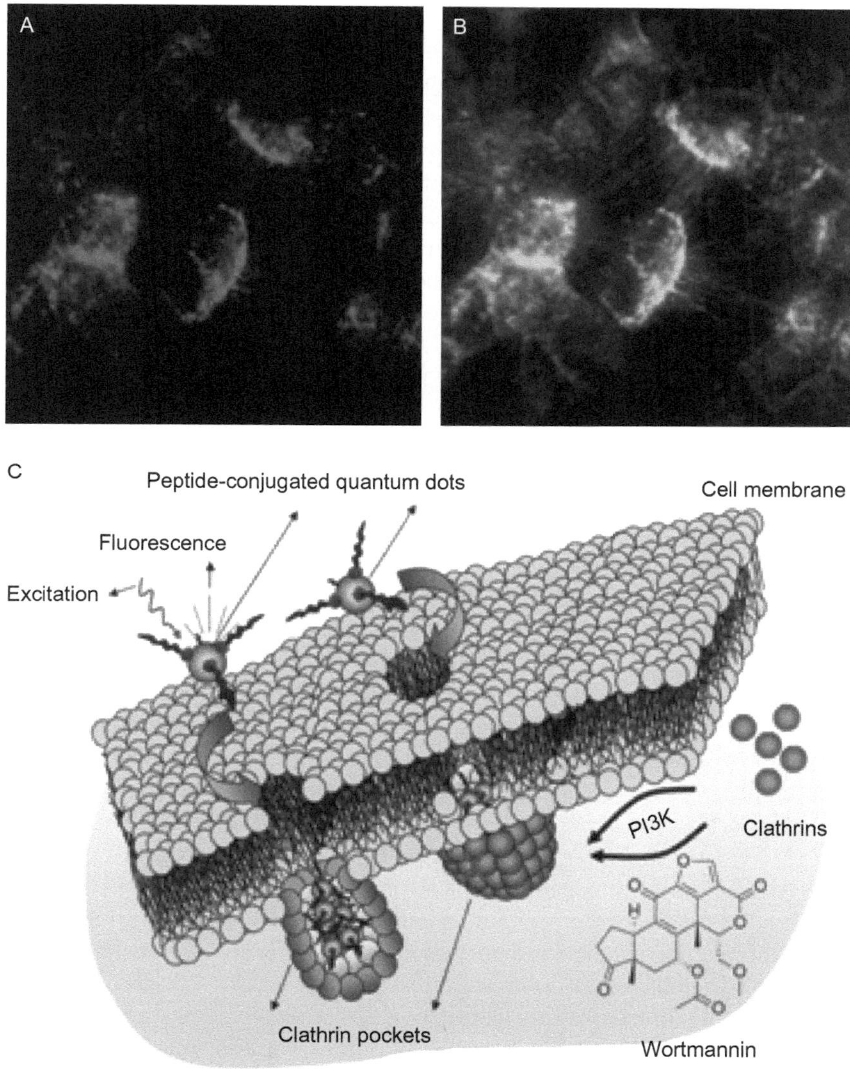

FIG. 7. Fluorescence images of A431 cells incubated with QD605–AST1 conjugate for 30 min, permeabilized using methanol, and subsequently incubated with QD565–CAb conjugate. (A) Image acquired through a filter for QD605 and (B) image acquired through a filter for QD565 and overlaid with A. (C) Schematic of clathrin-mediated endocytosis of QD–AST1. Reprinted with permission from Ref. 84; copyright (2009), American Chemical Society. (See Color Insert.)

endocytosis was shown by incubating the cells at 4 °C. The efficiencies at 41 °C were ~87% for A431 and ~55% for 3T3 cells, suggesting that direct cell penetration, the other ~13% for A431 and the other ~45% for 3T3, is operative in the delivery of QD–AST1. From the authors' work[83,84] and also from the publications by Ruan *et al.*[78] and Chen *et al.*,[80] multiple pathways are apparently involved in the delivery of QD–peptide conjugates.

Versatility of peptide-mediated delivery of QDs in cells was tested for several other natural and synthetic peptides. For example, Walther *et al.* employed calcitonin, a linear polypeptide hormone found in most animals and man, as a carrier peptide for QDs and RNA in HeLa and HEK-293 cells.[85] QD–calcitonin conjugates were endocytosed mainly through lipid raft-dependent pathways; however, the contributions of clathrin-mediated endocytosis and macropinocytosis were moderate. However, calcitonin is the ligand for G-protein-coupled calcitonin receptor in osteoclasts. Thus, receptor-mediated endocytosis will be operative for QD–calcitonin conjugates in osteoclasts. Maurocalcine, a 33-amino acid peptide toxin found in scorpion venom, is another natural peptide investigated for the delivery of QDs. Jayagopal *et al.* accomplished biotinylation of this peptide, conjugation of the biotinylated peptide to streptavidin-coated QD585 and QD655, and efficient delivery of QD–maurocalcine conjugates in immunomagnetically isolated monocytes and T-lymphocytes.[86] The examples in Section IV.A.3 show that the extent and the mode of delivery of QD–peptide conjugates depend on various factors, such as the amino acid sequence of the peptide, charge of the peptide, and the type of cells.

4. Proteins

Proteins, such as collagen, serum albumin, and virus capsid proteins, are cost-effective agents for the nonspecific delivery of QDs in cells. Parak *et al.* utilized collagen-coated QDs for evaluating the motility of cancer cells.[87] They seeded human mammary epithelial tumor cells (MDA-MB-231) and nontumor cells (MCF-10A) on plates coated with collagen and silica-shelled CdSe/ZnS QDs. Although both cancer cells and normal cells engulfed the QD–collagen complex, the uptake was prominent in cancer cells due to their motility. Thus, the motility of cancer cells was determined from nonfluorescent tracks made by engulfing QD–collagen complexes. Serum albumin is also cost-effective and a widely available protein for the delivery of QDs. Hanaki *et al.* first examined the efficiency in the delivery of QD–serum albumin complexes into the cytosol.[88] They found that the QD–serum mixture is efficiently endocytosed by Vero cells. One of the advantages of the QD–serum complex over other QD conjugates is that serum albumin renders QDs nontoxic.[89] Proteins derived from virus capsids are favorable for the delivery of QDs into cells through caveolae-dependent pathway. This pathway is effective in avoidance of endolysosomal trapping and degradation of QDs. Li *et al.* accomplished efficient delivery of CdSe/ZnS QDs in Vero cells by complexing QDs with capsid protein from SV40 virus.[90] They have constructed a chimeric virus-like nanostructure by encapsulating TGA-capped QDs in the capsid protein from SV40 virus.

Resembling the endocytosis of virus particles, the QD–protein assembly was delivered through caveolar endocytosis. This pathway was confirmed by colocalizing QDs with caveolin-1 fused with CFP. On the other hand, clathrin-mediated endocytosis was ruled out simply because transferrin was not colocalized with QDs. The endocytosed nanoparticles were transported in the cytosol along microtubules and accumulated in the endoplasmic reticulum, a process that mimics the early stage of viral infection.

5. Small Molecules

Nonspecific endocytosis of QDs by small molecules such as TGA,[29,44] mercaptopropionic acid (MPA),[30,91] DHLA,[29] and trimethoxysilylpropyl urea[92] was relevant only until bioconjugate chemistry of QD was established. QDs conjugated with such small molecules were utilized in control experiments or experiments related to occupational safety. Bruchez *et al.* tested the first biological application of QDs by delivering silica-shelled and trimethoxysilylpropyl urea-conjugated CdSe/CdS QD in 3T3 cells.[92] They found that the internalized QDs can be utilized for staining the nucleus. They also successfully detected specific F-actin filaments in 3T3 cells by treating the cells successively with biotinylated phalloidin, streptavidin, and QD-biotin. Jaiswal *et al.* utilized DHLA-capped CdSe/ZnS QDs for labeling of HeLa cells.[29] They have evaluated the endocytosis of the QD–DHLA conjugate by colocalizing QDs with an endosome-related protein, to which CFP was fused. Sun *et al.* successfully detected nonspecific uptake of TGA- or MPA-conjugated CdSe/ZnSe/ZnS QDs by human ovarian cancer cells (SKOV-3), multiple myeloma cells (RPMI 8226), and pancreatic cancer cells (AsPc-1), but not NIH3T3 cells.[91] Xu *et al.* found that a combination of UV light and the membrane-permeable dye 40,6-diamino-2-phenylindole facilitates delivery of QDs in cytoplasm and nuclei.[93] These reports indicate that nonspecific endocytosis of QDs depends on both the molecules on the surface of QDs and the type of cells.

6. Carbohydrates

Coating or conjugation of carbohydrates to QDs offers QDs water solubility and biocompatibility. In addition, complexes between QDs and carbohydrates showed substantial delivery in a variety of cells. For example, Hasegawa *et al.* successfully delivered QDs coated with a thick nanogel shell composed of pollulan ($CHPNH_2$), a polysaccharide, functionalized by cholesterol and primary amine.[94] They assembled the QD–nanogel complex by coating negatively charged QDs with positively charged $CHPNH_2$. The efficiency in delivery of the QD–nanogel complex is determined by the number of amino groups in pollulan. For example, QD–nanogel containing 9 amino groups per 100 glucose units provided efficiency in the delivery equal to that of QDs encompassed in lipofectamine. With increasing amino groups per 100 glucose units from 9 to 15, the efficiency in delivery was increased by a factor of 3.4.

These results indicate that positive charge of the nanogel from amino groups has considerably contributed to the delivery of QDs. Similarly, Xie *et al.* and de Farias *et al.* successfully delivered QDs coated with chitosan[95] or glucose[96] in yeast cells. Although a precise mechanism underlying the delivery of the QD–carbohydrate system is still open, the uptake of QD–glucose conjugate by yeast cells is attributable to an active transport by transmembrane proteins.

B. Specific Delivery

Specific delivery of QDs in cells is essential in extended imaging of the structures of subcellular organelles and the functions of molecules in cells. Receptor-mediated endocytosis is the main strategy for specific delivery of QDs. Membrane receptors tagged by QD–antibody or QD–ligand conjugates will be rapidly endocytosed by the activation of receptors. In such cases, QD–antibody and QD–ligand conjugates are taken by receptor-mediated endocytosis. The QDs trapped in endosomes will be transported by microtubules and actin filaments to the lysosome organization center, and degraded under enzymatic reactions or acidic conditions in the endolysosomal compartments. Thus, endosome-disrupting coating will be necessary for specific delivery. Here, one will find various antibodies and ligands used for specific delivery of QDs.

1. Antibodies

Antibodies to membrane receptors and subcellular organelles have extensively been utilized for both delivery of QDs in cells and subcellular labeling. While membrane permeability is indispensable for the delivery of QDs conjugated with antibodies directed against cell proteins, membrane receptors labeled with QD–antibody conjugates are taken by living cells through receptor-mediated endocytosis. Wu *et al.* elegantly demonstrated the delivery of bioconjugated multicolor CdSe/ZnS QDs in human epithelial cells or 3T3 cells, and in their nuclei, by sequentially incubating fixed cells with primary antibody, biotinylated secondary antibody, and QD–streptavidin conjugate.[44] Microtubules in 3T3 cells were labeled by sequentially incubating the cells with monoclonal anti-α-tubulin antibody, biotinylated antimouse IgG, and streptavidin–QD conjugate. Similarly, nucleus antigen in human epithelial cells and 3T3 cells was labeled with an antibody to human antinuclear antigens, biotinylated antihuman IgG, and QD–streptavidin conjugates by sequentially incubating fixed cells. Yezhelyev *et al.* utilized QD–antibody conjugates for the simultaneous detection of cytoplasmic and nuclear marker proteins, such as the target of rapamycin (mTOR), estrogen receptor (ER), and progesterone receptor (PR) in MCF 7, BT-474, and MDA-MB-231 cells.[46] Owing to the brightness and photostability of QDs, they detected low levels of mTOR located in the cytoplasm and ER and PR located in the nucleus, both with high sensitivity.

In contrast to the examples by Wu *et al.*[44] and Yezhelyev *et al.*[46] for labeling of fixed cells, QD–antibody conjugates have been extensively utilized for labeling of living cells through receptor-mediated endocytosis. For example, Tan *et al.* accomplished nonviral delivery of siRNA using chitosan nanoparticles conjugated with anti-Her2 antibody and doped with CdSe/ZnS QD.[97] Similarly, Tada *et al.* accomplished delivery of QD800–herceptin conjugate in Her2 positive human breast carcinoma cells (KPL-4).[98] They also utilized QD–herceptin conjugate for *in vitro* and *in vivo* detection of single-molecule diffusion of QD–herceptin conjugate along the cell membrane and into the perinuclear space. Zhang *et al.* successfully delivered CdTe QDs conjugated with an antibody (AVE-1642) to IGF1R in MCF 7 cells.[50] They found that the QD–antibody conjugate is first bound on the cell membrane, and then enters the cytosol and nucleus. Cambi *et al.* introduced a somewhat different approach to the delivery of QDs using virus-mimicking nanoparticles labeled with dendritic cell-specific pathogen-uptake receptor (DC-SIGN).[99] The nanoparticles were prepared by conjugating biotinylated anti-DC-SIGN antibody to streptavidin-coated QD525, QD605, or QD655. The QD–antibody conjugates readily bound on CHO cells and were gradually delivered into the cytosol by DC-SIGN-mediated endocytosis. In addition, contributions by clathrin- and caveolae-dependent endocytosis of the conjugates were evidenced by colocalizing the QD–nanoparticle assembly with labels for clathrin-coated pits and caveolin-1. Prasad and coworkers utilized Claudin 4 and PSCA, overexpressed antigens in pancreatic cancer cells (MiaPaCa, CoLo-357/Panc-1), as targets for receptor-mediated endocytosis of QDs.[100,101] They found that CdSe/CdS/ZnS QDs conjugated with anticlaudin 4 antibody are efficiently delivered in MiaPaCa, CoLo-357, and Panc-1 cells.[100] Here, receptor-mediated endocytosis of the conjugate was identified by colocalizing the conjugate with QD–transferrin conjugate. They extended investigations to labeling of pancreatic cancer cells by replacing CdSe/CdS/ZnS QDs with less toxic InP/ZnS QDs.[101] Conjugates of InP/ZnS QDs and anticlaudin 4 antibody or anti-PSCA antibody were efficiently delivered in MiaPaCa cells, in which the corresponding antigens are overexpressed.

Other antibodies tested for the delivery of QDs in cells are antihuman α-fetoprotein (AFP) antibody and αCD3 antibody. AFP is a plasma protein produced by the liver, yolk sac, and gastrointestinal tract of human fetus; it is a marker for liver cancer in adults. Chen *et al.* accomplished selective delivery of CdSe/ZnS QDs in human metastatic hepatocarcinoma cells (HCCLM6) by conjugating QDs with antihuman AFP antibody.[102] Bottini *et al.* utilized αCD3 antibody for the delivery of QD-doped silica nanoparticles in human Jurkat T cells.[103] The QD-silica–antibody conjugates were prepared by biotinylating the amino groups in both silica nanoparticles and αCD3 antibody, and then bridging them together through neutravidin. Receptor-mediated endocytosis of the conjugates was confirmed by the fluorescence in Jurkat T cells at 37 °C. On the other hand, the conjugates were not delivered in Jurkat T cells at 4 °C or αCD3-negative HeLa cells at 37 °C.

2. Ligands

Ligands for membrane proteins and receptors are the potential for preparation of QD bioconjugates and receptor-mediated delivery of the conjugates in cells. Activation of receptors by corresponding ligands results in the clustering of receptors followed by clathrin-mediated endocytosis of receptor–ligand complexes. Thus, ligands, such as transferrin, EGF, nerve growth factor (NGF), folic acid, wheat germ agglutinin (WGA), and certain sugars, have been utilized for the delivery of QDs. Chan and Nie accomplished for the first time specific delivery of CdSe/ZnS QDs conjugated to transferrin,[12] which is a carrier protein for iron. The QD–transferrin conjugates were efficiently taken by HeLa cells by transferrin receptor-mediated endocytosis. On the other hand, control QDs conjugated with TGA remained stuck in the cell membrane. Transferrin receptor functions to maintain cellular iron homeostasis through the delivery of transferrin–iron complex. Thus, transferrin receptor-mediated endocytosis is common to most cells. The endocytosis of transferrin receptor is well characterized. Thus, QD–transferrin conjugates are expected to be used as a standard endosome marker.[77,101,104,105]

Like transferrin receptor-mediated endocytosis of QD–transferrin conjugates, EGFR-mediated endocytosis is expected to be useful for the delivery of QD–EGF conjugates in almost all cells. For example, Derfus *et al.* accomplished efficient delivery of QD–EGF conjugates in rat hepatocytes by EGFR-mediated endocytosis.[89] The endocytosed QD–EGF conjugates were continuously tracked for analyzing the migration of micropatterned hepatocytes. The extended photostability of QDs allows one to follow receptor-mediated endocytosis of QD–EGF–EGFR complexes and tracking of endosomes encompassing the complex in cells. In addition to the delivery of QD–EGF conjugates in cells, Kim *et al.* advanced a step by accomplishing endosomal escape of QD–EGF, using an endosome-disrupting poly(D,L-lactide-co-glycolide) coat on QD (QDNC).[52] The NC coat facilitated pH-dependent endosomal escape of the QD–EGF conjugate in SK-BR-3 cells.

Delivery of QDs in nerve cells is accomplished by conjugating QDs to NGF and subsequent endocytosis mediated by TrkA or p75.[106] Furthermore, endocytosed QD–NGF conjugates were utilized for extended imaging of neuronal differentiation, neurite growth, and receptor tracking in cells. However, a marked decrease in the density of the QDs was detected 18 h after incubation. This observation is due to either exocytosis or photobleaching of QDs. Moreover, active transport of QD–NGFTrkA complex along microtubules to far-reaching subcellular regions and newly created regions was detected. Cui *et al.* showed that QD–NGF conjugates are delivered in a compartmentalized culture of rat dorsal root ganglion and utilized for tracking retrograde axonal transport of vesicles.[106] Single-molecule experiments in this report showed that only a single NGF dimer is sufficient to sustain signaling during the axonal transport. Also, they found that vesicle transport along axons takes place in a stop-and-go mode, without being affected by the concentration of QD–NGF.

Folate receptor, overexpressed in many cancers including the cancers of the ovary, prostate, brain, column, nose, and throat, is an ideal target for both receptor-mediated endocytosis of QD–folic acid conjugates and the detection of cancer cells. Bharali *et al.* accomplished efficient delivery of folic acid-conjugated InP/ZnS QDs in KB cells through folate receptor-mediated endocytosis.[23] On the other hand, cancer cells such as A549 and MCF 7, which are folate receptor-negative, were not labeled with QD–folic acid conjugates. Gao *et al.* showed that CdSe/ZnS QDs conjugated with WGA are delivered in heterogeneous human epithelial colorectal adenocarcinoma cells (Caco-2) by WGA receptor-mediated endocytosis.[107] The QD–WGA conjugates were taken by caveolae- and clathrin-dependent endocytosis. This observation was verified by inhibiting clathrin-mediated endocytosis with chlorpromazine and caveolae-mediated endocytosis with filipin. The endocytosed QD–WGA conjugates were further transported by microtubules and actin filaments to the Golgi apparatus and lysosome. Kikkeri *et al.* found that D-galactose effectively delivers CdSe/ZnS QDs into the cytosol of hepatocellular carcinoma (HepG2) cells by asialoglycoprotein receptor-mediated endocytosis.[41]

3. Arg-Gly-Asp Peptides

Arg-Gly-Asp, or RGD peptides are ligands for $\alpha_v\beta_3$ and $\alpha_v\beta_5$ integrins. Thus, peptides carrying RGD sequences were utilized for integrin-mediated delivery of QDs in cells.[63,105,108] Between these two integrins, $\alpha_v\beta_3$ is overexpressed in the membrane of angiogenetic endothelial cells and many cancer cells, in particular, metastatic cancer cells. Mulder *et al.* conjugated RGD peptide to CdSe/ZnS QDs coated with paramagnetic ligands and PEGylated lipids, and used the conjugates as a bimodal probe for imaging $\alpha_v\beta_3$ integrin in HUVEC cells.[63] The conjugates are initially bound on the membrane of proliferating HUVEC cells, and then internalized. This pathway was experimentally identified by fluorescence and magnetic resonance imaging. Lieleg *et al.* found that the efficiency of delivery of QD–RGD conjugates depends on the spacer in between QD and RGD.[108] Koole *et al.* utilized RGD peptide for the delivery of microemulsions encompassing silica/PEG-shelled CdSe/CdS/CdZnS/ZnS QDs in HUVEC cells.[105] They found that the efficiency in the delivery is improved by conjugating multiple copies of RGD to QDs.

4. Toxins

Bacterial toxins, such as Shiga toxin, cholera toxin B (CTB), and a plant toxin ricin, are bound on certain receptors in mammalian cells and facilitate receptor-mediated endocytosis of QDs. Tekle *et al.* found that subunit B in Shiga toxin or ricin toxin is an ideal tool for the delivery of QDs in HeLa cells.[104] Conjugates of QDs with these toxins initially bind on corresponding membrane receptors in HeLa cells and are then taken by receptor-mediated endocytosis. Despite their efficient delivery, these conjugates showed poor tracking and recycling in the cells, warning that QD–toxin conjugates delivered in the cytosol cause unfavorable physiological consequences. Chakraborty *et al.* accomplished efficient delivery of CdSe/ZnS QDs in many cell lines, such as NIH3T3, mouse muscle-derived cells, human mesenchymal stem cells, human melanoma cells (M21), and mouse tetra carcinoma cells (MH15) by conjugating QDs with CTB.[109]

Like proteins, amino groups in CTB are conjugated to carboxylic acid-functionalized QDs by simple carbodiimide cross-linking. The QD–CTB conjugate initially is bound on gangliosides composed of a glycosphingolipid and exists in lipid rafts in the cell membrane of all mammalian cells, and then taken by caveolae-dependent endocytosis. They utilized QD–CTB conjugates for extended imaging of mesenchymal stem cells, and identified that the endocytosed QD does not affect the differentiation potential of the cells. The advantages of QD–CTB conjugates are twofold: the extent of their aggregation in cells is less than that of QDs conjugated with several other peptides, and endolysosomal degradation of QDs is bypassed by caveolae-dependent endocytosis.

V. Photodynamic Therapy by Quantum Dots

The use of PDT supported by fluorescence imaging advances clinical trials for efficient detection and treatment of cancers. Fluorescence imaging of cancer is facilitated by targeting tumors using fluorescent dyes conjugated with anticancer antibodies and exciting the dyes with visible or NIR light. In PDT, cancers are treated by applying a photosensitizing (PS) drug and irradiation of light. The principle underlying PDT is that a photoactivated PS drug transfers energy or electrons to oxygen or other molecules, and creates reactive oxygen species (ROS), which immediately react with and damage vital biomolecules in cell organelles, resulting in cell death. The main advantage of fluorescence imaging over other biomedical imaging techniques, such as X-rays, computed tomography (CT), and positron emission tomography (PET), is that visible and NIR excitation in fluorescence imaging is nonionizing and less hazardous. The main advantage of PDT over chemotherapy and radiation therapy is that site-specific photoactivation of targeted PS drugs using visible or NIR light offers selective therapy, leaving the immune system and normal cells intact. However, fluorescence imaging and PDT of cancer are challenging owing to the limited availability of photostable and NIR dyes as PS drug. This section overviews PDT from its fundamentals to recent developments involving the use of QDs.

A. ROS, Key to PDT

The basic requirement for fluorescence imaging and PDT of cancer is the selective delivery of fluorescent dyes and PS drugs in tumors. The basic principle underlying PDT is photoactivation of a PS drug, which results in the formation of ROS, such as singlet oxygen ($^1O_2$), hydroxyl radical ($^{\bullet}OH$), super oxide anion ($^{\bullet}O_2^-$), and hydrogen peroxide (H_2O_2) through a series of energy and electron transfer reactions between PS and oxygen ($^3O_2$).[110,111] Figure 8 shows various photophysical and photochemical processes involved in PDT. Briefly, photoactivation of a PS drug places it at the lowest excited singlet

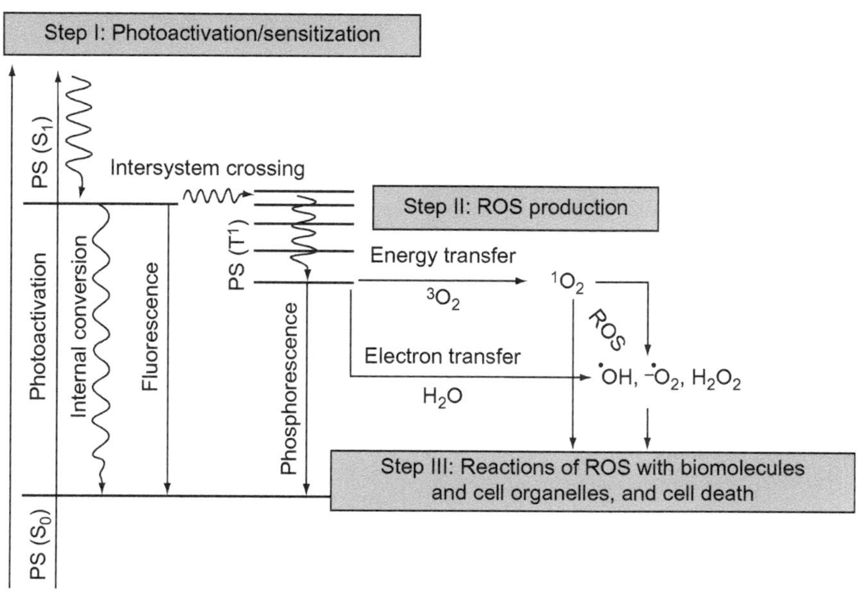

FIG. 8. Photophysical and photochemical processes in PDT.

(S_1) and triplet (T_1) states. The lifetime of the T_1 states for most PS drugs ranges from several hundred nanoseconds to milliseconds, much longer than the S_1 lifetime.

A PS drug in the T_1 state either relaxes to the ground state (S_0) by transferring excess energy to molecular oxygen or transfers an electron (also, at S_1 state) to oxygen, water, or a proximal molecule and enters into a series of photochemical reactions.[110,111] By the energy transfer from a PS to $^3O_2$, an electron in the πx^* and πy^* orbitals of $^3O_2$ changes its spin quantum number and forms $^1O_2$. The energy required for the formation is only 94.3 kJ/mol. $^1O_2$ is an unstable species and it reacts with water, generating a sequence of ROS such as $^\bullet OH$, $^\bullet O_2^-$, and H_2O_2. On the other hand, electron transfer from a PS drug directly produces ROS. However, electron transfer creates the cation radical of a PS, which irreversibly reacts with other molecule and results in the chemical transformation of PS.[112] Furthermore, photosensitized production of ROS through energy transfer is a renewable process. Thus, energy transfer is preferred over electron transfer for the durability of PS drugs. In both mechanisms, cell death is initiated by the photochemical reactions of ROS with biomolecules and cell organelles such as amino acids, endoplasmic reticulum, mitochondrion, lysosomes, and Golgi apparatus.

Early PS drugs for PDT are porphyrins, phthalocyanines, and chlorine derivatives. In the earlier days, a mixture of porphyrins, called the first-generation PS drugs, was used for PDT. For example, breast cancer is successfully healed in a mouse model by applying hematoporphyrin derivatives as PS drugs.[113] Later, with the introduction of purified PS drugs, also called the second-generation PS drugs, such as porphyrins, phthalocyanines, and chlorine derivatives, research on PDT has been advanced into clinical trials. For example, superficial bladder cancer was treated with nonspecific administration of photofrin as a PS drug, followed by illuminating the bladder with red light.[114] However, the patients suffered serious side effects due to nonspecific delivery and photoactivation of the drug. With the advancements, such as synthesis of new-generation PS drugs, targeted drug delivery, image-guided PDT, and introduction of tunable and fiber-optic lasers, imaging and PDT of cancer have become more popular methods for skin cancers, Barrett's esophagus, bronchial cancers, head and neck cancer, lung cancer, prostate cancer, and bladder cancer.

B. QDs for PDT of Cancer

Metal, semiconductor, polymer, and ceramic nanoparticles have attracted much attention in the imaging and PDT of cancer.[115] Polymer and ceramic nanoparticles have been widely employed as drug carriers, whereas metal and semiconductor nanoparticles act as probes for imaging and therapy. Among various nanoparticles, QDs attracted much attention as probes for bioimaging.

Photoactivated QDs and QD–PS conjugates and QD–PS mixtures produce $^1O_2$. When a QD is directly photoactivated, radiative carrier relaxation in the band edge dominates energy transfer to oxygen, and results in a poor efficiency for ROS production. On the other hand, energy transfer processes from QDs to PS drugs and PS drugs to oxygen dominate in QD–PS conjugates, providing a higher efficiency for ROS production. Advantages of QD–PS conjugates over conventional PS drugs are many; for example, indirect photoactivation of PS drugs using photostable QDs offers prolonged imaging and PDT without photobleaching, the large surface area of QDs allows space for conjugating multiple PS and cancer markers for efficient and targeted cancer imaging and PDT, and broad absorption band and large two-photon absorption cross section of QDs are advantageous for unrestricted and NIR photoactivation. Here, the synthesis of biocompatible QDs, *in vitro* and *in vivo* targeting, and imaging of cancer cells and tumors using QDs, photosensitization properties of QDs and QD–PS conjugates, prospects and remaining issues of QDs and QD–PS conjugates for imaging and PDT of cancer are summarized.

The quality of a PS drug for PDT depends on its efficiency for energy and/or electron transfer to molecular oxygen and the subsequent production of ROS. Compared with electron transfer, energy transfer is desirable for PDT because electron transfer products such as cation and anion radicals undergo irreversible chemical transformations. The irreversible nature prevents subsequent photoactivation of a PS drug and continuous generation of ROS. Samia *et al.* proposed the concept "QDs for PDT" and first investigated the use of QDs in PDT.[116] Unparalleled photostability of QDs is the most suitable property for PDT. Additionally, broad absorption band and large two-photon absorption cross section of QDs are advantages for photoactivation using various visible and NIR light. Despite these advantages, photosensitized production of ROS at high efficiency is the primary requirement for a standard PS drug. Although targeted delivery of QDs in cancer cells and tumors using anticancer antibodies and other biomolecules has become possible, compared with conventional PS drugs such as porphyrins and phthalocyanines, the efficiency of QDs to produce ROS under direct photoactivation is low. Thus, preparation of conjugates between QDs and conventional PS drugs, investigation of energy transfer efficiencies from QDs to PS drugs, and ROS production by the conjugates are being widely investigated.

C. QDs Versus Conventional PS Drugs for PDT

Samia *et al.* found that direct photoactivation of QDs produces $^1O_2$ through energy transfer from the dark exciton state of QDs to $^3O_2$.[116] In this investigation, they employed hydrophobic-capped CdSe QDs (photoluminescence quantum efficiency ~65%) dissolved in oxygen-saturated toluene, and estimated the $^1O_2$ efficiency at ~5% by recording and analyzing the steady-state and time-resolved phosphorescence of $^1O_2$ at $\sim$1270 nm. From femtosecond time-resolved pump-probe measurements, they found that the low quantum efficiency for $^1O_2$ is ascribed to ultrafast carrier trapping and nonradiative carrier relaxation. In other words, ultrafast carrier relaxation processes in QDs compete with relatively slow energy transfer to $^3O_2$. This process requires the formation of a collision complex. Another reason for the low efficiency for energy transfer is the large hydrodynamic size of QDs. On the other hand, conventional PS drugs based on porphyrins and phthalocyanines having small hydrodynamic size and long-living triplet state (submicroseconds to milliseconds) produce ROS at much higher efficiencies ($>$75%).

Despite the low efficiency for $^1O_2$, QDs offer prolonged photoactivation and persistent production of $^1O_2$ and other ROS owing to the incredible photostability. Thus, in contrast to conventional PS drugs that are less photostable, QDs offer cumulative effects in PDT. For example, Anas *et al.* found that prolonged photoactivation of a QD–plasmid DNA conjugate at 512 nm results in the breakage and damage of DNA.[117] The breakage and damage of DNA were induced by the photosensitized production of ROS. The product was identified as the ROS scavenger using nitroblue tetrazolium (NBT) chloride. Also, the strand breakage of DNA was characterized by atomic force microscopy imaging; nucleobase damage was characterized by gel electrophoresis and base excision repair enzyme assays. ROS such as hydroxyl radical abstract hydrogen atoms form the bases or pyranose ring and create radical centers in DNA. Subsequent rearrangement of free radicals in DNA results in the strand breakage and nucleobase damage in DNA. The photosensitized strand breakage and nucleobase damage of DNA suggest that QDs are emerging PS drugs for nucleus-targeted PDT if combined with intranuclear delivery of QDs in cancer cells. Also, Liang *et al.* reported that UV illumination of a mixture of calf thymus DNA and CdSe QDs results in DNA nicking.[118] This change is attributed to the reactions of DNA with ROS. Clarke *et al.* also reported that photoactivation of the QD–dopamine complex internalized in A9 cells results in DNA damage due to the production of $^1O_2$.[119] However, the production of $^1O_2$ comes from electron transfer from QD to dopamine followed by the oxidation of dopamine. Juzenas *et al.* investigated the potential of QDs as PS drugs for PDT.[120] They found that NIR photoactivation of QDs in cancer cells results in the production of ROS and reactive nitrogen species (RNS) such as superoxide and peroxynitrite. They employed dihydrorhodamine 123 as a sensor for the oxidation, and found that RONS generated by QDs results in the breakage of lysosomes. In contrast to the reports by Samia *et al.*[116] and Anas *et al.*,[117] specific tests made by Juzenas *et al.* using 9,10-dimethylanthracene, a $^1O_2$ scavenger, and $^1O_2$ Sensor Green™ indicated that $^1O_2$ is not produced by QDs under direct photoactivation.[120,121]

The properties of QDs, such as photostability, photosensitized production of ROS and RNS, and damage and breakage of DNA and lysosomes, show the potential of QDs for PDT. However, cytotoxicity of QDs due to photo-oxidation and chemical degradation should be resolved. For example, Derfus *et al.* found that CdSe QDs release toxic levels of cadmium ions inside cells and result in cell death.[89] Similarly, Cho *et al.* found that human breast cancer cells MCF-7 treated with cysteine or mercaptoacetic acid-capped CdTe QDs results in serious mitochondrial impairment, cell death due to both the release of cadmium ions through surface-etching, and the production of superoxide through electron transfer.[122]

D. QD and Photosensitizer Hybrids as Drugs for PDT

There are several advantages and limitations for both conventional PS drugs and QDs when individually applied for PDT. For example, the properties of QDs such as NIR absorption, large two-photon absorption cross section, broad absorption band, and photostability are favorable for PDT. In contrast to these unique optical properties of QDs, narrow absorption band, poor photostability, visible light absorption, and small two-photon absorption cross section of conventional PS drugs are less attractive for PDT. However, the efficiency (N75%) for ROS production by PS drugs is superior to that by QDs ($< 5\%$). In other words, the advantages and limitations of QDs and PS drugs complement each other. Thus, to utilize the photostability of QDs and improve the production of $^1O_2$, several conjugates and hybrids of QDs and conventional PS drugs were investigated as new-generation drugs for PDT. In such hybrid QD–PS pairs, the excited singlet ($^1PS^*$) and triplet ($^3PS^*$) states of PS drugs are indirectly generated by nonradiative energy transfer, also called Förster resonance energy transfer (FRET) from photoactivated QDs (QD*). The indirect photoactivation minimizes photobleaching of PS drugs. Also, the large surface area and biocompatibility of QDs allow them to be conjugated with hydrophobic multiple PS molecules. The indirectly excited PS drugs form collision complexes (QD-$^3PS^*$-$^3O_2$) with oxygen, transfer energy to $^3O_2$ and generate $^1O_2$ and other ROS.

Samia *et al.* first demonstrated the concept of FRET-based production of $^1O_2$ by QD–PS pairs by preparing a noncovalent mixture composed of CdSe QDs and a silicon phthalocyanine (Pc4).[116] They selected Pc4 for its high $^1O_2$ efficiency (43%) under direct photoactivation. In the QD–Pc4 pair, QD acts as the energy donor to Pc4, and Pc4 acts as both an energy acceptor from QD and an energy donor to $^3O_2$. Thus, high quantum efficiency in generating $^1O_2$ and stability of the hybrid pair were anticipated. However, according to the principle underlying FRET, the energy transfer efficiency inversely varies with the sixth power of the distance between a donor and an acceptor. Thus, close conjugation, typically within 10 nm, of PS drugs to QDs is necessary for efficient energy transfer and ROS production. Medintz and Mattoussi described the standards for the construction of energy donor–acceptor QD–PS pairs (Ref. 123, and the reference therein). Following the first investigation of QD–PS pair,[116] many researchers were attracted to the energy transfer properties of covalent and noncovalent QD–PS pairs composed of CdSe, CdSe/CdS/ZnS, CdSe/ZnS, and CdTe QDs as energy donors and various chromophores such as porphyrins, phthalocyanines, inorganic complexes, and other organic dyes as energy acceptors. Depending on the energy acceptor, the QD–PS pairs can be classified into QD–phthalocyanines, QD–porphines, QD–organic dyes, and QD–inorganic dyes.

1. QD–Phthalocyanine Conjugates for FRET and Singlet Oxygen Production

Phthalocyanine-conjugated QDs (QD-Pc) were widely investigated for energy transfer and $^1O_2$ production because of the high triplet quantum efficiency and long-living triplet state of Pc. Burda and coworkers extended investigations of energy transfer and $^1O_2$ production into many QD–Pc conjugates as functions of donor–acceptor distance, the relative number of QDs and Pcs, terminal functional group in Pc, bulk nature of spacers between donors and acceptors, the mode of binding between QD and Pc, and the size and surface states of QDs. For example, they used fluorescence up-conversion and transient absorption techniques. These techniques are valuable for characterizing the energy transfer kinetics from various exciton states in photoactivated QDs, and investigated energy transfer from CdSe QDs to silicon Pc molecules bearing one or two axial functional groups, such as thiol, hydroxyl, tertiary alkyl, and tertiary amine.[124,125] Figure 8 shows examples of Pc molecules that were used as energy acceptors in QD–Pc pairs. For these molecules, the energy transfer efficiency decreased with increasing length and the bulkiness of spacers between QD and Pc.[124,125] In addition, functional groups such as amine and thiol in Pc were found to play important roles in both QD to Pc bonding and quenching of the excited state of QDs. In particular, the energy transfer efficiency was found higher when Pc molecules were linked to QDs through two axial amine or thiol groups. Dayal *et al.* detected up to 70% efficiency for energy transfer from QDs to a primary amine-terminated Pc.[124] Also, quenching of QD's photoluminescence was effective for 1:1 and 1:2 conjugates between QD and Pc; however, the energy transfer efficiency decreased with increasing Pc per QD owing to self-absorbance,[124] indicating that many PS on the surface of a QD will be less attractive for PDT. One of the reasons for different energy transfer efficiencies for QD–Pc pairs linked through bulky or amine/thiol/alkyl functional groups was different electronic coupling between the donor and acceptor. Another important factor in the energy transfer efficiency is the surface states of QDs. The factor was identified from the nonlinear relationship between spectral overlap integral and energy transfer efficiency for QD–Pc pairs.[125] In short, Burda and coworkers have concluded that 1:1 or 1:2 complexes between QDs and PS molecules bearing two axial amine or thiol functional groups and non-bulky and short spacers would be ideal QD–PS donor–acceptor pairs for efficient energy transfer and $^1O_2$ production.

In preparation of QD–Pc pairs, energy transfer from QD to Pc and the generation of $^1O_2$ were further extended into construction of complexes between CdTe QDs and tetrasulfonated aluminum Pc (AlTSPc) pairs.[126–128] Here, Nyokong and coworkers prepared CdTe–AlTSPc mixtures by adding solutions of AlTSPc with various concentrations to solutions of CdTe QDs tethered with mercaptocarboxylic acids, such as thioglycolic acid (TDA), 3-MPA, and L-lysine.[126] In this mixture, the excited state of QDs was quenched and resulted in an increase in the triplet yield for AlTSPc along with fluorescence emission from AlTSPc. Among the CdTe QDs with TDA, MPA, and L-lysine, MPA-capped CdTe QDs provided long-living triplet state of AlTSPc. This character was attributed to the strong binding between AlTSPc and MPA. Later, they found that the CdTe–AlTSPc complex produces $^1O_2$ at 9.5–15% yield that was determined using phosphorescence decay of $^1O_2$ in the presence and absence of sodium azide, a $^1O_2$ scavenger.[127] They extended energy transfer investigations to various metallophthalocyanines (TSPc) linked to CdTe QDs through sulfonic acid, carboxylic acid, and pyridinium group.[128] By varying the metal ion and the functional groups in Pc, they obtained QD–Pc pairs with high triplet yields and energy transfer efficiencies (up to 80%). The most important properties of the CdTe–TSPc pairs are their water solubility and photosensitized production of $^1O_2$. However, the mode of binding between CdTe QDs and sulfonated Pcs, correlation between the quenching of QD's excited state and the formation of both the triplet and singlet states of TSPcs, toxicity due to cadmium, and the potential of QD–Pc pairs for *in vitro* and *in vivo* PDT need further attention.

2. QD–Porphine Conjugates for FRET and Singlet Oxygen Production

Porphines are classical photosensitizers clinically applied for PDT of various cancers due to their high triplet yields and high efficiencies for ROS production. However, as with most phthalocyanines, poor water solubility, inadequate mechanism for selective delivery in tumors, and lack of NIR absorption are major drawbacks of porphines for PDT. Tsay *et al.* eliminated most of these drawbacks by coating Chlorin e6 on the surface of CdSe/CdS/ZnS QDs either non-covalently binding an alkyl amine linker or covalently binding a lysine-terminated peptide linker.[129] They found that the photoluminescence lifetime of QDs decreased as a result of energy transfer from QDs to

Chlorin e6. Also, in contrast to the previous report,[124] Tsay *et al.* found that the energy transfer efficiency from QD to Chlorin e6 increased with increasing Chlorin e6 molecules attached to a single QD.[129] The QD–Chlorin e6 conjugate provided $^1O_2$ at 31% efficiency. Another example for a water-soluble QD–porphine pair for $^1O_2$ production is CdTe QDs electrostatically coated by a meso-tetra(4-sulfonatophenyl)porphine (TSPP), investigated by Shi *et al.*[130] The CdTe–TSPP composite produced $^1O_2$ at 43% efficiency when photoactivated at 355 nm. At this wavelength, both the donor and acceptor were directly photoactivated. Thus, the quantum efficiency for FRET-based $^1O_2$ production was probably overrated. However, based on an assumption that QDs quench the directly excited triplet state of an acceptor, Tsay *et al.* ruled out the production of $^1O_2$ through direct photoactivation of an acceptor in the proximity of a QD.[129] Moreover, in contrast to the production of $^1O_2$ and other ROS by CdSe QDs as reported by Samia *et al.*[116] and Anas *et al.*,[117] $^1O_2$ production was not detected for CdTe QDs alone, indicating that QD–PS pairs are an ideal tool for PDT compared with QDs alone. Despite the above two reports on QD–porphine pairs for energy transfer and $^1O_2$ production, systematic investigations of the relations among energy transfer, donor–acceptor distance, size of QDs, dielectric constant of the medium, and the efficiency for $^1O_2$ production remain open.

3. QD–Organic and Inorganic Dye Pairs for FRET and Singlet Oxygen Production

Organic and inorganic dyes having high triplet quantum efficiencies are potential energy acceptors for the construction of QD–PS pairs for $^1O_2$ and other ROS production and PDT. Tsay *et al.* investigated conjugates of Rose Bengal on the surface of CdSe/CdS/ZnS QDs through a lysine-terminated peptide linker.[129] In the conjugates, the photoluminescence lifetime of the QDs was considerably reduced, indicating efficient FRET from QD to Rose Bengal. Furthermore, they investigated the production of $^1O_2$ by recording the steady-state absorption spectrum of anthracene dipropionic acid, a widely known $^1O_2$ scavenger, and the phosphorescence spectrum of $^1O_2$ at 1270 nm. The $^1O_2$ quantum efficiency for QD–Rose Bengal conjugate excited at 355 nm was 17%. Here, the production of $^1O_2$ through direct excitation of the acceptor was ruled as stated in the previous section. The low quantum efficiency for $^1O_2$ production was attributed to inefficient energy transfer because of poor donor–acceptor spectral overlap integral. Interestingly, by selecting Chlorin e6 as the energy acceptor, they achieved 31% quantum efficiency for $^1O_2$ owing to better overlap between the photoluminescence spectrum of QDs and the absorption spectrum of Chlorin e6. Other examples of organic dyes for the preparation of QD–PS pairs are Merocyanine 540 (MC540) and Toluidine Blue O (TBO).[131,132] From steady-state and time-resolved fluorescence measurements, efficient FRET from CdSe/ZnS QDs to MC540 was detected,[131] a chemotherapeutic drug. Here, FRET efficiency was determined from the quenching of the steady-state and time-resolved photoluminescence of QDs. Narband *et al.* utilized the advantages of QD–PS pairs for photodynamic killing of bacteria by applying a mixture of NIR QD and TBO. Photoactivation of QDs resulted in FRET from QD to TBO and the production of $^1O_2$.[132] Here, the high molar extinction coefficient of QDs in the short-wavelength region and efficient overlap between the photoluminescence spectrum of QDs and the absorption spectrum of TBO were advantageous for the generation of various cytotoxic species including $^1O_2$. Energy transfer, $^1O_2$ production, and bactericidal action for TBO and QD mixtures were discussed in terms of ionic interactions between QD and TBO.

Covalent conjugates and physical mixtures between QDs and inorganic dyes are another class of donor–acceptor pairs with potentials for PDT. For example, Hsieh *et al.* conjugated iridium complexes with CdSe/ZnS QDs and prepared covalent donor– acceptor pairs.[133] Photoactivation of a deoxygenated solution of the conjugate resulted in a weak phosphorescence with a 2.1-μs decay component from the Ir complex. The phosphorescence disappeared when the solution was aerated. Here, the excited state of the Ir complex was generated through FRET from QDs. The disappearance of the phosphorescence during aeration was due to the quenching of the excited state of Ir complex by $^3O_2$ and the formation of $^1O_2$. Although high quantum efficiency (97%) for $^1O_2$ production was estimated for the QD–Ir complex, the roles of nonradiative relaxations of QDs and the Ir complex, spectral overlap integral, and donor–acceptor distance are yet to be addressed. Another example for QD–inorganic dye pairs was investigated by preparing a physical mixture between CdSe/ZnS QDs and trans-Cr(cyclam)Cl_2.[134] Here, the excited state of QD was quenched by the Cr complex, evidenced from a nonlinear Stern–Volmer quenching kinetics and a decrease in the photoluminescence lifetime of QDs with increasing concentration of Cr complex. The spectral overlap integral for the QD–Cr complex was ideal for efficient FRET.

Preparation of QD–PS pairs such as noncovalent and covalent assemblies between QDs and organic chromophores as well as investigation of energy transfer and $^1O_2$ production are emerging research topics with great potentials for environment and health management. The significance of QD–PS pairs compared to conventional PS is that the unparalleled photostability of QDs offers durability. Despite the reports by Tsay *et al.*[129] and other authors,[131–134] systematic investigations of the donor–acceptor distance, donor–acceptor spectral overlap integral, donor–acceptor orientation, efficiency of $^1O_2$ production, toxicity of the donor–acceptor pairs, and *in vitro* and *in vivo* PDT are important issues remaining to be investigated. In particular, QD–PS pairs composed of QDs with small size and without heavy metals would bring radical changes to PDT of cancer.

VI. Concluding Remarks

The important property of QDs in bioimaging is its bright and stable fluorescence of choice. This chapter introduces a variety of methods for interfacing QDs with cells, and the versatility of QDs conjugated with polymers, proteins, peptides, antibodies, ligands, and PS drugs for specific and nonspecific imaging of cells, subcellular organelles, and single molecules as well as PDT. The potential of QDs for cancer imaging was tested for *in vitro* and *in vivo* imaging and PDT of cancer. Now, synthesis, optical properties, and surface chemistry of QDs are all well optimized; therefore, the preparation of bioconjugated QDs is not a problem anymore. Also, ready-to-label QD bioconjugates of any color are now commercially available. Having such a well-built background for QDs, it is expected that the information in this chapter will be useful for starting *in vitro* and *in vivo* biological as well as medical applications of QDs, utilizing QD bioconjugates for investigating complex biophysical and biochemical problems, and developing novel hybrid nanomaterials by interfacing QDs and bioconjugated QDs with organic, semiconductor, and metal nanoparticles. Although the toxicity of QDs is a matter of considerable discussion, the focus of QDs in biology has been changing to multiplexed and multimodal imaging *in vivo*, in particular, in the field of noninvasive imaging of cancers. To raise the issue of high-quality QDs and develop nontoxic QDs to offer all the advantages of QDs to biomedical imaging and therapy, many researchers have invested invaluable efforts.

References

1. Brus LE. Electron-electron and electron-hole interactions in small semiconductor crystallites: the size dependence of the lowest excited electronic state. *J Chem Phys* 1984;**80**:4403–9.
2. Murray CB, Norris DJ, Bawendi MG. Synthesis and characterization of nearly monodisperse CdE (E = S, Se, Te) semiconductor nanocrystallites. *J Am Chem Soc* 1993;**115**:8706–15.
3. Biju V, Itoh T, Anas A, Sujith A, Ishikawa M. Semiconductor quantum dots and metal nanoparticles: syntheses, optical properties, and biological applications. *Anal Bioanal Chem* 2008;**391**:2469–95.
4. Han MY, Gao XH, Su JZ, Nie S. Quantum-dot-tagged microbeads for multiplexed optical coding of biomolecules. *Nat Biotechnol* 2001;**19**:631–5.
5. Burda C, Chen X, Narayanan R, El-Sayed MA. Chemistry and properties of nanocrystals of different shapes. *Chem Rev* 2005;**105**:1025–102.
6. Nirmal M, Dabbousi BO, Bawendi MG, Macklin JJ, Trautman JK, Harris TD, et al. Fluorescence intermittency in single cadmium selenide nanocrystals. *Nature* 1996;**383**:802–4.
7. Hohng S, Ha T. Near-complete suppression of quantum dot blinking in ambient conditions. *J Am Chem Soc* 2004;**126**:1324–5.
8. Hamada M, Nakanishi S, Itoh T, Ishikawa M, Biju V. Blinking suppression in CdSe/ZnS single quantum dots by TiO_2 nanoparticles. *ACS Nano* 2008;**8**:4445–54.

9. Wang X, Ren X, Kahen K, Hahn MA, Rajeswaran M, Maccagnano-Zacher S, et al. Non-blinking semiconductor nanocrystals. *Nature* 2009;**459**:686–9.
10. Larson DR, Zipfel WR, Williams RM, Clark SW, Bruchez MP, Wise FW, et al. Water-soluble quantum dots for multiphoton fluorescence imaging in vivo. *Science* 2003;**300**:1434–6.
11. Michalet X, Pinaud FF, Bentolila LA, Tsay JM, Doose S, Li JJ, et al. Quantum dots for live cells, in vivo imaging, and diagnostics. *Science* 2005;**307**:538–44.
12. Gao XH, Cui YY, Levenson RM, Chung LWK, Nie S. In vivo cancer targeting and imaging with semiconductor quantum dots. *Nat Biotechnol* 2004;**22**:969–76.
13. Cai WB, Shin DW, Chen K, Gheysens O, Cao QZ, Wang SX, et al. Peptide-labeled near infrared quantum dots for imaging tumor vasculature in living subjects. *Nano Lett* 2006;**6**:669–76.
14. Akerman ME, Chan WCW, Laakkonen P, Bhatia SN, Ruoslahti E. Nanocrystal targeting in vivo. *Proc Natl Acad Sci USA* 2002;**99**:12617–21.
15. Biju V, Mundayoor S, Omkumar RV, Anas A, Ishikawa M. Bioconjugated quantum dots for cancer research: present status, prospects and remaining issues. *Biotechnol Adv* 2010;**28**:199–213.
16. Katari JEB, Colvin VL, Alivisatos AP. X-Ray photoelectron-spectroscopy of CdSe nanocrystals with applications to studies of the nanocrystal surface. *J Phys Chem* 1994;**98**:4109–17.
17. Peng ZA, Peng XG. Formation of high-quality CdTe, CdSe, and CdS nanocrystals using CdO as precursor. *J Am Chem Soc* 2001;**123**:183–4.
18. Talapin DV, Rogach AL, Kornowski A, Haase M, Weller H. Highly luminescent monodisperse CdSe and CdSe/ZnS nanocrystals synthesized in a hexadecylamine-trioctylphosphine oxide-trioctylphospine mixture. *Nano Lett* 2001;**1**:207–11.
19. Vossmeyer T, Katsikas L, Giersig M, Popovic IG, Diesner K, Chemseddine A, et al. CdS nanoclusters—synthesis, characterization, size-dependent oscillator strength, temperature shift of the excitonic-transition energy, and reversible absorbency shift. *J Phys Chem* 1994;**98**:7665–73.
20. Qu LH, Peng ZA, Peng XG. Alternative routes toward high quality CdSe nanocrystals. *Nano Lett* 2001;**1**:333–7.
21. Peng ZA, Peng XG. Nearly monodisperse and shape-controlled CdSe nanocrystals via alternative routes: nucleation and growth. *J Am Chem Soc* 2002;**124**:3343–53.
22. Gaponik N, Talapin DV, Rogach AL, Hoppe K, Shevchenko EV, Kornowski A, et al. Thiol-capping of CdTe nanocrystals: an alternative to organometallic synthetic routes. *J Phys Chem B* 2002;**106**:7177–85.
23. Bharali DJ, Lucey DW, Jayakumar H, Pudavar HE, Prasad PN. Folate-receptor-mediated delivery of InP quantum dots for bioimaging using confocal and two-photon microscopy. *J Am Chem Soc* 2005;**127**:11364–71.
24. Reiss P, Protiere M, Li L. Core/shell semiconductor nanocrystals. *Small* 2009;**5**:154–68.
25. Chan WCW, Nie S. Quantum dot bioconjugates for ultrasensitive nonisotopic detection. *Science* 1998;**281**:2016–8.
26. Bentzen EL, Tomlinson ID, Mason J, Gresch P, Warnement MR, Wright D, et al. Surface modification to reduce nonspecific binding of quantum dots in live cell assays. *Bioconjug Chem* 2005;**16**:1488–94.
27. Walling MA, Novak JA, Shepard JRE. Quantum dots for live cell and in vivo imaging. *Int J Mol Sci* 2009;**10**:441–91.
28. Smith AM, Duan HW, Mohs AM, Nie S. Bioconjugated quantum dots for in vivo molecular and cellular imaging. *Adv Drug Deliv Rev* 2008;**60**:1226–40.
29. Jaiswal JK, Mattoussi H, Mauro JM, Simon SM. Long-term multiple color imaging of live cells using quantum dot bioconjugates. *Nat Biotechnol* 2003;**21**:47–51.

30. Gomez N, Winter JO, Shieh F, Saunders AE, Korgel BA, Schmidt CE. Challenges in quantum dot-neuron active interfacing. *Talanta* 2005;**67**:462–71.
31. Breus VV, Heyes CD, Tron K, Nienhaus GU. Zwitterionic biocompatible quantum dots for wide pH stability and weak nonspecific binding to cells. *ACS Nano* 2009;**3**:2573–80.
32. Murcia MJ, Minner DE, Mustata GM, Ritchie K, Naumann CA. Design of quantum dot-conjugated lipids for long-term, high-speed tracking experiments on cell surfaces. *J Am Chem Soc* 2008;**130**:15054–62.
33. Ryman-Rasmussen JP, Riviere JE, Monteiro-Riviere NA. Surface coatings determine cytotoxicity and irritation potential of quantum dot nanoparticles in epidermal keratinocytes. *Nano Lett* 2007;**7**:1344–8.
34. Mortensen LJ, Oberdorster G, Pentland AP, DeLouise LA. In vivo skin penetration of quantum dot nanoparticles in the murine model: the effect of UVR. *Nano Lett* 2008;**8**:2779–87.
35. Gopalakrishnan G, Danelon C, Izewska P, Prummer M, Bolinger PY, Geissbuhler I, et al. Multifunctional lipid/quantum dot hybrid nanocontainers for controlled targeting of live cells. *Angew Chem Int Ed* 2006;**45**:5478–83.
36. Chen F, Gerion D. Fluorescent CdSe/ZnS nanocrystal-peptide conjugates for long-term, nontoxic imaging and nuclear targeting in living cells. *Nano Lett* 2004;**4**:1827–32.
37. Ballou B, Ernst LA, Andreko S, Harper T, Fitzpatrick JAJ, Waggoner AS, et al. Sentinel lymph node imaging using quantum dots in mouse tumor models. *Bioconjug Chem* 2007;**18**:389–96.
38. Gussin HA, Tomlinson ID, Little DM, Warnement MR, Qian HH, Rosenthal SJ, et al. Binding of muscimol-conjugated quantum dots to GABA(c) receptors. *J Am Chem Soc* 2006;**128**:15701–13.
39. Jayagopal A, Russ PK, Haselton FR. Surface engineering of quantum dots for in vivo vascular imaging. *Bioconjug Chem* 2007;**18**:1424–33.
40. Liu W, Howarth M, Greytak AB, Zheng Y, Nocera DG, Ting AY, et al. Compact biocompatible quantum dots functionalized for cellular imaging. *J Am Chem Soc* 2008;**130**:1274–84.
41. Kikkeri R, Lepenies B, Adibekian A, Laurino P, Seeberger PH. In vitro imaging and in vivo liver targeting with carbohydrate capped quantum dots. *J Am Chem Soc* 2009;**131**:2110–1.
42. Winter JO, Liu TY, Korgel BA, Schmidt CE. Recognition molecule directed interfacing between semiconductor quantum dots and nerve cells. *Adv Mater* 2001;**13**:1673–7.
43. Howarth M, Liu WH, Puthenveetil S, Zheng Y, Marshall LF, Schmidt MM, et al. Monovalent, reduced-size quantum dots for imaging receptors on living cells. *Nat Methods* 2008;**5**:397–9.
44. Wu XY, Liu HJ, Liu JQ, Haley KN, Treadway JA, Larson JP, et al. Immunofluorescent labeling of cancer marker Her2 and other cellular targets with semiconductor quantum dots. *Nat Biotechnol* 2003;**21**:41–6.
45. Weng KC, Noble CO, Papahadjopoulos-Sterberg B, Chen FF, Drummond DC, Kirpotin DB, et al. Targeted tumor cell internalization and imaging of multifunctional quantum dot-conjugated immunoliposomes in vitro and in vivo. *Nano Lett* 2008;**8**:2851–7.
46. Yezhelyev MV, Al-Hajj A, Morris C, Marcus AI, Liu T, Lewis M, et al. In situ molecular profiling of breast cancer biomarkers with multicolor quantum dots. *Adv Mater* 2007;**19**:3146–51.
47. Huang DH, Su L, Peng XH, Zhang HZ, Khuri FR, Shin DM, et al. Quantum dot-based quantification revealed differences in subcellular localization of EGFR and E-cadherin between EGFR-TKI sensitive and insensitive cancer cells. *Nanotechnology* 2009;**20**:225102.
48. Barat B, Sirk SJ, McCabe KE, Li JQ, Lepin EJ, Remenyi R, et al. Cys-diabody quantum dot conjugates (immunoQdots) for cancer marker detection. *Bioconjug Chem* 2009;**20**:1474–81.
49. Lim YT, Cho MY, Noh YM, Chung JW, Chung BH. Near-infrared emitting fluorescent nanocrystals-labeled natural killer cells as a platform technology for the optical imaging of immunotherapeutic cells-based cancer therapy. *Nanotechnology* 2009;**20**:475102.

50. Zhang H, Sachdev D, Wang C, Hubel A, Gaillard-Kelly M, Yee D. Detection and down regulation of type I IGF receptor expression by antibody-conjugated quantum dots in breast cancer cells. *Breast Cancer Res Treat* 2009;**114**:277–85.
51. Kawashima N, Nakayama K, Itoh K, Itoh T, Ishikawa M, Biju V. Reversible dimerization of EGFR revealed by single-molecule fluorescence imaging using quantum dots. *Chem Eur J* 2010;**16**:1186–92.
52. Kim BYS, Jiang W, Oreopoulos J, Yip CM, Rutka JT, Chan WCW. Biodegradable quantum dot nanocomposites enable live cell labeling and imaging of cytoplasmic targets. *Nano Lett* 2008;**8**:3887–92.
53. Dahan M, Levi S, Luccardini C, Rostaing P, Riveau B, Triller A. Diffusion dynamics of glycine receptors revealed by single-quantum dot tracking. *Science* 2003;**302**:442–5.
54. Bouzigues C, Morel M, Triller A, Dahan M. Asymmetric redistribution of GABA receptors during GABA gradient sensing by nerve growth cones analyzed by single quantum dot imaging. *Proc Natl Acad Sci USA* 2007;**104**:11251–6.
55. Chen HF, Titushkin I, Stroscio M, Cho M. Altered membrane dynamics of quantum dot-conjugated integrins during osteogenic differentiation of human bone marrow derived progenitor cells. *Biophys J* 2007;**92**:1399–408.
56. Howarth M, Takao K, Hayashi Y, Ting AY. Targeting quantum dots to surface proteins in living cells with biotin ligase. *Proc Natl Acad Sci USA* 2005;**102**:7583–8.
57. Chen I, Choi YA, Ting AY. Phage display evolution of a peptide substrate for yeast biotin ligase and application to two-color quantum dot labeling of cell surface proteins. *J Am Chem Soc* 2007;**129**:6619–25.
58. Roullier V, Clarke S, You C, Pinaud F, Gouzer G, Schaible D, et al. High-affinity labeling and tracking of individual histidine-tagged proteins in live cells using Ni^{2+} tris-nitrilotriacetic acid quantum dot conjugates. *Nano Lett* 2009;**9**:1228–34.
59. Bonasio R, Carman CV, Kim E, Sage PT, Love KR, Mempel TR, et al. Specific and covalent labeling of a membrane protein with organic fluorochromes and quantum dots. *Proc Natl Acad Sci USA* 2007;**104**:14753–8.
60. Kim J, Park HY, Kim J, Ryu J, Kwon DY, Grailhe R, et al. Ni-nitrilotriacetic acid-modified quantum dots as a site-specific labeling agent of histidine-tagged proteins in live cells. *Chem Commun* 2008;1910–2.
61. Dif A, Boulmedais F, Pinot M, Roullier V, Baudy-Floc'h M, Coquelle FM, et al. Small and stable peptidic pegylated quantum dots to target polyhistidine-tagged proteins with controlled stoichiometry. *J Am Chem Soc* 2009;**131**:14738–46.
62. So MK, Yao HQ, Rao JH. HaloTag protein-mediated specific labeling of living cells with quantum dots. *Biochem Biophys Res Commun* 2008;**374**:419–23.
63. Mulder WJM, Koole R, Brandwijk RJ, Storm G, Chin PTK, Strijkers GJ, et al. Quantum dots with a paramagnetic coating as a bimodal molecular imaging probe. *Nano Lett* 2006;**6**:1–6.
64. Pinaud F, King D, Moore HP, Weiss S. Bioactivation and cell targeting of semiconductor CdSe/ZnS nanocrystals with phytochelatin-related peptides. *J Am Chem Soc* 2004;**126**:6115–23.
65. Derfus AM, Chan WCW, Bhatia SN. Intracellular delivery of quantum dots for live cell labeling and organelle tracking. *Adv Mater* 2004;**16**:961–4.
66. Hsieh SC, Wang FF, Lin CS, Chen YJ, Hung SC, Wang YJ. The inhibition of osteogenesis with human bone marrow mesenchymal stem cells by CdSe/ZnS quantum dot labels. *Biomaterials* 2006;**27**:1656–64.
67. Lagerholm BC, Wang MM, Ernst LA, Ly DH, Liu HJ, Bruchez MP, et al. Multicolor coding of cells with cationic peptide coated quantum dots. *Nano Lett* 2004;**4**:2019–22.

68. Zhang TT, Stilwell JL, Gerion D, Ding LH, Elboudwarej O, Cooke PA, et al. Cellular effect of high doses of silica-coated quantum dot profiled with high throughput gene expression analysis and high content cellomics measurements. *Nano Lett* 2006;**6**:800–8.
69. Duan H, Nie S. Cell-penetrating quantum dots based on multivalent and endosome-disrupting surface coatings. *J Am Chem Soc* 2007;**129**:3333–8.
70. Mok H, Park JW, Park TG. Enhanced intracellular delivery of quantum dot and adenovirus nanoparticles triggered by acidic pH via surface charge reversal. *Bioconjug Chem* 2008;**19**:797–801.
71. Yezhelyev MV, Qi LF, O'Regan RM, Nie S, Gao XH. Proton-sponge coated quantum dots for siRNA delivery and intracellular imaging. *J Am Chem Soc* 2008;**130**:9006–12.
72. Nair A, Shen JH, Thevenot P, Zou L, Cai T, Hu ZB, et al. Enhanced intratumoral uptake of quantum dots concealed within hydrogel nanoparticles. *Nanotechnology* 2008;**19**:485102.
73. Qi LF, Gao XH. Quantum dot-amphipol nanocomplex for intracellular delivery and real-time imaging of siRNA. *ACS Nano* 2008;**2**:1403–10.
74. Silver J, Ou W. Photoactivation of quantum dot fluorescence following endocytosis. *Nano Lett* 2005;**5**:1445–9.
75. Delehanty JB, Medintz IL, Pons T, Brunel FM, Dawson PE, Mattoussi H. Self-Assembled quantum dot-peptide bioconjugates for selective intracellular delivery. *Bioconjug Chem* 2006;**17**:920–7.
76. Zhang Y, So MK, Rao JH. Protease-modulated cellular uptake of quantum dots. *Nano Lett* 2006;**6**:1988–92.
77. Medintz IL, Pons T, Delehanty JB, Susumu K, Brunel FM, Dawson PE, et al. Intracellular delivery of quantum dot-protein cargos mediated by cell penetrating peptides. *Bioconjug Chem* 2008;**19**:1785–95.
78. Ruan G, Agrawal A, Marcus AI, Nie S. Imaging and tracking of tat peptide-conjugated quantum dots in living cells: new insights into nanoparticle uptake, intracellular transport, and vesicle shedding. *J Am Chem Soc* 2007;**129**:14759–66.
79. Lei Y, Tang H, Yao L, Yu R, Feng M, Zou B. Applications of mesenchymal stem cells labeled with tat peptide conjugated quantum dots to cell tracking in mouse body. *Bioconjug Chem* 2008;**19**:421–7.
80. Chen B, Liu QL, Zhang YL, Xu L, Fang XH. Transmembrane delivery of the cell-penetrating peptide conjugated semiconductor quantum dots. *Langmuir* 2008;**24**:11866–71.
81. Rozenzhak SM, Kadakia MP, Caserta TM, Westbrook TR, Stone MO, Naik RR. Cellular internalization and targeting of semiconductor quantum dots. *Chem Commun* 2005;2217–9.
82. Hoshino A, Manabe N, Fujioka K, Hanada S, Yasuhara M, Kondo A, et al. GFP expression by intracellular gene delivery of GFP-coding fragments using nanocrystal quantum dots. *Nanotechnology* 2008;**19**:495102.
83. Biju V, Muraleedharan D, Nakayama K, Shinohara Y, Itoh T, Baba Y, et al. Quantum dot-insect neuropeptide conjugates for fluorescence imaging, transfection, and nucleus targeting of living cells. *Langmuir* 2007;**23**:10254–61.
84. Anas A, Okuda T, Kawashima N, Nakayama K, Itoh T, Ishikawa M, et al. Clathrin mediated endocytosis of quantum dot-peptide conjugates in living cells. *ACS Nano* 2009;**2009**(3):2419–29.
85. Walther C, Meyer K, Rennert R, Neundorf I. Quantum dot-carrier peptide conjugates suitable for imaging and delivery applications. *Bioconjug Chem* 2008;**19**:2346–56.
86. Jayagopal A, Su YR, Blakemore JL, Linton MF, Fazio S, Haselton FR. Quantum dot mediated imaging of atherosclerosis. *Nanotechnology* 2009;**20**:165102.

87. Parak WJ, Boudreau R, Le Gros M, Gerion D, Zanchet D, Micheel CM, et al. Cell motility and metastatic potential studies based on quantum dot imaging of phagokinetic tracks. *Adv Mater* 2002;**14**:882–5.
88. Hanaki K, Momo A, Oku T, Komoto A, Maenosono S, Yamaguchi Y, et al. Semiconductor quantum dot/albumin complex is a long-life and highly photostable endosome marker. *Biochem Biophys Res Commun* 2003;**302**:496–501.
89. Derfus AM, Chan WCW, Bhatia SN. Probing the cytotoxicity of semiconductor quantum dots. *Nano Lett* 2004;**4**:11–8.
90. Li F, Zhang ZP, Peng J, Cui ZQ, Pang DW, Li K, et al. Imaging viral behavior in mammalian cells with self-assembled capsid-quantum-dot hybrid particles. *Small* 2009;**5**:718–26.
91. Sun YH, Liu YS, Vernier PT, Liang CH, Chong SY, Marcu L, et al. Photostability and pH sensitivity of CdSe/ZnSe/ZnS quantum dots in living cells. *Nanotechnology* 2006;**17**:4469–76.
92. Bruchez M, Moronne M, Gin P, Weiss S, Alivisatos AP. Semiconductor nanocrystals as fluorescent biological labels. *Science* 1998;**281**:2013–6.
93. Xu Y, Wang QB, He P, Dong QM, Liu F, Liu Y, et al. Cell nucleus penetration by quantum dots induced by nuclear staining organic fluorophore and UV-irradiation. *Adv Mater* 2008;**20**:3468–73.
94. Hasegawa U, Nomura SIM, Kaul SC, Hirano T, Akiyoshi K. Nanogel-quantum dot hybrid nanoparticles for live cell imaging. *Biochem Biophys Res Commun* 2005;**331**:917–21.
95. Xie M, Liu H, Chen MP, Zhang ZL, Wang XH, Xie ZX, et al. CdSe/ZnS-labeled carboxymethyl chitosan as a bioprobe for live cell imaging. *Chem Commun* 2005;5518–20.
96. de Farias PMA, Santos BS, Menezes FD, Brasil Jr. AG, Ferreira R, Motta MA, et al. Highly fluorescent semiconductor core-shell CdTe-CdS nanocrystals for monitoring living yeast cells activity. *Appl Phys A* 2007;**89**:957–61.
97. Tan WB, Jiang S, Zhang Y. Quantum-dot based nanoparticles for targeted silencing of HER2/neu gene via RNA interference. *Biomaterials* 2007;**28**:1565–71.
98. Tada H, Higuchi H, Watanabe TM, Ohuchi N. *In vivo* real-time tracking of single quantum dots conjugated with monoclonal anti-her2 antibody in tumors of mice. *Cancer Res* 2007;**67**:1138–44.
99. Cambi A, Lidke DS, Arndt-Jovin DJ, Figdor CG, Jovin TM. Ligand-conjugated quantum dots monitor antigen uptake and processing by dendritic cells. *Nano Lett* 2007;**7**:970–7.
100. Qian J, Yong KT, Roy I, Ohulchanskyy TY, Bergey EJ, Lee HH, et al. Imaging pancreatic cancer using surface-functionalized quantum dots. *J Phys Chem B* 2007;**111**:6969–72.
101. Yong KT, Ding H, Roy I, Law WC, Bergey EJ, Maitra A, et al. Imaging pancreatic cancer using bioconjugated InP quantum dots. *ACS Nano* 2009;**3**:502–10.
102. Chen LD, Liu J, Yu XF, He M, Pei XF, Tang ZY, et al. The biocompatibility of quantum dot probes used for the targeted imaging of hepatocellular carcinoma metastasis. *Biomaterials* 2008;**29**:4170–6.
103. Bottini M, D'Annibale F, Magrini A, Cerignoli F, Arimura Y, Dawson MI, et al. Quantum dot-doped silica nanoparticles as probes for targeting of T-lymphocytes. *Int J Nanomed* 2007;**2**:227–33.
104. Tekle C, van Deurs B, Sandvig K, Iversen TG. Cellular trafficking of quantum dot-ligand bioconjugates and their induction of changes in normal routing of unconjugated ligands. *Nano Lett* 2008;**8**:1858–65.
105. Koole R, van Schooneveld MM, Hilhorst J, Castermans JK, Cormode DP, Strijkers GJ, et al. Paramagnetic lipid-coated silica nanoparticles with a fluorescent quantum dot core: a new contrast agent platform for multimodality imaging. *Bioconjug Chem* 2008;**19**:2471–9.
106. Cui BX, Wu CB, Chen L, Ramirez A, Bearer EL, Li WP, et al. One at a time, live tracking of NGF axonal transport using quantum dots. *Proc Natl Acad Sci USA* 2007;**104**:13666–71.
107. Gao XL, Wang T, Wu BX, Chen J, Chen JY, Yue Y, et al. Quantum dots for tracking cellular transport of lectin-functionalized nanoparticles. *Biochem Biophys Res Commun* 2008;**377**:35–40.

108. Lieleg O, Lopez-Garcia M, Semmrich C, Auernheimer J, Kessler H, Bausch AR. Specific integrin labeling in living cells using functionalized nanocrystals. *Small* 2007;**3**:1560–5.
109. Chakraborty SK, Fitzpatrick JAJ, Phillippi JA, Andreko S, Waggoner AS, Bruchez MP, et al. Cholera Toxin B conjugated quantum dots for live cell labeling. *Nano Lett* 2007;**7**:2618–26.
110. Ochsner M. Photophysical and photobiological processes in the photodynamic therapy of tumours. *J Photochem Photobiol B* 1997;**39**:1–18.
111. Oleinick NL, Evans HH. The photobiology of photodynamic therapy: cellular targets and mechanisms. *Radiat Res* 1998;**150**:S146–56.
112. Lachheb H, Puzenat E, Houas A, Ksibi M, Elaloui E, Guillard C, et al. Photocatalytic degradation of various types of dyes (Alizarin S, Crocein Orange G, Methyl Red, Congo Red, Methylene Blue) in water by UV-irradiated titania. *Appl Catal B Environ* 2002;**39**:75–90.
113. Dougherty TJ, Grindey GB, Fiel R, Weishaupt KR, Boyle DG. Photoradiation therapy. II. Cure of animal tumors with hematoporphyrin and light. *J Natl Cancer Inst* 1975;**55**:115–21.
114. Nseyo UO, Dehaven J, Dougherty TJ, Potter WR, Merrill DL, Lundahl SL, et al. Photodynamic therapy (PDT) in the treatment of patients with resistant superficial bladder cancer: a long-term experience. *J Clin Laser Med Surg* 1998;**16**:61–8.
115. Brigger I, Dubernet C, Couvreur P. Nanoparticles in cancer therapy and diagnosis. *Adv Drug Deliv Rev* 2002;**54**:631–51.
116. Samia ACS, Chen XB, Burda C. Semiconductor quantum dots for photodynamic therapy. *J Am Chem Soc* 2003;**125**:15736–7.
117. Anas A, Akita H, Harashima H, Itoh T, Ishikawa M, Biju V. Photosensitized breakage and damage of DNA by CdSe–ZnS quantum dots. *J Phys Chem B* 2008;**112**:10005–11.
118. Liang J, He Z, Zhang S, Huang S, Ai X, Yang H, et al. Study on DNA damage induced by CdSe quantum dots using nucleic acid molecular "light switches" as probe. *Talanta* 2007;**71**:1675–8.
119. Clarke SJ, Hollmann CA, Zhang Z, Suffern D, Bradforth SE, Dimitrijevic NM, et al. Photophysics of dopamine-modified quantum dots and effects on biological systems. *Nat Mater* 2006;**5**:409–17.
120. Juzenas P, Generalov R, Asta J, Juzeniene A, Moan J. Generation of nitrogen oxide and oxygen radicals by quantum dots. *J Biomed Nanotechnol* 2008;**4**:450–6.
121. Juzenas P, Chen W, Sun Y-P, Coelho MAN, Generalov R, Generalova N, et al. Quantum dots and nanoparticles for photodynamic and radiation therapies of cancer. *Adv Drug Deliv Rev* 2008;**60**:1600–14.
122. Cho SJ, Maysinger D, Jain M, Röder B, Hackbarth S, Winnik FM. Long-term exposure to CdTe quantum dots causes functional impairments in live cells. *Langmuir* 2007;**23**:1974–80.
123. Medintz IL, Mattoussi H. Quantum dot-based resonance energy transfer and its growing application in biology. *Phys Chem Chem Phys* 2009;**11**:17–45.
124. Dayal S, Krolicki R, Luo Y, Qiu X, Berlin JC, Kenney ME, et al. Femtosecond time-resolved energy transfer from CdSe nanoparticles to phthalocyanines. *Appl Phys B* 2006;**84**:309–15.
125. Dayal S, Li J, Li Y-S, Wu H, Samia ACS, Kenney ME, et al. Effect of the functionalization of the axial phthalocyanine ligands on the energy transfer in QD-based donor–acceptor pairs. *Photochem Photobiol* 2008;**84**:243–9.
126. Idowu M, Chen J-Y, Nyokong T. Photoinduced energy transfer between water-soluble CdTe quantum dots and aluminium tetrasulfonated phthalocyanine. *N J Chem* 2008;**32**:290–6.
127. Moeno S, Nyokong T. The photophysical studies of a mixture of CdTe quantum dots and negatively charged zinc phthalocyanines. *Polyhedron* 2008;**27**:1953–8.
128. Moeno S, Nyokong T. Interaction of water-soluble CdTe quantum dots with octacarboxy met allophthalocyanines: a photophysical and photochemical study. *J Lumin* 2009;**129**:356–62.
129. Tsay JM, Trzoss M, Shi L, Kong X, Selke M, Jung ME, et al. Singlet oxygen production by peptide-coated quantum dot-photosensitizer conjugates. *J Am Chem Soc* 2007;**129**:6865–71.

130. Shi L, Hernandez B, Selke M. Singlet oxygen generation from water-soluble quantum dot-organic dye nanocomposites. *J Am Chem Soc* 2006;**128**:6278–9.
131. Narayanan SS, Sinha SS, Pal SK. Sensitized emission from chemotherapeutic drug conjugated to CdSe/ZnS QDs. *J Phys Chem C* 2008;**112**:12716–20.
132. Narband N, Mubarak M, Ready D, Parkin IP, Nair SP, Green MA, et al. Quantum dots as enhancers of the efficacy of bacterial lethal photosensitization. *Nanotechnology* 2008;**19**:445102.
133. Hsieh JM, Ho ML, Wu PW, Chou PT, Tsai TT, Chi Y. Iridium-complex modified CdSe/ZnS quantum dots; a conceptual design for bifunctionality toward imaging and photosensitization. *Chem Commun* 2006;615–7.
134. Neuman D, Ostrowski AD, Mikhailovsky AA, Absalonson RO, Strouse GF, Ford PC. Quantum dot fluorescence quenching pathways with Cr(III) complexes. Photosensitized NO production from trans-Cr(cyclam)(ONO)2. *J Am Chem Soc* 2008;**130**:168–75.

Fabrication of Dendrimers Toward Biological Application

Hsieh-Chih Tsai* and Toyoko Imae*,†

*Graduate Institute of Applied Science and Technology, National Taiwan University of Science and Technology, Taipei, Taiwan, ROC

†Department of Chemical Engineering, National Taiwan University of Science and Technology, Taipei, Taiwan, ROC

I. Introduction 102
II. Are Dendrimers Attractive in Biomedical Sciences? 103
A. *In Vitro* Cytotoxicity of Dendrimers 106
B. *In Vivo* Toxicity of Dendrimers 107
C. How to Solve the Problems of Biomedical Toxicity 108
III. Drug Delivery Systems of Dendrimers 110
A. Load of Drugs by Physical Doping in Dendrimers 110
B. Release of Drugs Loaded by Physical Doping in Dendrimers 113
C. Delivery of Drugs Loaded by Chemical Binding in Dendrimers 114
IV. Targeting Delivery of Dendrimers 116
A. Folic Acid Dendrimers for Cancer Targeting 117
B. Antibody-Dendrimer for Cancer Targeting 117
C. Glycodendrimer for Cancer Targeting 118
V. Dendrimers for Therapy 119
A. Dendrimers for Boron–Neutron Therapy 119
B. Dendrimers for Photodynamic Therapy 120
C. Dendrimers for Photothermal Therapy 122
D. Dendrimers for Gene Therapy 122
VI. Dendrimers as Imaging Agents for Inspection 126
A. Molecular Probes 126
B. Magnetic Resonance Imaging 128
C. X-ray Imaging and CT 129
D. Optical Fluorescence Imaging 131
VII. Conclusions 132
References 132

Dendrimers consisting of inner core and peripheral shell are well-designed branching architectures with abundant terminal groups. Therefore, the high level of control over dendritic architectures makes dendrimers ideal carriers in biomedical application. In addition, the toxicity of dendrimers mainly comes from the high cationic charge density in the periphery, where charges interact with biological cell membrane and then result in membrane disruption. Two strategies have been utilized to minimize the toxicity of dendrimers: first,

Progress in Molecular Biology
and Translational Science, Vol. 104
DOI: 10.1016/B978-0-12-416020-0.00003-6

selecting neutral or anionic biocompatible dendrimers, and second, masking of peripheral charge by chemical modification. Then the focus of this chapter shifts to biomedical application of biocompatible dendrimers, including drug delivery systems of dendrimers, targeting delivery by dendrimers, and applications of dendrimers for therapy and as imaging agents for inspection.

I. Introduction

Dendrimers are globular, highly branched macromolecules possessing a well-defined interior region and a large number of end groups. The unique structural features of dendrimers is that there are a number of spacers whose ends combine with a high degree of branching which leads to a variety of new physical properties, different from traditional linear polymers. Dendrimers consist of a core, several branch points, and outer surface moieties, as shown in Fig. 1. The size of dendrimers is defined by generations, each of which corresponds to the layer of branching. Dendrimers are generally synthesized by a sequential repeat of reaction steps and each additional step derives the higher-generation dendrimer. Each created layer (generation) accompanies the doubling of the surface end groups for the two-functional branch and the remarkable increase of the molecular weight.

The ultimate goal of nanotechnology is the handling of technology at the nanoscale (at the range of 1–100 nm), that is, the handling of the inorganic particles, macromolecules, and supramolecules. Especially, nanobiotechnology

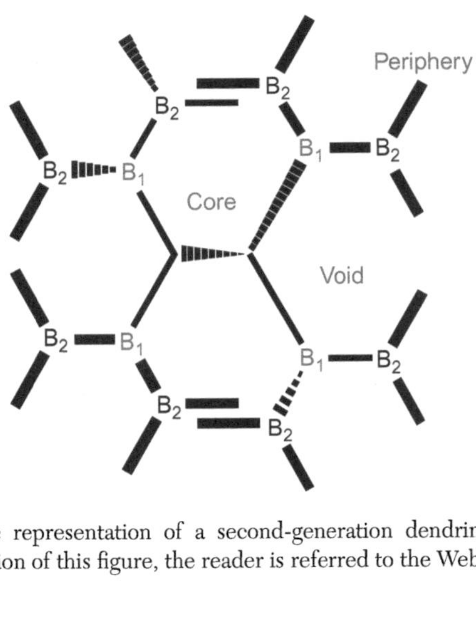

FIG. 1. Schematic representation of a second-generation dendrimer with two-functional branch. (For color version of this figure, the reader is referred to the Web version of this chapter.)

provides solutions for transformation of biosystems and offers a nanolevel control of techniques and processes on human life. The development of supramolecular nanostructures with well-defined particle size and shape are of prominent interest in the biological utilization such as delivery of pharmaceutical drugs and imaging agents or gene transfection. In general, architectures utilized as carriers in drug delivery should be in nanometer range and uniform in size to possess their ability for passing through cell membranes and to reduce the risk of undesired clearance from macrophage during blood circulation.

Dendrimers, one of such organic nanoparticles, provide a number of advantages compared to other polymeric nanoarchitectures which have been used for biological purposes. The structural regularity of dendrimers has motivated numerous studies with the aim of biomedical applications; for example, the enhancement of molecular effect or the creation of extremely high local concentration of drug, gene transfection, molecular label, or probe moiety. Moreover, the nanometer size of dendrimers and their narrow polydispersity can allow easier passage across biological barriers such as an extravasation through vascular endothelial tissue. The versatile poly(amido amine) (PAMAM) dendrimer and poly(propylenemine) (PPI) dendrimer as shown in Fig. 2 have been most widely investigated for these applications. The first- to tenth-generation PAMAM dendrimers with terminal groups such as —COOH, —NH_2, or —OH have been synthesized and then their size varied from 1.1 to 12.4 nm. In this chapter, the fabrication of dendrimers as biocompatible nanoparticles toward medical application is reviewed. After commenting on the overcoming of toxicity, the application to drug, and targeting delivery systems, therapy and inspection by imaging agents will be introduced.

II. Are Dendrimers Attractive in Biomedical Sciences?

Dendrimers have a three-dimensional structure of nanometer size without distribution of molecular weight. The unique characteristic properties of dendrimers are often compared with the globular shape of proteins and make this class of synthetic macromolecules suitable for the mimicry of proteins. In addition, dendrimers are on average less compact than proteins. The interior of dendrimers is not packed as efficiently as the typical proteins, and dendrimers possess a substantially large number of terminal functional groups superior to proteins of corresponding molecular weight.

It must be mentioned that the monodispersity is one of the important factors to launch dendrimers to the forefront of biomedical research. Monodispersity is a key, and it allows the investigation of explicit correlation between dendritic structure and biological system. By knowing the exact composition of

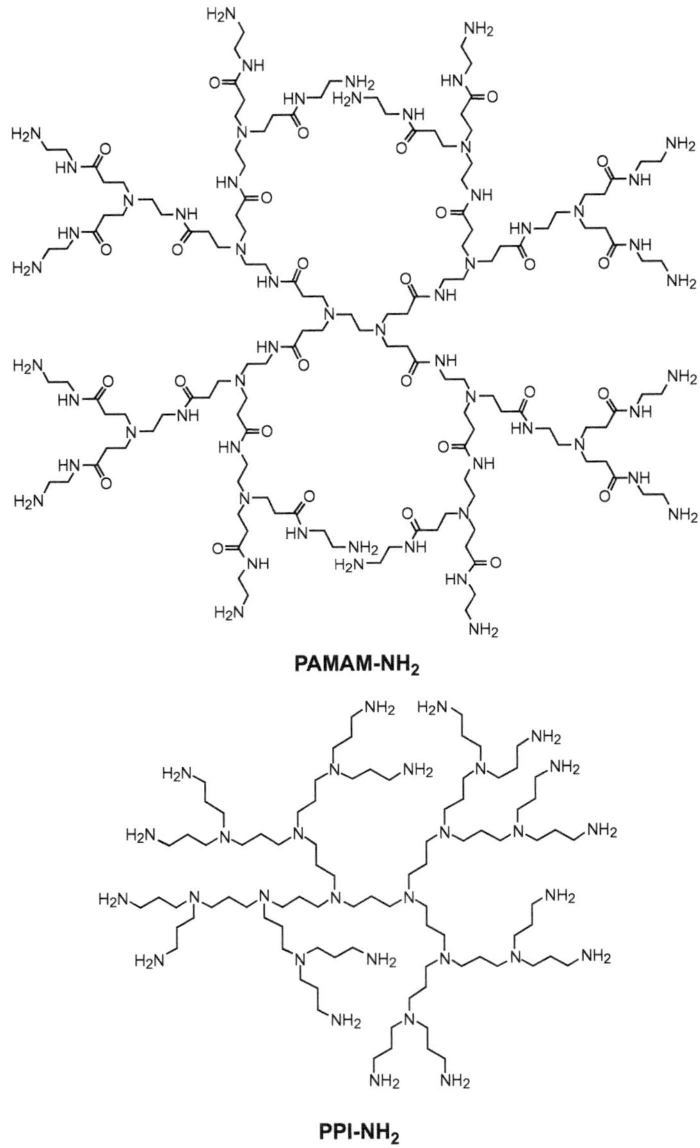

FIG. 2. Second-generation PAMAM dendrimer and PPI dendrimer.

macromolecules, scientists can determine the biological activity related to their specific configuration. This is a powerful tool for drug and medical systems. However, it must be noticed that dendrimers of high generation have some defects in the defined dendrimer structure because of a too crowded situation of peripheral functional moieties. Moreover, most dendrimers could be mounted with more than one type of ligand, such as targeting moiety, imaging moiety, and therapeutic moiety, all suitable to be conjugated in the periphery of one dendrimer.

Another important feature is the multivalent behavior of dendrimers. Multivalency leads the strong activity, as shown in Fig. 3, compared to the corresponding monomeric interaction. This synergistic enhancement of binding affinity from a monomeric to multimeric system could be ascribed as cluster effect or dendritic effect. The dendritic effect comes into working when simultaneous existence of N-binding sites in a ligand molecule results in a synergistic increase in affinity with a maximum binding affinity of N multiplication. Besides, the cluster effect can be usually observed for carbohydrate–protein receptors in a natural system, known as glycoside cluster effect.[1] Another important issue is that multivalency can also increase the specificity of a given interaction.[2] A factor that plays a role in the binding of ligands in a dendrimer is obviously the geometry of multivalently existing ligands and the flexibility of their attachment. Polyvalency of dendrimers orients bare reactive groups on the nanostructural exterior of dendrimers.

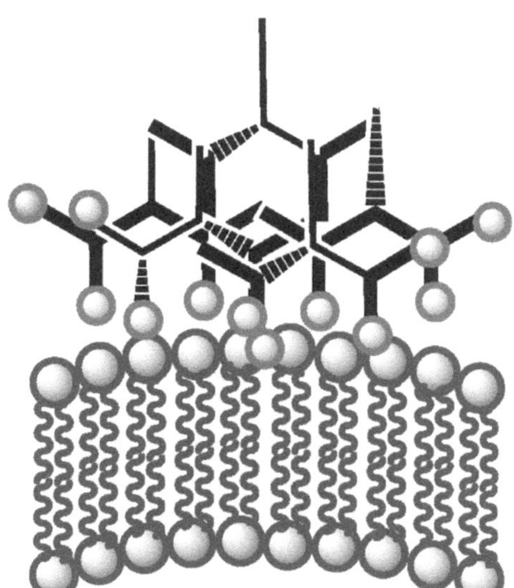

FIG. 3. Mutivalent behavior of a dendrimer on a cell membrane. (For color version of this figure, the reader is referred to the Web version of this chapter.)

A. *In Vitro* Cytotoxicity of Dendrimers

A major concern on the adoption of a new class of nanomaterials in biomedical systems is toward the biocompatibility of these materials. In order to be usable as drug delivery, gene delivery, and imaging agents, dendrimers have to be nontoxic and nonimmunogenic. The cytotoxicity of dendrimers has been evaluated *in vitro* and even *in vivo*. Cytotoxicity of dendrimers is dependent on their chemical structure but is most strongly influenced by the nature of their terminal groups. Depending on the modification of terminal groups on dendrimers, dendrimers carry a positive, neutral, or negative charge on their periphery. For example, among melamine-based dendrimers carrying amine, guanidine, carboxylate, sulfonate, phosphonate, or poly(ethylene glycol) (PEG) on the periphery, cationic dendrimers were much more cytotoxic (in MTT assay) than anionic or PEG dendrimers.[3] The positive charges on the dendrimer tend to destabilize cell membrane, cause cell lysis, and interact with blood components.[4–7]

The mechanism of membrane damage induced by a cationic PAMAM dendrimer was investigated by using 1,2-dimyristoyl-*sn*-glycero-3-phosphocholine liposome and human nasopharynx carcinoma (KB) and rat embryo fibroblast (Rat2) cells in culture.[5,8] Techniques such as atomic force microscopy, fluorescence microscopy, and confocal microscopy were used to visualize damage and release of enzyme lactate dehydrogenase to explore leak of cell membrane. Further, the cytotoxicity was found to be generation-dependent, that is, the higher generation of dendrimers provided stronger toxicity. The reason might be attributed to the higher-generation dendrimers attaining higher surface coverage on cells and liposome, which might induce cell death or membrane leakage. The mechanism of the interaction of dendrimers with the cell membrane is shown in Fig. 4.

PAMAM dendrimers with amino groups revealed the significant cytotoxicity on human intestinal adenocarcinoma (Caco-2) cells.[9,10] A generation dependency of amino-terminated PAMAM dendrimers was observed on the hemolytic effect with blood cells in a solution.[7] However, the related study has shown that amino-terminated PAMAM dendrimers exhibit lower toxicity than flexible linear polymers with amine groups. This should be ascribed to the low adherence of rigid globular dendrimers to the cell surface. The degree of substitution of amine as well as the type of amine functionality is also important, so that the primary amine is more toxic than the second or tertiary amine.[11] Similar cytotoxicity and hemolytic effects were found for amine-terminated PPI dendrimers, involving the increase of both effects depending on generation.[12] The cytotoxicity of dendrimers having terminal groups such as guanidine, carboxylate, sulphonate, or phosphonate besides amine is also reported.[3]

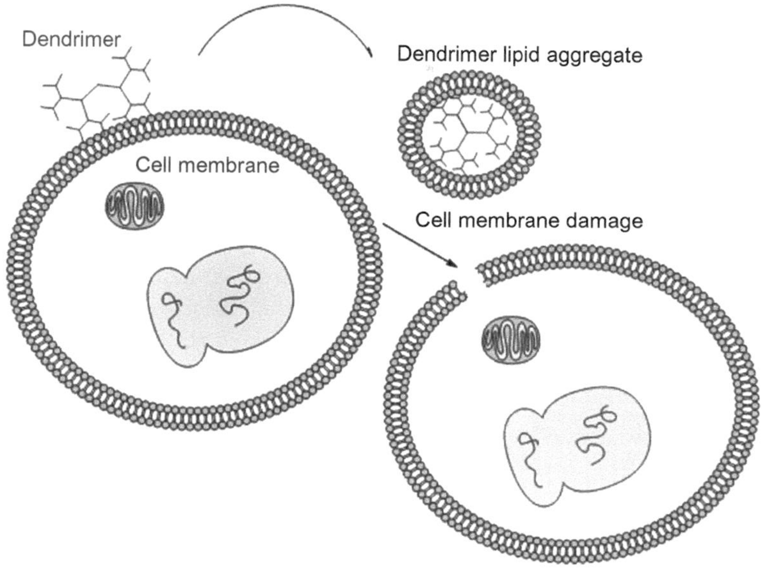

FIG. 4. Schematic representation of the interaction between dendrimer and cell membrane. (For color version of this figure, the reader is referred to the Web version of this chapter.)

B. *In Vivo* Toxicity of Dendrimers

Studies of *in vivo* toxicity basically prove the safety of any materials on biomedical application. PAMAM dendrimers have been injected into mice to evaluate the *in vivo* cytotoxicity. Results indicated that nontoxicity in mice was obtained for PAMAM dendrimers with both unmodified and modified amino-terminal groups.[13,14] *In vitro* and *in vivo* cytotoxicities of polyester dendrimers have been investigated.[15] Inhibition of cell growth was at the concentration of 40 mg/ml but no cell death occurred at the same concentration. Upon injection into mice, no acute or chronic toxicity was observed.

Recent reports suggest that higher-generation dendrimers exhibit greater hemolytic toxicity, which can be ascribed to the greater overall cationic charge. The cationic terminal groups of dendrimers interact with red blood cells. The polycationic nature of dendrimers also leads to hemolysis[16–18] and might influence hematological parameters attributing to the polycationic nature of unmodified dendrimers. The effect of cationic dendrimers, such as amino-terminated PAMAM and PPI dendrimers, on different blood components including white blood corpuscles (WBCs), red blood corpuscles (RBCs), hemoglobin (Hb), hematocrit (HCT), and mean corpuscular hemoglobin (MCH), could be determined by using a particle counter.[19] Significant decreases in contents of RBC, Hb, and MCH, a substantial increase in the count of WBCs, and a considerable difference of HCT values between the control and cationic dendrimers indicated that cationic dendrimers exhibit impairments in the hematological component. Some reports relating to the immunogenicity of dendrimers revealed no or only weak immunogenicity in mice treated with cationic dendrimers. In addition, from the investigation of the immunogenicity of peptide dendrimers in Bulb/C (a strain of inbred white mice that is readily developed for experimental myelomatosis) mice using the enzyme-linked immunosorbent assay for monitoring of an antibody, it has been observed that the multiple peptide dendrimers are unable to cause any detectable humoral immune response. This means that the dendrimers could be regarded as "pseudo-native" for the host immune system.

C. How to Solve the Problems of Biomedical Toxicity

There are two general ways to reduce the cytotoxicity and hemolytic toxicity of dendrimers. The first approach is the embedding of the biodegradable moieties in the interior of dendrimers. Monomers, which may be transferred to metabolic products via various biological pathways, are selected in the synthesis of dendrimers and hence they lead to the synthesis of biodegradable dendrimers. Monomers such as lactic acid, glycerol, succinic acid, and PEG have been utilized as good biocompatible and biodegradable component materials for the synthesis of dendrimers.[20–22] Another kind of biodegradable dendrimers was composed of different types of amino acids such as lysine, arginine, etc., and thus produced dendrimers promised to have drug carrier properties.[23–25] Peptide dendrimers with oxime, hydrazone, or thiazolidine linkage as a building block and a selective ligation between an aldehyde and a weak base were also synthesized.[26]

The second approach is the modification of the terminal amine/cationic groups of dendrimers with natural or anionic moiety. It is well documented that decreasing the surface charge of amine-terminated dendrimers toward neutral reduces their toxicity. The presence of multiple terminal sites makes possible the attachment of the moieties with various functionalities through the covalent or noncovalent bonding on the periphery of the dendrimers. In addition to the proper reduction of the toxicity of dendrimers, the functionalization also imparts some other properties, which are beneficial for their biological application including improvements of drug control and therapeutic potential.[18,27–30] It has been proven that hydrophilic PEG chains can enhance the loading efficiency of hydrophilic drugs and cause the stable enclosure of drugs within the dendrimers. In addition, PEG-modified dendrimers could heighten the circulation efficiency in blood, which allows the drug carrier by dendrimers to move toward the targeting specific tissue. Therefore, the PEGylation of dendrimers can effectively reinforce the therapeutic potentials of dendrimers alone.

It has been reported that the cytotoxicity related to amine-terminated PAMAM dendrimers for cancer therapy could be minimized by the amidation of terminal amine groups.[31] Additionally, the amidation of amines in dendrimers leads to a 10-fold reduction in cytotoxicity, while the ability of the cell transmission remains.[32] The partial amidation was effective in reducing the cytotoxicity, whereas the complete amidation of terminal amine groups in dendrimers evoked the absence of the cytotoxicity. The number of free amine groups was in linear relation to the cytotoxicity. It was also revealed that the toxicity of terminal-modified PAMAM dendrimers was similar to that of unmodified dendrimers.[32]

The modification of dendrimers with chemically inert PEGs or fatty acids is the other way to reduce the toxicity of cationic dendrimers.[6,9,10] PEGylation of dendrimers not only reduces the toxicity but also improves the drug loading and reduces the drug leakage.[30] Other purposes of PEGylation of dendrimers are improvement of biodistribution and pharmacokinetics, increase in solubility of dendrimers, sustained and controlled drug delivery, better transfection efficiency, and tumor localization.[18,29,33] Additionally, immunogenicity is one of the crucial *in vivo* biological properties of dendrimers. A lot of research revealed that the scarce or slight immunogenicity occurred in dendrimer materials. Other studies also indicated that the PEGylation of dendrimers resulted in relatively low immunogenicity, when compared with non-PEGylated ones. It should be noted that the PEGylated drug-loaded systems have a number of advantages including the retention of bioactivity and bioenvironmental protection. As a general rule, PEG with high molecular weight has been widely used as a drug carrier to improve the targeting and therapeutic efficacy.

Star dendrimers as drug carriers have been synthesized by combining dendrimers on the terminals of three-arm star PEGs.[34] The goal of this synthesis was a preparation of long-lived dendrimers for circulating *in vivo* to allow them to accumulate on the targeting tissue. In the design of drug carriers, the PEG polymer chains could prevent the recognition of macrophages and white cells during their circulation in the blood. Therefore, PEG-modified drug carriers have the long-lived circulating behavior when compared with unmodified ones. Further investigation has been reported to evaluate the biodistribution of a sixth-generation lysine dendrimer and two PEGylated derivatives in both normal and tumor-bearing mice.[15] The intact poly(lysine) dendrimer underwent the rapid clearance from the blood stream and the nonspecific accumulation in the liver and kidney. In contrast, the PEGylated poly(lysine) dendrimer with high-degree modification was accumulated effectively in the tumor tissue. This success of the effect suggested that the PEGylated poly(lysine) dendrimers should be useful materials for the tumor-targeting drug carrier.[15]

The toxicity of dendrimers could be related to the interaction of terminal cationic charge of dendrimers with negatively charged biological membranes *in vivo*. Such an interaction of dendrimers with biological membranes results in membrane disruption via nanohole formation, membrane thinning, and erosion. To minimize this toxicity, two strategies have been utilized: first, designing and synthesizing of biocompatible dendrimers, and second, masking of peripheral charges of dendrimers by surface engineering. In order to design an adequate dendrimer for biomedical application, one has to pay attention and realize what causes the toxicity of the dendrimer and how to reduce it.

III. Drug Delivery Systems of Dendrimers

Recently, developing nanotechnology-based efficient drug delivery systems for small molecule, protein, and DNA has attracted a great deal of attention. Dendrimers, unlike traditional polymers, can be obtained at the precise molecular weight even at high generation, which can provide reproducible pharmacokinetic results. This feature makes dendrimers an ideal candidate for drug delivery applications. In dendrimer-based delivery, drugs could be noncovalently encapsulated in the interior of dendrimers by the physical association or covalently conjugated to form prodrugs, as shown in Fig. 5.

A. Load of Drugs by Physical Doping in Dendrimers

Dendrimers consisting of a hydrophobic interior and a polar periphery may be referred to as monomolecular micelles without the limitation of the critical micelle concentration. Several therapeutic agents are examined as guest molecules for drug delivery. Hydrophobic anticancer drugs such as camptothecin are proven to damage DNA and lead to cell death. However, the extremely low water solubility of camptothecin limits its therapeutic efficiency. In order to solve this problem, the hydrophobic camptothecin has been utilized by encapsulating it in liposome, micelles, nanoparticles, and dendrimers.[35] A biocompatible polyether dendrimer has been modified to comprise glycerol and succinic acid to conjugate with 10-hydroxyl camptothecin.[18] The cytotoxicity of the dendrimer–drug complex showed a low IC_{50} (nmol/L) (half-maximal (50 %) inhibitory concentration) value; moreover, the measurement in human breast adenocarcinoma cells showed a 16-fold increase on cellular uptake and an increase on drug retention. Otherwise, dendrimers based on 1,3,5-triazine have been also utilized for the encapsulation of 10-hydroxyl camptothecin.[36] The building block, triazine, attracts a lot of attention due to not only its low cost but also the easy displacement of three chlorine atoms by amines to

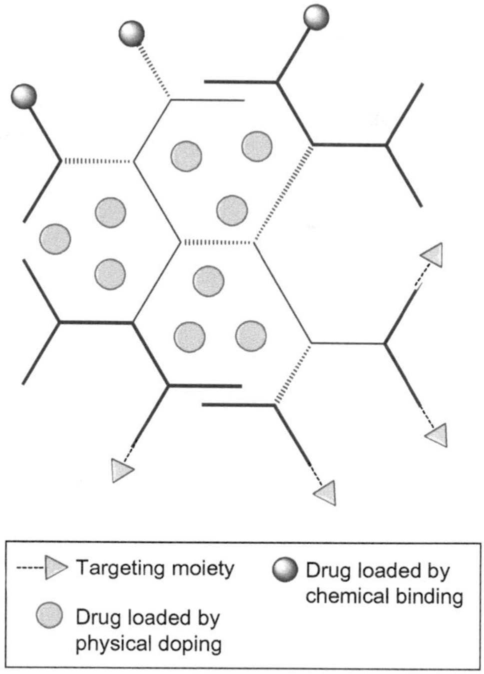

FIG. 5. Schematic representation of a functional dendrimer adapted as drug carrier. (For color version of this figure, the reader is referred to the Web version of this chapter.)

generate mono-, di-, and tri-substitute of 1,3,5-triazine. A second-generation triazine dendrimer has been scaled to kilogram quantities, which is an important issue in pharmaceutical applications.[37]

Another example of loading of drugs within dendrimers is the complexation of small molecular drugs. The small acidic molecules such as benzoate could bind within the interior of the PAMAM dendrimer via ion-pairing with protonated tertiary amine.[38,39] When benzoate is protonated in acidic condition, this nonionic guest suffers release and precipitation. Dendrimers often encapsulate small organic molecules (drugs) into the void space in the dendrimer's interior, while large guest molecules prefer to absorb onto the dendrimer periphery. Ibuprofen, an anti-inflammatory drug, was complexated and encapsulated into third- and fourth-generation PAMAM dendrimers.[40] A third-generation PAMAM dendrimer binds 32 molecules of ibuprofen, while a fourth-generation dendrimer with 64 terminal amine groups binds 78 molecules of ibuprofen.[41,42] It was found that ibuprofen is not only complexated with the terminal amine groups through electrostatic interaction but also encapsulated in the interior of larger dendrimers. In this study, ibuprofen was successfully transported to A549 human lung epithelial carcinoma cells by PAMAM dendrimers. The anti-inflammatory effect of the ibuprofen-complexated dendrimer system revealed more rapid suppression than the pure drug.

The electrostatic binding between the drug and the dendrimer was dependent on concentration and generation.[43] The presence of a PAMAM dendrimer enhanced the transdermal delivery of ketoprofen and increased 2.73-fold the bioavailability of a free drug. The solubility of ketoprofen in a PAMAM dendrimer increased linearly with the concentration of dendrimer under the fixed pH condition. The solubility of ketoprofen was highest at pH 6 and lowest at pH 3.[42] This result indicates that the drug would not be fully ionized at low pH and strongly interacted with a dendrimer. *In vivo* study of ketoprofen–PAMAM dendrimer complexes indicated the prolonged pharmacodynamic behavior of ketoprofen. This suggests that PAMAM dendrimers can act as potential drug carriers of ketoprofen with a sustained release behavior.[44]

One of the methods for physical drug loading is the increased density of terminal groups in higher-generation dendrimers. At a particular generation level, the terminal groups will reach the compact packing limitation, which is called "de Gennes's dense packing."[45,46] This situation will seal the interior void and prohibit drug penetration. However, the limitation of peripheral moieties depends on the strength of the intramolecular interaction between adjacent moieties, the peripheral shell thickness, and the environmental condition of dendrimer solutions such as temperature, pH, and polarity. Moreover, the dynamics of dendrimer chains is also one of the important factors.[47,48] These characteristics can promote the encapsulation and release of drugs in and from dendrimers, respectively.

The maximum amount of drug loading was directly related to the shape and size of internal cavities of dendrimers. Large and small molecules could be simultaneously entrapped within PPI dendrimers, which generally contain 4 large and 12 small cavities. PPI dendrimers could be opened under controlled conditions to release either some or all loaded molecules.[46,49] In a thermodynamic aspect, free drugs can be distinguished from this physical load as a complex with finite energy barriers.[50] This finite energy has a relation with the inlet and outlet of drugs in dendrimer cavities. A hydrophobic drug would be expected to be associated with the hydrophobic interior of a dendrimer. Further, the hydrophobic property of a drug should isolate itself from the polar periphery of a dendrimer and the polar media to keep the minimum contact with polar environments. Generally, when the drug molecule is significantly larger than the dendrimer cavity dimension, the complexation between drug and dendrimer will not occur.

In order to study the effect of ethylene glycol dendrimer on solubility enhancement of paclitaxel, poly[oligo(ethylene glycol) methacrylate] (poly(OEGM)), star-shaped poly(OEGM), and polyglycerol dendrimers (G3, G4, and G5) have been synthesized and evaluated for their drug-loading properties.[51] Poly(OEGM) increased the solubility of paclitaxel, but a significantly enhanced effect was observed in the star poly(POEGM) and polyglycerol dendrimers. The solubility of paclitaxel in water in the presence (10 wt%) of a star-shaped poly(OEGM) with five arms and polyglycerol dendrimers (G3, G4, G5) was 130-, 270-, 379-, and 434-fold greater than the solubility of paclitaxel in water. This result indicates that the solubility of paclitaxel is not enhanced by a molecular weight of PEG. It was strongly suggested that the high density of the ethylene glycol unit in the dendritic structure contributes to the physical loading of the hydrophobic paclitaxel drug.

B. Release of Drugs Loaded by Physical Doping in Dendrimers

Response to external stimuli is one of the most fundamental properties of the biological system. Indeed, bio-organisms exhibit very high sensitivity to their external environment, changing their shapes or releasing the chemicals from them as a response. Generally, the structure of dendrimers consists of three main components: (1) the core affects the three-dimensional shape of dendrimers, for example, spherical or cylindrical shape; (2) the interior has an influence on the guest–host properties of dendrimers; and (3) the periphery of dendrimers can be further polymerized or modified with functional groups. Dendrimers at low generation are usually flexible and open, while dendrimers at high generation are usually dense with three-dimensional shapes. The conformation of dendrimers could be affected by the ionic strength and pH in aqueous solutions.

The mechanism of drug release can be represented based on molecular conformation change in different pHs and ionic strengths. Conformation of amine-terminated PAMAM and PPI molecules is globular and compact at high pH than pK_a (=9.2).[52] At pH > 9, the folding of terminal groups occurs as a consequence of hydrogen bonding of the interior amide and/or tertiary amine with the primary amine, which results in a dense interior. The extended conformation dominates at lower pH than pK_a (=6.7) because of the electrostatic repulsions of protonated ternary amines in the interior. In the case of carboxyl acid-terminated dendrimers, at lower pH than pK_b (=4.2),[53] the dendrimer interior has the extended conformation due to the electrostatic repulsion of protonated tertiary amines. While at neutral pH ($4.2 < \text{pH} < 6.7$), the slight folding of terminals takes place as a result of attractive Coulomb's interaction between negatively charged terminal carboxyl groups and positively charged interior tertiary amines, at high pH (6.7<), the electrostatic repulsion between negatively charged terminal carboxyl groups generates an extended conformation due to highly expanded peripheral area.

When a PAMAM dendrimer physically loads a guest drug, the drug could be retained within the dendritic void by an electrostatic or hydrophobic interaction with the interior of the dendrimer. Then pH- and temperature-responsive drug carriers are known to be successful drug-releasing systems. A poly(*N*-isopropylacrylamide)-conjugated star-shaped dendrimer has been synthesized.[54] The aqueous solution of this dendrimer was thermo-sensitive, since the dendrimer achieved a lower critical solution temperature (LCST) around 32 °C. Moreover, the addition of methacrylic acid besides NIPAAM not only induces the shift of the LCST but also makes the dendrimer materials responsive to both temperature and pH.

C. Delivery of Drugs Loaded by Chemical Binding in Dendrimers

Dendrimer–drug conjugates generally consist of drugs covalently attached to the periphery of the dendrimer. This type of drug carrier offers different advantages over physically encapsulated drug carrier systems. Drugs can be conjugated with each dendrimer molecule, and the release of these therapeutic molecules could be controlled by the character of linkage. Drug molecule can also become an integral part of the dendritic carrier that is released through certain triggering events at the desired location, for example, a tumor site. The main advantage of these methods is that the drug–dendrimer conjugate diffuses slowly, different from the case where the free drug is absorbed on the specific interface. Therefore, cell targeting and controlled delivery are allowed. An additional benefit of this prodrug is in reproducible pharmacokinetics and pharmacodynamics. The covalent bond between dendrimer and drug also provides a stable structure for the internal encapsulation of drugs, which is less contributed by thermodynamics and physical factors.

The pharmacodynamic behaviors of two types of amide-bonded PAMAM dendrimer–methotrexate (MTX) drug carriers were evaluated.[55] One of the amide bonds was prepared by coupling a carboxyl-terminated G2.5 PAMAM dendrimer via two amine groups with an aromatic ring of MTX (drug carrier A). Another was synthesized by coupling an amine-terminated G3 PAMAM dendrimer with a γ-carboxyl group of MTX (drug carrier B). A drug carrier A with an MTX payload of 2.8 exhibited a significant anticancer activity toward MTX-resistance cells. However, the drug carrier B with an MTX payload of 22.4 revealed no cytotoxicity on the same cells. A chemical linkage of MTX with dendrimer is shown in Fig. 6. This result can be explained by the different ionic characters of these drug carriers, which results in the different residence time in the lysosomes. It was indicated that in the lysosomotropic effect, the displacement of small basic molecules from the lysosome by the positively charged dendrimer occurs with an increase in pH and eventually the disruption of the

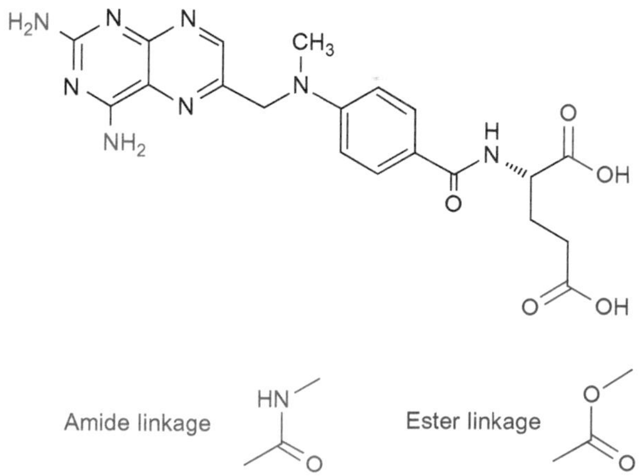

FIG. 6. Ester and amide linkages by MTX chemotherapeutic drug. (See Color Insert.)

lysosome. As a result, the conjugate linkage reduces the interaction with protease and retards the breaking of drug linkage. Therefore, the different behaviors of the intracellular drug release in drug carrier A and B lead to different cytotoxicities against the MTX-resistance cells.

A remarkable asymmetric architecture, where half of the dendritic drug carrier was modified with doxorubicin (DOX) and the other half with PEG, has been reported.[56] A pH-sensitive linkage between the drug and dendrimer was designed to release the drug once it is in cells. In the cell culture, the dendrimer–DOX prodrug was less than one-tenth toxic than free DOX on C-26 colon carcinoma cells after culturing for 72 h. Administration of DOX-conjugated dendrimer over 2 months caused complete tumor regression, while no cures were observed for mice treated only by the drug.

Drugs could attach via linkers or spacers to terminal groups of the dendrimer. Usually, ester or amide bonds are employed, since they can be hydrolyzed inside the cells by an endosomal or a lysosomal enzyme.[57,58] The drug release from a conjugate is controlled largely by the nature of a linking bond or a spacer between the drug and the target physiological domain for the intended release. Ester and amide bonds might be cleaved by enzymes or under hydrolytic condition. However, the ester cleavage is generally more facile than amide cleavage for the drug release.

The stability and release properties of both ester and amide bonds between Naproxen and PAMAM dendrimer were reported.[59,60] The amide bond conjugation withstood 80% the release in human plasma, indicating that the amide link conjugation exhibited stability in plasma. Meanwhile, the ester link conjugation suffered rapid esterase-catalyzed hydrolysis and the half time of cleavage cutoff was around 51 min, that is, the ester link conjugation releases the drug easily. In another experiment, a hydrazone bond and an amide bond between PEGylate PAMAM dendrimer and adriamycins have been studied with respect to their release properties.[61] The hydrazone bond of adriamycins was remarkably released at pH 5.0 for the endosome, which is comparable to the amide bond. Further, the conjugation by the hydrazone bond was seven times more efficient than by the amide bond.

Regarding the effect of molecular weight and size difference, the release properties of a high generation of dendrimer have been recently studied.[59] The three-dimensional nanostructure of dendrimers gives rise to sterical enzymatic release, different from the linear polymer system. Therefore, there should be space for the linkers and the specificity for enzymes on the dendrimer periphery. A drug carrier of tetra-peptide, Gly-Leu-Phe-Gly (GLPG), which is enzymatically cut off, has been applied for studying the drug release. For the GLPG peptide bond of ibuprofen by amide conjugation with PAMAM dendrimer, while the hydrolysis of ester conjugation showed a clear pH-dependent rate, the amide bond conjugation was very stable at all buffer pH. The linkage of peptide between drug and PAMAM dendrimer must rely on enzymatic breakage. The insertion of peptide spacer gives an access site of the enzymes in the presence of cathepsin B. As a result, the dendrimer–GLPG–ibuprofen released 40% pay loads of ibuprofen for 2 days, while the ester and amide bonds of ibuprofen remained at 95% conjugation in the dendrimer. Since steric hindrance played an important role in the enzymatic cutoff from the dendrimer conjugation, the appropriate choice of binding could raise a preferable effect on drug delivery.

IV. Targeting Delivery of Dendrimers

Multifunctional dendritic architecture allowed for the conjugation of both drug and targeting moiety such as folic acid, monoclonal antibody, and peptide to the periphery of the dendrimer for promoting the specific drug delivery (Fig. 5). In the field of oncology, the targeting delivery of chemotherapeutics to the tumor significantly reduces the side effect, compared with the nontargeting delivery, which might damage the healthy tissue such as liver, spleen, kidney, and heart with the accumulation of a toxic level of the drug. Nonspecific protease targeting or passive targeting to tumors could be achieved by PEGylation to increase the hydrodynamic radius of the drug carrier, leading to the accumulation of dendrimers in the tumor through the enhanced permeability retention (EPR) effect. The EPR effect caused the neovascular leak and defection with disorganized endothelial cells, which offers the accumulation and the retention of macromolecules in the tumor region. This may increase the effect of the antitumor drug up to 70-fold, when drug delivery systems are injected intravenously.[62] Specific targeting or active targeting relies on the conjugation of one or more targeting moieties in dendrimers to facilitate the receptor mediate endocytosis against cells.

A. Folic Acid Dendrimers for Cancer Targeting

The binding affinity of a multivalent targeting G5 PAMAM dendrimer containing a different number of folic acid moiety has been explicitly quantified.[63] The binding affinity to a folic acid receptor expressing cells increased with numbers of folic acid in the dendrimer up to 5–6 folic acids per dendrimer. However, the rate of internalization was not affected significantly at different numbers of folic acids. It was suggested that aggregates of 5–6 folic acid receptors are prerecognized on the cell membrane. The main discovery in this study indicated that the efficiency of the targeting cell is enhanced by the retention of receptors on the cell but not the rate of endocytosis. Several variations of folic acid-conjugated dendrimers for targeting drug delivery have been investigated.[64] Folic acid was conjugated on the G5 PAMAM dendrimer with remaining free amines, which were capped with glycidol for neutralization of positive charge on the dendrimer. Results in the drug release profile were compared between encapsulated MTX and covalently bound MTX. Physically loaded MTX showed the rapid release of the free drug (around 75% of initial load) within 2.5 h, while a slower release for chemical bond was only around 5% of the initial load in the same period. In addition, it was demonstrated that folic acid conjugated on the dendrimer was highly specific for KB cells (human carcinoma of the nasopharynx), overexpressing the folic acid receptor. Although both the free drug carrier and the actively targeting dendrimer-based drug carrier showed similar cytotoxicity, the targeting dendrimer lost the targeting effect when the receptor was blocked or underexpressed with free folic acid.

B. Antibody-Dendrimer for Cancer Targeting

Antibody-conjugated delivery systems are the novel strategy for delivery of therapeutic agent to specific cell type. Conjugation of monoclonal antibody to PAMAM dendrimer has been investigated for specific targeting of tumor cells that overexpress specific antigen.[65] Another targeting moiety J591, an anti-prostate specific antigen (PSA), was conjugated on a G5 PAMAM dendrimer and evaluated the cell binding and the internalization ability. The antibody-conjugated dendrimer has been found to bind specifically to PSA positive cells (LNCaP; human prostate cancer cell lines from lymph) but not to PSA negative cells (PC-3; human prostate cancer cell lines from bone). Further, the targeting efficiency of the antibody-conjugated dendrimer was evaluated from the image of a confocal microscope. The unconjugated dendrimer was not significantly taken in by a cell.

An anti-Her2 G5 PAMAM dendrimer tagged with AlexaFluro was synthesized for targeting HER2 expressing cells.[66] The conjugation of Anti-Her2 on the dendrimer was applied for the targeting of human growth factor receptor-2 overexpressing breast and ovarian cancer cells. Not only did the conjugation yield the preferable binding and affinity but also the *in vivo* result indicated that it targets the HER2 expressing tumors. The effect of conjugation on the immunoreactivity of antibody has been examined with respect to size and number of conjugation.[67] An anti-epidermal growth factor receptor antibody hMAb425 was conjugated with PAMAM dendrimers of different sizes containing up to 128 chelating moieties. Results in this study indicated that the number of derivated sites has a crucial effect on immunoreactivity, whereas the size of the dendrimer did not give a significant influence on activity.

C. Glycodendrimer for Cancer Targeting

The biological character of the carbohydrate receptor allows the carbohydrate cluster to be applied to a receptor with biologically meaningful affinity. Carbohydrate in some of the glycoconjugates played a key role in the process of cell-to-cell and cell-to-pathogen adhesion. Therefore, the glycodendrimers have several peripheral groups of carbohydrate residues accessible for multiple binding interactions. To evaluate such constructs, several T-antigens containing glycodendrimer clusters were synthesized for checking relative binding properties. Even though glycodendrimers may be less efficient than glycopolymers on the inhibition effect, the structure of the dendrimers is worthy of notice because of the lack of immunogenicity and the well-defined chemical pattern. Since glycodendrimers are chemically and geometrically well-defined monodispersed macromolecules, they are suitable as a tool for medical and pharmaceutical purposes. Glycodendrimers are a class of dendrimers that incorporate the sugar moiety such as glucose, galactose mannose,[68] and disaccharide.[69] A glycosylation of dendrimer was prepared and evaluated along an HeLa cell (cancer cell) line and a nontransformed mouse embryonic fibroblast cell (MEF, healthy cell) line.[70] While the glycopeptide dendrimers conjugated were not as antiproliferative as colchicine alone, the dendrimers were 20–100 times more effectively inhibiting the proliferation of HeLa cells than MEF cells. However, nonglycosylated dendrimers were 10-fold less selective for HeLa cells.

Chemically and geometrically well-defined T-Ag glycol PAMAM dendrimers with valences of 4, 8, 16, and 32 were synthesized with an amide linkage. Successive bioassays showed strong protein-binding properties, thus demonstrating an excellent cluster effect that is establishing these as a strong candidate for biological and immunochemical applications.[71]

V. Dendrimers for Therapy

A. Dendrimers for Boron–Neutron Therapy

Boron–neuron capture therapy (BNCT) is a binary approach to the treatment of cancer. Biologically targeted BNCT treatment is based on the production of radiation inside a tumor using ^{10}B and thermal neutron. For BNCT to be effective on a tumor, a minimum concentration of ^{10}B must be around 10–30 μg/g of the tumor. If ^{10}B can be delivered in sufficient quantities to the tumor tissue, the subsequent irradiation with thermal or epithermal neutrons produces a highly energetic α particle and $^{7}Li^{3+}$ ion that could damage the mitotic potential of the tumor cells.[72,73] A critical issue for BNCT is the precise localization of a boron therapeutic agent in the tumor. In this perspective, dendrimers are a very fascinating material for using as boron carriers due to their well-defined structure and multivalency as shown in Fig. 7.

In order to reach the above goal, in 1994, the starburst PAMAM dendrimer has been boronated by reacting them with an isocyanatopolyhedral borane ($Na(CH_3)_3NB_{10}H_8NCO$).[74] This compound was attached to monoclonal antibodies (IB16-6) to produce an immune conjugate for selective cell targeting. *In vivo* biodistribution studies revealed that this dendrimer–borane agent was localized with strong propensity in the liver and spleen. In subsequent studies, in order to achieve the effective brain tumor targeting, an epidermal growth factor (EGF) was attached to the boronated G4 PAMAM dendrimer, as gliomas express a level of epidermal growth factor receptor. It was found that the boronated G4 dendrimer was initially bound to the cell membrane and then endocytosed, resulting in an accumulation of boron in the cell lysosome, which was determined from the energy-filtered transmission electron microscopy known as electron spectroscopic imaging.[75] Another monoclonal antibody, cetuximab, which was directed against mutant isoforms of EGF, has been

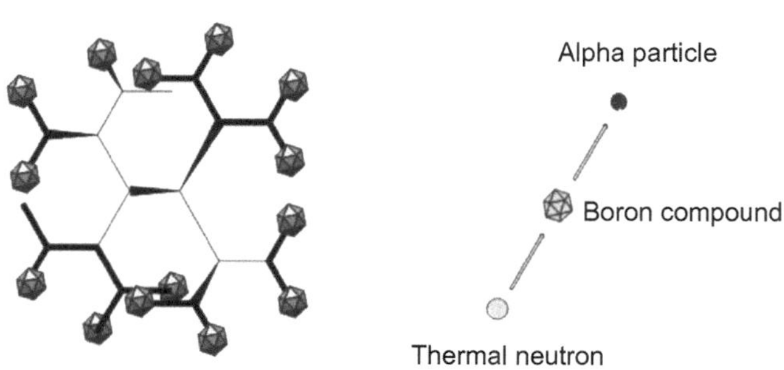

FIG. 7. An action of a dendrimer as a BNCT carrier. (For color version of this figure, the reader is referred to the Web version of this chapter.)

evaluated as a boron delivery agent for brain tumor. After 24 h of the administration of boronated cetuximab-conjugated dendrimer, the mean boron concentration in rats bearing gliomas was found to be 92.3 ± 23.3 μg/g. In contrast, the accumulation of the nontargeting boronated dendrimer was only 6.7 ± 3.6 μg/g in tumor. Moreover, the mean survival time of rats injected with a targeting boronated dendrimer was 45 ± 3 days as compared to 25 ± 3 days for nontargeting ones. These results demonstrated the efficacy of the boronated cetuximab on BNCT of an intracerebral glioma, and this examination can be regarded as a model for future studies.[76,77]

B. Dendrimers for Photodynamic Therapy

In photodynamic therapy (PDT), the drug becomes toxic upon *in situ* irradiation to generate the small amount of toxic singlet oxygen or radical oxygen, which has strong physiologically damaging effects. The photosensitive drug should be relatively nontoxic under nonirradiative conditions. According to the principle of light irradiation, the photosensitive drug is excited to the singlet state and then transformed to the long-lived triplet state. The triplet state of the photosensitive drug can transfer the excited energy to oxygen molecules to generate reactive oxidative species (ROS). Therefore, the selective photoirradiation of the target tissue could possibly minimize damage to the healthy tissue.

Since the initial approval of porphyrin as a photosensitizer for the cure of bladder cancer, the use of PDT for the treatment of cancer as well as the non-neoplastic lesion has increased dramatically, following advances in laser light and photosensitizer.[78] Until now, a number of these photosensitizers have gained regulatory authorization and some photosensitizers are under clinical evaluation.[79] Most of the photosensitizers show less solubility in water and poor selectivity for tumors. It is clear from the discussion above that dendrimers could be used for overcoming less solubility, nontargeting, toxicity, and other problems, since they are impeditive in the bioactivity of photosensitizers. Therefore, the dendritic carrier can be effectively manipulated to improve PDT.

Polyamide-based dendrimers with 5-aminolevulinic acid (ALA) have been prepared, and their activity has been evaluated on the photodynamic examination.[80] Protoporphyrin IX (PPIX) was formed from the substrate of ALA via the heme synthesis pathway in cells. Formation of PPIX in the cells can be used for tumor detection or for PDT. On the investigation of G0-ALA dendron with three ALA groups at the periphery, results revealed that similar PDT efficiency was obtained for G0-ALA when compared with free ALA. For G0-ALA in the cells, the ALA molecules are first cleaved from the dendron and then go within the cells through the heme synthesis pathway.[81,82] Lipophilicity and esterase (hydrolase enzyme) accessibility were also evaluated through a series of G0-ALA dendrons containing different focal points and linkers. Partition coefficient was calculated for the G0-ALA dendron with a nitro core and a propyl linker (3H-ALA), which was found to be the most lipophilic.[83] Analysis of the dendron structures indicated that both lipophilicity and accessibility to the ALA ester linkage might lead to the comparatively higher success of 3H-ALA.

A larger-generation (second-generation) ALA-based dendrimer with terminals of 18 ALAs (G2-18-ALA) was synthesized from a tripodent aromatic core by an ester linkage and its ability to deliver a high payload of ALA molecules was examined.[84] At low concentration, G2-18-ALA was superior to free ALA on the formation of PpIX. G2-18-ALA with propionamido linker was demonstrated to have slightly high efficacy, when compared with free ALA. This indicates that the existence of propionamido linker in the G2-18-ALA may raise a small steric advantage to make the ALA residue more accessible to enzymatic cleavage. In addition, G2-18ALA was found to be internalized via pinocytosis, while the G0-ALA goes through the cells by its active transport and passive diffusion.

Most photosensitizers are poorly soluble in water because of their large π-conjugation and hydrophobic characteristics. Therefore, photosensitizers easily form aggregates in water, which could result in the reduced ROS activity due to the self-quenching of the excited state.[85] Poly(benzyl ether) dendrimer with photosensitizer in the core has been synthesized.[86] The substitution of large dendritic wedges offered the isolation effect for photosensitizer and avoided the aggregation of photosensiziter as shown in Fig. 8. Cell viability of a dendrimer-photosensitizer system was evaluated for Lewis lung carcinoma after photoirradiation. The result showed that a dendrimer–porphyrin system is 10–100 times more photocytotoxic than a conventional photosensitizer PPIX. Polyion-complexified micelles could be prepared from ionic dendrimer–porphyrin and polyelectrolyte with counter ions for the examination of *in vitro* and *in vivo* PDT efficiency. High stability in the increasing ionic strength and further enhancement in photocytotoxicity were obtained in the

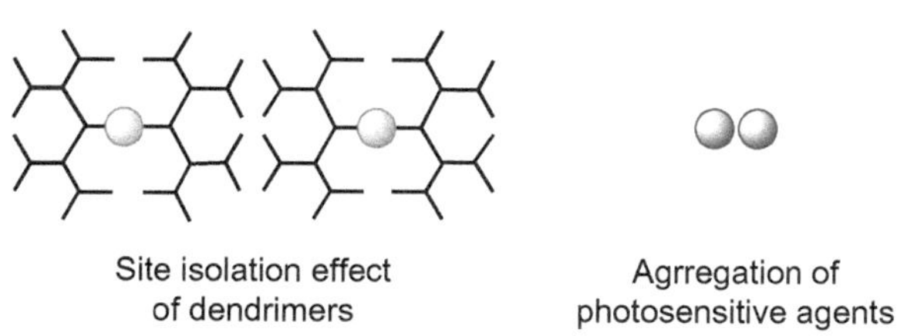

FIG. 8. Aspect of a drug in the core of a dendrimer for prevention of drug aggregation (left) in comparison with a free drug without a dendrimer (right). (For color version of this figure, the reader is referred to the Web version of this chapter.)

polyion-complexified micelles. In addition, the large hydrodynamic volume of micelles was expected to result in high localization into tumor tissue due to the enhanced permeation and retention effects.[87]

C. Dendrimers for Photothermal Therapy

Gold-based nanoparticles have been developed through different routes for a photo-induced therapy, called photothermal therapy. Photothermal therapy is unlike PDT, which could work in the target tissue or cells without oxygen. Gold nanoparticles strongly absorb light in the visible and near-infrared regions. Especially, near-infrared light can penetrate relatively deep into biological tissue. Therefore, gold nanoparticles can generate the lethal dose of heat at the site of the tumor.[88] Amine-terminated G5 PAMAM dendrimer-encapsulated gold nanoparticles were prepared and conjugated with fluorescein and folic acid for active drug carrier targeting.[89] The dendrimers displayed the specific binding with KB cells *in vitro* and were located in lysosomes within 2 h. The utility of these nanoparticles for target hyperthermia therapy and the *in vivo* performance research are still in progress.

Photothermal properties of PEGylated and non-PEGylated G4 PAMAM dendrimers, where gold nanoparticles are entrapped, have been investigated.[90] The gold nanoparticles were prepared by the reduction of $HAuCl_4$ in a dendrimer solution. The light irradiation produced an increase in temperature of the aqueous solution of the gold-encapsulated PEGylated dendrimer, indicating that gold particles retain photothermal properties even if they are encapsulated in a dendrimer. In addition, the gold-encapsulated PEGylated dendrimer was superior on photostability to the non-PEGylated one. *In vivo* results showed that the gold-encapsulated non-PEGylated dendrimer decreased excitation intensity within 3 days, but the intensity of the gold-encapsulated PEGylated dendrimer remained unchanged over 5 days. Future work for thermal therapy still needs to be stepped up, but gold nanoparticles encapsulated in a functional dendrimer could get additional beneficial contribution from the dendrimer, such as acting as a carrier to load chemotherapeutic drugs for double therapy.

D. Dendrimers for Gene Therapy

Gene therapy is emerging as a potential strategy for the treatment of genetic disease, cancer, cardiovascular disease, and infection disease. Delivery of plasma DNA or virus-based protein expression vector is considered a promising approach for the cure of hereditary disease by substituting defective genes. In case of DNA delivery, the nucleic acid needs to be introduced into the nucleus. However, a 21–23 basepair (bp) long double-stranded small interfering RNA (siRNA) could act in the cytosol, where a single strand of siRNA is integrated into the RNA-induced silencing complex.[91,92] Although

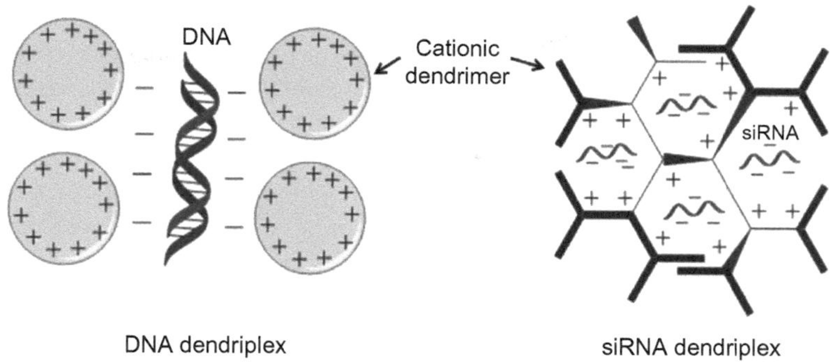

FIG. 9. Action of cationic dendrimers as gene carriers. (For color version of this figure, the reader is referred to the Web version of this chapter.)

other forms of siRNA such as short hairpin RNA (shRNA), microRNA, and so on are available, gene encoding for shRNA, like plasmid DNA, has to enter the nucleus, which causes the complexity in the gene delivery. Figure 9 represents the action of the dendrimer as a gene carrier for DNA and siRNA.

Free DNA or siRNA can be intravenously injected in the patients but they might be rapidly degraded by serum nucleases. As a result, vectors that can protect and downsize the nucleotides are vital. There are two kinds of vectors that could be used in gene delivery: viral and nonviral vectors.[93,94] Viral-mediated nucleotide vehicles play an important role in gene therapy but have a risk for safety such as immunogenicity and potential viral recombination. Therefore, more attention has been concentrated on nonviral vectors. Nonviral vectors such as cationic liposomes and polycations have attracted much attention due to their low immune response and safety. However, the low transfection efficiency greatly limits their clinical application. Scientists have been motivated to use polymers as gene-transfer systems for several reasons, including the fact that nucleic acids themselves are polymers with the nature of negative charges and spontaneously form complexes ("polyplexes") with cationic polymers.

The main restrictions for *in vitro* gene delivery can be categorized as follows: (1) complex stability, (2) lysosome escape, and (3) dissociation of gene from complex at target. Genetic materials are usually taken into the cell by the endocytosis pathway. At first it should be located at endosomes and then transported to lysosomes or cytosol. Thus the genetic materials reach endosomes in a mildly acidic condition and then end up in lysosomes, in which they face much lower pH condition and might be degraded by enzymes. Therefore, an important issue for the gene carrier is to escape from endosomes and to deliver it to cytosol, which is an extremely important process for effective gene therapy.

One reason for the high efficiency of PPI and PAMAM dendrimers in gene transfer is their stimulus behavior in response to the endosomal environment. These dendrimers are like sponges which can absorb protons. It means that the moderate protonation of nitrogen occurs at acidic pH, but the protonation of nitrogen would also arise with endosomal acidification. Finally, the increased density of positive charges causes an influx of chloride ion and water into the endosome.[95] In consequence, the endosome bursts due to the elevated osmotic pressure and the membrane-destabilizing effect of positively charged dendrimer (polymer), which is also called the "proton sponge effect." However, it still exhibits a bottleneck which has to be optimized for the delivery of therapeutic nucleic acid in efficient endosomal escape. Adequate amount of PAMAM dendrimer per DNA is preferable as the dendriplex for gene transfer,[96] where the word dendriplex describes the complex formed from the electrostatic interaction between cationic dendrimers and anionic DNA/RNA. Moreover, the gene transfer of PAMAM dendrimer with high generation is likely superior to that of low generation.[97–99] The transfection efficiency of the dendrimer shows linear increase with increase of generation up to 6, while the higher generation of PAMAM dendrimer ($G > 7$) reversely behaves in the transfection efficiency. These results could be ascribed to the augmented toxicity of the dendrimer as the generation increases. Therefore, the development of low generation of dendrimers ($G < 4$) in the gene carrier has attracted more attention due to the extremely low cytotoxicity.[100]

Different strategies have been used to improve the transfection efficiency and the biocompatibility of the dendrimer gene carrier. The enhanced gene transfer activity through the conjugation of α-cyclodextrin (α-CD) with the dendrimer could be ascribed to the disruption on endosome membrane.[101] Moreover, α-CD has also been reported as a good candidate for siRNA delivery.[102,103] The ternary complex of α-CD/PAMAM dendrimer/siRNA induced sequence-specific gene silence without off-target effect. In this system, α-CD was found to have the potent RNA interference effect compared to lipofectamine 2000 (L2000), a commercial agent for siRNA transfection.

To improve the targeting and the membrane penetration of the gene carrier, a cell-penetrating peptide conjugated in the dendrimer was developed: Modifying the amine groups of the dendrimer with arginine moiety could achieve the above goal.[104–106] Dendriplexes including arginine showed up to two orders of magnitude higher DNA transfection efficiency than dendriplexes without arginine in liver hepatocellular (HepG2), mouse neuroblastoma (Neuro2A), and smooth muscle cells. Gene transfection efficiency of the arginine-conjugated dendrimer could be improved by replacing the ester group in the amide bond. Confocal microscopic images indicated that the ester groups in arginines were cut off via the ester hydrolysis, since the hydroxyl-terminated PAMAM dendrimer did not interact with anionic protein in the cells. Studies also mentioned that high transfection efficiency could be further obtained by increasing the number of arginine in the periphery of the dendrimer. Di-arginine dendriplex could be found inside the nucleus, while mono-arginine dendriplex was mainly located inside the cytoplasm.

The targeting moiety has to attach onto the periphery of the dendrimer to increase the efficient delivery of the complex to the specific tissue.[107] *In vivo* targeted hepatic gene delivery was achieved by conjugating G3 PPI dendrimers with a galactose group. After intravenous (i.v.) injection of asialoglycoprotein receptor-targeting dendrimer complex with DNA into CD-1 mice, organs were harvested to check the luciferase expression. It was confirmed from the investigation of specific liver gene delivery that the luciferase expression occurred preferentially in the liver over other organs. The synergetic effect of targeting agents was evaluated from another *in vivo* study using a PPI dendrimer as a delivery agent. The siRNA was complexified with the G5 PPI dendrimer, following the cross-linking of the individual complexes by using dithiobispionimidate with reducible disulfide bond. Then the dendriplex was PEGylated to improve the stability and finally conjugated with the luteinizing hormone-releasing hormone (LHRH) peptide via maleimide building block. These targeting dendriplexes preferentially knocked down B-cell lymphoma 2 genes in positive LHRH cells *in vitro*. For the investigation of *in vivo* gene efficiency of the siRNA carrier, the results showed a predominance of labeled siRNA and dendrimer in tumor tissue. The ability of functionalization of the periphery of the dendrimer with exquisite control allows its use as a vector for both DNA and siRNA transfections. Cationic dendrimers typically produce cytotoxic effect but also possess the ability to replace cations on the dendrimer by conjugating neutral or hydrophobic groups and then to append a targeting agent for an active targeting gene delivery. However, the gene-silencing activity *in vivo* still needs to be proved.[108]

Promotion of the gene transfection is one of the key aspects in gene delivery. This can be accomplished by developing nontoxic gene deliveries that are controllably responsive. An innovative strategy based on photochemical internalization (PCI) gene delivery could provide time- and space-controlled endosomal escape of therapeutic molecules.[109] Ternary complexes with a size of around 100 nm were composed of cationic peptide ($C(YGRKKRRQRRRG)_2$) in the core that was enveloped in the second-generation aryl ether dendrimer (32 carboxyl groups at the periphery). The study of *in vitro* transfection denoted that this ternary system indicated the enhanced transgene express ($>$ 100-fold), when compared with the conventional reagents such as polyethyleneimine and lipofectamine. Fluorescence microscopic studies of rat conjunctival tissue revealed the positively induced gene delivery by PCI *in vivo*. This work demonstrated that the polyplex of polycations significantly affects both transfection efficiency and toxicity. Thus, PCI provides an opportunity to develop an efficient light-inducible gene delivery system. Photo-triggered gene transfection consisting of porphyrin and PAMAM dendrimer is also reported.[110] *In vitro* results indicated that PAMAM dendrimers with porphyrin core were able to make complexes with green fluorescent protein plasmid DNA and that the internalization of dendrimers in HeLa was concentration dependent. In addition, the conjugation of porphyrin in the core of the dendrimer did not exhibit photocytoxicity at a concentration below 20 μM. However, the expression of enhanced green fluorescent protein was enhanced in HeLa after PCI treatment.

VI. Dendrimers as Imaging Agents for Inspection

Polymer-based bioimaging probes for the diagnoses of different diseases were generated with the cooperation of polymer science and imaging science. The ultimate goal for *in vivo* bioimaging is to achieve highly sensitive and confident imaging systems for detecting and monitoring drug delivery. New probes with enhanced capabilities and performances should be developed based on nano-imaging technology. Key research of targeting *in vivo* bioimaging should address the design of nanostructures for extending the circulation time in the blood and preventing nanostructures from being recognized and cleared by macrophage before reaching the target cell. In addition, nanocarriers for imaging should selectively target diseased cells, tissues, and organs. Finally, new probes must exhibit compatibility with external activation of magnetic field, X-ray, or optics to trigger and enhance the monitoring ability.

A. Molecular Probes

Dendrimers are attractive molecules to be used as molecular probes due to their distinct morphology and unique properties. Large void volume and high density of functional groups in dendrimers make them very efficient to integrate molecular probes. Pt-coordinated complexes with peripheral functional groups were reported. The Pt complexes change to five-coordination Pt complexes via the electronic absorption of SO_2 adducts. The five-coordination Pt complexes could be easily recovered to the original state due to the steric repulsion between the coordination sites.[111] The reversible association of avidin on a biotin-functionalized dendrimer monolayer was reported, and its association behavior was monitored by cyclic voltammetry.[112]

Biosensors for DNA hybridization hold great promise for the rapid diagnosis of genetic diseases. Such sensors rely on the immobilization of single-stranded oligonucleotide probes that selectivity recognizes their complementary target tissue through hybridization. Most of the nucleotide-based and DNA-based dendrimers have been synthesized for signal hybridization. A fluorescence oligonucleotide dendrimer on the signal amplification system was modified by using microarray technology in order to improve signal detection on DNA microarrays.[113] The G4 dendrimer-conjugated 30 DNA single strands were immobilized onto a quartz crystal microbalance.[114] Bioactivity of waterborne pathogen cryptosporidium parvum could be detected through the mass sensitive piezoelectric transducers. A large resonant frequency was detected, while target DNA formed three-dimensional surface hybridization. Immobilized dendrimers on the molecular probe were found to contribute to a higher sensitivity and wider linear range for biosensor, when compared with a non-dendrimer-immobilized probe. Finally, dendrimers must be water-soluble and be protected from interaction with biomolecules in the blood. Therefore, it could be applied in dendritic sensors for *in vivo* application. The Pd complexes of tetrabenzoporphyrins with amino acid-based dendritic wedge-modified chain end with PEG have been synthesized for *in vivo* oxygen imaging.[115,116] These dendrimers have strong oxygen-dependent phosphorescence around 800 nm.

An amperometric enzyme electrode was developed for glucose sensor.[117] Highly sensitive electrodes were fabricated by immobilizing glucose oxidase onto carbon and platinum electrodes and then by modifying with ferrocene colbaltoceium dendrimers. Results indicated that higher-generation dendrimers were more efficient as electron transfer mediators and can be applicable as sensors. In another study, the core of a PPI dendrimer functionalized with octamethylferrocenyl promotes the redox reaction of hydrogen peroxide.[118] These dendrimer-based biosensors possess a good linear response of glucose detection, superior sensitivity, and data reproducibility as compared with other ferrocene-mediated glucose sensors.

Detection of human erythrocyte membrane acetylcholinesterase (AChE) was evaluated from the G4 PAMAM dendrimer-immobilized sensor.[119] Low concentration of dendrimer at 25 μM caused statistical enhancement in enzyme activity. Inhibition of AChE was observed at a concentration of dendrimer higher than 100 μM. These inhibitions were found for both amine and hydroxyl-terminated PAMAM dendrimers. One explanation is attributed to the change of AChE activity caused by the alteration of protein by the addition of dendrimer. Another study aimed to clarify whether AChE inhibition was a result of direct action of the dendrimer on the enzyme or direct change on lipid bilayer.[120] It was demonstrated that anionic and cationic dendrimers change the AChE conformation and the strongest effect was observed by carboxyl-terminated G3.5 PAMAM dendrimers. Therefore, the changes in AChE conformation and catalytic activity depended on not only the concentration of the dendrimer but also its type.

The binding of PAMAM dendrimer to sodium hyaluronate (NaHA) has been investigated.[121] Static light scattering denoted that when the dendrimer was bound, NaHA varied scarcely at low $[NH_2]/[COO—]$ ratio, but the structure of NaHA changed to rodlike at high $[NH_2]/[COO—]$ ratio. Electrostatic interaction between PAMAM dendrimers and NaHA is essential but not sufficient. The additional reason for effectively combining such a number of dendrimers on NaHA is the hydrogen bonding between dendrimer and NaHA.

B. Magnetic Resonance Imaging

Rapid diagnoses of genetic and pathogenetic diseases have been made possible by dendrimer-based biosensors. Complexes of PAMAM dendrimer loaded 111In or 153Gd have been used as anchoring agents for specific and radiolabeled monoclonal antibodies. Gadolinium is a Food and Drug Administration (FDA)-approved contrast agent for magnetic resonance imaging (MRI) and it could provide great contrast between normal tissue and abnormal tissue in the body. Gadolinium is safer than iodine-type contrast used in computed tomography (CT) scans and also is rapidly cleared by the kidney. MRI is one of the prominent noninvasive diagnostic tools for disease detection and is based on subtle difference of environmentally sensitive proton nuclear magnetic resonance (^{1}H NMR) in the living system. The image of MRI depends on the inhomogeneous relaxation time of protons in different tissues.[122]

The first imaging using the dendrimer-based MRI contrast agent for *in vivo* diagnosis was reported in 1994.[123] The dendrimer-based imaging agent exhibited blood-pooling properties and extraordinary relaxivity values when making chelate complex with gadolinium ion, Gd^{3+} (Gd(III)), in comparison with the commercially available small molecule agent.[124] Many early works of dendrimer MRI reagent focused on the PAMAM dendrimer.[125] G2 and G6 PAMAM dendrimers were conjugated with diethylene triamine petaacetic acid (DTPA), which is a commonly used gadolinium chelate agent, to create a chelate structure with 11 and 170 DTPA groups. Based on NMR studies, the chelates with G2 and G6 dendrimers increased longitudinal relaxivity (r1) to 21 and 34 $mM^{-1}s^{-1}$, respectively. These values were four- and sixfold higher ion relaxivities than that of free Gd(III) DTPA (5.4 $mM^{-1}s^{-1}$). The increase in r1 was attributed to the increased rotational correlation from chelating structure in dendrimer. There was significant enhancement in half-life from 24 min for Gd(III) DTPA to 40 and 200 min for G2 and G6 dendrimers, respectively. Generation dependency of dendrimer-based MRI reagents offered dramatic enhancement on MRI contrast properties, which is also called "dendritic effect." The limitation of relation to generation and r1 (an intrinsic property for evaluating the efficiency of MRI contrast agent) is up to G7 at a given field and temperature. Results indicated that the slow water exchange in higher-generation ($G > 7$) dendrimeric Gd(III) chelates displayed a limiting factor in r1.

The effects of scaffold rigidity on relaxivity enhancement have also been further investigated. Complex formation of chelating agent could increase relaxivity by reducing the internal motion of the dendrimer. A series of PPI Gd(III) DTPA structures with rigid linker groups was synthesized.[126] A G5 PPI dendrimer exhibited a relaxivity of 29 $mM^{-1}s^{-1}$. A comparable study was performed with a similar chelating dendrimer but modified with a flexible group.[127] The results indicated that the relaxivity was reduced to 29 $mM^{-1}s^{-1}$ due to larger Gd(III) movement allowedby a flexible linker. Physical properties of dendrimer based on contrast agents not only affect the imaging application but also influence the agent-related toxicity. The prolonged blood-pooling retention time of a dendrimer as an MR agent limited its clinical usage due to the increasing time for release of toxic Gd(III) ion from the dendrimer chelating agent. Although high-generation dendrimer-Gd(III) chelating revealed higher relaxivity, it exhibited higher contrast image. In addition, the smaller generation of dendrimer conjugates showed more rapid excretion from the body than the higher generation did. Therefore, it has been proposed that the low generation of MRI dendrimer agents is most suitable for clinical use due to their therapeutic lower risk of toxicity and their prolonged retention, compared with commercial MRI agents. Gadomer-17, which is currently in phase II clinical trial, is a polylysine dendrimer with a trimesic acid core and only 24 groups at their periphery for gadolinium-chelating conjugation.[128] Gadomer-17 is suitable for blood-pooling imaging and similar to the well-known Gd(III) DTPA poly(lysine), but showed a superior elimination rate, which could be attributed to the globular character of dendrimer derivatives. In a related study, the internal rotational flexibility of Gd(III) chelate in Gadomer-17 and the slow water exchange rate, where both reduce the proton relaxivity, are currently identified as the limiting factors for MRI application.[129]

C. X-ray Imaging and CT

X-ray CT is one of the medical diagnosis tools which could be applied to detect several diseases or organs, such as artioschlerotic vasculature, tumor, infarct, lung, and kidney. Similar to MRI reagents, high molecular weight contrast agents are better for quantitative detection of disease lesions. However, most of the commercial contrast agents are iodinated compounds. These small molecules could be quickly cleared and equilibrated between the intravascular and extracellular compartments in the body. Therefore, in order to achieve a clear contrast image, high concentration of contrast agents have to be injected into patients. This might give rise to other side effects because high dosages of iodine compounds could harm healthy cells or tissues. Consequently, to develop longer circulation time and higher contrast images at a low dosage becomes an important issue in the evolution of novel CT contrast agents. Dendrimers have attracted a lot of attention in association with X-ray contrast agents, since the agents (bismuth, tin, and iodine) could form organometallic complexes with the dendrimer.[130,131] Triiodobenzene derivative conjugated with the G4 dendrimer results in a water-soluble iodinated dendrimer particle for use as CT imaging agents.[132] Superior spatial resolution can be reached due to the high iodine load within the dendrimer.

More recently, gold nanoparticles have been considered possible materials for a CT contrast agent, since gold nanoparticles can attenuate X-rays.[133–137] It has been proved that contrast agents made from gold nanoparticles exhibited better contrast properties than commercially available iodine contrast agents did. Dendrimers could play a role in the control of the dispersion and the size of gold nanoparticles during the nanotemplating process. Various functional group-terminated dendrimers were evaluated in their applications for *in vivo* use. Gold nanoparticles as a CT contrast agent were loaded in a PEGylated dendrimer.[138] The contrast effect of this system was higher than that of a commercial iodated agent, which might be easily excreted from the body. The blood pool and the heart were enhanced in the CT image within 5 min after injection. Moreover, the biodistribution of gold nanoparticles could be monitored by CT and the further irradiation by light at the accumulation site induced the photothermal effect. Additionally, a new contrast agent was prepared by combining two contrast agent elements of gold and iodine within a PAMAM dendrimer nanodevice for CT imaging application.[139] Significant enhancing effects in X-ray attenuation intensity were obtained by combining multiple radio-dense elements in the dendrimer system.

A single photon emission computed tomography (SPECT) scan is another type of nuclear imaging test that reveals how the blood flows to tissues and organs. A SPECT scan integrates two technologies to view the body: (1) CT and (2) radioactive materials acting as a tracer. The tracer allows visualizing the flow of blood to tissue and organs. Technetium-99m (99mTc) is so far the most widely used radioactive tracer isotope in nuclear imaging. More than 80% of all usually used radiopharmaceuticals contain this short-lived metastable radionuclide. This is due to the highly characteristic physical properties of 99mTc, namely, short half-life (6.03 h) and γ photon emission of 140.5 keV, which is very important for both perspectives of effective imaging and patient safety. The 99mTc can be derived as a column eluted from a 99Mo/99mTc generator, which makes it readily available. Further, 99mTc possesses latent chemical properties, facilitating thereby the labeling of several types of kits for versatile diagnostic applications. In an approach to prepare well-defined poly(2,2-bis (hydroxymethyl)propanoic acid) dendron with reactive acid functionality at the core, it became possible to introduce a reactive ligand for 99mTc.[140] A radioactive ligand in the dendron core was considered to minimize the interaction with the biological environment during the blood and tissue circulation. It was found that the radiochemical yield decreased with increasing generation

(G5 > G6 > G7), indicating the elimination of the reactivity by the isolating effect of the dendron. *In vivo* results showed that G5–G7 dendrimers were rapidly and efficiently removed from the bloodstream via the kidney and excreted through the bladder within 15 min post injection.

D. Optical Fluorescence Imaging

Biocompatible fluorescent molecules could perfectly work in tumor biosensing by using fluorescence detection techniques. This kind of approach provided an additional advantage when compared with nonbiocompatible techniques such as radiation or chemical analysis. In a related project, PAMAM dendrimer-based sensors were targeted to tumor cells to monitor the anticancer activity of therapeutics.[141] Multifunctional folic acid-targeted PAMAM dendrimer delivery carriers were synthesized and covalently bound with apoptotic sensor, PhiPhiLux G1D2, in order to detect the extent of cell apoptosis. PhiPhiLux G1D2 is a caspase-specific Forster resonance energy transfer (FRET)-based agent that responds to the signal from an apoptosis-inducing agent. A fivefold increase in intracellular fluorescence intensity was detected when dendrimers were internalized within the Jurkat cells in the first 30 min of incubation. This demonstrated the potential applicability of targeting the apoptosis-measuring dendrimer nanodevice, which could be simultaneously used for monitoring the apoptotic delivery of a drug *in vivo*.

Although conventional fluorescence optical images are limited due to the absorption and scattering of light in tissue, the elimination of light scattering and absorption could be accomplished by using an optical fiber that is placed into a doubtful tumor tissue to identify the region. Several biosensors based on one-photon fluorescence have been developed for the quantification of fluorescence materials *in situ*. Recently, two-photon system have been used for the simultaneous detection of fluorophores with a broad range of excitation wavelength with a spatial resolution of only a few micrometers. In addition, two-photon systems employ near-infrared light for excitation, which minimizes tissue damage, photobleaching, and intrinsic tissue fluorescence. Dendrimers were conjugated with 6-TAMRA, which is a two-photon excited fluorescence-sensing agent, as a carrier to target xenograft tumors in mice. To quantify the concentration of a fluorescence agent in the tissue, a two-photon optical fiber could be inserted into tissue. This method has both advantages of being minimally invasive and being deeply sensitive in a live animal. The tumor fluorescence was observed in live mice at 0.5, 2, and 24 h with the help of a two-photon fluorescence probe. The results indicated that the dendrimer with folate targeting moiety showed selective accumulation in the tumor with maximum mean level at 673 ± 67 nM for 2 h after administration. The intensity of fluorescence for the non-targeting dendrimer as a control was at a level of 136 ± 28 nM for the same period and then decreased rapidly. Using the two-photon excitation fiber-optic probe for detecting the dendrimer nanoparticle conjugated with 6-TAMRA, tumors containing as little as 0.3% fluorescent protein cells could be identified.[142]

VII. Conclusions

The well-defined structure and multivalent periphery of dendrimers provide an excellent platform for attaching drugs, genes, targeting moieties, and imaging agents for biomedical and therapeutic applications. Although dendrimers have been expected to be a potential material in biomedical and pharmaceutical fields, the elimination of toxicity from dendrimers is an initial important issue. Nonselective interaction of cationic dendrimer systems caused membrane disruption and the erosion was followed by leakage of cytosolic enzymes and cell death. Reduction of the toxicity of dendrimers could be achieved by using the biocompatible anionic dendrimers or modifying the periphery of dendrimers.

The reactivity of dendrimers toward the terminal modification can achieve the high payload efficiency for bioactive agents. To avoid the drug release from physical loading, the drug must be conjugated on the periphery of dendrimers with the aim of preparing more stable drug delivery. Moreover, in order to apply to targeting therapy and diagnosis, anticancer bioactive and imaging agents are required for simple conjugation through chemical covalent bonding on the dendrimers. Dendrimers are also considered as an effective tool for various therapeutic applications such as neutron capture therapy, PDT, and gene therapy due to their unique structural architecture. In summary, dendrimers have achieved significant success in a variety of biomedical applications. Hopefully, this brief introduction of dendrimers toward biomedical applications clearly specifies perspectives for dendrimers in the emerging biomedical field.

References

1. Lundquist JJ, Toone EJ. The cluster glycoside effect. *Chem Rev* 2002;**102**:555–78.
2. Andre S, Ortega PJC, Perez MA, Roy R, Gabius HJ. Lactose containing starbust dendrimers: influence of dendrimer generation and binding site orientation of receptor (plant/animal lectins and immunoglobulins) on binding properties. *Glycobiology* 1999;**9**:1253–61.
3. Chen HT, Neerman MF, Parrish AR, Simanek EE. Cytotoxicity, hemolysis, and acute in vivo toxicity of dendrimers based on melamine, candidate vehicles for drug delivery. *J Am Chem Soc* 2004;**126**:10044–8.
4. Rittner K, Benavente A, Bompard-Sorlet A, Heitz F, Divita G, Brasseur R, et al. New basic membrane-destabilizing peptides for plasmid-based gene delivery in vitro and in vivo. *Mol Ther* 2002;**5**:104–14.

5. Hong SP, Bielinska AU, Mecke A, Keszler B, Beals JL, Shi XY, et al. Interaction of poly (amidoamine) dendrimers with supported lipid bilayers and cells: hole formation and the relation to transport. *Bioconjug Chem* 2004;**15**:774–82.
6. Stasko NA, Johnson CB, Schoenfisch MH, Johnson TA, Holmuhamedov EL. Cytotoxicity of polypropylenimine dendrimer conjugates on cultured endothelial cells. *Biomacromolecules* 2007;**8**:3853–9.
7. Malik N, Wiwattanapatapee R, Klopsch R, Lorenz K, Frey H, Weener JW, et al. Dendrimers: relationship between structure and biocompatibility in vitro, and preliminary studies on the biodistribution of I-125-labelled poly(amidoamine) dendrimers in vivo. *J Control Release* 2000;**65**:133–48.
8. Parimi S, Barnes TJ, Callen DF, Prestidge CA. Mechanistic insight into cell growth, internalization, and cytotoxicity of PAMAM dendrimers. *Biomacromolecules* 2010;**11**:382–9.
9. Jevprasesphant R, Penny J, Jalal R, Attwood D, McKeown NB, D'Emanuele A. The influence of surface modification on the cytotoxicity of PAMAM dendrimers. *Int J Pharm* 2003;**252**:263–6.
10. El-Sayed M, Ginski M, Rhodes C, Ghandehari H. Transepithelial transport of poly(amidoamine) dendrimers across Caco-2 cell monolayers. *J Control Release* 2002;**81**:355–65.
11. Fischer D, Li Y, Ahlemeyer B, Krieglstein J, Kissel T. In vitro cytotoxicity testing of polycations: influence of polymer structure on cell viability and hemolysis. *Biomaterials* 2003;**24**:1121–31.
12. Zinselmeyer BH, Mackay SP, Schatzlein AG, Uchegbu IF. The lower-generation polypropylenimine dendrimers are effective gene-transfer agents. *Pharm Res* 2002;**19**:960–7.
13. Bourne N, Stanberry LR, Kern ER, Holan G, Matthews B, Bernstein DI. Dendrimers, a new class of candidate topical microbicides with activity against Herpes simplex virus infection. *Antimicrob Agents Chemother* 2000;**44**:2471–4.
14. Roberts JC, Bhalgat MK, Zera RT. Preliminary biological evaluation of poly(amidoamine) (PAMAM) starburst dendrimers. *J Biomed Mater Res* 1996;**30**:53–65.
15. Padilla De Jesús OL, Ihre HR, Gagne L, Fréchet JMJ, Szoka Jr. FC. Polyester dendritic systems for drug delivery applications: in vitro and in vivo evaluation. *Bioconjug Chem* 2002;**13**:453–61.
16. Agrawal P, Gupta U, Jain NK. Glycoconjugated peptide dendrimers-based nanoparticulate system for the delivery of chloroquine phosphate. *Biomaterials* 2007;**28**:3349–59.
17. Asthana A, Chauhan AS, Diwan P, Jain NK. Poly(amidoamine) (PAMAM) dendritic nanostructures for controlled site-specific delivery of acidic anti-inflammatory active ingredient. *AAPS Pharm Sci Technol* 2005;**6**:E536–42.
18. Bhadra D, Bhadra S, Jain S, Jain NK. A PEGylated dendritic nanoparticulate carrier of fluorouracil. *Int J Pharm* 2003;**257**:111–24.
19. Agashe HB, Dutta TD, Garg M, Jain NK. Investigations on the toxicological profile of functionalized fifth-generation poly (propylene imine) dendrimer. *J Pharm Pharmacol* 2006;**58**:1491–8.
20. Carnahan MA, Grinstaff MW. Synthesis and characterization of polyether-ester dendrimers from glycerol and lactic acid. *J Am Chem Soc* 2001;**123**:2905–6.
21. Dhanikula RS, Hildgen P. Conformation and distribution of groups on the surface of amphiphilic polyether-co-polyester dendrimers: effect of molecular architecture. *J Colloid Interface Sci* 2007;**311**:52–8.
22. Hirayama Y, Nakamura T, Uehara S, Sakamoto Y, Yamaguchi K, Sei Y, et al. Synthesis and characterization of polyester dendrimers from acetoacetate and acrylate. *Org Lett* 2005;**7**:525–8.
23. Agrawal A, Asthana A, Gupta U, Jain NK. Tumor and dendrimers: a review on drug delivery aspects. *J Pharm Pharmacol* 2008;**60**:671–88.

24. Wegmann KW, Wagner CR, Whitham RH, Hinrichs DJ. Synthetic peptide dendrimers block the development and expression of experimental allergic encephalomyelitis. *J Immunol* 2008;**181**:3301–9.
25. Cubillos C, De la Torre BG, Jakab A, Clementi G, Borras E, Barcena J, et al. Enhanced mucosal immunoglobulin A response and solid protection against foot-and-mouth disease virus challenge induced by a novel dendrimeric peptide. *J Virol* 2008;**82**:7223–30.
26. Shao J, Tam JP. Unprotected peptides as building blocks for the synthesis of peptide dendrimers with oxime, hydrazone, and thiazolidine linkages. *J Am Chem Soc* 1995;**117**:3893–9.
27. Konda SD, Aref M, Wang S, Brechbiel M, Wiener EC. Specific targeting of folate dendrimer MRI contrast agents to the high affinity folate receptor expressed in ovarian tumor xenografts. *Magn Reson Mater Phys Biol Med* 2001;**12**:104–13.
28. Bhadra D, Yadav AK, Bhadra S, Jain NK. Glycodendrimeric nanoparticulate carriers of primaquine phosphate for liver targeting. *Int J Pharm* 2005;**295**:221–33.
29. Gajbhiye V, Kumar PV, Tekade RK, Jain NK. PEGylated PPI dendritic architectures for sustained delivery of H2 receptor antagonist. *Eur J Med Chem* 2009;**44**:1155–66.
30. Agrawal A, Gupta U, Asthana A, Jain NK. Dextran conjugated dendritic nanoconstructs as potential vectors for anti-cancer agent. *Biomaterials* 2009;**30**:3588–96.
31. Kukowska-Latallo JF, Candido KA, Cao Z, Nigavekar SS, Majoros IJ, Thomas TP, et al. Nanoparticle targeting of anticancer drug improves therapeutic response in animal model of human epithelial cancer. *Cancer Res* 2005;**65**:5317–24.
32. Kolhatkar RB, Kitchens KM, Swaan PW, Ghandehari H. Surface acetylation of polyamidoamine (PAMAM) dendrimers decreases cytotoxicity while maintaining membrane permeability. *Bioconjug Chem* 2007;**18**:2054–60.
33. Gajbhiye V, Kumar PV, Tekade RK, Jain NK. Pharmaceutical and biomedical potential of PEGylated dendrimers. *Curr Pharm Des* 2007;**13**:415–29.
34. Dufes C, Uchegbu IF, Schatzlein AG. Dendrimers in gene delivery. *Adv Drug Deliv Rev* 2005;**57**:2177–202.
35. Okuda T, Kawakami S, Akimoto N, Niidome T, Yamashita F, Hashida M. PEGylated lysine dendrimers for tumor-selective targeting after intravenous injection in tumor-bearing mice. *J Control Release* 2006;**116**:330–6.
36. Morgan MT, Nakanishi Y, Kroll DJ, Griset AP, Carnahan MA, Wathier M, et al. Dendrimer encapsulated camptothecins: increased solubility, cellular uptake, and cellular retention afford enhanced anticancer activity in vitro. *Cancer Res* 2006;**66**:11913–21.
37. Steffensen MB, Hollink E, Kuschel F, Bauer M, Simanek EE. Dendrimers based on [1,3,5]-triazines. *J Polym Sci A Polym Chem* 2006;**44**:3411–33.
38. Chouai A, Simanek EE. Kilogram-scale synthesis of a second-generation dendrimer based on 1,3,5-triazine using green and industrially compatible methods with a single chromatographic step. *J Org Chem* 2008;**73**:2357–66.
39. Twyman LJ, Beezer AE, Esfand R, Hardy MJ, Mitchell JC. The synthesis of water-soluble dendrimers, and their application as possible drug delivery systems. *Tetrahedron Lett* 1999;**40**:1743–6.
40. Beezer AE, King ASH, Martin IK, Mitchel JC, Twyman LJ, Wain CF. Dendrimers as potential drug carriers; encapsulation of acidic hydrophobes within water soluble PAMAM derivatives. *Tetrahedron* 2003;**59**:3873–80.
41. Kolhe P, Misra E, Kannan RM, Kannan S, Lieh-Lai M. Drug complexation, in vitro release and cellular entry of dendrimers and hyper-branched polymers. *Int J Pharm* 2003;**259**:143–60.
42. Cheng Y, Xu T. Dendrimers as potential drug carriers. Part I. Solubilization of non-steroidal anti-inflammatory drugs in the presence of polyamidoamine dendrimers. *Eur J Med Chem* 2005;**40**:1188–92.

43. Kannan S, Kolhe P, Raykova V, Glibatec M, Kannan RM, Lieh-Lai M, et al. Dynamics of cellular entry and drug delivery by dendritic polymers into human lung epithelial carcinoma cells. *J Biomater Sci Polym Ed* 2004;**15**:311–30.
44. Cheng Y, Xu T, Fu R. Polyamidoamine dendrimers used as solubility enhancers of ketoprofen. *Eur J Med Chem* 2005;**40**:1390–3.
45. Man N, Cheng Y, Xu T, Ding Y, Wang X, Li Z, et al. Dendrimers as potential drug carriers. Part II. Prolonged delivery of ketoprofen by in vitro and in vivo studies. *Eur J Med Chem* 2006;**41**:670–4.
46. Recker J, Tomcik DJ, Parquette JR. Folding dendrons: the development of solvent-, temperature-, and generation-dependent chiral conformational order in intramolecularly hydrogen-bonded dendrons. *J Am Chem Soc* 2000;**122**:10298–307.
47. Boas U, Karlsson AJ, de Waal BFM, Meijer EW. Synthesis and properties of new thiourea-functionalized poly(propylene imine) dendrimers and their role as hosts for urea functionalized guests. *J Org Chem* 2001;**66**:2136–45.
48. Funayama K, Imae T, Aoi K, Tsutsumiuchi K, Okada M, Furusaka M, et al. Small-angle neutron scattering investigations of layer-block dendrimers in aqueous solutions. *J Phys Chem B* 2003;**107**:1532–9.
49. Funayama K, Imae T, Seto H, Aoi K, Tsutsumiuchi K, Okada M, et al. Fast and slow dynamics of water-soluble dendrimers consisting of amido-amine repeating units by neutron spin echo. *J Phys Chem B* 2003;**107**:1353–9.
50. Kannaiyan D, Imae T. pH dependent encapsulation of pyrene in PPI-core: PAMAM-shell dendrimers. *Langmuir* 2009;**25**:5282–5.
51. Esfand R, Tomalia DA. Poly(amidoamine) (PAMAM) dendrimers: from biomimicry to drug delivery and biomedical applications. *Drug Discov Today* 2001;**6**:427–36.
52. Ooya T, Lee J, Park K. Effects of ethylene glycol-based graft, star-shaped, and dendritic polymers on solubilization and controlled release of paclitaxel. *J Control Release* 2003;**93**:121–7.
53. Leisner D, Imae T. Polyelectrolyte behavior of an interpolyelectrolyte complex formed in aqueous solution of a charged dendrimer and sodium poly(l-glutamate). *J Phys Chem B* 2003;**107**:13158–67.
54. Tamano K, Imae T. Investigation of luminescent poly(propylene imine) dendrimer. *J Nanosci Nanotechnol* 2008;**9**:4329–34.
55. Zheng Q, Pan CY. Preparation and characterization of dendrimer- star PNIPAAM using dithiobenzoate-terminated PPI dendrimer via RAFT polymerization. *Eur Polym J* 2006;**42**:807–14.
56. Gurdag S, Khandare J, Stapels S, Matherly LH, Kannan RM. Activity of dendrimer-methotrexate conjugates on methotrexate-sensitive and -resistant cell lines. *Bioconjug Chem* 2006;**17**:275–83.
57. Lee CC, Gillies ER, Fox ME, Guillaudeu SJ, Fréchet JM, Dy EE, et al. A single dose of doxorubicin-functionalized bow-tie dendrimer cures mice bearing C-26 colon carcinomas. *Proc Natl Acad Sci USA* 2006;**103**:16649–54.
58. Kolhe P, Khandare J, Pillai O, Kannan S, Lieh-Lai M, Kannan RM. Preparation, cellular transport, and activity of polyamidoamine based dendritic nanodevices with a high drug payload. *Biomaterials* 2006;**27**:660–9.
59. Kono K, Kojima C, Hayashi N, Nishisaka E, Kiura K, Watarai S, et al. Preparation and cytotoxic activity of poly(ethylene glycol)- modified poly(amidoamine) dendrimers bearing adriamycin. *Biomaterials* 2008;**29**:1664–75.
60. Najlah M, Freeman S, Attwood D, D'Emanuele A. Synthesis, characterization and stability of dendrimer prodrugs. *Int J Pharm* 2006;**308**:175–82.

61. Najlah M, Freeman S, Attwood D, D'Emanuele A. In vitro evaluation of dendrimer prodrugs for oral drug delivery. *Int J Pharm* 2007;**336**:183–90.
62. Kurtoglu YE, Mishra MK, Kannan S, Kannan RM. Drug release characteristics of PAMAM dendrimer–drug conjugates with different linkers. *Int J Pharm* 2010;**384**:189–94.
63. Duncan R. The dawning era of polymer therapeutics. *Nat Rev Drug Discov* 2003;**2**:347–60.
64. Hong S, Leroueil PR, Majoros IJ, Orr BG, Baker Jr. JR, Banaszak Holl MM. The binding avidity of a nanoparticle-based multivalent targeted drug delivery platform. *Chem Biol* 2007;**14**:107–15.
65. Chandrasekar D, Sistla R, Ahmad FJ, Khar RK, Diwan PV. The development of folate–PAMAM dendrimer conjugates for targeted delivery of anti-arthritic drugs and their pharmacokinetics and biodistribution in arthritic rats. *Biomaterials* 2007;**28**:504–12.
66. Patri AK, Myc A, Beals J, Thomas TP, Bander NH, Baker Jr. JR. Synthesis and in vitro testing of J591 antibody–dendrimer conjugates for targeted prostate cancer therapy. *Bioconjug Chem* 2004;**15**:1174–81.
67. Shukla R, Thomas TP, Peters JL, Desai AM, Kukowska-Latallo J, Patri AK, et al. HER2 specific tumor targeting with dendrimer conjugated anti-HER2 mAb. *Bioconjug Chem* 2006;**17**:1109–15.
68. Wangler C, Moldenhauer G, Eisenhut M, Haberkorn U, Mier W. Antibody-dendrimer conjugates: the number, not the size of the dendrimers, determines the immunoreactivity. *Bioconjug Chem* 2008;**19**:813–20.
69. Woller EK, Cloninger MJ. Mannose functionalization of a sixth generation dendrimer. *Biomacromolecules* 2001;**2**:1052–4.
70. Roy R, Baek MG. Glycodendrimers: novel glycotope isosteres unmasking sugar coding. Case study with T-antigen markers from breast cancer MUC1 glycoprotein. *J Biotechnol* 2002;**90**:291–309.
71. Lagnoux D, Darbre T, Schmitz ML, Reymond JL. Inhibition of mitosis by glycopeptide dendrimer conjugates of colchicines. *Chemistry* 2005;**11**:3941–50.
72. Gabius HJ, Schroter C, Gabius S, Brinck U, Tietze LF. Binding of T-antigen-bearing neoglycoprotein and peanut agglutinin to cultured tumor cells and breast carcinomas. *J Histochem Cytochem* 1990;**38**:1625–31.
73. Soloway AH, Tjarks W, Barnum BA, Rong FG, Barth RF, Codogni IM, et al. The chemistry of neutron capture therapy. *Chem Rev* 1998;**98**:1515–62.
74. Hawthorne MF, Maderna A. Applications of radiolabeled boron clusters to the diagnosis and treatment of cancer. *Chem Rev* 1999;**99**:3421–34.
75. Barth RF, Adams DM, Soloway AH, Alam F, Darby MW. Boronated starburst dendrimer-monoclonal antibody immunoconjugates: evaluation as a potential delivery system for neutron capture therapy. *Bioconjug Chem* 1994;**5**:58–66.
76. Capala J, Barth RF, Bendayan M, Lauzon M, Adams DM, Soloway AH, et al. Boronated epidermal growth factor as a potential targeting agent for boron neutron capture therapy of brain tumors. *Bioconjug Chem* 1996;**7**:7–15.
77. Wu G, Barth RF, Yang W, Chatterjee M, Tjarks W, Ciesielski MJ, et al. Site-specific conjugation of boron-containing dendrimers to anti-EGF receptor monoclonal antibody cetuximab (IMC-C225) and its evaluation as a potential delivery agent for neutron capture therapy. *Bioconjug Chem* 2004;**15**:185–94.
78. Barth RF, Wu G, Yang W, Binns PJ, Reley KJ, Patel H, et al. Neutron capture therapy of epidermal growth factor (+) gliomas using boronated cetuximab (IMC-C225) as a delivery agent. *Appl Radiat Isot* 2004;**61**:899–903.
79. Ackroyd R, Kelty C, Brown N, Reed M. The history of photodetection and photodynamic therapy. *Photochem Photobiol* 2001;**74**:656–69.

80. Najlah M, D'Emanuele A. Synthesis of dendrimers and drug-dendrimer conjugates for drug delivery. *Curr Opin Drug Discov* 2007;**10**:756–67.
81. Battah S, Chee CE, Nakanishi H, Gerscher S, Macrobert AJ, Edwards C. Synthesis and biological studies of 5-aminolevulinic acid-containing dendrimers for photodynamic therapy. *Bioconjug Chem* 2001;**12**:980–8.
82. Di Venosa GM, Casas AG, Battah S, Dobbin P, Fukuda H, Macrobert AJ, et al. Investigation of a novel dendritic derivative of 5- aminolaevulinic acid for photodynamic therapy. *Int J Biochem Cell Biol* 2006;**38**:82–91.
83. Battah S, O'Neill S, Edwards C, Balaratnam S, Dobbin P, MacRobert AJ. Enhanced porphyrin accumulation using dendritic derivatives of 5-aminolaevulinic acid for photodynamic therapy: an in vitro study. *Int J Biochem Cell Biol* 2006;**38**:1382–92.
84. Battah S, Balaratnam S, Casas A, O'Neill S, Edwards C, Batlle A, et al. Macromolecular delivery of 5-aminolaevulinic acid for photodynamic therapy using dendrimer conjugates. *Mol Cancer Ther* 2007;**6**:876–85.
85. Gerhardt SA, Lewis JW, Kliger DS, Zhang JZ, Simonis U. Effect of micelles on oxygen-quenching processes of triplet-state parasubstituted tetraphenylporphyrin photosensitizers. *J Phys Chem A* 2003;**107**:2763–7.
86. Zhang GD, Harada A, Nishiyama N, Jiang DL, Koyama H, Aida T, et al. Polyion complex micelles entrapping cationic dendrimer porphyrin: effective photosensitizer for photodynamic therapy of cancer. *J Control Release* 2003;**93**:141–50.
87. Matsumura Y, Maeda H. A new concept for macromolecular therapeutics in cancer chemotherapy: mechanism of tumoritropic accumulation of proteins and the antitumor agent smancs. *Cancer Res* 1986;**46**:6387–92.
88. Hirsch LR, Gobin AM, Lowery AR, Tam F, Drezek RA, Halas NJ, et al. Metal nanoshells. *Ann Biomed Eng* 2006;**34**:15–22.
89. Shi X, Wang S, Meshinchi S, Van Antwerp ME, Bi X, Lee I, et al. Dendrimer-entrapped gold nanoparticles as a platform for cancer-cell targeting and imaging. *Small* 2007;**3**:1245–52.
90. Haba Y, Kojima C, Harada A, Ura T, Horinaka H, Kono K. Preparation of poly(ethylene glycol)-modified poly(amido amine) dendrimers encapsulating gold nanoparticles and their heat-generating ability. *Langmuir* 2007;**23**:5243–6.
91. Elbashir SM, Harborth J, Lendeckel W, Yalcin A, Weber K, Tuschl T. Duplexes of 21-nucleotide RNAs mediate RNA interference in cultured mammalian cells. *Nature* 2001;**411**:494–8.
92. Fire A, Xu S, Montgomery MK, Kostas SA, Driver SE, Mello CC. Potent and specific genetic interference by double-stranded RNA in Caenorhabditis elegans. *Nature* 1998;**391**:806–11.
93. Cristiano RJ, Xu B, Nguyen D, Schumacher G, Kataoka M, Spitz FR, et al. Viral and nonviral gene delivery vectors for cancer gene therapy. *Cancer Detect Prev* 1998;**22**:445–54.
94. Ma H, Diamond SL. Nonviral gene therapy and its delivery systems. *Curr Pharm Biotechnol* 2001;**2**:1–17.
95. Sonawane ND, Szoka Jr. FC, Verkman AS. Chloride accumulation and swelling in endosomes enhances DNA transfer by polyamine-DNA polyplexes. *J Biol Chem* 2003;**278**:44826–31.
96. Mitra A, Imae T. Nanogel formation consisting of dna and poly(amido amine) dendrimer studied by static light scattering and atomic force microscopy. *Biomacromolecules* 2004;**5**:69–73.
97. Braun CS, Vetro JA, Tomalia DA, Koe GS, Koe JG, Middaugh CR. Structure/function relationships of polyamidoamine/DNA dendrimers as gene delivery vehicles. *J Pharm Sci* 2005;**94**:423–36.
98. Kukowska-Latallo JF, Bielinska AU, Johnson J, Spindler R, Tomalia DA, Baker Jr. JR. Efficient transfer of genetic material into mammalian cells using Starburst polyamidoamine dendrimers. *Proc Natl Acad Sci USA* 1996;**93**:4897–902.

99. Wang Y, Kong W, Song Y, Duan Y, Wang L, Steinhoff G, et al. Polyamidoamine dendrimers with a modified pentaerythritol core having high efficiency and low cytotoxicity as gene carriers. *Biomacromolecules* 2009;**10**:617–22.
100. Morgan DM, Larvin VL, Pearson JD. Biochemical characterisation of polycation-induced cytotoxicity to human vascular endothelial cells. *J Cell Sci* 1989;**94**:553–9.
101. Arima H, Motoyama K. Recent findings concerning PAMAM dendrimer conjugates with cyclodextrins as carriers of DNA and RNA. *Sensor* 2009;**29**:6346–61.
102. Tsutsumi T, Hirayama F, Uekama K, Arima H. Evaluation of polyamidoamine dendrimer/cyclodextrin conjugate (generation 3, G3) as a novel carrier for small interfering RNA (siRNA). *J Control Release* 2007;**119**:349–59.
103. Tsutsumi T, Hirayama F, Uekama K, Arima H. Potential use of polyamidoamine dendrimer/-cyclodextrin conjugate (generation 3, G3) as a novel carrier for short hairpin RNA expressing plasmid DNA. *J Pharm Sci* 2008;**97**:3022–34.
104. Choi JS, Nam K, Park J-Y, Kim J-B, Lee J-K, Park J-S. Enhanced transfection efficiency of PAMAM dendrimer by surface modification with L-arginine. *J Control Release* 2004;**99**:445–56.
105. Nam HY, Nam K, Hahn HJ, Kim BH, Lim HJ, Kim HJ, et al. Biodegradable PAMAM ester for enhanced transfection efficiency with low cytotoxicity. *Biomaterials* 2009;**30**:665–73.
106. Kim T-I, Bai CZ, Nam K, Park J-S. Comparison between arginine conjugated PAMAM dendrimers with structural diversity for gene delivery systems. *J Control Release* 2009;**136**:132–9.
107. Kim KS, Lei Y, Stolz DB, Liu D. Bifunctional compounds for targeted hepatic gene delivery. *Gene Ther* 2007;**14**:704–8.
108. Taratula O, Garbuzenko OB, Kirkpatrick P, Pandya I, Savla R, Pozharov VP, et al. Surface-engineered targeted PPI dendrimer for efficient intracellular and intratumoral siRNA delivery. *J Control Release* 2009;**140**:284–93.
109. Nishiyama N, Iriyama A, Jang WD, Miyata K, Itaka K, Inoue Y, et al. Light-induced gene transfer from packaged DNA enveloped in a dendrimeric photosensitizer. *Nat Mater* 2005;**4**:934–41.
110. Shieh MJ, Peng CL, Lou PJ, Chiu CH, Tsai TY, Hsu CY, et al. Non-toxic phototriggered gene transfection by PAMAM-porphyrin conjugates. *J Control Release* 2008;**129**:200–6.
111. Albrecht M, Gossage RA, Lutz M, Spek AL, van Koten G. Diagnostic organometallic and metallodendritic materials for SO2 gas detection: reversible binding of sulfur dioxide to arylplatinum(II) complexes. *Chem Eur J* 2000;**6**:1431–45.
112. Daniel MC, Ruiz J, Nlate S, Blais JC, Astruc D. Nanoscopic assemblies between supramolecular redox active metallodendrons and gold nanoparticles: synthesis, characterization, and selective recognition of H2PO4-, HSO4-, and adenosine-5′-triphosphate (ATP2-) anions. *J Am Chem Soc* 2003;**125**:2617–28.
113. Stears RL, Getts RC, Gullans SR. A novel, sensitive detection system for high-density microarrays using dendrimer technology. *Physiol Genomics* 2000;**3**:93–9.
114. Wang J, Jiang M, Nilsen TW, Getts RC. Dendritic nucleic acid probes for DNA biosensors. *J Am Chem Soc* 1998;**120**:8281–2.
115. Finikova O, Galkin A, Rozhkov V, Cordero M, Haegerhaell C, Vino-gradov S. Porphyrin and tetrabenzoporphyrin dendrimers: tunable membrane-impermeable fluorescent pH nanosensors. *J Am Chem Soc* 2003;**25**:4882–93.
116. Rietveld IB, Kim E, Vinogradov SA. Dendrimers with tetrabenzoporphyrin cores: near infrared phosphors for in vivo oxygen imaging. *Tetrahedron Lett* 2003;**59**:3821–31.
117. Alonso B, Armada PG, Losada J, Cuadrado I, González B, Casado CM. Amperometric enzyme electrodes for aerobic and anaerobic glucose monitoring prepared by glucose oxidase

immobilized in mixed ferrocene-cobaltocenium dendrimers. *Biosens Bioelectron* 2004;**19**:1617–25.

118. Armada MPG, Losada J, Zamora M, Alonso B, Cuadrado I, Casado CM. Electrocatalytical properties of polymethylferrocenyl den drimers and their applications in biosensing. *Bioelectrochemistry* 2006;**69**:65–73.
119. Klajnert B, Sadowska M, Bryszewska M. The effect of polyamidoamine dendrimers on human erythrocyte membrane acetylcholinesterase activity. *Bioelectrochemistry* 2004;**65**:23–6.
120. Shcharbin D, Jokiel M, Klajnert B, Bryszewska M. Effect of dendrimers on pure acetylcholinesterase activity and structure. *Bioelectrochemistry* 2006;**68**:56–9.
121. Imae T, Hirota T, Funayama K, Aoi K, Okada M. Binding of poly(amido amine) dendrimer to sodium hyaluronate in aqueous NaCl solution. *J Colloid Interface Sci* 2003;**263**:306–11.
122. Stiriba S, Frey H, Haag R. Dendritic polymers in biomedical applications: from potential to clinical use in diagnostics and therapy. *Angew Chem Int Ed* 2002;**41**:1329–34.
123. Wiener EC, Brechbiel MW, Brothers H, Magin RL, Gansow OA, Tomalia DA, et al. Dendrimer-based metal chelates: a new class of magnetic resonance imaging contrast agents. *Magn Reson Med* 1994;**31**:1–8.
124. Tomalia DA, Reyna LA, Svenson S. Dendrimers as multi-purpose nanodevices for oncology drug delivery and diagnostic imaging. *Biochem Soc Trans* 2007;**35**:61–7.
125. Wiener EC, Auteri FP, Chen JW, Brechbiel MW, Gansow OA, Schneider DS, et al. Molecular dynamics of ion–chelate complexes attached to dendrimers. *J Am Chem Soc* 1996;**118**:7774–82.
126. Kobayashi H, Kawamoto S, Jo S-K, Bryant Jr. HL, Brechbiel MW, Star RA. Macromolecular MRI contrast agents with small dendrimers: pharmacokinetic differences between sizes and cores. *Bioconjug Chem* 2003;**14**:388–94.
127. Langereis S, de Lussanet QG, van Genderen MHP, Backes WH, Meijer EW. Multivalent contrast agents based on gadolinium–diethylenetriaminepentaacetic acid-terminated poly (propylene imine) dendrimers for magnetic resonance imaging. *Macromolecules* 2004;**37**:3084–91.
128. Herborn CU, Barkhausen J, Paetsch I, Hunold P, Mahler M, Shamsi K, et al. Coronary arteries: contrast-enhanced MR imaging with SH L 643A—experience in 12 volunteers. *Radiology* 2003;**229**:217–23.
129. Nicolle GM, Toth E, Schmitt-Willich H, Raduchel B, Merbach AE. The impact of rigidity and water exchange on the relaxivity of a dendritic MRI contrast agent. *Chem Eur J* 2002;**8**:1040–8.
130. Schumann H, Wassermann BC, Schutte S, Velder J, Velder J, Aksu Y. Synthesis and characterization of water-soluble tin-based metallodendrimers. *Organometallics* 2003;**22**:2034–41.
131. Krause W, Hackmann-Schlichter N, Maier FK, Muller R. Dendrimers in diagnostics. *Top Curr Chem* 2000;**210**:261–308.
132. Fu Y, Nitecki DE, Maltby D, Simon GH, Berejnoi K, Raatschen HJ, et al. Dendritic iodinated contrast agents with PEG-cores for CT imaging: synthesis and preliminary characterization. *Bioconjug Chem* 2006;**17**:1043–56.
133. Hainfeld JF, Slatkin DN, Focella TM, Smilowitz HM. Gold nanoparticles: a new X-ray contrast agent. *Br J Radiol* 2006;**79**:248–53.
134. Kim D, Park S, Lee J, Jeong Y, Jon S. Antibiofouling polymer-coated gold nanoparticles as a contrast agent for in vivo x-ray computed tomography imaging. *J Am Chem Soc* 2007;**129**:7661–5.
135. Cai QY, Kim SH, Choi KS, Kim SY, Byun SJ, Kim KW, et al. Colloidal gold nanoparticles as a blood-pool contrast agent for X-ray computed tomography in mice. *Invest Radiol* 2007;**42**:797–806.

136. Xu C, Tung GA, Sun S. Size and concentration effect of gold nanoparticles on X-ray attenuation as measured on computed tomography. *Chem Mater* 2008;**20**:4167–9.
137. Jackson PA, Rahman WN, Wong CJ, Ackerly T, Geso M. Potential dependent superiority of gold nanoparticles in comparison to iodinated contrast agents. *Eur J Radiol* 2010;**75**:104–9.
138. Kojima C, Yasuhito U, Mikako O, Harada A, Yasuhiro M, Kono K. X-ray computed tomography contrast agents prepared by seeded growth of gold nanoparticles in PEGylated dendrimer. *Nanotechnology* 2010;**21**:1–6.
139. Guo R, Wang H, Peng C, Shen M, Zheng L, Zhang G, et al. Enhanced X-ray attenuation property of dendrimer-entrapped gold nanoparticles complexed with diatrizoic acid. *J Mater Chem* 2011;**21**:5120–7.
140. Parrott MC, Benhabbour SR, Saab C, Lemon JA, Parker S, Valliant JF, et al. Synthesis, radiolabeling, and bio-imaging of high-generation polyester dendrimers. *J Am Chem Soc* 2009;**131**:2906–16.
141. Myc A, Majoros IJ, Thomas TP, Baker Jr. JR. Dendrimer-based targeted delivery of an apoptotic sensor in cancer cells. *Biomacromolecules* 2007;**8**:13–8.
142. Thomas TP, Myaing MT, Ye JY, Candido K, Kotlyar A, Beals J, et al. Detection and analysis of tumor fluorescence using a two-photon optical fiber probe. *Biophys J* 2004;**86**:3959–65.

Carbohydrate-Based Nanoparticles for Potential Applications in Medicine

Marco Marradi,[*,†]
Isabel García,[*,†] and
Soledad Penadés[*,†]

[*]Laboratory of GlycoNanotechnology, Biofunctional Nanomaterials Unit, CIC biomaGUNE, Paseo Miramón 182, Parque Tecnológico de San Sebastián, Spain

[†]Biomedical Research Networking Center in Bioengineering, Biomaterials and Nanomedicine (CIBER-BBN), Paseo Miramón 182, San Sebastián, Spain

I. Introduction 142
II. Metallic Carbohydrate-Based Nanoparticles for *In Vitro* Detection of Cells and Pathogens 144
A. Detection of Cells 145
B. Detection of Pathogens 147
III. Metallic Carbohydrate-Based Nanoparticles for *In Vivo* Diagnosis by Magnetic Resonance Imaging 151
IV. Metallic Carbohydrate-Based Nanoparticles for *In Vivo* Diagnosis by Optical Imaging 156
V. Metallic Carbohydrate-Based Nanoparticles in Therapy 158
VI. NonMetallic Polysaccharide-Based Nanoparticles in Therapy 164
VII. Conclusions 165
References 166

Although carbohydrates are essential for life, the development of medical tools based on these important biomolecules is significantly slower than those based on proteins or nucleic acids. This chapter deals with the applications of nanoparticles decorated with carbohydrates and discusses the perspectives of their use in the field of medicine. The review is divided into two sections: diagnosis and therapy. Within these topics, the focus will be on two main types of systems: carbohydrate-coated metallic nanoparticles in which the carbohydrate ligands are "covalently" linked to a nanosized metallic cluster and polysaccharide-encapsulated metallic cores. The former glyconanoparticles (GNPs) represent a powerful chemical tool in the field of glycobiology as a specific carbohydrate can be selected to exert a concrete biological function and profile carbohydrate-based interactions. Up to now, these GNPs have been mainly used as potential anti-adhesion agents against pathogens.

Progress in Molecular Biology
and Translational Science, Vol. 104
DOI: 10.1016/B978-0-12-416020-0.00004-8

1877-1173/11 $35.00

The opportunity of inserting multifunctionality and changing the nucleus size/material is giving birth to new targeted systems for imaging and therapy. On the other hand, nonmetallic polysaccharide-based nanoparticles have been successfully used as drug delivery carriers in addition to molecular imaging.

I. Introduction

Bringing discoveries from bench to bedside is the key aim of biomedical "translational" research. The advent of nanotechnology is creating great expectations for the application of engineered and functional nanoparticles to different fields of medicine, including diagnosis and therapy.[1] The nanometric dimensions of the nanotechnology tools allow the study of and intervention in the cellular world and a better understanding of the biological processes at a molecular level. One of the main challenges in this field is the development of new therapeutic systems and diagnostic probes that ensure high stability, specificity, and low toxicity. An ideal therapeutic system should be effective at low doses, selective, and specific. To accomplish these requirements, one strategy is based on the conjugation of targeted biomolecules to nanoparticles. In addition to these characteristics, a diagnostic tool should give the opportunity to be used simultaneously in different imaging techniques with high sensitivity. To this aim, nanotechnology can contribute to enhance the performance of molecular probes due to the possibility of inserting magnetic and/or luminescent elements in the nanostructures and profit from quantum size effects. The design and synthesis of multifunctional and multimodal nanoclusters for biomedical applications[2] that include bioassays,[3] cancer therapy,[4–6] gene and drug delivery,[7] regenerative brain medicine,[8] optical biodiagnostics,[9] and multimodal imaging[10–13] have been extensively reviewed. Most of these bionanomaterials are based on the conjugation of nucleic acids, peptides, proteins, antibodies, and genetic material (DNA, RNA, etc.) to nanoparticles. However, the development of medical tools based on carbohydrates is significantly slower than that based on proteins or nucleic acids. Among biomolecules, carbohydrates have an important role in recognition and mediate a number of biological and pathological processes.[14,15] Carbohydrates (glycoproteins and glycolipids) are presented at the cell surfaces as multivalent domains. In order to mimic the presentation of glycoconjugates at cell surface, synthetic oligosaccharides functionalized with suitable spacers have been multimerized on a number of scaffolds including peptides and cyclopeptides,[16–19] dendrimers and dendrons,[20] polymers,[21–23] liposomes,[24–27] and calixarenes,[28–30] among others. These multivalent glycomaterials have been used to address basic studies, but also to face medical challenges with the aim of translating basic research to applicative fields.[31]

Metallic nanoparticles functionalized with carbohydrates (glyconanoparticles (GNPs)) are known for 10 years and have been used as a biomimetic model to study and to interfere with carbohydrate-mediated interactions.[32] The use of thiol-ended glycoconjugates to protect nascent gold nanoparticles, by the formation of carbohydrate self-assembled monolayers (SAMs) at gold surface,[33] was first presented by Penadés and coworkers.[34] The birth of gold GNPs was due to the interest in developing a new multivalent chemical system to investigate the existence of carbohydrate–carbohydrate interactions in water.[35] Since then, a plethora of different types of GNPs based on gold, but also on other noble metals, semiconductors, and/or magnetic elements has appeared in the literature.[32,36] Functionalizing metallic nanoparticles with glycoconjugates has in fact several advantages over other multivalent scaffolds. Their preparation is a simple, one-step process which allows the tuning of various physical and chemical properties such as colloid stability, water dispersibility, and biocompatibility. Furthermore, the same nanoparticle can host different molecules of interest (multifunctionality) which can be tailored on the metal surface by altering their ratios in a controlled way. However, the use of metallic GNPs to solve biomedical problems is still in a nascent state. Although the clinical translation of these new nanomaterials is difficult to foresee, some GNPs have been already applied in preclinical studies. Among the core materials which can be chosen to construct GNPs, gold nanoclusters have been extensively explored because of their relative inertness, low toxicity, and easy manipulation, and because the chemistry of their surface is easy to control.[37] In general, gold nanoparticles have been envisaged as promising candidates for *in vivo* applications such as drug delivery.[38] Concerns about nanotoxicity[39] and poor biodegradability suggest that their clinical use should be limited to out-of-body devices, such as ultrasensitive biosensors,[40] or to one-off/limited administration such as vaccination, diagnostics, and thermal cancer treatments. This is a general consideration for metallic nanoparticles that has to be taken into account especially for quantum dots (QDs).[41]

Polysaccharides have also been used in nanotechnology as they usually are biocompatible and can function as a protective coating for metallic cores. Most of the applications of polysaccharide-based nanoparticles deal with drug delivery and molecular imaging. The increase of circulation time in the body, avoiding nonspecific protein adsorption, and reduction of toxicity are some qualities that have allowed this type of nanoparticles to reach the market. Dextran-coated iron oxide nanoparticles are already in clinical use as contrast agents. The possibility of functionalizing the saccharidic units of this type of glycomaterials has been used to obtain biologically compatible systems which can also be targeted. These two types of systems (Fig. 1) present the great

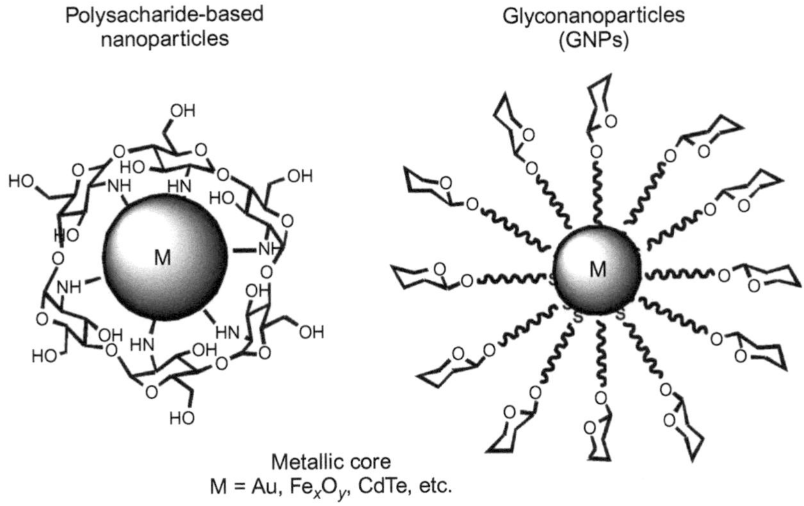

FIG. 1. Polysaccharide-encapsulated metallic nanoparticles (left) and carbohydrate-functionalized metallic nanoparticles (glyconanoparticles, GNPs).

advantage of high biocompatibility due to saccharide molecules, but the GNPs offer a greater versatility for chemical manipulation and control of the density and type of ligands than polysaccharide-encapsulated nanoparticles.

This chapter focuses on the application of these two types of carbohydrate-coated metallic nanoparticles (Fig. 1) to biomedical problems. First, the main examples are reported of the use of these glycomaterials in diagnostics (*in vitro* and *in vivo*) and then an overview given of their application as anti-adhesion and/or therapeutic agents stressing the few examples which can be translated to the clinic.

II. Metallic Carbohydrate-Based Nanoparticles for *In Vitro* Detection of Cells and Pathogens

There is a constant need to improve the performance of current diagnostic assays as well as to develop innovative testing strategies. Nanoparticles are promising tools for *in vitro* sensing and diagnosis.[42,43] Multifunctional GNPs able to specifically target carbohydrate-binding receptors and their integration into novel devices (GNP-based diagnostics) may open new frontiers for the detection of tumors, infectious diseases, and neurological diseases. More work is necessary to optimize the use of this type of nanoparticles for medical diagnosis, as this application is still in preliminary stages. Most of the applications of GNPs in the field of biosensing are "proofs of principle" based on model lectins (carbohydrate binding proteins) such as concanavalin A (ConA).[44–47] Many carbohydrate-based nanostructures have been used to evaluate carbohydrate–protein interactions.[32,36,48] The most popular detection technique is based on the surface plasmon resonance (SPR) of the metallic core (Au/Ag),[49,50] but other techniques have also been applied to detect protein biomarkers by using GNPs (carbohydrate–protein interactions).[51–53]

The following sections will review examples of the application of metallic carbohydrate-based nanoparticles in the detection of cells and pathogens. None of these examples are in clinical use, but some of them have a great potential to be translated to the clinic.

A. Detection of Cells

A number of reports have been proposed to detect cells with carbohydrate-coated gold, iron oxide, or semiconductor metallic nanoparticles. The combined use of lectins and GNPs for profiling sugar patterning on the cell surface has been reported. Ding *et al.* have recently developed a new biosensor platform that highlights the application of biofunctionalized GNPs as tools for *in vitro* diagnosis.[54] They designed a scanometric approach to detect cell surface carbohydrate that combines the aggregation of mannose-coated gold GNPs mediated by the recognition of ConA with the aggregation-regulated silver signal amplification and a spot test (Fig. 2).

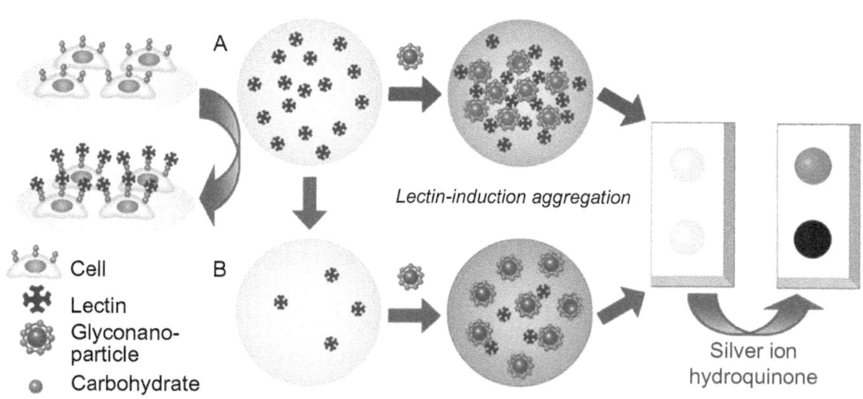

FIG. 2. Scanometric approach to detect mannose residues on living cells. Lectin-induction aggregation promoted by mannose GNPs of (A) control lectin solutions and (B) supernatants collected after the incubation of cells (BGC-823) with ConA. This aggregation was visualized by the gold-catalyzed silver enhancement process. Adapted from Ref. 54. (See Color Insert.)

The group of Huang has developed an effective method based on magnetic GNPs that may have great potential in *in vitro* cancer diagnosis by magnetic resonance imaging (MRI).[55] The method identifies and distinguishes the fine features of cancer cells by targeting endogenous lectins specifically expressed at their surface. Iron oxide core/silica shell magnetic nanoparticles bearing monosaccharide units (mannose, galactose, fucose, sialic acid, and glucosamine) were incubated with nine different types of cancer cells. When a particular magnetic GNP is specifically recognized by a type of cancer cell, aggregation of nanoparticles results in shorter T_2 (transverse relaxation time) and consequently a darker MRI image. Un-functionalized iron oxide magnetic nanoparticles were unable to generate significant T_2 changes and were used as controls. A range of cells including related isogenic tumor cells (cells with different metastatic potential) and malignant versus normal cells (breast cancer MCF-7/Adr-res cells versus normal breast endothelial 184B5 cells) were distinguished by MRI measurements. The interactions between glycoconjugates and endogenous lectins present on cancer cell surface are crucial for cancer and metastasis process. The importance of this work is the characterization of the glyco-code of individual cell lines. Recently, the same group has described the synthesis of hyaluronic acid (HA)-encapsulated iron oxide nanoparticles. HA is a polysaccharide constituted by a unit of the monosaccharide 2-acetamido-2-deoxy-D-glucuronic acid. These HA-encapsulated magnetic nanoparticles have been used *in vitro* to specifically detect macrophage-related atherosclerotic plaques.[56] The authors proposed the CD44 cell receptor as target protein. This receptor is the principal cell-adhesion receptor of HA and it is expressed on three cell types present in the atherosclerotic plaques, vascular endothelial cells, macrophages, and smooth muscle cells. They demonstrate that HA attached on the surface of the nanoparticles maintains binding with CD44 as determined through enzyme-linked immunosorbent assay (ELISA) and flow cytometry. Finally, *in vitro* phantom imaging of activated human acute monocytic leukemia cells (THP-1) showed significant changes in T_2-weighted MR images when the HA-modified magnetic nanoparticles are uptaken by the cells. HA-CD44 receptor–ligand interaction had been previously utilized to target *in vitro* cancer cells.[57] HA conjugated to dopamine was immobilized onto magnetic nanoparticles and their cancer-targeting capability was evaluated in $CD44^+$ cells (HCT116, human colon carcinoma cells) by means of T_2-weighted gradient-echo MR phantom imaging.

QDs (see the chapter "Luminescent Quantum Dots, Making Invisibles Visible in Bioimaging" by Ishikawa and Biju in this volume) are also interesting tools for imaging. CdSe/ZnS core shells QDs functionalized with β-*N*-acetylglucosamine (GlcNAc) or mannose (Man) were used as optical imaging probes in confocal microscopy to label mice, pigs, and sea-urchin live sperm.[58] Due to the different distribution of the GlcNAc and Man receptors on the sperm surface, acetyl glucosamine QDs were imaged at the sperm heads, while mannose QDs spread over the whole sperm body. The ability of the GlcNAc QDs to site-specifically bind sperm and to influence the acrosomal reaction opens up the opportunity to follow the dynamic changes in the fertilization process.

B. Detection of Pathogens

Nanotechnology-based approaches for the detection, identification, and/or isolation of pathogens are attracting much interest in current research.[59] As mentioned in Section I, the conjugation of nanoparticles with biomolecules not only ensures water dispersibility and biocompatibility but also allows selective targeting depending on the type of biofunctionalization. Furthermore, the opportunity to vary the inorganic core-material size, shape, and composition enables the use of a broad range of techniques to detect pathogens with high sensitivity (gold with biosensors, QDs in fluorescence imaging, and magnetic nanoparticles for relaxation-based measurements). These are the great advantages of nanotechnology with respect to traditional methods based on amplification and/or purification procedures. Especially SPR biosensors and magnetic- and fluorescent-based techniques have largely profited from nanoparticulate systems for quick, sensitive, and specific assays.

Carbohydrate-mediated interactions are involved in entry processes of pathogens.[60] The envelope of virus, bacteria, and parasites are glycoproteins, peptidoglycans, or unusual glycan structures that the pathogens use to target lectins of host cells. On the other hand, pathogens also profit from their carbohydrate binding regions to interact with surface glycoconjugates (glycocalix) of the host cells.[60] The use of carbohydrate-based nanoparticles in pathogen detection relies on the fact that the presentation of multiple copies of carbohydrates on their surface can mimic either the glycans of the pathogen envelop or the cell glycocalix. In addition, the multivalence of carbohydrates that specifically interact with a receptor in the target pathogen increases the low affinity of carbohydrate-mediated interactions. This chapter discusses the main applications of GNPs on the basis of the type of carbohydrate used for the *in vitro* detection of pathogens.

Some examples of how carbohydrate-based nanoparticles can address lectin-type receptors on the cell surface have been described.[55] Different types of bacteria, as well as toxins associated with pathogens, also express specific carbohydrate-binding domains that interact with glycoproteins and glycolipids at the host cell surface. For example, the Gram-negative bacterium *Escherichia coli* has many type 1 pili protein structures which contain a mannose-binding region at the tip of each pilus that allow for attachment to host endothelial cells. Different strains of this bacterium have been used for testing the detection capability of carbohydrate-based nanoparticles.

Gold nanoparticles coated with pentyl α-mannoside were used by Lin *et al.*[61] to target mannose-specific FimH adhesin of type 1 pili in *E. coli*. After incubation of the nanoparticles with different *E. coli* strains, transmission electron microscopy (TEM) images showed that mannose GNPs selectively bind the pili of the ORN178 strain which expresses the wild-type 1 pili, but not those of the ORN208 strain which lack the *fimH* gene and cannot mediate mannose-specific binding. About 2000-fold concentrations of methyl α-D-mannopyranoside were required to compete out ~90% of mannose GNPs binding to bacterial pili. Although TEM is not the most suitable technique to be translated into routine for detecting a pathogen, this is the first example of labeling a specific protein at pathogen surface by using carbohydrate-conjugated nanoparticles.

The detection and separation of *E. coli* from contaminated media was achieved by El-Boubbou *et al.*[62] by means of magnetic GNPs. After incubation of *E. coli* strain ORN178 with silica-coated magnetite nanoparticles functionalized with multiple copies of mannosides, a magnetic field was applied to separate the aggregates. *E. coli* could be detected within a limit of 10^4 cells mL^{-1} as measured by fluorescence microscopy imaging. Depending on the incubation time, the removal of *E. coli* from the medium ranged from 65% (5 min) up to 88% (45 min). The capture efficiency with magnetic mannose GNPs was higher than the one obtained with magnetic particles functionalized with antibodies or lectins. The identification of three *E. coli* strains was also proved by using mannose GNPs and galactose GNPs. They used the mannose-specific ORN178 strain, a mutant strain ORN208 with reduced mannose-binding affinity, and an environmental strain (ES) with unknown carbohydrate-binding specificity. The ability to distinguish pathogen strains is of great interest from the clinical perspective and the proposed application of magnetic-based separation of a pathogen from contaminated media can also allow the concentration and recovery of pathogens for further analyses by other techniques.

An attempt to detect *E. coli* by means of water-dispersible gold nanodots which exhibited fluorescent properties due to the nanometric size of the metallic cluster (<2 nm) was recently reported.[63] Incubation of mannose-protected gold nanodots (gold diameter ~1.8 nm and quantum yield 8.6%) with *E. coli* K12 wild-type strain ATCC 25404 caused the formation of brightly fluorescent cell clusters upon excitation at 365 nm. The fluorescence intensity (at 545 nm) versus *E. coli* concentration showed a linear response from 1.00×10^6 to 5.00×10^7 cells/mL and the detection limit of *E. coli* was 7.20×10^5 cells/mL. Although mannose GNPs are easier to prepare than silica-coated magnetic GNPs and no fluorescent dyes are needed for staining the bacteria, it seems that the sensitivity of these nanodots is lower than magnetic mannose GNP proposed by El-Boubbou *et al.*[62] Other mannose-based systems for selective *E. coli* detection based on micelles,[64] nanocapsules,[65] polymeric nanoparticles,[66] glycopolymers,[67] spindle-type hematite nanoparticles,[68] micrometric magnetic beads,[69] and QDs[70] have also been proposed. These few examples are part of a wider branch of research devoted to the development of nanotechnology-based rapid and ultrasensitive diagnostic systems, especially focused on the detection of infectious agents.[71] The application of glyconanomaterials in antimicrobial applications and biodefense is attracting much attention: an example is the use of carbohydrate-functionalized carbon nanotubes to selectively aggregate bacterial spores for diminishing danger in terms of lethal inhalation or cutaneous doses, as, for example, the case of anthrax infection.[72]

Polysaccharide-encapsulated metallic nanoparticles have been also used to intervene against pathogens. Recently, an assay based on dextran-coated gold nanoparticles was used to assess the conditions for preventing bacterial growth (antimicrobial susceptibility) in Mueller–Hinton broth medium by recording the changes in the UV–vis absorbance spectrum.[73] Relatively fast (3 h) and high-throughput assessment of the *E. coli* drug ampicillin resistance was achieved by using ConA-induced clustering of nanoparticles in starch-rich bacterial suspension in the presence of various concentrations of ampicillin. Under conditions of bacterial growth inhibition, addition of ConA caused starch-assisted nanoparticle aggregation, which resulted in wavelength red shifts of the gold surface plasmon band. Under conditions of increased bacterial growth, proliferating bacteria caused a reduction in starch amount in the solution and a corresponding decrease in the size of the aggregates (wavelength blue shifts in the plasmonic band) due to an increasing number of nanoparticles that bind to bacterial surface carbohydrates. In general, this type of methodology could be used to rapidly identify effective concentrations of antibacterial agents and speed up clinical decision-making for proper dose administration. Based on previous works of Weissleder and colleagues about magnetic nanoswitches,[74,75] Kaittanis *et al.* used iron oxide nanoparticles to identify the cellular cholera toxin B (CTB) receptor by MRI.[76] Galactose-conjugated nanoparticles interact with CTB, producing changes in T_2 relaxation times due to nanoparticle aggregation. The detection limit achieved was of 40 pM.

Pieters and collaborators explored the detection of the bacterial pathogen *Streptococcus suis* using magnetic GNPs. This bacterium contains an adhesion protein that binds to the disaccharide galabiose (α-Gal-(1 → 4)-Gal) and causes meningitis in humans.[77] After incubation of *S. suis* with magnetic GNPs, magnetic separation, and adenosine triphosphate (ATP) detection, bacterial levels down to 10^5 colony-forming unit (cfu) were detected. Submicrometer-sized particles (250 nm) with higher surface area were needed in bacterial detection, while larger nanoparticles (10 μm) failed in the detection.

In addition to detection of bacteria as a whole, the toxins secreted by certain types of bacteria (cholera toxin from *Vibrio cholerae* and shiga toxins from *E. coli*) that bind to glycolipid receptors on the surface of host cells have been also detected by means of GNPs. Lactose GNPs have been used for colorimetric detection of cholera toxin B subunit following the surface plasmon absorption changes upon aggregation.[78] Although it is known that cholera toxin uses the GM_1 ganglioside for binding receptors on gut cells and promoting the initial attachments of the B subunit to them, the disaccharide lactose (β-Gal-(1 → 4)-Glc) is also recognized through its galactose terminal moiety by this toxin.[79] The disadvantage of a lower and less specific interaction due to the use of lactose is compensated by the preparation simplicity of the GNP sensor system which is based on an accessible disaccharide instead of the pentameric unit of GM_1. Other works have focused on the detection of Shiga toxin type 1 (Stx1) released by *E. coli O157:H7* using gold GNPs coated with glycan-like Gb3 trisaccharides that mimic the glycolipid domain on host cells.[80,81] Kulkarni *et al.*[80] expanded their study to the more clinically relevant Shiga toxin type 2 (Stx2), demonstrating that inhibition is highly dependent on the structure and density of the glycans and the structure of the toxin (Fig. 3). GNPs bearing trisaccharide analogs of Gb3 were not cytotoxic to the Vero cells and

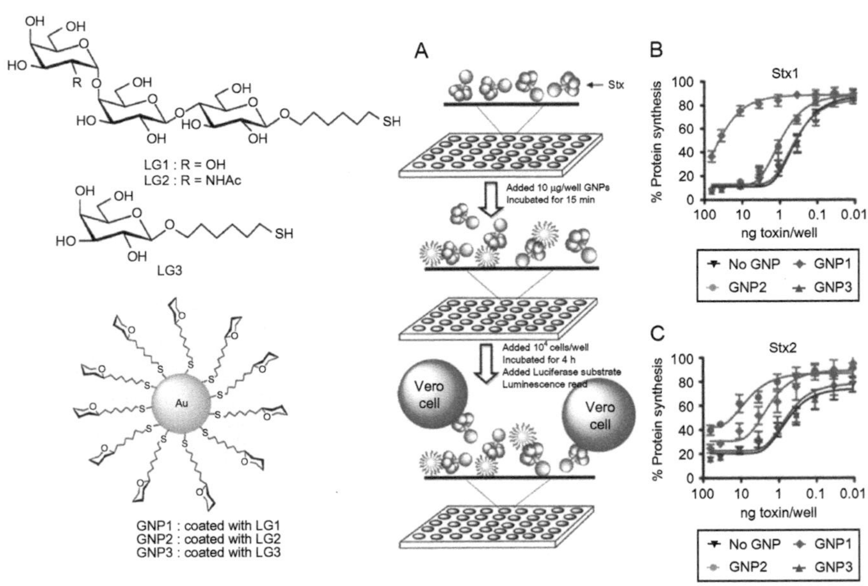

FIG. 3. GNPs coated with Gb3 trisaccharide analogs. (A) Cartoon presentation of the Vero cell toxin inhibition assay. GNPs were added to wells containing serial dilutions of toxin, followed by addition of 10^4 Vero cells expressing a destabilized form of luciferase, *luc2P*. Plates were incubated at 37 °C for 4 h. Luciferase substrate was added and luminescence was measured on a luminometer. (B) Inhibition of Stx1 using GNPs. (C) Inhibition of Stx2 using GNPs. (D) Glycan-like Gb3 trisaccharide gold nanoparticles. Adapted from Ref. 80. (See Color Insert.)

neutralized the toxin in a dose-dependent manner (Fig. 3). A few amino acid variations in the receptor-binding site of the B subunit or in the A subunit of the toxin (natural variants Stx2c and Stx2d) change the receptor specificity and the GNPs failed to inhibit the adhesion of the toxin to Vero cells.

An SPR competition binding assay was used by Chien *et al.*[81] to investigate the affinity of Gb3 globotriose-functionalized gold nanoparticles as multivalent probes for Stx1. Globotriose GNPs showed size- and linker length-dependent affinity for the toxin with affinity up to 10^8-fold than free globotriose. Due to this high affinity, GNPs have been used as probes for Shiga toxin purification. Interestingly, the authors also described in the same paper a chip-bioassay, similar to ELISA, based on a silver enhancement of these gold GNPs for the final detection of Stx. Although the method is not as sensitive as ELISA, it provides a rapid alternative for the detection of proteins.

So far the chapter has presented some examples of *in vitro* diagnostics based on carbohydrate-based metallic nanoparticles (Table I). The following section will focus on their *in vivo* applications to address the diagnosis of diseases.

III. Metallic Carbohydrate-Based Nanoparticles for *In Vivo* Diagnosis by Magnetic Resonance Imaging

Noninvasive imaging technologies such as MRI, positron emission tomography (PET), single-photon emission computed tomography (SPECT), and X-ray computed tomography (CT) for the diagnosis of diseases typically involve the administration of contrast agents for signal enhancement that facilitates the visualization of organs or the state of a particular disease. The development of nanoparticle-based probes for imaging is having a tremendous impact in the design of targeted and specific contrast agents with the potential to enhance the contrast between malignant and benign tissues.[82,83] Until now, only two types of molecular probes for MRI are in clinical use: T_1-contrast agents based on paramagnetic Gd-complex and T_2-contrast agents based on polysaccharide-encapsulated superparamagnetic iron oxide nanoparticles (Resovist®, Feridex®, Sinerem®). Dextran is the polysaccharide approved in clinical use, but chitosan-encapsulated iron oxide nanoparticles have also been developed for imaging, although the translation from animals to application in humans has not yet been achieved.[84] It seems that these nanoparticles could also provide an excellent template for new imaging nanoparticle formulations competing with other polysaccharide-coating nanomaterials such as dextran- and carboxydextran-coated magnetic nanoparticles. All of these contrast agents are nontargeted probes and their distribution *in vivo* cannot be controlled.

TABLE I
DIAGNOSIS *IN VITRO* WITH CARBOHYDRATE-BASED NANOPARTICLES

Nanoparticle type	Carbohydrate	Application	Reference
Gold GNPs	Glycosaminoglycans	Detection of disease-associated enzyme activity	51
Gold GNPs	Sialic acid	Electrochemical amyloid-beta detection	52
Superparamagnetic Fe_3O_4 NPs	Mannose, galactose, fucose, sialic acid, and glucosamine	*In vitro* detection and differentiation cancer cells by MRI	55
Superparamagnetic Fe_3O_4 NPs	Hyaluronic ancid (HA)	*In vitro* detection of macrophages	56
Gold GNPs	Mannose	*In vitro* detection of mannose motifs in cells	54
Gold GNPs	Dextran coated	*E. coli* growth inhibition (antimicrobial susceptibility)	73
Gold GNPs	Lactose	Detection of cholera toxin B-subunit by UV–vis spectroscopy	78
Superparamagnetic Fe_3O_4 NPs	Galactose	Detection of cholera toxin B using magnetic nanoprobes	74
Gold GNPs	Glycan-like Gb3 trisaccharides	Inhibition and neutralization studies of shiga toxins type 1 (Stx1) and shiga toxins type 1 (Stx2)	80
Gold GNPs	Gb3 trisaccharide	Studies of affinity bewteen globotriose GNPs and Stx1 by SPR and related ELISA techniques	81
Superparamagnetic Fe_3O_4 NPs	Galabiose (α-Gal-$(1\rightarrow4)$-Gal)	Effective magnetic separation and detection of *Streptococcus suis* pathogen	77

Targeted imaging probes offer the opportunity to substantially improve the specificity and sensitivity of diagnostic imaging.[85] Noninvasive and quantitative detection of specific molecular markers associated with the progression of diseases (mainly neural disorders and cancer) using targeted nanoprobes is currently being developed.[86] There are some examples where carbohydrates have been used as targeting or stabilizing agents for diagnostic purposes.[87–91] These carbohydrate-based MRI contrast agents were designed to specifically address a molecular target and allow an increase of local concentration of contrast agents via selective binding.

Neurodegenerative diseases (multiple sclerosis, brain tumors, and Alzheimer and Parkinson diseases) mainly affect developed countries. MRI-based molecular imaging methods are beginning to have an impact in the diagnosis of these neurological disorders. A growing number of molecular imaging agents have been synthesized and tested *in vitro*, but so far relatively few have been validated in the brains of live animals. The difficulty of delivering contrast agents through the blood–brain barrier (BBB) is one of the principal obstacles to achieve the imaging of the central nervous system.[92] The second issue to be addressed is to achieve highly sensitive contrast agents to reduce the doses for delivery, and to alleviate the side effects. In general, BBB breakdown reflects end-stage pathology. Therefore, one of the most important challenges in imaging of the brain is the detection of early-stage pathologies which imply that the BBB is still intact.

An excellent work on the development of dual specific targeting nanoprobes for selective accumulation in brain tumors across BBB has been described by Veiseh *et al.*[87] Their nanoprobes consist of iron oxide nanoparticles stabilized with a PEGylated chitosan-branched copolymer (NPCP) that incorporate a brain tumor-targeting ligand (chlorotoxin (CTX)) and a near-infrared region (NIR) fluorophore (Cy5.5) for multimodal optical imaging/MRI (Fig. 4A). PEG-chitosan acts as a stabilizer and cross-linking agent, and facilitates the transport across the BBB by triggering adsorption-mediated transport via electrostatic interaction between positively charged chitosan and negatively charged brain endothelium. Finally, *in vivo* MR and NIR-fluorescence imaging of brain tumors in genetically engineered ND2:SmoA1 mice were evaluated after tail vein nanoprobe injection (Fig. 4). Evans blue exclusion assay, immunostaining assessment, and Gd-DTPA contrast agent exclusion method confirmed the integrity of the BBB in the tumor before and after the crossing of the nanoprobes. Additionally, biodistribution study and safe assessment were also demonstrated.

Superparamagnetic (T_2 contrast agents) carbohydrate-coated nanoparticles designed to target relatively accessible physiological vascular epitopes of cerebral inflammation have been successfully used as *in vivo* contrast agents.[88] Cross-linked, amine-functionalized dextran-coated iron oxide nanoparticles were used as a platform for the conjugation of multiple copies of the oligosaccharide sialyl Lewis X (sLe^X) bearing an amine reactive *S*-cyanomethyl functional group. These sLe^X GNPs were used to visualize the cerebral endothelium activation by labeling the initial recruitment of leukocytes to the damaged site in multiple sclerosis, ischemic lesions, and tumors. This recruitment takes place in an intact, but activated brain endothelium (outside the BBB). The targeted sLe^X-dextran-encapsulated nanoparticles enabled the detection of E- and P-selectin upregulation on the surface of the endothelium. In this way, it was possible to obtain information of ongoing lesion activity

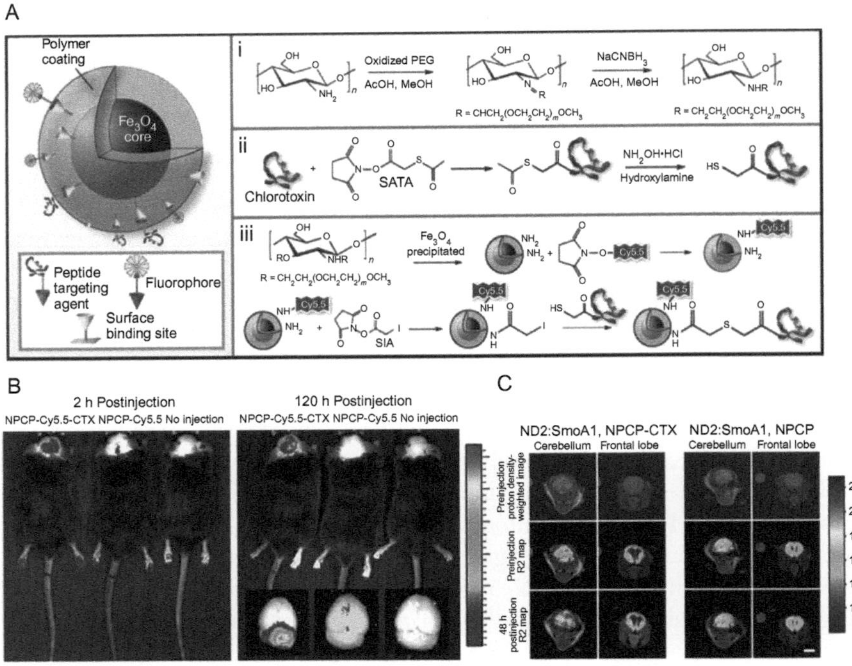

FIG. 4. (A) Synthesis and characterization of NPCP-Cy5.5-CTX nanoprobes. (B) *In vivo* NIRF imaging of autochthonous medulloblastoma tumors in genetically engineered ND2:SmoA1 mice injected with either NPCP-Cy5.5-CTX or NPCP-Cy5.5, or receiving no injection (left to right). Images were acquired at 2 h (right) and 120 h (left) post injection. *Ex vivo* fluorescence images of mice brains from the same mice following necropsy (inset, left image). The spectrum gradient bar (right) corresponds to the fluorescence intensity (p/s/cm2/sr) of the images. (C) *In vivo* MR images of ND2:SmoA1 acquired before and 48 h after administration of either NPCP-CTX or NPCP nanoprobes. Coronal cross sections of the frontal lobe (no tumor present) of the cerebral hemisphere and cerebellum (containing tumor tissues) were analyzed. Colorized R2 maps of the brain region were superimposed onto proton density-weighted images. Varying R2 values (s^{-1}) from low (blue) to high (red) were visually represented in colors generated from the gradient (right). Figure and caption reproduced by permission from Ref. 87. (See Color Insert.)

inside the brain by targeting inflammation outside the BBB. Unmodified and *N*-acetyl lactosamine-modified dextran-encapsulated nanoparticles were used as negative controls to confirm the selectivity of sLe^X GNPs. These data suggest that the magnetic sLe^X nanoparticles are highly sensitive biomarker ligands for MRI.

Superparamagnetic GNPs that combine carbohydrates and antibodies on their surface have been designed to obtain biocompatible and highly specific targeted agents for *in vivo* imaging (Fig. 5). These new targeted T_2-contrast

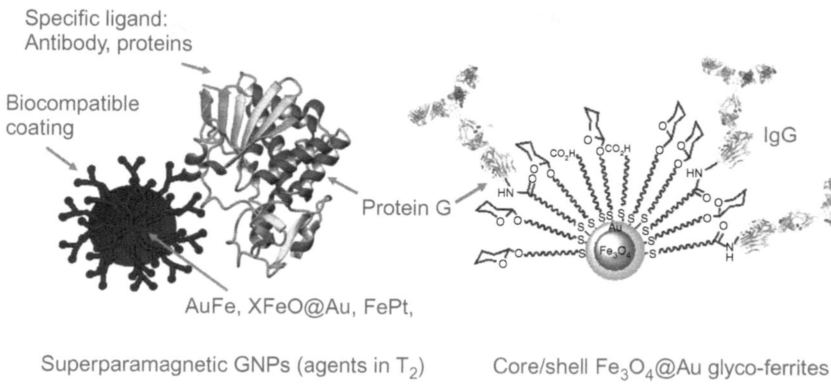

FIG. 5. Left: Schematic representation of targeted and biocompatible superparamagnetic glyconanoparticles. Right: Core/shell iron oxide/gold antibody-coated glyconanoparticles (*immuno*-glycoferrites). (See Color Insert.)

agents have been prepared from water-soluble core@shell Fe_3O_4@Au nanoferrites which are capped with carbohydrates and functional linkers (*glyco*-ferrites).[93] The glyo-ferrites were functionalized with Protein G to attach well-oriented antibodies (*immuno*-glycoferrites).[94] In this work, the biofuncionality of the antibody on the nanoparticles and their capability to specifically label cells have been demonstrated using immunocytochemistry techniques and MRI in different cell lines.

The efficiency of antibody-coated GNPs to label specifically cell populations within a complex biological medium as the whole human blood has been demonstrated.[94a] The success of *in vivo* targeting of specifically pathological sites via the recognition of biomarkers by antibodies involves the labeling of a specific cell population present at very low concentration in a multicellular environment.

In the field of contrast agents design, there is also a growing attention toward paramagnetic nanoparticles which incorporate Gd(III) ions.[10,95] Gold GNPs bearing thiol-ending glycoconjugates of glucose, galactose, or lactose, and an *N*-alkyl (mercaptopentyl or mercaptoundecanyl) tetraazacyclododecane triacetic acid (DO3A) derivative have been prepared in one step.[89] Subsequent incubation with a Gd(III) salt converted the GNPs in paramagnetic T_1 probes that were used for *in vivo* imaging of glioma in mice after studying their *in vitro* properties. Although lactose GNPs yielded the highest *in vitro* relaxivities ($\sim 20\ mM^{-1}\ s^{-1}$), six times the values of Magnevist®, only the glucose GNPs were able to enhance *in vivo* the contrast of the tumor zones better than the clinically used Magnevist®. Presumably, lactose GNPs did not reach the tumor at a proper concentration because of liver sequestration, although they highly enhanced the contrast outside the brain. Although this work is very preliminary and has some limitations from the perspective of clinical use, the system can be improved by changing the DO3A derivatives with DOTA to obtain more stable probes and by controlling the Gd(III) amounts in the nanoparticles.

IV. Metallic Carbohydrate-Based Nanoparticles for *In Vivo* Diagnosis by Optical Imaging

QDs and/or NIR-emitting fluorophores have also been prepared and applied to *in vivo* optical imaging.[96] Fluorescent microscopy is not the best choice for *in vivo* imaging, but achievable improvements of this technology may be translated in the future into clinical applications once the biocompatibility issue and detection in deep tissues are completely overcome. The application of QDs for optical imaging in humans has to face the great concern of metal toxicity, in addition to toxicity associated with the nano-dimensions of the particles.

Optical imaging has been achieved by means of semiconductor nanocrystals capped with different monosaccharides (glyco-QDs).[97] Injection of mannose and galactosamine QDs in mice resulted in specific liver sequestration presumably mediated by mannose receptor (highly expressed on Kupffer cells and sinusoidal endothelial cells in the liver) and asialoglycoprotein receptor (expressed predominantly on hepatocytes), respectively. These results were obtained by fluorescence microscopy after mice sacrifice (2 h after injection), liver perfusion, and preparation of paraffin sections of the livers. The authors used a polyethylene glycol-based spacer to conjugate the carbohydrates in order to further reduce the cytotoxicity of the colloidal CdSe/ZnS core of QDs. In this way, the carbohydrate assisted the water solubility and modulated the *in vivo* behavior of the QDs. In particular, the hepatotoxicity of these QDs in a model of liver injury was studied by co-injection with lipopolysaccharides (LPS). A significant increase in serum transaminases after i.v. injection of galactosamine QDs was recorded, indicating that this type of QDs selectively mediate liver injury, as expected by the fact that D-galactosamine is a well--known hepatotoxin. On the other hand, galactose QDs caused a slighter increase of transaminases, indicating that the QDs themselves are somewhat hepatotoxic possibly due to the release of the CdSe core. Scale-up preparation of these glyco-QDs by using microreactors has also been developed.[98]

Park *et al.*[91] have prepared biodegradable dextran-coated luminescent porous silicon nanoparticles (D-LPSiNPs) (Fig. 6A and B). Whole-body *in vivo* fluorescence imaging of nude mice (Fig. 6C) and of mice bearing MDA-MB-435 human carcinoma (Fig. 6D) was performed after intravenous injection. The functional role of the polysaccharide coating is to prevent rapid

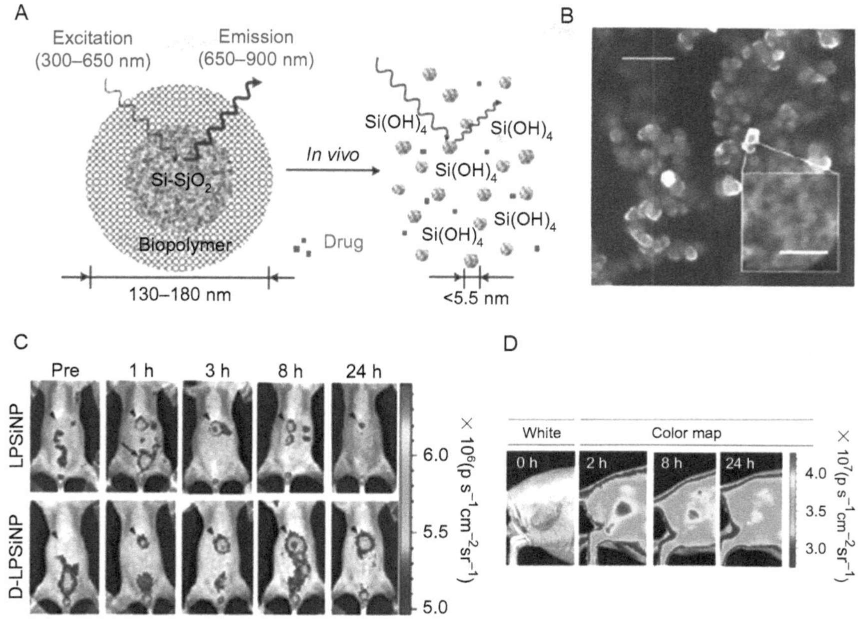

FIG. 6. (A) Schematic diagram depicting the structure and *in vivo* degradation process for the (dextran-coated) nanoparticles used in this study. (B) SEM image of LPSiNPs (the inset shows the porous nanostructure of one of the nanoparticles). The scale bar is 500 nm (50 nm for the inset). (C) *In vivo* images of LPSiNPs and D-LPSiNPs. The mice were imaged at multiple time points after intravenous injection of LPSiNPs and D-LPSiNPs (20 mg kg^{-1}). Arrowheads and arrows with solid lines indicate liver and bladder, respectively. (D) Lateral image of the same mice shown in (C) 8 h after LPSiNP or D-LPSiNP injection. Arrows with dashed lines indicate spleen. Figure and caption reproduced by permission from Ref. 91. (See Color Insert.)

degradation of the nanoparticles after injection and to increase their blood half-life, as confirmed by subsequent biodistribution and histological studies. This is the first study of tumor imaging by means of silicon nanoparticles, which are interesting tools for *in vivo* applications because of their low toxicity due to the favorable biodegradation pathway.

Lee *et al.*[90] conjugated NIR dye HiLyte Fluor 647-labeled HA to gold nanoparticles for noninvasive detection of reactive oxygen species (ROS) and hyaluronidase activity. NIR dyes assure the deep tissue imaging while the glycosaminoglycan HA provides the target for *in vivo* detection of arthritic inflammation and human ovarian carcinoma tumor in mice. Initial fluorescence quenching due to Au-fluorophore interaction is suppressed by hyaluronidase hydrolysis of HA. The detection of the HiLyte Fluor 647–HA conjugate emission by optical fluorescence imaging confirms the presence of the GNPs in the targeted zones.

TABLE II
DIAGNOSIS *IN VIVO* BY MRI AND OPTICAL IMAGING

Nanoparticle type	Carbohydrate	Application	Reference
Superparamagnetic Fe_3O_4 NPs	PEGylated chitosan	*In vivo* MR and NIRF imaging of brain tumors	87
Superparamagnetic Fe_3O_4 NPs	Sialyl Lewis-X	*In vivo* MR imaging of cerebral endothelium activation	88
Paramagnetic Gd-based Au NPs	Glucose, galactose, lactose	*In vivo* MR imaging of glioma	89
CdSe/ZnS quantum dots	Mannose, galactosamine	*Ex vivo* liver optical imaging and liver nanotoxicity	97
Porous silica-based NPs	Dextran	*In vivo* optical imaging of tumor	42
Gold nanoparticles	HiLyte-hyaluronic acid (HA)	*In vivo* optical detection of arthritic inflammation and human ovarian carcinoma tumor in mice	96
CdSe/CdS/ZnS quantum dots	Hyaluronicacid (HA)	*In vivo* optical imaging of lymphatic vessels in mice	99
Quantum dots	Hyaluronicacid (HA)	*In vivo* optical imaging of cirrhotic liver in mice	100

HA-encapsulated QDs (HA-QDs) have been also used to visualize *in vivo* lymphatic vessels[99] and liver in cirrhotic mice.[100] In the first case, positively charged CdSe/CdS/ZnS QDs were encapsulated with HA by means of electrostatic interactions.[99] In the other case, commercially available QDs were modified by amide bond formation between carboxylic acid groups of QDs and amine groups of adipic acid-modified HA.[100]

The examples summarized in Table II show that carbohydrate-based nanoparticles are good systems as detection probes for *in vivo* imaging.

V. Metallic Carbohydrate-Based Nanoparticles in Therapy

Several nanoparticle formulations are currently undergoing development for a range of therapeutic applications.[101] The first metallic nanomaterials in the pharmaceutical pipeline have appeared.[102] However, the application of carbohydrate-based nanoparticles in therapeutic treatments is still in an embryonic state. As far as is known, there are no examples of clinical application of metallic GNPs for therapy, but it is predictable that carbohydrate-based nanoparticles will increase their presence on the market. The metal cores allow different therapeutic applications such as photothermal-induced ablation/

hyperthermia by using suitable colloidal gold nanoparticles[103] or radiofrequency-mediated treatment with magnetic nanoparticles[104] in the fight against cancer. Nanoparticle-mediated hyperthermia in animal studies is particularly promising, and early clinical testing is currently under way. The systemic administration of metallic nanoparticles has to face many concerns such as retention in organs (especially liver) and nanotoxicity, which have still to be solved before the complete translation to clinic.[39]

The previous part of this review showed that metallic carbohydrate-based nanoparticles can be used in the detection and identification of microbes.[59] The fact that the adhesion of different types of pathogens to host cells occurs through carbohydrate-mediated interactions inspired the creation of antimicrobial agents which mimic the presentation of host or pathogen glycoconjugates (glycoproteins, glycolipids, and polysaccharides).[105] Polyvalent systems presenting analogs of these glycoconjugates are potential candidates as anti-adhesion agents against virus and bacteria and may afford a valid alternative to current treatments. This approach is also valid in the development of anticancer agents. It is in fact possible to use the specific glyco-code of tumor cells to create anti-adhesion systems against metastasis development.

Metallic GNPs have been used as multivalent tools that mimic glycosphingolipids at the cell surface. It has been demonstrated that GNPs can work as anti-metastatic agents depending on the carbohydrate (lactose or glucose) coating.[106] Pre-incubation of lactose GNPs with B16F10 melanoma cells before inoculation to mice prevented the development of lung metastasis *in vivo* up to 70%. On the contrary, the use of glucose GNPs did not inhibit the spread of metastasis. The hypothesis of this work was based on the fact that the expression of ganglioside GM3 on tumor cells and lactosylceramide, Gg3, and Gb4 on the lung endothelium is involved in the critical step of tumor cell adhesion to the endothelium before transmigration and metastasis. This work was the first demonstration of the potential application of gold GNPs in biomedicine.

Different types of pathogens, ranging from bacteria to viruses, express glycosylation patterns which can be tailored onto nanoparticles in order to mimic the pathogen itself and create anti-adhesion barriers. On the other hand, many pathogens profit from the glycocalix of host cells to enter the host and start the infection process. The human immunodeficiency virus (HIV) is a good example of a pathogen which can be targeted by mimicking the glycoclusters on its envelope glycoprotein gp120 or the carbohydrate patterns of some host cells which are responsible for the first contacts with the virus. In the last years, research efforts have been directed toward the development of new strategies against HIV in order to complement the highly active antiretroviral therapy (HAART),[107] which is the only approved way to treat infected people. Different types of microbicides[108] and/or vaccines[109] against HIV have been proposed in order to protect HIV-negative people from becoming infected or to treat HIV-positive people, but all attempts failed up to date. As carbohydrates are involved in the recognition and infection of host cells by HIV, sugar-based systems have been developed to get insights into the molecular biology of the virus and to develop potential anti-HIV agents. Mainly, two different approaches are being used to face this challenge. The first one is to target the virus with endogenous-like cell receptors which are involved in HIV/host interactions in order to avoid the viral adhesion process; the other one is based on mimicking the virus to block the receptors that HIV uses for entering and spreading out infection in the host.

Regarding the first approach, galactosyl ceramide (GalCer) and host-associated glycolipids may function as an alternative receptor of HIV for viral entry to host cells which do not express the receptor CD4.[110] Synthetic multivalent systems presenting GalCer have been prepared and tested as inhibitors in a system that mimics the binding of gp120 to GalCer clusters at the membrane of CD4-negative cells.[111,112] Another strategy based on this approach is to create nanoparticles which target different epitopes of the viral envelope. To this aim, multivalent gold nanoparticles coated with 50% density of sulfated ligands and 50% of glucose conjugates[113] have been used to target the positively charged amino acids in the V3 loop in the gp120. The sulphated ligands on the GNPs try to mimic the *O*-sulfated tyrosine residues of the *N*-terminus of the co-receptor CCR5.[114] These nanoparticles were able to intervene in the adsorption/fusion process of the virus infection and neutralize *in vitro* the HIV infection of T-cells at the nanomolar range.

Regarding the second approach, HIV-associated high-mannose glycans present on gp120 have been multimerized onto different synthetic scaffolds to produce viral mimics.[115] High-mannose glycans interact with lectins of host cells and thus have been used for developing anti-adhesive agents against HIV. Multivalent gold nanoparticles coated with different structural motifs of high-mannose-type oligosaccharides of gp120 (*manno*-GNPs) have been prepared.[116] The oligomannosides were functionalized with thiol-armed long amphiphilic linkers in order to ensure a suitable presentation of the antigens on the gold nanoclusters, while a glucoside bearing a short aliphatic spacer was used as the inner component. These GNPs were able to inhibit the lectin DC-SIGN (dendritic cell-specific intercellular adhesion molecule-grabbing non-integrin) mediated HIV-1 *trans*-infection of T-cells[117] at nanomolar concentrations in cellular models.[118] The experimental settings tried to mimic the natural route of virus transmission from dendritic cells (DCs) to T-lymphocytes. The combination of high local concentration of oligomannosides on the *manno*-GNPs and multivalent effects resulted in enhanced affinity of the carbohydrate antigens for DC-SIGN lectin and allowed the prevention of viral attachment to DC-SIGN-expressing cells. This study demonstrates the ability of *manno*-GNPs to interfere with the binding of HIV to DCs and to block viral dissemination to T-cells. *Manno*-GNPs can thus potentially work as an anti-adhesive barrier at an early stage of HIV infection (microbicides).

High-mannose clusters of gp120 also play a crucial role in viral immunity.[119] The broadly neutralizing human antibody 2G12 recognizes a cluster of these high-mannose glycans.[120] Some of the synthetic oligomannosides used to construct *manno*-GNPs showed high affinity for antibody 2G12, so that *manno*-GNPs were tested as 2G12 multivalent ligands.[121] SPR-based experiments and competition saturation transfer difference NMR (STD-NMR) spectroscopy demonstrated that *manno*-GNPs are synthetic mimics of the 2G12 epitope. Depending on the oligomannoside and its density on the gold surface, *manno*-GNPs were able to bind chip-immobilized 2G12 with high affinity and inhibit the 2G12/gp120 binding as measured by SPR. The affinities of *manno*-GNPs toward 2G12 were obtained in isotropic solution by STD-NMR competitive titrations.[121] The GNPs-mediated displacement of the monovalent aminoethyl oligomannosides from the binding sites of antibody 2G12 was followed by the decay of the STD signals of the oligomannosides. The inhibitory effect of selected *manno*-GNPs on 2G12-mediated neutralization of HIV-1 infection of TZM-bl, a luciferase reporter Hela cell line that stably expresses CD4, CCR5, and CXCR4 (viral receptor and co-receptors), was evaluated. After pre-incubation of serial dilutions of *manno*-GNPs with 2G12, addition of HIV-1 (NL4-3 strain), and incubation with TZM-bl cells, the decrease of 2G12 neutralization activity was assessed as the percentage of luciferase activity measured from cell lysates. In these experiments, nanoparticles bearing a linear tetramannoside induced a reproducible inhibition of 2G12 neutralization within the micromolar range.[121] These results indicate that *manno*-GNPs could also be considered as a platform to develop vaccine candidates against HIV. The opportunity of simultaneously tailoring on the same gold nanoparticle both oligomannosides and molecules which target different stages of the HIV life cycle makes nanoparticles a suitable scaffold to develop new multifunctional anti-HIV systems. Although carbohydrates are usually poorly immunogenic, examples of carbohydrate-based vaccines against a diverse set of diseases are already in clinical use or in development, indicating that the strategy of using carbohydrates to induce effective protection is viable.[122]

Another interesting example of GNPs against virus was recently reported.[123] The fusion protein hemagglutinin of influenza virus was targeted by the group of Papp using sialic acid-functionalized gold nanoparticles.[123] Two different sized gold nanoparticles were tested (2 and 14 nm). Electron microscopy techniques allowed the direct visualization of functionalized gold nanoparticles and their interaction with viral fusion proteins (Fig. 7).

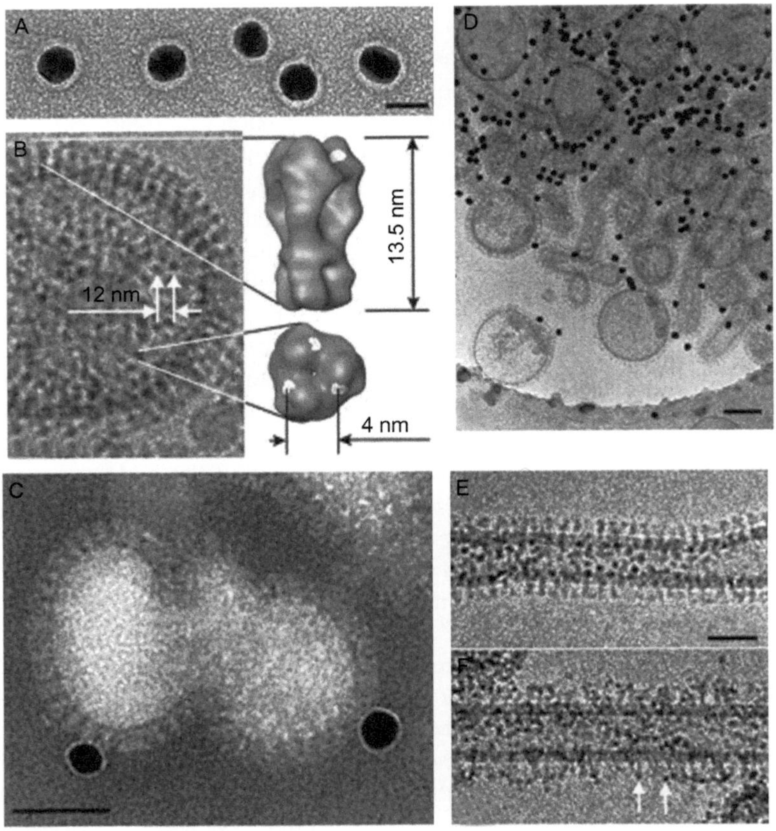

Fig. 7. (A) TEM image of AuNP (14 nm) showing the gold core (darkest spot) and the sialic acid organic shell (clearer halation). (B) Influenza A X31 virion showing the individual trimeric fusion protein hemagglutinin sticking out of the viral membrane. (C) Cryo-TEM preparation of influenza A virions after 60 min incubation with AuNP (14 nm). (D) Image showing multiple binding of individual functionalized AuNPs to viral hemagglutinin. (E) Cryo-TEM preparations of influenza A virions before (top) and after (bottom) 60 min incubation with AuNP (14 nm) at lower magnifications. Adapted from Ref. 123.

The ability of sialic acid GNPs to inhibit viral infection was evaluated preincubating the virus with the nanoparticles and subsequent exposing it to adherent Madin–Darby canine kidney (MDCK) cells. The expression of viral nucleoproteins in the cells was analyzed by flow cytometry using anti-nucleoprotein antibodies. Six hours post infection, gold nanoparticles of 14 nm reduced the viral infection by 40%, in contrast to those of 2 nm that led to a reduction of only 20%. Direct binding with hemagglutinin was also demonstrated using transfected chinese hamster ovary (CHO) cells with HA coupled to a yellow fluorescent protein (HA-YFP). These data indicate that multivalent interactions strongly depend on the size of the interacting particle surfaces.

There is also an increasing interest in exploring metallic nanoclusters as novel carriers to prepare fully synthetic carbohydrate-based vaccines. In the case of bacteria, tailoring the repeating units of bacterial capsular polysaccharides in the nanocluster surface is an alternative to the use of immunogenic protein carriers (conjugate vaccines). Among the polysaccharide-coated bacteria, *Neisseria meningitidis*[124] is a Gram-negative diplococcus that causes severe epidemics (meningitis and septicaemia) in the sub-Saharan region of Africa and Asia. The intent to explore gold GNPs as synthetic carbohydrate vaccines has been reported by Scrimin and collaborators.[125] They used gold GNPs coated with synthetic analogs of α-D-*N*-acetyl-mannosamine-(1 → 6)phosphate, the repeating unit of the capsular polysaccharide of type A bacterium, to inhibit the binding to its specific mouse polyclonal antibodies. In these synthetic analogs, the phosphodiester bridge connecting the glycosidic units of the capsular polysaccharide were substituted with a phosphonate as a hydrolytically more stable moiety. ELISA showed that 100% density of mannosaminide analogs of bacterium antigens on the nanoparticle surface performed slightly better than GNPs with lower percentages of mannosaminide compounds. Di- or trisaccharide phosphonate derivatives did not show better inhibition than *N*-acetyl mannosamine, indicating that the antibodies under study recognize preferentially the external mannoside units on the GNP surface as a whole.

A similar strategy has been followed by the authors' group to prepare vaccine candidates against *Streptococcus pneumoniae.* The bacterium is a major cause of invasive respiratory tract infections in both children and the elderly.[126] Multicomponent gold GNP coated with a synthetic tetrasaccharide epitope related to the type 14 capsular polysaccharide of the bacterium (Pn14PS), the T-helper $OVA_{323-339}$ peptide, and D-glucose were prepared and tested in mice.[126a] The GNP coated with the synthetically branched tetrasaccharide β-D-Gal*p*-(1 → 4)-β-D-Glc*p*-(1 → 6)-[β-D-Gal*p*-(1 → 4)-]β-D-Glc*p*NAc-(1→), which corresponds to a single repeating unit of Pn14PS, was able to trigger significant titre of specific anti-Pn14PS IgG antibodies. These results demonstrate the feasibility of a fully synthetic carbohydrate vaccine against *S. pneumoniae* based on gold nanoclusters. The usefulness of gold nanoparticles as a versatile carrier for the development of antibacterial vaccine may be extended to other pathogens.

The few examples given in Table III demonstrate the potential of carbohydrate-based nanoparticles as anti-adhesion agents and vaccines, and establish the basis for their future application in therapy.

TABLE III
GOLD GLYCONANOPARTICLES AS ANTIADHESION AGENTS

Nanoparticle type	Carbohydrate	Application	Reference
Gold GNPs	Lactose	Reduction of the progression of metastasis in mice	106
Gold GNPs	Galactose	Competitive adhesion assays: competition of GNPs for HIV gp120 reversibly bound to βGalCer-coated polystyrene plates	111
Gold GNPs	High-mannose-type oligosaccharides	Binding to DC-SIGN receptor and antibody 2G12 by SPR-bioassays, anti-HIV activity in *trans*-infection models, and 2G12 neutralization assays in cells	116,118,121
Gold GNPs	Sialic acid	Inhibition of Influenza A X31 virus infection in cells	123
Gold GNPs	Mannosaminides	Inhibition of specific polyclonal antibodies/*Neisseria meningitidis* binding	125

VI. NonMetallic Polysaccharide-Based Nanoparticles in Therapy

Contrary to metallic carbohydrate-based nanoparticles, which are still far away from clinical application, natural polysaccharides-based nanoparticles are already in a more advanced state toward clinical trials. Natural polysaccharides have received more and more attention as drug delivery systems.[127] Polysaccharides are the most promising materials in the preparation of nano-carriers. Delivery of complex molecules such as peptides, proteins, oligonucleotides, and plasmids or small drugs is one of the most intensively studied applications of these natural macromolecules. The use of carriers made of hydrophilic polysaccharides is a promising alternative to conventional formulations for improving the transport of hydrophobic drugs across biological barriers. From a biopharmaceutical point of view, polysaccharides have the special feature of adhering to a mucosal surface and act as good bioadhesive material that allows increasing drug loading and slow drug rate releasing. These interesting biopharmaceutical features are accompanied by its well-known biocompatibility and low cytotoxicity. For drug delivery application, chitosan has special interest and is widely used[128] because nanoparticles of this polysaccharide are biocompatible, degradable, and nontoxic drug carriers.[129] Chitosan is a natural cationic polysaccharide consisting of *N*-acetylglucosamine and D-glucosamine units. Its biopharmaceutically relevant property is the capacity for the transient opening of tight junctions, which could facilitate mucosal transport of bioactive drugs. The *in vivo* classical reported applications of chitosan-based nanoparticles are mainly as drug delivery vehicles[128,129] and as transfection agents.[130] One decade of research in the frame of ophthalmic applications confirms the high performance of chitosan-based nanoparticles to encapsulate and facilitate the penetration of drugs through internal epithelial ocular barriers.[131] An excellent review of Alonso and colleagues focuses on the use of polysaccharide-based nanocarriers for the nasal delivery of antigenic molecules (nasal vaccination).[132] In addition, alginate-based nanoparticles[133] and chitosan nanoparticles[134] are in preclinical development as carriers for drug delivery in therapy against cancer.

Polysaccharide-encapsulated metallic nanoparticles take advantage of the good features of polysaccharides as carriers and the singular physical properties of metallic cores (magnetic, fluorescent) at the nanosized scale. Intensive research on this type of nanomaterials can assure their future translation in medicine.

VII. Conclusions

This review has presented an overview of some biomedical applications of carbohydrate-based nanoparticles. Functional, magnetic, and fluorescent metallic carbohydrate-based nanoparticles with high selectivity and high sensitivity are promising tools in pathogen (bacteria, toxins, virus) and tumor cell detection. Metallic GNPs have shown to be interesting tools as anti-adhesion agents (metastasis, HIV, influenza virus) and vaccine candidates (*N. meningitidis*, *S. pneumoniae*). In all these processes, carbohydrate-mediated interactions are implicated at molecular level. Actually, the main focus of clinical applications is diagnosis and treatment of cancer. In this field, detection of molecular changes on cells as early as possible is a key requirement. To achieve this objective, extreme selectivity and highly sensitive techniques must be used. Carbohydrate antigens of glycoproteins and glycolipids are specific ligands of cellular receptors and their multivalent presentation on nanoparticles has the potential to target and detect single cells by using different techniques. There are already promising examples of tumor targeting and *in vivo* imaging using GNPs and polysaccharide-encapsulated metallic nanoparticles. However, important issues such as their toxicity and unpredictable behavior in complex biological environments in terms of biodistribution are still a current limitation that remains to be solved.[135,136]

Another interesting field of application of nanomaterials is regenerative medicine.[137] Much attention has been specifically directed to the tracking of stem cells which intervene in repairing injuries and degenerative diseases. Understanding the pathways of stem cells to injury sites by tracking them *in vivo* and guiding them to specific sites requires both the development of advanced techniques and new targeted and highly sensitive materials such as biofunctional magnetic nanoparticles and QDs.[138] Nanotechnology-based approaches can contribute to the development of carbohydrate-based nanoparticles in regenerative medicine.

The next generation of nanoparticles should be stimuli-responsive, and multifunctional and should comprise active targeting molecules on their surface to ensure high specificity and high selectivity. No clinically approved compound of this kind has been produced so far. It is foreseen that carbohydrate, in tandem with other (bio)molecules and materials, may be useful to obtain this type of smart nanoparticles that can enable the translation of basic research to clinic applications.

Acknowledgments

The authors thank the Spanish Ministry of Science and Innovation, the European Commission, and the Department of Industry of the Basque Government for their financial support.

References

1. Vogel V. Nanomedicine: the next waves of medical innovations. Introduction. In: Vogel V, editor. *Nanotechnology. Nanomedicine*, vol. 5. Weinheim: Wiley–VCH Verlag GmbH & Co. KGaA; 2009. pp. 1–16.
2. Katz E, Willner I. Integrated nanoparticle-biomolecule hybrid systems: synthesis, properties, and applications. *Angew Chem Int Ed Engl* 2004;**43**:6042–108.
3. Giljohann DA, Mirkin CA. Drivers of biodiagnostic development. *Nature* 2009;**462**:461–4.
4. Bhattacharyya S, Kudgus RA, Bhattacharya R, Mukherjee P. Inorganic nanoparticles in cancer therapy. *Pharm Res* 2011;**28**:237–59.
5. Petros RA, DeSimone JM. Strategies in the design of nanoparticles for therapeutic applications. *Nat Rev Drug Discov* 2010;**9**:615–27.
6. Kim DK, Dobson J. Nanomedicine for targeted drug delivery. *J Mater Chem* 2009;**19**:6294–309.
7. McBain SC, Yiu HP, Dobson J. Magnetic nanoparticles for gen and drug delivery. *Int J Nanomed* 2008;**3**:169–80.
8. Orive G, Anitua E, Pedraz JL, Emerich DF. Biomaterials for promoting brain protection, repair and regeneration. *Nat Rev Neurosci* 2009;**10**:682–92.
9. Shim S-Y, Lim D-K, Nam J-M. Ultrasensitive optical biodiagnostic methods using metallic nanoparticles. *Nanomedicine* 2008;**3**:215–32.

10. Kim J, Piao Y, Hyeon T. Multifunctional nanostructured materials for multimodal imaging, and simultaneous imaging and therapy. *Chem Soc Rev* 2009;**38**:372–90.
11. Sharma P, Brown S, Walter G, Santra S, Moudgil B. Nanoparticles for bioimaging. *Adv Colloid Interface Sci* 2006;**123**:471–85.
12. Jun YW, Lee JH, Cheon J. Chemical design of nanoparticle probes for high-performance magnetic resonance imaging. *Angew Chem Int Ed Engl* 2008;**47**:5122–35.
13. Cheon J, Lee JH. Synergistically integrated nanoparticles as multimodal probes for nano-biotechnology. *Acc Chem Res* 2008;**41**:1630–40.
14. Varki A, Cummings R, Esko J, Freeze H, Hart G, Marth J, editors. *Essentials of glycobiology*. Cold Spring Harbor, New York: Cold Spring Harbor Laboratory Press; 1999. p. 653.
15. Gabius H-J, editor. *The sugar code: fundamentals of glycosciences*. Wiley-VCH Verlag GmbH & Co. KGaA, Weinheim, Germany: Wiley-Blackwell; 2009. p. 569.
16. Herzner H, Reipen T, Schultz M, Kunz H. Synthesis of glycopeptides containing carbohydrate and peptide recognition motifs. *Chem Rev* 2000;**100**:4495–537.
17. Renaudet O. Recent advances on cyclopeptide-based glycoclusters. *Mini-Rev Org Chem* 2008;**5**:274–86.
18. Wang J, Li H, Zou G, Wang LX. Novel template-assembled oligosaccharide clusters as epitope mimics for HIV-neutralizing antibody 2G12. Design, synthesis, and antibody binding study. *Org Biomol Chem* 2007;**5**:1529–40.
19. Krauss IJ, Joyce JG, Finnefrock AC, Song HC, Dudkin VY, Geng X, et al. Fully synthetic carbohydrate HIV antigens designed on the logic of the 2G12 antibody. *J Am Chem Soc* 2007;**129**:11042–4.
20. Chabre YM, Roy R. Design and creativity in synthesis of multivalent neoglycoconjugates. *Adv Carbohydr Chem Biochem* 2010;**63**:165–393.
21. Kiessling LL, Gestwicki JE, Strong LE. Synthetic multivalent ligands as probes of signal transduction. *Angew Chem Int Ed Engl* 2006;**45**:2348–68.
22. Pearson S, Chen G, Stenzel MH. Synthesis of glycopolymers. In: Narain R, editor. *Engineered carbohydrate-based materials for biomedical applications. Polymers, surfaces, densdrimers, nanoparticles, and hydrogels*. Hoboken, NJ: John Wiley & Sons, Inc; 2011. pp. 1–118. Chapter 1.
23. Otsuka I, Blanchard B, Borsali R, Imberty A, Kakuchi T. Enhancement of plant and bacterial lectin binding affinity by three-dimensional organized cluster glycosides constructed on helical poly(phenylacetylene) backbones. *Chembiochem* 2010;**11**:2399–408.
24. Yamazaki N, Kojima S, Bovin NV, André S, Gabius S, Gabius H-J. Endogenous lectins as targets for drug delivery. *Adv Drug Deliv Rev* 2000;**43**:225–44.
25. Torchilin VP. Recent advances with liposomes as pharmaceutical carriers. *Nat Rev Drug Discov* 2005;**4**:145–60.
26. Zhang H, Ma Y, Sun X-L. Recent developments in carbohydrate-decorated targeted drug/gene delivery. *Med Res Rev* 2010;**30**:270–89.
27. Marradi M, Chiodo F, García I, Penadés S. Glycoliposomes and metallic glyconanoparticles in glycoscience. In: Renaudet O, Spinelli N, editors. *Synthesis and biological applications of glycoconjugates*. Bentham e-Books; 2011. Chapter 10, pp. 164–202.
28. Baldini L, Casnati A, Sansone F, Ungaro R. Calixarene-based multivalent ligands. *Chem Soc Rev* 2007;**36**:254–66.
29. Dondoni A, Marra A. Calixarene and calixresorcarene glycosides: their synthesis and biological applications. *Chem Rev* 2010;**110**:4949–77.
30. Cecioni S, Lalor R, Blanchard B, Praly JP, Imberty A, Matthews SE, et al. Achieving high affinity towards a bacterial lectin through multivalent topological isomers of calix[4]arene glycoconjugates. *Chem Eur J* 2009;**15**:13232–40.

31. El-Boubbou K, Huang X. Glyco-nanomaterials: translating insights from the "sugar-code" to biomedical applications. *Curr Med Chem* 2011;**18**:2060–78.
32. Marradi M, Martín-Lomas M, Penadés S. Glyconanoparticles: polyvalent tools to study carbohydrate-based interactions. *Adv Carbohydr Chem Biochem* 2010;**64**:211–90.
33. Love JC, Estroff LA, Kriebel JK, Nuzzo RG, Whitesides GM. Self-assembled monolayers of thiolates on metals as a form of nanotechnology. *Chem Rev* 2005;**105**:1103–69.
34. de la Fuente JM, Barrientos AG, Rojas TC, Rojo J, Cañada J, Fernández A, et al. Gold glyconanoparticles as water-soluble polyvalent models to study carbohydrate interactions. *Angew Chem Int Ed Engl* 2001;**40**:2258–61.
35. de la Fuente JM, Penadés S. Understanding carbohydrate-carbohydrate interactions by means of glyconanotechnology. *Glycoconj J* 2004;**21**:149–63.
36. de la Fuente JM, Penadés S. Glyconanoparticles: types, synthesis and applications in glycoscience, biomedicine and material science. *Biochim Biophys Acta Gen Subj* 2006;**1760**:636–51.
37. Boisselier E, Astruc D. Gold nanoparticles in nanomedicine: preparations, imaging, diagnostics, therapies and toxicity. *Chem Soc Rev* 2009;**38**:1759–82.
38. Han G, Ghosh P, Rotello VM. Functionalized gold nanoparticles for drug delivery. *Nanomedicine* 2007;**2**:113–23.
39. Hall JB, Dobrovolskaia MA, Patri AK, McNeil SE. Characterization of nanoparticles for therapeutics. *Nanomedicine* 2007;**2**:789–803.
40. White KA, Rosi NL. Gold nanoparticle-based assays for the detection of biologically relevant molecules. *Nanomedicine* 2008;**3**:543–53.
41. García I, Marradi M, Penadés S. Glyconanoparticles: multifunctional nanomaterials for biomedical applications. *Nanomedicine* 2010;**5**:777–92.
42. Perez-Lopez B, Merkoci A. Nanoparticles for the development of improved (bio)sensing systems. *Anal Bioanal Chem* 2011;**399**:1577–90.
43. Liu SQ, Tang ZY. Nanoparticle assemblies for biological and chemical sensing. *J Mater Chem* 2010;**20**:24–35.
44. Hone DC, Haines AH, Russell DA. Rapid, quantitative colorimetric detection of a lectin using mannose-stabilized gold nanoparticles. *Langmuir* 2003;**19**:7141–4.
45. Lin CC, Yeh YC, Yang CY, Chen GF, Chen YC, Wu YC, et al. Quantitative analysis of multivalent interactions of carbohydrate-encapsulated gold nanoparticles with concanavalin A. *Chem Commun* 2003;**23**:2920–1.
46. Halkes KM, Carvalho de Souza A, Maljaars CEP, Gerwig GJ, Kamerling JP. A facile method for the preparation of gold glyconanoparticles from free oligosaccharides and their applicability in carbohydrate–protein interaction studies. *Eur J Org Chem* 2005;**17**:3650–9.
47. Thygesen MB, Sauer J, Jensen KJ. Chemoselective capture of glycans for analysis on gold nanoparticles: carbohydrate oxime tautomers provide functional recognition by proteins. *Chem Eur J* 2009;**15**:1649–60.
48. Jayaraman N. Multivalent ligand presentation as a central concept to study intricate carbohydrate-protein interactions. *Chem Soc Rev* 2009;**38**:3463–83.
49. Jiang XZ, Housni A, Gody G, Boullanger P, Charreyre MT, Delair T, et al. Synthesis of biotinylated alpha-D-mannoside or N-acetyl beta-D-glucosaminoside decorated gold nanoparticles: study of their biomolecular recognition with Con A and WGA lectins. *Bioconj Chem* 2010;**21**:521–30.
50. Wang X, Ramstrom O, Yan M. Quantitative analysis of multivalent ligand presentation on gold glyconanoparticles and the impact on lectin binding. *Anal Chem* 2010;**82**:9082–9.
51. Ban Z, Bosques CJ, Sasisekharan R. A simple assay to probe disease-associated enzyme activity using glycosaminoglycan-assisted synthesized gold nanoparticles. *Org Biomol Chem* 2008;**6**:4290–2.

52. Chikae M, Fukuda T, Kerman K, Idegami K, Miura Y, Tamiya E. Amyloid-beta detection with saccharide immobilized gold nanoparticle on carbon electrode. *Bioelectrochemistry* 2008;**74**:118–23.
53. La Belle JT, Gerlach JQ, Svarovsky S, Joshi L. Label-free impedimetric detention of glycan-lectin interactions. *Anal Chem* 2007;**79**:6959–64.
54. Ding L, Qian R, Xue Y, Cheng W, Ju H. In situ scanometric assay of cell surface carbohydrate by glyconanoparticle-aggregation-regulated silver enhancement. *Anal Chem* 2010;**82**:5804–9.
55. El-Boubbou K, Zhu DC, Vasileiou C, Borhan B, Prosperi D, Li W, et al. Magnetic glyco-nanoparticles: a tool to detect, differentiate, and unlock the glyco-codes of cancer via magnetic resonance imaging. *J Am Chem Soc* 2010;**132**:4490–9.
56. Medha K, El-Boubbou K, Zhu DC, Lansdell T, Lu X, Li W, et al. Hyaluronic acid immobilized magnetic nanoparticles for active targeting and imaging of macrophages. *Bioconj Chem* 2010;**21**:2128–35.
57. Lee Y, Lee H, Kim YB, Kim J, Hyeon T, Park HW, et al. Bioinspired surface immobilization of hyaluronic acid on monodisperse magnetite nanocrystals for targeted cancer imaging. *Adv Mater* 2008;**20**:4154–7.
58. Robinson A, Fang J-M, Chou P-T, Liao K-W, Chu R-M, Lee S-J. Probing lectin and sperm with carbohydrate-modified quantum dots. *Chembiochem* 2005;**6**:1899–905.
59. Luo PG, Stutzenberger FJ. Nanotechnology in the detection and control of microorganisms. *Adv Appl Microbiol* 2008;**63**:145–81.
60. Varki A, Freeze HH, Gagneux P. Evolution of glycan diversity. In: Varki A, Cummings RD, Esko JD, Freeze HH, Stanley P, Bertozzi CR, Hart GW, Etzler ME, editors. *Essentials of Glycobiology.* New York: Cold Spring Harbor; 2009. pp. 281–92, Chapter 19; Nizet V, Esko JD. Bacterial and viral infections. In: Varki A, Cummings RD, Esko JD, Freeze HH, Stanley P, Bertozzi CR, Hart GW, Etzler ME, editors. *Essentials of glycobiology.* New York: Cold Spring Harbor; 2009. pp. 537–52, Chapter 39; Cummings RD, Turco S. Parasitic infections. In: Varki A, Cummings RD, Esko JD, Freeze HH, Stanley P, Bertozzi CR, Hart GW, Etzler ME, editors. *Essentials of glycobiology.* New York: Cold Spring Harbor; 2009. pp. 553–6, Chapter 40.
61. Lin C-C, Yeh Y-C, Yang C-Y, Chen C-L, Chen G-F, Chen C-C, et al. Selective binding of mannose-encapsulated gold nanoparticles to type 1 pili in *Escherichia coli*. *J Am Chem Soc* 2002;**124**:3508–9.
62. El-Boubbou K, Grunden C, Huang X. Magnetic glyco-nanoparticles: a unique tool for rapid pathogen detection, decontamination, and strain differentiation. *J Am Chem Soc* 2007;**129**:13392–3.
63. Huang C-C, Chen C-T, Shiang Y-C, Lin Z-H, Chang H-T. Synthesis of fluorescent carbohydrate-protected Au nanodots for detection of concanavalin A and *Escherichia coli*. *Anal Chem* 2009;**81**:875–82.
64. Joralemon MJ, Murthy SK, Remsen EE, Becker ML, Wooley KL. Synthesis, characterization, and bioavailability of mannosylated shell cross-linked nanoparticles. *Biomacromolecules* 2004;**5**:903–13.
65. Kim BS, Yang WY, Ryu JH, Yoo YS, Lee M. Carbohydrate-coated nanocapsules from amphiphilic rod–coil molecule: binding to bacterial type 1 pili. *Chem Commun* 2005;**41**:2035–7.
66. Qu LW, Luo PG, Taylor S, Lin Y, Huang WJ, Anyadike N, et al. *J Nanosci Nanotechnol* 2005;**5**:319–22.
67. Xue C, Velayudham S, Johnson S, Saha R, Smith A, Brewer W, et al. Highly water-soluble, fluorescent, conjugated fluorene-based glycopolymers with poly(ethylene glycol)-tethered spacers for sensitive detection of *Escherichia coli*. *Chem Eur J* 2009;**15**:2289–95.

68. Liu L-H, Dietsch H, Schurtenberger P, Yan M. Photoinitiated coupling of unmodified monosaccharides to iron oxide nanoparticles for sensing proteins and bacteria. *Bioconj Chem* 2009;**20**:1349–55.
69. Hatch DM, Weiss AA, Kale RR, Iyer SS. Biotinylated bi- and tetra-antennary glycoconjugates for *Escherichia coli* detection. *Chembiochem* 2008;**9**:2433–42.
70. Mukhopadhyay B, Martins MB, Karamanska R, Russell DA, Field RA. Bacterial detection using carbohydrate-functionalised CdS quantum dots: a model study exploiting *E. coli* recognition of mannosides. *Tetrahedron Lett* 2009;**50**:886–9.
71. Kim J, Yoon M-Y. Recent advances in rapid and ultrasensitive biosensors for infectious agents: lesson from *Bacillus anthracis* diagnostic sensors. *Analyst* 2010;**135**:1182–90.
72. Luo PG, Wang H, Gu L, Lu F, Lin Y, Christensen KA, et al. Selective interactions of sugar-functionalized single-walled carbon nanotubes with Bacillus spores. *ACS Nano* 2009;**3**:3909–16.
73. Nath S, Kaittanis C, Tinkham A, Perez JM. Dextran-coated gold nanoparticles for the assessment of antimicrobial susceptibility. *Anal Chem* 2008;**80**:1033–8.
74. Sun EY, Weissleder R, Josephson L. Continuous analyte sensing with magnetic nanoswitches. *Small* 2006;**2**:1144–7.
75. Koh I, Hong R, Weissleder R, Josephson L. Sensitive NMR sensors detect antibodies to influenza. *Angew Chem Int Ed Engl* 2008;**47**:4119–21.
76. Kaittanis C, Banerjee T, Santra S, Santiesteban OJ, Teter K, Perez JM. Identification of molecular-mimicry-based ligands for cholera diagnostics using magnetic relaxation. *Bioconj Chem* 2011;**22**:307–14.
77. Parera Pera N, Kouki A, Haataja S, Branderhorst HM, Liskamp RMJ, Visser GM, et al. Detection of pathogenic *Streptococcus suis* bacteria using magnetic glycoparticles. *Org Biomol Chem* 2010;**8**:2425–9.
78. Schofield CL, Field RA, Russell DA. Glyconanoparticles for the colorimetric detection of cholera toxin. *Anal Chem* 2007;**79**:1356–61.
79. Vrasidas I, de Mol NJ, Liskamp RMJ, Pieters RJ. Synthesis of lactose dendrimers and multivalency effects in binding to the cholera toxin B subunit. *Eur J Org Chem* 2001;**24**:4685–92.
80. Kulkarni AA, Fuller C, Korman H, Weiss A, Iyer SS. Glycan encapsulated gold nanoparticles selectively inhibit Shiga Toxins 1 and 2. *Bioconj Chem* 2010;**21**:1486–93.
81. Chien Y-Y, Jan M-D, Adak AK, Tzeng H-C, Lin Y-P, Chen Y-J, et al. Globotriose-functionalized gold nanoparticles as multivalent probes for Shiga-like toxin. *Chembiochem* 2008;**9**:1100–9.
82. Mulder JM, Strijkers GJ, VanTilborg GAF, Cormode DP, Fayad ZA, Nicolay K. Nanoparticulate assemblies of amphiphiles and diagnostically active materials for multimodality imaging. *Acc Chem Res* 2009;**42**:904–14.
83. Gao J, Gu H, Xu B. Multifunctional magnetic nanoparticles: design, synthesis, and biomedical applications. *Acc Chem Res* 2009;**42**:1097–107.
84. Agrawal P, Strijkers GJ, Nicolay K. Chitosan-based systems for molecular imaging. *Adv Drug Deliv Rev* 2010;**62**:42–58.
85. Terreno E, Dastru W, Castelli DD, Gianolio E, Crich SG, Longo D, et al. Advances in metal-based probes for MR molecular imaging applications. *Curr Med Chem* 2010;**17**:1052–63.
86. Hahn MA, Singh AK, Sharma P, Brown SC, Moudgil BM. Nanoparticles as contrast agents for in-vivo bioimaging: current status and future perspectives. *Anal Bioanal Chem* 2011;**399**:3–27.

87. Veiseh O, Sun C, Fang C, Bhattarai N, Gunn J, Kievit F, et al. Specific targeting of brain tumors with an optical/magnetic resonance imaging nanoprobe across the blood-brain barrier. *Cancer Res* 2009;**69**:6200–7.
88. Van Kasteren SI, Campbell SJ, Serres S, Anthony DC, Sibson NR, Davis BG. Glyconanoparticles allow pre-symptomatic *in vivo* imaging of brain disease. *Proc Natl Acad Sci USA* 2009;**106**:18–23.
89. Marradi M, Alcántara D, de la Fuente JM, García-Martín ML, Cerdán S, Penadés S. Paramagnetic Gd-based gold glyconanoparticles as probes for MRI: tuning relaxivities with sugars. *Chem Commun* 2009;**29**:3922–4.
90. Lee H, Lee K, Kim IK, Park TG. Synthesis, characterization, and in vivo diagnostic applications of hyaluronic acid immobilized gold nanoprobes. *Biomaterials* 2008;**29**:4709–18.
91. Park J-H, Gu L, Von-Maltzahn G, Ruoslahti E, Bhatia SN, Sailor MJ. Biodegradable luminescent porous silicon nanoparticles for in vivo applications. *Nat Mater* 2009;**8**:331–6.
92. Neuwelt EA, Bauer B, Fahlkr C, Fricker G, Ladecola C, Janigro D, et al. Engaging neuroscience to advance translational research in brain barrier biology. *Nat Rev Neurosci* 2011;**12**:169–82.
93. Gallo J, García I, Padro D, Arnáiz B, Penadés S. Water-soluble magnetic glyconanoparticles based on metal-doped ferrites coated with gold: synthesis and characterization. *J Mater Chem* 2010;**20**:10010–20.
94. García I, Gallo J, Genicio N, Padro D, Penadés S. Magnetic glyconanoparticles as a versatile platform for selective immunolabeling and imaging of cells. *Bioconj Chem* 2011;**22**:264–73.
94a. Gallo J, García I, Genicio N, Padro D, Penades S. Specific labeling of cell populations in blood with targeted immuno-fluorescent/magnetic glyconanoparticles. Biomaterials, In press, doi:10.1016/j.biomaterials.2011.09.010.
95. Na HB, Song IC, Hyeon T. Inorganic nanoparticles for MRI contrast agents. *Adv Mater* 2009;**21**:1–16.
96. Gao X, Cui Y, Levenson RM, Chung LWK, Nie S. *In vivo* cancer targeting and imaging with semiconductor quantum dots. *Nat Biotechnol* 2004;**22**:969–76.
97. Kikkeri R, Lepenies B, Adibekian A, Laurino P, Seeberger PH. *In vitro* imaging and *in vivo* liver targeting with carbohydrate capped quantum dots. *J Am Chem Soc* 2009;**131**:2110–2.
98. Kikkeri R, Laurino P, Odedra A, Seeberger PH. Synthesis of carbohydrate-functionalized quantum dots in microreactors. *Angew Chem Int Ed Engl* 2010;**49**:2054–7.
99. Bhang SH, Won N, Lee T-J, Jin H, Nam J, Park J, et al. Hyaluronic acid-quantum dot conjugates for *in vivo* lymphatic vessel imaging. *ACS Nano* 2009;**3**:1389–98.
100. Kim K, Hur W, Park S-J, Hong SW, Choi JE, Goh EJ, et al. Bioimaging for targeted delivery of hyaluronic acid derivatives to the livers in cirrhotic mice using quantum dots. *ACS Nano* 2010;**4**:3005–14.
101. Sakamoto JH, van de Ven AL, Godin B, Blanco E, Serda RE, Grattoni A, et al. Enabling individualized therapy through nanotechnology. *Pharm Res* 2010;**62**:57–89.
102. Dobrovolskaia MA, McNeil SE. Immunological properties of engineered nanomaterials. *Nat Nanotechnol* 2007;**2**:469–78.
103. Kennedy LC, Bickford LR, Lewinski NA, Coughlin AJ, Hu Y, Day ES, et al. New era for cancer treatment: gold-nanoparticle-mediated thermal therapies. *Small* 2011;**7**:169–83.
104. Jain KK. Advances in the field of nanooncology. *BMC Med* 2010;**8**:83.
105. Sharon N. Carbohydrates as future anti-adhesion drugs for infectious diseases. *Biochim Biophys Acta* 2006;**1760**:527–37.
106. Rojo J, Diaz V, de la Fuente JM, Segura I, Barrientos AG, Riese HH, et al. Gold glyconanoparticles as new tools in antiadhesive therapy. *Chembiochem* 2004;**5**:291–7.
107. De Clercq E. Antiretroviral drugs. *Curr Opin Pharmacol* 2010;**10**:507–15.

108. Balzarini J, Van Damme L. Microbicide drug candidates to prevent HIV infection. *Lancet* 2007;**369**:787–97.
109. Barouch DH. Challenges in the development of an HIV-1 vaccine. *Nature* 2008;**455**:613–9.
110. Harouse JM, Bhat S, Spitalnik SL, Laughlin M, Stefano K, Silberberg DH, et al. Inhibition of entry of HIV-1 in neural cell lines by antibodies against galactosyl ceramide. *Science* 1991;**253**:320–3.
111. Nolting B, Yu J-J, G-y Li, Cho S-J, Kauzlarich S, Gervay-Hague J. Synthesis of gold glyconanoparticles and biological evaluation of recombinant gp120 interactions. *Langmuir* 2003;**19**:6465–73.
112. McReynolds KD, Gervay-Hague J. Chemotherapeutic interventions targeting HIV interactions with host-associated carbohydrates. *Chem Rev* 2007;**107**:1533–52.
113. Di Gianvincenzo P, Marradi M, Martínez-Ávila OM, Bedoya LM, Alcamí J, Penadés S. Gold nanoparticles capped with sulfate-ended ligands as anti-HIV agents. *Bioorg Med Chem Lett* 2010;**20**:2718–21.
114. Huang C-c, Lam SN, Acharya P, Tang M, Xiang S-H, Shahzad-ul Hussan S, et al. Structures of the CCR5 N terminus and of a tyrosine-sulfated antibody with HIV-1 gp120 and CD4. *Science* 2007;**317**:1930–4.
115. Wang LX. Carbohydrate-based vaccines against HIV/AIDS. In: Klyosov AA, Witczak ZJ, Platt D, editors. *Carbohydrate drug design*. Washington, DC, USA: ACS Symposium Series; 2006. pp. 133–60, No. 932.
116. Martínez-Ávila O, Hijazi K, Marradi M, Clavel C, Campion C, Kelly C, et al. Gold manno-glyconanoparticles: multivalent systems to block HIV-1 gp120 binding to the lectin DC-SIGN. *Chem Eur J* 2009;**15**:9874–88.
117. Geijtenbeek TBH, Kwon DS, Torensma R, van Vliet SJ, van Duijnhoven GCF, Middel J. DC-SIGN, a dendritic cell-specific HIV-1-binding protein that enhances trans-infection of T cells. *Cell* 2000;**100**:587–97.
118. Martínez-Ávila O, Bedoya LM, Marradi M, Clavel C, Alcamí J, Penadés S. Multivalent manno-glyconanoparticles inhibit DC-SIGN-mediated HIV-1 trans-infection of human T cells. *Chembiochem* 2009;**10**:1806–9.
119. Calarese DA, Scanlan CN, Zwick MB, Deechongkit S, Mimura Y, Kunert R, et al. Antibody domain exchange is an immunological solution to carbohydrate cluster recognition. *Science* 2003;**300**:2065–71.
120. Scanlan CN, Pantophlet R, Wormald MR, Ollmann Saphire E, Stanfield R, Wilson IA, et al. The broadly neutralizing anti-human immunodeficiency virus type 1 antibody 2G12 recognizes a cluster of $\alpha 1 \rightarrow 2$ mannose residues on the outer face of gp120. *J Virol* 2002;**76**:7306–21.
121. Marradi M, Di Gianvincenzo P, Enríquez-Navas PM, Martínez-Ávila OM, Chiodo F, Yuste E, et al. Gold nanoparticles coated with oligomannosides of HIV-1 glycoprotein gp120 mimic the carbohydrate epitope of antibody 2G12. *J Mol Biol* 2011;**410**:798–810.
122. Astronomo RD, Burton DR. Carbohydrate vaccines: developing sweet solutions to sticky situations? *Nat Rev Drug Discov* 2010;**9**:308–24.
123. Papp I, Sieben C, Ludwig K, Roskamp M, Böttcher C, Schlecht S, et al. Inhibition of Influenza virus infection by multivalent sialic-acid-functionalized gold nanoparticles. *Small* 2010;**6**:2900–6.
124. Shafer WM, Folster JP, Nicholas RA. Molecular mechanisms of antibiotic resistance expressed by the pathogenic *Neisseria*. In: Genco C, Wetzler L, editors. *Neisseria: molecular mechanisms of pathogenesis*. Norfolk, UK: Caister Academic Press; 2010. Chapter 13, pp. 245–70.

125. Manea F, Bindoli C, Fallarini S, Lombardi G, Polito L, Lay L, et al. Multivalent, saccharide-functionalized gold nanoparticles as fully synthetic analogs of type A *Neisseria meningitidis* antigens. *Adv Mater* 2008;**20**:4348–52.
126. Snippe H, Jansen WTM, Kamerling JP. Immunology of experimental synthetic carbohydrate-protein conjugate vaccines against *Streptococcus pneumoniae* serotypes. In: Roy R, editor. *Carbohydrate-based vaccines*, vol. 989. Washington, DC, USA: ACS Symposium Series, American Chemical Society; 2008. pp. 85–104.
126a. Safari D, Marradi M, Chiodo F, Dekker HA Th, Shan Y, Adamo R, et al. Gold nanoparticles as carriers for a synthetic *S. pneumoniae* type 14 conjugate vaccine. *Nanomedicine* 2011; (accepted for publication).
127. Liu Z, Jiao Y, Wang Y, Zhou C, Zhang Z. Polysaccharides-based nanoparticles as drug delivery systems. *Adv Drug Deliv Rev* 2008;**60**:1650–62.
128. Duceppe N, Tabrizian M. Advances in using chitosan-based nanoparticles for *in vitro* and *in vivo* drug and gene delivery. *Expert Opin Drug Deliv* 2010;**7**:1191–207.
129. Wang JJ, Zeng ZW, Xiao RZ, Xie T, Zhou GL, Zhan XR, et al. Recent advances of chitosan nanoparticles as drug carriers. *Int J Nanomed* 2011;**6**:765–74.
130. Mao S, Sun W, Kissel T. Chitosan-based formulations for delivery of DNA and siRNA. *Adv Drug Deliv Rev* 2010;**62**:12–27.
131. de la Fuente M, Raviña M, Paolicelli P, Sanchez A, Seijo B, Alonso MJ. Chitosan-based nanostructures: a delivery platform for ocular therapeutics. *Adv Drug Deliv Rev* 2010;**62**:100–17.
132. Csaba N, Garcia-Fuentes M, Alonso MJ. Nanoparticles for nasal vaccination. *Adv Drug Deliv Rev* 2009;**61**:140–57.
133. Chavanpatil MD, Khdair A, Panyam J. Surfactant–polymer nanoparticles: a novel platform for sustained and enhanced cellular delivery of water-soluble molecules. *Pharm Res* 2007;**24**:803–10.
134. Park JH, Kwon S, Lee M, Lee M, Chung H, Kim J-H, et al. Self-assembled nanoparticles based on glycol chitosan bearing hydrophobic moieties as carriers for doxorubicin: in vivo biodistribution and anti-tumor activity. *Biomaterials* 2006;**27**:119–26.
135. Dobrovolskaia MA, Germolec DR, Weaver JL. Evaluation of nanoparticle immunotoxicity. *Nat Nanotechnol* 2009;**4**:411–4.
136. Khlebtsov N, Dykman L. Biodistribution and toxicity of engineered gold nanoparticles: a review of *in vitro* and *in vivo* studies. *Chem Soc Rev* 2011;**40**:1647–71.
137. Verma S, Domb AJ, Kumar N. Nanomaterials for regenerative medicine. *Nanomedicine* 2011;**6**:157–81.
138. Solanki A, Kim JD, Lee K-B. Nanotechnology for regenerative medicine: nanomaterials for stem cell imaging. *Nanomedicine* 2008;**3**:567–78.

Carbon Nanotubes: Engineering Biomedical Applications

Diego A. Gomez-Gualdrón,
Juan C. Burgos, Jiamei Yu, and
Perla B. Balbuena

Department of Chemical Engineering and Materials Science and Engineering Program, Texas A&M University, College Station, Texas, USA

I.	Introduction	176
II.	Characterization	176
	A. Carbon Nanotube Structure	177
	B. Electrical and Optical Properties	181
	C. Solubility, Toxicity, and Distribution in Biological Systems	186
III.	Synthesis	194
	A. Arc Discharge	195
	B. Laser Ablation	199
	C. Chemical Vapor Deposition	202
	D. Chirality Control	205
IV.	Post-Synthesis	207
	A. Purification Methods	208
	B. Separation Methods	211
	C. Functionalization of Carbon Nanotubes	217
V.	Biomedical Applications	223
	A. Applications of Carbon Nanotubes in Diagnostics	224
	B. Applications of Carbon Nanotubes in Therapeutics	229
	C. Drug and Gene Delivery and Targeting	231
	D. Tissue Regenerations	232
VI.	Summary and Outlook	233
	References	234

Carbon nanotubes (CNTs) are cylinder-shaped allotropic forms of carbon, most widely produced under chemical vapor deposition. They possess astounding chemical, electronic, mechanical, and optical properties. Being among the most promising materials in nanotechnology, they are also likely to revolutionize medicine. Among other biomedical applications, after proper functionalization carbon nanotubes can be transformed into sophisticated biosensing and biocompatible drug-delivery systems, for specific targeting and elimination of tumor cells. This chapter provides an introduction to the chemical and electronic structure and properties of single-walled carbon nanotubes, followed by a description of the main synthesis and post-synthesis methods. These sections allow the reader to become familiar with the specific characteristics of these materials and the manner

Progress in Molecular Biology
and Translational Science, Vol. 104
DOI: 10.1016/B978-0-12-416020-0.00005-X

1877-1173/11 $35.00

in which these properties may be dependent on the specific synthesis and post-synthesis processes. The chapter ends with a review of the current biomedical applications of carbon nanotubes, highlighting successes and challenges.

I. Introduction

Since their discovery in 1991,[1] carbon nanotubes (CNTs) became one of the main exponents of the nanotechnology era.[2,3] Their special architecture grants them outstanding physical–chemical properties that have captured the imagination of researchers in nearly every field of science.[4–10] As the use of nanotechnology in medicine—*nanomedicine*—promises to revolutionize the way human diseases are detected and treated,[11–13] CNTs have become logical building block candidates of new detection systems[14] and therapeutic strategies.[15] Recent efforts in the research community have shown that CNTs have the potential to be transformed into sophisticated biocompatible systems for bio-detection[16] and drug delivery.[17] For instance, a CNT can be functionalized to specifically bind a biomolecule and allow its detection through changes in the CNT fluorescence spectra.[18] Also, a CNT functionalized with a small drug can become a delivery agent after being endocitozed by a target cell.[19]

The realization of CNT-based technology in nanomedicine, as in other fields, depends on the commercial availability of CNTs with desired specifications at a reasonable cost. In this regard, several nanomedical applications require semiconductor nanotubes with a specific band gap.[20,21] However, relatively large-scale synthesis processes, such as chemical vapor deposition (CVD), normally produce a mixture of metallic and semiconducting CNTs of different band gaps.[22–24] Thus, the adoption of an affordable strategy to either separate the CNTs or produce them selectively is of critical importance. In the following sections of this chapter, various nanotube properties and their role in nanomedicine are introduced, followed by a discussion of the most important synthesis processes emphasizing the issues and challenges of selective nanotube production. Then, an overview is presented of the main post-synthesis treatments aimed to prepare nanotubes for their nanomedical use, such as functionalization. The chapter ends with a selection of potential nanomedical applications of CNTs.

II. Characterization

The last two decades have witnessed the scientific community making outstanding efforts in research and development of CNT technology. The following section contains a description of the CNTs' structural characteristics and main properties as well as the connection of these properties with their potential uses in nanomedicine.

A. Carbon Nanotube Structure

From a conceptual standpoint, CNTs are allotropic forms of carbon obtained by rolling up graphene sheets into seamless cylindrical shapes. Depending on the number of sheets rolled into concentric cylinders, there are two broad categories, namely, single-wall carbon nanotubes (SWCNTs) and multi-wall carbon nanotubes (MWCNTs). MWCNTs usually have larger diameters (from 10 to 20 nm) than SWCNTs, whereas SWCNTs can be as small as 1 nm. As a comparison, cell sizes vary on the order of 10^4–10^5 nm. However, the CNT lengths vary greatly with the synthesis conditions and can reach values as high as several centimeters,[25] thus having extremely high aspect (length-to-diameter) ratios.

Similar to graphene, the carbon atoms constituting CNT walls possess sp^2 hybridization, which generates a delocalized electron cloud along the wall. Such an electron cloud is responsible for the π–π interactions existing between adjacent cylindrical layers in MWCNTs, resembling the interactions between adjacent layers in graphite, although a fraction of inter-wall sp^3 bonds can also be found.[26] The sp^2 hybridization also facilitates the formation of nanotube bundles often observed in experiments, through π–π interactions of electron clouds from neighboring CNTs. This kind of interaction plays a role in the formation of complex CNT structures, such as "ropes," "forests," as well as in non-covalent functionalization of CNTs for nanomedical applications, which is discussed in Section IV.

Perhaps the most important structural feature of SWCNTs is their *chirality*, since it is closely connected with the nanotube's electronic and optical properties. The concept of chirality is widespread in areas such as biochemistry, wherein otherwise identical molecules are non-superimposable mirror images of each other.[27] For instance, amino acids such as alanine show both *right-handed* and *left-handed* versions of the molecule, whose difference lies on the different spatial arrangement of substituent groups around a *chiral* carbon. In SWCNTs, the chirality comes from global arrangements of atoms in the structure, and not from a particular chiral center.[28]

Thus, building on the geometrical concept of rolling graphene into a cylinder, it is possible to understand SWCNT chirality as a result of the rolling direction, as shown in Fig. 1.

Rolling the graphene sheet by bringing points A and B together creates a SWCNT with some C–C bonds along its axis direction, namely, a *zigzag* (zz) SWCNT. Similarly, bringing points A and C together creates a SWCNT with some C–C bonds perpendicular to its axis direction, namely, an *armchair* (ac) SWCNT. Notice that a vector drawn from point A to B (or C) can be expressed in terms of the unit vectors $\hat{a}_1$ and $\hat{a}_2$, as $n\hat{a}_1 + m\hat{a}_2$, where n and m are positive integers. Although the SWCNTs created when $n = m$ (zz) and $m = 0$ (ac)

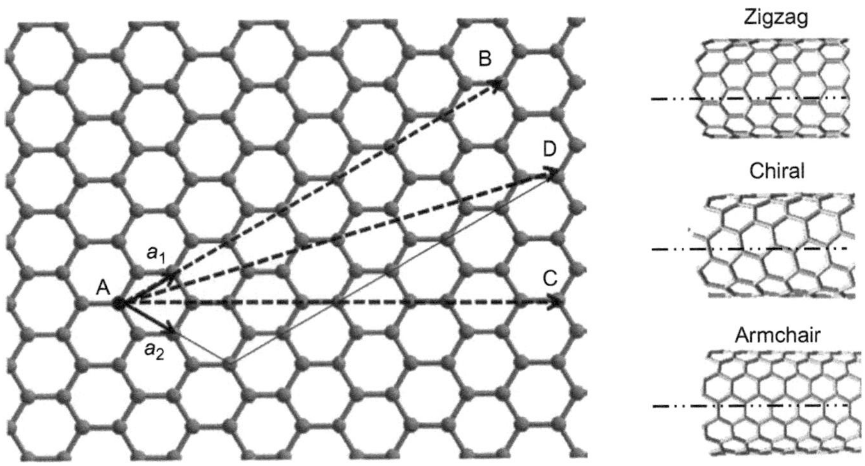

FIG. 1. Rolling the graphene sheet by bringing together points A and B creates a zigzag nanotube, bringing together points A and C creates an armchair nanotube, and bringing together points A and D creates a chiral nanotube. The vector connecting points A and B (or C, or D) is the chiral vector, which can be expressed as $na_1 + ma_2$. For instance, the chiral vector AD is expressed as $6a_1 + 2a_2$; hence the corresponding chiral indexes are (6,2).

are symmetric, the ones created when $n \neq m$ (e.g., bringing together points *A* and *D*) feature a helical pattern in their wall hexagonal lattices, making them non-superimposable on their mirror images, resulting in right-handed and left-handed SWCNTs. Such SWCNTs are said to be chiral, and the indexes (n,m) are used to express the SWCNT chirality, or lack thereof. For instance, a (6,6)-CNT is achiral, whereas the (6,5) and (5,6) ones are chiral, and they are mirror images of each other.

The (n,m) indexes can be used to determine other SWCNT characteristics, such as their diameter d and chiral angle θ, by using Eqs. (1) and (2). The chiral angle expresses the degree of helicity of the SWCNTs, ranging from 0° to 30° due to the sixfold symmetry of carbon rings constituting their walls. It can be seen from Eqs. (1) and (2) that it is possible to have SWCNTs with identical chiral angle and different diameter, or SWCNTs with identical diameter and different chiral angle. This characteristic is relevant to the efforts for designing chiral-selective CNT synthesis processes. It can also be inferred from Eq. (1) that for MWCNTs, the increase in diameter from the inner to the outer cylinder results in different chiral indexes for each of the concentric cylinders, whereas the chiral angle is not necessarily different, or equal.

$$d = \left(\frac{a}{\pi}\right)\left(n^2 + nm + m^2\right)^{1/2} \tag{1}$$

$$\theta = \tan^{-1}(\sqrt{3}m/(m + 2n) \tag{2}$$

Another important structural feature is the CNT ending. After synthesis, CNTs can have one of their ends open or closed.[29] If the end is open, each carbon atom in the open end is bonded to two other carbon atoms, that is, one less atom than usual; therefore, the nanotube rim constitutes a structural defect in an otherwise perfect wall lattice (Fig. 2).

On the other hand, if the end is closed, such an end adopts a hemi-spherical carbon structure known as the *cap*; thus each carbon atom maintains its three bonds. The cap structure is reminiscent of the carbon allotropes known as fullerenes.[2] Although in general there are many possible cap structures corresponding to a given chirality, in all of them the hemi-spherical character of the cap is given by the presence of *six* pentagonal rings, which create the needed curvature, according to Euler's theorem.[30] Because of these pentagonal rings and the pronounced curvature, the cap region is also a CNT structural defect, albeit an important one.

It was mentioned that for a given SWCNT of (n,m) chiral indexes, there may be a number of compatible cap structures. However, for a given cap, there is only one compatible (n,m)-SWCNT structure. Thus, chiral indexes (n,m) are often assigned to the caps, according to the (n,m)-SWCNT they correspond to. Therefore, cap formation has been suggested as a key process for the control of the nanotube structure (i.e., chirality) during synthesis.[31,32] Also, as any structural defect, the cap structure is capable of modifying certain properties in comparison to those of the infinite-length defect-free SWCNT, thus

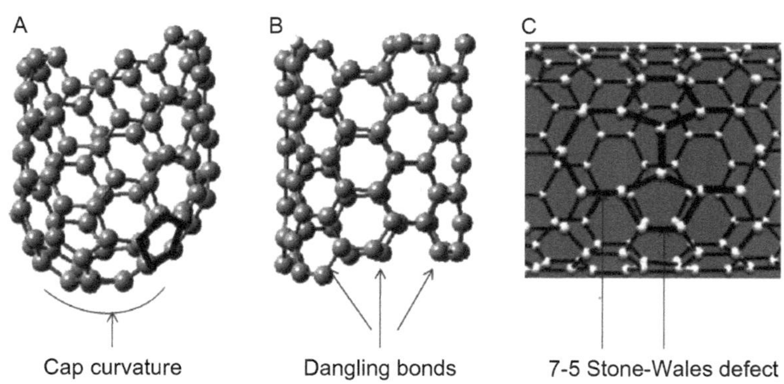

FIG. 2. Defects in carbon nanotubes: (A) A capped end of an SWCNT, wherein the spherical curvature is introduced by the presence of pentagonal rings. (B) An open end of an SWCNT, wherein the rim atoms have one less bond (dangling bond) than atoms in the body. (Adapted from *J. Phys. Chem. C* 2008;**112**(34):13175.) (C) A Stone–Wales defect within the SWCNT wall, wherein the curvature effects of adjacent pentagonal and heptagonal rings are canceled out. (Adapted from *Compos Struct* **92**(7):1701.)

becoming a relevant factor for several applications. For instance, the SWCNT field emission properties relevant to medical imaging have been shown to be affected by the presence of the cap.[33]

Realistically, the CNT structure is likely to have a number of defects after synthesis. Although depending on the target application, CNTs may be treated—*or not*—to become open-ended, they are usually produced as capped. Also, the CNT wall may feature imperfections such as substitutional defects produced by replacing some carbon atoms with other elements such as nitrogen or boron,[34] as well as structural defects such as the so-called *Stone–Wales* defect[35] (Fig. 2). In substitutional defects, foreign elements are incorporated into the CNT structure, sometimes intentionally, to fine-tune its properties.[36] In the Stone–Wales defect, the bonds of two pairs of adjacent hexagonal rings rearrange themselves to form a pentagon–heptagon pair that can move along the nanotube wall in a similar fashion to a dislocation defect. This defect is important for the nanotube structure because it can lead to a change of either the chiral angle or the diameter along the CNT,[37] and even if this is not the case, its presence can impact its electronic, mechanical, and thermal properties.

The description in the above paragraphs corresponds to pristine CNTs. However, after post-synthesis treatments discussed below in Section IV, CNTs can be found in functionalized form (FCNTs). In fact, such functionalization is critical for most of their nanomedical applications, wherein the CNT can be "coated" with a number of biomolecules or polymeric molecules of interest.[38] It is at this point that the particular geometry of CNTs is advantageous. For pristine SWCNTs, the totality of their atoms makes up their surface, resulting in a specific surface area around 1300 m^2/g.[39] Therefore, there is a vast surface available to accommodate more than one type of molecule, conferring *polyfunctionality* to the CNT.[40] Similarly, because of their high length-to-diameter ratio, such polyfunctionality can be varied along the CNT length, which increases their interaction versatility; for instance, simultaneous interaction with two different biomolecules on different cellular sites, or two different *biomarkers* in the blood stream. If the functionalizing molecules have a polycyclic aromatic region, they can accommodate on the CNT surface through π–π interactions with the CNT wall, in similar fashion to the interwall interactions in MWCNTs, corresponding to the non-covalent functionalization mentioned earlier.[38] Alternatively, covalent sp^3 bonds can be created between the CNT wall and the functionalizing molecule,[41] although this can affect the CNT properties because of the disruption of its wall lattice.

Although the chapter has focused on the structure of individual CNTs, when they are in proximity to other CNTs, there is a clear CNT trend to self-assemble in bundles,[42] ropes, or other structures,[2] due to their inter-wall interactions. Since such proximity is usually favored by synthesis conditions, CNTs are seldom found in individual form. The basic difference between bundles and ropes is the higher organization of the latter, wherein the axes of the individual CNTs are aligned in the same direction. Sometimes, CNTs are synthesized on a flat support, in which depending on the synthesis conditions they can grow either vertically or horizontally, forming densely packed structures called *forests*[43] or *carpets*,[44] respectively. Although these structures are attractive for several applications, in the nanomedicine field—*wherein the CNTs will be often delivered into the body*—individual CNTs are preferred in order to fully exploit their properties, hence the importance of post-synthesis CNT separation.

B. Electrical and Optical Properties

At a fundamental level, the properties of a material depend on its atomic level structure. For instance, electrical and optical properties of SWCNTs depend on their chirality, because different rolling patterns of the graphene sheet lead to different electronic wave functions.[45] The electronic wave function of graphene describes its delocalized electron cloud, resulting in its metallic character (i.e., good electrical conductivity). On the other hand, the graphene wave function is modified in the rolled-up SWCNT. As a result, two-thirds of the chiralities correspond to semiconducting SWCNTs, whereas only one-third correspond to metallic SWCNTs.[2,46] Theoretical calculations predict that if $(n - m)/3 = k$, wherein k is an integer number, then the SWCNT is of metallic character.[47–49] However, deviations to this rule can occur due to factors such as curvature effects, finite length, structural defects, doping, and/or cap structure.

The metallic or semiconductor character of a material may be evaluated through its *density of states* (DOS), which is the number of energy states corresponding to a given energy level. A material of infinite dimensions features a continuous DOS, in contrast with a discrete one for a finite-size material (Fig. 3). In the first case, the DOS of a metallic material appears continuous between energy regions corresponding to occupied (valence) and unoccupied (conduction) states (bands). Such continuity implies that electrons can easily occupy the conduction band (Fig. 3). On the other hand, a semiconducting material has a (band) gap—*an intermediate region where the DOS is zero*—between the occupied and unoccupied states, which means that electrons need to gain a specific amount of energy to reach the conduction band (Fig. 3). In a finite-size material, the metallic or semiconducting character may be characterized by the separation between the highest occupied state (HOMO) and the lowest unoccupied state (LUMO) (Fig. 3). The amount of energy needed to excite electrons from occupied to unoccupied states (bands) may be reflected in the material optical properties as well.

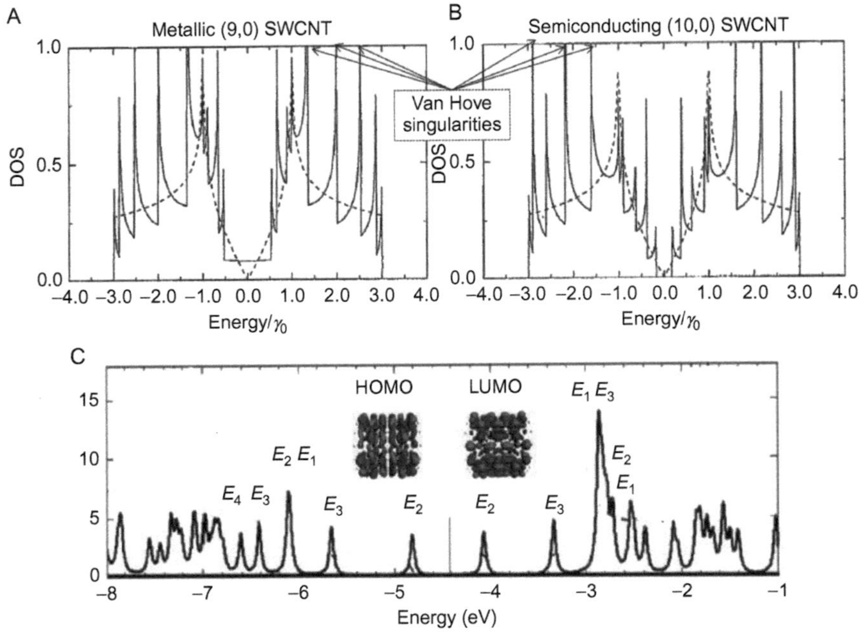

FIG. 3. Density of states (DOS) of infinite-length SWCNTs: (A) Metallic (9,0) SWCNT ($n - m/3 = k$; wherein k is an integer) features a DOS that never drops to zero. (B) Semiconducting (10,0) SWCNT features an intermediate forbidden region where the DOS drops to zero; it possesses a band gap. The characteristic peaks of the DOS are known as Van Hove singularities. Optical and electronic transitions involve excitations from one singularity to another. (Adapted from *Appl Phys Lett* 1992;**60** (18):2204.) (C) DOS of a finite-length (14,0) SWCNT, a Lorentzian convolution is made around the discrete levels (hence the DOS is zero in several regions). Optical and electronic transitions involve excitations from orbitals below the HOMO to orbitals above the LUMO. (Adapted from *Phys Rev B* 2008:78, 035432.) (See Color Insert.)

The high aspect ratio of SWCNTs makes them one-dimensional materials with a one-dimensional DOS (Fig. 3). A typical SWCNT feature is the sudden rise and fall of their DOS, resulting in the appearance of peaks in both occupied and unoccupied regions known as *Van Hove singularities* (VHS).[50] The location (energy) of these singularities depends on the SWCNT structure, giving a characteristic DOS for each chirality.[48] While the metallic or semiconducting character of each SWCNT depends on whether the band gap exists, the optical behavior depends on electron transitions from one VHS to another one.[51,52] In semiconducting SWCNTs, the band gap corresponds to the gap between the highest occupied (v_1) and the lowest occupied (c_1) VHS, hence the relevance of the SWCNT band gap for nanomedical applications based on the optical properties of SWCNTs.[53]

Light is absorbed by CNTs when its photons possess the exact energy to excite electrons from an occupied state to an unoccupied one, as long as such *optical* transition is not forbidden.[54] Conversely, the reverse de-excitation process results in the emission of photons, whose energy corresponds to the energy liberated by the de-excited electrons. Because of the distinctive sharp DOS existing for each SWCNT chirality, the energies of absorbed and emitted photons of *individual* SWCNTs are very specific to each chirality. In comparison to other materials, the specific photon absorption or emission of CNTs is easier to identify against the background noise of detecting devices.[9] The energy values relevant to the optical transitions of SWCNTs correspond to photons in the near-infrared region (NIR).[55,56] Biological tissues do not interact (absorb) with photons whose wavelengths are in the 800–2000 nm range, constituting the so-called *biological transparency window.*[57]

Excited electrons do not necessarily return to their original state, but rather to an intermediate state. Therefore, the energy of the absorbed and emitted photons is different and the CNT is said to *fluoresce.*[58] This fluorescence can make possible their detection in biological systems, if properly stimulated. Semiconducting SWCNTs normally fluoresce with higher intensity than their metallic counterparts[59]; therefore, they are preferred for nanomedical applications. An additional advantage is that the fluorescence stability in SWCNTs allows them to glow for months, whereas other fluorescent structures such as proteins and quantum dots fade within few hours or days. As the SWCNTs emission properties depend on their DOS and band gap, they will be impacted by any structural modification that changes any of the latter. The importance of the band gap energy led to the creation of the *Kataura plot*,[60] which allows reading the band gap energy for a particular SWCNT diameter (Fig. 4).

In MWCNTs the inter-wall interactions between the nested SWCNTs can alter the original DOS, thus hindering optical and electrical properties, as occurs with the formation of CNT bundles. Moreover, the nested SWCNTs could consist of different proportions of metallic and semiconducting chiralities of different band gaps contributing to the overall character of the MWCNTs.[61] However, an interesting feature of MWCNTs is the change of the DOS with the rotation of the nested SWCNTs with respect to each other.[62] The band gap increases or decreases, because as the rotation angle changes, the inter-wall interaction changes accordingly. Similarly, interactions between the wall—*either for MWCNTs or SWCNTs*—and functionalizing molecules can also modify the DOS, resulting in the alteration of the optical and electrical behavior.[63] The stronger the interaction, the larger is the effect on the DOS; thus, covalently bonded molecules are expected to have a greater impact than non-covalently bonded ones.

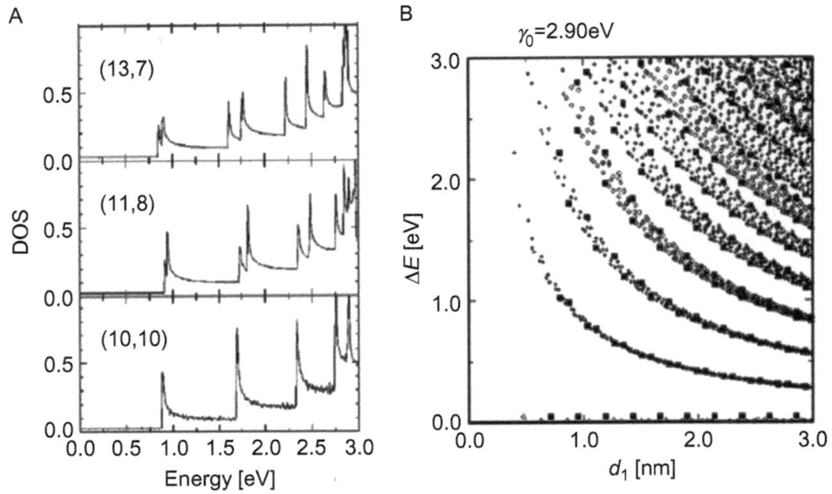

FIG. 4. (A) DOS as a function of SWCNT chirality, inducing variations of optical and electronic properties with chirality. (B) Kataura plot. (Adapted from *Phys Rev B* 2000:**61**;2981 and Kataura *et al.*, *Synth Met* 1999:**103**;2555–2558.)

Alterations in the DOS of the CNTs can be used as an indication of the adsorption of biomolecules on their wall, because such changes are reflected in modifications of their electrical and optical behavior.[64,65] Moreover, under appropriate conditions, FCNTs are sensible enough for such modifications to occur when biomolecules adsorb on the functionalizing molecule, but not directly on the CNT wall. For instance, molecular adsorption can alter the fluorescence characteristics of glowing CNTs, such as bright intensity and emitted wavelength.[65] Similarly, the amount of current running through a CNT at a particular applied voltage can be expected to increase or decrease as the amount of adsorbed molecules increases.[64] It is possible that a minimum number of molecules have to be adsorbed before a measurable change is detected. Such a number depends on the characteristics of the CNT, the molecules adsorbed, those functionalizing the CNT, and the conditions of the surroundings. The different CNT responses to molecular adsorption, such as the ones previously described, can be exploited to create CNT-based biosensors with potentially lower detection limits and higher sensitivity.

The crystallinity of the SWCNT wall results in the quantization of its vibrational modes in *phonons*.[66] If incident photons interact only with the SWCNT electrons, the excitation and de-excitation of the latter among VHSs can result in fluorescence, as long as the photons have the correct energy.[58] On the other hand, if incident photons of any energy excite the SWCNT electrons to a short-lived virtual state, the interactions between electrons and phonons lead to electron de-excitation from the virtual state to an *excited vibrational* state.[66] This results in *Raman* scattering,[66] wherein the energy (frequency) difference between the incident and emitted photon (light) corresponds to the energy of a vibrational state (mode). The *radial breathing mode* (RBM) and the *tangential mode* (G-band) are the most significant modes for Raman detection/characterization of defect-free pristine SWCNTs (Fig. 5), whereas a *disorder-induced mode* (D-band) appears due to defects created by either imperfect growth or SWCNT functionalization.[67]

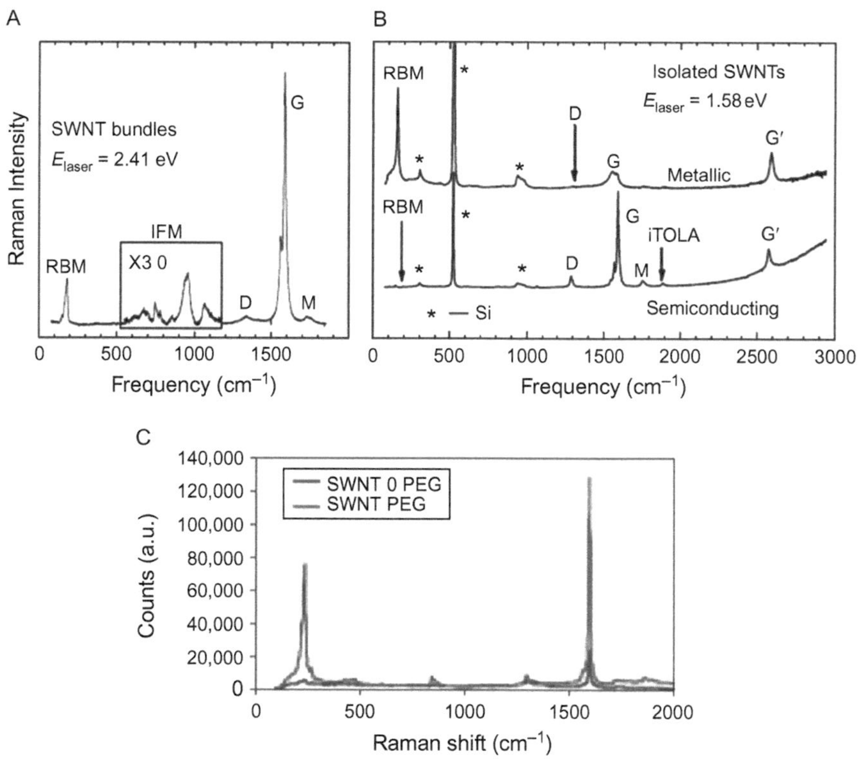

FIG. 5. (A) Raman spectrum of a sample of SWCNT bundles showing the RBM, D, G, and M peaks. (Adapted from *Phys Rep* 2005;**409**:47–99.) (B) Raman spectra of individual metallic (top) and semiconducting (bottom) SWCNTs (notice the different G-peak shape). The individual tubes are deposited on silicon, hence the appearance of peaks corresponding to such material. (Adapted from *Phys Rep* 2005;**409**:47–99.) (C) Raman spectra of covalently functionalized (SWNT-O-PEG) and non-covalently functionalized (SWNT PEG) nanotubes, wherein the peaks of the former are reduced in intensity due to disruption of the wall. (Adapted from *Nat Nanotechnol* 2008;**3**:216.) (See Color Insert.)

The RBM corresponds to the synchronized out-of-plane vibration of the wall atoms, with a frequency ($150 \text{ cm}^{-1} < \omega_{RBM} < 300 \text{ cm}^{-1}$) correlated to the inverse of the SWCNT diameter. Thus, the RBM can be used to identify an SWCNT of a given diameter. The G-band corresponds to in-plane vibrations of the wall atoms at a frequency $\omega_G \sim 1600 \text{ cm}^{-1}$, wherein the peak shape can be used to identify a semiconducting SWCNT against a metallic one. The D-band ($1300 \text{ cm}^{-1} < \omega_{RBM} < 1400 \text{ cm}^{-1}$) is attributed to bad resonance, but there is no clear correlation with the number of defects or amount of functionalization[68] (Fig. 5).

The energy (frequency) of the incident photons (light) does not affect the frequencies where the RBM and the G-band peaks appear in the Raman spectrum of a particular SWCNT; however, specific energy values of the incident photons can intensify the signal due to resonance.[67] Resonance occurs when the excited virtual state correspond to a VHS.[68] Since the VHS locations change with the DOS—*therefore with SWCNT chirality*—a particular SWCNT chirality can be identified from a heterogeneous mixture by selecting the appropriate incident light frequency.[67] For nanomedical applications, such light should be in the biological transparent window. Additionally, SWCNTs are also susceptible to *surface-enhanced Raman spectroscopy* (SERS),[69] wherein the intensity of their Raman signal is enhanced by the proximity of a metallic surface, or even by functionalization with metallic nanoparticles.

C. Solubility, Toxicity, and Distribution in Biological Systems

A material dissolves into another when the change in free energy ΔG_{mixing} associated with the process is negative. In a first approximation, it can be said that such a change is favored by a negative change in enthalpy (ΔH_{mixing}) and/or a positive change in entropy (ΔS_{mixing}). Since the introduction of a new component to a molecular system normally increases the number of possible "molecular arrangements," ΔS_{mixing} is usually positive.[70] Therefore, ΔH_{mixing} normally determines the sign of ΔG_{mixing}, and consequently is a first indicator of the favorable dissolution—or not—of a material.[70] The change in enthalpy depends on intermolecular interactions. Thus, the behavior of CNTs in solvents is influenced by the relative strengths of solute–solute, solvent–solvent, and solute–solvent interactions.[70]

A more familiar interpretation of the previous thermodynamic concepts is that dissolution occurs if the solute–solvent interactions allow the solute molecules to disrupt the solvent structure, while allowing the solvent to *disperse* the solute molecules. This is expected to occur when solute–solute and solvent–solvent interactions are similar (i.e., like dissolves like). The strength of attraction between CNT walls has been reported to be 500 eV per each micrometer of CNT contact,[71] which along with their specific area and aspect ratio favors the formation of bundles, as mentioned in a previous subsection. The sp^2 bonding on the surface of the CNT wall makes it highly hydrophobic; therefore, the interaction between CNTs and aqueous solvents is not strong enough to induce dispersion.[72] Moreover, most conventional solvents, either aqueous or organic, are unable to disperse and dissolve pristine CNTs.[73]

The insolubility of pristine CNTs is an obstacle for their nanomedical applications, because aggregation masks the outstanding CNT properties. For instance, their fluorescence is diminished due to the interaction between CNTs that alters their DOS.[74] Additionally, applications inside the body require their dissolution in a biologically compatible saline or buffer solution.[9] Moreover, an adequate CNT biodistribution inside the body is unlikely to be achieved if the CNTs aggregate as a result of hydrophobic forces. Although oxidized CNTs are able to dissolve in water, they aggregate in saline biological solutions.[75] Therefore, methods have been developed in response to the necessity of dispersing and dissolving the CNTs, wherein CNT functionalization plays a significant role.[41,76]

Functionalization of CNTs with molecules possessing selected moieties introduces the required interactions with the solvent promoting CNT dispersion in either aqueous or organic solutions. Since it is possible to functionalize CNTs in either a covalent or non-covalent fashion, a similar classification applies for their dissolution methods. The interaction between functionalizing molecules and the CNT is stronger with the covalent method, making the FCNT stable at a variety of conditions. This gives more robustness to the covalent method, wherein the dispersion of CNTs is more resilient to condition changes.[75] On the other hand, the non-covalent method relies on comparatively weaker interactions (e.g., hydrophobic π–π interactions) for the accommodation of surfactant molecules on the CNT wall, making the FCNT stable for a narrow set of conditions.[76] Nonetheless, it is important to recall that with the non-covalent method the original CNT properties are usually maintained.

A surfactant is an amphiphilic molecule, wherein its hydrophobic tail can be directed toward the CNT wall, while the hydrophilic head is directed toward the solvent. Thus, the dispersion mechanism is related to *micelle* formation around the CNTs,[77] which can be described by classical colloidal chemistry for dispersion of solid particles (Fig. 6). Accordingly, the *critical micelle concentration* (CMC) plays a significant role in the stability of the CNT dispersion, because if the surfactant concentration decreases below the CMC, then the micelles are destabilized.[78] In nanomedical applications, high surfactant concentrations are not desirable, because it may have negative effects, such as protein denaturation. Thus, the use of traditional surfactants such as sodium dodecylsulfate (SDS) and Triton X-100 is not always recommended.[75]

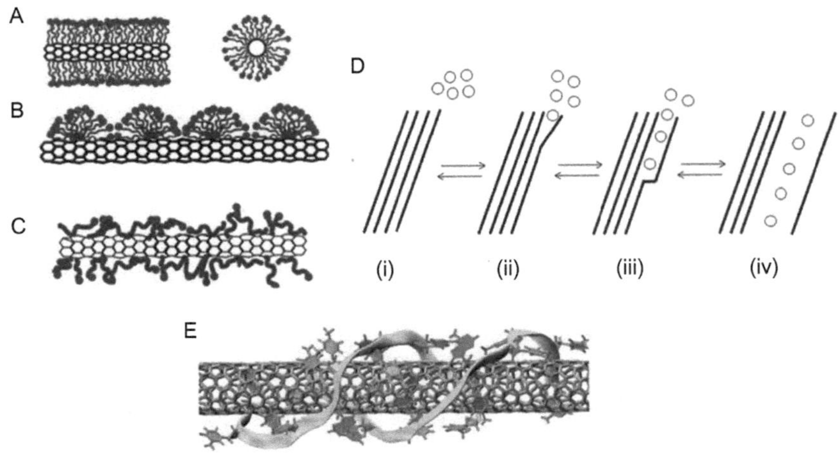

FIG. 6. Top left: several geometries of micelle formation. Surfactant hydrophilic heads are represented by dots, and hydrophobic tails by lines. (A) Cylindrical micellar adsorption; (B) hemimicellar adsorption; (C) random micellar adsorption. (Adapted from *JACS* 2004,**126**(32):9902–9903.) Top right: mechanism of separation of a SWCNT from the bundle assisted by surfactant molecules and ultrasonication. (Adapted from *J Nanosci Nanotechol* 2003;**3**(1/2): 81–86.) (E) Adsorption of a macromolecule (DNA) around a SWCNT facilitates its solubilization by reducing the hydrophobic interface. (Adapted from *Nature* 2009;**460**:50.) (See Color Insert.)

It has been hypothesized[79] that once the surfactant molecules begin to accumulate on a CNT, the latter may begin to separate from the bundle due to steric effects and electrostatic interactions. In order to facilitate the access of surfactant molecules between CNTs, the dispersion process is generally assisted by ultrasonication—*the mechanical vibration of the solution at ultrasonic frequencies*—to induce mechanical breaking of CNT inter-wall interactions.[79] Nevertheless, excessive mechanical stimulation may result in CNT damage, which demands a careful control of the ultrasonication process (Fig. 6), even though some extent of CNT length shortening is practically unavoidable.[80] The medium targeted for CNT dissolution usually determines the nature of the surfactant. For instance, aqueous solvents call for ionic surfactants, and organic solvents for nonionic ones, whereas biological media usually require biocompatible, *bioconjugable*, nontoxic surfactants that are stable in serums with high salt and protein content.[9,75]

Adsorption of ionic surfactants, such as dodecyl benzene sodium sulfonate (NaDDBS), on the CNT wall is promoted by electrostatic interactions.[81] Thus, the surface charge of the CNT in the target media plays an important role in the dispersion stability. However, the cationic or anionic character of the surfactants has not been clearly correlated to their performance. On the other hand, the bond unsaturation of certain hydrophobic tails appears to affect the dispersion performance of both ionic and nonionic surfactants, because it determines the stability of the interaction of the surfactant with the CNT wall.[82] For instance, the nonionic surfactant SPAN-85 (having three double bonds in its tail) has been shown to be superior for dispersing CNTs than other surfactants within the SPAN family.[82]

Alternatively, both block copolymers and single-stranded DNA (ss-DNA) have been shown to improve CNT dispersion in both aqueous and organic solvents (Fig.6). It has been suggested that these macromolecules wrap the CNTs, decreasing the hydrophobic interface with the solvent, hence promoting dispersion.[83] Such wrapping is driven by selective interactions with specific regions of the macromolecules, such as hydrophobic blocks of the copolymers, or bases in the ss-DNA, while leaving exposed other regions with higher affinity with the solvent, such as hydrophilic blocks of the copolymers, or the ss-DNA phosphate backbone.[83,84] Interestingly, in some cases, the affinity of the wrapping macromolecule with the CNT has been shown to be dependent on the CNT chirality, being potentially useful for chiral separation of the CNTs.[84]

Enclosing of the CNTs by either wrapping macromolecules or forming a micellar structure of surfactant molecules around them effectively refrains the CNTs from directly interacting with the media.[77,83] Therefore, such methods also facilitate the potential use of CNTs in nanomedicine, because they reduce their potential toxicity.[9] Along this line, the emergence of CNTs as one of the most promising materials in the nanotechnology era has been accompanied by increasing concerns about their toxicity. Most of these concerns were raised because of the similarities between pristine CNTs and asbestos fibers, whose inhalation is known to produce serious health problems. Subjects exposed to asbestos usually develop a rare cancer known as mesothelioma, originated in a type of protective membrane known as mesothelium.[85] In consequence, initial studies on the toxicology of CNTs have been focused on the response of pleural tissue to their inhalation.[86–88]

All the initial toxicological studies focusing on the inhalation of pristine CNTs indicate an inflammatory response of the pleural tissues along with the formation of granulomas.[86] Some of these studies in animals reveal that a short time after exposure, CNTs are found in several lung structures, which triggers an immune response as suggested by the counting of mononuclear cell aggregates such as macrophages.[86] These macrophages aggregate, forming granulomas in an attempt to engulf the CNTs, because their remarkable stability prevents their elimination by the immune system. It has been also reported that the severity of the immune response increases with the number of walls and the length of the CNTs, because the larger dimensions deter their engulfing by the macrophages.[86] Therefore, it is generally accepted that the prolonged accumulation of CNTs larger than 10 μm should be avoided, even though there is no consensus on the relevance of the dosages.[9] Nevertheless, further studies have shown that the accumulation of CNTs previously biodegraded using the human myeloperoxidase (hMPO) does not result in the above discussed inflammatory response.[89] Therefore, this is still an open topic where a lot of research efforts are being invested.

Additional studies have addressed the toxicity of CNTs in cell cultures, indicating the decrease in viability (Fig. 7) due to the presence of pristine CNTs in the culture media.[90,91] For instance, one study demonstrates the "germicide" effect of pristine CNTs in *Escherichia coli*, wherein staining assays reveal the higher percentage of dead cells in cultures containing CNTs.[91] Since the CNTs were carefully purified and stripped of any metal traces, the toxic effect can be attributed exclusively to the CNTs. Similarly, another study on human fibroblast cultures suggests the toxic effect of the hydrophobic surface of carbonaceous materials, wherein SWCNTs were found to have a higher impact on cell survival than MWCNTs.[90] Although it is not specified whether the cell dies via necrosis or apoptosis, it is suggested that the CNTs activates protein signaling ultimately leading to cell death.[90] Nonetheless, up to date there is no consensus on the toxicological impact of the degree of dispersion of the CNTs.

The toxicological effects found for pristine CNTs are attributed to interactions with their hydrophobic surface, being rather representative of accidental exposure to CNTs in the workplace.[9] As already mentioned, CNT functionalization—*which effectively isolates it from the biological milieu*—is found to reduce their toxicity allowing their deliberate insertion into the patient's body.[92] For instance, the functionalization of the CNT with a polymer such as polyethylene glycol (PEG)—*or some derivative*—effectively reduces CNT toxicity, because the exposed PEG moiety has been shown to be biocompatible, nontoxic, and non-immunogenic.[92] Therefore, the toxicity of FCNTs is not correlated to the toxicity of pristine CNTs, and must be evaluated independently.

Since the CNT functionalization results in higher versatility of the surface exposed for interaction with the biological milieu, the use of different groups is likely to result in varied toxicity and biodistribution of the CNT inside the body. Therefore, the corresponding toxicological studies have been focused on determining the biodistribution and adverse effects of FCNTs (in solution) intravenously applied to animal test subjects.[9] Thus, along with the toxicity evaluation, the organs where CNTs preferably accumulate—*if any*—and the preferred excretion routes are determined through different tracking methods.[9] Methods such as Raman spectroscopy, or NIR fluorescence, take advantage of some of the optical CNT properties discussed in the two previous sections.

A common conclusion of *in vivo* toxicological studies in FCNTs is that after administration, FCNTs have a tendency to accumulate in the reticulo-endothelial system (RES).[9] Thus, FCNTs behave similarly to other nanoparticles that are also found to accumulate in RES organs such as the liver. Most importantly, no apparent adverse effects were found among the injected animal

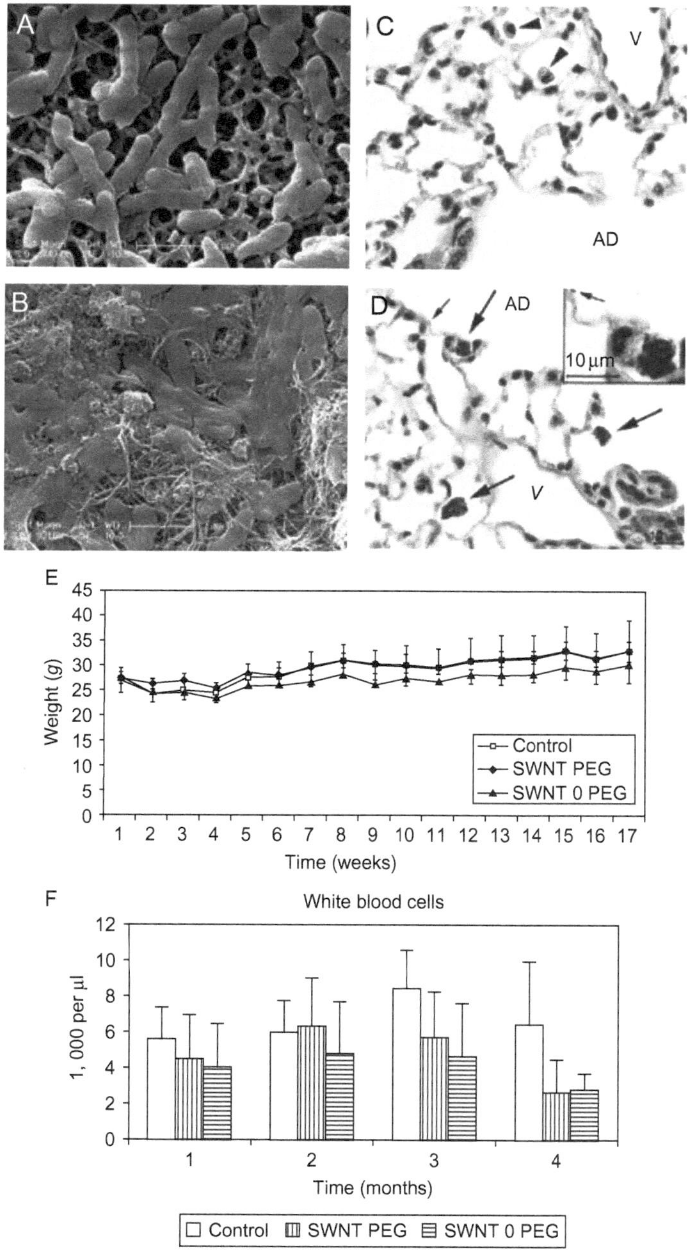

FIG. 7. Left images show viability of *Escherichia coli* cultures. (A) *E. coli* culture in the absence of SWCNTs. (B) *E. coli* culture in the presence of SWCNTs. (Adapted from *Langmuir* 2007;23:8671.) Center images show pulmonary tissue in control and MWCNT-exposed rats:

subjects.[9] However, desorption of functionalizing molecules due to the interactions with the biological milieu has also been observed, which could lead to the exposure of the "toxic" hydrophobic surface of the CNT. Hence, a careful control of the "coating" stability must be assured.

While *in vitro* studies have assessed the viability of cell cultures in the presence of FCNTs to evaluate their cytotoxicity, *in vivo* studies assess a wider spectrum of physiological variables in the animal subject. *In vitro* studies have shown some extent of cytotoxicity and oxidative stress for FCNTs (even though less severe than those found for pristine ones), whereas *in vivo* studies have not revealed overall adverse effects on the living subject due to the FCNTs.[9] In a pilot study, cardiac rhythm, respiratory patterns, body weight, red cell counting, blood pressure, tissue histology, and behavioral patterns were analyzed in mice upon the injection of either a control saline solution or one containing pegylated CNTs (non-covalently or covalently bonded). After monitoring for 4 months, there were not statistically significant variations respecting to the control subjects (Fig. 7).[93] Nevertheless, the nonbiodegradability of the FCNTs is evidenced, raising uncertainty on longer-term effects.

It is apparent that there are no clearly established protocols and toxicological models to obtain a unified evaluation of the CNT toxicity. For instance, reported studies usually differ in the methods and materials used, leading to partial disagreement among some of them.[9,17] Furthermore, it is uncertain if current studies realistically represent exposure conditions, because there is no consensus on the dosage and formulation of the tested CNTs, the duration and route of exposure, the tested biological subject, and the establishment of standards and references that permit the creation of a scale of toxicity and risks. Additionally, there is also uncertainty on the final environmental fate of the CNTs after leaving the body, because CNTs are durable and practically nonbiodegradable.[94]

In several instances, questions about whether CNTs are toxic or not are raised because their small size, as discussed earlier in this section, could facilitate the cellular uptake of the CNTs.[95] Accordingly, several studies have been addressed to determine if the CNTs are able to cross cellular membranes (Fig. 8), with affirmative results.[95–97] Therefore, it is possible that once the FCNTs reach certain organs in accordance to their characteristic

(C) Unremarkable tissue in control rat. (D) Formation of macrophages engulfing MWCNTs in exposed rats. (Adapted from *Toxicol Sci* 2007;**100**(1):203–214.) On the right, some statistical data are given for *in vivo* studies in mice injected with functionalized nanotubes (FCNTs), wherein statistically significant differences in weight evolution and white blood counts were not found between exposed and control subjects. (Adapted from *Nature Nanotechnol* 2008;**3**:216.) (For color version of this figure, the reader is referred to the Web version of this chapter.)

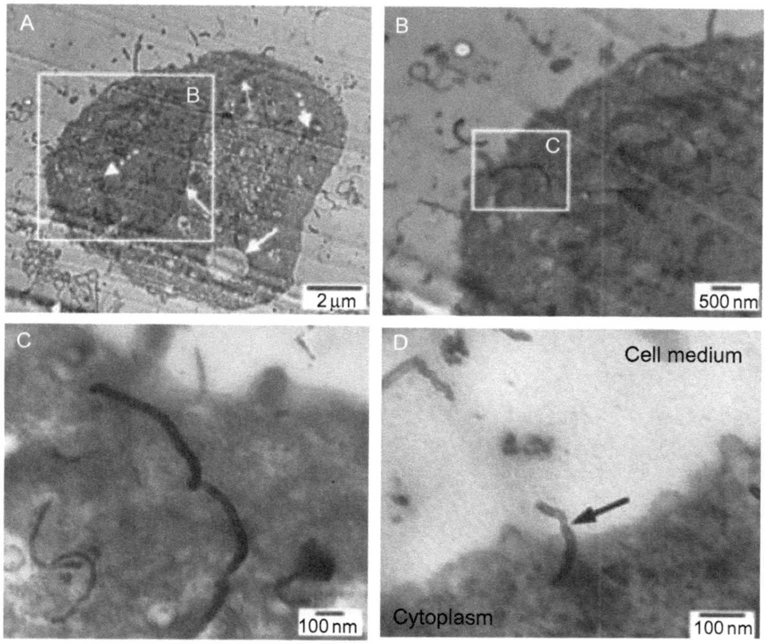

FIG. 8. Panels showing ultrathin section of cells treated with functionalized nanotubes, at different amplifications. (A) Whole cell. (B) and (C) Subsections of the cells. (D) Amplified image showing a nanotube trespassing the cell membrane. (Adapted from *Ang Chem Int Ed* 2004;**43** (39):5242–5246.)

biodistribution, crossing of the cellular membranes occurs, and the FCNTs are internalized inside the cell. There are two proposed mechanisms that explain internalizations of the FCNTs, wherein the two of them are not necessarily exclusive.[98] One is cellular uptake via endocytosis,[99] and the other one is insertion and diffusion mechanism.[100] Since experimental evidence supporting either mechanism has been found, it is possible that the occurrence of either internalization mechanism depends on several factors such as diameter and length of the FCNTs, nature and number of functional groups, and cellular environment variables such as temperature and ATP reserves.[17]

The endocytosis mechanism proposes that once the FCNT reaches the outside of the cellular membrane, the cell uses a small region of the latter to encapsulate the FCNT inside a vesicle.[17] Then, the cellular membrane rearranges to bring the vesicle inside the cytoplasm, wherein it merges with an endosome, which eventually carries it into a lysosome.[17] On the other hand, the insertion mechanism is not fully understood, experimental observations show that FCNTs appear inside the cell oriented perpendicularly to the membrane.[101] Molecular simulations have been used to elucidate the mechanism, wherein it was shown that the FCNT is adsorbed parallel to the membrane, and then assisted by the phospholipid molecules to reorient transversally to the membrane and proceed with the insertion.[102] It is important to remark that whether either mechanism applies, the internalization of the FCNTs is a process that can be taken advantage of for the design of advance drug-delivery systems.[17,103]

III. Synthesis

The production of CNTs consists in the transformation of a carbon source into nanotubes, usually at high temperature and low pressure, wherein the synthesis conditions influence the characteristics of the final product. For instance, the synthesis conditions can be modified to obtain either MWCNTs or SWCNTs to satisfy different needs, although the properties of the latter make them more attractive for a number of nanomedical applications. Nonetheless, it is also important to control the formation of undesired products, as well as the quality of the synthesized nanotubes. The formation of undesired products such as amorphous carbon, fullerenes, or graphite is reflected in the *selectivity*.

Selectivity toward CNT formation is usually given by the ratio between the mass of produced CNTs and the total mass recovered. For instance, in a process featuring a selectivity of 90%, 9 g of nanotubes can be extracted for every 10 g recovered from the reactor, using a suitable purification process. The production cost of CNTs can be decreased by improving the selectivity, because this results in reduction of purification costs. Additionally, the term selectivity can also be used to quantify the formation of specific types of nanotubes in a given sample. Some authors denominate *selectivity type I* as the ratio between the mass of SWCNTs produced and the total mass of CNTs produced,[104] and *selectivity type II* as the ratio between the mass of SWCNTs produced with a specific chirality and the total mass of SWCNTs produced.[105]

Achieving a high selectivity type II—*chiral selectivity*—constitutes one of the most challenging goals in the nanotube synthesis. Currently, both experimentalists and theoretical researchers are making significant efforts in disclosing the key to attain chiral selectivity. Designing a synthesis with a high chiral selectivity would allow producing batches of SWCNTs with one dominant chirality and nanotubes having specific properties such as diameter, conductance, band gap that best suit a particular application. Although attaining chiral selectivity still remains elusive, some success has been achieved in designing synthesis processes that produce majority of either metallic or semiconducting nanotubes.[106,107] The effectiveness of these processes is quantified by either the *metallic* or *semiconducting selectivity*, wherein the ratio between the mass of metallic—*or semiconducting*—nanotubes and the total mass of nanotubes is calculated.[106,107]

Similar to other chemical syntheses, the effectiveness of the production of CNTs can be assessed by the *relative yield*, which correlates with the amount of product obtained from the corresponding growth reaction. Accordingly, the relative yield can be defined as the ratio between the mass of the product and the mass of carbon available for reaction. However, for reaction systems using a catalyst, researchers usually define an alternative yield as the ratio between the mass of the product and the mass of the catalyst used for such purposes. Thus, it is possible to evaluate the performance of the catalyst before its deactivation. Significant efforts are continuously made by the research community to improve both selectivity and yield of the nanotube synthesis processes so large-scale production becomes economically feasible. There are three main nanotube synthesis processes (a few example products shown in Fig. 9): arc discharge (AD), laser ablation (LA), and CVD.[2]

A. Arc Discharge

AD stands out as the synthesis method used by Iijima in 1991 when CNTs were first identified.[1] The method was originally intended for the synthesis of fullerenes, the allotropic form of carbon discovered in 1985.[108] Briefly, the method relies on the electrical breakdown of gas molecules located in the gap between two carbon electrodes. As a result, a continuous electric discharge from one electrode to the other occurs through the normally nonconductive gas. Such an extreme condition turns the gas into plasma—*a fluid of negatively and positively charged particles*—generating a temperature high enough to vaporize most materials. Accordingly, the electric arc vaporizes carbon, which reassembles in a variety of allotropic forms, including nanotubes. Notwithstanding energy and scale requirements, its conceptual simplicity results in AD being perhaps the easiest method to produce CNTs.[2]

Typically, the two carbon electrodes (Fig. 10) are separated approximately by 1 mm, wherein a *direct current* in the range of 50–100 A (i.e., high current arc) is driven by an applied voltage of approximately 20 V between electrodes.[109]

The AD apparatus is usually filled with inert gases such as helium and argon at low pressures (50–700 mbar), although other gases such as hydrogen or methane can be added as well to modify the synthesis conditions.[110,111] The temperature in the plasma can be as high as 4000 K, although existing models predict that the CNTs form in the region of the plasma where the temperature is relatively low (i.e., 1300–1800 K).[112] The anode is hotter than the cathode due to electron collisions, so carbon vaporization occurs in the former. This results in dimensional changes in the anode, requiring its continuous

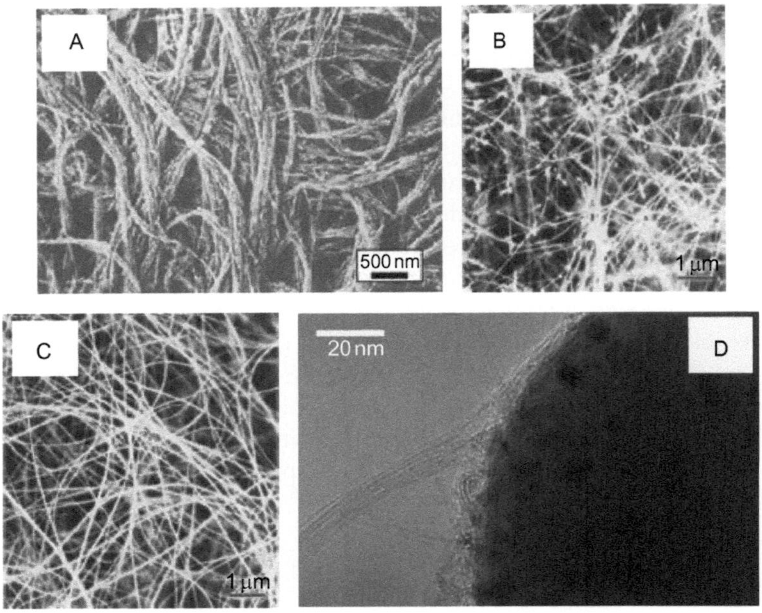

FIG. 9. TEM images of (A) bundles of MWCNTs produced by AD (adapted from Ohkohchi *et al.*, *Jpn J Appl Phys* 1993;**32**:L1248) Copyright *J Appl Phys* 1993, The Japan Society of Applied Physics; (B) and (C) SWCNTs produced by AD in the presence of hydrogen before (B) and after (C) purification (adapted from Ando *et al.*, *Jpn J Appl Phys* 1998;**37**: L61) Copyright *J Appl Phys* 1998, The Japan Society of Applied Physics; (D) SWCNTs produced on supported CoMoCAT catalyst nanoparticles (from http://www.swentnano.com/tech/what_is_comocat.php). (For color version of this figure, the reader is referred to the Web version of this chapter.)

adjustment to ensure that the separation with the cathode is maintained constant. Additionally, both the cathode and the usually smaller anode are water-cooled during the synthesis.

The synthesized material can be typically found deposited in the cathode after several minutes of anode evaporation. The appearance of the as-grown material is that of carbonaceous powder, but inspection using *scanning electron microscopy* (SEM) reveals the presence of needle-like structures (CNTs) in the carbon soot.[109] The characteristics of the produced CNTs and their proportion in the carbon soot can be modified by altering the synthesis conditions. For instance, changing the proportion of helium and argon used in the apparatus has been shown to alter the diameter distribution of the produced CNTs.[113] More importantly, the selectivity type I can be altered depending on the exact composition of the anode, wherein the presence of metal atom has been shown to be indispensable for the synthesis of SWCNTs.[114]

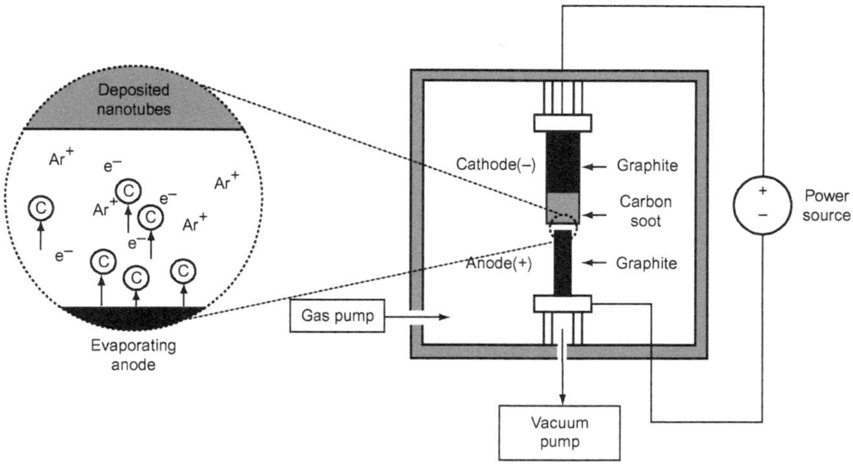

FIG. 10. Schematic (non-scaled) of the AD process based on Zhao *et al.*, *Bull Res Inst Meijo Univ* 1996;**1**:7. A schematic of the conditions in the gap between electrodes (circled region) is shown on the left.

Using pure graphite electrodes in a variety of conditions produces a majority of MWCNTs, as shown by several experiments in the early 1990s. For instance, the observed diameter distribution (4–30 nm) in the samples reported by Iijima[1] suggested the formation of MWCNTs. Similarly, Ebbesen and Ajayan[113] observed nanotubes in the form of MWCNTs in samples produced in a variety of inert gas compositions, pressures, and dimensions of the electrodes. In one instance, the selectivity was found to monotonically increase with pressure up to 67 kPa. Further pressure increase up to 333 kPa maintained constant the selectivity, while decreasing the relative yield. Alternatively, Ohkohchi *et al.*[115] demonstrated that doping the graphite electrodes with rare earth metals such as, yttrium, scandium, and lanthanum, increases the relative yield possibly due to catalytic effects of the molten metal. On the other hand, Ando *et al.*[110] demonstrated that the addition of methane improved the selectivity of MWCNTs by inhibiting the formation of fullerene, and reducing the number of other carbon species.

In 2003, a modified method was proposed to increase the economical viability of the AD synthesis of MWCNTs.[116] Briefly, the carbon electrodes are submerged in liquid nitrogen within a container, where the nitrogen eliminates the need of creating vacuum conditions, setting a cooling system for the electrodes, and using expensive inert gases. In similar fashion to normal AD, the product is deposited in the cathode, which is separated by approximately 1 mm from the anode. The applied voltage is in the range of 20–27.5 V, resulting in a current of up to 80 A.[116] Inspection of the nanotubes reveals high-quality structures and no traces of nitrogen. The selectivity was found to be 70%, which facilitates further purification of the sample. The purification of the MWCNTs is also simplified by the absence of catalyst. Accordingly, modified AD synthesis appears as a potential alternative for scaled-up production of MWCNTs.

Increase in the quality of the MWCNTs can be achieved by introducing a magnetic field in AD synthesis.[117] However, although large-diameter MWCNTs are attractive as a host of pharmaceutical species to be delivered at a specific location, SWCNTs posses a broader spectrum of potential nanomedical applications, hence the importance of synthesizing the latter as well. The incorporation of a metal catalyst in either the cathode or the anode is mandatory for the synthesis of SWCNTs using AD. The first SWCNTs synthesized by this method used iron or cobalt as catalyst, but only a low yield was achieved.[114,118] In one instance, a small piece of iron incorporated in the anode melted into a droplet and then vaporized to further deposit as iron carbide onto the cathode.[118] Additionally, the presence of argon and methane was required for the production of SWCNTs, which appeared as bundles with a wide diameter distribution. A difference with the formation of MWCNTs is that SWCNTs do not appear in the normal deposit on the cathode, but rather in other parts of the chamber.[118]

The yield of SWCNTs has been shown to improve by using bimetallic combinations such as nickel and yttrium.[119] Also, modifications in the setup, such as positioning the electrodes at an angle other than 180° are able to reduce the deposition in the cathode, while increasing the deposition in locations where SWCNTs form, up to a rate of 1 g/min.[120] This modified method is named arc plasma jet (APJ) due to the characteristic arc plasma generated. The presence of metal nanoparticle heavily coated with amorphous carbon makes the purification of the sample more tedious. To reduce the amount of amorphous carbon, hydrogen can be added to the chamber, resulting in a web structure containing SWCNT bundles and metal catalyst thinly covered with amorphous carbon.[120] Since the addition of hydrogen somewhat destabilizes the arc plasma, heavy inert gases such as krypton and xenon may be added to optimize the process.[121]

Inspection of the SWCNTs produced with AD typically reveals the presence of tubes with diameters between 1.2 and 1.4 nm. However, the diameter can be moderately controlled by altering the inert gas nature, which changes the thermal and mass diffusion rates.[122] It was shown that increasing the argon fraction in the inert gas mixture resulted in decreasing the nanotube diameter.[122] Similarly, control on the diameter of the SWCNT has been attempted by the use of different bimetallic combinations, which changes the diameter of the nanoparticles created by metal vaporization. For example, the use of a mixture of cobalt and molybdenum proved to result in thinner SWCNTs in the range of 0.6–1.2 nm.[114]

Raman spectroscopy reveals that AD synthesis results in SWCNTs of high quality in terms of crystallinity and density of structural defects. On the other hand, although the relative yield has been improved using bimetallic catalysts, or changing the geometrical arrangement of the electrodes (i.e., angle and distance), the amount of nanotubes produced is relatively small.[114,118] Including a magnetic field during AD synthesis has been shown to improve the quality, length, purity, and quantity of SWCNTs produced due to a change in diffusional patterns driven by changes in plasma density.[112]

Further improvement is needed for AD to be able to produce nanotubes at a large scale. However, the AD method is widely used to provide nanotubes in the quantities required for laboratory research, wherein their high quality is well regarded. Therefore, since nanotubes produced through AD usually require extensive purification, it is important to design a purification strategy wherein the quality of the nanotube is maintained during the elimination of species such as amorphous carbon and metallic nanoparticles, among others.[109]

B. Laser Ablation

LA was the second process utilized for the synthesis of CNTs. The process was developed by Smalley in 1995, with the intention of improving the purity of the final product.[123] Briefly, the method relies on focusing an energetic laser beam on an ablation target that is majorly made of carbon and small amounts of transition metals, which results in a laser plume as a result of the target vaporization.[123] Thus, the laser plume contains vaporized carbon and metallic nanoparticles that lead to the reassembling of carbon in the form of nanotubes, in similar fashion to AD. Nonetheless, a major difference is that AD synthesis has been shown to produce both MWCNTs and SWCNTs, whereas LA synthesis has only been shown to produce the latter. The formation of metallic catalyst nanoparticles appears to play a fundamental role in the growth mechanism of nanotubes produced using LA. Therefore, the laser-induced vaporization of a pure carbon target *does not* yield nanotubes at all. On the other hand, as discussed in relation to AD synthesis, SWCNTs were favored when metal was involved in the synthesis. Accordingly, SWCNTs are expected to be preferred over MWCNTs whenever LA synthesis is utilized.[124]

In the original design (Fig. 11), concentric tubes of 25 and 56 mm of diameter are placed inside a furnace, wherein the whole system is heated up to 1500 K.[123] An inert gas flows through the small tube, with the ablation target located in front of it.[123] The laser beam travels through the tubes and over the target, resulting in the formation of the LA plume. The relative positions of the smaller tube and the target allow the spatial confinement of the plume, reportedly improving the conditions for nanotube formation. Subsequent designs are derived from the original one, wherein geometrical or operational

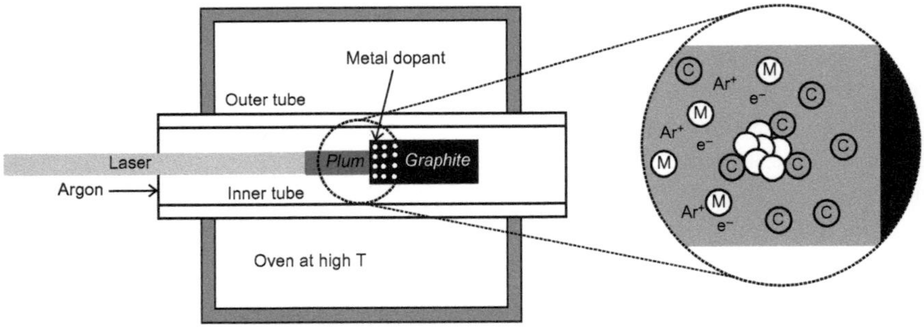

FIG. 11. Schematic (non-scaled) of the LA process based on Smalley *et al.*, *Appl Phys A* 2001;**72**:573–580. A schematic of the conditions in the laser plume (circled region) is shown on the right.

modifications can be introduced. For instance, a particular wavelength for the laser may be preferred, as well as the use of pulsed laser over a continuous one.[125,126]

The time scales for the different events involved in nanotube growth depend on whether a continuous or a pulsed laser is utilized. For instance, the selection of a continuous laser results in a synthesis controlled by heat and mass transfer processes, in similar fashion to AD. Accordingly, reaction conditions that increase the relative yield in AD have been shown to do so for LA with a continuous laser.[125] On the other hand, experiments using a single pulsed laser—typically a *Nd:YAG* (neodymium-doped yttrium aluminum garnet) laser—have shown that the yield increases with the frequency of the pulse, perhaps because higher frequencies increase the average energy input resulting in higher ablation rates.[126] In 2001, Zhang *et al.*[127] studied the impact of the laser power on the nanotube yield. A continuous CO_2 infrared laser was used to ablate a carbon–nickel/cobalt target in an argon atmosphere. It was shown that a minimum power of 500 W was needed to vaporize the target. Additionally, power increases above 800 W resulted in a significant increase of the yield.[127]

There is no conclusive evidence regarding the effect of the operation mode (continuous or pulsed), frequency, wavelength, or power of the laser on the nanotube diameter. Variation of the frequency between 0.01 and 10 Hz of the Nd:YAG laser does not significantly alter the nanotube size.[126] Similarly, increasing the power from 600 to 700 W of the CO_2 laser only increases the diameter from 1.2 to 1.4 nm.[127] On the other hand, independent LA experiments differing in the laser wavelength used also differ in the nanotube diameter obtained, but it is possible that additional differences in the experimental setups also contribute to modify the diameter.[127–129] Thus, the relation between wavelength and nanotube diameter is deemed unclear. Nonetheless, the temperature of the oven has been shown to affect the nanotube diameter, wherein larger oven temperatures produce nanotubes of larger diameter.[130]

Similarly to AD synthesis, the pressure and composition of the inert gas play a significant role in LA synthesis. Successful formation of nanotubes has been reported with argon and nitrogen atmospheres at different pressures, but growth with a helium atmosphere was reported to be unsuccessful.[131] A high yield was obtained when any of the two former gases was used in the pressure range of 27–53 kPa.[131] Specific conditions of temperature and plume density are necessary to promote nanotube growth, wherein nanotubes have been shown in the plume high-density/high-temperature zone nearby the target.[124] Thus, the gas pressure affects the yield due to the effect on the plume shape and density, wherein elongated and diffuse plumes resulting from a low gas pressure do not favor a large production of nanotubes.[131]

The development of nanotube synthesis by LA brought about the development of theoretical models to increase the understanding of nanotube growth. The earliest model proposed by Smalley *et al.* was the *scooter model*.[71,123] It was proposed that metal atoms attached at the end of growing fullerene structures to prevent the closure of the structure, which is in agreement with the observed formation of fullerenes for the ablation of the pure target. A particular atom is envisioned to *scoot* around the open end, adsorbing carbon atoms and incorporating them into a tubular structure, forming SWCNTs as observed for the ablation of a metal-doped target. Since doping the target with two metals results in a yield increase, it suggested an increased mobility of the scooting atoms due to the presence of two different metallic species. However, further studies have casted doubts on the scooter model.[124]

Later, another model was proposed, wherein the vaporized carbon and metal atoms aggregate into molten carbon–metal droplets, based on the eutectic diagram of binary carbon–metal systems.[130] The nanotubes grow from the segregation of carbon atoms in the supersaturated droplets at ablation temperatures. The size of the droplets can be relatively large (10–50 nm), which allows the growth of several nanotubes from a single droplet.[132] Therefore, there is no evident connection between the catalyst size and the diameter of nanotubes produced under LA. Thus, the most frequently found nanotube diameter ($\sim$ 1.4 nm) is favorably grown as a balance between strain and curvature energy.[133] There are large and transient temperature and concentration gradients in the plume, wherein the windows for effective growth conditions are relatively narrow. In consequence, the catalyst only performs effectively for a relative short time, resulting in relatively short nanotubes.[124]

Similarly to AD, LA can be used to produce high-quality nanotubes in quantities suitable for research work. Further efforts to scale up the process are necessary to achieve synthesis of larger quantities. Nonetheless, it is possible that the cost of operating the laser at larger scales becomes prohibitive rendering the scale-up unfeasible.[2]

C. Chemical Vapor Deposition

Historically, CVD had been widely used to catalytically decompose hydrocarbon gases such as methane and ethylene to obtain different carbon products such as carbon fibers.[134] Currently, CVD has become one of the preferred methods for production of CNTs since many researchers regard CVD as the only viable approach for mass production.[135,136] Briefly, a carbon-containing precursor gas is catalytically decomposed on metallic nanoparticles. Residual gas diffuses away, whereas freed carbon atoms dissolve into the nanoparticles and then segregate to the catalyst surface to form nanotubes. Since it is not necessary to vaporize a solid carbon source, the temperature at which CVD occurs (900–1200 K) is lower than that for AD and LA.[23,24,137] It is possible to obtain either SWCNTs or MWCNTS using CVD, depending on the reaction conditions and the catalyst preparation.[104]

Catalyst preparation precedes CVD synthesis, playing an important role in the growth reaction since it affects the structure of the nanoparticle before the reaction commences. There is a connection between the nanoparticle diameter and both the diameter and type of nanotubes produced. Usually, nanoparticles smaller than 3 nm lead to the formation of SWCNTs, whereas nanoparticles larger than 3 nm lead to the formation of MWCNTs.[105,136] Additionally, during CVD synthesis each nanoparticle gives rise to only one nanotube; therefore, the larger the nanoparticle size, the larger is the diameter of the produced nanotubes. The catalyst can be prepared to perform the CVD with a floating catalyst or a supported one, while targeting a given diameter for the synthesis.[138]

Methane, acetylene, ethanol, and carbon monoxide are commonly used as precursors.[23,24,139] Iron, cobalt, and nickel[24,139] are among the most common catalytic materials, although copper, gold, and silver are also used under the right conditions.[137] Similarly, silicon oxide, magnesium oxide, and aluminum oxide are typically used to support the catalyst.[24,139] A particular metal catalyst can be applied on the support through metal sputtering or by impregnating the support with a precursor saline solution containing the metal.[2] Then, the formation of nanoparticles is induced through chemical or thermal annealing, which can include oxidation or reduction steps in the presence of oxygen or hydrogen followed by heating in the presence of an inert gas atmosphere.

Among other factors, the nanoparticle size depends on the nature of the selected metal(s), composition and pressure of chamber gases, and nature of the support, as well as rate, duration, and temperature of heating. Thus, careful control of these variables can significantly control the diameter of the catalyst, and hence the diameter of the produced nanotubes. The diameter of the nanoparticles for CVD using the floating catalysts can be controlled during their fabrication through gas-phase synthesis before the CVD.[140] Notice that according to Eq. (1), a moderate control of the chirality can be achieved by controlling the nanotube diameter.

Two possible mechanisms occur for the nanotube growth, namely, tip-growth and base-growth mechanisms.[141,142] In the tip-growth mechanism, carbon precipitates from the nanoparticle to form a nanotube cap. However, the cap does not lift up from the nanoparticle during growth. Nonetheless, the carbon atoms are still incorporated into the nanotube structure, thus increasing the nanotube length, and consequently displacing the nanoparticle away from the support.[141] It appears that the tip-growth mechanism is favored for unsupported nanoparticles and for supported large nanoparticles weakly adsorbed on the support.[143] On the other hand, a small nanoparticle strongly adsorbed on the support is held in place during nanotube growth favoring the base-growth mechanism.[142] In the base-growth mechanism, carbon precipitates to form a nanotube cap, similar to the tip-growth mechanism, but the cap detaches from the nanoparticle while carbon is added to the nanotube "open" end at the metal–nanotube interface.[142]

Nanotube CVD synthesis with a supported catalyst is a heterogeneous process that can be performed either using a fixed-bed or fluidized-bed reactor. Typically, materials such as zeolites, mesoporous silica, and graphite are used to disperse the catalyst in the fixed-bed method.[144–146] Additionally, more sophisticated synthesis can be performed by lithographic patterning of the catalyst on high-quality silicon, quartz, or sapphire wafers to produce either vertically or horizontally aligned nanotubes.[44,146,147] On the other hand, in the fluidized-bed method, the oxide support is a fine powder impregnated with the catalyst, and suspended due to the upward gas flow in a vertical tubular reactor.[148] Although it is possible to obtain either SWCNTs or MWCNTs with both the fixed-bed and fluidized-bed methods, it is easier to produce SWCNTs with the former.[149] However, an advantage of the latter method is the large space for nanotube growth, and the large specific area for catalysis to occur.[149]

Nanotube CVD synthesis with a floating catalyst is a homogeneous process typically performed in a quartz tube vertically placed in an oven, wherein a first region of the reactor is kept at a lower temperature while the region where nanotubes grow is maintained at a higher temperature (900–1200 K).[136] A gaseous current with the precursor gas and inert gases flows through the reactor. The catalyst can be added to the process through a gas precursor, such as ferrocene,[150] or alternatively through injection of nanoparticles previously prepared, which enables a stricter control of the nanoparticle diameter.[140] A high ratio between the catalyst and the carbon precursor favors the production of SWCNTs, whereas the production of MWCNTs is favored otherwise.[22,135,151,152] Additionally, the yield can be increased through the use of additives such as sulfur compounds, as well as the introduction of oxygen through the use of carbon precursors such as alcohols.[153–155]

Sometimes the CVD process can be modified in an effort to make the nanotube synthesis more efficient. For example, a previously prepared wafer supporting a catalyst is placed on a grounded electrode, whereas a second electrode is placed at a certain distance above the wafer.[156] A carbon precursor gas is added to the chamber where it flows between the electrodes. An external power source induces electric discharge between the electrodes, resulting in plasma formation, which is utilized for the nanotube growth.[156] This plasma-enhanced CVD (PECVD) possesses higher diffusion rates than thermal CVD, resulting in preferential formation of MWCNTs. Similarly, a laser can be introduced to the synthesis in a laser-assisted CVD (LCVD), wherein a continuous CO_2 laser is directed on an oxide wafer.[157] The laser performs pyrolysis of the carbon and metal precursor gases, which leads to the formation of metallic nanoparticles and subsequently to the formation of nanotubes. Similarly to the floating catalyst method, the formation of either SWCNTs (0.7–2.5 nm) or MWCNTs (30–80 nm) can be controlled by tuning the ratio between the carbon and metal precursor gases.[157]

A widely known method for producing nanotubes is the *high-pressure CO conversion* (HiPCO) developed by Smalley *et al.*[22,151] Carbon monoxide (CO) and iron pentacarbonyl ($Fe(CO)_5$) are used as carbon and metal precursor gases respectively. The gas continuously flows through the heated reactor (Fig. 12), where iron nanoparticles are formed and catalyze the growth of SWCNTs.[22,151] Diameter distribution is moderately controlled through adjustment of the CO pressure, wherein nanotubes as small as 0.7 nm can be produced. A yield superior to 70%, and a selectivity toward SWCNT formation (purity) superior to 96% can be achieved through HiPCO synthesis.[22] Therefore, HiPCO is one of the most attractive alternatives for mass production.

The relatively low temperature in CVD synthesis generally results in nanotubes of lower quality than those obtained through LA and AD synthesis. Nevertheless, during CVD, the conditions can be adjusted in order to reduce poisoning and increase the catalyst life, resulting in conditions favorable to mass production of nanotubes. Additionally, CVD can produce longer nanotubes with a more effective diameter control than AD and LA synthesis. Furthermore, both vertically and horizontally aligned patterned growth is possible under the appropriate conditions,[43,44,147] having the potential for in situ fabrication of biosensors and similar devices.[158] Thus, CVD has a high prospective for dominating the commercial production of nanotubes, wherein perhaps the remaining obstacle is the control of the chirality during synthesis.

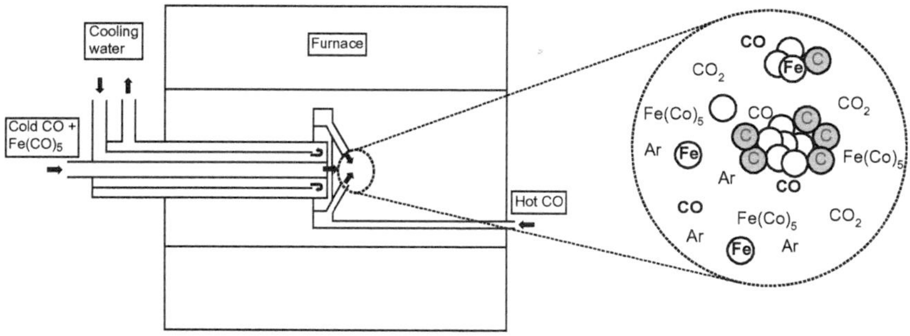

FIG. 12. Schematic (non-scaled) of a CVD process corresponding to the HiPCO process, based on Smalley *et al.*, *Chem Phys Lett* 1999;**313**:91–97. A schematic of the conditions in the reactive zone is shown on the right.

D. Chirality Control

Since there is a connection between nanotube properties and chirality, selectively synthesizing nanotubes of a particular chirality (n,m) is equivalent to selectively synthesizing nanotubes with particular properties. Therefore, chiral selectivity is a desirable feature to achieve in a synthesis process. As previously mentioned, controlling the diameter introduces a moderate chirality control, because it restricts the possible chiralities resulting from the nanotube growth. Thus, since CVD appears to have a superior control of the catalyst nanoparticles than AD and LA, it also has a superior potential for chirality control.[24,107,140]

Interestingly, a glance through some of the most important CVD techniques reveals a tendency to produce nanotubes with a high chiral angle (near-armchair).[22–24] It has been suggested that near-armchair tubes are kinetically favored to increase their length (i.e., formation of new rings in the nanotube wall), because they possess more active sites for the incorporation of carbon dimers—thus creating new rings—than near-zigzag tubes.[32] Furthermore, the growth rate may be proportional to the chiral angle as indicated from the application of crystal growth theory to the nanotube case.[159] Because of the connection between cap structure and nanotube chirality (discussed in Section I), it has been proposed that the nanotube chirality is set during nucleation of the cap structure.[32,160,161]Therefore, the thermodynamic stability of near-armchair caps may play a role in the abundance of near-armchair tubes.[32,162]

Performing CVD with a cobalt–molybdenum catalyst, in the so-called *CoMoCAT synthesis*, it has been possible to produce the near-armchair (6,5) nanotube with a chiral selectivity higher than 70%.[24] In the CoMoCAT synthesis, CO is decomposed by fine cobalt nanoparticles resulting from the preparation (Section III) of a mixture of cobalt and excess molybdenum.

Although the chiralities (9,1), (10,2), (7,5), and (6,5) are all consistent with the observed nanoparticle diameter, only (6,5)—and to a lesser extent, (7,5)—are preferentially synthesized.[24] It could be speculated that the diameter control together with the preferential growth of near-armchair chirality results in the preferential growth of (6,5) and (7,5); however, the selection mechanism is not clearly understood.[24] Interestingly, a selective synthesis of the larger (9,8) nanotube—another near-armchair tube—has been reported recently using a cobalt-based catalyst.[163]

Nonetheless, recent studies have suggested the possibility of achieving chirality control through a template effect (Fig. 13) between the catalyst and the nanotube.[107,140,164] In 2008, Zhu *et al.*[165] using high-resolution transmission electron microscopy (TEM) reported a correlation between the facet orientation of a cobalt nanoparticle and the chirality of the corresponding nanotube. In 2009, Chiang *et al.*[140] reported a floating catalyst synthesis obtaining different chiral distributions. Changes of the chiral distributions resulted from varying the composition of nickel–iron nanoparticles of *identical*

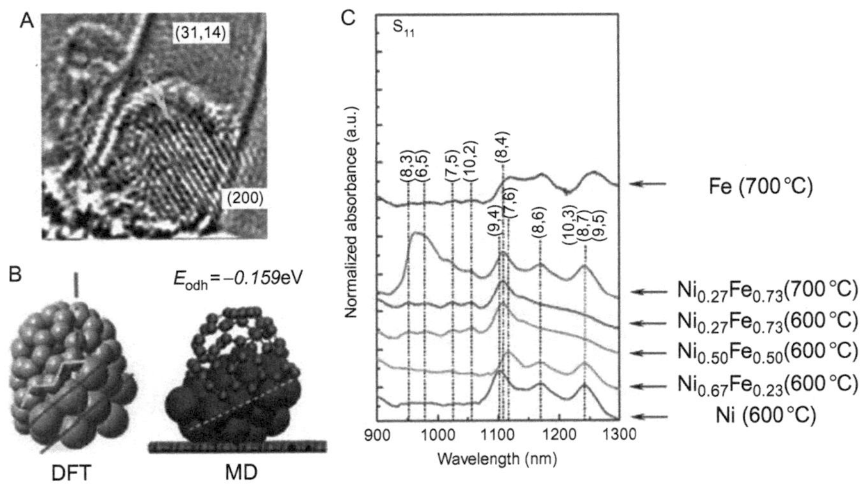

FIG. 13. (A) TEM image showing the observed correlation between the orientation of nanoparticle crystallographic planes and the nanotube chirality. (Zhu *et al.*, *J Crystal Growth* 2008;**310**:5473–5476.) (B) On the left, geometry of a nanotube cap on a metallic nanoparticle optimized using density functional theory showing a similar correlation between nanoparticle structure and nanotube chirality; on the right, snapshot of a reactive molecular dynamics simulation shows a similar trend, with the nanoparticle structure influenced by the catalyst support. (Gomez-Gualdron *et al.*, *J Chem Phys* 2011;**134**:014705.) (C) Variation in chiral distribution on CVD-produced nanotubes upon nanoparticle structural changes (achieved through compositional changes) shown by variations in the Raman signal of the samples. (Chiang and Sankaran, *Nat Mater* 2009;**8**:882–886.) (See Color Insert.)

diameter, wherein the variations are connected to structural changes detected in the nanoparticles.[140] In the same year, Harutyunyan *et al.*[107] reported the preferential synthesis of metallic nanotubes (> 90% selectivity) as a result of a catalyst treatment modifying the morphology of iron nanoparticles. Iron nanoparticles with sharp edges and steps resulted in metallic nanotubes, whereas rounded ones did not result in selective growth.

Some theoretical calculations support the possibility of inducing a template effect on the nucleating nanotube.[164,166] Since step formation has been observed on the nanoparticle during nanotube growth, stepped surfaces can be used to model the growth of the nanotube at the metal–carbon interface. Norskov *et al.*[166] showed the stabilization of graphene in stepped metallic surfaces, as well as the epitaxial effect induced by the latter. Similarly, Balbuena *et al.*[164] demonstrated the preferential formation of either armchair or zigzag structures as a result of different surface structures. Additionally, it was suggested that for the template effect to be possible at the synthesis temperature (1000 K), the support may help stabilize the nanoparticle structure.[164] It was shown that different adsorption strengths between the nanoparticle and the support preserve the nanoparticle structure to different extents.[164]

It has also been suggested that different reaction environments may enhance the production of particular chiralities. Liu *et al.*[106] have reported the preferential synthesis of semiconducting nanotubes (> 90% selectivity), wherein appears an apparent correlation between the mechanism of reaction and the chiral distribution. Namely, the presence of methanol in the gas mixture used for the growth of horizontally aligned nanotubes results in a selective synthesis.[106] It was hypothesized that the methanol decomposition gives rise to hydroxyl radicals that etch the nascent metallic nanotubes, resulting in semiconducting tubes outgrowing the metallic ones.[106]

Systematic chirality control remains elusive. However, the tremendous benefits of synthesizing nanotubes with particular properties at large scale warrants continued research in this field. Completely understanding the different growth mechanisms is a necessary stepping stone in the quest of chiral selectivity. Since the details of these mechanisms are still under debate, there is a long road ahead before the chirality can be easily manipulated. Meanwhile, obtaining specific chiralities for particular applications depends on separation methods wherein a specific chirality can be withdrawn from a heterogeneous nanotube mixture.

IV. Post-Synthesis

In the previous sections, several SWCNTs synthesis methods were described in detail, and it was emphasized that despite all the efforts carried out for improving and optimizing the growth processes, complete control over important structural variables such as chirality, diameter, and length has not yet been reached. Biomedical applications require availability of highly purified and monodisperse SWCNTs, particularly semiconducting nanotubes with specific band gaps which make them potentially useful for biosensors and other biological uses. Furthermore, compatibility of SWCNTs with living organism environments is only achieved after a successful functionalization process that reduces the hydrophobicity of nanotube side wall surfaces and therefore their toxicity. The following section discusses the most effective methods employed to purify and separate SWCNTs according to their conductivity, chirality, and diameter, as well as the most remarkable functionalization reactions known so far.

A. Purification Methods

Although some processes have reached a high level of selectivity toward SWCNTs, further purification processes are still needed in order to satisfy the requirements of highly pure SWCNTs for the majority of applications. All the synthesis processes employed for mass production of SWCNTs involve catalytic decomposition of precursor carbon compounds. Multiwalled carbon nanotubes, graphite, and carbonaceous material are examples of a wide variety of carbon species beside SWCNTs that may grow on the catalyst particle, and in addition to the presence of the metallic catalyst they become potential impurities for SWCNTs. Thus, all these residual impurities remaining from the growth process must be removed; however, the type of impurities and their abundance depend upon the synthesis process used, and therefore they require the use of specific purification techniques.[167]

Although AD methods are capable of synthesizing MWCNTs without being assisted by a metallic catalyst, nucleation of SWCNTs through these methods requires the presence of catalyst particles which together with unwanted carbonaceous products would become relevant impurities to be removed after synthesis. Thus, two main approaches have been developed to purify CNTs synthesized from AD. On the one hand, *gas phase oxidation* is employed for purifying MWCNTs by means of a chemical attack with oxygen to the capped ends of the nanotubes.[167] Capped ends are the CNT segments most susceptible to be chemically attacked due to the presence of carbon pentagons which introduce high curvature to the ends and therefore high reactivity. Addition of oxygen or CO_2 at high temperatures on the order of 700 °C leads to a complete oxidation of the caps as well as that of other undesired carbon species, followed by a thinning process which results from detaching of outer layers from the multiwalled nanotubes.[168,169] Unfortunately, this oxidation process ends up eliminating about 99% of the total weight of CNT samples. On the other hand, *liquid phase oxidation* appears to increase the purification yields and is regarded as the most important approach to successfully purify

SWCNTs produced by AD methods.[170] The action of strong liquid oxidants like potassium permanganate, nitric acid, sulfuric acid, or potassium permanganate solutions in acid may reduce the weight losses from 99% to 60%. Further purification treatments must be carried out after oxidation in order to remove the residual oxidant; generally, filtration and washing with water and hydrochloric acid are employed for complete purification. Oxidation in liquid phase is useful for purification of SWCNTs. For instance, chemical action of nitric acid refluxed more than once is able to remove the metallic catalyst particle attached to the tip of SWCNTs produced by AD. This method was optimized by Dillon *et al.*,[171] achieving a high level of purification above 98%, even though the long times of acid exposure resulted in a great proportion of damaged SWCNTs. However, caution should be taken when extracting impurities through oxidation since undesirable carboxylic functional groups might be produced and adhere to the CNT side wall defects and open ends, representing a relevant issue for biomedical applications of SWCNTs, as will be shown in a later section.

SWCNTs synthesized by LA methods come generally accompanied by carbon encapsulated nanoparticles, carbon nanospheres, fullerenes, and metallic nanoparticles. Thus, multistep purification methods have arisen to fractionate SWCNTs, separating them from their side products.[172,173] Most of the processes proposed for purifying laser vaporization SWCNTs involve the renowned sonication process, which consists in the application of ultrasound waves capable of breaking non-covalent bonds, separating metallic particles and carbon nanospheres from carbon nanotubes. In addition, sonication necessarily requires the assistance of a microfiltration procedure where the nanotubes are collected in the filter whereas metal and metal carbide nanoparticles are filtered through the micropores of a permeable membrane. Although the above-described process is the basis for purification of LA SWCNTs, variations may be introduced in order to improve yields. Thus, some techniques introduce a first stage of filtration before sonication where a specific solvent is employed to separate insoluble nanotubes and nanoparticles from other soluble fullerenes, with toluene and CS_2 among the most common solvents.[172] Other purification procedures take advantage of the liquid acid oxidation followed by subsequent centrifugations and decantations for removing the residual acid, and before performing the sonically assisted filtration.[174] Thus, SWCNTs obtained after applying these purification procedures can be above 90 wt% of purity. Nevertheless, filtration procedures require a large number of repetitions in order to reach these high purity levels, but several consecutive filtrations unavoidably lead to material loss. Therefore, other non-oxidant methods appear to separate SWCNTs, MWCNTs, metallic particles, and other carbon forms formed after synthesis, minimizing the amount of material wasted during the process. Size exclusion chromatography (SEC) is a well-known method useful for the separation of biological macromolecules also applied to fractionate MWCNTs by length.[175,176] Moreover, it was found that SEC is able to fractionate heavier SWCNTs from lighter fullerenes and catalyst particles, although a further centrifugation step is generally needed to completely remove the metallic particles attached to the nanotube.[177]

As discussed in Section III, CVD growth processes have reached the highest selectivities toward SWCNTs. However, despite the efforts in optimization and the significant improvements achieved in selectivity, purification processes are still needed. Growth of SWCNTs through CVD methods involves assistance of catalytic particles yielding similar side products as those produced by LA or AD methods. Therefore, the purification processes for CVD-grown SWCNTs are quite similar to those described above. They consist basically of one of the four main procedures: gas phase oxidation, acid oxidation, ultrasonically assisted filtration, chromatography, or combinations of some of them. Moreover, washing steps are frequently introduced before and after filtration stages, leading to high levels of purification, with toluene and acids as washing liquids at initial stages and deionized water at final stages. Li *et al.* proposed a multistep process for purifying SWCNTs.[178] This method involves two of the four main purification procedures: acid oxidation and filtration assisted by sonication. Initially, the raw sample of SWCNTs is washed in benzene where benzene-soluble compounds such as fullerenes and catalyst particles are extracted. This step is followed by subsequent sonically assisted filtrations supported by dissolution of catalyst particles in hydrochloric acid, thus removing catalyst particles and amorphous carbon from nanotubes. To break some residual carbon nanofibers and MWCNTs, the sample is quenched at low temperatures in liquid nitrogen between both filtrations steps. Finally the sample is washed in deionized water and SWCNTs of 95 wt% of purity are collected. Alternatively, other methods suggest a phase gas oxidation in Ar/O_2 at high temperatures at the initial step followed by an acid oxidation.[179] The presence of the catalyst is found to catalyze not only the growth of CNTs, but also the oxidation of carbon at the cap–metal interface. Thus, the carbon shell is broken apart and the hydrochloric acid is now able to oxidize and remove the catalyst particle. Thermogravimetric analysis provides useful information about the purification mechanism. The TGA plot shown in Fig. 14 illustrates the time/temperature dependence of the percentage weight loss.

Oxidation of the metal particle is related to the weight increase during the initial minutes; however, a continuous decrease of the temperature-dependent weight is observed afterwards as a result of detachment of carbon shells. The subsequent action of the HCl-driven oxidation can be alternated with ultrasonication to remove the oxidized catalyst particles followed by microfiltration and washing with deionized water. Finally, further approaches have been suggested to purify SWCNTs produced by catalytic decomposition of hydrocarbons. For instance, HiPCO nanotubes may be purified through organic

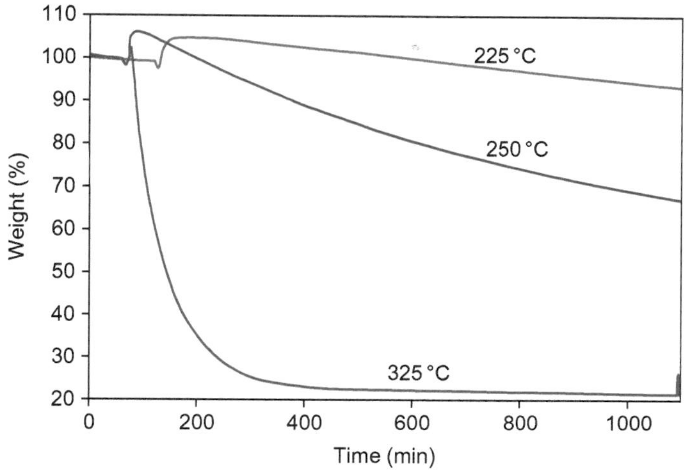

FIG. 14. Thermogravimetric analysis of HiPCO nanotubes treated at different temperatures in a wet Ar/O_2 environment. (Adapted from Chiang *et al.*, *J Phys Chem B* 2001;**105**:8297–8301.) (For color version of this figure, the reader is referred to the Web version of this chapter.)

functionalization of raw SWCNTs side walls and solubilization in organic solvents such as dimethylformamide.[180] The catalyst particles remain insoluble in this kind of solvent whereas solubilized nanotubes must be subjected to a final stage of elimination of the organic functional groups that adhere to the nanotube walls. On the other hand, other methods recall the widely used gas phase oxidation, however, introducing microwave local heating aiming at preferential oxidation of carbon covering catalyst particles, and therefore avoiding SWCNT damage.[181]

Purified SWCNTs are generally characterized by TEM, Raman spectra, and UV–vis-NIR spectra among other techniques. Figure 15 shows a comparison between as-received CNTs and purified CNT Raman curves. Intense Raman active modes at 1500–1600 cm^{-1} are related to the presence of symmetrical tubes, mostly armchair, and zigzag in a lower proportion. The notorious reduction of peak intensity at $\sim$1300 cm^{-1} after purification procedures is associated with the removal of amorphous carbon and graphitic structures during the process.

B. Separation Methods

Since many biological applications require specific types of SWCNTs, a further fractionation is eventually needed in order to obtain high concentrations of tubes with defined characteristics. A correct application of the above-described purification methods achieves the separation of SWCNTs

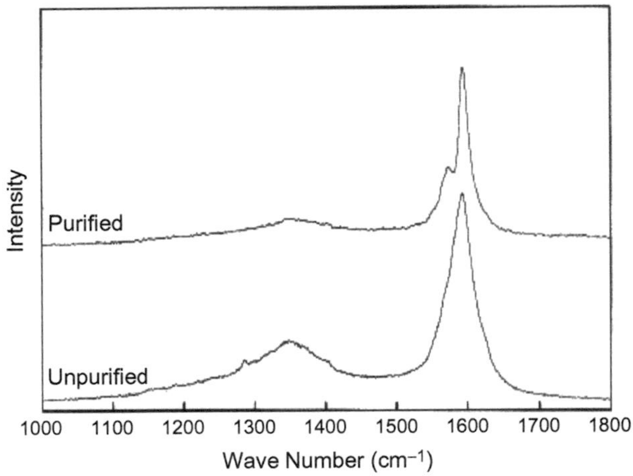

FIG. 15. Raman spectra of SWCNTs before and after purification with a laser wavelength of 514.5 nm. (Reprinted from Li *et al.*, *Carbon* 2000;**38**:2041–2045.)

from other carbon species and catalyst particles. However, even a collection of 100% purified carbon nanotubes may not find useful applications since the growth methods known to date are not highly selective toward specific structures, and the physical and chemical properties of SWCNTs are known to be mainly dependent on chirality and diameter, though the length represents a critical issue for some biomedical applications. Furthermore, availability of a mixture of SWCNTs with a wide chirality and diameter distribution makes the presence of both semiconducting and metallic tubes within the bundle highly probable. As shown later, some biological applications require semiconducting nanotubes exclusively, and even sometimes with specific band gaps. Thus, sorting SWCNTs according to their principal structural variables and/or electronic behavior constitutes a key step toward the final adaptation of SWCNTs to biological systems. Electrophoresis, chromatography, and density gradient, which are among the most efficient methods known to achieve these objectives,[182] are discussed in detail in the following section.

1. Separation by Electrophoresis

Detection of specific segments of DNA fragments for structural studies,[183] separation of proteins,[184] and detection of polymorphisms in DNA[185] are some of the applications of electrophoresis in the biochemistry area. In a similar fashion as electric fields are applied to fractionate DNA sections by size, SWCNTs can be separated by diameter, length, and electronic type by means of electrophoretic methods. Electrophoretic separation of SWCNTs can be approached dissimilarly according to the kind of electric field applied and solvent medium, and the desired type of fractionation (e.g., by diameter, length, or electronic type) will depend on the selected approach. Thus, alternating current (ac) dielectrophoresis method is usually employed to sort individual SWCNTs by electronic type.[186] As discussed in Section III, SWCNTs can be classified into three main groups: metallic, semiconducting with small band gaps or quasimetallic ($n - m =$ multiple of 3), and semiconducting with large band gaps ($n - m \neq$ multiple of 3). Application of a variable electric field to SWCNTs in aqueous solution is found to induce a dielectrophoretic force capable of exerting a translational movement on certain types of SWCNTs. The dielectrophoretic force is a function of the dielectric constant of the solvent and directly proportional to the dielectric constant of nanotubes, whereas the latter is in turn inversely proportional to the square of the band gaps of semiconducting nanotubes. Hence, as the nanotube band gap approaches zero, the dielectric force becomes more positive and the nanotubes are strongly attracted to the microelectrodes of the electrophoretic dispositive; this is the case of metallic and quasimetallic SWCNTs. Conversely, semiconducting nanotubes with large band gaps register small dielectric constants and negative dielectrophoretic forces, and consequently they remain suspended in aqueous solution. The effectiveness of this separation technique is quantified through the analysis of Raman spectra of electrophoretically deposited nanotubes, and nanotubes without any sorting method applied. The relative height of the peaks and the characteristics of the RBMs and G-modes, before and after electrophoresis, are irrefutable evidence of a metallic enrichment in the sample.[186] This method achieves the deposition of samples of up to 80% of metallic nanotubes, and this percentage can be enhanced if the frequency of the electric field is increased. Nevertheless, it must be noted that the method only applies to individual SWCNTs, not to those arranged in bundles, because the presence of a single metallic tube in a semiconducting bundle can provoke the electrophoretic deposition of the whole bundle on the electrode.

On the other hand, application of a direct current (dc) electric field to SWCNTs in a capillary, gel, or solution medium can achieve the sorting of SWCNTs by diameter and/or length, or even separate single nanotubes from bundles. Capillary electrophoresis (CE) is carried out to obtain the fractionation of SWCNTs by length as a result of differences in mobility in solution when the dc electric field is applied.[187] The SWCNTs pristine sample is prepared by ultrasonication and suspended in a solution of SDS. SDS provides negative charges to the hydrophobic side walls of the nanotubes and the concentration of the negative charges is directly proportional to the surface area of the tubes, with the surface area in turn being directly proportional to the tube length and diameter. Therefore, when the sample is passed through the electrophoretic device, the flow (mobility) of negatively charged nanotubes is delayed (becomes higher) and the shorter nanotubes elute first, whereas the longer SWCNTs elute later due to their high concentration of negative charges on the side walls. Bundles are also separated from individual nanotubes by this method,[188] since they exhibit higher diameters than individual tubes of the same length and hence larger surface areas and larger amount of negative charges. Alternatively, agarose gel electrophoresis is applied to separate SWCNTs by diameter and length.[189] This method is assisted by a diameter-selective ultrasonication which produces longer nanotubes when the diameter is small, whereas shorter nanotubes are deposited when the diameter is large. The electrophoretic mobility of the tubes through the gel depends on the size of the nanotubes and therefore this method leads to separation by length that necessarily implies a diameter fractionation. Other approaches include the formation of nucleic acids/SWCNTs complexes,[190] where electrophoretic mobility in the gel depends on diameter and length among other parameters.

2. Separation by Chromatography

As mentioned previously, SEC has become a useful tool to purify SWCNTs from their side products in addition to its main application, which is sorting MWCNTs and SWCNTs by length.[176,177] SWCNTs produced by either LA or AD, and initially treated by ultrasonication, are passed through a chromatographic column. The smallest tubes flow faster through a porous glass bed, while the rest of the tubes are fractionated along the column according to their size. Since the small carbon nanoparticles, fullerenes, and catalyst particles flow faster, and the smallest nanotubes and all the other carbon compounds not removed during sonication remain suspended at the top of the column, this method becomes applicable for purification of SWCNTs.

In a similar way as nucleic acids (DNA or RNA) encapsulate SWCNTs to assist the agarose gel electrophoresis, single-strand DNA (ssDNA) adsorbed on SWCNT walls is capable of assisting the diameter and electronic type separation through ion exchange chromatography (IEC).[191,192] Before performing the chromatography, ssDNA is added to an aqueous solution of nanotube bundles. The flexibility of the DNA macromolecules enables them to wrap around the nanotube's body in a helical shape, and the strong binding energy to side walls compete with the Van der Waals interactions between nanotubes separating them apart into individual SWCNTs. Once the DNA/SWCNTs remain individually suspended in the aqueous solution, IEC is carried out to sort the nanotubes according to their electronic state. ssDNA phosphate groups provide an effective negative surface charge to the SWCNTs walls which is a function of their electronic state. Metallic tubes are found to have less charge density than semiconducting ones; thus, metallic tubes become less electrostatically attracted to a positively charged resin and elute first in the chromatographic column than the semiconducting tubes.

Reduced negative charge densities in metallic nanotubes induce positive screening images charge on the ssDNA/SWCNT hybrid surface. This screening process is enhanced by high salt concentrations in the eluting solution which boost the rapid elution of metallic tubes. In such a way, the electronic state fractionation is accomplished although the wide length distribution of pristine SWCNTs represents a remarkable interference for the electronic separation because short semiconducting tubes may elute together with the metallic ones.[192] In the specific case of semiconducting nanotubes, the elevated negative charge density is due to the low polarizability of their surface in aqueous solution, which is in turn diameter dependent. In consequence, semiconducting ssDNA/SWCNTs hybrid can be also sorted by diameter through IEC because this kind of nanotubes will be differently attracted to the anodal resin according to their diameters.[191] The sequential action of SEC and IEC assisted by DNA has been found to mitigate the effects of length distribution on the diameter and electronic type separation. In addition, an initial length fractionation by SEC has shown to enable a later chirality-specific separation of CoMoCAT nanotubes by ion exchange chromatography.[193]

3. Separation by Density Gradient Ultracentrifugation (DGU)

To date, the most proficient method to separate SWCNTs by either diameter or electronic type is the DGU, reaching up to 99% of electronic-type or diameter-specific nanotubes.[182] Ultracentrifugation methods sorting macromolecules by sedimentation rates have been widely used for biochemical and pharmaceutical purposes.[194] Determination of high-density cholesterol in serum[195] and quantification of human immunodeficiency virus type-1 (HIV-1) in plasma[196] are good examples of biomedical procedures assisted by ultracentrifugation. For separation of SWCNTs, the experimental setup of a DGU process basically consists in a column filled with an aqueous solution with a defined linear density gradient. First, nanotubes are individually suspended in aqueous solution after an ultrasonication process that breaks the bundle. As illustrated in Fig. 16A and B, the buoyant density of the SWCNTs in aqueous solution becomes diameter dependent after the addition of specific surfactants such as sodium cholate (SC) which encapsulate the SWCNTs, reducing their hydrophobicity and enabling diameter-specific sedimentation of SWCNTs along the column after accelerated centrifugation.[197]

Initially, instead of amphiphilic surfactants (like SC), hybrids formed by ssDNA wrapping SWCNTs have been used to perform DGU. The sedimentation constant or buoyant density of ssDNA/SWCNT is diameter dependent and do not follow a simple inverse relation to the nanotube diameter as may be expected. The augmentation in diameter of an ssDNA/SWCNT hybrid will significantly increase the mass respect to the volume increase, contrary to what would happen in the case of an unwrapped SWCNT. Thus, the buoyant density

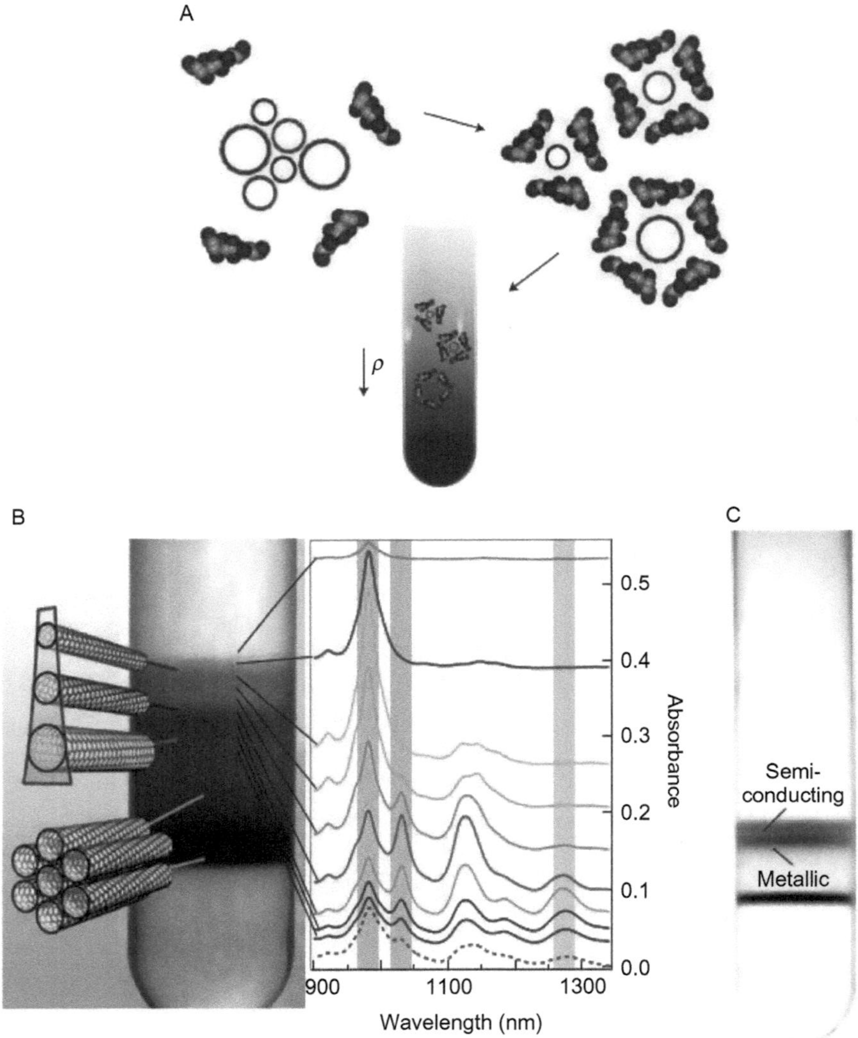

FIG. 16. Sorting SWCNTs by diameter and electronic nature using density gradient ultra centrifugation. (A) Schematic of SWCNTs wrapped by surfactants. (B) Formation of colored bands indicates diameter separation of SWCNTs wrapped by SC surfactant. The optical absorbance spectra demonstrate the high concentration of large-diameter nanotubes at larger densities. (A and B reprinted from Arnold *et al.*, *Nat Nanotechnol* 2006;**1**:60–65.) (C) Electronic nature separation by means of mixture of SC and SDS surfactant mixtures wrapping SWCNTs. (Reprinted from Hersam *Nat Nanotechnol* 2008;**3**:387–394.) (See Color Insert.)

of the hybrid is directly proportional to its diameter and hence a diameter specific sedimentation can be achieved.[198] However, most of the methods involving DNA wrapping, including ion exchange chromatography, have the same problems. Wrapping irreversibility, incapability of sorting tubes of diameters larger than 1.2 nm, low stability in aqueous solutions with density gradients, and expensiveness, constitute the principal disadvantages.

In consequence, other amphiphilic surfactants are preferred over ssDNA for DGU methods because they overcome the limitations just mentioned. Also, if a correct selection or combination of surfactants is made, electronic, specific band gap, and specific chirality fractionation with high yields can be achieved. In addition to SC, other surfactant-built salts have been tested. It is found that although an amphiphilic surfactant such as sodium dodecylbenzene sulphonate (SDBS) is incapable of achieving a diameter fractionation, band gap separation can be obtained by using SDBS encapsulations and carrying out a later ultracentrifugation.[199] Furthermore, the repeated action of the centrifugation process increases the levels of separations of metallic/semiconducting tubes sorting up to 97% of semiconducting nanotubes with a narrow diameter distribution within a range of 0.2 Å. Moreover, an adequate mixture of surfactants optimizes the electronic-type separation (Fig. 16C), as well as being able to achieve chirality separation as explained below. The large polarizability of metallic tubes leads to different interactions with the surfactant in comparison to the semiconducting ones; on the other hand, the hydration of the surfactant is found to be electronic nature dependent. All of this is reflected in variations in the sedimentation constant which allows the electronic-type separation by DGU.

Alternatively, chirality-specific purification of SWCNTs was obtained by combining SC and SDS surfactants in conventional ultracentrifugation extraction. Particular combination ratios of the mentioned surfactants produced selective binding to SWCNTs of specific chiralities, allowing the selective extraction of nanotubes of specific band gaps.[200] The advantages of this method, and amphiphilic surfactant methods in general, include high levels of separation, easy elimination of surfactants through basic washing steps, low cost, and scalability to mass production levels, which have made SWCNTs obtained by DGU commercially available for a variety of applications.

C. Functionalization of Carbon Nanotubes

After purification and separation of carbon nanotubes, functionalization is the next and final step in the treatment to make them useful for nanomedical and biological applications such as drug delivery into cells and sensing of proteins, DNA, and other biological molecules. Functionalization reduces the toxicity caused by the nanotubes' highly hydrophobic surface. Addition of functional groups to the CNT side walls makes the tubular structures soluble in aqueous solutions, less harmless to the organism cells, and therefore biocompatible, allowing the advantage of their excellent physical properties to be used within biological environments. The chemistry of CNTs, and specifically SWCNTs, allows their functionalization by means of either one of the following pathways: covalent side wall functionalization, covalent chemistry at side wall defects or open ends, non-covalent functionalization at side walls, non-covalent polymer wrapping, and molecular insertion inside SWCNTs (Fig. 17).[182]

In the following sections, covalent and non-covalent functionalization approaches are described as they represent the most selective processes in terms of diameter and electronic type, though some of them were already introduced in the previous section because of their important roles in separation methods. For instance, ssDNA-assisted chromatography involves a polymer-wrapping functionalization, whereas SDS encapsulation of SWCNTs is basically a non-covalent functionalization prior to the DGU process.

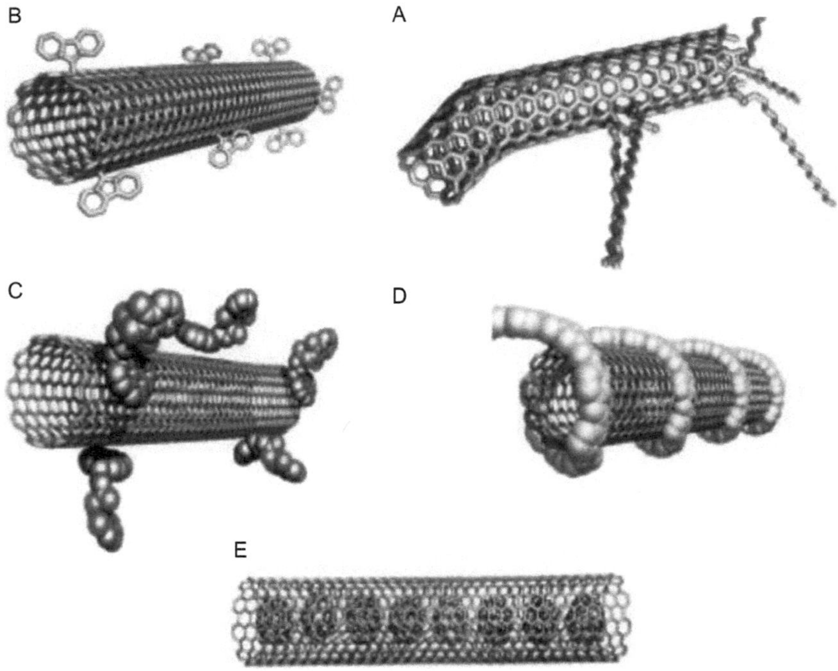

FIG. 17. Possible ways of functionalization. (A) Covalent functionalization at defects or open ends. (B) Covalent side wall chemistry. (C) Non-covalent surfactant encapsulation. (D) Non-covalent functionalization through polymer wrapping. (E) Molecular insertion inside SWCNTs walls. (Reproduced from Hirsch *Ang Chem Int Ed* 2002;**41**:1853–1859.) (For color version of this figure, the reader is referred to the Web version of this chapter.)

1. Covalent Functionalization

Addition of functional groups to the side walls of carbon nanotubes is frequently carried out by chemical reactions between functional organic molecules and the SWCNTs' surface. The strong curvature experienced by the SWCNTs' carbon lattice make some carbon atoms belonging to the nanotube body more susceptible to chemical attacks, thus favoring the attachment of functional groups to the nanotube wall. For all fullerenes in general, the reactivity of the surface carbon increases with the curvatures of their lattice, as demonstrated in investigations performed over C_{70}.[201] Additionally, the presence of five- and seven-member rings constitutes an important source of chemical activity in the side walls as well as the open ends of SWCNTs that may have resulted from oxidation or ultrasonication during the purification process. In fact, covalent functionalization is observed after the purification process by oxidation in the liquid phase, where the chemical attack of HNO_3 generally removes the catalyst particles, leaving carboxylic groups attached to the open ends and defective sites on the nanotubes' surface, as described by the chemical reaction in Fig. 18A.[202] A different approach that suggests the addition of

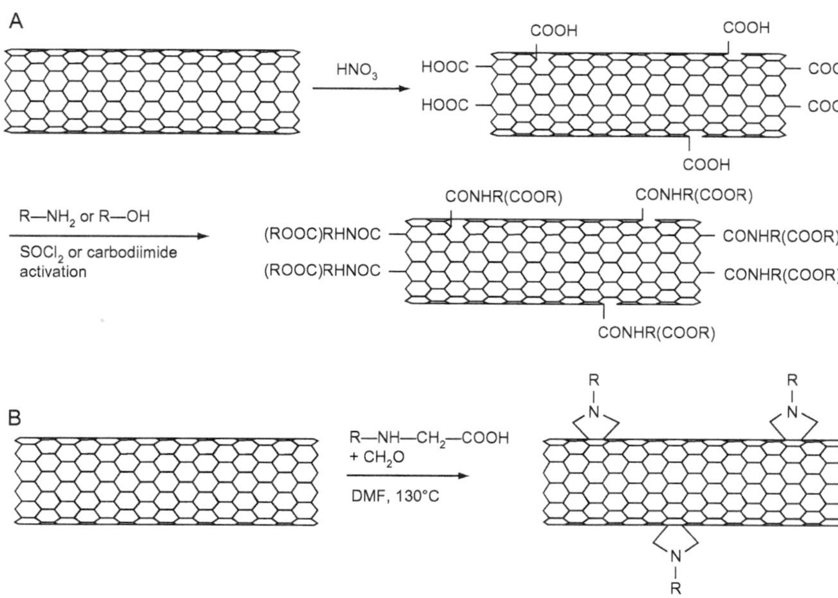

Fig. 18. Covalent functionalization of SWCNTs. (A) SWCNTs functionalized at side wall defects and open ends during purification with HNO_3 and further treatment with hydrophilic polymers to reduce toxicity in biological environments. (B) Addition of pyrene functional groups on non-defective sites of the SWCNT side walls by 1,3 dipolar cycloaddition. (Adapted from Liu *et al.*, *Nano Res* 2009;**2**:85–120.)

amide functional groups to the open ends of the nanotube has been also demonstrated to favor their solubility. The attachment of the amide groups is reached after reacting long amine molecules such as octadecylamine (ODA) to purified SWCNTs previously functionalized with thionychloride at the open ends.[203] However, despite the fact that carboxylic functional groups attached to the SWCNTs side walls and open ends reduce their hydrophobicity and increase their solubility in aqueous solutions, their applicability in biological applications becomes restricted by the high concentration of salt aggregates when functionalized nanotubes are dissolved in biological aqueous solutions. Nonetheless, hydrophilic polymers have been adhered to the oxidized SWCNTs, reducing the formation of salty aggregates in the solutions, and reducing the toxicity of functionalized SWCNTs within cell environments where they may accomplish several tasks as drug and protein delivery, and even tumor imaging.[93] However, despite the several benefits of covalent functionalization, the distribution of the functional groups along the carbon framework of the nanotube depend on the concentration of defective sites and non-six-membered rings in the side walls, which represents a notorious limitation.

For covalent functionalization of non-defective sites of SWCNT side walls, another type of chemical reaction different than oxidation usually take place over less reactive six-membered rings of the side wall. Organic functionalization of SWCNTs with carbenes and nitrenes accomplishes this goal, augmenting the solubility of nanotubes in organic solvents. As an example, the reaction of dichlorocarbene with SWCNTs is observed to functionalize the non-defective sites of the nanotubes' side walls by disruption of the double bonds between carbon atoms. This reaction facilitates solubility of SWCNTs in organic solvents by taking advantage of their solution-phase wall chemistry. On the other hand, the 1,3-dipolar cycloaddition of azomethine ylides has been found to introduce pyrene → (pyrrolidine) functional groups, improving in such a way the solubility of SWCNTs principally in water as well as in other polar solvents whereas it remains insoluble in nonpolar solvents,[204] illustrated in Fig. 18B. The specific functionalization of SWCNTs always aims to enhance their solubility in a class of solvents. For instance, the attack of SWCNTs by fluorine provokes the formation of C–F covalent bonds on the tube side walls, which improves the solubility of the nanotube in alcoholic solvents. Nevertheless, the temperature of the fluorination must be carefully controlled since a functionalization reaction carried out at high temperatures could instead produce fluorolysis of SWCNT walls. Other significant disadvantages of the fluorine functionalization are the negative consequences on the electrical behavior of nanotubes. The adhesion of fluorine groups to the side walls at certain temperature conditions is accompanied by the formation of a band gap at the Fermi level of metallic nanotubes, turning them into insulators. In addition to these limitations, the covalent functionalization of

SWCNTs may negatively impact other SWCNT physical properties. For example, photoluminescence and Raman-scattering properties are significantly affected due to the presence of functional groups along the carbon nanotube body.

2. Non-covalent Functionalization

The previous section discussed how wrapping of hydrophilic polymers around the carbon graphitic carbon network of a nanotube would reduce the formation of salty aggregates of the oxidized nanotubes during its covalent functionalization. The addition of hydrophilic polymers and other surfactants represents other feasible ways of functionalization of SWCNTs where chemical reactions and all their disadvantages are not present while the same goal is reached: the dissolution of nanotubes in organic solutions and their compatibility with biological environments. In the case of surfactants, non-covalent hybrids are formed after molecules with high surface activity are aggregated to the hydrophobic side walls of the SWCNTs. The adhesion between surfactants and nanotube walls becomes very strong due to the π–π-stacking interactions resulted from the attachment of aromatic groups of the amphiphile surfactant to the aromatic network of the nanotube side walls, as evidenced in the case of adhesion of *N*-succinimidyl-1-pyrenebutanoate. In general, pyrene molecules or derivatives favor the non-covalent functionalization of SWCNTs. The aromatic groups of the pyrene aggregates introduce hydrophobicity to the surfactant that make them highly compatible with the hydrophobic side walls of the nanotube, favoring π–π-stacking interactions between aromatic groups,[201] as illustrated schematically by Fig. 19A. But the SWCNTs solubilization is not the only performance improvement achieved by surfactant functionalization. The excellent electrical properties of SWCNTs together with the specific immobilization of biological macromolecules onto their functionalized surface make them potentially useful for biosensing applications. For instance, bifunctionalization with pyrene and ester groups was known to immobilize proteins on the SWCNT surface. The nucleophilic reaction of amine constituents, present in most proteins, with the ester groups of the functionalized nanotube surface allows the immobilization of the proteins with a high level of specificity.[205] SDS and benzylalkonium chloride are other good examples of surfactants non-covalently aggregated to the nanotube side walls that facilitate dissolution of carbon nanotubes in water, besides providing an effective charge to the nanotube surface which enables their later purification as previously mentioned. Additional to SDS, ssDNA molecules have been also proposed to functionalize SWCNTs with the final objective of separating them according to their electronic structure. Functionalization is a consequence of the presence of aromatic groups in DNA backbone that permits the π–π-stacking interaction with the aromatic side wall of the nanotube.[192] However, functionalizations with DNA are accompanied by one of the most undesirable effects. The instability of

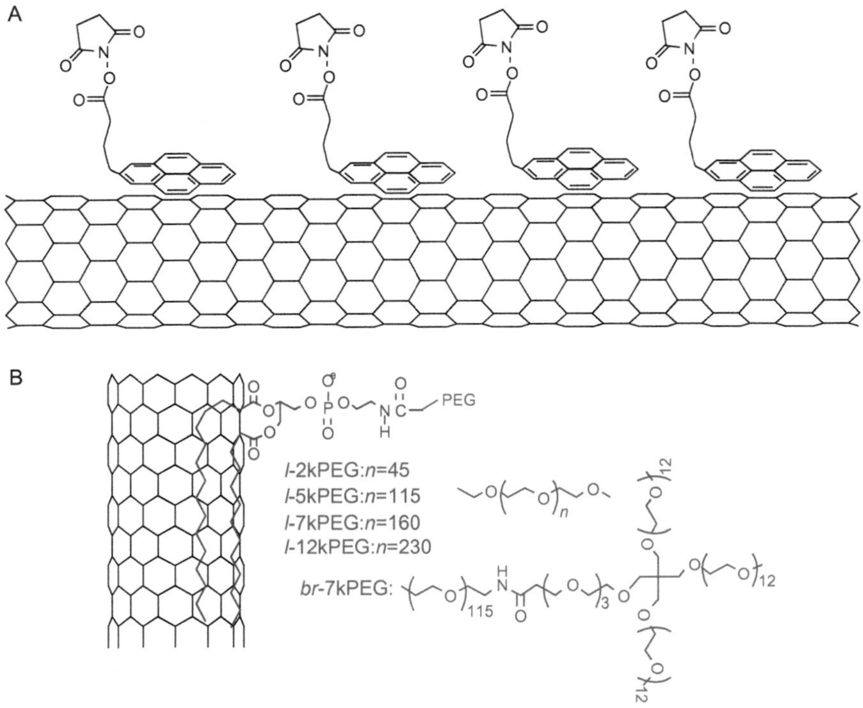

FIG. 19. Non-covalent functionalization of SWCNTs. (A) Functionalization by pyrene groups directly added to the nanotube surface due to π–π-stacking interactions. (B) Scheme of functionalization by linear or branched PEGylated phospholipids. (Adapted from Liu *et al.*, *Nano Res* 2009;**2**:85–120.) (For color version of this figure, the reader is referred to the Web version of this chapter.)

ssDNA/SWCNTs hybrids in biological solutions and the subsequent separation of the DNA molecule from the nanotube surface in the presence of nucleases in serum make this functionalization less applicable for biomedical purposes.

On the other hand, non-covalent polymer wrapping of SWCNTs arises as an alternative for nontoxic stable functionalization. Besides the required stability of the surfactant/SWCNTs hybrids, the nontoxicity and biocompatibility are the other prerequisites for a suitable SWCNTs functionalization for biological applications. Thus, polymer wrapping of SWCNTs seems to accomplish each one of these requirements. As in the case of surfactant functionalization, solubility and conductivity properties are appreciably modified after wrapping the nanotube surface with polymers. For example, polyvinylpyrrolidone (PVP), having polar sides along its long chain, assists the dissolution of PVP/SWCNT aggregates in polar solvents. Moreover, the conductivity of SWCNTs is found to be eight times enhanced when the nanotube is in the form of a complex aggregate with poly(mphenylene-co-2,5-dioctoxy-p-phenylenevinylene) (PmPV) in organic solvents.[206] The most important characteristic required for biological applications is the combination of solubility, stability, and biocompatibility, and PEGylated phospholipid is a polymer capable of accomplishing the functionalization requirements just mentioned[207] (Fig. 19B). The hydrophilic chain of the polyethyleneglycol (PEG) favors the water solubility, whereas the phospholipids assure the safety use in biological systems. PEGylated phospholipids remain stable within harsh biological environments, even remaining immune to the attack of nucleases in serum, unlike ssDNA and other surfactants. The reduction in hydrophobicity of the nanotube walls is intrinsically related to the increase in selective adsorption of proteins. This fact, together with its stability, and biocompatibility, make the PEGylated phospholipid/SWCNTs complex an alternative for immobilization and biosensing of proteins and other biological molecules on nanotubes surface. Drug delivery and imaging are also boosted by non-covalent functionalization with PEGylated phospholipid. All these applications take advantage, of course, of the high surface area, semiconductivity, and band gap fluorescence of SWCNTs. Thus, post-synthesis treatments collectively, including purification, separation, and functionalization of SWCNTs, complement each other to reach the common goal, which is obtaining high-quality SWCNTs, with definite characteristics, and specifically functionalized for the application of SWCNTs complexes in biological environments.

V. Biomedical Applications

Unique intrinsic physical and chemical properties of CNTs, including needle-like shape, large surface areas, flexible configurations, specific surface functionalization, ability to penetrate biological membranes, relatively low toxicity, special electrical, thermal, and spectroscopic properties, and others, allow them to be promising in medical applications from diagnosis to treatment. Especially, functionalized CNTs are very attractive as a new nanomedicine platform due to their extensively tailorable possibility and thereby excellent biocompatibility. The diagnostic and therapeutic applications of nanotubes are intensively explored by both academic and industrial laboratories, in parallel with their toxicological and other potential pathogenic reactions. Despite being in its infancy, recent research in CNTs for their medicinal applications has produced large numbers of publications containing several excellent reviews.[17,207–218] The following section briefly outlines the research advances of CNTs in several medicine-related topics, including imaging, electrochemical sensors, drug delivery, gene therapy, and tissue regeneration.

A. Applications of Carbon Nanotubes in Diagnostics

1. Imaging

As noted in Section II, semiconducting SWCNTs exhibit photoluminescence in the NIR range. The emission ranges of ~800–2000 nm[219] cover the biological tissue transparency window, which is therefore suitable for the imaging of biological systems. For example, Weisman and coworkers[220] tested the NIR fluorescence microscopy of SWCNTs in phagocytic cells. The samples with an average uptake of ~70 000 nanotubes per cell were used to record fluorescence microscopic images of the macrophages. It was found that controlling samples without nanotubes showed no detectable emission, whereas all samples incubated with nanotubes gave emission. Figure 20 shows a false-color-coded fluorescence image of one cell, which was taken near the cell's vertical center. This image shows many localized intracellular regions of NIR light emission. Authors interpret these as nanotubes located in small

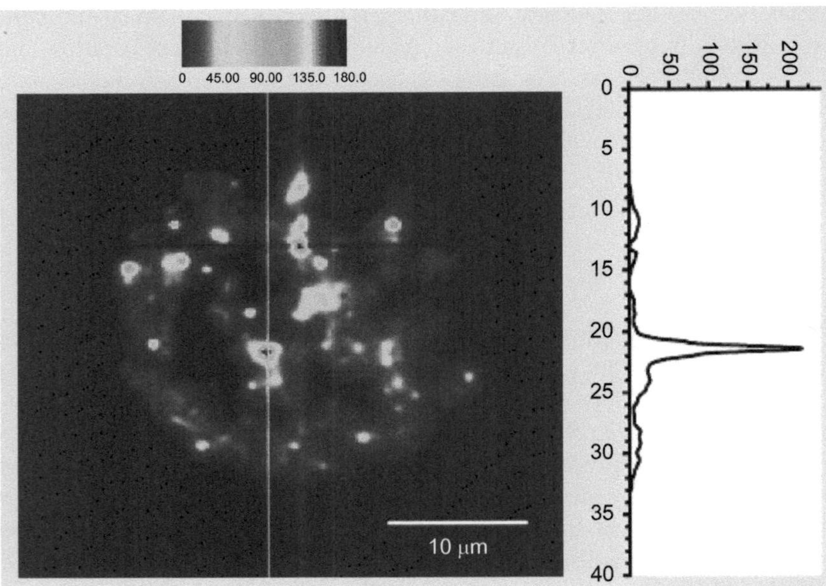

FIG. 20. Fluorescence image of one macrophage-like cell incubated with SWCNTs, showing emission detected from 1125 to 1600 nm with excitation at 660 nm. Intensities are coded with false color, and the image was obtained from a *z*-axis series by deconvolution processing. Intensity along the yellow vertical line is plotted on the graph to the right, showing high image contrast and localized emission sources. (Reprinted from Cherukuri *et al.*, *JACS* 2004;**126**:15638–15639.) (For interpretation of the references to color in this figure legend, the reader is referred to the Web version of this chapter.)

phagosomes and suggest that nanotubes were actively ingested through phagocytosis. Combining other related characterizations with these results showed that macrophage cells can actively ingest SWCNTs without showing toxic effects. The ingested nanotubes can be imaged through NIR fluorescence microscopy.

The same group also reported the mammalian pharmacokinetics of SWCNTs using intrinsic NIR fluorescence.[221] In this work, individualized SWCNTs were intravenously administered to rabbits and monitored through their characteristic NIR fluorescence. The experiments showed that the nanotube coating with synthetic surfactant molecules was displaced by blood proteins within seconds. The nanotube concentration in the blood serum decreased exponentially with a half-life of about 1.0 h. Behavior or pathological examination demonstrated no adverse effects from low-level nanotube exposure. After 24 h, significant concentrations of nanotubes were found only in the liver. These results demonstrate that SWCNTs are high-contrast NIR fluorophores in mammalian tissues that can be tracked using optical methods.

In addition to the direct measurement of SWCNT fluorescence in immunoassay formats, Strano and coworkers[18] have used band gap modulation and charge transfer effects via photoluminescence in the transduction and quantification of biomolecules to detect DNA hybridization. In their work, SWCNTs were non-covalently functionalized by 24-mer DNA and subsequent hybridized by DNA in proximity to the SWCNT surface. These modifications changed the dielectric constant at the SWCNT surface, and produced a 2-meV increase in band gap energy observed as a blue shift in emission. As a result, this method yielded a theoretical detection limit of 6 nmol/L for 24-mer DNA.

It should be pointed out that the fluorescent quantum yield of those SWCNTs in the NIR is dependent upon their chemical environment, thus it is required to avoid quenching and maximize quantum yield during processing. Furthermore, signal transduction suffers from the limits of spectral resolution and photoluminescence intensity, which restricts the utility of these methods to analytes at higher concentration.

Raman spectroscopy has also been explored for biomedical applications in diagnostics because it can provide detailed information on the chemical composition of cells and tissues. Compared with fluorescence methods, the sharp and narrow Raman peaks of SWCNTs make it easily differentiable from the autofluorescence background. The Raman excitation and scattering photons of SWCNTs are both in the NIR region, which is a transparent optical window for *in vivo* imaging.[222]

As a typical example, Gambhir and coworkers[223] studied noninvasive deep-tissue molecular images in a living subject using Raman spectroscopy. Surface-enhanced Raman-scattering SWCNTs with high Raman intensities were used to obtain whole-body Raman imaging and *in vivo* tumor targeting.

Subsequently, the same group further investigated the *in vivo* tumor targeting and localization of SWCNTs in mice.[224] Integrin $\alpha_v\beta_3$ and an arginine–glycine–aspartic acid (RGD) peptide acting as targeting molecule and ligand, respectively, were used in these studies. After 3 days, Raman imaging demonstrated a higher accumulation of SWCNT-RGD in the U87MG human glioblastoma tumor (integrin $\alpha_v\beta_3$-positive) than the plain nontargeted SWCNTs.

Recently, Dai and coworkers reported multiplexed Raman imaging of live cells with isotopically modified SWCNTs.[225] In this work, SWCNTs with different isotope (^{12}C and/or ^{13}C) compositions (which have distinct Raman G-band peaks) were studied, and different cancer cells combining different surface receptors were used for receptor-specific targeting. As shown in Fig. 21, three SWCNTs with different colors were each conjugated with a different targeting ligand. Each targeting ligand can recognize one specific receptor, which resulted in the multicolor Raman imaging of cells in a multiplexed manner. In another report, antibody-functionalized SWCNTs were used as multicolor Raman tags for multiplexed protein detection.[226] Combined with SERS substrates, the strong Raman intensity of SWCNTs resulted in the protein detection sensitivity down to the 1 fM (or 0.15 pg/ml) of analyte. Thus, when pure ^{12}C- and ^{13}C-SWCNTs are conjugated by different antibodies, multiplexed Raman detection of different proteins can be demonstrated.

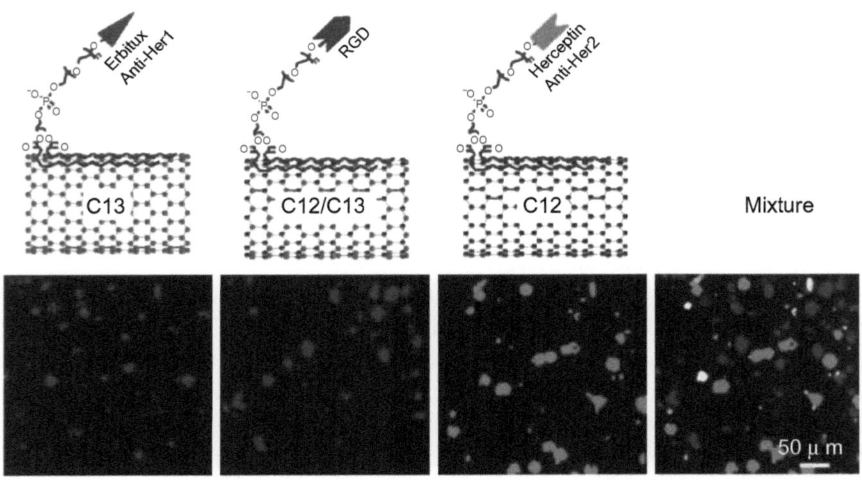

FIG. 21. Deconvoluted confocal Raman spectroscopy images of a mixture of three cell lines after incubation with a mixture of three different, molecularly targeted SWCNT conjugates. Each of the SWCNT conjugates has a different Raman signature (due to different carbon isotope composition) and was conjugated with a targeting moiety for a different molecular target. (Reprinted from Liu *et al.*, *JACS* 2008;**130**:13540.) (See Color Insert.)

CNT-based field emission X-ray source technology has also been investigated for diagnostic imaging applications. Cao and coworkers[227] have demonstrated the ability of this technology for high-resolution prospective-gated cardiac micro-computed tomography (CT) imaging under free-breathing condition. Their results showed that high-resolution CT images of beating mouse hearts can be obtained at 15 ms temporal resolution and 6.2 lp/mm spatial resolution at 10% of system MTF.

In addition, magnetic carbon nanotube labeling for hematopoietic stem/progenitor cell (HSPC) tracking has been reported by Chen and coworkers.[228] In this chapter, the uptake efficiency of fluorescein-isothiocyanate-labeled mCNT (FITC-mCNT) into HSPC and their effect on the cytotoxicity and differentiation of HSPC has been investigated. The result showed that the cellular uptake of FITC-mCNT was concentration- and time-dependent and the FITC-mCNT uptake has no adverse effect on the cell viability, cytotoxicity, and differentiation of HSPC.

2. Electrochemical Sensors

Carbon nanotubes have been incorporated in electrochemical sensors to detect proteins, neurotransmitters, and small bio-molecules for medical applications. Several electrochemical methods are used in these sensors including direct electrochemical detection of DNA biosensors, indirect detection of an oxidation product using enzyme sensors, and detection of conductivity changes using CNT field effect transistors (FETs). A discussion can be found in a recent comprehensive review contributed by Venton and coworkers.[211] Two examples are given herein to get a brief overview of experimental explorations in this topic.

Rusling and coworkers[229] reported an ultrasensitive electrochemical immunosensor for human interleukin-6 (IL-6), which is a suitable biomarker overexpressed by several types of cancers, such as head and neck squamous cell carcinoma (HNSCC). The measurement of biomarker for early cancer detection and treatment monitoring is significant and challenging. In this work,[229] antibodies (Ab_1) attached SWCNT forests with enzyme label horseradish peroxidase (HRP) were used in an electrochemical sandwich immunoassay protocol to measure the levels of IL-6. It was found that Ab_2-MWCNT–HRP bioconjugated immunosensors with 106 HRP labels per 100 nm gave the highest sensitivity of 19.3 nA mL cm^{-2} (pg IL-6)$^{-1}$ and the best detection limit of 0.5 pg mL^{-1} (25 fM) for IL-6 in 10 μL of calf serum. As shown in Fig. 22A, the steady-state sensor current increased with IL-6 concentration in serum over a range of ~0.5–30 pg mL^{-1}.The sensitivity, reflected by the slope of the calibration plot in the nearly linear region increased about fivefold compared with the Ab_2-biotin-streptavidin-HRP system. Furthermore, the small error bars implied the good device-to-device reproducibility.

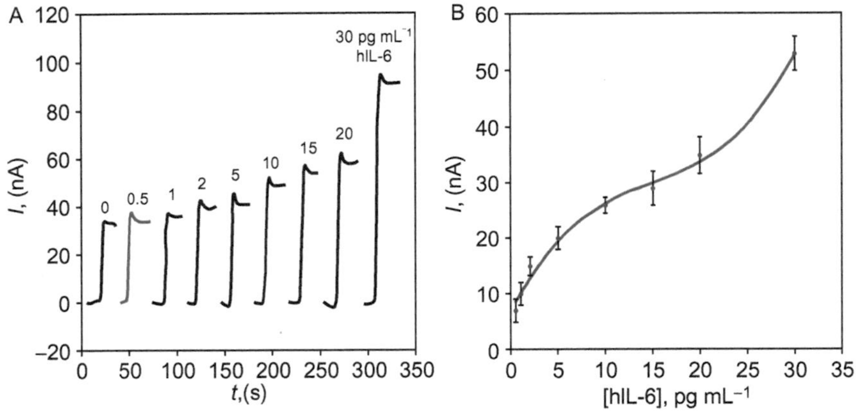

FIG. 22. Amperometric response for SWCNT immunosensors incubated with IL-6 in 10 μL of calf serum, then biotin-Ab2-MWNT-HRP: (A) current at − 0.3 V and rotating the electrode at 3000 rpm in PBS buffer containing 1 mM hydroquinone mediator and then injecting H_2O_2 to 0.04 mM to develop the signal; (B) influence of IL-6 concentration on steady-state current for immunosensor using the bioconjugate. (Reprinted from Malhotra *et al.*, *Analyt Chem* 2010;**82**:3118–3123.)

The secreted IL-6 in a wide range of HNSCC cells can thus be accurately measured by these immunosensors, indicating that the combination of SWCNT immunosensors and multilabel detection has great potential for detecting IL-6 for practical applications.

In a second example, a proof of concept of the electronic detection of stanozolol (Stz) and methylboldenone (MB) was carried out using CNT-FETs with two specific antibodies.[230] The results showed that CNT-FETs bonded to the antibodies are able to detect the presence of Stz or MB through the changes in the threshold voltage and drain current which were registered in the transistors. In addition, it was found that the CNT-FETs cannot detect other steroids, showing the high selectivity.

It should be pointed out that, to date, although various CNT-FETs and other CNT-based electrochemical devices have been explored for selective detection of medicine-related bio-molecules, the sensitivity and utility of these devices in the detection are limited by several issues. The detection is related to the charge screening of both ions and CNT-bound ligands in physiological buffers. The incompatibility of common aqueous buffers with these devices is one of the problems encountered for FET applications. For example, the high ionic strength of buffer systems is often required to retain the structure and function of bio-molecules such as proteins; protection of FET devices from charge effects of these bio-molecules significantly complicates the systems and limits the utility of these detections.[207]

B. Applications of Carbon Nanotubes in Therapeutics

1. Photothermal Treatments

CNTs exhibit strong absorbance in the NIR, which makes this a promising tool in photothermal therapy because NIR radiation could destroy the endosome and cells. As a typical example, Dai and coworkers have shown that, when irradiated by NIR light, CNTs that were internalized in cells can trigger endosomal rupture and cell death without harming normal cells.[231] This observation led to the development of functionalization schemes of SWCNTs with specific ligands for recognizing and targeting tumor cells. They exploited FRs that are common tumor markers by making highly water-soluble specific SWCNTs non-covalently functionalized by PL-PEG-FA (Fig. 23A). During studies of this system, both normal cells and FR-positive HeLa cells (FR^+ cells) with overexpressed FRs on the cell surfaces obtained by culturing cells in FA-depleted cell medium were exposed to PL-PEG-FA-SWCNTs for ~ 12–18 h, washed, and then radiated by a 808-nm laser (1.4 W/cm^2) for 2 min. After the NIR radiation, extensive cell death for the FR^+ cells was confirmed by cell morphology changes, whereas the normal cells remained intact as shown in Fig. 23B and C, respectively. The selective destruction of FR^+ cells implied that PL-PEG-FA-SWCNTs were efficiently internalized inside cells, which was confirmed by fluorescence shown in Fig. 23D for SWCNTs with FA cargo and FITC labels. The lack of fluorescence as shown in Fig. 23E implied that functionalized SWCNTs were not uptaken.

In another two examples, Choi and coworkers[232] demonstrated that NIR irradiation effect of PEG-SWCNTs can lead to eradication of tumors without recovery over 6 months in a human epidermoid mouth carcinoma model. Gmeiner and coworkers[233] showed that DNA encapsulated by CNT can lead to improved heat emission efficacy. When injecting this DNA with CNT into PC3 xenograft tumors and irradiated, the resulting conjugates can complete the tumor eradication.

In addition, Dai and coworkers[234] reported the first dual application of intravenously injected SWCNTs as photoluminescent agents for *in vivo* tumor imaging and as NIR absorbers and heaters for photothermal tumor elimination at the low injected dose and laser irradiation power. It was found that complete tumor elimination can be achieved for the photothermally treated mice without any toxic effects after more than 6 months. Furthermore, side-by-side comparison experiments showed that SWCNTs have higher effective tumor elimination with ten times lower injected doses and lower irradiation powers than those of gold nanorods (AuNRs). These results suggest the significant potential of utilizing biocompatible SWCNTs for combined cancer imaging and therapy.

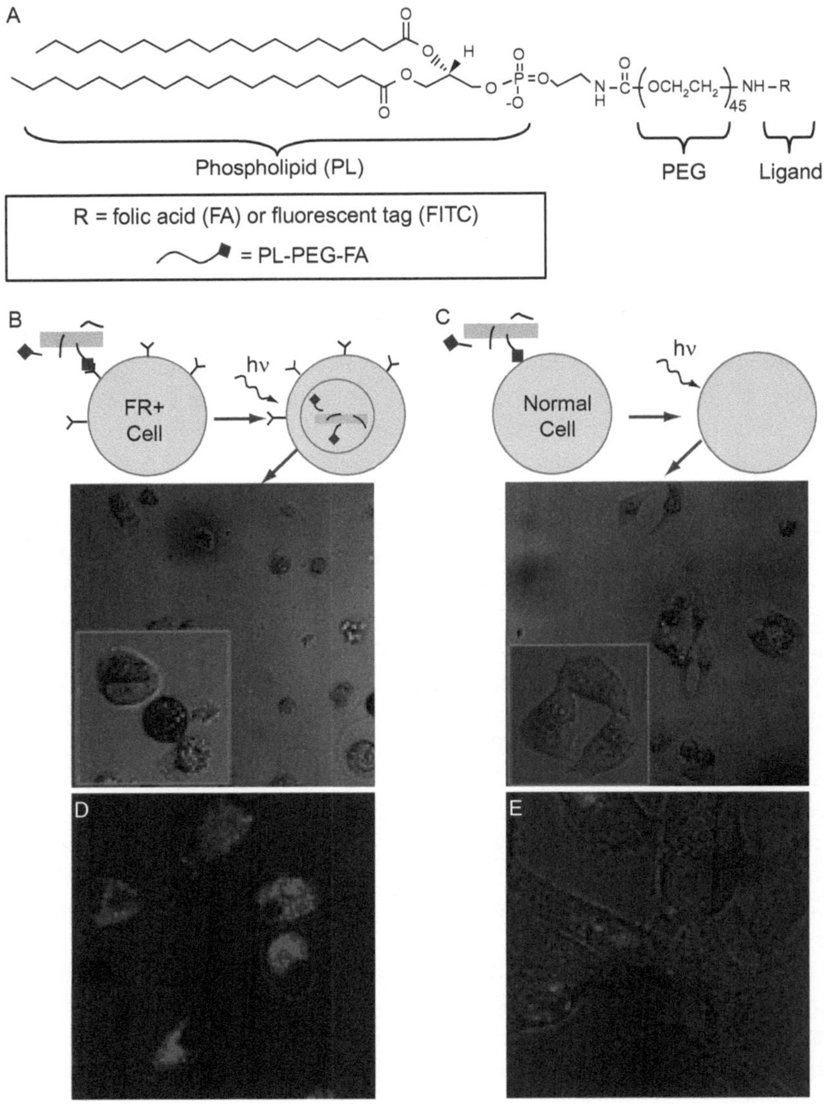

FIG. 23. Selective targeting and killing of cancer cells. (A) Chemical structure of PL-PEG-FA and PL-PEG-FITC used for solubilizing individual SWCNTs. (B) (Upper) Schematic of selective internalization of PL-PEG-FA-SWCNTs into FR^+ cells via receptor binding and then NIR 808 nm laser radiation; (Lower) image showing death of FR^+ cells after the process in Upper. (C) (Upper) Schematic of no internalization of PL-PEG-FA-SWCNTs into normal cells without available FRs; (Lower) image showing normal cells with no internalized SWCNTs are unharmed by the same laser radiation condition as in (B). (C) Confocal image of FR^+ cells after incubation in a solution of SWCNTs with two cargoes (PL-PEG-FA and PL-PEG-FITC). (E) The same as (D) for normal cells without abundant FRs on cell surfaces. (Reprinted from Kam *et al.*, *Proc Natl Acad Sci USA* 2005;**102**:11600–1605. Copyright (2005) *National Academy of Sciences, USA.*) (See Color Insert.)

C. Drug and Gene Delivery and Targeting

Another important application of CNTs is the delivery of gene, small drug molecules, and biomacromolecules to a desirable group of cells with the purpose of transporting a proper amount of drugs to the desirable sites, such as tumors, diseased tissues, etc., while minimizing side effects of the drugs on other tissues. There are several reviews which outline the progress in this topic.[17,207,209,216,218]

A specific application for drug delivery is in cancer therapy where the drugs require to be directed to the targeted tissues. In a typical example reported by Dai and coworkers,[235] SWCNTs were conjugated with paclitaxel (PTX) and showed high efficacy in inhibiting tumor growth in mice. In this work, SWCNTs were first functionalized with branched PEG chains and then PTX was linked to the PEG terminal (Fig. 24). It was demonstrated that this resulting SWCNT–PTX conjugate can stay in the blood for longer times than the Taxol (clinical drug formulation of PTX) and PTX coated with PEG. This prolonged circulation time and the enhanced permeability and retention (EPR) effect from SWCNT resulted in a tenfold higher accumulation of the drug at tumor sites than Taxol and a sixfold higher accumulation than PEG–PTX after 2 h of injection, and six- and fourfold higher, respectively, after 24 h of injection. The increased delivery efficacy of PTX by SWCNT led to the tumor inhibition of SWCNT–PTX (5 mg/kg PTX) for the 4T1 tumor model.

To deliver genes to cells is another issue of disease therapy. Guo and coworkers conjugated amino-functionalized MWCNTs (NH_2-MWCNTs) to plasmid DNA and successfully achieved the delivery of a green fluorescence

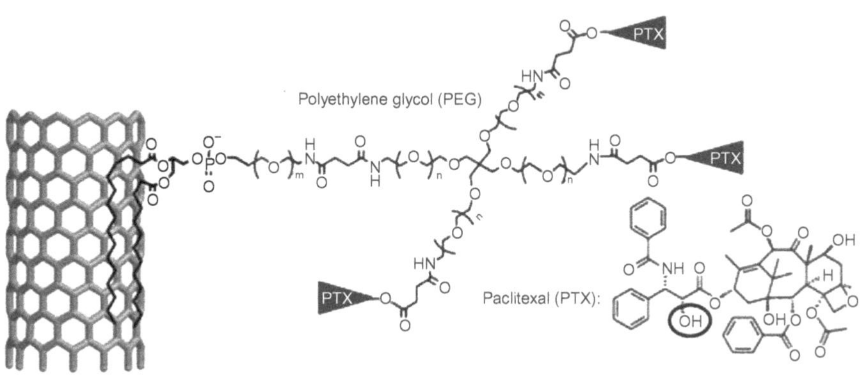

FIG. 24. Schematic representation of a CNT conjugate for PTX delivery. SWCNT functionalized by phospholipids with branched PEG chains which was linked with PTX. The blue ring indicates where the PTX reacted to attach to PEG. (Reprinted from Liu *et al.*, *Cancer Res* 2008;**68**:6652–6660.) (For interpretation of the references to color in this figure legend, the reader is referred to the Web version of this chapter.)

protein (GFP) gene into cultured human umbilical vein endothelial cells.[236] In another example reported by Prato and coworkers,[97] DNA was linked to functionalized CNTs via electrostatic interactions and delivered to cells. It was found that the gene expression levels were up to ten times higher than those achieved with DNA alone.

D. Tissue Regenerations

CNTs are also explored for applications in the field of regenerative medicine, such as for bone regeneration and neural regeneration.[17] In the applications acting as scaffolds for bone regeneration, CNTs can be ideal candidates due to their excellent mechanical properties and special thin linear shape with suitable length close to the size of the triple helix of collagen fibrils. Although pristine SWCNTs cannot support new bone growth, after being conjugated with functional groups that attract calcium cations, CNTs become effective in this activity as demonstrated in a mimic example reported by Haddon and coworkers.[237] In their studies, SWCNTs (CNT–COOH) were functionalized by poly(aminobenzene sulfonic acid) (PABS) using amidation of the carboxylic acid groups on SWCNTs with amines (PABS). Thin films of either SWCNT or SWCNT–COOH functionalized by PABS were deposited on glass slides. After 7 days of soaking the films in a solution of $CaCl_2$ and Na_2HPO_4, a large amount of plate-shaped hydroxyapatite (HA) crystals were formed on the surface of SWCNT–PABS films. The thickness of the HA crystal layer was about 2.4 μm after 7 days of mineralization and grew up to 3.5 μm thick after 14 days. However, the SWCNT–COOH thin films showed no HA crystals formed. This result demonstrated a method to engineer CNTs for supporting the mineralization of HA through functionalization with special groups. These functionalized CNTs can be used as scaffolds for bone regeneration.

In neural tissue regeneration, Mattson and coworkers[238] showed for the first time that functionalized MWCNTs can be used to support neuronal cell attachment and growth. Furthermore, Parpura and coworkers[239] revealed that CNTs functionalized with various bioactive molecules can improve neural regeneration activity including outgrowth, neurite branching, and attachment of growth cones. In addition, it was revealed that the high electrical conductivity of CNTs can enhance neuronal circuit activities. As a special example, Hanein and coworkers[240] designed a series of experiments to get an insight into the activities. First, they fabricated patterned substrates composed of CNT islands and then cultured neuronal cells from the cortices of 1-day-old Charles River rats on the CNT island substrates to get a well-organized neural network as shown in Fig. 25. Clearly, the cells preferentially aggregated to the CNT islands. Then, neuronal cells between neighboring islands grew axons and dendrites to connect with each other (Fig. 25B). Finally, organized neuronal networks formed on the CNT islands.

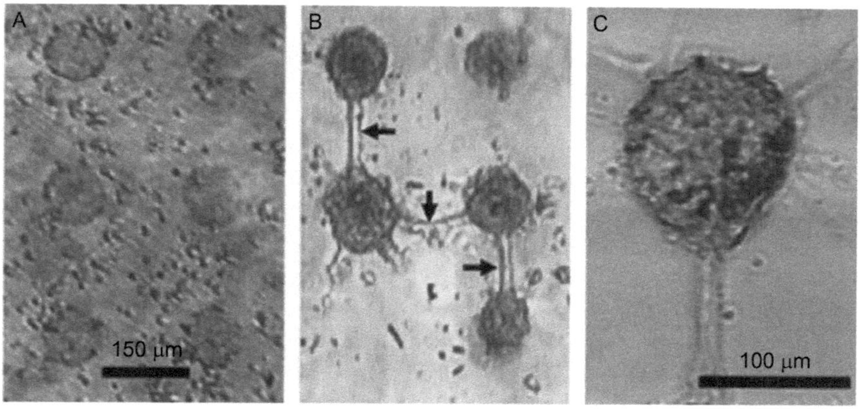

FIG. 25. Inverted microscope images of self-organized neuronal networks with predesigned geometry on CNT islands. (A) Neuron cells randomly attached on the substrates after 1 h of culture. (B) Cell aggregates on CNT islands formed interconnected neuronal networks between neighboring islands after 96 h of culture. (C) Neuron clusters formed on a CNT island. Arrows point to the connections formed between two neighboring neuronal clusters. (Reprinted from Gabay *et al. Phys A* 2005;**350**:611–621.) (See Color Insert.)

Overall, CNTs with unique physical and surface features have significant application potential in nanomedicine. Although it was demonstrated that they are not toxic in most cases, their nonbiodegradability remains a problem to be challenged to date. In addition, a standard protocol to prepare CNTs with high purity and at large scale is lacking, which is mandatory for clinical applications.

VI. Summary and Outlook

As stated in the Introduction, carbon nanotubes hold promises to become extremely useful materials for biomedical applications. Among their most important properties, their absorption and fluorescence as well as their electronic characteristics make CNTs excellent materials for diagnosis and treatment of diseases. However, after synthesis, they need to be properly treated to enhance their solubility in aqueous media and therefore their biocompatibility, and to minimize their potential toxicity. Synthesis methods are constantly being revised and investigated, aiming toward selective processes producing specific type of carbon nanotubes instead of a mixture of them. This selectivity goal has not been achieved yet, and significant research efforts are currently invested in developing better understanding and refinement of the processes that have shown at least partial selectivity, as well as searching for additional approaches aiming in this direction.

Biomedical applications require availability of highly purified and monodisperse SWCNTs, particularly semiconducting nanotubes with specific band gaps which make them potentially useful for biosensors and other biological uses. Furthermore, compatibility of SWCNTs with living organism environments is only achieved after a successful functionalization process that reduces the hydrophobicity of the nanotubes' side wall surfaces and therefore their toxicity. Thus, further research related to purification and separation and functionalization methods is crucial for improved biomedical applications. This chapter ends with a brief overview of the current most significant applications of CNTs in biomedicine, such as imaging and sensing devices as tools for diagnosis, and the uses of CNTs in drug delivery for therapeutical applications. The field is definitely in its infancy in many aspects; however, the high potentiality of these materials as highlighted in this report deserves and most likely will generate huge scientific and technological investments and consequently discoveries that definitely will benefit society.

References

1. Iijima S. Helical microtubules of graphitic carbon. *Nature* 1991;**354**:56–8.
2. Dresselhaus MS, Avouris P. *Carbon nanotubes: synthesis, structure, properties and applications.* Berlin: Springer; 2001.
3. Baughman RH, Zakhidov AA, de Heer WA. Carbon nanotubes–the route toward applications. *Science* 2002;**297**:787–92.
4. Hafner JH, Cheung CL, Lieber CM. Growth of nanotubes for probe microscopy tips. *Nature* 1999;**398**:761–2.
5. Rueckes T, Kim K, Joselevich E, Tseng GY, Cheung C-L, Lieber CM. Carbon nanotube-based nonvolatile random access memory for molecular computing. *Science* 2000;**289**:94–7.
6. Collins PG, Bradley K, Ishigami M, Zettl A. Extreme oxygen sensitivity of electronic properties of carbon nanotubes. *Science* 2000;**287**:1801–4.
7. Thostenson ET, Ren Z, Chou T-W. Advances in the science and technology of carbon nanotubes and their composites: a review. *Compos. Sci. Technol.* 2001;**61**:1899–912.
8. Avouris P. Carbon nanotube electronics. *Chem. Phys.* 2002;**281**:429–45.
9. Kostarelos K, Bianco A, Prato M. Promises, facts and challenges for carbon nanotubes in imaging and therapeutics. *Nat Nanotechnol* 2009;**4**:627–33.
10. Xu M, Futaba DN, Yamada T, Yumura M, Hata K. Carbon nanotubes with temperature-invariant viscoelasticity from − 196° to 1000 °C. *Science* 2010;**330**:1364–8.
11. Davis ME, Chen Z, Shin DM. Nanoparticle therapeutics: an emerging treatment modality for cancer. *Nat Rev Drug Discov* 2008;**7**:771–82.
12. Scheinberg DA, Villa CH, Escorcia FE, McDevitt MR. Conscripts of the infinite armada: systemic cancer therapy using nanomaterials. *Nat Rev Clin Oncol* 2010;**7**:266–76.
13. Strano MS. Biomedical materials: nanoscale radiosurgery. *Nat Mater* 2010;**9**:467–8.
14. Sorgenfrei S, Chiu Cy, Gonzalez RL, Yu YJ, Kim P, Nuckolls C, et al. Label-free single-molecule detection of DNA-hybridization kinetics with a carbon nanotube field-effect transistor. *Nat Nanotechnol* 2011;**6**:126–32.

15. Gannon CJ, Cherukuri P, Yakobson BI, Cognet L, Kanzius JS, Kittrell C, et al. Carbon nanotube-enhanced thermal destruction of cancer cells in a noninvasive radiofrequency field. *Cancer* 2007;**110**:2654–65.
16. Wang J. Carbon-nanotube based electrochemical biosensors: a review. *Electroanalysis* 2005;**17**:7–14.
17. Tran PA, Zhang L, Webster TJ. Carbon nanofibers and carbon nanotubes in regenerative medicine. *Adv Drug Deliv Rev* 2009;**61**:1097–114.
18. Jeng ES, Moll AE, Roy AC, Gastala JB, Strano MS. Detection of DNA hybridization using the near-infrared band-gap fluorescence of single-walled carbon nanotubes. *Nano Lett* 2006;**6**:371–5.
19. Kam NWS, Dai H. Carbon nanotubes as intracellular protein transporters: generality and biological functionality. *J Am Chem Soc* 2005;**127**:6021–6.
20. Heller DA, Baik S, Eurell TE, Strano MS. Single-walled carbon nanotube spectroscopy in live cells: towards long-term labels and optical sensors. *Adv Mater* 2005;**17**:2793–9.
21. Welsher K, Liu Z, Daranciang D, Dai H. Selective probing and imaging of cells with single walled carbon nanotubes as near-infrared fluorescent molecules. *Nano Lett* 2008;**2**:586–90.
22. Bachilo SM, Strano MS, Kitrell C, Hauge RH, Smalley RE, Weisman RB. Structure-assigned optical spectra of single-walled carbon nanotubes. *Science* 2002;**298**:2361–6.
23. Miyauchi Y, Chiashi S, Murakami Y, Hayashida Y, Maruyama S. Fluorescence spectroscopy of single-walled carbon nanotubes syntheszied from alcohol. *Chem Phys Lett* 2004;**387**:198–203.
24. Lolli G, Zhang L, Balzano L, Sakulchaicharoen N, Tan Y, Resasco DE. Tailoring (n, m) structure of single-walled carbon nanotubes by modifying reaction conditions and the nature of the support of CoMo catalysts. *J Phys Chem B* 2006;**110**:2108–15.
25. Wang X, Li Q, Xie J, Jin Z, Wang J, Li Y, et al. Fabrication of ultralong and electrically uniform single-walled carbon nanotubes on clean substrates. *Nano Lett* 2009;**9**:3137–41.
26. Xia ZH, Guduru PR, Curtin WA. Enhancing mechanical properties of multiwall carbon nanotubes via sp^3 interwall bridging. *Phys Rev Lett* 2007;**98**:245501.
27. Compton RN, Pagni RM. The chirality of biomolecules. In: Benjamin B, Herbert W, editors. *Advances in atomic, molecular, and optical physics*, vol. 48. Academic Press; 2002. pp. 219–61.
28. Dalla Cort A, Mandolini L, Pasquini C, Schiaffino L. “Inherent chirality” and curvature. *New J. Chem.* 2004;**28**:1198–9.
29. Iijima S. Growth of carbon nanotubes. *Mater Sci Eng B* 1993;**19**:172–80.
30. Ashrafi AR, Hamadanian M, Tavangar Z, Sabzyan H. Symmetry of a capped nanotube. *Digest J Nanomat Biostru* 2009;**4**:319–22.
31. Reich S, Li L, Robertson J. Control the chirality of carbon nanotubes by epitaxial growth. *Chem. Phys. Lett.* 2006;**421**:469–72.
32. Gómez-Gualdrón DA, Balbuena PB. The role of cap chirality in the mechanism of growth of single-wall carbon nanotubes. *Nanotechnology* 2008;**19**:485604.
33. Tanaka H, Akita S, Pan L, Nakayama Y. Comparison of capped carbon nanotube with open-ended one for field emission. *Jpn J Appl Phys* 2004;**43**:L427–9.
34. Stephan O, Ajayan PM, Colliex C, Redlich P, Lambert JM, Bernier P, et al. Doping graphitic and carbon nanotube structures with boron and nitrogen. *Science* 1994;**266**:1683–5.
35. Stone AJ, Wales DJ. Theoretical studies of icosahedral C_{60} and some related species. *Chem. Phys. Lett.* 1986;**128**:501–3.
36. Wang R, Zhang D, Zhang Y, Liu C. Boron-doped carbon nanotubes serving as a novel chemical sensor for formaldehyde. *J Phys Chem B* 2006;**110**:18267–71.
37. Zhu W, Rosen A, Bolton K. Changes in single-walled carbon nanotube chirality during growth and regrowth. *J Chem Phys* 2008;**128**:124708-6.
38. Bianco A, Kostarelos K, Partidos CD, Prato M. Biomedical applications of functionalised carbon nanotubes. *Chem Commun* 2005;**5**:571–7.

39. Peigney A, Laurent C, Flahaut E, Bacsa RR, Rousset A. Specific surface area of carbon nanotubes and bundles of carbon nanotubes. *Carbon* 2001;**39**:507–14.
40. Holzinger M, Baur J, Haddad R, Wang X, Cosnier S. Multiple functionalization of single-walled carbon nanotubes by dip coating. *Chem Commun* 2011;**47**:2450–2.
41. Balasubramanian K, Burghard M. Chemically functionalized carbon nanotubes. *Small* 2005;**1**:180–92.
42. Ajayan PM, Banhart F. Nanotubes: strong bundles. *Nat Mater* 2004;**3**:135–6.
43. Yamada T, Namai T, Hata K, Futaba DN, Mizuno K, Fan J, et al. Size-selective growth of double-walled carbon nanotube forests from engineered iron catalysts. *Nat Nanotechnol* 2006;**1**:131–6.
44. Yuan D, Ding L, Chu H, Feng Y, McNicholas TP, Liu J. Horizontally aligned single-walled carbon nanotube on quartz from a large variety of metal catalysts. *Nano Lett* 2008;**8**:2576–9.
45. Venema LC, Wildöer JWG, Janssen JW, Tans SJ, Tuinstra HLJT, Kouwenhoven LP, et al. Imaging electron wave functions of quantized energy levels in carbon nanotubes. *Science* 1999;**283**:52–5.
46. Ebbesen TW, Lezec HJ, Hiura H, Bennett JW, Ghaemi HF, Thio T. Electrical conductivity of individual carbon nanotubes. *Nature* 1996;**382**:54–6.
47. Hamada N, Sawada Si, Oshiyama A. New one-dimensional conductors: graphitic microtubules. *Phys Rev Lett* 1992;**68**:1579.
48. Saito R, Fujita M, Dresselhaus G, Dresselhaus MS. Electronic structure of chiral graphene tubules. *Appl Phys Lett* 1992;**60**:2204–6.
49. Saito R, Fujita M, Dresselhaus G, Dresselhaus MS. Electronic structure of graphene tubules based on C_{60}. *Phys Rev B* 1992;**46**:1804.
50. Saito R, Dresselhaus G, Dresselhaus MS. Trigonal warping effect of carbon nanotubes. *Phys Rev B* 2000;**61**:2981.
51. Saito R, Grüneis A, Samsonidze GG, Dresselhaus G, Dresselhaus MS, Jorio A, et al. Optical absorption of graphite and single-wall carbon nanotubes. *Appl Phys A* 2004;**78**:1099–105.
52. Iakoubovskii K, Minami N, Kim Y, Miyashita K, Kazaoui S, Nalini B. Midgap luminescence centers in single-wall carbon nanotubes created by ultraviolet illumination. *Appl Phys Lett* 2006;**89**:173108-3.
53. Liu Z, Yang K, Lee S-T. Single-walled carbon nanotubes in biomedical imaging. *J Mater Chem* 2011;**21**:586–98.
54. Miyauchi Y, Oba M, Maruyama S. Cross-polarized optical absorption of single-walled nanotubes by polarized photoluminescence excitation spectroscopy. *Phys Rev B* 2006;**74**:205440.
55. O'Connell MJ, Bachilo SM, Huffman CB, Moore VC, Strano MS, Haroz EH, et al. Band gap fluorescence from individual single-walled carbon nanotubes. *Science* 2002;**297**:593–6.
56. Hartschuh A, Pedrosa HN, Novotny L, Krauss TD. Simultaneous fluorescence and raman scattering from single carbon nanotubes. *Science* 2003;**301**:1354–6.
57. Wray S, Cope M, Delpy DT, Wyatt JS, Reynolds EOR. Characterization of the near infrared absorption spectra of cytochrome aa3 and haemoglobin for the non-invasive monitoring of cerebral oxygenation. *Biochim Biophys Acta* 1988;**933**:184–92.
58. Lakowicz JR. *Principles of fluorescence spectroscopy, vol. 1*. 3rd ed. New York: Springer; 2006.
59. Crochet J, Clemens M, Hertel T. Quantum yield heterogeneities of aqueous single-wall carbon nanotube suspensions. *J Am Chem Soc* 2007;**129**:8058–9.
60. Kataura H, Kumazawa Y, Kojima N, Maniwa Y, Umezu I, Masubuchi S, et al. *International winter school on electronic properties of novel materials*. NY: Woodbury; 1999.
61. Forró L, Schönenberger C. Physical properties of multi-wall nanotubes. In: Dresselhaus M, Dresselhaus G, Avouris P, editors. *Carbon nanotubes*, vol. 80. Heidelberg: Springer Berlin; 2001. pp. 329–91.

62. Miyamoto Y, Saito S, Tomanek D. Electronic interwall interactions and charge redistribution in multiwall nanotubes. *Phys Rev B* 2001;**65**:041402.
63. Strano MS, Dyke CA, Usrey ML, Barone PW, Allen MJ, Shan H, et al. Electronic structure control of single-walled carbon nanotube functionalization. *Science* 2003;**301**:1519–22.
64. Star A, Gabriel J-CP, Bradley K, Grüner G. Electronic detection of specific protein binding using nanotube FET devices. *Nano Lett* 2003;**3**:459–63.
65. Lee AJ, Wang X, Carlson LJ, Smyder JA, Loesch B, Tu X, et al. Bright fluorescence from individual single-walled carbon nanotubes. *Nano Lett* 2011;**11**:1636–40.
66. Dresselhauss MS, Eklund PC. Phonons in carbon nanotubes. *Adv Phys* 2000;**49**:705–814.
67. Jorio A, Saito R, Hafner JH, Liebner CM, Hunter M, McClure T, et al. Structural (n, m) determination of isolated single-wall carbon nanotubes by resonant Raman scattering. *Phys Rev Lett* 2001;**86**:1118–21.
68. Dresselhaus MS, Dresselhaus G, Saito R, Jorio A. Raman spectroscopy of carbon nanotubes. *Phys Rep* 2005;**409**:47–99.
69. Lefrant S. Raman and SERS studies of carbon nanotube systems. *Curr Appl Phys* 2002;**2**:479–82.
70. Bergin SD, Nicolosi V, Streich PV, Giordani S, Sun Z, Windle AH, et al. Towards solutions of single-walled carbon nanotubes in common solvents. *Adv Mater* 2008;**20**:1876–81.
71. Thess A, Lee R, Nikolaev P, Dai H, Petit P, Robert J, et al. Crystalline ropes of metallic carbon nanotubes. *Science* 1996;**273**:483–7.
72. Tasis D, Tagmatarchis N, Georgakilas V, Prato M. Soluble carbon nanotubes. *Chemistry* 2003;**9**:4000–8.
73. Bahr JL, Mickelson ET, Bronikowski MJ, Smalley RE, Tour JM. Dissolution of small diameter single-wall carbon nanotubes in organic solvents? *Chem Commun* 2001;**2**:193–4.
74. Amiot CL, Xu S, Liang S, Pan L, Zhao JX. Near-infrared fluorescent materials for sensing of biological targets. *Sensors* 2008;**8**:3082–105.
75. Liu Z, Tabakman S, Welsher K, Dai H. Carbon nanotubes in biology and medicine: in vitro and in vivo detection, imaging and drug delivery. *Nano Res* 2009;**2**:85–120.
76. Vaisman L, Wagner HD, Marom G. The role of surfactants in dispersion of carbon nanotubes. *Adv Colloid Interface Sci* 2006;128–30 37–46.
77. Yurekli K, Mitchell CA, Krishnamoorti R. Small-angle neutron scattering from surfactant-assisted aqueous dispersions of carbon nanotubes. *J Am Chem Soc* 2004;**126**:9902–3.
78. Christian SD, Scamehorn JF. *Solubilization in surfactants aggregates.* New York: Marcel Dekker; 1995.
79. Strano MS, Moore VC, Miller MK, Allen MJ, Haroz EH, Kittrell C, et al. The role of surfactant adsorption during ultrasonication in the dispersion of single-wall carbon nanotubes. *J Nanosci Nanotechol* 2003;**3**:81–6.
80. Lu KL, Lago RM, Chen YK, Green MLH, Harris PJF, Tsang SC. Mechanical damage of carbon nanotubes by ultrasound. *Carbon* 1996;**34**:814–6.
81. Islam MF, Rojas E, Bergey DM, Johnson AT, Yodh AG. High weight fraction surfactant solubilization of single-wall carbon nanotubes in water. *Nano Lett* 2003;**3**:269–73.
82. Vaisman L, Marom G, Wagner HD. Dispersions of surface-modified carbon nanotubes in water-soluble and water-insoluble polymers. *Adv Funct Mater* 2006;**16**:357–63.
83. Kang Y, Taton TA. Micelle-encapsulated carbon nanotubes: a route to nanotube composites. *J Am Chem Soc* 2003;**125**:5650–1.
84. Tu X, Manohar S, Jagota A, Zheng M. DNA sequence motifs for structure-specific recognition and separation of carbon nanotubes. *Nature* 2009;**460**:250–3.
85. Mossman BT, Craighead JE. Mechanisms of asbestos carcinogenesis. *Environ Res* 1981;**25**:269–80.

86. Mitchell LA, Gao J, Wal RV, Gigliotti A, Burchiel SW, McDonald JD. Pulmonary and systemic immune response to inhaled multiwalled carbon nanotubes. *Toxicol Sci* 2007;**100**:203–14.
87. Chou C-C, Hsiao H-Y, Hong Q-S, Chen C-H, Peng Y-W, Chen H-W, et al. Single-walled carbon nanotubes can induce pulmonary injury in mouse model. *Nano Lett* 2008;**8**:437–45.
88. Shvedova AA, Kisin E, Murray AR, Johnson VJ, Gorelik O, Arepalli S, et al. Inhalation vs. aspiration of single-walled carbon nanotubes in C57BL/6 mice: inflammation, fibrosis, oxidative stress, and mutagenesis. *Am J Physiol Lung Cell Mol Physiol* 2008;**295**:L552–65.
89. Kagan VE, Konduru NV, Feng W, Allen BL, Conroy J, Volkov Y, et al. Carbon nanotubes degraded by neutrophil myeloperoxidase induce less pulmonary inflammation. *Nat Nanotechnol* 2010;**5**:354–9.
90. Tian F, Cui D, Schwarz H, Estrada GG, Kobayashi H. Cytotoxicity of single-wall carbon nanotubes on human fibroblasts. *Toxicol in Vitro* 2006;**20**:1202–12.
91. Kang S, Pinault M, Pfefferle LD, Elimelech M. Single-walled carbon nanotubes exhibit strong antimicrobial activity. *Langmuir* 2007;**23**:8670–3.
92. Liu Z, Tabakman SM, Chen Z, Dai H. Preparation of carbon nanotube bioconjugates for biomedical applications. *Nat Protocols* 2009;**4**:1372–81.
93. Schipper ML, Nakayama-Ratchford N, Davis CR, Kam NWS, Chu P, Liu Z, et al. A pilot toxicology study of single-walled carbon nanotubes in a small sample of mice. *Nat Nanotechnol* 2008;**3**:216–21.
94. Helland A, Wick P, Koehler A, Schmid K, Som C. Reviewing the environmental and human health knowledge base of carbon nanotubes. *Ciência Saúde Coletiva* 2008;**13**:441–52.
95. Kostarelos K, Lacerda L, Pastorin G, Wu W, Wieckowski Sebastie, Luangsivilay J, et al. Cellular uptake of functionalized carbon nanotubes is independent of functional group and cell type. *Nat Nanotechnol* 2007;**2**:108–13.
96. Pantarotto D, Briand JP, Prato M, Bianco A. Translocation of bioactive peptides across cell membranes by carbon nanotubes. *Chem Commun* 2004;**1**:16–7.
97. Pantarotto D, Singh R, McCarthy D, Erhardt M, Briand J-P, Prato M, et al. Functionalized carbon nanotubes for plasmid DNA gene delivery. *Angew Chem Int Ed* 2004;**43**:5242–6.
98. Lacerda L, Raffa S, Prato M, Bianco A, Kostarelos K. Cell-penetrating CNTs for delivery of therapeutics. *Nano Today* 2007;**2**:38–43.
99. Shi Kam NW, Jessop TC, Wender PA, Dai H. Nanotube molecular transporters: internalization of carbon nanotube–protein conjugates into mammalian cells. *J Am Chem Soc* 2004;**126**:6850–1.
100. Wu W, Wieckowski S, Pastorin G, Benincasa M, Klumpp C, Briand J-P, et al. Targeted delivery of amphotericin B to cells by using functionalized carbon nanotubes. *Angew Chem Int Ed* 2005;**44**:6358–62.
101. Bianco A, Kostarelos K, Prato M. Applications of carbon nanotubes in drug delivery. *Curr Opin Chem Biol* 2005;**9**:674–9.
102. Lopez CF, Nielsen SO, Preston BM, Klein ML, DeGrado WF. Understanding nature's design for a nanosyringe. *Proc Natl Acad Sci USA* 2004;**101**:4431–4.
103. Menard-Moyon C, Venturelli E, Fabbro C, Samori C, Ros TD, Kostarelos K, et al. The alluring potential of functionalized carbon nanotubes in drug discovery. *Expert Opin. Drug Discov* 2010;**5**(7):691–707.
104. Herrera JE, Resasco DE. Role of Co–W interaction in the selective growth of single-walled carbon nanotubes from CO disproportionation. *J Phys Chem B* 2003;**107**:3738–46.
105. Herrera JE, Resasco DE. Loss of single-walled carbon nanotubes selectivity by disruption of the Co-Mo interaction in the catalyst. *J Catalysis* 2004;**221**:354–64.
106. Ding L, Tselev A, Wang J, Yuan D, Chu H, McNicholas TP, et al. Selective growth of well-aligned semiconducting single-walled carbon nanotubes. *Nano Lett* 2009;**9**:800–5.

107. Harutyunyan AR, Chen G, Paronyan TM, Pigos EM, Kuznetsov OA, Hewaparakrama K, et al. Preferential growth of single-walled carbon nanotubes with metallic conductivity. *Science* 2009;**326**:116–20.
108. Kroto HW, Heath JR, O'Brien SC, Curl RF, Smalley RE. C60: buckminsterfullerene. *Nature* 1985;**318**:162–3.
109. Ando Y. Carbon nanotube: the inside story. *J Nanosci Nanotechnol* 2010;**10**:3726–38.
110. Tai Y, Inukai K, Osaki T, Tazawa M, Murakami J, Tanemura S, et al. Identification of compounds produced through contact arc vaporization of graphite under CH4 ambience. *Chem. Phys. Lett.* 1994;**224**:118–22.
111. Zhao X, Ohkohchi M, Shimoyama H, Ando Y. Morphology of carbon allotropes prepared by hydrogen arc discharge. *J Crystal Growth* 1999;**198–199**:934–8.
112. Keidar M, Levchenko I, Arbel T, Alexander M, Waas AM, Ostrikov KK. Magnetic-field-enhanced synthesis of single-wall carbon nanotubes in arc discharge. *J Appl Phys* 2008;**103**:094318-8.
113. Ebbesen TW, Ajayan PM. Large-scale synthesis of carbon nanotubes. *Nature* 1992;**358**:220–2.
114. Journet C, Bernier P. Production of carbon nanotubes. *Appl Phys A* 1998;**67**:1–9.
115. Ohkohchi M, Ando Y, Bandow S, Saito Y. Formation of carbon nanotubes by evaporation of carbon rod containing scandium oxide. *Jpn J Appl Phys* 1993;**32**:L1248–51.
116. Jung SH, Kim MR, Jeong SH, Kim SU, Lee OJ, Lee KH, et al. High-yield synthesis of multi-walled carbon nanotubes by arc discharge in liquid nitrogen. *Appl Phys A* 2003;**76**:285–6.
117. Anazawa K, Shimotani K, Manabe C, Watanabe H, Shimizu M. High-purity carbon nanotubes synthesis method by an arc discharging in magnetic field. *Appl Phys Lett* 2002;**81**:739–41.
118. Iijima S, Ichihashi T. Single-shell carbon nanotubes of 1-nm diameter. *Nature* 1993;**363**:603–5.
119. Journet C, Maser WK, Bernier P, Loiseau A, de la Chapelle ML, Lefrant S, et al. Large-scale production of single-walled carbon nanotubes by the electric-arc technique. *Nature* 1997;**388**:756–8.
120. Ando Y, Zhao X, Hirahara K, Suenaga K, Bandow S, Iijima S. Mass production of single-wall carbon nanotubes by the arc plasma jet method. *Chem. Phys. Lett.* 2000;**323**:580–5.
121. Ando Y, Zhao X, Inoue S, Suzuki T, Kadoya T. Mass production of high-quality single-wall carbon nanotubes by H_2-N_2 arc discharge. *Diam Relat Mater* 2005;**14**:729–32.
122. Farhat S, de La Chapelle ML, Loiseau A, Scott CD, Lefrant S, Journet C, et al. Diameter control of single-walled carbon nanotubes using argon–helium mixture gases. *J Chem Phys* 2001;**115**:6752–9.
123. Guo T, Nikolaev P, Thess A, Colbert DT, Smalley RE. Catalytic growth of single-walled manotubes by laser vaporization. *Chem. Phys. Lett.* 1995;**243**:49–54.
124. Scott CD, Arepalli S, Nikolaev P, Smalley RE. Growth mechanisms for single-wall carbon nanotubes in a laser-ablation process. *Appl Phys A* 2001;**72**:573–80.
125. Maser WK, Muñoz E, Benito AM, Martínez MT, de la Fuente GF, Maniette Y, et al. Production of high-density single-walled nanotube material by a simple laser-ablation method. *Chem. Phys. Lett.* 1998;**292**:587–93.
126. Yudasaka M, Ichihashi T, Komatsu T, Iijima S. Single-wall carbon nanotubes formed by a single laser-beam pulse. *Chem. Phys. Lett.* 1999;**299**:91–6.
127. Zhang H, Chen K, He Y, Zhu Y, Chen Y, Wu C, et al. Formation and Raman spectroscopy of single wall carbon nanotubes synthesized by CO_2 continuous laser vaporization. *J Phys Chem Solids* 2001;**62**:2007–10.
128. Rao AM, Richter E, Bandow S, Chase B, Eklund PC, Williams KA, et al. Diameter-selective Raman scattering from vibrational modes in carbon nanotubes. *Science* 1997;**275**:187–91.

129. Sugano M, Kasuya A, Tohji K, Saito Y, Nishina Y. Resonance Raman scattering and diameter-dependent electronic states in single-wall carbon nanotubes. *Chem. Phys. Lett.* 1998;**292**:575–9.
130. Kokai F, Takahashi K, Yudasaka M, Yamada R, Ichihashi T, Iijima S. Growth dynamics of single-wall carbon nanotubes synthesized by CO_2 laser vaporization. *J Phys Chem B* 1999;**103**:4346–51.
131. Muñoz E, Maser WK, Benito AM, Martínez MT, de la Fuente GF, Maniette Y, et al. Gas and pressure effects on the production of single-walled carbon nanotubes by laser ablation. *Carbon* 2000;**38**:1445–51.
132. Saito Y. Nanoparticles and filled nanocapsules. *Carbon* 1995;**33**:979–88.
133. Tersoff J, Ruoff RS. Structural properties of a carbon-nanotube crystal. *Phys Rev Lett* 1994;**73**:676.
134. Zhao G, Giolando DM, Kirchhoff JR. Chemical vapor deposition fabrication and characterization of silica-coated carbon fiber ultramicroelectrodes. *Anal Chem* 1995;**67**:2592–8.
135. Resasco DE, Alvarez WE, Pompeo F, Balzano L, Herrera JE, Kitiyanan B, et al. A scalable process for production of single-walled carbon nanotubes (SWNTs) by catalytic disproportionation of CO on a solid catalyst. *J Nanopart Res* 2002;**4**:131–6.
136. Kumar M, Ando Y. Chemical vapor deposition of carbon nanotubes: a review on growth mechanism and mass production. *J Nanosci Nanotechnol* 2010;**10**:3739–58.
137. Takagi D, Homma Y, Hibino H, Suzuki S, Kobayashi Y. Single-walled carbon nanotube growth from highly activated metal nanoparticles. *Nano Lett* 2006;**6**:2642–5.
138. Balbuena P, Zhao J, Huang S, Wang Y, Sakulchaicharoen N, Resasco DE. Role of the catalyst in the growth of single-wall carbon nanotubes. *J Nanosci Nanotechnol* 2006;**6**:1247–58.
139. Cantoro M, Hofmann S, Pisana S, Scardaci V, Parvez A, Ducati C, et al. Catalytic chemical vapor deposition of single-wall carbon nanotubes at low temperatures. *Nano Lett* 2006;**6**:1107–12.
140. Chiang W-H, Mohan Sankaran R. Linking catalyst composition to chirality distributions of as-grown single-walled carbon nanotubes by tuning NixFe1-x nanoparticles. *Nat Mater* 2009;**8**:882–6.
141. Charlier JC, Amara H, Lambin P. Catalytically assisted tip growth mechanism for single-wall carbon nanotubes. *ACS Nano* 2007;**1**:202–7.
142. Gavillet J, Loiseau A, Journet C, Willaime F, Ducastelle F, Charlier JC. Root-growth mechanism for single-wall carbon nanotubes. *Phys Rev Lett* 2001;**87**:275504.
143. Bhushan B. *Handbook of nanotechnology*. 2nd ed. New York: Springer; 2007.
144. Li WZ, Xie SS, Qian LX, Chang BH, Zou BS, Zhou WY, et al. Large-scale synthesis of aligned carbon nanotubes. *Science* 1996;**274**:1701–3.
145. Fan S, Liang W, Dang H, Franklin N, Tombler T, Chapline M, et al. Carbon nanotube arrays on silicon substrates and their possible application. *Phys E* 2000;**8**:179–83.
146. Huang L, Cui X, White B, O'Brien SP. Long and oriented single-walled carbon nanotubes grown by ethanol chemical vapor deposition. *J Phys Chem B* 2004;**108**:16451–6.
147. Ismach A, Joselevich E. Orthogonal self-assembly of carbon nanotube crossbar architectures by simultaneous graphoepitaxy and field-directed growth. *Nano Lett* 2006;**6**:1706–10.
148. Liu B-C, Tang S-H, Liang Q, Gao L-Z, Zhang B-L, Qu M-Z, et al. Production of carbon nanotubes over pre-reduced $LaCoO_3$ by using fluidized-bed catalytic reactor. *Chin J Chem* 2001;**19**:983–6.
149. Baddour CE, Briens C. Carbon nanotube synthesis: review. *Int J Chem React Eng* 2005;**3**:1–20.
150. Moisala A, Nasibulin AG, Brown DP, Jiang H, Khriachtchev L, Kauppinen EI. Single-walled carbon nanotube synthesis using ferrocene and iron pentacarbonyl in a laminar flow reactor. *Chem Eng Sci* 2006;**61**:4393–402.

151. Nikolaev P, Bronikowski MJ, Bradley RK, Rohmund F, Colbert DT, Smith KA, et al. Gas-phase catalytic growth of single-walled carbon nanotubes from carbon monoxide. *Chem. Phys. Lett.* 1999;**313**:91–7.
152. Bai S, Li F, Yang Q, Cheng H-M, Bai J. Influence of ferrocene/benzene mole ratio on the synthesis of carbon nanostructures. *Chem. Phys. Lett.* 2003;**376**:83–9.
153. Han W-Q, Kohler-Redlich P, Seeger T, Ernst F, Ruhle M, Grobert N, et al. Aligned CN_x nanotubes by pyrolysis of ferrocene/C_{60} under NH_3 atmosphere. *Appl Phys Lett* 2000;**77**:1807–9.
154. Ci L, Rao Z, Zhou Z, Tang D, Yan X, Liang Y, et al. Double wall carbon nanotubes promoted by sulfur in a floating iron catalyst CVD system. *Chem. Phys. Lett.* 2002;**359**:63–7.
155. Maruyama S, Kojima R, Miyauchi Y, Chiashi S, Kohno M. Low-temperature synthesis of high-purity single-walled carbon nanotubes from alcohol. *Chem. Phys. Lett.* 2002;**360**:229–34.
156. Chen M, Chen C-M, Chen C-F. Preparation of high yield multi-walled carbon nanotubes by microwave plasma chemical vapor deposition at low temperature. *J Mater Sci* 2002;**37**:3561–7.
157. Alexandrescu R, Crunteanu A, Morjan RE, Morjan I, Rohmund F, Falk LKL, et al. Synthesis of carbon nanotubes by CO_2-laser-assisted chemical vapour deposition. *Infrared Phys Technol* 2003;**44**:43–50.
158. Chikkaveeraiah BV, Bhirde A, Malhotra R, Patel V, Gutkind JS, Rusling JF. Single-wall carbon nanotube forest arrays for immunoelectrochemical measurement of four protein biomarkers for prostate cancer. *Anal Chem* 2009;**81**:9129–34.
159. Ding F, Harutyunyan AR, Yakobson BI. Dislocation theory of chirality-controlled nanotube growth. *Proc Natl Acad Sci USA* 2009;**106**:2506–9.
160. Gómez-Gualdrón DA, Balbuena PB. Effect of metal cluster-cap interactions on the catalyzed growth of single-wall carbon nanotubes. *J Phys Chem C* 2008;**113**:698–709.
161. Gómez-Gualdrón DA, Balbuena PB. Growth of chiral single-walled carbon nanotube caps in the presence of a cobalt cluster. *Nanotechnology* 2009;**20**:215601.
162. Reich S, Li L, Robertson J. Structure and formation energy of carbon nanotube caps. *Phys Rev B* 2005;**72**:165423.
163. Wang H, Wang B, Quek X-Y, Wei L, Zhao J, Li L-J, et al. Selective synthesis of (9,8) single walled carbon nanotubes on cobalt incorporated TUD-1 catalysts. *J Am Chem Soc* 2010;**132**:16747–9.
164. Gomez-Gualdron DA, Zhao J, Balbuena PB. Nanocatalyst structure as a template to define chirality of nascent single-walled carbon nanotubes. *J Chem Phys* 2011;**134**:014705.
165. Zhu H, Suenaga K, Wei J, Wang K, Wu D. A strategy to control the chirality of single-walled carbon nanotubes. *J Crystal Growth* 2008;**310**:5473–6.
166. Saadi S, Abild-Pedersen F, Helveg S, Sehested J, Hinnemann B, Appel CC, et al. On the role of metal step-edges in graphene growth. *J Phys Chem C* 2010;**114**:11221–7.
167. Sinha N, Yeow JTW. Carbon nanotubes for biomedical applications. *IEEE Trans NanoBiosci* 2005;**4**:180–95.
168. Ebbesen TW, Ajayan PM, Hiura H, Tanigaki K. Purification of nanotubes. *Nature* 1994;**367**:519.
169. Ajayan PM, Ebbesen TW, Ichihashi T, Iijima S, Tanigaki K, Hiura H. Opening carbon nanotubes with oxygen and implications for filling. *Nature* 1993;**362**:522–5.
170. Hiura H, Ebbesen TW, Tanigaki K. Opening and purification of carbon nanotubes in high yields. *Adv Mater* 1995;**7**:275–6.
171. Dillon AC, Gennett T, Jones KM, Alleman JL, Parilla PA, Heben MJ. Simple and complete purification of single-walled carbon nanotube materials. *Adv Mater* 1999;**11**:1354–8.
172. Bandow S, Rao AM, Williams KA, Thess A, Smalley RE, Eklund PC. Purification of single-wall carbon nanotubes by microfiltration. *J Phys Chem B* 1997;**101**:8839–42.

173. Shelimov KB, Esenaliev RO, Rinzler AG, Huffman CB, Smalley RE. Purification of single-wall carbon nanotubes by ultrasonically assisted filtration. *Chem. Phys. Lett.* 1998;**282**:429–34.
174. Rinzler AG, Liu J, Dai H, Nikolaev P, Huffman CB, Rodriguez-Macías FJ, et al. Large-scale purification of single-wall carbon nanotubes: process, product, and characterization. *Appl Phys A* 1998;**67**:29–37.
175. Duesberg GS, Burghard M, Muster J, Philipp G. Separation of carbon nanotubes by size exclusion chromatography. *Chem Commun* 1998;435–6.
176. Duesberg GS, Blau W, Byrne HJ, Muster J, Burghard M, Roth S. Chromatography of carbon nanotubes. *Synth Met* 1999;**103**:2484–5.
177. Duesberg GS, Muster J, Krstic V, Burghard M, Roth S. Chromatographic size separation of single-wall carbon nanotubes. *Appl Phys A* 1998;**67**:117–9.
178. Li F, Cheng HM, Xing YT, Tan PH, Su G. Purification of single-walled carbon nanotubes synthesized by the catalytic decomposition of hydrocarbons. *Carbon* 2000;**38**:2041–5.
179. Chiang IW, Brinson BE, Huang AY, Willis PA, Bronikowski MJ, Margrave JL, et al. Purification and characterization of single-wall carbon nanotubes (SWNTs) obtained from the gas-phase decomposition of CO (HiPco process). *J Phys Chem B* 2001;**105**:8297–301.
180. Georgakilas V, Voulgaris D, Vázquez E, Prato M, Guldi DM, Kukovecz A, et al. Purification of HiPCO carbon nanotubes via organic functionalization. *J Am Chem Soc* 2002;**124**:14318–9.
181. Harutyunyan AR, Pradhan BK, Chang J, Chen G, Eklund PC. Purification of single-wall carbon nanotubes by selective microwave heating of catalyst particles. *J Phys Chem B* 2002;**106**:8671–5.
182. Hersam MC. Progress towards monodisperse single-walled carbon nanotubes. *Nat Nanotechnol* 2008;**3**:387–94.
183. Southern EM. Detection of specific sequences among DNA fragments separated by gel electrophoresis. *J Mol Biol* 1975;**98**:503–17.
184. Schägger H, von Jagow G. Tricine-sodium dodecyl sulfate-polyacrylamide gel electrophoresis for the separation of proteins in the range from 1 to 100 kDa. *Anal Biochem* 1987;**166**:368–79.
185. Orita M, Iwahana H, Kanazawa H, Hayashi K, Sekiya T. Detection of polymorphisms of human DNA by gel electrophoresis as single-strand conformation polymorphisms. *Proc Natl Acad Sci USA* 1989;**86**:2766–70.
186. Krupke R, Hennrich F, Löhneysen Hv, Kappes MM. Separation of metallic from semiconducting single-walled carbon nanotubes. *Science* 2003;**301**:344–7.
187. Doorn SK, Fields RE, Hu H, Hamon MA, Haddon RC, Selegue JP, et al. High resolution capillary electrophoresis of carbon nanotubes. *J Am Chem Soc* 2002;**124**:3169–74.
188. Doorn SK, Strano MS, O'Connell MJ, Haroz EH, Rialon KL, Hauge RH, et al. Capillary electrophoresis separations of bundled and individual carbon nanotubes. *J Phys Chem B* 2003;**107**:6063–9.
189. Heller DA, Mayrhofer RM, Baik S, Grinkova YV, Usrey ML, Strano MS. Concomitant length and diameter separation of single-walled carbon nanotubes. *J Am Chem Soc* 2004;**126**:14567–73.
190. Alexandre AV, et al. Fractionation of SWNT/nucleic acid complexes by agarose gel electrophoresis. *Nanotechnology* 2006;**17**:4263.
191. Zheng M, Jagota A, Strano MS, Santos AP, Barone P, Chou SG, et al. Structure-based carbon nanotube sorting by sequence-dependent DNA assembly. *Science* 2003;**302**:1545–8.
192. Zheng M, Jagota A, Semke ED, Diner BA, McLean RS, Lustig SR, et al. DNA-assisted dispersion and separation of carbon nanotubes. *Nat Mater* 2003;**2**:338–42.
193. Zheng M, Semke ED. Enrichment of single chirality carbon nanotubes. *J Am Chem Soc* 2007;**129**:6084–5.

194. Schuck P. Size-distribution analysis of macromolecules by sedimentation velocity ultracentrifugation and Lamm equation modeling. *Biophys J* 2000;**78**:1606–19.
195. Lopes-Virella M, Stone P, Ellis S, Colwell J. Cholesterol determination in high-density lipoproteins separated by three different methods. *Clin Chem* 1977;**23**:882–4.
196. Piatak M, Saag M, Yang L, Clark S, Kappes J, Luk K, et al. High levels of HIV-1 in plasma during all stages of infection determined by competitive PCR. *Science* 1993;**259**:1749–54.
197. Green AA, Hersam MC. Ultracentrifugation of single-walled nanotubes. *Mater Today* 2007;**10**:59–60.
198. Arnold MS, Stupp SI, Hersam MC. Enrichment of single-walled carbon nanotubes by diameter in density gradients. *Nano Lett* 2005;**5**:713–8.
199. Arnold MS, Green AA, Hulvat JF, Stupp SI, Hersam MC. Sorting carbon nanotubes by electronic structure using density differentiation. *Nat Nanotechnol* 2006;**1**:60–5.
200. Wei L, Wang B, Goh TH, Li L-J, Yang Y, Chan-Park MB, et al. Selective enrichment of (6,5) and (8,3) single-walled carbon nanotubes via cosurfactant extraction from narrow (n, m) distribution samples. *J Phys Chem B* 2008;**112**:2771–4.
201. Hirsch A. Functionalization of single-walled carbon nanotubes. *Angew Chem Int Ed* 2002;**41**:1853–9.
202. Liu Z, Tabakman S, Welsher K, Dai H. Carbon nanotubes in biology and medicine: in vitro and in vivo detection, imaging and drug delivery. *Nano Res* 2009;**2**:85–120.
203. Chen J, Hamon MA, Hu H, Chen Y, Rao AM, Eklund PC, et al. Solution properties of single-walled carbon nanotubes. *Science* 1998;**282**:95–8.
204. Georgakilas V, Kordatos K, Prato M, Guldi DM, Holzinger M, Hirsch A. Organic functionalization of carbon nanotubes. *J Am Chem Soc* 2002;**124**:760–1.
205. Chen RJ, Zhang Y, Wang D, Dai H. Noncovalent sidewall functionalization of single-walled carbon nanotubes for protein immobilization. *J Am Chem Soc* 2001;**123**:3838–9.
206. Curran SA, Ajayan PM, Blau WJ, Carroll DL, Coleman JN, Dalton AB, et al. A composite from poly(m-phenylenevinylene-co-2,5-dioctoxy-p-phenylenevinylene) and carbon nanotubes: a novel material for molecular optoelectronics. *Adv Mater* 1998;**10**:1091–3.
207. Liu Z, Tabakman S, Welsher K, Dai HJ. Carbon nanotubes in biology and medicine: in vitro and in vivo detection, imaging and drug delivery. *Nano Res* 2009;**2**:85–120.
208. Foldvari M, Bagonluri M. Carbon nanotubes as functional excipients for nanomedicines: I. pharmaceutical properties. *Nanomedicine* 2008;**4**:173–82.
209. Foldvari M, Bagonluri M. Carbon nanotubes as functional excipients for nanomedicines: II. Drug delivery and biocompatibility issues. *Nanomedicine* 2008;**4**:183–200.
210. Pascu SI, Arrowsmith RL, Bayly SR, Brayshaw S, Hu ZY. Towards nanomedicines: design protocols to assemble, visualize and test carbon nanotube probes for multi-modality biomedical imaging. *Phil Trans R Soc A* 2010;**368**:3683–712.
211. Jacobs CB, Peairs MJ, Venton BJ. Review: carbon nanotube based electrochemical sensors for biomolecules. *Anal Chim Acta* 2010;**662**:105–27.
212. Movia D, Del Canto E, Giordani S. Purified and oxidized single-walled carbon nanotubes as robust near-IR fluorescent probes for molecular imaging. *J Phys Chem C* 2010;**114**:18407–13.
213. Kostarelos K, Bianco A, Prato M. Promises, facts and challenges for carbon nanotubes in imaging and therapeutics. *Nat Nanotechnol* 2009;**4**:627–33.
214. Xie J, Lee S, Chen XY. Nanoparticle-based theranostic agents. *Adv Drug Deliv Rev* 2010;**62**:1064–79.
215. Zhang Y, Bai YH, Yan B. Functionalized carbon nanotubes for potential medicinal applications. *Drug Discov Today* 2010;**15**:428–35.
216. Cheung W, Pontoriero F, Taratula O, Chen AM, He HX. DNA and carbon nanotubes as medicine. *Adv Drug Deliv Rev* 2010;**62**:633–49.

217. Thakare VS, Das M, Jain AK, Patil S, Jain S. Carbon nanotubes in cancer theragnosis. *Nanomedicine* 2010;**5**:1277–301.
218. Menard-Moyon C, Venturelli E, Fabbro C, Samori C, Da Ros T, Kostarelos K, et al. The alluring potential of functionalized carbon nanotubes in drug discovery. *Expert Opin Drug Discov* 2010;**5**:691–707.
219. O'Connell MJ, Bachilo SM, Huffman CB, Moore VC, Strano MS, Haroz EH, et al. Band gap fluorescence from individual single-walled carbon nanotubes. *Science* 2002;**297**:593–6.
220. Cherukuri P, Bachilo SM, Litovsky SH, Weisman RB. Near-infrared fluorescence microscopy of single-walled carbon nanotubes in phagocytic cells. *J Am Chem Soc* 2004;**126**:15638–9.
221. Cherukuri P, Gannon CJ, Leeuw TK, Schmidt HK, Smalley RE, Curley SA, et al. Mammalian pharmacokinetics of carbon nanotubes using intrinsic near-infrared fluorescence. *Proc Natl Acad Sci USA* 2006;**103**:18882–6.
222. Hong H, Gao T, Cai W. Molecular imaging with single-walled carbon nanotubes. *Nano Today* 2009;**4**:252–61.
223. Keren S, Zavaleta C, Cheng Z, de la Zerda A, Gheysens O, Gambhir SS. Noninvasive molecular imaging of small living subjects using Raman spectroscopy. *Proc Natl Acad Sci USA* 2008;**105**:5844–9.
224. Zavaleta C, de la Zerda A, Liu Z, Keren S, Cheng Z, Schipper M, et al. Noninvasive Raman spectroscopy in living mice for evaluation of tumor targeting with carbon nanotubes. *Nano Lett* 2008;**8**:2800–5.
225. Liu ZA, Li XL, Tabakman SM, Jiang KL, Fan SS, Dai HJ. Multiplexed multicolor Raman imaging of live cells with isotopically modified single walled carbon nanotubes. *J Am Chem Soc* 2008;**130**:13540.
226. Chen Z, Tabakman SM, Goodwin AP, Kattah MG, Daranciang D, Wang X, et al. Protein microarrays with carbon nanotubes as multicolor Raman labels. *Nat Biotechnol* 2008;**26**:1285–92.
227. Cao GH, Burk LM, Lee YZ, Calderon-Colon X, Sultana S, Lu JP, et al. Prospective-gated cardiac micro-CT imaging of free-breathing mice using carbon nanotube field emission X-ray. *Med Phys* 2010;**37**:5306–12.
228. Gul H, Lu WB, Xu P, Xing J, Chen J. Magnetic carbon nanotube labelling for haematopoietic stem/progenitor cell tracking. *Nanotechnology* 2010;**21**:1–9.
229. Malhotra R, Patel V, Vaque JP, Gutkind JS, Rusling JF. Ultrasensitive electrochemical immunosensor for oral cancer biomarker IL-6 using carbon nanotube forest electrodes and multilabel amplification. *Anal Chem* 2010;**82**:3118–23.
230. Martinez MT, Tseng YC, Salvador JP, Marco MP, Ormategui N, Loinaz I, et al. Electronic anabolic steroid recognition with carbon nanotube field-effect transistors. *ACS Nano* 2010;**4**:1473–80.
231. Kam NWS, O'Connell M, Wisdom JA, Dai H. Carbon nanotubes as multifunctional biological transporters and near-infrared agents for selective cancer cell destruction. *Proc Natl Acad Sci USA* 2005;**102**:11600–5.
232. Moon HK, Lee SH, Choi HC. In vivo near-infrared mediated tumor destruction by photothermal effect of carbon nanotubes. *ACS Nano* 2009;**3**:3707–13.
233. Ghosh S, Dutta S, Gomes E, Carroll D, D'Agostino R, Olson J, et al. Increased heating efficiency and selective thermal ablation of malignant tissue with DNA-encased multiwalled carbon nanotubes. *ACS Nano* 2009;**3**:2667–73.
234. Robinson JT, Welsher K, Tabakman SM, Sherlock SP, Wang HL, Luong R, et al. High performance in vivo near-IR (> 1 mu m) imaging and photothermal cancer therapy with carbon nanotubes. *Nano Res* 2010;**3**:779–93.
235. Liu Z, Chen K, Davis C, Sherlock S, Cao Q, Chen X, et al. Drug delivery with carbon nanotubes for in vivo cancer treatment. *Cancer Res* 2008;**68**:6652–60.

236. Gao L, Nie L, Wang T, Qin Y, Guo Z, Yang D, et al. Carbon nanotube delivery of the GFP gene into mammalian cells. *Chembiochem* 2006;**7**:239–42.
237. Zhao B, Hu H, Mandal SK, Haddon RC. A bone mimic based on the self-assembly of hydroxyapatite on chemically functionalized single-walled carbon nanotubes. *Chem Mater* 2005;**17**:3235–41.
238. Mattson M, Haddon R, Rao A. Molecular functionalization of carbon nanotubes and use as substrates for neuronal growth. *J Mol Neurosci* 2000;**14**:175–82.
239. Hu H, Ni Y, Mandal SK, Montana V, Zhao B, Haddon RC, et al. Polyethyleneimine functionalized single-walled carbon nanotubes as a substrate for neuronal growth. *J Phys Chem B* 2005;**109**:4285–9.
240. Gabay T, Jakobs E, Ben-Jacob E, Hanein Y. Engineered self-organization of neural networks using carbon nanotube clusters. *Phys A* 2005;**350**:611–21.

Engineered Biological Entities for Drug Delivery and Gene Therapy: Protein Nanoparticles

JOAN DOMINGO-ESPÍN,[*,†,‡] UGUTZ UNZUETA,[*,†,‡] PAOLO SACCARDO,[*,†,‡] ESCARLATA RODRÍGUEZ-CARMONA,[*,†,‡] JOSÉ LUÍS CORCHERO,[*,†,‡] ESTHER VÁZQUEZ,[*,†,‡] AND NEUS FERRER-MIRALLES[*,†,‡]

[*]*Institute for Biotechnology and Biomedicine, Universitat Autònoma de Barcelona, Bellaterra, Barcelona, Spain*

[†]*Department of Genetics and Microbiology, Universitat Autònoma de Barcelona, Bellaterra, Barcelona, Spain*

[‡]*CIBER de Bioingeniería, Biomateriales y Nanomedicina (CIBER-BBN), Bellaterra, Barcelona, Spain*

I. Introduction 248
II. Protein Nanoparticle Formulation and Biological Barriers 249
  A. Interaction with Drugs and Nucleic Acids 250
  B. Protein Stability in Serum 254
  C. Defeating Biological Barriers 254
  D. Ways to Get Over the BBB 257
III. Multifunctional Proteins 257
  A. Protein Engineering: Direct Evolution, *De Novo* Synthesis, and Rational Design 257
  B. Designing a Protein Nanoparticle 259
  C. Production of Protein Nanoparticles 261
  D. Physicochemical Characterization 262
IV. Natural Self-Assembling Protein Nanoparticles: VLPs 264
  A. Structure of Protein Self-Assembled Nanovehicles 265
  B. Characteristic Features of VLPS and Their Limitations 265
  C. Tuning VLPS (Chemically or Genetically) for Their Uses/ Applications in Gene Therapy and Drug Delivery 268
V. Nonviral Self-Assembling Proteins 270
VI. Medical Applications of Protein Nanoparticles 274
  A. Therapeutic Protein Nanoparticles Currently in the Market 274
  B. Therapeutic Protein Nanoparticles Currently in Clinical Trials 276

Progress in Molecular Biology
and Translational Science, Vol. 104
DOI: 10.1016/B978-0-12-416020-0.00006-1

C. Therapeutic Protein Nanoparticles in Preclinical Models of Human Diseases 277
VII. Conclusions 281
References 282

The development of genetic engineering techniques has speeded up the growth of the biotechnological industry, resulting in a significant increase in the number of recombinant protein products on the market. The deep knowledge of protein function, structure, biological interactions, and the possibility to design new polypeptides with desired biological activities have been the main factors involved in the increase of intensive research and preclinical and clinical approaches. Consequently, new biological entities with added value for innovative medicines such as increased stability, improved targeting, and reduced toxicity, among others have been obtained. Proteins are complex nanoparticles with sizes ranging from a few nanometers to a few hundred nanometers when complex supramolecular interactions occur, as for example, in viral capsids. However, even though protein production is a delicate process that imposes the use of sophisticated analytical methods and negative secondary effects have been detected in some cases as immune and inflammatory reactions, the great potential of biodegradable and tunable protein nanoparticles indicates that protein-based biotechnological products are expected to increase in the years to come.

I. Introduction

The design of new chemical entities (NCE) for diagnosis and treatment of human diseases has relied on the discovery of active chemical drugs from a diverse library of compounds or from naturally occurring molecules.[1,2] Further chemical modifications improve pharmacokinetic properties to obtain a final product with a known mechanism of action and decreased toxicity.[3] Nonetheless, using such approaches, the final products present low specificity for their target molecules, interacting with many other molecules and accumulating in some tissues, disturbing the correct homeostasis of the system. In some cases, the adverse effects of drug administration exceed pharmacological effect and despite the concise mechanism of action of the drug over the target molecule representing an improvement in the patient's state, the treatment has to be prevented or discontinued.[4] In fact, although a maintained steady increase in the number of launched NCE has been observed in the last years, the question arises whether this classical approach has already exhausted the discovery of innovative molecules.[5]

On the other hand, macromolecular new biological entities (NBE) have been used to supplement cellular deficiencies or to inhibit cellular pathways exploiting their relatively specific mode of action. Proteins and peptides have been obtained first from their natural source or produced as recombinant versions after the development of genetic engineering techniques in the late 1970s. However, the delivery of biological entities is sometimes hampered by its low half-life in the bloodstream by unspecific degradation, resulting in an expensive and ineffective process. Nevertheless, some solutions have already been explored for biopharmaceuticals to increase solubility and stability and to reduce immunogenicity including postranslational modifications such as glycosylation and covalent conjugation of polyethylene glycol.[6]

Thus, one of the main objectives in the use of drugs (for either NCE or NBE) is the need to optimize the delivery system to reduce the pharmacological dose which would consequently represent a concomitant reduction in toxicity and cost. In that scenario, new delivery approaches have been implemented using biological interactions such as antigen–antibody binding (immunoliposomes)[7] or more sophisticated interactions including the binding between nutrient concentrator SPARC (secreted protein acidic and rich in cysteine) and albumin in the treatment of some types of cancer (Abraxane®).[8,9]

Proteins can be then used for their targeting qualities as molecular delivery vehicles both for the specific delivery of drugs or nucleic acids in gene therapy approaches and by themselves as therapeutic molecules. One of the interesting characteristics of proteins is their ability to form intermolecular driven complexes as sophisticated and structurally perfect as in the case of viral capsids. In addition, through the use of genetic engineering, recombinant proteins can be tuned to include additional properties to optimize drug delivery and nucleic acid delivery in gene therapy.

In this chapter, the main available strategies to develop protein-based nanovehicles or biopharmaceuticals will be described. In this context, several parameters will be defined such as proper formulation, stability, immunogenicity, and delivery to the correct cell type and cell compartment. Modular protein engineering, virus-like particles (VLPs), and other self-assembling entities are envisioned as modulatable novel protein nanoparticles able to include many desirable properties in the correct delivery of drugs and nucleic acids. Finally, some successful examples of protein nanoparticles on the market will be described in addition to protein products currently in clinical trials and under preclinical research in order to envision which type of protein nanoparticles will be available soon on the market.

II. Protein Nanoparticle Formulation and Biological Barriers

When a protein-only nanoparticle is meant to be used as a vector to deliver therapeutic nucleic acid, drug, or peptide, there are several steps that the nanoparticle has to perform to successfully get inside the target cell. In the first instance, it is necessary to obtain the proper formulation of the complex with the therapeutic molecule to generate a vehicle capable of being transported in the blood if a systemic administration is needed and retaining a significant stability before reaching the target cell.[10,11] In addition, the biological system poses specific barriers that have to be overcome such as membranes (cytoplasmic, endocytic, and nuclear), degradation (protease degradation induced by acid denaturalization in lysosomes, cytosolic proteosomes, and nucleases), cytosolic transport, and nuclear entry if necessary.[12,13] For central nervous system therapies, the blood–brain barrier (BBB) represents the main bottleneck, and for that, a specific strategy has to be designed.[14] Furthermore, the therapeutic complex has to be flexible enough in order to release the therapeutic molecule in the specific cell compartment.

Thus, several protein motifs have been described to overcome each and every process described earlier so that a modular multifunctional protein can be generated including those modules that are necessary to achieve its goal. In order to get a rational construction of the multifunctional vector, each step has to be carefully taken into account so as to overcome every step which is needed to achieve its final goal (Table I).

A. Interaction with Drugs and Nucleic Acids

The DNA/RNA condensation or drug interaction with the protein vector is a critical step in the formulation of protein nanoparticles for gene therapy. They have to remain attached to the vector during the whole transport process through the body and the cell until it can be released in the desired localization within the target cell. Highly positively charged peptides containing a large number of arginines or polylysines have been used to promote electrostatic interactions since nucleic acids are highly negatively charged molecules.[15–22] Natural DNA-condensing proteins as nuclear histidines or protamines can also be used to bind nucleic acids.[22–25] Protamine, which is the protein that replaces histidines during the spermatogenesis process, is a sperm chromatin component and just as the histidines do, it has very high DNA condensation ability to protect nucleic acids form cytosolic endonucleases.[23,26] In addition, as soon as the complex reaches the cellular nucleus, protamine is degraded by chromatin-remodeling proteins, releasing the transported DNA allowing its expression.[15,23] In contrast, polycationic DNA condensation modules such as polylysines and polyarginines—even they can present higher DNA condensation ability depending on the polycationic chain length—usually present lower DNA-releasing ability, interfering negatively with the accessibility of cellular transcription factors and DNA expression capacity.[15]

All these DNA condensation modules described above interact with any DNA that is incubated in an unspecific way. However, there are proteins such as GAL4 that are able to recognize specific DNA sequences[27–29] and that permit to bind and condensate specific DNA sequences in the final vector.[30,31]

TABLE I

Selection of Peptide Motifs Used in Gene Therapy and Drug Delivery to Improve Protein Nanovehicle Performance

Peptide motif	Sequence	References
Nucleic acid condensation peptides		
Polylysine	$(KKKKKKKKKKKKKKKKKKKK)_n$	16–18
Polylysine containing peptides	YKAKKKKKKKKWK and derivatives	19–22
Salmon protamine	PRRRRSSSRPVRRRRRPRVSRRRRRRGGRRRR	23–25
GAL4	MKLLSSIEQACDICRLKKLKCSKEKPKCAKCLKNNWECRYSPK	30,31
Blood–brain barrier (BBB) peptides		
g7	H2N-Gly-L-Phe-D-Thr-Gly-L-Phe-L-Leu-L-Ser(O-β-D-glucose)-$CONH_2$	107
RVG	YTIWMPENPRPGTPCDIFTNSRGKRASNG	56
Tat	YGRKKRRQRRR	108
R9	RRRRRRRRR	14
Cell-penetrating peptides (CPP)		
Tat	GRKKRRQRRPPQ	36–41
R9	RRRRRRRRR	42
Penetratin	CRQIKIWFQNRRMKWKK	43,44
bPrPp	MVKSKIGSWILVLFVAMWSDVGLCKKRPKP	43
Transportan	CLIKKALAALAKLNIKLLYGASNLTWG	44–46
Receptor-specific ligands (ligand/receptor)		
RGD/integrins (mainly αvβ3)	GRGDSP	47,48
CXCL12/CXCR4	KPVSLSYRCPCRFFESHVARANVKHLKILNTPNCALQIVARLKNNNRQVCIDPKLKWIQEYLEKALN	49,50
Transferrin receptor ligand (12Aa)/transferrin receptor	THRPPMWSPVWP	51,52
EGF/EGF receptor	NPVVGYIGERPQYRDL	53,54

(Continues)

TABLE I (*Continued*)

Peptide motif	Sequence	References
Asioaloglycoprotein/ asioaloglycoprotein receptor		55
RVG/acetil-colin receptor	YTIWMPENPRPGTPCDIFTNSRGKRASNG	56
PLAEIDGIELTY/integrin a9b1	PLAEIDGIELTY	57
Molossin (RGD)/integrin	ICRRARGDNPDDRCT	58
Secretin/Secretin receptor	HSDGTFTSELSRLRDSARLQRLLQGLV	59
NL4 (loop 4 of nerve growth factor)/TrkA	CTTTHTFVKALTMDGKQAAWRFIRIDTAC	60
Neurotensin/Neurotensin receptor (NTRH)	ELYENKPRRPYIL	61
LSIPPKA, FQTPPQL, LTPATAI/LOX-1	LSIPPKA, FQTPPQL, LTPATAI	62
Monoclonal Abs/antigen recognized by the antibody	–	63–65
Endosomal escape fusiogenic peptides		
HA-2	GLFGAIAGFIENGWEGMIDGWYG	12,69,70
GALA	WEAALAEALAEALAEHLAEALAEALEALAA	12,71–74
KALA	WEAKLAKALAKALAKHLAKALAKALKACEA	12,75
JTS-1	GLFEALLELLESLWELLLEA	12,19
ppTG20	GLFRALLRLLRSLWRLLLRA	12,76
PPTG1	GLFKALLKLLKSLWKLLLKA	12,76
Melittin	GIGAVLKVLTTGLPALISWIKRKRQQ	12,77
Tat	GRKKRRQRRRPPQ	12,39,40
Penetratin	RQIKIWFQNRRMKWKK	12,78,79
Transportant	GWTLNSAGYLLGKINLKALAALAKKIL	12,45,46
INF 7	GLFEAIEGFIENGWEGMIDGWYG	12,80

Endosomal escape histidine-rich peptides		
CHK6HC	CHKKKKKKHC	12,22
H5WYG	GLFHAIAHFIHGGWHGLIHGWYG	12,80–82
LAH4	KKALLALALHHLAHLALHLALALKKA	12,83
Nuclear import peptides		
SV40 large T antigen	PKKKRKV	36
Tat	VIH transcription factor	37
EBNA-1	Epstein–Barr virus	91
Melittin	Honeybee venom (*Apis mellifera*)	77
M1 (c-myc transcription factor)	PAAKRVKLD	92
M2 (c-myc transcription factor)	RQRRNELKRSP	92
GAL4 amino terminal domain	Transcription factor	93
Protamines	Sperm DNA condensation protein	23
Histone H1	Nuclear DNA condensation protein	94,95
M9 (heterogeneous nuclear ribonucleoprotein A1 (hnRNP A1)	NQSSNFGPMKGGNFGGRSSGPYGGGGQYFAKPRNQGGY	96
Vp3	SV40 estructural protein Vp3	97
Adenovirus E1 protein C-terminus	KRPRP	98
Xenopus N1 protein	VRKKRKTEEESPLKDKDAKKSKQE	99
Fibroblast growth factor 3 (FDF3)	RLRRDAGGRGGVYEHLGGAPRRRK	100
Poly ADP-ribose polymerase (PARP)	KRKGDEVDGVDECAKKSKK	101
Xenopus protein nucleoplasmine	KRPAATKKAGQAKKKK	102

B. Protein Stability in Serum

In many cases, the multifunctional protein vector is *in vivo* administrated by the systemic route in order to travel in the blood and reach the target cells. That exposes the vector to all blood components, making it susceptible to be degraded. Thus, it is completely necessary that the vector remain in the blood long enough to be able to reach the target cells. It has also been described that naked DNA has an estimated half-life in blood of minutes[10]; so protein nano-vehicles in gene therapy, among other properties, are intended to protect nucleic acids from degradation.

One important factor when the vector is exposed to the blood is that it can be recognized by the immune system components and produces an immune response against the vector. Thus, it is also very important to try to make the vector as less antigenic as possible in order to avoid being degraded or even being toxic to the organism.[32]

C. Defeating Biological Barriers

1. Cell Binding and Internalization

Peptide uptake or internalization involves a step before the protein binding to the cell surface. This attachment can be either specific or unspecific but in all cases the promotion of its internalization is required.[33]

Positively charged peptides usually bind the cellular surface by unspecific electrostatic interactions with the negatively charged cell surface proteoglicans. This kind of peptides can be used in the multifunctional protein if specific targeting is not required.[33] Cell-penetrating peptides (CPPs) have been widely described as unspecific cell-binding and internalization peptides[34–46] (see also the chapter "Peptide Nanoparticles for Oligonucleotide Delivery" by Lehto *et al.* in this volume). However, specific interactions can be obtained by incorporating cell receptor ligands if cell or tissue targeting is required for the therapeutic action. Moreover, some of those ligand–receptor interactions promote the ligand–receptor complex internalization. Many peptides have been described in the literature as receptor-specific ligands so any of them can be added to the multifunctional proteins in order to confer them cell specificity.[47–62] The most natural specific ligands that can also be used for cell targeting are monoclonal antibodies.[32,63–65] In addition, if no specific peptides are available for an intended target, new specific binding peptides can be found by using phage display[66] or combinatorial chemistry.[67]

2. Endosomal Escape

Several internalization pathways are possible depending on the vector properties,[27,33] including endocytosis (clathrin/caveolae-mediated, clathrin/caveolae-independent), macropinocytosis, and non-endocytic pathways.

It is known that more than one internalization pathway can be performed at the same time but usually the peptide-based vector uses endocytic pathways.[68] Moreover, it seems that proteins that interact with a specific cellular receptor are internalized by the clathrin-mediated endocytic pathway.[33] Most of the generated endosomal vesicles will converge to late endosomes that eventually will fuse with cellular lysosomes.[15,33] Remaining in the cellular endosomes, the multifunctional protein will be degraded, so it is strictly necessary that the internalized multifunctional proteins be released into the cellular cytoplasm escaping from degradation.

Several peptides have been described that are able to promote endosomal escape and can be classified into two types depending on their escape mechanism: fusiogenic peptides and histidine-rich peptides.[36] The fusiogenic peptides are small peptides that have hydrophobic amino acids (Aa-s) interspersed at constant intervals with negatively charged Aa-s.[12,19,39,40,45,46,69–80] Thus, when early endosomes become late endosomes, their low pH induces a conformational change in the peptide, which adopts a alpha-helix structure, in an amphipathic structure able to fuse with the endosomal membrane, leading to pore formation and releasing all the endosomal content into the cell cytoplasm.[36] The histidine-rich peptides are small peptides with a high histidine content whose endosmolytic activity is mediated by a mechanism called "proton sponge".[12,22,80–83] When the endosomal pH becomes low in late stages, the imidazole groups of the histidines are protonated and attract endosomal Cl^- ions, buffering against the proton pump. Thus, the endosomes collapse by an osmolytic swelling process and the endosomal content is released to the cell cytoplasm.[36] Further details are given in the chapter "Peptide Nanoparticles for Oligonucleotide Delivery" by Lehto *et al.* in this volume.

3. Vector Stability in the Cytosol

Once the protein has achieved the cellular cytosol, it can be degraded by cellular proteases or by the cellular proteosome system.[84] It is important to avoid this process, especially if the protein has to reach the cellular nucleus. If the final target of the nanoparticle is the cellular cytoplasm, it is necessary that it remain there at least long enough to perform its therapeutical action.

Several peptide proteosome inhibitors have been described that are able to avoid this type of protein degradation. By adding these peptides to the final protein vector it is possible to protect it and enhance cytoplasmatic stability. Epstein–Barr virus nuclear antigen 1 (EBNA1) contains a proteosome inhibitor consisting of glycine–alanine repeats able to prevent proteosomal proteolysis. It has been shown that a minimum of 4 Aa-s Gly-Ala repeats are necessary to achieve such protective activity.[85–87] If the protein vector is carrying nucleic acids (DNA or RNA), degradation by the cytosolic endonucleases has to be taken into account, so it is also very important to protect this nucleic acid in order to maintain its integrity. Some DNA/RNA condensing peptides as protamines also protect the DNA against cytoplasmic endonucleases and enhance its stability as has been described above.[15]

4. Intracytosolic Mobility

The cellular cytoplasm is a very crowded and compartmentalized environment where cellular organelles and cytoskeleton make the free diffusion of macromolecules such as protein vectors difficult. However, cytoskeleton elements such as microtubules are used by endosomes and other cytosolic macromolecules for intracytosolic mobility.[33] Dyneins have been described as being capable of carrying those macromolecules and endosomes along the microtubules in a retrograde transport toward the nucleus. Some small peptides that are able to bind dyneins have been identified. They can be added to the multifunctional protein vector in order to mediate an intracytosolic mobility toward the cellular nucleus.[36] Several dynein-binding proteins have been identified in viruses that are able to use this transport system. Comparing those protein sequences, a consensus peptide sequence (KSTQT) that is able to bind to the dynein LC8 light chain has been identified.[88]

5. Nuclear DNA Delivery and Expression

Molecules lower than 45 kDa/10–30 nm are able to enter in the cellular nucleus by passive diffusion. However, macromolecules higher than 45 kDa/ 10–30 nm generally require an active transport system through the nuclear pore system. This transport mechanism generally requires a specific targeting signal peptide named nuclear localization signal (NLS). These signaling peptides are usually rich in basic Aa-s, which are recognized by the cellular importines and actively transported through the nuclear pore.[15,89] Monopartite or bipartite NLS sequences which are NLS peptides that have one or two NLS recognized sequences respectively have been described.[12] Thus, these peptidic sequences can be added into the final multifunctional protein if nuclear localization is required in order to express a carried DNA. It has been reported that a single NLS sequence is sufficient to transport the vector to the nucleus and that a large number of NLS sequences can result in inhibition of its activity.[90]

One of the most used NLS signal peptides are fragments derived from the 111–135 Aa-s of the simian virus SV40 large tumor antigen (T-ag). Other NLS sequences can be found in GAL4, protamines, or Tat.[23,36,37,77,91–102]

It is important that when the transported DNA reaches the cellular nucleus, it has to be released in order to be accessible to the nuclear transcription factors and achieve the desired expression level. Thus, while designing the multifunctional protein vector, this aspect has to be taken into account.

Once the DNA has been released in the cell nucleus, it will be necessary to control its expression level depending on which therapeutic action is being promoted. When the goal is to kill a cell as in cancer therapies, the uncontrolled DNA expression levels would not be a problem. However, when a specific protein expression level is required, achieving good control is very important.[13] Some expression systems have been developed that can be pharmacologically regulated by oral drug formulation.[103] Cell-specific promoters and enhancers can be also used in order to confer high cell specificity to the therapy.[104,105]

D. Ways to Get Over the BBB

The BBB is a hermetic barrier that only allows nonlipophilic molecules smaller than 400 Da to cross it. However, some human proteins such as insulin, transferrin, insulin-like growth factor, or leptins are able to go across it by receptor-mediated transporters. Thus, the most important factor limiting central nervous system-targeting therapeutics is the presence of the BBB.[106] Finding the way to cross it will be the main challenge.

Some peptides have been described that are able to reach the brain crossing the BBB. Moreover, it has been seen that they can be associated with another molecule and transported through the barrier. Thus, they could be interesting candidates to be included in the multifunctional vectors if central nervous system targeting is required.[14,56,107,108]

Antibodies have also been described that bind transferrin and insulin receptors and that are able to cross the BBB efficiently. They can be conjugated with large molecules, allowing its translocation to the central nervous system.[63,64,109–111]

III. Multifunctional Proteins

A. Protein Engineering: Direct Evolution, *De Novo* Synthesis, and Rational Design

The development of genetic engineering techniques has increased the natural repertoire of proteins for the design of useful and/or valuable proteins with the aim to obtain new proteins with desired functions. There are three main strategies leading to the construction of engineered proteins: (a) direct evolution, (b) *de novo* protein design, and (c) rational design.

Directed evolution has developed quickly to become a method of choice for protein engineers in order to create enzymes having desired properties for all kind of processes. Over the past decade, this technique has become a daily part of the molecular toolbox of every biochemist. This is emphasized by the increasing number of publications about the subject.[112]

In nature, evolution and creation of new functionalities is achieved by mutagenesis, recombination, and survival of the fittest. Directed evolution mimics this and is a process of iterative cycles of producing mutants and finding the mutant with the desired properties. Mutations can be introduced at specific places using site-directed mutagenesis or throughout the gene by random mutagenesis. Several mutagenesis techniques have been developed in order to avoid codon bias.[113,114] The first technique used to mimic evolution was DNA shuffling.[115] This method is based on the mixing and subsequent joining of different related small DNA fragments in order to form a complete new gene. In the process of shuffling, the recombination frequency is dependent on the degree of homology. A high level of recombination is important to get all possible combinations of mutations. Since recombination can be biased, several methods to overcome problems arising from the use of shuffling in the early years were tackled by novel strategies, all having their own advantages and disadvantages.[112] The products obtained by these methods have to be screened for desired qualities and not all of them can be easily screened.

De novo protein design offers the broadest possibility for new structures. It is based on searches for amino acid sequences that are compatible with a three-dimensional protein backbone template using *in silico* techniques. Several research groups in the field have applied *in silico* methods to design the hydrophobic cores of proteins, with the novel sequences being validated with experimental data.[116] *In silico* protein design has allowed novel functions on templates originally lacking those properties, modifying existing functions, and increasing protein stability or specificity. Beyond any doubt, intense research activities are ongoing in the field, the potential of which is simply enormous.[117] So far there have been numerous examples of full sequences designed "from scratch" that were confirmed to fold into the target three-dimensional structures by experimental data.[118] The zinc-finger protein designed by Dahiyat and Mayo[119] was the first one to appear by this method.

Rational design of proteins is based on the modification or insertion of selected amino acids or domains in a polypeptide chain backbone to obtain proteins with new or altered biological functions. When using that strategy, a detailed knowledge of the structure and function of the backbone protein is needed to make desired changes. This generally has the advantage of being inexpensive and technically feasible. However, a major drawback of this approach is that detailed structural knowledge of a protein is often unavailable or it can be extremely difficult to predict the effects of various mutations. Modular engineering enables, by using simple DNA recombinant techniques, the construction of chimerical polypeptides in which selected domains, potentially from different origins, provide the required activities. An equilibrate combination and spatial distribution of such partner elements has generated promising prototypes, able to deliver expressible DNA or molecules to tissue culture but also to specific cell types in whole organisms.[120] Modular fusion proteins that combine distinct functions required for cell type-specific uptake and intracellular delivery of DNA or drugs present an attractive approach for the development of self-assembling vectors for targeted gene or drug delivery.[121] One of the first examples was described by the group of Uherek *et al.* They combined a cell-specific target module (antibody fragment specific for the tumor-associated ErbB2 antigen), a DNA-binding domain (Gal4), and a translocation domain for endosomal escape.[121]

In this context, many strategies for the construction of safer vehicles are being explored and the number of nonviral prototype vectors for gene and drug delivery is noticeably increasing. Here, the common steps that an approach like this might explore are presented (Fig. 1).

B. Designing a Protein Nanoparticle

When designing a new protein for drug or gene delivery there are many critical aspects, namely (a) design of the vehicle itself, required functions, stability, etc.; (b) production of the protein, suitable expression system, purification procedure, scaling up process, etc.; (c) characterization of the vehicle by physicochemical and functional tests; and finally (d) the administration route and regulatory guidance for biological products. Although all these aspects belong to different disciplines, they have to be overviewed together. Here, the major needs of a modular protein for gene and drug delivery are presented.

To enhance the physicochemical stability of the cargo molecules and their resistance to nuclease/protease-mediated degradation, protein vehicles should ideally exhibit, like their natural counterparts (viruses), nucleic-acid binding and condensing properties.[27] Such abilities are, in general, conferred by cationic segments of the main scaffold molecules that interact with nucleic acids, mainly through electrostatic interactions. In addition, such complexes need to efficiently release the nucleic acid in the nucleus (if the cargo is a therapeutic gene), for which endosomal escape is required. Such functions have been found in some peptides in many natural molecules and they are suitable for functionalizing protein vehicles.

The ability to bind a particular cell type with high specificity is especially significant in a systemic delivery in which appropriate biodistribution and tissue targeting are essential.[122] For nuclear targeting, only naked short nucleic acids can freely enter the nucleus of nondividing cells via free diffusion through the nuclear pore. Large molecules require active transport mediated by NLSs that are often found in viral proteins. Because the molecular mass of plasmidic DNA varies from to 2 to 10 MDA, DNA that is to be expressed, and essentially

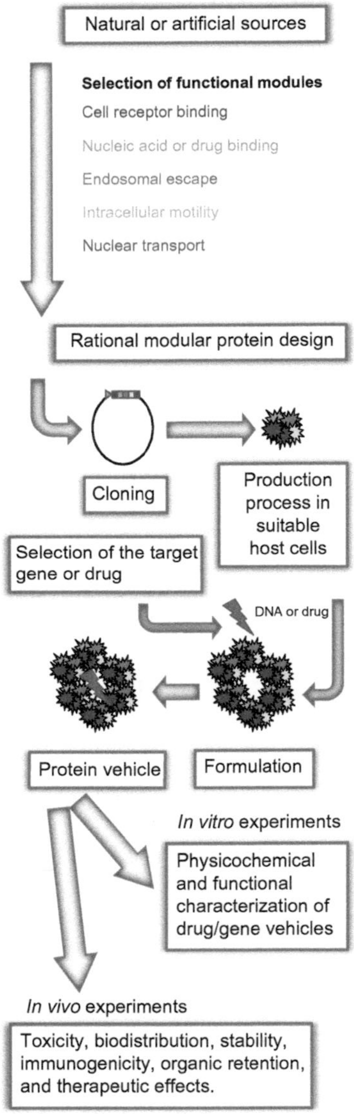

FIG. 1. Scheme in the development of protein nanoparticles for drug delivery and gene therapy. (See Color Insert.)

any macromolecular complex for nucleic acid delivery, requires NLSs.[123] The role and types of functional modules peptides used for all these purposes will be discussed in depth in the following sections.

Finally, which protein or peptide is better for a given cargo is to be determined empirically and only few rules can be taken literally.[38,124]

C. Production of Protein Nanoparticles

Some steps in the production of a protein-based vehicle after molecular cloning such as protein production and protein purification[125] might be experimentally labor intense with a variable success rate. For that reason, when small proteins are needed, solid-phase peptide synthesis[44] guarantees the process. However, the classical procedure of biological production allows scaling up the process in most of the cases and the production of larger polypeptides and full-length proteins.

Generally, in protein nanoparticle approaches, the protein is composed by different modules of natural sources such as the cell-penetrating peptide transactivator of transcription (TAT) derived from the TAT of the human immunodeficiency virus (HIV)[126] or artificial sequences not present in any organism such as the polylysine DNA-condensing sequence.[127]

Once it has been defined which modules will be part of the protein, it is important to define the order they will have in the final construct. It has been demonstrated by Boekle and coworkers using melittin conjugated to polyethylenimine (PEI) that depending on the side of the linkage (C- or N-terminus), the lytic activity could be changed. Some other modules have the need to be in a determined position for its correct function.[128]

When producing a protein for gene or drug delivery, it is important to know the origin of its domains to choose the most suitable expression system for its production. For instance, if any module naturally carries a posttranslational modification that is essential for its biological function, the expression system chosen will have to be able to reproduce the same crucial modification.

The main biological production systems for protein drugs are described below.

Escherichia coli is the most widely used prokaryotic organism for the expression of recombinant proteins.[129] The use of this host is relatively simple and inexpensive.[130] Added advantages include its short duplication time, growth to high cell densities, ease of cultivation, and high yields of the recombinant product. However, since it lacks fundamental prerequisites for efficient secretion, recombinant proteins manufactured by *E. coli* systems are mainly produced as inclusion bodies.[125,131] Moreover, posttranscriptional modifications are not achieved with this system. There are many examples of proteins for gene delivery produced in *E. coli* with probed efficiency.[132,133]

Like *E. coli*, yeasts can be grown cheaply and rapidly and are amenable to high-cell-density fermentations. Besides possessing complex posttranslational modification pathways, they offer the advantage of being neither pyrogenic nor pathogenic and are able to secrete more efficiently. Species established in industrial production procedures are *Saccharomyces cerevisiae*, *Kluyveromyces lactis*, *Pichia pastoris*, and *Hansenulapolymorpha*. *S. cerevisiae* is the best genetically characterized eukaryotic organism among them all and is still the prevalent yeast species in pharmaceutical production processes.[131] In spite of their physiological advantageous properties and natively high expression and secretion capacity, the employability of yeasts in some cases, however, might reach a limit, particularly when the pharmacological activity of the product is impaired by the glycosylation pattern. In such cases, either a postsynthetic chemical modification has to be considered or the employment of more highly developed organisms. Most examples of nanoparticles produced in yeast are for VLPs.[134]

Animal cell expression systems show the highest similarity to human cells regarding the pattern and capacity of posttranslational modifications and the codon bias. However, their culture is more complicated and costlier and usually yields lower product titers. Among the known systems, insect cells infected by baculovirus vectors have reached popularity since they are considered to be more stress-resistant, easier to handle, and more productive compared with mammalian systems and are thus frequently employed for high-throughput protein expression. For commercial application, scale-up related questions have to be solved.[135–137] Preferably applied in pharmaceutical production processes are mammalian systems like chinese hamster ovary (CHO) cells and baby hamster kidney (BHK) cells. These systems are genetically more stable and easier to transform and handle in scale-up processes, to grow faster in adherent and submerged cultures, and to be more similar to human cells and more consistent in their complete spectrum of modification.[138] In some cases, mammalian cell systems can be the only choice for the preparation of correctly modified proteins.

Peptides, being complex and unique complex molecules with regard to its chemical and physical properties, can be produced synthetically by the solid-phase method.[139,140] This technology can be used to avoid problems related to biological production. General advantages of synthetic peptides are that they are very stable compounds, solid-phase chemistry produces highly standardized peptides, and the crucial polycation component is provided by a "natural" polycation, thus minimizing toxicity.[141]

However, some disadvantages related to synthetic peptides have been reported such as the difficulty to synthesize long and well-folded oligopeptides, peptides with multiple cysteine, methionine, arginine, and tryptophan residues due to technical limitations or production cost.[141]

D. Physicochemical Characterization

When working with protein nanoparticles, it is very important to characterize them physically and functionally in order to understand their behavior.

The size and charge of protein/cargo particles are crucial properties which influence rates of diffusion, binding to polyanionic components of connective tissues, transversal of anatomical barriers, binding of serum proteins, attachment to cells, and mechanisms of endocytosis, among other factors. Stability in physiological salt solutions is a key issue for *in vivo* delivery, as salt is found everywhere in the body.[141] Mixing a multivalent polycation and DNA results in electrostatic binding of both molecules, with charge neutralization of DNA and a particle formation named conjugate. Charge neutralization can be easily seen by retardation gel assays and particle formation by dynamic light scattering (DLS). DLS is a good method to see particle formation but not to quantify relative number of particles of different sizes.[142]

To visualize particles, many groups have used transmission electron microscopy (TEM)[15,143] with good results while others have used fluid particle image analyzer (FPIA) to photograph individual particles in physiological solutions.[58]

The net charge of protein/cargo particles is an important variable. Generally, optimal gene delivery for cell lines requires a net positive charge but, as stated previously, it has to be determined empirically. One of the best techniques to determine the net charge is by calculating the Zeta potential that measures the electrophoretic mobility of particles.[144]

Despite the fact that physical characterization is a key element, understanding and testing the functionality and pharmacokinetics of a gene or drug is the most important part of its development process. Most of the initial tests are done using cell lines in *in vitro* experiments using reporter genes, RNA, or drugs.[145,146] Quantifying the percentage of transfected cells or drug-induced changes is a very valuable tool to evaluate nanoparticle performance in both nuclear and cytoplasmic delivery, respectively. In addition, *in vitro* experiments may be designed to select a candidate for the *in vivo* experiments from a group of possible therapy vectors.

The quantitative kinetics of particle binding, the molecular basis of particle interactions with target cell membranes, the efficiency of particle internalization, and endosomal escape are all poorly understood.[141]

Interaction of particles with plasma membranes prior to protein internalization can be either unspecific or specific. Untargeted delivery normally is the consequence of electrostatic interactions between anionic ligands in the cell surface and cationic components of the vehicle. On the other hand, targeted delivery to specific membrane molecules is a more sophisticated approach. It aims to improve cell specificity and efficiency, by directing to molecules, only expressed or overexpressed in a particular cell type, that initiate internalization by endocytosis. Targeting moieties include many types of molecules and is discussed afterwards.

Internalization of particles, its mechanisms, and kinetics are not well known and most studies about nanoparticle delivery do not focus on this aspect. There are several endocytic pathways each initiated by different ligands.[147] Enhancing the delivery by addition of chloroquine, a synthetic molecule used primarily for the prophylaxis and treatment of malaria that disrupts endosomes,[148] is an accepted parameter to demonstrate endosomal localization of particles.

Endosomal escape is the area most intensively investigated but is poorly understood. An important practical point to note is that some reagents that are used can be toxic.[141] To enhance this step, anionic fusogenic peptides can be used. These peptides fuse to membranes in an acidic-dependent manner causing its disruption.[149]

In gene delivery approaches, translocation of DNA expression plasmids into the cell nucleus involves an active, energy-dependent process through the nuclear pore complex.[150] Directly injected DNA into the cytosol is usually, but not always, poorly transferred to the nucleus[150,151] and because of that, the use of proteins carrying cationic nuclear-localizing sequences (such as that of SV40 large T antigen) has been widely used to overcome this step.[143]

IV. Natural Self-Assembling Protein Nanoparticles: VLPs

Ideal drug delivery and gene therapy vehicles must accomplish some desired features such as appropriate packaging size for its cargo, target cell-specificity, safe and efficient cargo delivery, and protection against immune recognition, or capability to escape immune recognition. Moreover, these vehicles must avoid inflammatory toxicity and rapid clearance.[152]

In this context, viral vectors have been exploited as one of the vehicles of choice. Viruses are nano-sized (15–400 nm) supramolecular nucleoprotein-based entities, covered or not with a lipid bilayer (enveloped/nonenveloped viruses) that satisfy, into relatively simple structures, outstanding properties and functions that are relevant to drug and gene delivery. Viruses are able to recognize and interact specifically with cells by receptor-mediated binding, internalize, escape from endosomes, and uncoat and release nucleic acids in different cellular compartments. They are also capable of transcribing and translating their viral proteins to self-assemble into new infectious virus particles and exit the host cell.[120,153–155]

Despite all these relevant properties of viral vectors or some other rising vehicles in drug and gene delivery such as cationic liposomes, their therapeutic use presents some limitations and risks because of the complexity of production, limited packaging capacity, insertional mutagenesis and gene inactivation, low probability of integration, reduced efficacy of repeat administration or reduced expression overtime, unfavorable immunological recognition or strong immune response against vehicle and transgene, inflammatory toxicity, and rapid clearance.[120,152] In this context, virus capsids or VLPs, produced by recombinant capsid proteins but lacking the viral genome, have noticeably emerged as a safer alternative to viral vectors.

A. Structure of Protein Self-Assembled Nanovehicles

VLPs are classically described as self-assembling, nonreplicative and non-pathogenic, highly organized supramolecular multiprotein nanoparticles (coats) (ranging from 20 to 100 nm) that can be formed from the minimal spontaneous self-assembling of one or more viral structural capsid proteins. It has been described that the self-assembling process of the structural viral proteins for VLP formation involves both spontaneous assembly, under favorable experimental conditions, and the requirement of scaffold proteins as catalysts.[156,157] Therefore, VLPs are considered protein "coats", "shells", or "boxes" that lack the viral genome, still conserve the structure, morphology, and some properties of viruses. Some of these properties such as cellular tropism and uptake, intracellular trafficking, membrane translocation, and transfer of nucleic acids or molecules across the cytoplasmic, endosomal, and nuclear membranes are important for drug delivery and gene therapy.[120,153,155,158–160] Usually, the degree of similarity of VLPs and their viruses depends on the number of proteins incorporated into the constructs.[161,162]

Since the first description in 1983 of the viral DNA packaging into mouse polyomavirus (MPyV) VLPs and its transduction *in vitro*,[163] VLPs of different viruses such as papillomaviruses,[164–166] hepatitis B, C, and E viruses,[167–169] polyomaviruses,[163,170–179] lentivirus,[180,181] rotavirus,[145,182] parvovirus,[183,184] and norovirus[185] have been generated.

B. Characteristic Features of VLPS and Their Limitations

VLPs offer some structure, dynamics, characteristic features, and functions that make them appealing bionanomaterials to be exploited in the biomedicine arena as drug and gene delivery vehicles and are discussed in detail afterward.

On the one hand, viral coat proteins have the ability to spontaneously self-assemble, which ensures the formation of highly organized, regular, repetitive structurally stable, and very low morphological polydisperse particles that provide useful properties to be used as scaffolds for bioimaging, synthesis of bionanomaterials, and as nanocarriers in drug and gene therapy.[186] In addition, homogeneity of particle size and composition is a desired production factor when developing therapeutic molecules. The overexpression of structural viral proteins in a convenient expression system renders recombinant proteins capable of being folded and assembled in discrete organized nanoparticles with a defined size corresponding to the natural capsid geometry.[187–189] Moreover, even though VLPs are structurally stable particles, some biochemical and structural studies have observed that viral capsids and bacteriophages may show some structurally dynamic properties varying in shape, size, or rearrangements of the coat proteins, in response to different factors such as pH.[190–193]

On the other hand, VLPs are considered biologically safe nanostructures since they are not infectious (lack of viral genome) and do not replicate, representing a safer alternative to viral vectors.[160,194–197] However, they can elicit immune and inflammatory responses, especially when repeated administration is needed.[152] It has to be also noted that when used in vaccination, VLPs could show excellent adjuvant properties and the majority of VLPs stimulate strong cellular and humoral immune responses as direct immunogens.[198] It has been suggested that recombinant VLPs derived from infection of insect cells with baculovirus or even those derived from prokaryotic systems could be contaminated with different residual components of these host cells, contributing those impurities to the adjuvant properties.[153]

One interesting property of VLPs is that coat viral proteins present an enormous elasticity and adaptability to be modified chemically and/or by protein genetic engineering[154,160,199] to incorporate multiple directed functionalities, in order to be addressed in biomedical applications such as drug delivery or gene therapy. It has been recently reviewed that chemically and/or genetically modified VLPs, including CPMV, CCMV, MS2, M13 bacteriophages, and other virus-based nanoparticles,[155,186] could maintain their structural integrity and improve their physical stability[154] and, moreover, these modifications could also confer desired cell-targeting properties to the nanovehicle.[153–155,186,200,201] VLPs can be successfully engineered with spatial precision to incorporate (attached or genetically displayed on the surface) targeting tissue-specific ligands such as epidermal growth factor (EGFR) and antibodies, or other molecules such as oligonucleotides, peptides, gold, and other metals, target proteins, carbohydrates, polymers, fluorophores, quantum dots, drugs, or small molecules.[152,154,155] Moreover, one of the potential benefits of such modifications is that the specific geometric rearrangement confers precise recognition patterns.[200,201]

Furthermore, accessibility of the materials carried within the particle and the ability of inclusion and separation of nucleic acids, small molecules, and unusual cargoes with appropriate charge is another outstanding feature and key advantage of VLPs that has also made them excellent vessels for gene and drug delivery.[152,195] As described above, VLPs can be used as empty nanocarriers to transport molecules chemically attached on their surface or can be loaded *ex vivo* with therapeutic small molecules such as drugs, DNAs, mRNAs, siRNAs, oligonucleotides, quantum dots, magnetic nanoparticles, or proteins.[155,157,160] VLPs of different papillomavirus and polyomavirus have been widely characterized and used for directed delivery in biomedical applications.[132,165,173,174,194,202] Osmotic shock and *in vitro* self-assembling of VLP subunits in the presence of the cargo have been the two main strategies used to packaged nucleic acid or other small molecules. It has to be taken into account that some attachment of the cargo on the VLP surface can occur.[195] Besides, diversity of natural tropism including liver for hepatitis B VLPs, spleen for some papillomavirus and polyomavirus VLPs, antigen-presenting cells for certain papillomavirus VLPs, and glial cells for human polyomavirus JC (JCV) VLPs, among others[152] is one of the key advantages offered by VLPs providing a wide spectrum of specific targeting and distribution profiles depending on the directed application. Although each VLP has its own characteristic receptors, entry pathway, and intracellular trafficking, it has been demonstrated that tropism of VLPs could be customized, modifying the residues identified as ligands of the cellular receptor on VLPs' surface or even varying the delivery routes.[155,189,203]

Another key advantage of VLPs is that they can be easily produced by using a wide range of hosts and expression systems, each of them with its own conditionings.[162] In the past years, there has been an increasing need to improve and optimize efficient large-scale production systems, process control and monitoring, and up- and down-streaming processes.[153,157,159,204] Production of VLPs usually involves transfection of the cell host expression system of choice with a plasmid encoding one or more viral structural proteins, further and rigorous purification for the removal of immunogenic cellular contaminants, and quality control of the produced VLP and encapsulation of the cargo *ex vivo* before administration.[152,158] The most frequent and convenient expression systems, adaptable to large-scale processes are (1) yeast cells[176], (2) mammalian cells, (3) insect cells infected with recombinant baculovirus[205,206], (4) bacteria[204,207], (5) green plants infected with modified viruses[208,209], and (6) cell-free systems.[163,204] The preparative and large-scale manufacture of VLPs in some of these hosts has been reviewed by Pattenden *et al.* and can be classified into two main methods of bioprocessing: *in vivo* and *in vitro* systems.[157] In addition, the capability of *in vitro* dissociation and reassociation of VLPs contribute to the application of easy and more accurate purification methods than those of viral vectors.[152,157] Furthermore, depending on the expression system, the resulting VLP might be significantly different even though expressing the same viral proteins. Thus, a broad spectrum of VLPs could be customized depending on the VLP type, the number of proteins needed for VLP assembling, and the targeted final application.[158,210]

As described above, VLPs have great potential as nanocarriers in drug and gene delivery. At the same time, although there is an increasing flow of developments in this area, these vehicles also present some limitations that should be addressed and taken into account, such as residual cellular components, variable yield of functional VLPs after disassembly/reassembly process, immunostimulation and unsuitability for repeated administration, tolerance to the transgene, ineffective therapeutic molecule loading, and low transfection rates.[152]

C. Tuning VLPS (Chemically or Genetically) for Their Uses/Applications in Gene Therapy and Drug Delivery

Due to their versatile nanoparticulate structure and morphology, and non-replicative and noninfecting nature combined with their natural immunogenic properties and ease production, VLPs have principally emerged as an excellent alternative tool to attenuate viruses for vaccination.[152,153,204,210] There are currently commercialized upon the US Food and Drug Administration (FDA) approvals of some VLP-based vaccines that effectively protect humans from hepatitis B virus (HBV) (GlaxoSmithKline's Engerix® and Merck and Co., Inc.'s Recombivax HB®) and human papillomavirus (HPV) (Cervarix®, an HPV 16/18 VLP vaccine developed by GlaxoSmithKline's and Gardasil® developed by Merck against types 6, 11, 16, and 18 HPV). Other immunogenic VLP-based vaccines are already under clinical trials, preclinical test, or basic investigation including HBV,[211,212] HIV,[180,213] influenza virus,[214] parvovirus,[159] Norwalk virus,[185] rotavirus,[182] and Ebola virus.[215,216]

Although VLP-based vaccines have been primarily developed for their use against the corresponding virus, in the last decades genetic engineering or chemical modifications have been applied in order to generate chimeric VLPs. Thus, on the one hand, commonly short heterologous peptide epitopes or full proteins that are unable to form VLPs or that are unsafe for vaccination have been presented on surface-exposed loops or fused to N- or C-exposed termini of structural viral capsid proteins on VLPs.[154,161,210] Different HPV,[217–219] HBV,[220,221] parvovirus,[222,223] and chimeric polyoma VLPs have been engineered[170,175] and tested for different applications including vaccination against viral or bacterial diseases, against virus-induced tumors, and more recently, for immunotherapy of nonviral cancer.[161,210] On the other hand, chemical bioconjugation for covalent coupling of protein epitopes and small molecules to lysines, cysteines, or tyrosine residues of VLP surfaces has been applied in viral or cancer vaccines.[200] Chackerian *et al.* have demonstrated the efficient induction of protective autoantibodies using self-antigen conjugation to HPV VLPs.[224]

It is important to point out that VLPs can also be engineered to incorporate heterologous cell-specific ligands to cell receptors, thus altering their cellular tropism.[154,155,186,201] This great convertibility and flexibility of VLPs to be modified (chemically and/or genetically), their high stability, natural and diverse tropism, their nanocontainer properties, and their ability to enter in the cell and incorporate, bind, and deliver nucleic acids and small molecules have positioned VLPs as appealing entities not only for vaccination applications but also for a broad spectrum of other diverse and emerging applications in nanomedicine and nanotechnology such as immunotherapy against cancer,[210,225] gene therapy delivery of therapeutic genes into specific cells,[161,165,171,184,226,227] and targeted delivery of drugs and small molecules using VLPs as nanocarriers.[174,196]

Although there is no commercial VLP as vector in gene therapy, since the initial work in 1970 of uncoating polyoma pseudovirus in mouse embryo cells as gene delivery vector[228] and the establishment in 1983 of the viral DNA packaging into MPyV VLPs and its transduction *in vitro,*[163] different VLPs such as HBV and hepatitis E virus,[229] HPV and polyomavirus nanoparticles[172,178,229] have been modified toward the specific delivery of therapeutic genes and proteins in different target cells, organs, and tissues *in vitro* and *in vivo* by systemic injection[229] or oral administration.[230] For example, recombinant VP1-based polyomavirus VLPs can encapsulate *in vitro* exogenous DNA, and deliver it by cell surface sialic acid residues to human brain cells and fetal kidney epithelial cells.[178] Furthermore, VLPs have recently emerged as novel nanocarriers or nanocontainers to store unnatural cargos, deliver modified oligonucleotides,[154] synthetic small interfering RNAs, and plasmids expressing short hairpin RNAs as therapy to downregulate gene expression.[171,231] In this context, Chou *et al.* have recently described the use of JCV VLPs as an efficient vector for delivering RNAi *in vitro* using murine macrophage RAW 264.7 cells and *in vivo* using BALB/c mice in silencing the cytokine gene of IL-10 without significant cytotoxicity for systemic lupus erythematosus gene therapy.[171]

One of the key aspects in targeted gene and drug delivery is cell-specific delivery. It is important to point out that VLPs are tunable nanoparticles that can also be chemically or genetically engineered to modify their natural cellular tropism in order to diversify the range of therapeutic applications in targeted gene or drug delivery.[154,201] Some effective approaches to modify the natural cellular tropism include:

(1) Genetic engineering of VLP chimeras incorporating heterologous cell-specific short peptides that contain recognition sites of target cell receptors.[232] In this context, polyoma and papillomavirus, with solved atomic structures of their major structural capsid proteins, have been extensively used to obtain chimeric VLPs as delivery vector systems.[165,233] However, this approach has some bioprocessing limitations such as low production levels as a consequence of VLP modification, alterations of size and properties of the VLPs that could affect the structural interactions and conformations for VLP assembly, disassembly and packaging, and low transduction efficiencies.[157]

(2) Chemical bioconjugation of purified VLPs with epitope-containing peptides[234,235] or a wide range of small molecules conferring cell-specific targeting such as transferrins, folic acid, or other targeting molecules. As an example, CMPV VLPs have been successfully conjugated with Tfn using "click" chemistry[236] and with NHS-ester-derivatized folic acid, demonstrating both as internalized into HeLa cells and KB cells, respectively.[183,184]

(3) High-throughput library and directed evolution method is a rational approach that has been recently used to engineer viral vectors with the desired tropism properties.[237]
(4) Pseudotyping, which consists of replacing the envelope protein of one virus species by the envelope protein of another virus species.[238]
(5) Modification of the delivery route of the VLPs. It has been shown that the levels of expression of β-galactosidase in heart, lung, kidney, spleen, liver, and brain are different depending on the delivery route of polyomavirus VP1 VLPs.[203]

The great accessibility and reactivity showed by VLPs, as well as their ability to serve as nanocarriers, which made them suitable to be exploited in gene therapy, have also been applied to targeted drug delivery.[195] Genetic modification and/or chemical functionalization of exposed amino acid residues on the capsid surface in order to attach small molecules, such as markers or bioactives molecules, is one of the most common approaches applied to target drug delivery.[174,239] As an example, canine parvovirus (CPV) VLPs produced in a baculovirus expression system and exhibiting natural tropism to transferrin receptors (TfRs) were chemically modified on accessible lysines of the capsid surface with fluorescent dye molecules and delivered to tumor cells. Derivatization of CPV-VLPs did not interfere with the binding and internalization into tumor cells.[183,184]

One limitation of VLPs in gene therapy is the low efficiency of gene transduction due to inefficient DNA packaging. However, a recent study presented a novel *in vivo* DNA packaging of JCV VLPs in *E. coli* that effectively reduced human colon carcinoma volume in a nude mouse model. In this study, the exogenous plasmid DNA was transformed into the JCV VP1 expressing *E. coli*. The packaging of the second plasmid occurs simultaneously as the *in vivo* assembly of the JCV VLP. Even though it is still not clear how the plasmid DNA molecules are encapsidated in the VLP, the authors showed that gene transduction efficiency by their *in vivo* package system was about 80% in contrast to the 1–2% of gene transduction efficiency achieved by the *in vitro* osmotic shock system.[226] In addition, the administration of exogenous proteins may induce the immune system response, reducing therapy effectiveness or causing undesirable secondary effects, albeit immunological response of protein nanoparticles can be modulated.[240]

V. Nonviral Self-Assembling Proteins

Spontaneous protein self-assembly to form ordered oligomers is a common event in biology. It can prove advantageous in terms of genome-size minimization, formation of large structures, stabilization of complexes, and inclusion of functional features.[241] It has been widely documented that cellular oligomer proteins as well as viral capsids are stabilized by several weak noncovalent interactions as hydrophobic interaction, electrostatic energy, and Van der Waals forces.[242–244] These interactions result in a complex quaternary structure described by three symmetry point groups named cyclic (Cn), dihedral (Dm), and cubic (T, O, I).[245,246]

The development of computational techniques to predict protein–protein interactions using solved 3D protein structures makes it possible to predict and/or strengthen experimental data performing in *in silico* approaches.[247] Furthermore, its use opens up the possibility to design proteins not only displaying specific biological functions but also interesting intermolecular interactions to obtain increased multivalency in the resulting complexes. Moreover, it should be considered that not only whole proteins can self-assemble in smart nanoparticles; oligopeptides are also capable of forming organized structures. Many applications are possible due to the enormous quantity of different combinations and features that can be exploited with peptides.[248,249]

Furthermore, protein–protein interactions are not the unique parameters involved in particle formation, nucleic acid–peptide interactions, salt concentration, order of mix, and ratio between nucleic acid and protein can also strongly influence the condensation process.[250,251]

Due to their natural tendency to self-assemble forming highly ordered structures, viruses provide a wide variety of scaffold proteins which are used as gene/drug carriers. Among them, VLPs have been reviewed in the previous section. However, simple bacterial proteins can be also utilized as carriers for gene delivery. For example, heat shock proteins (HSP) from hyperthermophilic archeaon *Methanococcus jannaschii* can assemble in a small structure of 24 subunits having an octahedral symmetry. These 12 nm structures are stable at high temperature, up to 70 °C, and wide range of pH. Residue modifications are allowed to elicit specific attachment of small molecules.[186,252]

In bacteria, bacterial microcompartments (BMC) which are intracellular organelles consisting of enzymes encapsulated within polyhedral, protein-only shells, somewhat similar to viral capsids, have been described. BMCs are composed of a few thousand copies of a few repeated protein species (including one or more enzymes involved in specific metabolic pathways), and with sizes of around 100–150 nm in cross section. The general role of BMCs is to confine toxic or volatile metabolic intermediates, while allowing enzyme substrates, products, and cofactors to pass.

The first described BMC, the carboxysome, was isolated in the early 1970s[253,254] and has been found to contain both CO_2-fixing ribulose bisphosphate carboxylase/oxygenase (RuBisCO)[253,254] and carbonic anhydrase[255–257] enzymes. Carboxysomes' function is to enhance autotrophic CO_2 fixation at low CO_2 levels.

Other BMCs were later identified in cyanobacteria and some chemoautotroph bacteria. Among them, BMC proteins have been later found to be encoded in the propanediol utilization operon (pdu) of the heterotroph Salmonella[258] and by an operon for metabolizing ethanolamine (eut) in enteric bacterial species, including *Salmonella* and *Escherichia.*[259] *Salmonella enterica* forms a polyhedral organelle during growth on 1,2-propanediol (1,2-PD) as a sole carbon and energy source, but not during growth on other carbon sources.[260,261]

The pdu organelles' function is to minimize the harmful effects of a toxic intermediate of 1,2-PD degradation (propionaldehyde).[261–263] Other studies have shown that a polyhedral organelle is involved in ethanolamine utilization (eut) by *S. enterica.*[259] The function of the eut microcompartment is to metabolize ethanolamine without allowing the release of acetaldehyde into the cytosol, therefore minimizing the potentially toxic effects of excess aldehyde in the bacterial cytosol[264–266] and also preventing volatile acetaldehyde from diffusing across cell membrane.[267]

So far, about 1700 proteins containing BMC domains have been identified, covering at least 10 different bacterial phyla. The typical BMC protein consists of approximately 90 amino acids, with an alpha/beta fold pattern.[268,269] Some individual BMC proteins self-assemble to form hexamers, which further assemble side by side to form the flat facets of the shell.[268,270,271] The formation of icosahedral, closed shells from such flat layers was elucidated in part by structural studies in carboxysomes: some BMC proteins assemble to form pentamers, which are located at and form the vertices of the icosahedral shell.[270]

Mechanisms directing enzyme encapsulation within protein-based BMCs have been studied during the last years. It has been described that, in some carboxysomes, protein CcmM is used as a scaffold to form interactions between both shell proteins and enzymes,[272,273] through a CcmM C-terminal region with homology to the small subunit of RuBisCO.[274] Other studies revealed that pdu shells can self-assemble without needing interior enzymes[275] and that carboxysomes can self-assemble *in vivo* when RuBisCO has been deleted.[276]

Regarding properties of the encapsulated enzymes, in the pdu BMC some of the internal enzymes are encapsulated by specific N-terminal targeting sequences.[275,277] In this line, Sutter and colleagues[278] described a conserved C-terminal amino acid sequence that mediates the physical interaction of an iron-dependent peroxidase (DyP) or a protein closely related to ferritin (Flp) with a specific type of BMC (encapsulins).

In another example, an icosahedral enzyme complex, lumazine synthase (AaLS) from *Bacillus subtilis* and *Aquifex aeolicus*, was engineered to encapsulate target molecules by means of charge complementarity and can also be modified to give different characteristics to the assembled structure.[279,280]

Moreover, enzymatic subunits, like E2 of pyruvate dehydrogenase from *Bacillus stearotermophilus*, can be modified to be used in gene delivery. E2 peptides naturally form a dodecahedron of 60 subunits of 24 nm in diameter allowing modification for drug-like accommodation. The assembling/disassembling of these structures can be modulated by changing the operative pH in the experimental environment. These nanoparticles can also be functionalized with antigens for vaccine development.[281,282]

According to these results, specific targeting sequences could be of use in biotechnological applications to package proteins inside the stable self-assembled icosahedral shell of BMCs, offering appealing opportunities to manipulate in the laboratory such nanocages to fill them with therapeutic molecules. The simplicity of this system makes it very attractive for engineering studies to design, mimicking nature, new applications in biotechnology, providing a new, intriguing platform of microbial origin for drug delivery.

Bovine serum albumin (BSA) is able to form microspheres after sonochemical treatment in aqueous medium. Chemical effects of ultrasound radiation and coupling with an anticancer drug such as Taxol (paclitaxel) led to the assembling of a spherical carrier with an average diameter of 120 nm. BSA particles resulting from S–S bonds, due to HO_2 radical formation, are able to release the encapsulated Taxol in cancer tissue with best results if compared with mere Taxol treatment. This drug for breast cancer treatment is commercially available.[283,284]

Also little cationic peptides can lead to self-assembling particles. Among others, arginine-rich cationic peptides are widely known as good tools for gene delivery. For example, purified R9-tailored GFP in solution is described to form nanodisk particles 20 nm in diameter. This structure is proved to be induced by the 9 arg tails and is able to bind and condense DNA. These nanodisks are also able to deliver DNA toward the nucleus where the reporter gene is expressed.[285]

On the other hand, the expression of recombinant proteins over physiological rates can cause a bad functioning of cellular quality control system, leading to self-organizing, pseudo-spherical, protein aggregates known as inclusion bodies. These mechanically stable nanoparticles, ranging from 50 to 500 nm in diameter, were considered for a long time as undesired bio-products. Recently, it became clearer that they are suitable for medical approaches when utilized as scaffold surface to promote cellular proliferation.[286–288]

One of the most difficult goals for a foreign gene delivery is to reach the nucleus. An approach to overpass this obstacle is by fusing an NLS in a nonessential position of a DNA-binding protein. Such type of modification has been described for a tetracycline repressor protein (TetR) fused with an SV40 NLS. The TetR–NLS affinity and specificity to TetO DNA sequence is exploited to form spontaneous protein–DNA complexes which allow an enhancing of DNA transportation into the nucleus and subsequent expression of foreign genes, combining the two peculiar characteristics of each fusion component.[289]

VI. Medical Applications of Protein Nanoparticles

There is still a tremendous gap between progresses made in protein-based nanoparticle research for drug delivery and clinical reality. Hundreds of publications in basic research describe the combination of two or more functional elements in a single protein nanoparticle, by which the delivery of a carried drug is enhanced. These agents act by improving critical steps in the drug delivery process, such as increasing the systemic stability or tissue specificity, favoring internalization, endosomal escape, and entry into the nucleus, or transporting therapeutic material through the BBB, in *in vitro* and *in vivo* studies.

Besides the human recombinant therapeutic proteins currently on the market (or functional segments of them), there are also some fusion proteins approved for clinical use (most by incorporating an antibody fragment or a ligand to enhance cell specificity). Sadly no gene therapy trials have so far used full protein carriers *in vivo*, but rather peptide-functionalized vehicles.

Bottlenecking the gap between research and clinical application, the US FDA/European Medicines Agency (EMEA) only approves human proteins, to avoid the risk of an immune response that could affect not only the effectiveness of the nanoparticle but also challenge patients' health. Another critical factor is the administration route, where the protein is degraded before arriving at the target; this problem could be solved or minimized by the use of protein D-isomers, PEGylation, or the design of protecting groups for labile sites. Despite the current situation mentioned above, there are many good examples of multifunctional modular proteins that, when carrying therapeutic material, can improve the prognosis *in vivo* in animal models for different diseases. These examples are reviewed below, along with those few protein nanoparticles that are currently on the market or in clinical trials.

A. Therapeutic Protein Nanoparticles Currently in the Market

Albumin is a natural protein transporter of hydrophobic molecules throughout plasma that has been approved by the FDA to reversibly bind water-insoluble anticancer agents, as is the case of albumin-bound (nab) paclitaxel, Abraxane®. This albumin-nab technology-based drug is in use in patients with metastatic breast cancer who have failed combination therapy, and it is the first protein nanoparticle approved by the FDA. Albumin potentiates paclitaxel concentration within the tumor by increasing paclytaxel endothelial transcytosis through caveolae formation. It also contributes to the fact that tumors secrete an albumin-binding protein SPARC (also called BM-40) to attract and keep albumin-bound nutrients inside the tumor cell.[290] The albumin–paclitaxel complex was not formally considered a nanoparticle in the United States (due to an average size of 130 nm) but only so in Europe.

Apart from whole recombinant therapeutic proteins being currently commercialized, there are also some examples of vehicles formed by chimerical proteins with target ligands already in the market. DAB389IL-2 (denileukin diftitox or Ontak) is a fusion of Diphtheria toxin catalytic and translocation domains for lethal effect and interleukin-2 (IL-2) to gain cell specificity in the treatment of persistent or recurrent T-cell lymphoma. Belatacept (BMS-224818) is a CTLA4-Ig fusion protein formed by the cytotoxic T-lymphocyte-associated antigen 4 joined to an immunoglobulin G1 Fc fragment fusion protein, developed by Bristol–Miers–Squibb. Etanercept (Enbrel) fusion tumor necrosis factor receptor (TNFR), which binds and inhibits specifically TNF activity, to an immune globulin G1 Fc, to prevent inflammation mediated by TNF in autoimmune diseases like arthritis and psoriasis.

On the other hand, fusion proteins which include an antihuman epidermal growth factor receptor 2 (HER2) monoclonal antibody that binds tumor cell surfaces, among them the so-called "trastuzumab" (commercialized as Herceptin by Roche), associated to DM-1, an antimitotic drug, aimed at improving the treatment of breast cancer.

Finally, VLPs, that is, empty viral entities formed by the self-assembly of a viral capsid protein, are the only truly protein nanoparticles (architectonically speaking) which are currently used in clinical practice. HBsAg recombinant protein of HBV expressed in yeast and the capsid L1 recombinant protein of HPV (types 6, 11, 16, and 18) administered currently as vaccines tend to form spontaneously VLPs that elicit T and B immune response. Recently, there have been preclinical and clinical trials to test the security and efficacy of VLP vaccines against Chikungunya[291] and seasonal influenza virus (http://www.medpagetoday.com/MeetingCoverage//ICAAC/22129), respectively. Influenza VLP vaccines have proven to provide complete protection against H1N1 2009 flu pandemics,[292] within a record preparation time when compared to 9 months for traditional vaccines. The use of VLPs as a delivery system for drugs or nucleic acids in gene therapy is still under investigation.[194]

Drugs and proteins may be transformed through pegylation, a process that can assist them in overcoming some of the potential problems that delay the adoption of protein nanoparticles for clinical use. The covalent attachment of PEG can reduce immunogenicity and antigenicity by hiding the particle from the immune system, can increase the circulating time by reducing renal clearance, and can also improve the water solubility of a hydrophobic particle. The use of pegylation has been approved for commercial use by the FDA and EMEA, and some examples of pegylated protein products are Adagen® (PEG-bovine adenosine deaminase), the first pegylated protein approved by the FDA in 1990, Pegasys® (PEG-interferon alpha), and Oncaspar® (PEG-L-asparaginase).

B. Therapeutic Protein Nanoparticles Currently in Clinical Trials

The majority of protein nanoparticles studied in clinical trials (http://clinicaltrias.gov) are fusion proteins composed of a therapeutic protein/peptide and a target cell-specific ligand. An example is ALT-801, a biologic compound composed of IL-2 genetically fused to a humanized soluble T-cell receptor directed against the p53-derived antigen. The clinical trials evaluated whether directing IL-2 activity using ALT-801 to the patient's tumor sites that overexpress p53 results in clinical benefits (NCT01029873, NCT00496860). Another ligand joined to IL-2 is L19, a tumor-targeted immunocytokine constituted of a single chain fragment variable (scFv) directed against the ED-B domain of fibronectin, one of the most important markers for neoangiogenesis. L19–IL-2 is in a Phase I/II study for patients with solid tumors and renal cell carcinoma (RCC) (NCT01058538). L19 has also been fused to TNFα with the intention to target TNFα directly to tumor tissues resulting in high and sustained intralesional bioactive TNFα concentrations. The L19TNFα is under clinical trial using isolated inferior limb perfusion (ILP) with the standard treatment with melphalan 10 mg/l limb volume in subjects affected by stage III/IV limb melanoma (NCT01213732). NGR-hTNF is another bifunctional protein which combines a tumor-homing peptide (NGR) that selectively binds to amino peptidase N/CD13 highly expressed on tumor blood vessels, thus affecting tumor vascular permeability, and hTNF, with direct anticancer activity. NGR-hTNF is undergoing 14 clinical trials as a single agent to treat different cancers, as well as in combination with chemotherapy agents.

Another strategy to direct a therapeutic protein to the target cell is through fusion to a growth factor receptor ligand. An example is TP-38, a recombinant chimerical protein composed of the EGFR binding ligand (TGF-α) and a genetically engineered form of the Pseudomonas exotoxin, PE-38, to treat recurrent grade IV malignant brain tumors (NCT00071539).

Many clinical trials are based on a therapeutic protein fused to a targeting antibody, as is the case of APC8015. This drug stimulates the immune system and stops cancer cells from growing by the combination of biological therapies with Bevacizumab®, an already approved monoclonal antibody that locates tumor cells and kills them in a specific way (NCT00849290). There are also many putative protein drugs against cancer which include antibodies anti-integrins (e.g., cilengitide and IMGN388), sometimes in combination with classical therapies. A recently developed tool, the nanobodies or single domain antibodies,[293] have several advantages: small size (only 12–15 KDa), which lowers the possibility of triggering immune response, safety in clinical trials (NCT01020383), and is easy to be joined to different kinds of compounds. All these features make nanobodies competent drugs against different diseases, and have been tested *in vivo* as bifunctional proteins associated to a prodrug, very efficient in mice cancer xenografts.[294]

Even though CPPs are very useful tools to deliver drugs and in gene therapy (see the chapter "Peptide Nanoparticles for Oligonucleotide Delivery" by Lehto *et al.* in this volume), their toxicity and endosomal entrapment slows their inclusion for systemic delivery in clinical trials. Nevertheless, there are a few examples of use to prevent undesirable cell proliferation in coronary artery bypass grafts, as is the case of a CPP $(R\text{-}Ahx\text{-}R)_4$AhxB–PMO conjugate targeted to human c-myc to be applied *ex vivo*. The trial, in phase II, has been completed in 2009 (NCT00451256). Another case is PsorBan®, a product patented for the treatment of psoriasis based on a cyclosporine–polyarginine conjugate of local application, which circumvents the specificity problem of intravenous (i.v.) application. It is in clinical trial phase III, but not yet in the market. Finally, KAI-9803, a PKCδ inhibitor peptide conjugated to Tat to function as an intravenous drug for the treatment of acute myocardial infarction, is currently in phase 2b clinical trial (NCT00785954, KAI pharmaceuticals).

C. Therapeutic Protein Nanoparticles in Preclinical Models of Human Diseases

There are many proteins, often organized as nanoparticles, that when associated to a drug, therapeutic protein, peptide, or nucleic acid increase the therapeutic efficacy of a cargo alone in the treatment of various diseases. Some of them proved effective in animal models, which are discussed in more detail in this section, with relevant examples listed in Table II.

These nanoparticles may simply be (a) a CPP to promote nonspecific internalization,[295–300] (b) a peptide to confer cargo specificity by joining a receptor distinctive of a cell type, including scFvs or peptides obtained by phage display,[301] and (c) a mixture of both,[302] since as observed in several studies the CPP does not reduce ligand specificity and increases nanoparticle potency.[303–305] Complex and multifunctional vehicles including endosomal escape peptides enhance the therapeutic potency of the complex, or other domains that allow their selective activation in certain contexts.[306,307]

Apart from the cases listed in Table II, the spectrum of additional examples of multidomain protein nanoparticles tested *in vivo* is wide, and a considerable proportion of them include CPPs, mainly Tat and polyarginines. A classical Tat fusion protein is the transducible D-isomer RI-TATp53C′′ fusion protein that

TABLE II

Representative Examples of Protein Nanoparticles That, Acting as Carriers, Improve The Efficiency of Cargo Alone in the Treatment of Diseases Using *In Vivo* models

Carrier	Cargo	Administration route	Disease	References
VP-22	Gata4	Transplant of transfected cells	Myocardial infartion	295
$(RXR)_4XB$	Dystrophin exon skipping PMO	i.p.[a]	Duchenne muscular dystrophy	296
Tat-ErbB2	STAT3BP	i.p.	Breast cancer xenograft	302
Penetratin	scFVs-radionuclide	i.v.[b]	Colon carcinoma xenograft	305
8R	Taxol	i.p.	i.p. tumor xenografts	297
Penetratin	Caveolin-1	i.p.	Inflammation models	298
Tat-HA	Bcl-xL	i.p.	Cerebral ischemia	313
Protamine-Erb2 Ab Fab	c-myc, MDM2, VEGF-siRNA	i.v.	Breast cancer	301
9-D-arginine-RVG	JEV-siRNA	i.v.	JEV infection	56
Pegylated Pep-3	Cyclin B1 -PNA	i.v.	Human prostate carcinoma xenograft	299
Chol-MPG-8	Cyclin B1-siRNA	i.v.	Prostate and lung cancer xenografts	300
Tat	pVHL	i.p.	Mice with renal tumors dorsally implanted	310
Tat	MHC class I antigens	s.c.[c]	Dendritic cell vaccine for tumor regresion	311

[a]Intraperitoneal.
[b]Intravenous.
[c]Subcutaneous.

activates p53 protein in cancer cells, but not in normal cells. RI-TATp53C′ treatment in terminal peritoneal carcinomatosis and peritoneal lymphoma preclinical models results in significant increases in life span (higher than sixfold) and full recovery from the disease.[308] There are also several studies *in vivo* using Tat-fused therapeutic proteins which have proven effective in treating tumors[309–311] and cerebral ischemia[312,313] when applied intraperitoneally (i.p.).

Regarding polyarginines, Kumar and colleagues have presented two different models in which a bifunctional peptide formed by nine arginines (9R) and a specific ligand constitute an effective siRNA vehicle. In the first model, a chimerical peptide derived from rabies virus glycoprotein (to confer neuronal specificity) fused to 9- D-arginines (RVG-9R), was able to transport si-RNA across the BBB and silence specific gene expression in the brain when applied intravenously.[56] In the second model, a CD7-specific single-chain antibody was conjugated to oligo-9-arginine peptide (scFvCD7-9R) for T cell-specific antiviral siRNA delivery in humanized mice reconstituted with human lymphocytes. In HIV-infected humanized mice, this treatment controlled viral replication and prevented the disease-associated CD4 T cell loss. Moreover, it effectively suppressed viremia in infected mice.[314]

Some other examples of polyarginines in tumor models are 9-D-arginines fused to a tumor-suppressor peptide, which stopped tumor growth in hepatocellular carcinoma-bearing mice when applied intraperitoneally, and also colesteryl oligoarginines carrying VEGF siRNA, which inhibited tumor growth in colon adenocarcinoma after local application.[315] Another BBB-crossing peptide is g7, which is able to transport nanoparticles loaded with Loperamide.[107]

In general, the partner fusion peptide can confer specificity instead of penetrability, as is the case of EGFR Fab fragment associated to liposomes that contain anticancer drug, which increases efficiency of anticancer effect in EGF overexpressing xenograft tumors[316]; in addition, RGD-4 C-doxorubicin in human breast xenografts increases efficacy and diminishes toxicity.[317]

In many conjugates, the therapeutic peptide of the chimerical proteins is a toxin. Anthrax lethal toxin has been modified to be activated by methaloproteases, and it has probed to be effective for human xenografted tumors such as melanoma, lung, and colorectal cancer.[318] Anthrax toxin has also been associated to antibodies or growth factors for lethal effects specifically on cancer cells.[319] The specific cytotoxicity desired to treat a tumor might derive from a tissue factor, which promotes clotting to restrict blood supply in tumor vessels, fused to peptides that provide specificity, like V-CAM antibodies, fibronectin, and integrin ligands.[320]

Eventually, drug activity may decrease when conjugated to a carrier protein, although if the entry of the drug is favored, the overall balance of activity can be much more efficient.[321] On the other hand, the use of noncovalent bond drug carrier could avoid interfering with the activity of the drug.

An important issue in a preclinical study to be considered for a clinical trial is the administration route. In *in vivo* experiments, most of the protein nanoparticles are administered by local or intraperitoneal injection, avoiding systemic spreading and clearance in the vascular system, in a way very similar to *in vitro* experiments. The FDA and EMEA, on the other hand, will preferentially approve i.v. and oral administrations rather than intraperitoneal or local injections except for very accessible tissues. Another relevant issue is the number of active domains to be included in a therapeutic protein carrier, an issue that seems to be relevant for the functionality of the construct. For example, the CPP neutralization of a ligand may depend on the CPP/ligand ratio that is in the vehicle.[322] It has also been observed that the integrin binding power of RGD-containing motives increases with the number of RGD domains over the monomer until a maxim of four moieties.[323] Another example is Tat activity empowerment when attached to molecules that form tetramers, such as beta-galactosidase[108] and p-53.[324]

Some multidomain protein carriers allow the drug entrance only in selected target cells by tailored smart selective mechanisms.[325] For instance, CPPs neutralized by polyanions are activated and enter the cells when they are released by metalloproteases[326] or by lowering the pH,[327] both situations being very common in tumors.

CPP-morpholino oligomer (PMO) nanoparticles have also shown their effectiveness in treating viral infections by inhibiting viral replication, as demonstrated with the carrier (R-ahx-R) 4AhxB-PMO administered i.v. in animal models infected with picornaviruses, i.p. in mice infected with coronaviruses and flaviviruses, and the carrier R9F2C-PMO administered also i.p. in mice infected with Ebola virus. Furthermore, it has also been shown in some of these studies that the efficacy of the treatment is dependent on the incorporation of arginine-rich peptides in the nanoparticle.[328]

A good example of how a CPP can improve the internalization of a therapeutic protein is the case of insulin. The instability and low absorption in the digestive tract of insulin prevents its oral administration, even though it would be very convenient for a daily administrated drug. In recent studies, noncovalent conjugation of insulin to different CPPs enhances its absorption without toxic intestinal effect, L-penetratin being the most efficient as insulin carrier.[329]

Among the protein nanoparticles tested *in vivo*, it is worth making special mention of Trojan horses generated in Pardridge's laboratory to cross the BBB, through a strategy of fusing within a chimerical peptide the therapeutic protein which has to reach the CNS to a monoclonal antibody against the human insulin receptor (HIRMAb). This Trojan horse is very potent for humans and primates, and has proven effective to transport β-glucuronidase, α-L-iduronidase, GDNF, Abeta amyloid peptides, paroxonase, etc., with potential benefits in diseases like mucopolysaccharidosis type VII, Hurler syndrome, Parkinson, Alzheimer, and organophosphates toxicity, respectively.[330]

There are also promising results when protein nanoparticles have been tested as carriers for gene therapy *in vivo*, some examples being listed in Table II. In this regard, the use of modular proteins generated by insertional mutagenesis of β-galactosidase condensing the SOD gene are able to protect neurons against ischemic injury[133]; a bifunctional galactosylated polylysine is able to conjugate plasmid DNA and to differentially promote expression in hepatocytes that display asialoglycoprotein receptor[331]; a suicide multidomain protein particle formed by herpes simplex virus thymidine kinase (HSV-TK) conjugated to transferrin (Tf) by a biotin-streptavidin bridging, which, administered i.v. in K562 massively metastasized nude mice, was able to reduce tumor size and to increase mouse survival.[332]

VII. Conclusions

In this chapter, proteins and peptides have been envisioned as potent biotechnological tools for the development of new biocompatible biological entities that can be used as therapeutic agents by themselves or as nanovehicles for the delivery of associated drugs. Proteins are nanostructures that can form complex high-order entities such as VLPs, resulting in appropriate cages for the internalization of therapeutic molecules. In addition, the design of modular proteins displaying selected functions has been possible by using *in silico* approximations to the feasibility of recombinant protein production. This approach has demonstrated the versatility of such molecules in the generation of novel delivery nanovehicles opening up the possibility of new functional combinations to enhance the specific interaction with the target tissue. Such tunable specificity in the delivery of drugs, nucleic acids, or other proteins is one of the main properties that make multifunctional proteins appealing as more rational delivery vehicles.

The presence on the market of such complex entities, which started with the approval of Insulin for the treatment of diabetes, has been increasing over the past years, and this tendency is expected to continue. In fact, there are some products in clinical trials that will probably end up being approved and some more are being explored in preclinical experiments which might enter in clinical trials.

Acknowledgments

The authors appreciate the financial support received through grants BFU2010-17450 from MICINN, PS0900165 from FISS, and 2009SGR-108 from AGAUR. The authors also acknowledge the support of the CIBER de Bioingeniería, Biomateriales y Nanomedicina (CIBER-BBN), an initiative funded by the VI National R&D&i Plan 2008–2011, Iniciativa Ingenio 2010, Consolider Program, CIBER Actions and financed by the Instituto de Salud Carlos III with assistance from the European Regional Development Fund.

References

1. Shun TY, Lazo JS, Sharlow ER, Johnston PA. Identifying actives from HTS data sets: practical approaches for the selection of an appropriate HTS data-processing method and quality control review. *J Biomol Screen* 2011;**16**:1–14.
2. Vuorela P, Leinonen M, Saikku P, Tammela P, Rauha JP, Wennberg T, et al. Natural products in the process of finding new drug candidates. *Curr Med Chem* 2004;**11**:1375–89.
3. Zhao HY, Akritopoulou-Zanze I. When analoging is not enough: scaffold discovery in medicinal chemistry. *Expert Opin Drug Discovery* 2010;**5**:123–34.
4. Adamina M, Joerger M. Dose-toxicity models in oncology. *Expert Opin Drug Metab Toxicol* 2011;**7**:201–11.
5. Schmid EF, Smith DA. Is declining innovation in the pharmaceutical industry a myth? *Drug Discov Today* 2005;**10**:1031–9.
6. Veronese FM, Mero A. The impact of PEGylation on biological therapies. *BioDrugs* 2008;**22**:315–29.
7. Raja SM, Clubb RJ, Ortega-Cava C, Williams SH, Bailey TA, Duan L, et al. Anticancer activity of celastrol in combination with ErbB2-targeted therapeutics for treatment of ErbB2-overexpressing breast cancers. *Cancer Biol Ther* 2011;**11**:263–76.
8. Desai N, Trieu V, Damascelli B, Soon-Shiong P. SPARC expression correlates with tumor response to albumin-bound paclitaxel in head and neck cancer patients. *Transl Oncol* 2009;**2**:59–64.
9. Desai NP, Trieu V, Hwang LY, Wu R, Soon-Shiong P, Gradishar WJ. Improved effectiveness of nanoparticle albumin-bound (nab) paclitaxel versus polysorbate-based docetaxel in multiple xenografts as a function of HER2 and SPARC status. *Anticancer Drugs* 2008;**19**:899–909.
10. Houk BE, Martin R, Hochhaus G, Hughes JA. Pharmacokinetics of plasmid DNA in the rat. *Pharm Res* 2001;**18**:67–74.
11. Wang W. Instability, stabilization, and formulation of liquid protein pharmaceuticals. *Int J Pharm* 1999;**185**:129–88.
12. Martin ME, Rice KG. Peptide-guided gene delivery. *AAPS J* 2007;**9**:E18–29.
13. Mastrobattista E, van der Aa MA, Hennink WE, Crommelin DJ. Artificial viruses: a nanotechnological approach to gene delivery. *Nat Rev Drug Discov* 2006;**5**:115–21.
14. Gabathuler R. Approaches to transport therapeutic drugs across the blood-brain barrier to treat brain diseases. *Neurobiol Dis* 2010;**37**:48–57.
15. Glover DJ, Ng SM, Mechler A, Martin LL, Jans DA. Multifunctional protein nanocarriers for targeted nuclear gene delivery in nondividing cells. *FASEB J* 2009;**23**:2996–3006.
16. Tiera MJ, Winnik FO, Fernandes JC. Synthetic and natural polycations for gene therapy: state of the art and new perspectives. *Curr Gene Ther* 2006;**6**:59–71.
17. Mannisto M, Vanderkerken S, Toncheva V, Elomaa M, Ruponen M, Schacht E, et al. Structure-activity relationships of poly(L-lysines): effects of pegylation and molecular shape on physicochemical and biological properties in gene delivery. *J Control Release* 2002;**83**:169–82.
18. Ward CM, Read ML, Seymour LW. Systemic circulation of poly(L-lysine)/DNA vectors is influenced by polycation molecular weight and type of DNA: differential circulation in mice and rats and the implications for human gene therapy. *Blood* 2001;**97**:2221–9.
19. Gottschalk S, Sparrow JT, Hauer J, Mims MP, Leland FE, Woo SL, et al. A novel DNA-peptide complex for efficient gene transfer and expression in mammalian cells. *Gene Ther* 1996;**3**:448–57.
20. Plank C, Tang MX, Wolfe AR, Szoka Jr. FC. Branched cationic peptides for gene delivery: role of type and number of cationic residues in formation and in vitro activity of DNA polyplexes. *Hum Gene Ther* 1999;**10**:319–32.

21. McKenzie DL, Collard WT, Rice KG. Comparative gene transfer efficiency of low molecular weight polylysine DNA-condensing peptides. *J Pept Res* 1999;**54**:311–8.
22. McKenzie DL, Smiley E, Kwok KY, Rice KG. Low molecular weight disulfide cross-linking peptides as nonviral gene delivery carriers. *Bioconjug Chem* 2000;**11**:901–9.
23. Sorgi FL, Bhattacharya S, Huang L. Protamine sulfate enhances lipid-mediated gene transfer. *Gene Ther* 1997;**4**:961–8.
24. Brewer LR, Corzett M, Balhorn R. Protamine-induced condensation and decondensation of the same DNA molecule. *Science* 1999;**286**:120–3.
25. Masuda T, Akita H, Harashima H. Evaluation of nuclear transfer and transcription of plasmid DNA condensed with protamine by microinjection: the use of a nuclear transfer score. *FEBS Lett* 2005;**579**:2143–8.
26. Balhorn R. The protamine family of sperm nuclear proteins. *Genome Biol* 2007;**8**:227.
27. Ferrer-Miralles N, Vazquez E, Villaverde A. Membrane-active peptides for non-viral gene therapy: making the safest easier. *Trends Biotechnol* 2008;**26**:267–75.
28. Chan CK, Jans DA. Enhancement of MSH receptor- and GAL4-mediated gene transfer by switching the nuclear import pathway. *Gene Ther* 2001;**8**:166–71.
29. Fominaya J, Wels W. Target cell-specific DNA transfer mediated by a chimeric multidomain protein: novel non-viral gene delivery system. *J Biol Chem* 1996;**271**:10560–8.
30. Box M, Parks DA, Knight A, Hale C, Fishman PS, Fairweather NF. A multi-domain protein system based on the HC fragment of tetanus toxin for targeting DNA to neuronal cells. *J Drug Target* 2003;**11**:333–43.
31. Baleja JD, Thanabal V, Wagner G. Refined solution structure of the DNA-binding domain of GAL4 and use of 3J(113Cd,1H) in structure determination. *J Biomol NMR* 1997;**10**:397–401.
32. Bessis N, GarciaCozar FJ, Boissier MC. Immune responses to gene therapy vectors: influence on vector function and effector mechanisms. *Gene Ther* 2004;**11**(Suppl. 1):S10–7.
33. Vazquez E, Ferrer-Miralles N, Villaverde A. Peptide-assisted traffic engineering for nonviral gene therapy. *Drug Discov Today* 2008;**13**:1067–74.
34. Richard JP, Melikov K, Vives E, Ramos C, Verbeure B, Gait MJ, et al. Cell-penetrating peptides: a reevaluation of the mechanism of cellular uptake. *J Biol Chem* 2003;**278**:585–90.
35. Said HF, Saleh AF, Abes R, Gait MJ, Lebleu B. Cell penetrating peptides: overview and applications to the delivery of oligonucleotides. *Cell Mol Life Sci* 2010;**67**:715–26.
36. Vazquez E, Ferrer-Miralles N, Mangues R, Corchero JL, Schwartz S, Villaverde A. Modular protein engineering in emerging cancer therapies. *Curr Pharm Des* 2009;**15**:893–916.
37. Rudolph C, Plank C, Lausier J, Schillinger U, Muller RH, Rosenecker J. Oligomers of the arginine-rich motif of the HIV-1 TAT protein are capable of transferring plasmid DNA into cells. *J Biol Chem* 2003;**278**:11411–8.
38. Fawell S, Seery J, Daikh Y, Moore C, Chen LL, Pepinsky B, et al. Tat-mediated delivery of heterologous proteins into cells. *Proc Natl Acad Sci USA* 1994;**91**:664–8.
39. Brooks H, Lebleu B, Vives E. Tat peptide-mediated cellular delivery: back to basics. *Adv Drug Deliv Rev* 2005;**57**:559–77.
40. Vives E, Brodin P, Lebleu B. A truncated HIV-1 Tat protein basic domain rapidly translocates through the plasma membrane and accumulates in the cell nucleus. *J Biol Chem* 1997;**272**:16010–7.
41. Vives E. Cellular uptake [correction of utake] of the Tat peptide: an endocytosis mechanism following ionic interactions. *J Mol Recognit* 2003;**16**:265–71.
42. Wender PA, Mitchell DJ, Pattabiraman K, Pelkey ET, Steinman L, Rothbard JB. The design, synthesis, and evaluation of molecules that enable or enhance cellular uptake: peptoid molecular transporters. *Proc Natl Acad Sci USA* 2000;**97**:13003–8.
43. Lundberg P, El-Andaloussi S, Sutlu T, Johansson H, Langel U. Delivery of short interfering RNA using endosomolytic cell-penetrating peptides. *FASEB J* 2007;**21**:2664–71.

44. Muratovska A, Eccles MR. Conjugate for efficient delivery of short interfering RNA (siRNA) into mammalian cells. *FEBS Lett* 2004;**558**:63–8.
45. Pooga M, Hallbrink M, Zorko M, Langel U. Cell penetration by transportan. *FASEB J* 1998;**12**:67–77.
46. Pooga M, Kut C, Kihlmark M, Hallbrink M, Fernaeus S, Raid R, et al. Cellular translocation of proteins by transportan. *FASEB J* 2001;**15**:1451–3.
47. Hart SL. Integrin-mediated vectors for gene transfer and therapy. *Curr Opin Mol Ther* 1999;**1**:197–203.
48. Kessler T, Bieker R, Padro T, Schwoppe C, Persigehl T, Bremer C, et al. Inhibition of tumor growth by RGD peptide-directed delivery of truncated tissue factor to the tumor vasculature. *Clin Cancer Res* 2005;**11**:6317–24.
49. Amara A, Gall SL, Schwartz O, Salamero J, Montes M, Loetscher P, et al. HIV coreceptor downregulation as antiviral principle: SDF-1alpha-dependent internalization of the chemokine receptor CXCR4 contributes to inhibition of HIV replication. *J Exp Med* 1997;**186**:139–46.
50. Liang X. CXCR4, inhibitors and mechanisms of action. *Chem Biol Drug Des* 2008;**72**:97–110.
51. Daniels TR, Delgado T, Helguera G, Penichet ML. The transferrin receptor part II: targeted delivery of therapeutic agents into cancer cells. *Clin Immunol* 2006;**121**:159–76.
52. Joung I, Harber G, Gerecke KM, Carroll SL, Collawn JF, Engler JA. Improved gene delivery into neuroglial cells using a fiber-modified adenovirus vector. *Biochem Biophys Res Commun* 2005;**328**:1182–7.
53. Liu X, Tian PK, Ju DW, Zhang MH, Yao M, Cao XT, et al. Systemic genetic transfer of p21WAF-1 and GM-CSF utilizing of a novel oligopeptide-based EGF receptor targeting polyplex. *Cancer Gene Ther* 2003;**10**:529–39.
54. Wolschek MF, Thallinger C, Kursa M, Rossler V, Allen M, Lichtenberger C, et al. Specific systemic nonviral gene delivery to human hepatocellular carcinoma xenografts in SCID mice. *Hepatology* 2002;**36**:1106–14.
55. Wu YT, Jiaang WT, Lin KG, Huang CM, Chang CH, Sun YL, et al. A new N-acetylgalactosamine containing peptide as a targeting vehicle for mammalian hepatocytes via asialoglycoprotein receptor endocytosis. *Curr Drug Deliv* 2004;**1**:119–27.
56. Kumar P, Wu H, McBride JL, Jung KE, Kim MH, Davidson BL, et al. Transvascular delivery of small interfering RNA to the central nervous system. *Nature* 2007;**448**:39–43.
57. Schneider H, Harbottle RP, Yokosaki Y, Kunde J, Sheppard D, Coutelle C. A novel peptide, PLAEIDGIELTY, for the targeting of alpha9beta1-integrins. *FEBS Lett* 1998;**429**:269–73.
58. Collins L, Fabre JW. A synthetic peptide vector system for optimal gene delivery to corneal endothelium. *J Gene Med* 2004;**6**:185–94.
59. McKay T, Reynolds P, Jezzard S, Curiel D, Coutelle C. Secretin-mediated gene delivery, a specific targeting mechanism with potential for treatment of biliary and pancreatic disease in cystic fibrosis. *Mol Ther* 2002;**5**:447–54.
60. Zeng J, Too HP, Ma Y, Luo ES, Wang S. A synthetic peptide containing loop 4 of nerve growth factor for targeted gene delivery. *J Gene Med* 2004;**6**:1247–56.
61. Martinez-Fong D, Navarro-Quiroga I, Ochoa I, varez-Maya I, Meraz MA, Luna J, et al. Neurotensin-SPDP-poly-L-lysine conjugate: a nonviral vector for targeted gene delivery to neural cells. *Brain Res Mol Brain Res* 1999;**69**:249–62.
62. White SJ, Nicklin SA, Sawamura T, Baker AH. Identification of peptides that target the endothelial cell-specific LOX-1 receptor. *Hypertension* 2001;**37**:449–55.
63. Pardridge WM, Buciak JL, Friden PM. Selective transport of an anti-transferrin receptor antibody through the blood-brain barrier in vivo. *J Pharmacol Exp Ther* 1991;**259**:66–70.

64. Boado RJ, Zhang Y, Zhang Y, Pardridge WM. Humanization of anti-human insulin receptor antibody for drug targeting across the human blood-brain barrier. *Biotechnol Bioeng* 2007;**96**:381–91.
65. Jeong JH, Lee M, Kim WJ, Yockman JW, Park TG, Kim YH, et al. Anti-GAD antibody targeted non-viral gene delivery to islet beta cells. *J Control Release* 2005;**107**:562–70.
66. Balestrieri ML, Napoli C. Novel challenges in exploring peptide ligands and corresponding tissue-specific endothelial receptors. *Eur J Cancer* 2007;**43**:1242–50.
67. Aina OH, Liu R, Sutcliffe JL, Marik J, Pan CX, Lam KS. From combinatorial chemistry to cancer-targeting peptides. *Mol Pharm* 2007;**4**:631–51.
68. Lundberg M, Wikstrom S, Johansson M. Cell surface adherence and endocytosis of protein transduction domains. *Mol Ther* 2003;**8**:143–50.
69. Wagner E, Plank C, Zatloukal K, Cotten M, Birnstiel ML. Influenza virus hemagglutinin HA-2 N-terminal fusogenic peptides augment gene transfer by transferrin-polylysine-DNA complexes: toward a synthetic virus-like gene-transfer vehicle. *Proc Natl Acad Sci USA* 1992;**89**:7934–8.
70. Plank C, Oberhauser B, Mechtler K, Koch C, Wagner E. The influence of endosome-disruptive peptides on gene transfer using synthetic virus-like gene transfer systems. *J Biol Chem* 1994;**269**:12918–24.
71. Subbarao NK, Parente RA, Szoka Jr. FC, Nadasdi L, Pongracz K. pH-dependent bilayer destabilization by an amphipathic peptide. *Biochemistry* 1987;**26**:2964–72.
72. Parente RA, Nir S, Szoka Jr. FC. Mechanism of leakage of phospholipid vesicle contents induced by the peptide GALA. *Biochemistry* 1990;**29**:8720–8.
73. Parente RA, Nadasdi L, Subbarao NK, Szoka Jr. FC. Association of a pH-sensitive peptide with membrane vesicles: role of amino acid sequence. *Biochemistry* 1990;**29**:8713–9.
74. Li W, Nicol F, Szoka Jr. FC. GALA: a designed synthetic pH-responsive amphipathic peptide with applications in drug and gene delivery. *Adv Drug Deliv Rev* 2004;**56**:967–85.
75. Wyman TB, Nicol F, Zelphati O, Scaria PV, Plank C, Szoka Jr. FC. Design, synthesis, and characterization of a cationic peptide that binds to nucleic acids and permeabilizes bilayers. *Biochemistry* 1997;**36**:3008–17.
76. Rittner K, Benavente A, Bompard-Sorlet A, Heitz F, Divita G, Brasseur R, et al. New basic membrane-destabilizing peptides for plasmid-based gene delivery in vitro and in vivo. *Mol Ther* 2002;**5**:104–14.
77. Ogris M, Carlisle RC, Bettinger T, Seymour LW. Melittin enables efficient vesicular escape and enhanced nuclear access of nonviral gene delivery vectors. *J Biol Chem* 2001;**276**:47550–5.
78. Derossi D, Joliot AH, Chassaing G, Prochiantz A. The third helix of the Antennapedia homeodomain translocates through biological membranes. *J Biol Chem* 1994;**269**:10444–50.
79. Derossi D, Chassaing G, Prochiantz A. Trojan peptides: the penetratin system for intracellular delivery. *Trends Cell Biol* 1998;**8**:84–7.
80. Pichon C, Goncalves C, Midoux P. Histidine-rich peptides and polymers for nucleic acids delivery. *Adv Drug Deliv Rev* 2001;**53**:75–94.
81. Midoux P, Kichler A, Boutin V, Maurizot JC, Monsigny M. Membrane permeabilization and efficient gene transfer by a peptide containing several histidines. *Bioconj Chem* 1998;**9**:260–7.
82. Midoux P, LeCam E, Coulaud D, Delain E, Pichon C. Histidine containing peptides and polypeptides as nucleic acid vectors. *Somat Cell Mol Genet* 2002;**27**:27–47.
83. Kichler A, Leborgne C, Danos O, Bechinger B. Characterization of the gene transfer process mediated by histidine-rich peptides. *J Mol Med* 2007;**85**:191–201.
84. Kim J, Chen CP, Rice KG. The proteasome metabolizes peptide-mediated nonviral gene delivery systems. *Gene Ther* 2005;**12**:1581–90.

85. Leonchiks A, Stavropoulou V, Sharipo A, Masucci MG. Inhibition of ubiquitin-dependent proteolysis by a synthetic glycine-alanine repeat peptide that mimics an inhibitory viral sequence. *FEBS Lett* 2002;**522**:93–8.
86. Sharipo A, Imreh M, Leonchiks A, Imreh S, Masucci MG. A minimal glycine-alanine repeat prevents the interaction of ubiquitinated I kappaB alpha with the proteasome: a new mechanism for selective inhibition of proteolysis. *Nat Med* 1998;**4**:939–44.
87. Sharipo A, Imreh M, Leonchiks A, Branden C, Masucci MG. cis-Inhibition of proteasomal degradation by viral repeats: impact of length and amino acid composition. *FEBS Lett* 2001;**499**:137–42.
88. Martinez-Moreno M, Navarro-Lerida I, Roncal F, Albar JP, Alonso C, Gavilanes F, et al. Recognition of novel viral sequences that associate with the dynein light chain LC8 identified through a pepscan technique. *FEBS Lett* 2003;**544**:262–7.
89. Melchior F, Gerace L. Mechanisms of nuclear protein import. *Curr Opin Cell Biol* 1995;**7**:310–8.
90. Zanta MA, Belguise-Valladier P, Behr JP. Gene delivery: a single nuclear localization signal peptide is sufficient to carry DNA to the cell nucleus. *Proc Natl Acad Sci USA* 1999;**96**:91–6.
91. Fischer N, Kremmer E, Lautscham G, Mueller-Lantzsch N, Grasser FA. Epstein-Barr virus nuclear antigen 1 forms a complex with the nuclear transporter karyopherin alpha2. *J Biol Chem* 1997;**272**:3999–4005.
92. Dang CV, Lee WM. Identification of the human c-myc protein nuclear translocation signal. *Mol Cell Biol* 1988;**8**:4048–54.
93. Fominaya J, Uherek C, Wels W. A chimeric fusion protein containing transforming growth factor-alpha mediates gene transfer via binding to the EGF receptor. *Gene Ther* 1998;**5**:521–30.
94. Bauerle M, Doenecke D, Albig W. The requirement of H1 histones for a heterodimeric nuclear import receptor. *J Biol Chem* 2002;**277**:32480–9.
95. Baake M, Bauerle M, Doenecke D, Albig W. Core histones and linker histones are imported into the nucleus by different pathways. *Eur J Cell Biol* 2001;**80**:669–77.
96. Subramanian A, Ranganathan P, Diamond SL. Nuclear targeting peptide scaffolds for lipofection of nondividing mammalian cells. *Nat Biotechnol* 1999;**17**:873–7.
97. Gharakhanian E, Takahashi J, Kasamatsu H. The carboxyl 35 amino acids of SV40 Vp3 are essential for its nuclear accumulation. *Virology* 1987;**157**:440–8.
98. Lyons RH, Ferguson BQ, Rosenberg M. Pentapeptide nuclear localization signal in adenovirus E1a. *Mol Cell Biol* 1987;**7**:2451–6.
99. Kleinschmidt JA, Seiter A. Identification of domains involved in nuclear uptake and histone binding of protein N1 of Xenopus laevis. *EMBO J* 1988;**7**:1605–14.
100. Kiefer P, Acland P, Pappin D, Peters G, Dickson C. Competition between nuclear localization and secretory signals determines the subcellular fate of a single CUG-initiated form of FGF3. *EMBO J* 1994;**13**:4126–36.
101. Schreiber V, Molinete M, Boeuf H, de Murcia G, Menissier-de MJ. The human poly(ADP-ribose) polymerase nuclear localization signal is a bipartite element functionally separate from DNA binding and catalytic activity. *EMBO J* 1992;**11**:3263–9.
102. Robbins J, Dilworth SM, Laskey RA, Dingwall C. Two interdependent basic domains in nucleoplasmin nuclear targeting sequence: identification of a class of bipartite nuclear targeting sequence. *Cell* 1991;**64**:615–23.
103. Rivera VM, Gao GP, Grant RL, Schnell MA, Zoltick PW, Rozamus LW, et al. Long-term pharmacologically regulated expression of erythropoietin in primates following AAV-mediated gene transfer. *Blood* 2005;**105**:1424–30.

104. Wang X, Qiu R, Tsark W, Lu Q. Rapid promoter analysis in developing mouse brain and genetic labeling of young neurons by doublecortin-DsRed-express. *J Neurosci Res* 2007;**85**:3567–73.
105. Ishiwata A, Mimuro J, Mizukami H, Kashiwakura Y, Takano K, Ohmori T, et al. Liver-restricted expression of the canine factor VIII gene facilitates prevention of inhibitor formation in factor VIII-deficient mice. *J Gene Med* 2009;**11**:1020–9.
106. Pardridge WM. Blood-brain barrier delivery. *Drug Discov Today* 2007;**12**:54–61.
107. Tosi G, Costantino L, Rivasi F, Ruozi B, Leo E, Vergoni AV, et al. Targeting the central nervous system: in vivo experiments with peptide-derivatized nanoparticles loaded with Loperamide and Rhodamine-123. *J Control Release* 2007;**122**:1–9.
108. Schwarze SR, Ho A, Vocero-Akbani A, Dowdy SF. In vivo protein transduction: delivery of a biologically active protein into the mouse. *Science* 1999;**285**:1569–72.
109. Boado RJ, Zhang Y, Zhang Y, Pardridge WM. Genetic engineering, expression, and activity of a fusion protein of a human neurotrophin and a molecular Trojan horse for delivery across the human blood-brain barrier. *Biotechnol Bioeng* 2007;**97**:1376–86.
110. Boado RJ, Zhang Y, Zhang Y, Xia CF, Wang Y, Pardridge WM. Genetic engineering of a lysosomal enzyme fusion protein for targeted delivery across the human blood-brain barrier. *Biotechnol Bioeng* 2008;**99**:475–84.
111. Boado RJ, Zhang Y, Zhang Y, Wang Y, Pardridge WM. GDNF fusion protein for targeted-drug delivery across the human blood-brain barrier. *Biotechnol Bioeng* 2008;**100**:387–96.
112. Otten LG, Quax WJ. Directed evolution: selecting today's biocatalysts. *Biomol Eng* 2005;**22**:1–9.
113. Lutz S, Patrick WM. Novel methods for directed evolution of enzymes: quality, not quantity. *Curr Opin Biotechnol* 2004;**15**:291–7.
114. Neylon C. Chemical and biochemical strategies for the randomization of protein encoding DNA sequences: library construction methods for directed evolution. *Nucleic Acids Res* 2004;**32**:1448–59.
115. Stemmer WP. Rapid evolution of a protein in vitro by DNA shuffling. *Nature* 1994;**370**:389–91.
116. Dahiyat BI, Mayo SL. Protein design automation. *Protein Sci* 1996;**5**:895–903.
117. Floudas CA, Fung HK, McAllister SR, Mönnigmann M, Rajgaria R. Advances in protein structure prediction and de novo protein design: a review. *Chem Eng Sci* 2006;**61**:966–88.
118. Walsh ST, Cheng H, Bryson JW, Roder H, DeGrado WF. Solution structure and dynamics of a de novo designed three-helix bundle protein. *Proc Natl Acad Sci USA* 1999;**96**:5486–91.
119. Dahiyat BI, Mayo SL. De novo protein design: fully automated sequence selection. *Science* 1997;**278**:82–7.
120. Aris A, Villaverde A. Modular protein engineering for non-viral gene therapy. *Trends Biotechnol* 2004;**22**:371–7.
121. Uherek C, Fominaya J, Wels W. A modular DNA carrier protein based on the structure of diphtheria toxin mediates target cell-specific gene delivery. *J Biol Chem* 1998;**273**:8835–41.
122. Li SD, Huang L. Gene therapy progress and prospects: non-viral gene therapy by systemic delivery. *Gene Ther* 2006;**13**:1313–9.
123. Chan CK, Jans DA. Using nuclear targeting signals to enhance non-viral gene transfer. *Immunol Cell Biol* 2002;**80**:119–30.
124. Dietz GP, Bahr M. Delivery of bioactive molecules into the cell: the Trojan horse approach. *Mol Cell Neurosci* 2004;**27**:85–131.
125. Dietz GP, Bahr M. Synthesis of cell-penetrating peptides and their application in neurobiology. *Methods Mol Biol* 2007;**399**:181–98.

126. Ignatovich IA, Dizhe EB, Pavlotskaya AV, Akifiev BN, Burov SV, Orlov SV, et al. Complexes of plasmid DNA with basic domain 47–57 of the HIV-1 Tat protein are transferred to mammalian cells by endocytosis-mediated pathways. *J Biol Chem* 2003;**278**:42625–36.
127. Aris A, Villaverde A. Molecular organization of protein-DNA complexes for cell-targeted DNA delivery. *Biochem Biophys Res Commun* 2000;**278**:455–61.
128. Carrio MM, Villaverde A. Role of molecular chaperones in inclusion body formation. *FEBS Lett* 2003;**537**:215–21.
129. Baneyx F. Recombinant protein expression in Escherichia coli. *Curr Opin Biotechnol* 1999;**10**:411–21.
130. Sorensen HP, Mortensen KK. Advanced genetic strategies for recombinant protein expression in Escherichia coli. *J Biotechnol* 2005;**115**:113–28.
131. Schmidt FR. Recombinant expression systems in the pharmaceutical industry. *Appl Microbiol Biotechnol* 2004;**65**:363–72.
132. Ou WC, Wang M, Fung CY, Tsai RT, Chao PC, Hseu TH, et al. The major capsid protein, VP1, of human JC virus expressed in Escherichia coli is able to self-assemble into a capsid-like particle and deliver exogenous DNA into human kidney cells. *J Gen Virol* 1999;**80**:39–46.
133. Peluffo H, Acarin L, Aris A, Gonzalez P, Villaverde A, Castellano B, et al. Neuroprotection from NMDA excitotoxic lesion by Cu/Zn superoxide dismutase gene delivery to the postnatal rat brain by a modular protein vector. *BMC Neurosci* 2006;**7**:35.
134. Sasnauskas K, Buzaite O, Vogel F, Jandrig B, Razanskas R, Staniulis J, et al. Yeast cells allow high-level expression and formation of polyomavirus-like particles. *Biol Chem* 1999;**380**:381–6.
135. Birch JR, Froud SJ. Mammalian cell culture systems for recombinant protein production. *Biologicals* 1994;**22**:127–33.
136. Ikonomou L, Schneider YJ, Agathos SN. Insect cell culture for industrial production of recombinant proteins. *Appl Microbiol Biotechnol* 2003;**62**:1–20.
137. Reiter M, Bluml G. Large-scale mammalian cell culture. *Curr Opin Biotechnol* 1994;**5**:175–9.
138. Gervais A, Hammel YA, Pelloux S, Lepage P, Baer G, Carte N, et al. Glycosylation of human recombinant gonadotrophins: characterization and batch-to-batch consistency. *Glycobiology* 2003;**13**:179–89.
139. Collins L, Asuni AA, Anderton BH, Fabre JW. Efficient gene delivery to primary neuron cultures using a synthetic peptide vector system. *J Neurosci Methods* 2003;**125**:113–20.
140. Kim ES, Yang SW, Hong DK, Kim WT, Kim HG, Lee SK. Cell-penetrating DNA-binding protein as a safe and efficient naked DNA delivery carrier in vitro and in vivo. *Biochem Biophys Res Commun* 2010;**392**:9–15.
141. Fabre JW, Collins L. Synthetic peptides as non-viral DNA vectors. *Curr Gene Ther* 2006;**6**:459–80.
142. Parker AL, Collins L, Zhang X, Fabre JW. Exploration of peptide motifs for potent non-viral gene delivery highly selective for dividing cells. *J Gene Med* 2005;**7**:1545–54.
143. Aris A, Villaverde A. Engineering nuclear localization signals in modular protein vehicles for gene therapy. *Biochem Biophys Res Commun* 2003;**304**:625–31.
144. Chen L, McCrate JM, Lee JC, Li H. The role of surface charge on the uptake and biocompatibility of hydroxyapatite nanoparticles with osteoblast cells. *Nanotechnology* 2011;**22**:105708.
145. Cortes-Perez NG, Sapin C, Jaffrelo L, Daou S, Grill JP, Langella P, et al. Rotavirus-like particles: a novel nanocarrier for the gut. *J Biomed Biotechnol* 2010;**2010**:317545.
146. Alcala P, Feliu JX, Aris A, Villaverde A. Efficient accommodation of recombinant, foot-and-mouth disease virus RGD peptides to cell-surface integrins. *Biochem Biophys Res Commun* 2001;**285**:201–6.
147. Marsh M, Helenius A. Virus entry: open sesame. *Cell* 2006;**124**:729–40.

148. Luthman H, Magnusson G. High efficiency polyoma DNA transfection of chloroquine treated cells. *Nucleic Acids Res* 1983;**11**:1295–308.
149. Vaccaro L, Cross KJ, Kleinjung J, Straus SK, Thomas DJ, Wharton SA, et al. Plasticity of influenza haemagglutinin fusion peptides and their interaction with lipid bilayers. *Biophys J* 2005;**88**:25–36.
150. Ludtke JJ, Zhang G, Sebestyen MG, Wolff JA. A nuclear localization signal can enhance both the nuclear transport and expression of 1 kb DNA. *J Cell Sci* 1999;**112**:2033–41.
151. Wilke M, Fortunati E, van den Broek M, Hoogeveen AT, Scholte BJ. Efficacy of a peptide-based gene delivery system depends on mitotic activity. *Gene Ther* 1996;**3**:1133–42.
152. Seow Y, Wood MJ. Biological gene delivery vehicles: beyond viral vectors. *Mol Ther* 2009;**17**:767–77.
153. Ludwig C, Wagner R. Virus-like particles-universal molecular toolboxes. *Curr Opin Biotechnol* 2007;**18**:537–45.
154. Mateu MG. Virus engineering: functionalization and stabilization. *Protein Eng Des Sel* 2011;**24**:53–63.
155. Singh P, Gonzalez MJ, Manchester M. Viruses and their uses in nanotechnology. *Drug Dev Res* 2006;**67**:23–41.
156. Dokland T. Scaffolding proteins and their role in viral assembly. *Cell Mol Life Sci* 1999;**56**:580–603.
157. Pattenden LK, Middelberg AP, Niebert M, Lipin DI. Towards the preparative and large-scale precision manufacture of virus-like particles. *Trends Biotechnol* 2005;**23**:523–9.
158. Scheerlinck JPY, Greenwood DLV. Virus-sized vaccine delivery systems. *Drug Discov Today* 2008;**13**:882–7.
159. Roldao A, Mellado MC, Castilho LR, Carrondo MJ, Alves PM. Virus-like particles in vaccine development. *Expert Rev Vaccines* 2010;**9**:1149–76.
160. Sanchez-Rodriguez SP, Munch-Anguiano L, Bustos-Jaimes I. Advances in the development of virus-like particles as tools in medicine and nanoscience. *Curr Chem Biol* 2010;**4**:231–43.
161. Ramqvist T, Andreasson K, Dalianis T. Vaccination, immune and gene therapy based on virus-like particles against viral infections and cancer. *Expert Opin Biol Ther* 2007;**7**:997–1007.
162. Roy P, Noad R. Virus-like particles as a vaccine delivery system: myths and facts. *Hum Vaccin* 2008;**4**:5–12.
163. Slilaty SN, Aposhian HV. Gene transfer by polyoma-like particles assembled in a cell-free system. *Science* 1983;**220**:725–7.
164. Lenz P, Day PM, Pang YYS, Frye SA, Jensen PN, Lowy DR, et al. Papillomavirus-like particles induce acute activation of dendritic cells. *J Immunol* 2001;**166**:5346–55.
165. Xu YF, Zhang YQ, Xu XM, Song GX. Papillomavirus virus-like particles as vehicles for the delivery of epitopes or genes. *Arch Virol* 2006;**151**:2133–48.
166. Buonaguro FM, Tornesello ML, Buonaguro L. Virus-like particle vaccines and adjuvants: the HPV paradigm. *Expert Rev Vaccines* 2009;**8**:1379–98.
167. Brandenburg B, Stockl L, Gutzeit C, Roos M, Lupberger J, Schwartlander R, et al. A novel system for efficient gene transfer into primary human hepatocytes via cell-permeable hepatitis B virus-like particle. *Hepatology* 2005;**42**:1300–9.
168. Li TC, Takeda N, Miyamura T, Matsuura Y, Wang JC, Engvall H, et al. Essential elements of the capsid protein for self-assembly into empty virus-like particles of hepatitis E virus. *J Virol* 2005;**79**:12999–3006.
169. Xiang J, Wunschmann S, George SL, Klinzman D, Schmidt WN, LaBrecque DR, et al. Recombinant hepatitis C virus-like particles expressed by baculovirus: utility in cell-binding and antibody detection assays. *J Med Virol* 2002;**68**:537–43.

170. Andreasson K, Tegerstedt K, Eriksson M, Curcio C, Cavallo F, Forni G, et al. Murine pneumotropic virus chimeric Her2/neu virus-like particles as prophylactic and therapeutic vaccines against Her2/neu expressing tumors. *Int J Cancer* 2009;**124**:150–6.
171. Chou MI, Hsieh YF, Wang M, Chang JT, Chang D, Zouali M, et al. In vitro and in vivo targeted delivery of IL-10 interfering RNA by JC virus-like particles. *J Biomed Sci* 2010;**17**:51.
172. Gleiter S, Lilie H. Cell-type specific targeting and gene expression using a variant of polyoma VP1 virus-like particles. *Biol Chem* 2003;**384**:247–55.
173. Goldmann C, Petry H, Frye S, Ast O, Ebitsch S, Jentsch KD, et al. Molecular cloning and expression of major structural protein VP1 of the human polyomavirus JC virus: formation of virus-like particles useful for immunological and therapeutic studies. *J Virol* 1999;**73**:4465–9.
174. Goldmann C, Stolte N, Nisslein T, Hunsmann G, Luke W, Petry H. Packaging of small molecules into VP1-virus-like particles of the human polyomavirus JC virus. *J Virol Methods* 2000;**90**:85–90.
175. Lawatscheck R, Aleksaite E, Schenk JA, Micheel B, Jandrig B, Holland G, et al. Chimeric polyomavirus-derived virus-like particles: the immunogenicity of an inserted peptide applied without adjuvant to mice depends on its insertion site and its flanking linker sequence. *Viral Immunol* 2007;**20**:453–60.
176. Sasnauskas K, Bulavaite A, Hale A, Jin L, Knowles WA, Gedvilaite A, et al. Generation of recombinant virus-like particles of human and non-human polyomaviruses in yeast Saccharomyces cerevisiae. *Intervirology* 2002;**45**:308–17.
177. Voronkova T, Kazaks A, Ose V, Ozel M, Scherneck S, Pumpens P, et al. Hamster polyomavirus-derived virus-like particles are able to transfer in vitro encapsidated plasmid DNA to mammalian cells. *Virus Genes* 2007;**34**:303–14.
178. Krauzewicz N, Stokrova J, Jenkins C, Elliott M, Higgins CF, Griffin BE. Virus-like gene transfer into cells mediated by polyoma virus pseudocapsids. *Gene Ther* 2000;**7**:2122–31.
179. Clark B, Caparros-Wanderley W, Musselwhite G, Kotecha M, Griffin BE. Immunity against both polyomavirus VP1 and a transgene product induced following intranasal delivery of VP1 pseudocapsid-DNA complexes. *J Gen Virol* 2001;**82**:2791–7.
180. Young KR, McBurney SP, Karkhanis LU, Ross TM. Virus-like particles: designing an effective AIDS vaccine. *Methods* 2006;**40**:98–117.
181. Muratori C, Bona R, Federico M. Lentivirus-based virus-like particles as a new protein delivery tool. *Methods Mol Biol* 2010;**614**:111–24.
182. Istrate C, Hinkula J, Charpilienne A, Poncet D, Cohen J, Svensson L, et al. Parenteral administration of RF 8-2/6/7 rotavirus-like particles in a one-dose regimen induce protective immunity in mice. *Vaccine* 2008;**26**:4594–601.
183. Singh P, Destito G, Schneemann A, Manchester M. Canine parvovirus-like particles, a novel nanomaterial for tumor targeting. *J Nanobiotechnology* 2006;**4**:2.
184. Singh P. *Tumor targeting using canine parvovirus nanoparticles*. Berlin: Springer-Verlag; 2009 pp. 123–141.
185. Herbst-Kralovetz M, Mason HS, Chen Q. Norwalk virus-like particles as vaccines. *Expert Rev Vaccines* 2010;**9**:299–307.
186. Manchester M, Singh P. Virus-based nanoparticles (VNPs): platform technologies for diagnostic imaging. *Adv Drug Deliv Rev* 2006;**58**:1505–22.
187. Pease III LF, Lipin DI, Tsai DH, Zachariah MR, Lua LH, Tarlov MJ, et al. Quantitative characterization of virus-like particles by asymmetrical flow field flow fractionation, electrospray differential mobility analysis, and transmission electron microscopy. *Biotechnol Bioeng* 2009;**102**:845–55.
188. Stehle T, Harrison SC. High-resolution structure of a polyomavirus VP1-oligosaccharide complex: implications for assembly and receptor binding. *EMBO J* 1997;**16**:5139–48.

189. Neu U, Stehle T, Atwood WJ. The Polyomaviridae: contributions of virus structure to our understanding of virus receptors and infectious entry. *Virology* 2009;**384**:389–99.
190. Johnson JE, Chiu W. Structures of virus and virus-like particles. *Curr Opin Struct Biol* 2000;**10**:229–35.
191. Zhang S. Fabrication of novel biomaterials through molecular self-assembly. *Nat Biotechnol* 2003;**21**:1171–8.
192. Ivanovska IL, de Pablo PJ, Ibarra B, Sgalari G, MacKintosh FC, Carrascosa JL, et al. Bacteriophage capsids: tough nanoshells with complex elastic properties. *Proc Natl Acad Sci USA* 2004;**101**:7600–5.
193. Bothner B, Taylor D, Jun B, Lee KK, Siuzdak G, Schultz CP, et al. Maturation of a tetravirus capsid alters the dynamic properties and creates a metastable complex. *Virology* 2005;**334**:17–27.
194. Petry H, Goldmann C, Ast O, Luke W. The use of virus-like particles for gene transfer. *Curr Opin Mol Ther* 2003;**5**:524–8.
195. Garcea RL, Gissmann L. Virus-like particles as vaccines and vessels for the delivery of small molecules. *Curr Opin Biotechnol* 2004;**15**:513–7.
196. Georgens C, Weyermann J, Zimmer A. Recombinant virus like particles as drug delivery system. *Curr Pharm Biotechnol* 2005;**6**:49–55.
197. Noad R, Roy P. Virus-like particles as immunogens. *Trends Microbiol* 2003;**11**:438–44.
198. Grgacic EV, Anderson DA. Virus-like particles: passport to immune recognition. *Methods* 2006;**40**:60–5.
199. Carrasco C, Castellanos M, de Pablo PJ, Mateu MG. Manipulation of the mechanical properties of a virus by protein engineering. *Proc Natl Acad Sci USA* 2008;**105**:4150–5.
200. Fischlechner M, Donath E. Viruses as building blocks for materials and devices. *Angew Chem Int Ed Engl* 2007;**46**:3184–93.
201. Lee LA, Wang Q. Adaptations of nanoscale viruses and other protein cages for medical applications. *Nanomedicine* 2006;**2**:137–49.
202. Adjei AA. Blocking oncogenic Ras signaling for cancer therapy. *J Natl Cancer Inst* 2001;**93**:1062–74.
203. May T, Gleiter S, Lilie H. Assessment of cell type specific gene transfer of polyoma virus like particles presenting a tumor specific antibody Fv fragment. *J Virol Methods* 2002;**105**:147–57.
204. Liew MW, Rajendran A, Middelberg AP. Microbial production of virus-like particle vaccine protein at gram-per-litre levels. *J Biotechnol* 2010;**150**:224–31.
205. Kost TA, Condreay JP. Recombinant baculoviruses as mammalian cell gene-delivery vectors. *Trends Biotechnol* 2002;**20**:173–80.
206. Maranga L, Cruz PE, Aunins JG, Carrondo MJ. Production of core and virus-like particles with baculovirus infected insect cells. *Adv Biochem Eng Biotechnol* 2002;**74**:183–206.
207. Lee CD, Yan YP, Liang SM, Wang TF. Production of FMDV virus-like particles by a SUMO fusion protein approach in Escherichia coli. *J Biomed Sci* 2009;**16**:69.
208. Santi L, Huang Z, Mason H. Virus-like particles production in green plants. *Methods* 2006;**40**:66–76.
209. Santi L, Batchelor L, Huang Z, Hjelm B, Kilbourne J, Arntzen CJ, et al. An efficient plant viral expression system generating orally immunogenic Norwalk virus-like particles. *Vaccine* 2008;**26**:1846–54.
210. Ramqvist T, Dalianis T. Immunotherapeutic polyoma and human papilloma virus-like particles. *Immunotherapy* 2009;**1**:303–12.
211. McAleer WJ, Buynak EB, Maigetter RZ, Wampler DE, Miller WJ, Hilleman MR. Human hepatitis B vaccine from recombinant yeast. *Nature* 1984;**307**:178–80.

212. Pumpens P, Razanskas R, Pushko P, Renhof R, Gusars I, Skrastina D, et al. Evaluation of HBs, HBc, and frCP virus-like particles for expression of human papillomavirus 16 E7 oncoprotein epitopes. *Intervirology* 2002;**45**:24–32.
213. Hammonds J, Chen X, Zhang X, Lee F, Spearman P. Advances in methods for the production, purification, and characterization of HIV-1 Gag-Env pseudovirion vaccines. *Vaccine* 2007;**25**:8036–48.
214. Peek LJ, Middaugh CR, Berkland C. Nanotechnology in vaccine delivery. *Adv Drug Deliv Rev* 2008;**60**:915–28.
215. Sun Y, Carrion Jr. R, Ye L, Wen Z, Ro YT, Brasky K, et al. Protection against lethal challenge by Ebola virus-like particles produced in insect cells. *Virology* 2009;**383**:12–21.
216. Warfield KL, Bosio CM, Welcher BC, Deal EM, Mohamadzadeh M, Schmaljohn A, et al. Ebola virus-like particles protect from lethal Ebola virus infection. *Proc Natl Acad Sci USA* 2003;**100**:15889–94.
217. Freyschmidt EJ, Alonso A, Hartmann G, Gissmann L. Activation of dendritic cells and induction of T cell responses by HPV 16 L1/E7 chimeric virus-like particles are enhanced by CpG ODN or sorbitol. *Antivir Ther* 2004;**9**:479–89.
218. Kaufmann AM, Nieland JD, Jochmus I, Baur S, Friese K, Gabelsberger J, et al. Vaccination trial with HPV16 L1E7 chimeric virus-like particles in women suffering from high grade cervical intraepithelial neoplasia (CIN 2/3). *Int J Cancer* 2007;**121**:2794–800.
219. Kaufmann AM, Nieland J, Schinz M, Nonn M, Gabelsberger J, Meissner H, et al. HPV16 L1E7 chimeric virus-like particles induce specific HLA-restricted T cells in humans after in vitro vaccination. *Int J Cancer* 2001;**92**:285–93.
220. Cheong WS, Reiseger J, Turner SJ, Boyd R, Netter HJ. Chimeric virus-like particles for the delivery of an inserted conserved influenza A-specific CTL epitope. *Antiviral Res* 2009;**81**:113–22.
221. Yin Y, Zhang J, Dong D, Liu S, Guo Q, Song X, et al. Chimeric hepatitis B virus core particles carrying an epitope of anthrax protective antigen induce protective immunity against Bacillus anthracis. *Vaccine* 2008;**26**:5814–21.
222. Amexis G, Young NS. Parvovirus B19 empty capsids as antigen carriers for presentation of antigenic determinants of dengue 2 virus. *J Infect Dis* 2006;**194**:790–4.
223. Ogasawara Y, Amexis G, Yamaguchi H, Kajigaya S, Leppla SH, Young NS. Recombinant viral-like particles of parvovirus B19 as antigen carriers of anthrax protective antigen. *In Vivo* 2006;**20**:319–24.
224. Chackerian B, Lowy DR, Schiller JT. Conjugation of a self-antigen to papillomavirus-like particles allows for efficient induction of protective autoantibodies. *J Clin Invest* 2001;**108**:415–23.
225. Cornuz J, Zwahlen S, Jungi WF, Osterwalder J, Klingler K, Van MG, et al. A vaccine against nicotine for smoking cessation: a randomized controlled trial. *PLoS One* 2008;**3**:e2547.
226. Chen LS, Wang M, Ou WC, Fung CY, Chen PL, Chang CF, et al. Efficient gene transfer using the human JC virus-like particle that inhibits human colon adenocarcinoma growth in a nude mouse model. *Gene Ther* 2010;**17**:1033–41.
227. Ramsey JD, Vu HN, Pack DW. A top-down approach for construction of hybrid polymer-virus gene delivery vectors. *J Control Release* 2010;**144**:39–45.
228. Osterman JV, Waddell A, Aposhian HV. DNA and gene therapy: uncoating of polyoma pseudovirus in mouse embryo cells. *Proc Natl Acad Sci USA* 1970;**67**:37–40.
229. Yamada T, Iwasaki Y, Tada H, Iwabuki H, Chuah MK, VandenDriessche T, et al. Nanoparticles for the delivery of genes and drugs to human hepatocytes. *Nat Biotechnol* 2003;**21**:885–90.
230. Takamura S, Niikura M, Li TC, Takeda N, Kusagawa S, Takebe Y, et al. DNA vaccine-encapsulated virus-like particles derived from an orally transmissible virus stimulate mucosal and systemic immune responses by oral administration. *Gene Ther* 2004;**11**:628–35.

231. Kimchi-Sarfaty C, Brittain S, Garfield S, Caplen NJ, Tang Q, Gottesman MM. Efficient delivery of RNA interference effectors via in vitro-packaged SV40 pseudovirions. *Hum Gene Ther* 2005;**16**:1110–5.
232. Asokan A, Conway JC, Phillips JL, Li C, Hegge J, Sinnott R, et al. Reengineering a receptor footprint of adeno-associated virus enables selective and systemic gene transfer to muscle. *Nat Biotechnol* 2010;**28**:79–82.
233. Windram OP, Weber B, Jaffer MA, Rybicki EP, Shepherd DN, Varsani A. An investigation into the use of human papillomavirus type 16 virus-like particles as a delivery vector system for foreign proteins: N- and C-terminal fusion of GFP to the L1 and L2 capsid proteins. *Arch Virol* 2008;**153**:585–9.
234. Gleiter S, Lilie H. Coupling of antibodies via protein Z on modified polyoma virus-like particles. *Protein Sci* 2001;**10**:434–44.
235. Stubenrauch K, Gleiter S, Brinkmann U, Rudolph R, Lilie H. Conjugation of an antibody Fv fragment to a virus coat protein: cell-specific targeting of recombinant polyoma-virus-like particles. *Biochem J* 2001;**356**:867–73.
236. Sen GS, Kuzelka J, Singh P, Lewis WG, Manchester M, Finn MG. Accelerated bioorthogonal conjugation: a practical method for the ligation of diverse functional molecules to a polyvalent virus scaffold. *Bioconj Chem* 2005;**16**:1572–9.
237. Schaffer DV, Koerber JT, Lim KI. Molecular engineering of viral gene delivery vehicles. *Annu Rev Biomed Eng* 2008;**10**:169–94.
238. Kobinger GP, Weiner DJ, Yu QC, Wilson JM. Filovirus-pseudotyped lentiviral vector can efficiently and stably transduce airway epithelia in vivo. *Nat Biotechnol* 2001;**19**:225–30.
239. Qu Q, Sawa H, Suzuki T, Semba S, Henmi C, Okada Y, et al. Nuclear entry mechanism of the human polyomavirus JC virus-like particle: role of importins and the nuclear pore complex. *J Biol Chem* 2004;**279**:27735–42.
240. Raja KS, Wang Q, Gonzalez MJ, Manchester M, Johnson JE, Finn MG. Hybrid virus-polymer materials. 1. Synthesis and properties of PEG-decorated cowpea mosaic virus. *Biomacromolecules* 2003;**4**:472–6.
241. Marianayagam NJ, Sunde M, Matthews JM. The power of two: protein dimerization in biology. *Trends Biochem Sci* 2004;**29**:618–25.
242. Tu RS, Tirrell M. Bottom-up design of biomimetic assemblies. *Adv Drug Deliv Rev* 2004;**56**:1537–63.
243. Leckband D. Measuring the forces that control protein interactions. *Annu Rev Biophys Biomol Struct* 2000;**29**:1–26.
244. Meyer EE, Rosenberg KJ, Israelachvili J. Recent progress in understanding hydrophobic interactions. *Proc Natl Acad Sci USA* 2006;**103**:15739–46.
245. Schneider HJ. Binding mechanisms in supramolecular complexes. *Angew Chem Int Ed Engl* 2009;**48**:3924–77.
246. Goodsell DS, Olson AJ. Structural symmetry and protein function. *Annu Rev Biophys Biomol Struct* 2000;**29**:105–53.
247. Eriksson J, Fenyo D. Modeling experimental design for proteomics. *Methods Mol Biol* 2010;**673**:223–30.
248. Kopecek J. Smart and genetically engineered biomaterials and drug delivery systems. *Eur J Pharm Sci* 2003;**20**:1–16.
249. Zhao X, Zhang S. Molecular designer self-assembling peptides. *Chem Soc Rev* 2006;**35**:1105–10.
250. Domingo-Espin J, Vazquez E, Ganz J, Conchillo O, Garcia-Fruitos E, Cedano J, et al. Nanoparticulate architecture of protein-based artificial viruses is supported by protein-DNA interactions. *Nanomedicine* 2011;**6**:1047–61.

251. Tsumoto K, Luckel F, Yoshikawa K. Giant DNA molecules exhibit on/off switching of transcriptional activity through conformational transition. *Biophys Chem* 2003;**106**:23–9.
252. Flenniken M, Willits D, Brumfield S, Young MJ, Douglas T. The small heat shock protein cage from Methanococcus jannaschii is a versatile nanoscale platform for genetic and chemical modification. *Nano Lett* 2003;**3**:1573–6.
253. Shively JM, Decker GL, Greenawalt JW. Comparative ultrastructure of the thiobacilli. *J Bacteriol* 1970;**101**:618–27.
254. Shively JM, Ball F, Brown DH, Saunders RE. Functional organelles in prokaryotes: polyhedral inclusions (carboxysomes) of Thiobacillus neapolitanus. *Science* 1973;**182**:584–6.
255. Price GD, Coleman JR, Badger MR. Association of carbonic anhydrase activity with carboxysomes isolated from the Cyanobacterium Synechococcus PCC7942. *Plant Physiol* 1992;**100**:784–93.
256. So AK, Espie GS, Williams EB, Shively JM, Heinhorst S, Cannon GC. A novel evolutionary lineage of carbonic anhydrase (epsilon class) is a component of the carboxysome shell. *J Bacteriol* 2004;**186**:623–30.
257. Yu JW, Price GD, Song L, Badger MR. Isolation of a putative carboxysomal carbonic anhydrase gene from the Cyanobacterium Synechococcus PCC7942. *Plant Physiol* 1992;**100**:794–800.
258. Chen P, Andersson DI, Roth JR. The control region of the pdu/cob regulon in Salmonella typhimurium. *J Bacteriol* 1994;**176**:5474–82.
259. Kofoid E, Rappleye C, Stojiljkovic I, Roth J. The 17-gene ethanolamine (eut) operon of Salmonella typhimurium encodes five homologues of carboxysome shell proteins. *J Bacteriol* 1999;**181**:5317–29.
260. Bobik TA, Havemann GD, Busch RJ, Williams DS, Aldrich HC. The propanediol utilization (pdu) operon of Salmonella enterica serovar Typhimurium LT2 includes genes necessary for formation of polyhedral organelles involved in coenzyme B(12)-dependent 1, 2-propanediol degradation. *J Bacteriol* 1999;**181**:5967–75.
261. Havemann GD, Sampson EM, Bobik TA. PduA is a shell protein of polyhedral organelles involved in coenzyme B(12)-dependent degradation of 1,2-propanediol in Salmonella enterica serovar typhimurium LT2. *J Bacteriol* 2002;**184**:1253–61.
262. Havemann GD, Bobik TA. Protein content of polyhedral organelles involved in coenzyme B12-dependent degradation of 1,2-propanediol in Salmonella enterica serovar Typhimurium LT2. *J Bacteriol* 2003;**185**:5086–95.
263. Leal NA, Havemann GD, Bobik TA. PduP is a coenzyme-a-acylating propionaldehyde dehydrogenase associated with the polyhedral bodies involved in B12-dependent 1,2-propanediol degradation by Salmonella enterica serovar Typhimurium LT2. *Arch Microbiol* 2003;**180**:353–61.
264. Rondon MR, Horswill AR, Escalante-Semerena JC. DNA polymerase I function is required for the utilization of ethanolamine, 1,2-propanediol, and propionate by Salmonella typhimurium LT2. *J Bacteriol* 1995;**177**:7119–24.
265. Rondon MR, Kazmierczak R, Escalante-Semerena JC. Glutathione is required for maximal transcription of the cobalamin biosynthetic and 1,2-propanediol utilization (cob/pdu) regulon and for the catabolism of ethanolamine, 1,2-propanediol, and propionate in Salmonella typhimurium LT2. *J Bacteriol* 1995;**177**:5434–9.
266. Sampson EM, Bobik TA. Microcompartments for B12-dependent 1,2-propanediol degradation provide protection from DNA and cellular damage by a reactive metabolic intermediate. *J Bacteriol* 2008;**190**:2966–71.
267. Penrod JT, Roth JR. Conserving a volatile metabolite: a role for carboxysome-like organelles in Salmonella enterica. *J Bacteriol* 2006;**188**:2865–74.

268. Kerfeld CA, Sawaya MR, Tanaka S, Nguyen CV, Phillips M, Beeby M, et al. Protein structures forming the shell of primitive bacterial organelles. *Science* 2005;**309**:936–8.
269. Yeates TO, Crowley CS, Tanaka S. Bacterial microcompartment organelles: protein shell structure and evolution. *Annu Rev Biophys* 2010;**39**:185–205.
270. Tanaka S, Kerfeld CA, Sawaya MR, Cai F, Heinhorst S, Cannon GC, et al. Atomic-level models of the bacterial carboxysome shell. *Science* 2008;**319**:1083–6.
271. Tanaka S, Sawaya MR, Yeates TO. Structure and mechanisms of a protein-based organelle in Escherichia coli. *Science* 2010;**327**:81–4.
272. Cot SS, So AK, Espie GS. A multiprotein bicarbonate dehydration complex essential to carboxysome function in cyanobacteria. *J Bacteriol* 2008;**190**:936–45.
273. Long BM, Badger MR, Whitney SM, Price GD. Analysis of carboxysomes from Synechococcus PCC7942 reveals multiple Rubisco complexes with carboxysomal proteins CcmM and CcaA. *J Biol Chem* 2007;**282**:29323–35.
274. Price GD, Howitt SM, Harrison K, Badger MR. Analysis of a genomic DNA region from the cyanobacterium Synechococcus sp. strain PCC7942 involved in carboxysome assembly and function. *J Bacteriol* 1993;**175**:2871–9.
275. Parsons JB, Frank S, Bhella D, Liang M, Prentice MB, Mulvihill DP, et al. Synthesis of empty bacterial microcompartments, directed organelle protein incorporation, and evidence of filament-associated organelle movement. *Mol Cell* 2010;**38**:305–15.
276. Menon BB, Dou Z, Heinhorst S, Shively JM, Cannon GC. Halothiobacillus neapolitanus carboxysomes sequester heterologous and chimeric RubisCO species. *PLoS One* 2008;**3**:e3570.
277. Fan C, Cheng S, Liu Y, Escobar CM, Crowley CS, Jefferson RE, et al. Short N-terminal sequences package proteins into bacterial microcompartments. *Proc Natl Acad Sci USA* 2010;**107**:7509–14.
278. Sutter M, Boehringer D, Gutmann S, Gunther S, Prangishvili D, Loessner MJ, et al. Structural basis of enzyme encapsulation into a bacterial nanocompartment. *Nat Struct Mol Biol* 2008;**15**:939–47.
279. Seebeck FP, Woycechowsky KJ, Zhuang W, Rabe JP, Hilvert D. A simple tagging system for protein encapsulation. *J Am Chem Soc* 2006;**128**:4516–7.
280. Zhang X, Konarev PV, Petoukhov MV, Svergun DI, Xing L, Cheng RH, et al. Multiple assembly states of lumazine synthase: a model relating catalytic function and molecular assembly. *J Mol Biol* 2006;**362**:753–70.
281. Dalmau M, Lim S, Chen HC, Ruiz C, Wang SW. Thermostability and molecular encapsulation within an engineered caged protein scaffold. *Biotechnol Bioeng* 2008;**101**:654–64.
282. Dalmau M, Lim S, Wang SW. pH-triggered disassembly in a caged protein complex. *Biomacromolecules* 2009;**10**:3199–206.
283. Grinberg O, Hayun M, Sredni B, Gedanken A. Characterization and activity of sonochemically-prepared BSA microspheres containing Taxol–an anticancer drug. *Ultrason Sonochem* 2007;**14**:661–6.
284. Rivkin I, Cohen K, Koffler J, Melikhov D, Peer D, Margalit R. Paclitaxel-clusters coated with hyaluronan as selective tumor-targeted nanovectors. *Biomaterials* 2010;**31**:7106–14.
285. Vazquez E, Roldan M, Diez-Gil C, Unzueta U, Domingo-Espin J, Cedano J, et al. Protein nanodisk assembling and intracellular trafficking powered by an arginine-rich (R9) peptide. *Nanomedicine* 2010;**5**:259–68.
286. Diez-Gil C, Krabbenborg S, Garcia-Fruitos E, Vazquez E, Rodriguez-Carmona E, Ratera I, et al. The nanoscale properties of bacterial inclusion bodies and their effect on mammalian cell proliferation. *Biomaterials* 2010;**31**:5805–12.
287. Garcia-Fruitos E, Rodriguez-Carmona E, Diez-Gil C, Ferraz RM, Vazquez E, Corchero JL, et al. Surface cell growth engineering assisted by a novel bacterial nanomaterial. *Adv Mater* 2009;**21**:4249.

288. Rodriguez-Carmona E, Villaverde A. Nanostructured bacterial materials for innovative medicines. *Trends Microbiol* 2010;**18**:423–30.
289. Vaysse L, Harbottle R, Bigger B, Bergau A, Tolmachov O, Coutelle C. Development of a self-assembling nuclear targeting vector system based on the tetracycline repressor protein. *J Biol Chem* 2004;**279**:5555–64.
290. Fukunaga-Kalabis M, Herlyn M. Unraveling mysteries of the multifunctional protein SPARC. *J Invest Dermatol* 2007;**127**:2497–8.
291. Akahata W, Yang ZY, Andersen H, Sun S, Holdaway HA, Kong WP, et al. A virus-like particle vaccine for epidemic Chikungunya virus protects nonhuman primates against infection. *Nat Med* 2010;**16**:334–8.
292. Pushko P, Kort T, Nathan M, Pearce MB, Smith G, Tumpey TM. Recombinant H1N1 virus-like particle vaccine elicits protective immunity in ferrets against the 2009 pandemic H1N1 influenza virus. *Vaccine* 2010;**28**:4771–6.
293. Harmsen MM, De Haard HJ. Properties, production, and applications of camelid single-domain antibody fragments. *Appl Microbiol Biotechnol* 2007;**77**:13–22.
294. Cortez-Retamozo V, Backmann N, Senter PD, Wernery U, De BP, Muyldermans S, et al. Efficient cancer therapy with a nanobody-based conjugate. *Cancer Res* 2004;**64**:2853–7.
295. Bian J, Popovic ZB, Benejam C, Kiedrowski M, Rodriguez LL, Penn MS. Effect of cell-based intercellular delivery of transcription factor GATA4 on ischemic cardiomyopathy. *Circ Res* 2007;**100**:1626–33.
296. Fletcher S, Honeyman K, Fall AM, Harding PL, Johnsen RD, Steinhaus JP, et al. Morpholino oligomer-mediated exon skipping averts the onset of dystrophic pathology in the mdx mouse. *Mol Ther* 2007;**15**:1587–92.
297. Dubikovskaya EA, Thorne SH, Pillow TH, Contag CH, Wender PA. Overcoming multidrug resistance of small-molecule therapeutics through conjugation with releasable octaarginine transporters. *Proc Natl Acad Sci USA* 2008;**105**:12128–33.
298. Bucci M, Gratton JP, Rudic RD, Acevedo L, Roviezzo F, Cirino G, et al. In vivo delivery of the caveolin-1 scaffolding domain inhibits nitric oxide synthesis and reduces inflammation. *Nat Med* 2000;**6**:1362–7.
299. Morris MC, Gros E, drian-Herrada G, Choob M, Archdeacon J, Heitz F, et al. A non-covalent peptide-based carrier for in vivo delivery of DNA mimics. *Nucleic Acids Res* 2007;**35**:e49.
300. Crombez L, Morris MC, Dufort S, drian-Herrada G, Nguyen Q, Mc MG, et al. Targeting cyclin B1 through peptide-based delivery of siRNA prevents tumour growth. *Nucleic Acids Res* 2009;**37**:4559–69.
301. Song E, Zhu P, Lee SK, Chowdhury D, Kussman S, Dykxhoorn DM, et al. Antibody mediated in vivo delivery of small interfering RNAs via cell-surface receptors. *Nat Biotechnol* 2005;**23**:709–17.
302. Tan M, Lan KH, Yao J, Lu CH, Sun M, Neal CL, et al. Selective inhibition of ErbB2-overexpressing breast cancer in vivo by a novel TAT-based ErbB2-targeting signal transducers and activators of transcription 3-blocking peptide. *Cancer Res* 2006;**66**:3764–72.
303. Myrberg H, Zhang L, Mae M, Langel U. Design of a tumor-homing cell-penetrating peptide. *Bioconj Chem* 2008;**19**:70–5.
304. Laakkonen P, Porkka K, Hoffman JA, Ruoslahti E. A tumor-homing peptide with a targeting specificity related to lymphatic vessels. *Nat Med* 2002;**8**:751–5.
305. Jain M, Chauhan SC, Singh AP, Venkatraman G, Colcher D, Batra SK. Penetratin improves tumor retention of single-chain antibodies: a novel step toward optimization of radioimmunotherapy of solid tumors. *Cancer Res* 2005;**65**:7840–6.
306. Vocero-Akbani AM, Heyden NV, Lissy NA, Ratner L, Dowdy SF. Killing HIV-infected cells by transduction with an HIV protease-activated caspase-3 protein. *Nat Med* 1999;**5**:29–33.

307. Bidwell III GL, Fokt I, Priebe W, Raucher D. Development of elastin-like polypeptide for thermally targeted delivery of doxorubicin. *Biochem Pharmacol* 2007;**73**:620–31.
308. Snyder EL, Meade BR, Saenz CC, Dowdy SF. Treatment of terminal peritoneal carcinomatosis by a transducible p53-activating peptide. *PLoS Biol* 2004;**2**:E36.
309. Harada H, Hiraoka M, Kizaka-Kondoh S. Antitumor effect of TAT-oxygen-dependent degradation-caspase-3 fusion protein specifically stabilized and activated in hypoxic tumor cells. *Cancer Res* 2002;**62**:2013–8.
310. Datta K, Sundberg C, Karumanchi SA, Mukhopadhyay D. The 104–123 amino acid sequence of the beta-domain of von Hippel-Lindau gene product is sufficient to inhibit renal tumor growth and invasion. *Cancer Res* 2001;**61**:1768–75.
311. Shibagaki N, Udey MC. Dendritic cells transduced with protein antigens induce cytotoxic lymphocytes and elicit antitumor immunity. *J Immunol* 2002;**168**:2393–401.
312. Bright R, Raval AP, Dembner JM, Perez-Pinzon MA, Steinberg GK, Yenari MA, et al. Protein kinase C delta mediates cerebral reperfusion injury in vivo. *J Neurosci* 2004;**24**:6880–8.
313. Cao G, Pei W, Ge H, Liang Q, Luo Y, Sharp FR, et al. In vivo delivery of a Bcl-xL fusion protein containing the TAT protein transduction domain protects against ischemic brain injury and neuronal apoptosis. *J Neurosci* 2002;**22**:5423–31.
314. Kumar P, Ban HS, Kim SS, Wu H, Pearson T, Greiner DL, et al. T cell-specific siRNA delivery suppresses HIV-1 infection in humanized mice. *Cell* 2008;**134**:577–86.
315. Kim WJ, Christensen LV, Jo S, Yockman JW, Jeong JH, Kim YH, et al. Cholesteryl oligoarginine delivering vascular endothelial growth factor siRNA effectively inhibits tumor growth in colon adenocarcinoma. *Mol Ther* 2006;**14**:343–50.
316. Mamot C, Drummond DC, Noble CO, Kallab V, Guo Z, Hong K, et al. Epidermal growth factor receptor-targeted immunoliposomes significantly enhance the efficacy of multiple anticancer drugs in vivo. *Cancer Res* 2005;**65**:11631–8.
317. Arap W, Pasqualini R, Ruoslahti E. Cancer treatment by targeted drug delivery to tumor vasculature in a mouse model. *Science* 1998;**279**:377–80.
318. Liu S, Wang H, Currie BM, Molinolo A, Leung HJ, Moayeri M, et al. Matrix metalloproteinase-activated anthrax lethal toxin demonstrates high potency in targeting tumor vasculature. *J Biol Chem* 2008;**283**:529–40.
319. Frankel AE, Powell BL, Duesbery NS, Vande Woude GF, Leppla SH. Anthrax fusion protein therapy of cancer. *Curr Protein Pept Sci* 2002;**3**:399–407.
320. Hu P, Yan J, Sharifi J, Bai T, Khawli LA, Epstein AL. Comparison of three different targeted tissue factor fusion proteins for inducing tumor vessel thrombosis. *Cancer Res* 2003;**63**:5046–53.
321. Lindgren M, Rosenthal-Aizman K, Saar K, Eiriksdottir E, Jiang Y, Sassian M, et al. Overcoming methotrexate resistance in breast cancer tumour cells by the use of a new cell-penetrating peptide. *Biochem Pharmacol* 2006;**71**:416–25.
322. Anderson DC, Nichols E, Manger R, Woodle D, Barry M, Fritzberg AR. Tumor cell retention of antibody Fab fragments is enhanced by an attached HIV TAT protein-derived peptide. *Biochem Biophys Res Commun* 1993;**194**:876–84.
323. Dijkgraaf I, Kruijtzer JA, Liu S, Soede AC, Oyen WJ, Corstens FH, et al. Improved targeting of the alpha(v)beta (3) integrin by multimerisation of RGD peptides. *Eur J Nucl Med Mol Imaging* 2007;**34**:267–73.
324. Kawamura KS, Sung M, Bolewska-Pedyczak E, Gariepy J. Probing the impact of valency on the routing of arginine-rich peptides into eukaryotic cells. *Biochemistry* 2006;**45**:1116–27.
325. Vives E, Schmidt J, Pelegrin A. Cell-penetrating and cell-targeting peptides in drug delivery. *Biochim Biophys Acta* 2008;**1786**:126–38.

326. Jiang T, Olson ES, Nguyen QT, Roy M, Jennings PA, Tsien RY. Tumor imaging by means of proteolytic activation of cell-penetrating peptides. *Proc Natl Acad Sci USA* 2004;**101**:17867–72.
327. Sethuraman VA, Bae YH. TAT peptide-based micelle system for potential active targeting of anti-cancer agents to acidic solid tumors. *J Control Release* 2007;**118**:216–24.
328. Lebleu B, Moulton HM, Abes R, Ivanova GD, Abes S, Stein DA, et al. Cell penetrating peptide conjugates of steric block oligonucleotides. *Adv Drug Deliv Rev* 2008;**60**:517–29.
329. Kamei N, Morishita M, Eda Y, Ida N, Nishio R, Takayama K. Usefulness of cell-penetrating peptides to improve intestinal insulin absorption. *J Control Release* 2008;**132**:21–5.
330. Pardridge WM. Biopharmaceutical drug targeting to the brain. *J Drug Target* 2010;**18**:157–67.
331. Perales JC, Ferkol T, Beegen H, Ratnoff OD, Hanson RW. Gene transfer in vivo: sustained expression and regulation of genes introduced into the liver by receptor-targeted uptake. *Proc Natl Acad Sci USA* 1994;**91**:4086–90.
332. Sato Y, Yamauchi N, Takahashi M, Sasaki K, Fukaura J, Neda H, et al. In vivo gene delivery to tumor cells by transferrin-streptavidin-DNA conjugate. *FASEB J* 2000;**14**:2108–18.

Polymer Nanoparticles

Xiao-Yun Lu,[*] Dao-Cheng Wu,[*]
Zheng-Jun Li,[†] and
Guo-Qiang Chen[‡]

[*]*School of Life Science and Technology, Xi'an Jiaotong University, Xi'an, China*

[†]*College of Life Science and Technology, Beijing University of Chemical Technology, Beijing, China*

[‡]*School of Life Sciences, Tsinghua University, Beijing, China*

I. Definition and General Ingredients of Polymer Nanoparticles 300
II. Production Procedures 302
A. The Solvent Evaporation Method 302
B. Emulsification–Solvent Diffusion Method 302
C. The Solvent Displacement Method 303
D. Monomer Polymerization 304
III. Characterization, Quality, and Stability of Polymer Nanoparticles 305
A. Particle Size, Size Distribution, and Morphology 305
B. Drug Loading and Drug Encapsulation Efficiency 306
C. Drug Release 307
D. *In Vitro* Stability Performances 307
IV. Applications of Microbial PHA-Based Nanoparticles 308
A. PHA Nanoparticles as Drug Delivery Carriers 308
B. PHA Nanoparticles Designed for Targeted Drug Delivery Carriers 311
C. Functionalization of PHA Nanoparticles 313
D. PHA Biopolymer Granules Production from GRAS Bacteria 318
V. Summary and Outlook 318
References 320

Over the past few decades, interest in designing and developing polymeric nanoparticles has undergone considerable explosion. Indeed, these nanoparticulated polymer-based systems provide potential solution to improve therapeutic efficacy and diagnosis sensitivity. In this chapter, general properties, production, and characterization of polymer nanoparticles are introduced. Specifically, the development and application of polyhydroxyalkanoate (PHA)-based nanoparticles are emphasized because of the good biocompatible, biodegradable properties, as well as their mechanical flexibility. These PHAs nanoparticles can serve as targeting drug delivery carriers and protein purification and immobilization matrices. The perspective outlook in the last section highlights the future application of polymer nanoparticles in translational science.

Progress in Molecular Biology
and Translational Science, Vol. 104
DOI: 10.1016/B978-0-12-416020-0.00007-3

1877-1173/11 $35.00

I. Definition and General Ingredients of Polymer Nanoparticles

Polymer nanoparticles may be defined as being colloidal systems, which are usually from around 5–10 nm to an upper size limit of 1000 nm, although the range generally obtained is 100–500 nm.[1] The term "polymer nanoparticles" is a collective term used for any type of polymer nano-sized particles, but specifically for polymer nanospheres and nanocapsules. Polymer nanospheres are matrix particles, that is, particles whose entire mass is solid. They can further serve as carriers for other biologically active molecules which may be either adsorbed at the sphere surface or encapsulated within the particles. Here, biologically active materials include drugs, genes, nucleic, fluorescence, and other functional materials. Unlike polymer nanospheres, polymer nanocapsules are vesicular systems in which the bioactive agents are confined to an aqueous core and surrounded by the polymeric shell around them. A schematic representation of polymer nanoparticles is shown in Fig. 1.

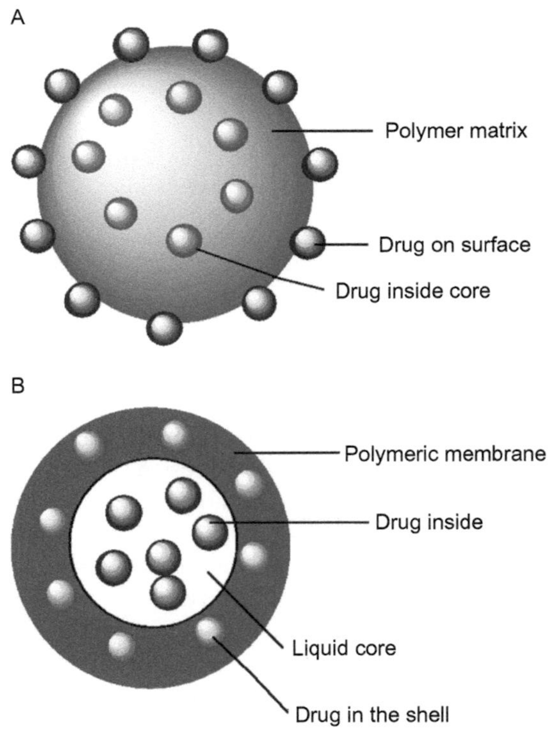

FIG. 1. Polymer nanoparticle stuctures. (A) Polymer nanoparticles or polymer nanospheres. (B) Polymer nanocapsules.

In the preparation process, two types of polymers, both natural and derived from synthesis, can be involved. Generally, most of the natural polymers are biodegradable while some of the synthesis polymers are not. Thus, natural polymers are preferred for the preparation of polymer nanoparticles as *in vivo* drug delivery systems. Typical natural polymers for the preparation of polymer nanoparticles include starch, polypeptides, albumin, sodium alginate, chitin, gelatin, cellulose, and polyhydroxyalkanoates (PHAs), while polyethylene glycol (PEG), poly-lactic-acid-*co*-glycolic-acid (PLGA), polyvinyl alcohol (PVA), polyvinyl pyrrolidone (PVP), polyethylene (PE), polyanhydrides, and polyorthoesters are common synthesis polymer materials. The surface of polymer nanoparticles may be functionalized with a range of metal ions, small molecules, surfactants, or polymers, for purposes of better targeting, binding with ligands, and avoiding immunological reactions.[1]

The historical development of polymer nanoparticles was started by Paul Ehrlich with the first experimental attempts by Ursula Scheffel. Extensive works were conducted by the group of Peter Speiser at ETH Zürich in the late 1960s and early 1970s.[2,3] Nanoparticles have since been extensively studied as drug carriers in the pharmaceutical field. Generally, polymer nanoparticles are suitable for nearly all administration routes including intravenous or intramuscular injections, dermatic or nasal absorptions, oral administration, and so on. The first commercial polymer nanoparticle drug delivery system termed Abraxane, which was human serum albumin nanoparticles containing paclitaxel, appeared for clinics at the beginning of 2005. Poly(isohexyl cyanoacrylate) nanoparticles named doxorubicin-Trans-drug, loaded with doxorubicin, is presently being developed by Bio Alliance in Paris for cancer treatment, and Phase I/II clinical trials have been successfully conducted.[4]

The advantages of polymer nanoparticles as active substance delivery systems include high drug encapsulation efficiency, higher intracellular uptake than other particulate drug delivery systems, higher stability of encapsulating active substances, and biocompatibility with tissue and cells when prepared from biopolymers that are biocompatible or biodegradable. Importantly, polymer nanoparticles can be designed to effectively deliver the drug to a target site and thus increase therapeutic results, minimizing the side effects. As an outstanding drug delivery system, polymer nanoparticles have the merits of diversity, maneuverability, and multifunctional potentials. Among the diverse polymer nanoparticles emerging rapidly, several are commercially available for biological and medical applications. New developments focus on new functions, novel polymer material, improved supporting functions, and promotion of drug performances.[4,5]

Disadvantages of polymer nanoparticles include possible nonbiodegradability, fragileness, higher manufacturing costs, toxic solvent residuals among others. It is imperative to have a better understanding of basic principles involved in designing and applying polymer nanoparticles for diagnosis, therapeutical treatment, or a combination of imaging and therapy in different clinical situations. Many factors need to be optimized to design advanced polymer nanoparticles for molecular imaging of drug delivery, among which are biocompatibility, pharmacokinetics, *in vivo* targeting efficacy, and cost-effectiveness. Foremost, minimizing the potential toxicity of polymer nanoparticles is critically important. In order to validate the potential of therapeutic drug delivery with polymer nanoparticles, *in vitro* cytotoxicity resulting from polymer nanoparticles must be carefully studied.[6,7]

II. Production Procedures

Polymer nanoparticles can be conveniently prepared either from dispersion of preformed polymers or by direct polymerization of monomers using classical polymerization. For dispersion of preformed polymers, several methods have been suggested to prepare biodegradable polymer nanoparticles made of PLA, PLG, PLGA, and poly(ε-caprolactone).[8]

A. The Solvent Evaporation Method

The polymer is first dissolved in an organic solvent such as dichloromethane, chloroform, or ethyl acetate to become a solution, while the drug is dissolved or dispersed into this preformed polymer solution. The two phases are emulsified to make an oil (O) in water (W) emulsion using surfactant/emulsifying agents like dioctyl sodium sulfosuccinate (AOT), polysorbate-80, poloxamer-188, etc. By continuous stirring for a certain period of time, the stable emulsion is formed, and the residual organic solvent can be either evaporated by increasing the system temperature/under pressure or by continuous stirring for some time (Fig. 2). Subsequently, the solidified polymer nanoparticles can be collected by ultracentrifugation and washed with distilled water to remove impurities such as surfactants or residual additives. Finally, the product is lyophilized and stored prior to use. The US Food and Drug Administration (FDA) and other administrations have specified the residual amount of organic solvents in injectable polymer nanoparticle systems.[1]

B. Emulsification–Solvent Diffusion Method

Emulsification–solvent diffusion technique (Fig. 3) is another method that has been developed for polymer nanoparticle preparation. The encapsulating polymer is first dissolved in a partially water-soluble solvent such as propylene carbonate saturated with water. To produce the polymer precipitation and the resulting nanoparticles, the solvent of the dispersed phase is diluted with an excess of water or with another organic solvent. Subsequently, the polymer–water-saturated solvent phase is emulsified in an aqueous solution containing a stabilizer, leading to

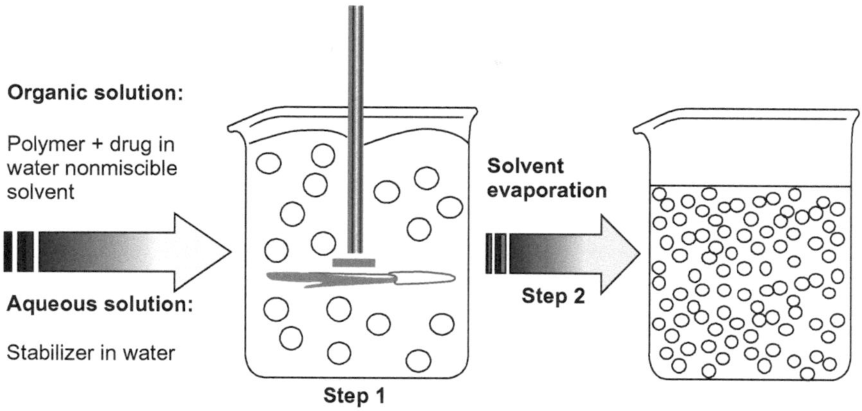

FIG. 2. The procedure of solvent evaporation method.[8]

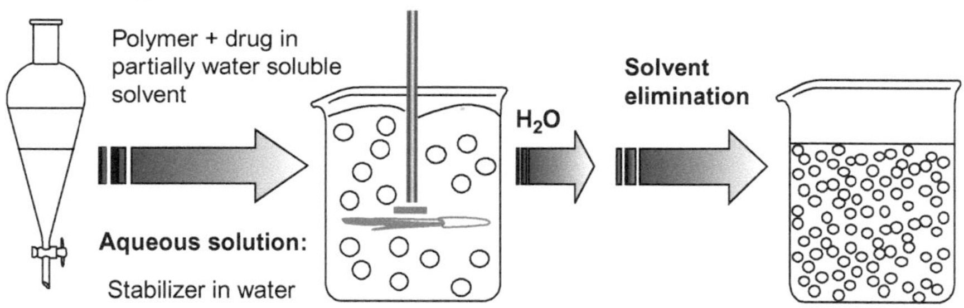

FIG. 3. The procedure of emulsification–solvent diffusion method.[8] A partially water-soluble solvent dissolved with the encapsulating polymer is saturated with water and then emulsified in an aqueous solution containing stabilizer. The formation of nanospheres or nanocapsules can be resulted from the solvent diffusion to the external phase.

solvent diffusion to the external phase and the formation of nanospheres or nanocapsules. The solvent is eliminated by evaporation.[1,8] This technique presents several advantages including high encapsulation efficiencies for lipophilic drugs, high batch-to-batch reproducibility, easy scale-up, simplicity, and narrow size distribution. Disadvantages are the large amount of water to be removed from the suspension and the leakage of water-soluble drug into the saturated-aqueous external phase during emulsification, reducing encapsulation efficiency.

C. The Solvent Displacement Method

The basic principle of this technique is based on the interfacial deposition of a polymer by an organic solvent or salts; rapid diffusion of the solvent into a nonsolvent phase results in the decrease of interfacial

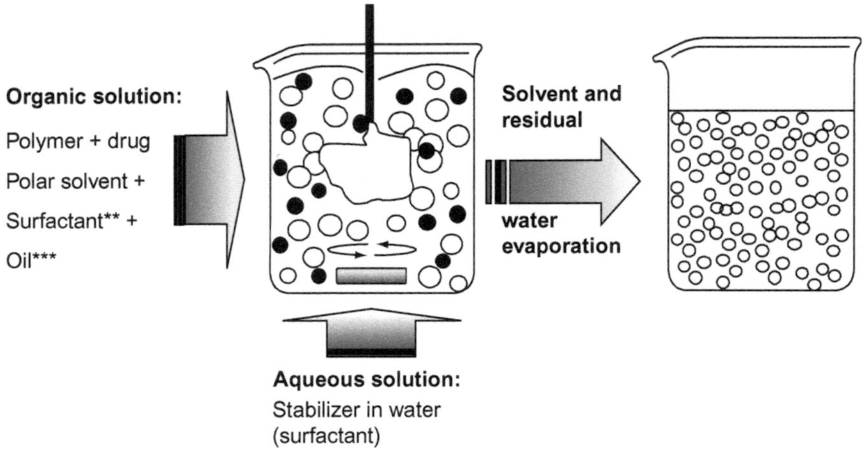

FIG. 4. The solvent displacement method.[8]

tension between the two phases, which can increase the surface area, and leads to the formation of nano-sized particles (Fig. 4). This method is widely used for the preparation of polymer nanoparticles such as PLA, PLGA, PCL, and so on.[1,6]

D. Monomer Polymerization

Monomer polymerization uses the following ingredients including water, a monomer of low water solubility, water-soluble initiator, and a surfactant. They are mixed at a certain temperature and stirred for an extended period of time to proceed with the polymerization. At the end of the reaction, polymer nanoparticles typically 10–150 nm in size are formed generally with some surfactant trapped in the polymer particles, as their long aliphatic chains either become entangled in the polymer network or they become grafted onto the polymer by side reactions with the radicals. They are therefore charge-stabilized. Removal of surfactants is a hard and time-consuming process that increases the cost of production.[1,2]

Interfacial polymerization also used for the preparation of polymer nanoparticles involves the process polymerization of two reactive monomers or agents which are dissolved respectively in two phases (i.e., continuous and dispersed phase); the reaction takes place at the interface of the two liquids. No purification process is needed and high efficiency of drug encapsulation can be reached. The organic solvent is an external phase that could have toxicity.

Drugs including insulin, calcitonin, octreotide, darodipine, indomethacin, and photoactivatable cytotoxic compounds are efficiently encapsulated in these nanoparticles formed by this polymerization process.[1–3]

III. Characterization, Quality, and Stability of Polymer Nanoparticles

Most frequently sought properties of polymer nanoparticles are sizes, zeta potential, encapsulation efficiency, drug-release behavior, stability, and *in vitro* and *in vivo* performances in relation to their clinic applications.

A. Particle Size, Size Distribution, and Morphology

Particle size and size distribution are the key parameters used for evaluating the physical properties of polymer nanoparticles. Photon correlation spectroscopy (PCS), also known as dynamic light scattering (DLS), laser diffraction (LD), and coulter counter are commonly used to measure the size and size distribution of polymer nanoparticles. PCS/DLS is widely used to determine the size and size distribution of nano-sized particles suspended in liquid medium. The mean particle size and size distribution indicated as polydispersity index (PDI) is the typical measured parameter for this technique. A PDI value range from 0.1 to 0.25 indicates a narrow size distribution, while a PDI more than 0.5 refers to a broad distribution. This technique is not useful for measuring the size of dry powders of polymer nanoparticles. Although PCS, LD, and coulter counter techniques provide rapid measurement of particle size and size distribution, they cannot be used to evaluate particle morphology. In this case, scanning electron microscopy (SEM), transmission electron microscopy (TEM), and atomic force microscopy (AFM) are widely used for the observation of particle morphology. AFM can provide a three-dimensional surface profile. Additionally, samples viewed by AFM do not require any special treatments such as metal/carbon coatings that would irreversibly change or damage the sample; most of the AFM modes can work perfectly well in ambient air or even a liquid environment. Importantly, AFM provides higher resolution than SEM does (Fig. 5).[9]

The zeta potential of polymer nanoparticles is commonly measured by laser Doppler electrophoresis, which evaluates electrophoretic mobility of suspended particles in the medium. It is a rule of thumb that an absolute value of zeta potential above 60 mV yields excellent stability, while 30, 20, and less than 5 mV generally result in good stability, acceptable short-term stability, and

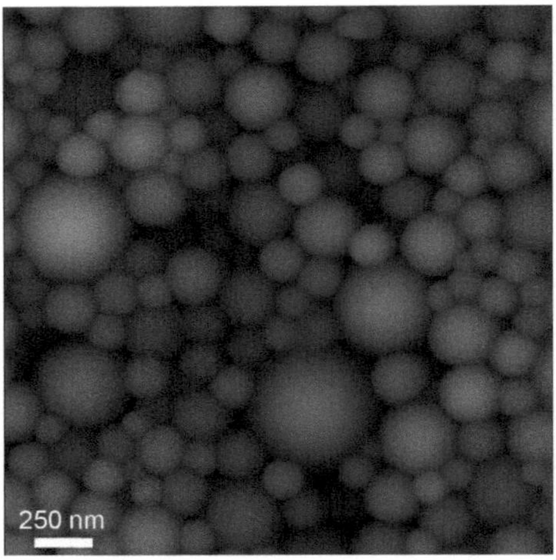

FIG. 5. AFM topography images of siRNA-loaded polymer nanoparticles with two different scan sizes.[9] (For color version of this figure, the reader is referred to the Web version of this chapter.)

fast particle aggregation, respectively, for pure electrostatic stabilization or in combination with low-molecular weight surfactants, but not for high-molecular weight stabilizers present in the medium.

B. Drug Loading and Drug Encapsulation Efficiency

One of the most noted translational medicine applications of polymer nanoparticles is as drug carriers. Drug loading and drug encapsulation efficiency are two important nanoparticle quality control keys, parameters with important impacts in their stability and application. Successful polymer nanoparticles have a high loading capacity to reduce the quantity of the carriers required for administration.[5,6,8]

Drug loading into polymer nanoparticles is achieved by incorporating the drug at the time of polymer nanoparticle production, or by adsorbing the drug after the formation of polymer nanoparticles through their incubation in the drug solution. Generally, drug loading and drug encapsulation efficiency of polymer nanoparticles were calculated as follows:

$$\text{Encapsulation efficiency}(\%) = \frac{\text{Concentration of drug input} - \text{Concentration of free drug}}{\text{Concentration of drug input}} \times 100\%, \quad (1)$$

$$\text{Drug Loading}(\%) = \frac{\text{Concentration of drug input} - \text{Concentration of free drug}}{\text{Concentration of nanoparticle}} \times 100\%. \quad (2)$$

Adsorption isotherms for the polymer nanoparticles/drug delivery system give vital information on the best possible formulation, drug binding capacity onto the surface of polymer nanoparticles, and the amount of drug adsorbed.

C. Drug Release

Drug-release behavior is an important factor for polymer nanoparticle application, directly related to drug stability and therapeutic results, as well as formulation development.[3,5] The drug-release rates from polymer nanoparticles depend on (1) desorption of the surface-bound/adsorbed drug; (2) diffusion from the polymer nanoparticles; (3) polymer nanoparticle erosion; and a combined erosion/diffusion process. Thus, diffusion and biodegradation govern the process of drug release. In most cases, rapid drug release from polymer nanoparticles, called " burst release", can be observed initially. It is reported that the release profiles of the drugs from polymer nanoparticles depend upon the nature of the delivery system. In the case of a polymer nanoparticle matrix, the drug is uniformly distributed/dissolved in the matrix and the release occurs by diffusion or erosion of the matrix. If the diffusion of the drug is faster than matrix degradation, the mechanism of drug release occurs mainly by diffusion. Rapid initial release or burst release is attributed to the fraction of the drug which is adsorbed or weakly bound to the large surface area of the polymer nanoparticles rather than to the drug incorporated in polymer nanoparticles.[5]

D. *In Vitro* Stability Performances

In vitro stability performances indicate the drug-loaded particle stability after storage for a certain period of time.[8] Generally, polymer nanoparticles are stored for 6 or 12 months at room temperature or 4 °C; then, parameters in relation to stability are measured and compared with those of samples before storage. The stability of polymer nanoparticles is mainly affected by their preparation method, particle size and distribution, morphology, and drug loading and drug encapsulation efficiency described earlier. Dosage form, delivery route, and nature of drug (small molecules vs. large biomolecules) can also affect the *in vitro* stability performances. The stability of drug-loaded polymer nanoparticles remains a very challenging issue for drug carrier development. The addition of a stabilizer or combination of stabilizers is the most commonly used and preferred approach to enhance the stability of polymer nanoparticle suspensions. Development of self-stabilized polymer nanoparticle suspensions is expected to grow with the continuing advancement in the field of polymer particle engineering.

IV. Applications of Microbial PHA-Based Nanoparticles

Polymer-based drug delivery and sustained drug-release systems are of high interest for both academic research and clinical applications. Over the past decades, many drug carriers based on polymer nanoparticles have been developed and a variety of polymer nanoparticles based on homo- and copolymers of lactate, glycolate, and PHAs have been developed with a focus on the design and development of nanocarriers for therapeutic and diagnostic purposes.[10–14] Indeed, these nanoparticulated systems provide a potential solution to improve therapeutic efficacy and diagnosis sensitivity.

Polymeric nanoparticles have been extensively studied as drug carriers.[15,16] Several polymer-based, nanoparticulated, antidrug formulations have been approved by the FDA for clinical applications.[17] Extensive reviews were published concerning the formation mechanisms of polymer nanoparticle[8,15,18] and their pharmaceutical applications.[17,19–21] The following section focuses on nanoparticles obtained from microbial polyesters PHAs as the novel multifunctional nanoparticles which have promising applications in translational science.

A. PHA Nanoparticles as Drug Delivery Carriers

PHAs are linear biopolyesters produced as energy- and carbon-storage materials by many bacteria (Fig. 6). By adjusting the contents and structures of hydroxyalkanoate (3HA) monomers, PHA demonstrates various physicochemical properties suitable for different applications. The medical applications of PHA have been extensively explored in recent years for implant biomedical applications including sutures, nerve conduits, patches, slings, cardiovascular patches, stents, guided tissue repair/regeneration devices, bone marrow scaffolds, and so on.[22–24] PHA nanoparticles were also developed as controlled drug-release vectors due to their biocompatibility and biodegradability.[22,25–27]

In the early 1990s, PHA, mainly poly(3-hydroxybutyrate) (PHB) and poly (3-hydroxybutyrate-*co*-3-hydroxyvalerate) (PHBV), became a candidate for drug delivery due to its biodegradability and biocompatibility.[28] The matrices produced by direct compression of PHBV for oral administration was proven with benefits of simplified processing over alternative sustained-release technologies.[29] PHA molecular weights had significant effect on the rate of sulfamethizole release from PHA-based carriers. Increasing the molecular weight led to enhanced release rate, which is diffusion-controlled.[30] PHA was first aimed as anticancer drug-release carriers compared with PLA.[23,24] It was found that the anticancer drug was released from the PHB microspheres faster than that from PLA microspheres.

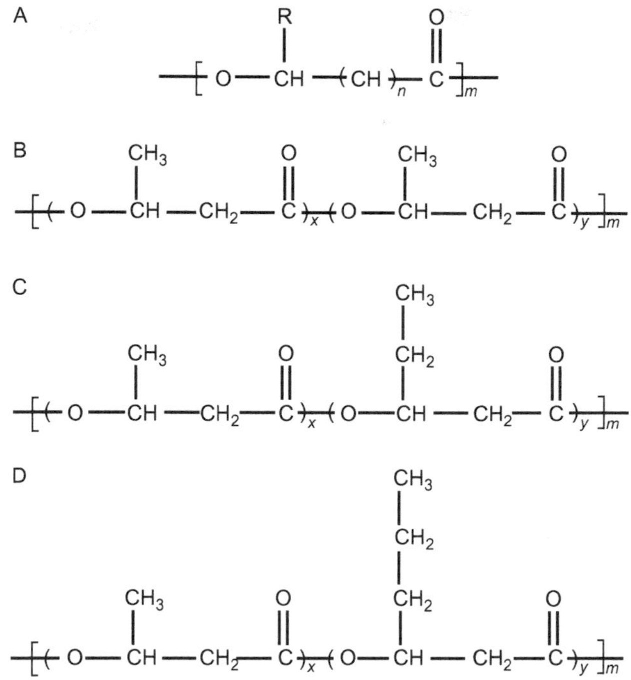

FIG. 6. Chemical structures of (A) PHA, (B) PHB, (C) PHBV, and (D) PHBHHx. R: alkyl chain of 1–11 carbon atoms; $n = 1$–3; $m > 100$; $x \geq 0$; $y \geq 0$.

Recently, a new PHA family member poly(3-hydroxybutyrate-*co*-3-hydroxyhexanorate) (PHBHHx) was intensively studied as a candidate of sustained drug-release carriers. Xiong *et al.* developed rhodamine B isothiocyanate (RBITC)-loaded PHA nanoparticles to study the macrophage-targeted intracellular sustained drug-release behaviors.[26] Figure 7 shows the SEM images of RBITC-loaded PHA nanoparticles.[25] They found that PHA nanoparticles could be endocytosed by macrophages effectively during the first 3–12 h and the RBITC-loaded PHA nanoparticles released the drug continuously over more than 20 days. Both the size of nanoparticles and the composition of polymers affected its sustained-release properties. PHB and PHBHHx nanoparticles with a small diameter demonstrated a faster drug release compared with that of a large one. Notably, PHBHHx nanoparticles always released drug faster than that those made of PHB. The amorphous nature of PHBHHx may be the reason for the faster release compared with the compact crystal PHB particles. This phenomenon was confirmed in another recent study in which a hydrophobic phosphoinositide 3-kinase (PI3K) inhibitor, TGX221, was used as a model drug. Lu *et al.* investigated the sustained release of TGX221 from

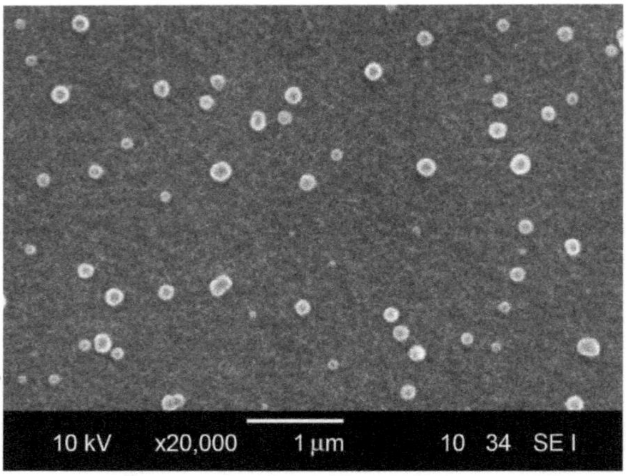

FIG. 7. SEM images of RBITC-loaded PHA nanoparticles.[25]

PHB, PHBHHx, and PLA nanoparticles. All nanoparticles showed similar profiles of an initial burst release followed by a sustained discharge process. The rate of TGX221 released from PHB nanoparticles was slower than that from PLGA ones, while PHBHHx nanoparticles appeared to have the fastest release rate of TGX221 under the same conditions. About 80% of the entrapped drug was released from PHBHHx nanoparticles at the end of initial burst-release stage, while only about 42% and 54% of entrapped TGX221 were released from PHB and PLGA nanoparticles, respectively, under the same conditions.[27] The sustained release of TGX221 from PHB nanoparticles could last for at least 6 days. About 65% of entrapped TGX221 was released at the end of the 6-day term and the sustained-release stage was expected to continue for a longer period of time.

TGX221 is described as a PI3K p110β-selective inhibitor, which has been widely used as one of the very few p110β-selective inhibitors which failed in different preclinical trials due to their poor solubility and very short circulating half-life. PHA nanoparticles showed a higher entrapment efficiency of hydrophobic compound and less cytotoxicity compared with PLGA nanoparticles. The entrapment of TGX221 into PHA nanoparticles could sufficiently extend its half-life and thus enhance its bioavailability. TGX221 could be sustained released from PHA-based nanoparticles and the drug-loaded nanoparticles significantly slowed down the growth of cancer cell lines compared with the free TGX221 alone.[27] Since hydrophobicity and poor bioavailability are common features of many anticancer kinase inhibitors, these results highlight the bright future of PHA nanoparticles in cancer treatments.

B. PHA Nanoparticles Designed for Targeted Drug Delivery Carriers

PHA-based nanoparticles can be further applied for targeting delivery when combined with specific targeting ligands. Yao *et al.* reported a specific PHA-based drug targeting system in which PHA nanoparticles containing the lipid-soluble colorant RBITC were employed as drug carriers (Fig. 8).[25] PHA nanoparticles can specifically target macrophages or hepatocellular carcinoma cells depending on the loading ligand of human α1-acid glycoprotein (rhAGP) or human epidermal growth factor (rhEGF), respectively. The ligand rhAGP bound to the mannose receptor of macrophages, and rhEGF targeted to EGF receptors on carcinoma cells were fused with the PHA granule-binding protein phasin (PhaP) to form fusion proteins rhAGP–PhaP and rhEGF–PhaP, which were expressed by *Pichia pastoris* GS115 and *Escherichia coli* BL21 (DE3), respectively. PhaP has been proven to have affinity to PHA,[31,32] allowing its stable binding to PHA drug carriers. The purified fusion proteins rhAGP–PhaP and rhEGF–PhaP were bound *in vitro* on the surface of PHA nanoparticles through the hydrophobic interaction between PhaP and PHA nanoparticles, generating rhAGP–PhaP-loaded PHA nanoparticles and rhEGF–PhaP-loaded PHA nanoparticles. These PHA nanoparticles were taken up specifically by macrophages or hepatocellular carcinoma cells under *in vitro* conditions (Fig. 9).

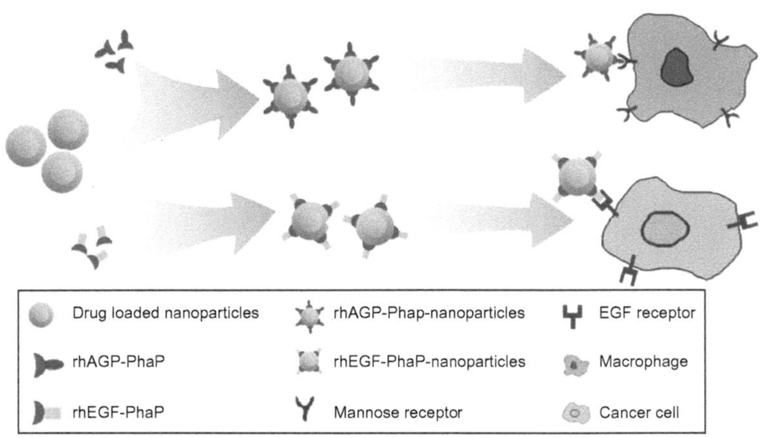

FIG. 8. A specific drug-targeting system based on PhaP fused with targeted cell ligands.[25] Drug-loaded PHA nanoparticles were coated with hypermannosylated rhAGP–PhaP or rhEGF–PhaP fusion protein to become targeting drug delivery vectors. The two vectors were specifically bound to the surface of macrophages and hepatocellular carcinoma cells, respectively, and were taken up by both types of the cells via a receptor-mediated endocytosis process. (For color version of this figure, the reader is referred to the Web version of this chapter.)

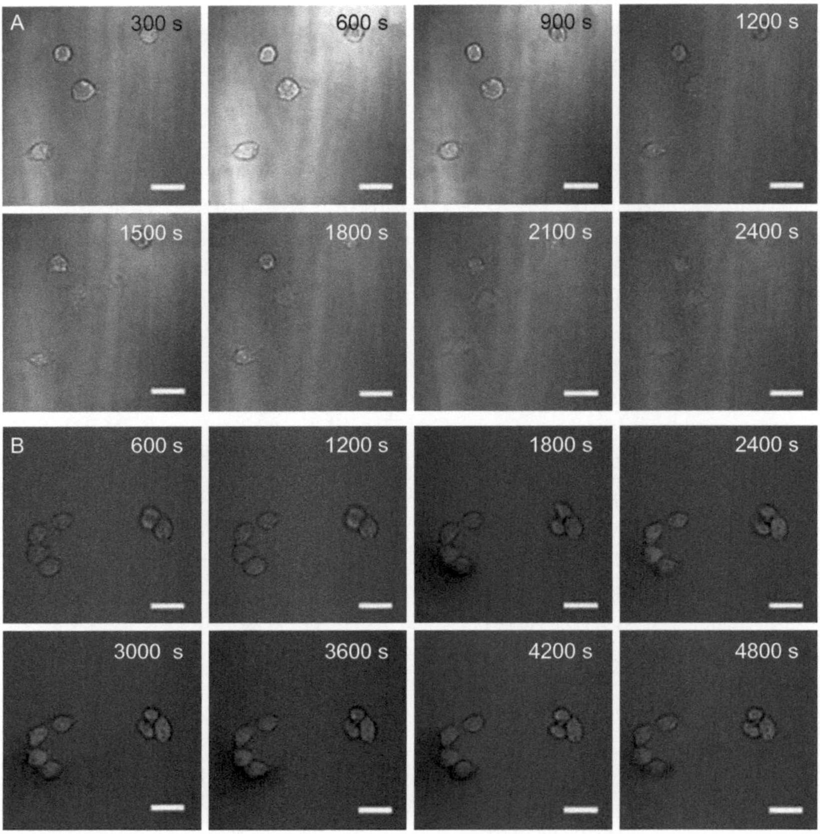

FIG. 9. Dynamic analysis of targeting drug delivery PHA nanoparticles uptake by macrophages and Bel7402 cells, respectively.[25] (A) Dynamics of rhAGP–PhaP nanoparticles incubated with macrophages. (B) Dynamics of rhEGF–PhaP nanoparticles incubated with Bel7402 cells. (See Color Insert.)

The *in vivo* studies employed an EGFR-overexpressed xenograft mouse model which clearly showed that rhEGF–PhaP-loaded PHA nanoparticles were able to achieve specific accumulation in the tumor tissue with only very little rhEGF–PhaP-loaded PHA nanoparticles observed in liver and lung tissues, indicating the high efficiency of the PHA nanoparticle-based targeting delivery system.[25] The advantage of PhaP-mediated surface functionalization is obvious since there is no cross-linking reaction during the ligand loading and thus no risk of any chemical contamination.

Although PHA nanoparticles have potential for targeted drug release, many studies remain to be done before the clinical application can be initiated. For example, phagocytic clearance of PHA, their degradation, and release products during blood circulation lead to their systematic spread which can cause toxic side effects as well as reduced therapeutic benefits resulting from drug dilution by body fluids. It is also expected that other PHA family members with diverse properties will bring more controlled-release properties for the drug-release field. This is still an area that remains to be exploited.

C. Functionalization of PHA Nanoparticles

As mentioned earlier, the surface of PHA-based nanoparticles can be further modified by many different types of signaling molecules that can be thus used for purposes other than targeted delivery. This can also be achieved by the engineering of not only PhaP but also many other PHA surface-associated proteins including PhaC, PhaR, and PhaZ (Fig. 10). The PHA synthase (PhaC) was proved to be covalently linked to the growing polyester chain and catalyzed the 3-hydroxyacyl-CoA polymerization reaction.[31] The phasin PhaP is a small amphiphilic protein that can associate with most of the hydrophobic polymer granule surfaces via strong hydrophobic interaction.[32–37] PhaP is abundant on the surface of natural PHA granules, and it could reach approximately 5% (wt/wt) of the total cell proteins in PHA-accumulating bacteria.[33] On the other hand, the PHA depolymerase (PhaZ) and PHA accumulation regulator (PhaR) were also found to attach on the granule surface. These proteins provide various targets for displaying various molecules on polymer nanoparticle surfaces. Thus, multifunctional polymer nanoparticles could be obtained through protein engineering and applied in versatile situations (Fig. 11).

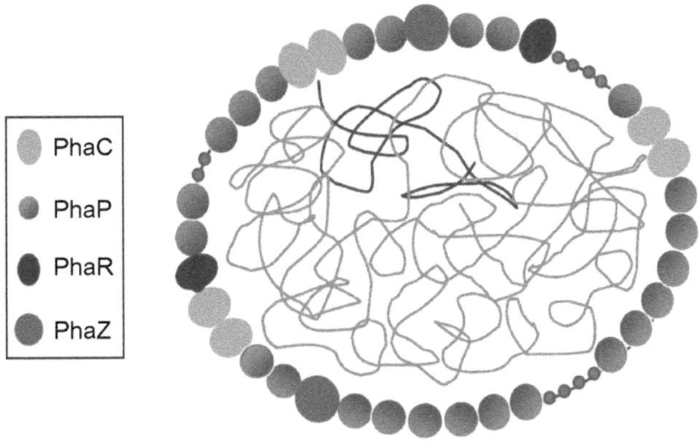

FIG. 10. PHA granule-associated proteins. (See Color Insert.)

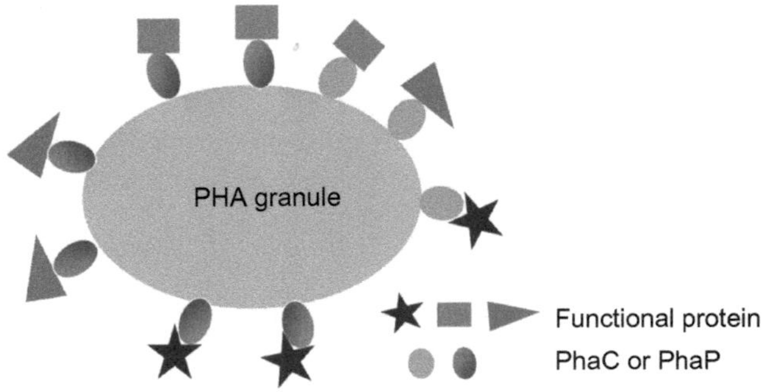

FIG. 11. Functional PHA nanoparticles through protein engineering. (See Color Insert.)

1. Functional Fusion Protein Via PHA Polymerase (PhaC)

PHA synthase (PhaC) was reported to be covalently attached to the PHA granule surface and provides another choice for protein immobilization and display. When PhaC was fused with foreign peptides, a functional PHA nanoparticle could also be obtained via the polymer synthesis process. During PHA synthesis, foreign peptides can be introduced on the polymer surface; thus, the functional nanoparticles can be produced in one step. Several interesting works in this area were reported.[34–38]

a. Immunoglobulin (IG) Purification: PhaC Fused with ZZ Domain of Protein A from Staphylococcus aureus. Staphylococcal protein A (SpA) is a type I membrane protein from the bacterium *Staphylococcus aureus*. The N-terminal half of the protein is a ZZ domain specifically binding the Fc fragment of IG. The IG-binding ZZ domain of protein A from *S. aureus* was fused to the N terminus of the PhaC from *Ralstonia eutropha* and displayed on the PHA granule surface.[34] The IGs were purified from human serum by using the ZZ–PHA granules displaying the ZZ domain as part of the PHA synthase on the surfaces of the PHA granules.[34] The IGs eluted from the engineered ZZ–PHA granules showed a high degree of purity, demonstrating that the engineered ZZ–PHA granules performed equally to commercial protein A-Sepharose beads with respect to immunoglobulin G (IgG) purification. Based on this result, an efficient and simple method for purifying IgG from mouse hybridoma supernatants was then designed using the ZZ polyester beads.[35] Three different subclasses—IgG1, IgG2a, and IgG2b—were isolated with acceptable purity. Very importantly, the ZZ polyester beads could be repeatedly used and maintained heat stability, indicating a promising commercial application for IG purification.

b. Streptavidin Beads Preparation: PhaC Fused with Streptavidin. Streptavidin is purified from the bacterium *Streptomyces avidinii.*[36] It has an extraordinarily high affinity for biotin and is used extensively in molecular biology and bionanotechnology as a high-affinity biotin-binding agent which is also resistant to extreme pH, temperature, organic solvents, denaturants, detergents, and proteolytic enzymes. Streptavidin beads were applied to display immobilized biotinylated antibodies to detect the existence of certain antigen. Generally, streptavidin beads preparation required protein production, purification, and chemical cross-linking processes.[36] PHA nanoparticle-based streptavidin presentation provided a simplified process for streptavidin beads preparation. The C terminus of streptavidin was fused to the N terminus of PHA polymerase to generate the fusion protein, which catalyzed the polyester polymerization and displayed streptavidin on the PHA granule surface at the same time[36] (Fig. 12). Several applications of the engineered polymer beads with streptavidin in enzyme-linked immunosorbent assay (ELISA), DNA purification, enzyme immobilization, and flow cytometry processes were successfully developed. Moreover, streptavidin beads constructed in this way could be effectively produced by engineered microorganisms, which simplified the process and lowered the production cost.

c. Enzyme Immobilization: PhaC Fused with α-Amylase. Industrial enzymes, for example, amylases and lipases, are often required to be immobilized onto various solid supports to facilitate repeated use and to improve

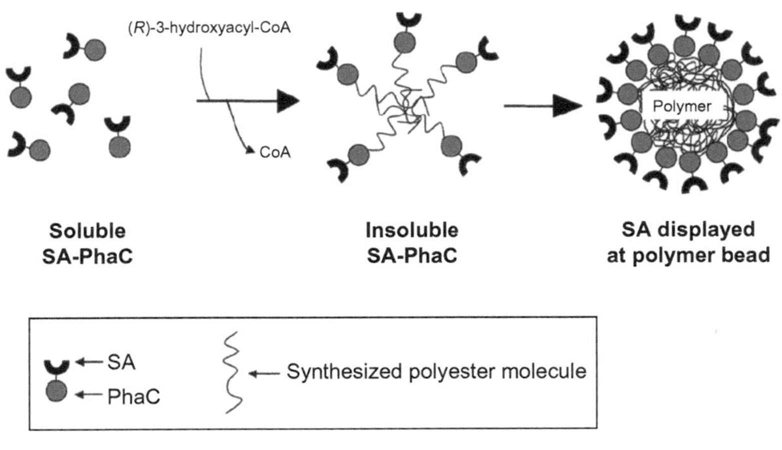

FIG. 12. Schematic view of the *in vivo* streptavidin beads assembly. SA, streptavidin; PhaC, PHA polymerase.[36]

catalytic performances. PHA granules and the granule-associated proteins provide an ideal system for protein immobilization. The industrial enzymes could be displayed on the PHA granule surface via co-expression with PhaP or PhaC, which realized enzyme expression and immobilization in one step.

A highly thermostable α-amylase (BLA) from *Bacillus licheniformis* has been successfully co-expressed with PHB synthesis operon and functionally displayed on the polymer beads in recombinant *E. coli*, in which PhaC was employed as the fusion target.[37] Sodium dodecyl sulfate polyacrylamide gel electrophoresis (SDS-PAGE) results indicated that the fusion protein was overexpressed at the polyester surface. The BLA beads exhibited α-amylase activity and the immobilized BLA indicated Michaelis–Menten enzyme kinetics, with higher binding affinity than that of free or chemically cross-linked cellulose beads of BLA.[37]

d. Protein Fused with the C Terminal of PhaC. The N terminus of PHA polymerase was believed to be a highly variable region but was not essential for the catalysis activity; thus, it was first chosen to be fused with foreign peptides for surface display. However, the displayed protein may require free C terminus for its activity. Therefore, the relatively conserved C terminus region of PhaC was investigated for foreign peptide translational fusion.[38]

The maltose-binding protein (MalE) and green fluorescent protein (GFP) were employed as reporter proteins and translationally fused to the C terminus of PhaC from *R. eutropha*. For MalE fusion, when a linker was inserted between PhaC and MalE, PhaC could maintain its activity, while MalE was inactive. Meanwhile, a 458-amino acid residues extension of PhaC N terminus enabled MalE activity. For GFP fusion, both PhaC and GFP could function well without a linker and/or N-terminal extension. Then the IgG-binding ZZ domain of protein A was fused to the N terminus of PhaC–GFP to generate a tripartite fused protein ZZ–PhaC–GFP, which displayed two functional protein domains on the polymer granule surface. All partners of the tripartite fused protein functioned well, indicating the prospect for various peptide displays via PhaC.[38]

2. Functional Fusion Protein Via Phasins

The phasins (PhaP) with high affinity to hydrophobic materials are accumulated at high levels in PHA-producing bacteria,[39] which offers a simple and low-cost approach for protein purification and surface modification.[40,41] As mentioned above, PhaP can be fused with targeted cell ligands and applied in drug-targeting delivery system.[25] In addition, various peptides were translationally fused to PhaP and improved the polymer performances. Recently,

PhaP was fused with cell adhesion motif RGD for PhaP–RGD attached on polymer surface; better cell attachment and growth were observed compared with uncoated polymer surfaces.

The Arg-Gly-Asp (RGD) sequence is a cell-adhesion-promoting peptide relating to cell attachment, and it can improve cell survival, attachment, motility, and proliferation. Thus, RGD or RGD-containing peptides have been employed for improving cell adhesion on artificial surfaces. The PhaP–RGD could bind to hydrophobic material surfaces without introducing any cross-linking agents. Dong *et al.* and You *et al.* reported the translational medical application of this functionalized PhaP–RGD with fibroblasts L929, NIH/3T3, and human bone marrow mesenchymal stem cells (hBMSCs), respectively.[40,41] Cell attachment studies using fibroblasts L929 and NIH/3T3 revealed that PhaP–RGD-coated material surfaces showed better cell attachment and proliferation than that of PhaP-coated and uncoated materials[40] (Fig. 13). When hBMSCs were inoculated on/in the scaffolds coated by PhaP–RGD, more homogeneous spread of cells, better cell adhesion, proliferation, and chondrogenic differentiation were observed than those of PhaP-coated or uncoated scaffolds. Furthermore, the RGD sequence showed obvious positive effect on extracellular matrix production, cartilage-specific extracellular substances

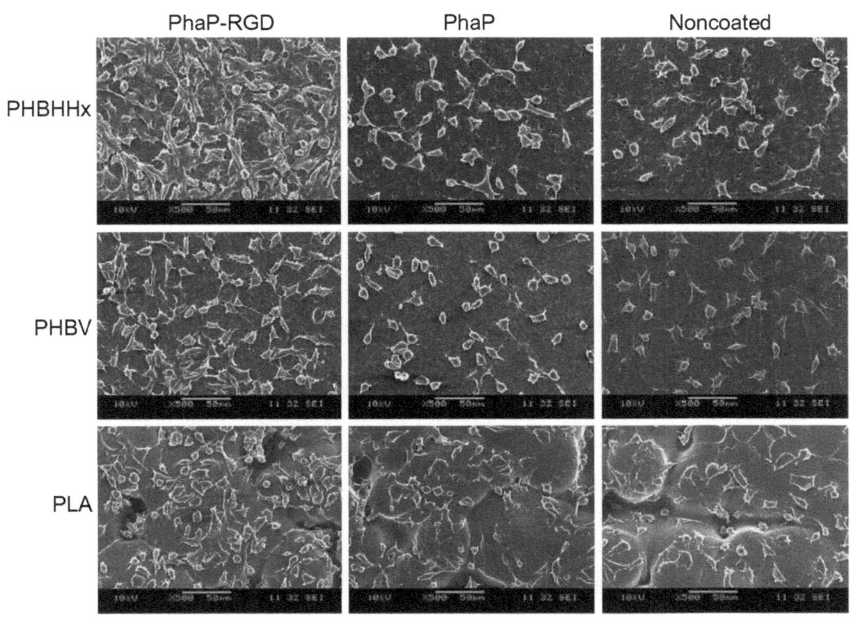

FIG. 13. SEM study of fibroblasts L929 grown for 72 h on film surfaces of PHBHHx, PHBV, and PLA coated without and with PhaP–RGD and PhaP.[40] The scale bar is 500 mm.

sulphated glycosaminoglycans (sGAG), and total collagen content.[41] These studies demonstrated that PhaP–RGD-coated polymers showed better performance than uncoated ones in tissue engineering applications, and the fused RGD sequence on PhaP brought in new properties to the polymers.

D. PHA Biopolymer Granules Production from GRAS Bacteria

Generally, biopolymers with functional proteins were produced by recombinant *E. coli* and other gram-negative bacteria containing lipopolysaccharide (LPS) endotoxins. Therefore, the polymer nanoparticles cannot be applied for *in vivo* human application directly unless LPS is completely removed. To solve the problem and obtain polymer nanoparticles suitable for biomedical applications *in vivo*, lactic acid bacterium *Lactococcus lactis* was employed to produce functional PHA nanoparticles free of LPS.[42]

L. lactis has been widely used in the food industry, and was a generally regarded-as-safe (GARS) host to produce a heterologous protein for medical applications. The PHB synthesis operon *phbCAB* from *R. eutropha* was codon-optimized for better transcription in *L. lactis* and expressed by nisin-controlled gene expression system. The highest PHB content was about 6% (wt/wt) of cell dry weight, and PhaC was found to be the dominant protein associated with polymer granules. When ZZ domain of protein A from *S. aureus* was fused to the N terminus of the PHA synthase, the fusion protein ZZ–PhaC was successfully displayed on the granule surface, exhibiting similar performance compared with that of *E. coli.*[42] Thus, *L. lactis* could be developed to become an efficient platform to produce various polymer particles suitable for medical applications.

V. Summary and Outlook

Polymer nanoparticles continue to generate excitement that can greatly accelerate the development of their applications and introduce new opportunities for biotechnological and biomaterial evolution. The rapid advancement of this technology over the past decade has allowed the design of specific and multifunctional polymer nanoparticles for medical and pharmaceutical applications such as molecular imaging, drug delivery, and targeting cancer research. The future of polymer nanoparticle will depend on the rational design of nanomaterials and the understanding of complex biological processes, whose development will lead to different categories of materials as discussed in the following sections.

a. Long-Circulating Stealth Nanoparticles. Although the interest in designing and developing nano-sized drug delivery systems has been growing, their clinical applications are still limited, owing to their rapid elimination from the blood circulation, and a nonspecific uptake by the mononuclear phagocyte system (MPS).[43,44] Consequently, in order to overcome the opsonization of the nanoparticles, effects have been made to render nanoparticles "invisible" by the immune system, creating long-circulating nanoparticles, known as stealth nanoparticles. This could be achieved by coating the nanoparticulate surface with PEG, which is referred to as a PEGylation process, resulting in the PEGylated nanoparticles that exhibit decreased levels of uptake by the MPS and an increased circulation time in the bloodstream. For PHA-based nanoparticles, PEGylation could be easily done during the fermentation process in which PEG was supplied to proper PHA synthesis microorganisms.[45,46] The resulting copolymers were end-caped with PEG fragments and can be applied as the matrix for stealth nanoparticles.

b. Environment-Sensitive Nanoparticles. Some polymer nanoparticles have been designed to have significant sensitivity to environmental variations including pH, redox, temperature, light, and ultrasound for controlled-release delivery systems. Some environmental changes are generated by abnormalities in pathological sites, which provide opportunities for the specific release of drugs in the disease site. For example, a decreased pH environment at a tumor site and an inflammation of cell cytoplasm or endosome could trigger the release of drugs from pH-sensitive polymer nanoparticles at these sites.

Pyridylhydrazone nanoparticles prepared from polylysine and propionaldehyde-PEG was shown to have the greatest acid-dependent hydrolysis. By using a polyplex RNAi nanovector shielded with bioreversible PEG conjugates, a 100-fold higher *in vitro* gene expression was achieved than in polyplexes with the analogous stable PEG shields.[47] The "proton-sponge" effect related to pH sensitivity is necessary for polynucleotide-containing polymer nanoparticles to escape from the endosome if cells have taken them up through clathrin-assisted endocytosis.

In addition to pH-sensitive carriers, polymer nanoparticles sensitive to other environmental factors, such as temperature, ion strength, and redox systems, are also under investigation.[48] Ultrasound has also been applied to enhance drug-release efficiency via microbubbles. Recent studies on cationic lipids and polymers indicated that they could also be triggered by ultrasound. Significant evidence showed that ultrasonic cavitation events create transient holes in the cell membranes, increasing the passive diffusion of both nanoparticles and drugs into the cells in the case of polymer nanoparticle-based drug delivery.

c. Multifunctional Polymer Nanoparticles. Ideal multifunctional polymer nanoparticle should be able to (1) target delivery of drug to designated cells, (2) track locations of polymer nanoparticles, (3) validate targeting result when a target is hit, and (4) indicate efficiency of usage. Multifunctional polymer nanoparticles show prospects for overcoming current barriers. The joint multi-functions of polymer nanoparticle vector and drug protection, tissue and cell targeting, imaging and external energy-assisted approach, and controlled release of drug would significantly improve therapeutic efficacy. Multifunctional polymer nanoparticles would also be powerful tools for gaining insight into the drug therapy or gene modulation process.[47,48] Several multifunctional polymer nanoparticles have been designed to solve specific problems, but the optimum organization of functional moieties requires delicate design since some of these functional parts might interact with each other and affect their functions. The multifunctional polymer nanoparticles have advantages for preoperative tumor assessment and intraoperative surgical guidance for tumor tissue resection. While successful experiments were conducted *in vitro*, progress in practical applications of these multifunctional polymer nanoparticles may be slow, particularly due to the unknown effects in humans. The development of novel multifunctional polymer nanoparticles and their performance characteristics open the way to new drug delivery systems in the future. It is a highly multidisciplinary area, requiring a range of scientific knowledge, from polymer chemistry involved in the preparation of different types of polymer nanoparticles through chemistry and biological sciences to allow for their functionalization and, of course, the basic physics, optics, and magnetic materials for special use. It is believed that polymer nanoparticles could play a very important role in the development of translational medical science.

Acknowledgments

This research was financially supported by National High Tech 863 Grants (Project No. 2010AA101607) and 973 Basic Research Fund (Grant No. 2007CB707807). CGQ was supported by Li Ka-Shing foundation to complete some of the PHA research.

References

1. Quintanar-Guerrero D, Allemann E, Fessi H, Doelker E. Preparation techniques and mechanisms of formation of biodegradable nanoparticles from preformed polymers. *Drug Dev Ind Pharm* 1998;**24**:1113–28.
2. Khanna SC, Speiser P. Epoxy resin beads as a pharmaceutical dosage form. I. Method of preparation. *J Pharm Sci* 1969;**58**:1114–7.

3. Khanna SC, Jecklin T, Speiser P. Bead polymerisation technique for sustained release dosage form. *J Pharm Sci* 1970;**59**:614–8.
4. Kreuter J. Nanoparticles—a historical perspective. *Int J Pharm* 2007;**331**:1–10.
5. Hoffman AS. The origins and evolution of "controlled" drug delivery systems. *J Control Release* 2008;**132**:153–63.
6. Singh S, Nalwa HS. Nanotechnology and health safety—toxicity and risk assessments of nanostructured materials on human health. *J Nanosci Nanotechnol* 2007;**7**:3048–70.
7. Do JH, An J, Joun YS, Chung DJ, Kim JH. Cellular-uptake Behavior of Polymer Nanoparticles into Consideration of Biosafety. *Macromol. Res.* 2008;**16**:695–703.
8. Reis CP, Neufeld RJ, Ribeiro AJ, Veiga F. Nanoencapsulation I. Methods for preparation of drug-loaded polymeric nanoparticles. *Nanomedicine* 2006;**2**:8–21.
9. Cun D, Jensen DK, Maltesen MJ, Bunker M, Whiteside P, Scurr D, et al. High loading efficiency and sustained release of siRNA encapsulated in PLGA nanoparticles: quality by design optimization and characterization. *Eur J Pharm Biopharm* 2011;**77**:26–35.
10. Xiong YC, Yao YC, Zhan XY, Chen GQ. Application of polyhydroxyalkanoates nanoparticles as intracellular sustained drug-release vectors. *J Biomater Sci Polym Ed* 2010;**21**:127–40.
11. Xiao RZ, Zeng ZW, Zhou GL, Wang JJ, Li FZ, Wang AM. Recent advances in PEG-PLA block copolymer nanoparticles. *Int J Nanomedicine* 2010;**5**:1057–65.
12. Patil Y, Panyam J. Polymeric nanoparticles for siRNA delivery and gene silencing. *Int J Pharm* 2009;**367**:195–203.
13. Khemtong C, Kessinger CW, Gao JM. Polymeric nanomedicine for cancer MR imaging and drug delivery. *Chem Commun* 2009;**24**:3497–510.
14. Mora-Huertas CE, Fessi H, Elaissari A. Polymer-based nanocapsules for drug delivery. *Int J Pharm* 2010;**385**:113–42.
15. Letchford K, Burt H. A review of the formation and classification of amphiphilic block copolymer nanoparticulate structures: micelles, nanospheres, nanocapsules and polymersomes. *Eur J Pharm Biopharm* 2007;**65**:259–69.
16. Chaubal MV. Application of formulation technologies in lead candidate selection and optimization. *Drug Discov Today* 2004;**9**:603–9.
17. Wang M, Thanou M. Targeting nanoparticles to cancer. *Pharmacol Res* 2010;**62**:90–9.
18. Moinard-Checot D, Chevalier Y, Briancon S, Beney L, Fessi H. Mechanism of nanocapsules formation by the emulsion-diffusion process. *J Colloid Interface Sci* 2008;**317**:458–68.
19. Moinard-Checot D, Chevalier Y, Briancon S, Fessi H, Guinebretiere S. Nanoparticles for drug delivery: review of the formulation and process difficulties illustrated by the emulsion-diffusion process. *J Nanosci Nanotechnol* 2006;**6**:2664–81.
20. Rajesh Singh JWLJ. Nanoparticle-based targeted drug delivery. *Exp Mol Pathol* 2009;**86**:215–23.
21. Patil Y, Panyam J. Polymeric nanoparticles for siRNA delivery and gene silencing. *Int J Pharm* 2009;**367**:195–203.
22. Valappil SP, Misra SK, Boccaccini AR, Roy I. Biomedical applications of polyhydroxyalkanoates, an overview of animal testing and in vivo responses. *Expert Rev Med Devices* 2006;**3**:853–68.
23. Wu Q, Wang Y, Chen GQ. Medical application of microbial biopolyesters polyhydroxyalkanoates. *Artif Cells Blood Substit Biotechnol* 2009;**37**:1–12.
24. Chen GQ, Wu Q. The application of polyhydroxyalkanoates as tissue engineering materials. *Biomaterials* 2005;**26**:6565–78.
25. Yao YC, Zhan XY, Zhang J, Zou XH, Wang ZH, Xiong YC, et al. A specific drug targeting system based on polyhydroxyalkanoate granule binding protein PhaP fused with targeted cell ligands. *Biomaterials* 2008;**29**:4823–30.

26. Xiong YC, Yao YC, Zhan XY, Chen GQ. Application of polyhydroxyalkanoates nanoparticles as intracellular sustained drug-release vectors. *J Biomater Sci Polym Ed* 2010;**21**:127–40.
27. Lu XY, Ciraolo E, Stefenia R, Chen GQ, Zhang YL, Hirsch E. Sustained release of PI3K inhibitor from PHA nanoparticles and in vitro growth inhibition of cancer cell lines. *Appl Microbiol Biotechnol* 2011;**89**:1423–33.
28. Koosha F, Muller RH, Davis SS. Polyhydroxybutyrate as a drug carrier. *Crit Rev Ther Drug Carrier Syst* 1989;**6**:117–30.
29. Gould PL, Holland SJ, Tighe BJ. Polymers for biodegradable medical devices. 4. Hydroxybutyrate valerate copolymers as nondisintegrating matrices for controlled-release oral dosage forms. *Int J Pharm* 1987;**38**:231–7.
30. Brophy MR, Deasy PB. Application of the Higuchi model for drug release from dispersed matrices to particles of general shape. *Int J Pharm* 1987;**37**:41–7.
31. Wieczorek R, Pries A, Steinbuchel A, Mayer F. Analysis of a 24-kilodalton protein associated with the polyhydroxyalkanoic acid granules in alcaligenes-eutrophus. *J Bacteriol* 1995;**177**:2425–35.
32. Potter M, Madkour MH, Mayer F, Steinbuchel A. Regulation of phasin expression and polyhydroxyalkanoate (PHA) granule formation in Ralstonia eutropha H16. *Microbiology* 2002;**148**:2413–26.
33. Dennis D, Sein V, Martinez E, Augustine B. PhaP is involved in the formation of a network on the surface of polyhydroxyalkanoate inclusions in *Cupriavidus necator* H16. *J Bacteriol* 2008;**190**:555–63.
34. Brockelbank JA, Peters V, Rehm BHA. Recombinant *Escherichia coli* strain produces a ZZ domain displaying biopolyester granules suitable for immunoglobulin G purification. *Appl Environ Microbiol* 2006;**72**:7394–7.
35. Lewis JG, Rehm BHA. ZZ polyester beads: an efficient and simple method for purifying IgG from mouse hybridoma supernatants. *J Immunol Methods* 2009;**346**:71–4.
36. Peters V, Rehm BHA. Protein engineering of streptavidin for in vivo assembly of streptavidin beads. *J Biotechnol* 2008;**134**:266–74.
37. Rasiah IA, Rehm BHA. One-step production of immobilized alpha-amylase in recombinant *Escherichia coli*. *Appl Environ Microbiol* 2009;**75**:2012–6.
38. Jahns AC, Rehm BHA. Tolerance of the *Ralstonia eutropha* class I polyhydroxyalkanoate synthase for translational fusions to its C terminus reveals a new mode of functional display. *Appl Environ Microbiol* 2009;**75**:5461–6.
39. Pötter M, Madkour MH, Mayer F, Steinbüchel A. Regulation of phasin expression and polyhydroxyalkanoate (PHA) granule formation in *Ralstonia eutropha* H16. *Microbiology-Sgm* 2002;**148**:2413–26.
40. Dong Y, Li P, Chen CB, Wang ZH, Ma P, Chen GQ. The improvement of fibroblast growth on hydrophobic biopolyesters by coating with polyhydroxyalkanoate granule binding protein PhaP fused with cell adhesion motif RGD. *Biomaterials* 2010;**31**:8921–30.
41. You ML, Peng GF, Li JA, Ma P, Wang ZH, Shu WL, et al. Chondrogenic differentiation of human bone marrow mesenchymal stem cells on polyhydroxyalkanoate (PHA) scaffolds coated with PHA granule binding protein PhaP fused with RGD peptide. *Biomaterials* 2011;**32**:2305–13.
42. Mifune J, Grage K, Rehm BHA. Production of functionalized biopolyester granules by recombinant *Lactococcus lactis*. *Appl Environ Microbiol* 2009;**75**:4668–75.
43. Passirani C, Benoit J-P. Complement activation by injectable colloidal drug carriers. In: *Biomaterials for Delivery and Targeting of Proteins and Nucleic Acids*. Boca Raton, FL, USA: CRC Press; 2005. pp. 187–230.
44. Moghimi SM, Hunter AC, Murray JC. Long-circulating and target-specific nanoparticles: theory to practice. *Pharmacol Rev* 2001;**53**:283–318.

45. Foster LJR, Sanguanchaipaiwong V, Gabelish CL, Hook J, Stenzel M. A natural-synthetic hybrid copolymer of polyhydroxyoctanoate-diethylene glycol: biosynthesis and properties. *Polymer* 2005;**46**:6587–94.
46. Zanzig J, Scholz C. Effects of poly(ethylene glycol) on the production of poly(beta-hydroxybutyrate) by Azotobacter vinelandii UWD. *J. Polym. Environ.* 2003;**11**:145–54.
47. Walker GF, Fella C, Pelisek J, Fahrmeir J, Boeckle S, Ogris M, et al. Toward synthetic viruses: endosomal pH-triggered deshielding of targeted polyplexes greatly enhances gene transfer in vitro and in vivo. *Mol Ther* 2005;**11**:418–25.
48. Liu YC, Le Ny ALM, Schmidt J, Talmon Y, Chmelka BF, Lee Jr. CT. Photo-assisted gene delivery using light-responsive catanionic vesicles. *Langmuir* 2009;**25**:5713–24.

Nanobiotechnology and Personalized Medicine

K.K. Jain

Jain PharmaBiotech, Basel, Switzerland

I. Introduction ... 326
II. Role of Nanodiagnostics in Personalized Medicine ... 326
A. Cantilever Arrays for Personalized Medical Diagnostics ... 328
B. Nanotechnology-Based Biochips ... 328
C. Nanobiotechnology for Detection of Single Nucleotide Polymorphisms ... 329
D. Role of Nanobiotechnology in Sequencing ... 330
III. Nanobiotechnology, Biomarkers, and Personalized Medicine ... 332
IV. Nanoparticle-Based Combination of Diagnostics and Therapeutics ... 333
A. Nanobiotechnology and Pharmacogenetics ... 334
V. Discovery, Development, and Delivery of Personalized Nanomedicines ... 334
A. Pharmacogenomics and Nanobiotechnology ... 335
B. Pharmacoproteomics and Nanobiotechnology ... 335
C. Nanobiotechnology-Based Drug Discovery ... 336
D. Nanoparticle-Based Drug Development ... 337
E. Nanoparticle-Based Drug Delivery ... 338
F. Perfluorocarbon Nanoparticles for Imaging and Targeted Drug Delivery ... 340
G. Nanoparticles for Imaging and Intracellular Delivery of Nucleic Acids ... 340
VI. Personalized Nanooncology ... 341
A. Nanobiotechnology for Early Detection of Cancer to Improve Treatment ... 341
B. Nanobiotechnology for Combination of Cancer Diagnostics with Therapeutics ... 342
C. Radiolabeled Carbon Nanotubes for Tumor Imaging and Targeting ... 343
D. QD Conjugates Combine Cancer Imaging, Therapy, and Sensing ... 343
E. Gold Nanoshells for Thermal Ablation of Cancer ... 343
F. Nanobombs for Cancer ... 345
G. Nanoparticles, MRI, and Thermal Ablation of Tumors ... 345
H. Gold Nanoparticle Plus Bombesin for Imaging and Therapy of Cancer 346
I. Role of Nanoparticle-Based Imaging in Oncology Clinical Trials ... 346
J. Concluding Remarks ... 347
VII. Personalized Nanoneurology ... 347
A. Nanoparticles for Neuroprotection ... 347
B. Peptide Nanostructures for Repair of the CNS ... 348
C. Nanobiotechnology for Brain Tumor Management ... 349
VIII. Personalized Nanocardiology ... 349
A. Gene–Environment Interactions in Personalized Cardiovascular Medicine ... 350
B. Use of Magnetic Nanoparticles as MRI Contrast Agents for Cardiac Disorders ... 350

Progress in Molecular Biology
and Translational Science, Vol. 104
DOI: 10.1016/B978-0-12-416020-0.00008-5

1877-1173/11 $35.00

C. Nanoparticles for Cardiovascular Imaging and Targeted Drug Delivery........ 350
IX. Future Prospects of Personalized Nanomedicine........ 351
References........ 352

This chapter will start with a definition and scope of personalized medicine and describe how various nanobiotechnologies will contribute to its development. Nanodiagnostics and its combination with therapeutics as well as nanoparticle-based drug delivery will play an important role. The most important applications of nanobiotechnology will be personalized management of cancer, neurological disorders, and cardiovascular diseases.

I. Introduction

Personalized medicine simply means the prescription of specific treatments and therapeutics best suited for an individual taking into consideration genomic, genetic, metabolic, environmental, and other factors that influence response to therapy.[1] Advances in biotechnology have facilitated the development of personalized medicine. The concept of personalized medicine is the best approach to integration of new technologies for improving the understanding of the pathomechanism of diseases and management of patients.

Among the new technologies, nanotechnology has made a significant contribution to personalized medicine.[2] Nanotechnology is the popular term for the construction and utilization of functional structures with at least one characteristic dimension measured in nanometers ($1\ \mathrm{nm} = 10^{-9}\ \mathrm{m}$). Given the inherent nanoscale functional components of living cells, it was inevitable that nanotechnology would be applied in biotechnology, giving rise to the term nanobiotechnology. Nanomedicine is the application of nanoparticles and other nanobiotechnologies in health care.[3] Relationships of nanobiotechnologies to personalized medicine are shown in Fig. 1.

II. Role of Nanodiagnostics in Personalized Medicine

Clinical application of molecular technologies to elucidate, diagnose, and monitor human diseases is referred to as molecular diagnosis, which is an important component of personalized medicine. Nanomolecular diagnostics is the use of nanobiotechnology in molecular diagnostics and can be termed "nanodiagnostics."[4] The chemical modification and global amplification of the nucleic acid samples are achieved by polymerase chain reaction (PCR), which can introduce artifacts caused by the preferential amplification of certain

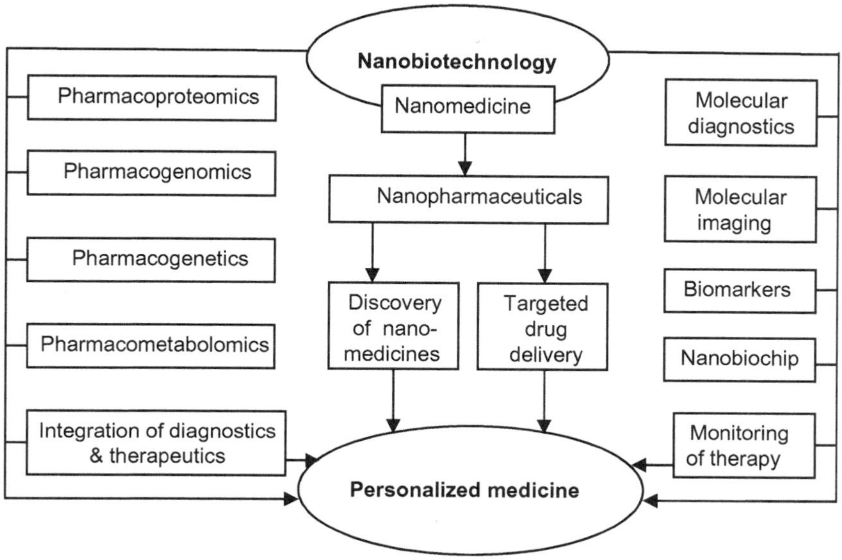

FIG. 1. Relationship of nanobiotechnology and personalized medicine.

sequences. Alternative label-free methods for avoiding the artifacts include surface plasmon resonance and quartz crystal microbalance, which rely on mass detection. Nanotechnologies also provide label-free detection.

Because of the small dimension, most of the applications of nanobiotechnology in molecular diagnostics fall under the broad category of biochips/microarrays but are more correctly termed nanobiochips and nanoarrays. Nanotechnology-on-a-chip is a general description that can be applied to several methods although some of these do not use nanotechnologies but only have the capability to analyze nanoliter amounts of fluids.

Nanobiotechnologies have refined and extended the limits of molecular diagnostics. Numerous nanodevices and nanosystems for sequencing single molecules of DNA are feasible. Given the inherent nanoscale of receptors, pores, and other functional components of living cells, the detailed monitoring and analysis of these components will be made possible by the development of a new class of nanoscale probes. Biological tests measuring the presence or activity of selected substances become quicker, more sensitive, and more flexible when certain nanoscale particles are put to work as tags or labels. Nanobiotechnology will improve the sensitivity and integration of analytical methods to yield a more coherent evaluation of life processes.[5]

Nanomaterials can be assembled into massively parallel arrays at much higher densities than is achievable with current sensor array platforms and in a format compatible with current microfluidic systems. Currently, quantum dot

(QD) technology is the most widely employed nanotechnology for diagnostic developments. This will have applications in genomic analysis, proteomics, and molecular diagnostics. Nanotechnology has potential advantages in applications of molecular diagnostics that are relevant for the development of personalized medicine: (i) point-of-care (POC) diagnosis, which also includes diagnosis on patient's bedside and self-diagnostics for use in the home; (ii) genetic screening; (iii) sequencing; and (iv) integration of diagnostics with therapeutics.

A. Cantilever Arrays for Personalized Medical Diagnostics

Among the recently emerging technologies, cantilevers are the most promising. Short complementary nucleic acid segments (sensors) are attached to silicon cantilevers which are 450 nm thick and therefore react with extraordinary sensitivity. Binding of the targeted gene transcript to its matching counterpart on one of the cantilevers results in optically measurable mechanical bending. This technology complements and extends current DNA and protein microarray methods, because nanomechanical detection requires no labels, optical excitation, or external probes and is rapid, highly specific, sensitive, and portable. An example is the method for rapid and sensitive detection of active genes relevant to diseases and their treatment by direct measurement of their transcripts (mRNA), which represent the intermediate step and link to protein synthesis.[6]

Another application is the observation of differential gene expression of the gene 1-8U, a potential biomarker for cancer progression or viral infections, in a complex background. The measurements provide results within minutes at the picomolar level without target amplification, and are sensitive to base mismatches. An array of different gene transcripts can even be measured in parallel by aligning appropriately coated cantilevers alongside each other like the teeth of a comb. The new method complements current molecular diagnostic techniques such as the gene chip and real-time PCR. It could be used as a real-time sensor for continuously monitoring various clinical parameters or for detecting rapidly replicating pathogens that require prompt diagnosis. These findings qualify the technology as a rapid method to validate biomarkers that reveal disease risk, disease progression, or therapy response. Cantilever arrays have potential as a tool to evaluate treatment response efficacy for personalized medical diagnostics.

B. Nanotechnology-Based Biochips

Nanotechnology on a chip, also referred to as a nanobiochip, is a new paradigm for total chemical analysis systems. The ability to make chemical and biological information much cheaper and easier to obtain is expected to fundamentally change health care.

Large numbers of structural variations have been discovered in chromosomal DNA ranging from kilobase to megabase pairs in length that account for much of the variance among individual human genomes. Structural variations are also associated clinically with many diseases and conditions, such as Crohn's disease, autism, schizophrenia, morbid obesity, and cancers. However, current tools to detect and analyze structural variations, including sequencing technologies that rely on short DNA fragments, complex analysis, and assembly approaches, are inadequate. To address this problem and in a radical departure from other genomic analysis platforms, nanoAnalyzer® technology (BioNanomatrix) is able to unravel, sort, and confine native-state, long genomic DNA fragments into an orderly, linear format. The technology does not require front-end amplification or shearing of the sample DNA into small fragments, thus preserving potentially clinically valuable genomic structural information such as copy number variations (CNVs).

C. Nanobiotechnology for Detection of Single Nucleotide Polymorphisms

Small stretches of DNA that differ in only one base are called single nucleotide polymorphisms (SNPs) and serve to distinguish one individual's genetic material from that of another. SNPs comprise some 80% of all known polymorphisms. Millions of SNPs have been discovered in humans and are available in public databases. SNPs are related to an individual's disease susceptibility and drug response and can be used for prediction of adverse effects of drugs. SNPs can be used as biomarkers to segregate individuals with different levels of response to treatment (beneficial or adverse) in clinical settings for developing personalized medicine.

The NanoChip System (Nanogen Inc), by utilizing electronically enhanced hybridization of complementary DNA strands, integrates advanced microelectronics and molecular biology into a platform technology with broad commercial applications in the fields of genomic diagnostics and has achieved 100% accuracy in the detection of SNPs. This technology enables "multiplex'" assays (the ability to run assays that determine the presence or absence of multiple genetic mutations at the same time and on the same chip). The technology is being developed further to allow amplification of DNA material directly on the NanoChip cartridge, which eliminates a time-consuming preparatory step and folds it into a single, simplified detection procedure. NanoChip System has been used at the Mayo Clinic (Rochester, MN) to examine SNPs in the gene coding for thiopurine methyl transferase (TPMT), an enzyme found in humans. Polymorphisms affecting TPMT alter therapeutic response to thiopurine drugs, which are used to treat leukemia, Crohn's disease, dermatological disorders, and as immunosuppressants.

ClearRead™ (Nanosphere Inc) nanoparticle-based technology enables a microarray-based method for multiplex SNP genotyping in total human genomic DNA without the need for target amplification.

D. Role of Nanobiotechnology in Sequencing

The term DNA sequencing refers to methods for determining the exact order of the 3 billion nucleotide bases—adenine, guanine, cytosine, and thymine—that make up the DNA of the 23 pairs of human chromosomes. In de novo sequencing, short DNA fragments purified from individual bacterial colonies are individually sequenced and assembled electronically into one long, contiguous sequence. This method does not require any preexisting information about the sequence of the DNA. Resequencing, using next-generation technologies, means determination of variations of DNA sequence in an organism where the nominal sequence is already known.

Among various technologies, sequencing will play an important role in the development of personalized medicine. The role of sequencing in personalized medicine is supported by the recent efforts in personal genome sequencing, for example, Quake's genome sequence in 2010, which was undertaken as an integrated analysis of a complete human genome in a clinical context on a person with a family history of vascular disease and early sudden death.[7] Genetic analysis included the development of novel methods for the integration of whole genome and clinical risk. Disease and risk analysis focused on prediction of genetic risk of variants associated with Mendelian disease, recognized drug responses, and pathogenicity for novel variants. Personalized genome sequencing would become an integral part of personalized medicine as the cost comes down. Sequencing will also lead to the development of many diagnostic assays that will contribute to personalized medicine. Numerous technologies and next-generation sequencers are available or in development.[8]

An efficient, nanoliter-scale microfabricated bioprocessor integrating all three Sanger sequencing steps—thermal cycling, sample purification, and capillary electrophoresis—has been developed and evaluated.[9] The performance of this miniaturized DNA sequencer provides a benchmark for predicting the ultimate cost and efficiency limits of Sanger sequencing. The goal now is to make DNA sequencing feasible as a routine part of health care. Traditional DNA sequencing techniques are being blended with cutting-edge nanotechnology to develop faster and less costly methods of analysis.

Direct observation of single molecules is the most elegant sequencing technology because it enables rapid sequencing of even small amounts of DNA. Also referred to by the term "single-molecule genomics," it includes a group of molecular methods in which single molecules are detected or sequenced. Single molecule sequencing (SMS) enables analysis of genomic information without the need for cloning or amplification, enabling data generation in only a few hours versus days or weeks with current systems. This is an important requirement for eventual use in a clinical setting. Although technically challenging, the analysis of single molecules has the potential to play a major role in the delivery of truly personalized medicine.[10] The two main subgroups of single molecule genomic methods are single-molecule digital PCR and SMS. Single-molecule PCR has a number of advantages over competing technologies, including improved detection of rare genetic variants and more precise analysis of CNVs, and is more easily adapted to the often small amount of material that is available in clinical samples. Most of the SMS techniques in commercial development are based on nanobiotechnology. Personal Genome Machine (Life Technologies Corporation) is a nonoptical chip-based bench-top sequencer based on nanobiotechnology.

Carbon nanotubes with unique arrangements of carbon atoms (see the chapter by Gomez-Gualdrón and coauthors in this volume) exhibit many special physical and chemical properties, which have been used for DNA sequencing. Single-stranded DNA (ssDNA) can translocate through a single-walled carbon nanotube (SWCNT) with a diameter of 1–2 nm.[11] Devices in which one SWCNT spans a barrier between two fluid reservoirs enable direct electrical measurement of ion transport through the tube. A fraction of the tubes pass anomalously high ionic currents. Electrophoretic transport of small ssDNA oligomers through these tubes is marked by large transient increases in ion current and confirmed by PCR analysis. SWCNTs simplify the construction of nanopores, permit new types of electrical measurements, and may open avenues for control of DNA translocation. It is possible to slow the rate of translocation to a speed where reading the sequence may actually be possible.

BASE™ (Oxford Nanopore Technology) has been used for accurate and continuous identification of DNA bases using nanopores. As single DNA bases pass through the nanopore, each base causes a characteristic disruption of current that allows the molecule to be identified. A study has shown that this system can also directly identify methylated cytosine, which can be distinguished from the four DNA bases using the same method.[12] This is important in the study of cancer, where genome methylation is implicated but existing study techniques require complex labeling of the DNA sample. Hybridization-assisted nanopore sequencing (NABsys) platform combines nanopore sequencing with sequencing by hybridization. Unlike other nanopore-based sequencing approaches, the NABsys platform does not depend on single-base resolution of the nanopore detector in order to obtain accurate sequence information. Although the DNA sequence information obtained from nanopores comes from the signal collected during DNA translocation, the throughput of the method is determined by the rate at which molecules arrive and thread into the pores. The process of DNA capture into nanofabricated pores of molecular dimensions has been investigated.[13] For fixed analyte concentrations there was an increase in capture rate as the DNA length increases from 800 to 8000 base pairs, a length-independent capture rate for longer molecules, and increasing capture rates when ionic gradients are established across the pore. Mycobacterium smegmatis porin A (MspA) has been used for nanopore sequencing.[14] Based on the short, narrow nature of the MspA pore, it is considered better than commonly used alpha-hemolysin protein for nanopore sequencing applications. It is just short enough to accommodate one or two nucleotides.

III. Nanobiotechnology, Biomarkers, and Personalized Medicine

A biomarker is a characteristic that can be objectively measured and evaluated as an indicator of a physiological as well as a pathological process or pharmacological response to a therapeutic intervention. Any specific molecular alteration of a cell on DNA, RNA, metabolite, or protein level can be referred to as a molecular biomarker. In the era of molecular biology, biomarkers usually mean molecular biomarkers and can be divided into three broad categories[15]:

1. Those that track disease progression over time and correlate with known clinical measures
2. Those that detect the effect of a drug
3. Those that serve as surrogate end points in clinical trials.

While researchers are studying all three categories, biotechnology and pharmaceutical companies favor using biomarkers as drug discovery tools—not only to detect biological responses to experimental drugs but also to aid in the discovery of new targets for therapeutic intervention. A biomarker can be as simple as a laboratory test or as complex as a pattern of genes or proteins. From a practical point of view, the biomarker would specifically and sensitively reflect a disease state and could be used for diagnosis as well as for disease monitoring during and following therapy.

Nanobiotechnology has refined and extended the limits of detection of biomarkers. Water-soluble, biocompatible, fluorescent, and stable silver/dendrimer nanocomposites have been synthesized that exhibit a potential for labeling cells in vitro as cell biomarkers.[16]

The longevity of QDs (see the chapter by Ishikawa and Biju in this volume) has also made them suitable molecular labels enabling scientists to study the earliest signs of cancer and track the effectiveness of pharmaceuticals that target the cellular underpinnings of disease. PEG-coated QDs can be used for in vivo imaging of breast and prostate cancers, which could be detected at very early stages and treated effectively.

Bioconjugated QDs, collections of different-sized nanoparticles embedded in tiny polymer beads, provide a new class of biological labels for evaluating biomarkers on intact cells and tissue specimens. In particular, the use of multicolor QD probes in immunohistochemistry is considered one of the most important and clinically relevant applications. The medical use of QD-based immunohistochemistry has been limited by the lack of specific instructions, protocols, for clinicians. Preliminary results and detailed protocols for QD–antibody conjugation, tissue specimen preparation, multicolor QD staining, image processing, and biomarker quantification have been published.[17] The results demonstrate that bioconjugated QDs can be used for multiplexed profiling of biomarkers, and ultimately for correlation with disease progression and response to therapy. This will increase the clinician's ability to predict the likely outcomes of drug therapy in a personalized approach to disease management. Bioinformatics and systems biology is used to link each individual's molecule profile with disease diagnosis and treatment decisions. The usefulness of these protocols was demonstrated by simultaneously identifying multiple biomarkers in prostate cancer tissue. In general, QD bioconjugation is completed within 1 day, and multiplexed molecular profiling takes 1–3 days depending on the number of biomarkers and QD probes used.

Development of personalized medicine is closely linked to biomarkers, which may serve as the basis for diagnosis, drug discovery, and monitoring of diseases. Nanobiotechnology, by improving biomarker detection, enhances their value for personalized medicine. Moreover, biomarkers may serve as common denominators for the diagnosis as well as treatment of a disease.

IV. Nanoparticle-Based Combination of Diagnostics and Therapeutics

Integration of diagnostics with therapeutics is an important feature of personalized medicine. A related term is "companion diagnostic," which is used when a diagnostic test is linked with a therapeutic product. Hercep test (DAKO), an immunohistochemistry assay, was used to identify patients with HER2-positive metastatic breast cancer when it was discovered that patients with HER2 amplification responded better to Herceptin (Genentech/Roche). The field of companion diagnostics is growing as an increasing number of biomarkers are being discovered and validated. Nanobiotechnology has contributed to the improvement of companion diagnostics. Currently there are approximately 40 drugs in the United States that have companion diagnostic tests associated with them, either as a requirement to their being prescribed (only 5 of 40), or a recommendation for use, or label information that lists genetic susceptibility relating to efficacy or dose. For example, warfarin labeling now includes FDA-recommended genotyping for mutations in two genes that cause increased susceptibility to bleeding, but the label does not require it. Testing for safety of drug use and reduction of adverse effects is discussed further under pharmacogenetics.

A. Nanobiotechnology and Pharmacogenetics

Pharmacogenetics, a term recognized in pharmacology in the pregenomic era, is the study of influence of genetic factors on action of drugs as opposed to genetic causes of disease. In the genomic era, it is the study of the linkage between the individual's genotype and the individual's ability to metabolize a foreign compound. The pharmacological effect of a drug depends on pharmacodynamics (interaction with the target or the site of action) and pharmacokinetics (absorption, distribution, and metabolism). It also covers the influence of various factors on these processes. Drug metabolism is one of the major determinants of drug clearance and the factor that is most often responsible for interindividual differences in pharmacokinetics. Pharmacogenetics links genotype and phenotype.

In the area of pharmacogenetic diagnostics, gold nanoparticle probe technology is used to develop leading-edge, clinical tests that enable rapid, multi-target detection of SNPs and similar DNA sequence variations that alter an individual's metabolism of specific drugs.

The Verigene Warfarin Metabolism Nucleic Acid Test (Nanosphere Inc), approved by the FDA, detects variants of CYP2C9 and VKORC1 genes, responsible for sensitivity to the anticoagulant warfarin. Also approved is the Verigene® F5/F2/MTHFR nucleic acid test, which detects disease-associated gene mutations that can contribute to blood coagulation disorders and difficulties metabolizing folate.

V. Discovery, Development, and Delivery of Personalized Nanomedicines

Personalized medicine started with individualizing the use of already established drugs to improve their efficacy and safety. The role of nanobiotechnology was to improve drug delivery. This process extended to drugs in clinical trials where biomarkers and molecular diagnostics/genotyping were used to identify the subgroups where the drugs would be most effective. Eventually the concept has extended to personalization of drugs at discovery stage and this is applicable to nanomedicines as well. Nanobiotechnology has been used for facilitating drug discovery and some nanoparticles are drug candidates as well. The basic part of personalized medicine relevant to pharmaceuticals is covered by the terms pharmacoproteomics and pharmacogenomics, which overlaps with pharmacogenetics.

A. Pharmacogenomics and Nanobiotechnology

Pharmacogenomics implies the use of genetic sequence and genomics information in patient management to enable therapy decisions. The genetic sequence and genomics information can be that of the host (normal or diseased) or of the pathogen. Pharmacogenomics will have an impact on all phases of drug development, from drug discovery to clinical trials. It will also apply to a wide range of therapeutic products including bioengineered proteins, cell therapy, antisense therapy, and gene therapy. These treatments are also subject to constraints and complexities engendered by individual variability. Various technologies, including nanobiotechnology, enable the analysis of these complex multifactorial situations to obtain individual genotypic and gene expression information. Nanobiotechnology facilitates the understanding of disease mechanism, which is an important component of personalized medicine along with pharmacogenomics, and is an important consideration in drug discovery and development.

B. Pharmacoproteomics and Nanobiotechnology

The term "proteomics" indicates PROTEins expressed by a genOME and is the systematic analysis of protein profiles of tissues. Proteomics parallels the related field of genomics. Now that the human genome has been sequenced, science is faced with the greater challenge of making use of this information for improving health care and discovering new drugs. There is an increasing interest in proteomics technologies now because DNA sequence information provides only a static snapshot of the various ways in which the cell might use its proteins whereas the life of the cell is a dynamic process. A detailed discussion of proteomics is given in a special report on this topic.[18] The role of proteomics in drug development can be termed "pharmacoproteomics." Proteomics-based characterization of multifactorial diseases may help to match a particular target-based therapy to a particular marker in a subgroup of patients. The industrial sector is taking a lead in developing this area. Individualized therapy may be based on differential protein expression rather than a genetic polymorphism. Nanobiotechnology has refined technologies used in proteomics, leading to the introduction of the term nanoproteomics.

Nanoproteomics improves on most current protocols including protein purification/display and automated identification schemes that yield unacceptably low recoveries with reduced sensitivity and speed while requiring more starting material. Low abundant proteins and proteins that can only be isolated from limited source material (e.g., biopsies) can be subjected to nanoscale protein analysis, that is, nanocapture of specific proteins and complexes, and optimization of all subsequent sample handling steps leading to mass analysis of peptide fragments. This is a focused approach, also termed targeted proteomics, and involves examination of subsets of the proteome, for example, those proteins that are either specifically modified, or bind to a particular DNA sequence, or exist as members of higher-order complexes, or any combination thereof. Some examples of applications of nanoproteomics are listed below:

- Nanofiber structures of some peptides and proteins as biological materials are being studied including their molecular mechanism of self-assembly and reassembly. The reassembly process is important for the construction of new scaffolds for 3D cell culture, tissue repair, and regenerative medicine.
- An ion mobility technology, high-field asymmetric waveform ion mobility mass spectrometry, uses ion separation to improve detection limits of peptide ions when used in conjunction with electrospray and nanoelectrospray mass spectrometry. This will play an important role in drug discovery and biomarker programs for monitoring of disease progression and drug efficacy.
- An intein-based method for site-specific conjugation of QDs to target proteins in vivo has been described, which enables the covalent conjugation of any nanostructure and/or nanodevice to any protein and thus the targeting of such material to any intracellular compartment or signaling complex within the cells.[19]

C. Nanobiotechnology-Based Drug Discovery

Current drug discovery process needs improvement in several areas. Although many targets are being discovered through genomics and proteomics, the efficiency of screening and validation processes need to be increased. Microfluidics has already proven useful for drug discovery. Through further miniaturization, nanotechnology will improve the ability to fabricate massive arrays in small spaces using nanofluidics and the time efficiency. This would enable direct reading of the signals from nanofluidic circuits in a manner similar to a microelectronics circuit where one does not require massive instrumentation. This would increase the ability to do high-throughput drug screening. QDs, gold nanoparticles, lipoparticles, nanobarcodes, micelles, dendrimers, and carbon nanotubes have received a considerable attention with their unique properties for potential use in drug discovery.[20] Ligand-conjugated nanocrystals can be used to target cell surface receptors, ion channels, and transporters. They are also versatile probes of membrane proteins in living cells. Several nanodevices that are commonly used for research in life sciences are also useful in drug discovery, for example, atomic force microscopy, nanolasers, and nanobiosensors.

Among various nanoparticles, the use of QDs for drug discovery has been explored extensively. Cellular receptors are a critical target for drug discovery, but older imaging tools such as fluorescent dyes are too unstable to effectively perform single-molecule tracking, which can be accomplished by QD conjugates that produce photo resolutions up to eight times more detailed than the older imaging tools. QDs carrying a surface-immobilized antagonist remain with nanomolar affinity on the cell surface, and particles carrying an agonist are internalized upon receptor binding. The receptor functions like a gate that grants cell access only to those particles that carry a receptor ligand and where the ligand is an agonist.[21] G-protein-coupled receptors (GPCRs) are the largest protein superfamily in the human genome; they comprise 30% of current drug targets and regulate diverse cellular signaling responses. The role of endosomal trafficking in GPCR signaling regulation is significant but this process remains difficult to study due to the inability to distinguish among many individual receptors because of simultaneous trafficking within multiple endosomal pathways. Accurate measurement of the internalization and endosomal trafficking of single groups of serotonin (5-hydroxytryptamine, 5-HT) receptors was shown by using single QD probes and quantitative colocalization.[22] Single-molecule imaging of endosomal trafficking will significantly impact the understanding of cellular signaling and provide powerful tools to elucidate the actions of GPCR-targeted therapeutics. Another advantage of QDs is that they enable the combination of diagnostics with targeted therapies for cancer.

D. Nanoparticle-Based Drug Development

Nanoparticles such as dendrimers and fullerenes are drug candidates whereas nanobodies, the smallest fragments of naturally occurring heavy-chain antibodies, can be developed as therapeutics.

Dendrimers are a novel class of 3D nanoscale, core-shell structures that can be precisely synthesized for a wide range of applications. Specialized chemistry techniques allow for precise control over the physical and chemical properties of the dendrimers. They are most useful in drug delivery but can also be used for the development of new pharmaceuticals with novel activities. Polyvalent dendrimers interact simultaneously with multiple drug targets. They can be developed into novel targeted cancer therapeutics. Polymer–protein and polymer–drug conjugates can be developed as anticancer drugs. These have the following advantages:

- Tailor-made surface chemistry
- Nonimmunogenic
- Inherent body distribution enabling appropriate tissue targeting
- Possibly biodegradable

Dendrimer conjugation with low-molecular weight drugs has been of increasing interest recently for improving pharmacokinetics, targeting drugs to specific sites, and facilitating cellular uptake. Opportunities for increasing the performance of relatively large therapeutic proteins such as streptokinase using dendrimers have been explored in one study.[23]

A key attribute of the fullerene molecules is their numerous points of attachment, allowing for precise grafting of active chemical groups in 3D orientations. This attribute, the hallmark of rational drug design, enables positional control in matching fullerene compounds to biological targets. In concert with other attributes of the fullerene molecules, that is, size, redox potential, and relative inertness in biological systems, it is possible to tailor requisite pharmacokinetic characteristics to fullerene-based compounds and optimize their therapeutic effect.

Like conventional antibodies, Nanobodies (Ablynx) show high target specificity and low inherent toxicity; however, like small molecule drugs, they can inhibit enzymes and can access receptor clefts. They combine the advantages of conventional antibodies with important features of small molecule drugs. An example of the use of nanobodies as novel drugs is nanobody-conjugated human trypanolytic factor for the treatment of human African trypanosomiasis.

E. Nanoparticle-Based Drug Delivery

Drug delivery is one of the important considerations in drug development and therapeutics. New technologies are applied for constructing innovative formulations and delivering them. The focus is on targeted drug delivery, which is desirable for personalized medicine. There are several requirements for developing a device small enough to efficiently leave the vasculature and enter cells for performing multiple, smart tasks. However, size is a critical factor. Vascular pores limit egress of therapeutics to materials less than approximately 50 nm in diameter, and cells will not internalize materials much greater than 70 nm. Another problem faced by systemic delivery of nanoparticle therapeutics is trapping by the reticuloendothelial system or excretion in urine by the kidneys.

One of the major problems with delivery of drugs to the human body is solubility, which is an essential factor for drug effectiveness, independent of administration route. It is also a major challenge for pharmaceutical companies developing new products since nearly half of the new chemically based drugs are insoluble, or poorly soluble, in water. Many, otherwise promising, compounds never reach the market. Others reach the market but in a suboptimal formulation, possibly providing low or unpredictable bioavailability, or posing an increased side effect risk. A technology to enhance solubility can be used to reformulate such drugs and increase their commercial potential. Some drugs are available as nanoformulations whereas others require nanoparticle-based delivery technologies. Nanobiotechnology provides the following solutions to the problems of drug delivery:

- Particle size is reduced to nanometer size range to increase the surface area, thereby increasing the rate of dissolution, for example, Nanoedge technology (Baxter).
- Improving solubilization of the drug.
- Using noninvasive routes of administration eliminates the need for administration of drugs by injection.
- Development of novel nanoparticle formulations with improved stabilities and shelf-lives.
- Development of nanoparticle formulations for improved absorption of insoluble compounds and macromolecules enable improved bioavailability and release rates, potentially reducing the amount of dose required and increasing safety through reduced side effects.
- Manufacture of nanoparticle formulations with controlled particle sizes, morphology, and surface properties would be more effective and less expensive than other technologies.
- Nanoparticle formulations that can provide sustained-release profiles up to 24 h can improve patient compliance with drug regimens.
- Direct coupling of drugs to targeting ligands restricts the coupling capacity to a few drug molecules but coupling of drug carrier nanosystems to ligands allows the import of thousands of drug molecules by means of one receptor-targeted ligand. Nanosystems offer opportunities to couple drugs with newly discovered disease-specific targets.

The unique properties of dendrimers such as high degree of branching and well-defined molecular weight make them ideal scaffolds for drug delivery. Well-characterized, commercially available dendritic polymers have been subjected to functionalization for preparing drug delivery systems of low toxicity, high loading capacity, ability to target specific cells, and transport through their membranes. This has been achieved by surface targeting ligands, which render the carriers specific to certain cells and PEG, securing water solubility, stability, and prolonged circulation. Moreover, transport agents facilitate transport through cell membranes while fluorescent probes detect their intracellular localization. A common feature of surface groups is multivalency, which considerably enhances their binding strength with complementary cell receptors. To these properties, one should also add the property of attaining high loading of active ingredients coupled with controlled and/or triggered release.[24] Dendrimers are particularly useful for the delivery of anticancer drugs such as cisplatin and doxorubicin. They are also agents for boron neutron capture therapy and photodynamic therapy for cancer. By adding stimuli-responsive properties to the dendrimers, dendritic polymers capable of controlled release can be produced.[25] These stimuli-responsive dendrimers are potential next-generation drug carriers.

F. Perfluorocarbon Nanoparticles for Imaging and Targeted Drug Delivery

Perfluorocarbon (PFC) nanoparticles are approximately 200 nm in diameter and are encapsulated in a phospholipid shell, which provides an ideal surface for the incorporation of targeting ligands, imaging agents, and drugs. PFC nanoparticles can serve as a platform technology for molecular imaging and targeted drug delivery applications. For molecular imaging, PFC nanoparticles can carry very large payloads of gadolinium to detect pathological biomarkers with magnetic resonance imaging (MRI). A variety of different epitopes, including $\alpha v \beta 3$, tissue factor, and fibrin, have been imaged using nanoparticles formulated with appropriate antibodies or peptidomimetics as targeting ligands. Lipophilic drugs can also be incorporated into the outer lipid shell of nanoparticles for targeted delivery. Upon binding to the target cell, the drug is exchanged from the particle surfactant monolayer to the cell membrane through a novel process called "contact-facilitated drug delivery." By combining targeted molecular imaging and localized drug delivery, PFC nanoparticles provide diagnosis and therapy with a single agent and would facilitate the development of personalized medicine.[26]

G. Nanoparticles for Imaging and Intracellular Delivery of Nucleic Acids

Polymer beacons have been developed that enable the delivery of nucleic acids to be visualized at nanoscale.[27] The polycations have been designed to contain repeated oligoethyleneamines, for binding and compacting nucleic acids into nanoparticles, and lanthanide chelates (either luminescent europium Eu^{3+} or paramagnetic gadolinium Gd^{3+}). The chelated Lns allow the visualization of the delivery vehicle both on the nm/μm scale via microscopy and on the sub-mm scale via MRI. These delivery beacons effectively bind and compact plasmid DNA (pDNA) into nanoparticles and protect nucleic acids from nuclease damage. These delivery beacons efficiently deliver pDNA into cultured cells and do not exhibit toxicity. Micrographs of cultured cells exposed to the nanoparticle complexes formed with fluorescein-labeled pDNA and the europium-chelated polymers reveal effective intracellular imaging of the delivery process. MRI of bulk cells exposed to the complexes formulated with pDNA and the gadolinium-chelated structures show bright image contrast, allowing visualization of effective intracellular delivery. Because of their versatility, these delivery beacons possess remarkable potential for tracking and understanding nucleic acid transfer in vitro, and have promise as in vivo vectors for gene therapy and agents for combining diagnostics and therapeutics.

VI. Personalized Nanooncology

Application of nanotechnology in cancer can be termed nanooncology and includes both diagnostics and therapeutics.[28] Nanobiotechnology plays an important role in the discovery of biomarkers of cancer as well as diagnosis and drug delivery. Several drugs in development for cancer are based on nanobiotechnology and a few of these are already approved. Nanobiotechnology-based devices are in development as aids to cancer surgery. Finally nanobiotechnology is playing an important role in the personalized therapy of cancer, an important feature in early detection and combination of diagnostics with delivery of therapeutics.

A. Nanobiotechnology for Early Detection of Cancer to Improve Treatment

Cancer is easier to treat and less likely to develop drug resistance when treatment is started very early. Cancer cells in very early stages are less likely to have mutations that make them resistant to treatment. Cancer cells themselves may be difficult to detect at an early stage but they leave a fingerprint, that is, a pattern of change in biomarker proteins that circulate in the blood. Multiple biomarkers may require as many as 500 measurements, all of which should be made from a drop of blood obtained by a pinprick. Thus nanoscale diagnostics will play an important role in this effort. Nanowire sensors are in development at California Institute of Technology (Pasadena, CA) for very early diagnosis of cancer, when there are just a few thousand cells. Nanowires can electronically detect a few proteins molecules along with other biochemical markers that are early signs of cancer. Nanowires in a set are coated with different compounds; each of these binds to a particular biomarker and changes the conductivity of the nanowire that can be measured. Thousands of such nanowires are combined on a single chip that enables identification of the type of cancer. Currently such a chip can detect between 20 and 30 biomarkers and is being used for the early diagnosis of brain cancer.

An automated gold nanoparticle bio-barcode assay probe has been described for the detection of prostate-specific antigen (PSA) at 330 fg/mL, along with the results of a clinical pilot study designed to assess the ability of the assay to detect PSA in the serum of 18 men who have undergone radical prostatectomy for prostate cancer.[29] Available PSA immunoassays are often not capable of detecting PSA in the serum of men after radical prostatectomy. This new bio-barcode PSA assay is approximately 300 times more sensitive than commercial immunoassays and all patients in this study had a measurable serum PSA level after radical prostatectomy. Because the patient outcome depends on the level of PSA, this ultrasensitive assay enables (i) informing patients, who have undetectable PSA levels with conventional assays but detectable and nonrising levels with the barcode assay, that their cancer will not recur; (ii) early detection of recurrence earlier because of the ability to measure increasing levels of PSA before conventional tools can make such assignments; and (iii) use of PSA levels, which would otherwise not be detectable with conventional assays, to follow the response of patients to treatment.

B. Nanobiotechnology for Combination of Cancer Diagnostics with Therapeutics

Use of the same nanoparticle for diagnosis as well as therapy enables integration of two important facets of cancer management. Dendrimers can be used as advanced contrast agents for imaging techniques such as MRI that can be targeted specifically to cancer cells. Dendrimers can also be used to deliver a variety of cancer therapies to improve their safety and efficacy. For example, applications of dendrimers in photodynamic therapy, boron neutron capture therapy, and gene therapy for cancer are being investigated.[30]

A biocompatible, multimodal iron oxide nanoparticle has been synthesized for targeted cancer therapy, and optical imaging as well as MRI. A modified solvent diffusion method is used for the co-encapsulation of both an anticancer drug and NIR dyes.[31] The resulting folate-derived nanoparticles combining diagnostic and therapeutic properties can be used for imaging as well as targeted killing of folate-expressing cancer cells.

Magnetic nanoparticles have shown promise for targeted drug delivery, hyperthermia, and MRI in cancer. Aptamer-conjugated magnetic nanoparticles controlled by an externally applied 3D rotational magnetic field have been developed as a nanosurgical approach for the removal of cancerous cells selectively from the interior of an organ or tissue without any collateral damage.[32] This system could be upgraded for the selective removal of complex cancers from diverse tissues by incorporating various target-specific ligands on magnetic nanoparticles.

C. Radiolabeled Carbon Nanotubes for Tumor Imaging and Targeting

SWCNTs with covalently attached multiple copies of tumor-specific MAbs, radiometal–ion chelates, and fluorescent probes can target lymphomas and deliver both imaging and therapeutic molecules to these tumors.[33] Each nanotube, which contained approximately 6 antibody molecules and 114 radioactive atoms, proved to be stable in human plasma for at least 96 h and was able to bind to targeted tumor cells. Most importantly, the chemical linkages binding the radioactive element indium-111 was completely stable in human plasma for the entire 4-day experiment. Tests using a mouse model of human lymphoma showed that this nanotube constructs successfully targeted tumors while avoiding healthy cells. The ability to specifically target tumors with prototype-radiolabeled or fluorescent-labeled antibody-appended SWCNT constructs was encouraging and suggested further investigation of these as diagnostic combined with drug delivery for cancer.

D. QD Conjugates Combine Cancer Imaging, Therapy, and Sensing

The specificity and sensitivity of a QD–aptamer–doxorubicin (QD–Apt–Dox) conjugate as a targeted cancer imaging, therapy, and sensing system has been demonstrated in vitro.[34] By functionalizing the surface of fluorescent QD with an RNA aptamer, which recognizes the extracellular domain of the prostate-specific membrane antigen (PSMA), the system is capable of differential uptake and imaging of prostate cancer cells that express the PSMA. The intercalation of Dox, an anticancer drug with fluorescent properties, in the double-stranded stem of the aptamer results in a targeted conjugate with reversible self-quenching properties based on a Bi-FRET mechanism. A donor-acceptor model FRET between QD and Dox and a donor-quencher model FRET between Dox and aptamer result when Dox is intercalated within the aptamer. This simple multifunctional nanoparticle system can deliver Dox to the targeted prostate cancer cells and sense the delivery of Dox by activating the fluorescence of QD, which concurrently images the cancer cells.

E. Gold Nanoshells for Thermal Ablation of Cancer

Metal nanoshells belong to a class of nanoparticles with tunable optical resonances that have been used for thermal ablative therapy for cancer. Nanoshells can be tuned to strongly absorb light in the near-infrared (NIR), where optical transmission through tissue is optimal. Nanoshells placed at depth in tissues can be used to deliver a therapeutic dose of heat by using moderately low exposures of extracorporeally applied NIR. Exposure to low doses of NIR in solid tumors treated with metal nanoshells can reach temperatures capable of inducing irreversible tumor destruction within minutes. Gold nanoshells are $\sim$120 nm in diameter and a cancer cell is 170 times bigger. Therefore, nanoshells can penetrate the tumor capillaries and lodge in the tumor. Application of NIR light, which passes through the skin harmlessly, heats the nanoshells and kills the tumor cells. Since no drug is used, the cancer cells are unlikely to develop drug resistance.

The ability to control both wavelength-dependent scattering and absorption of nanoshells offers the opportunity to design nanoshells which provide both diagnostic and therapeutic capabilities in a single nanoparticle. A nanoshell-based all-optical platform technology can integrate cancer imaging and therapy applications. Immunotargeted nanoshells are engineered to both scatter light in the NIR range enabling optical molecular cancer imaging and absorb light to enable selective destruction of targeted carcinoma cells through photothermal therapy. In a proof of principle experiment, dual imaging/therapy immunotargeted nanoshells were used to detect and destroy breast carcinoma cells that overexpress HER2, a clinically relevant cancer biomarker. This approach has some significant advantages over alternatives that are under development.[35] For example, optical imaging is much faster and less expensive than other medical imaging techniques. Gold nanoparticles are also more biocompatible than other types of optically active nanoparticles, such as QDs. A commercial preparation of nanoshell, AuroShell™ (Nanospectra Biosciences Inc), is in development. AuroShell™ microparticles are injected intravenously and specifically collect in the tumor through the associated leaky vasculature. After the particles accumulate in a tumor, the area is illuminated with an NIR laser at wavelengths chosen to allow the maximum penetration of light through tissue. Unlike solid metals and other materials, AuroShell™ microparticles are designed to specifically absorb this wavelength, converting the laser light into heat. This results in the rapid destruction of the tumor along its irregular boundaries. The advantages of nanoshell-based tumor cell ablation include the following:

- Targeting to specific cells and tissues to avoid damage to surrounding tissue
- Less adverse effects than targeted chemotherapeutic agents or photodynamic therapy
- Repeatability because of lack of "tissue memory" as in radiation therapy and biocompatibility
- Ability to treat malignancies such as glioblastoma multiforme, metastases, and inoperable tumors

F. Nanobombs for Cancer

Nanobombs are nanoscale bombs, which infiltrate into tumors in a minimally invasive manner and then explode on exposure to physical or chemical triggers. Various nanomaterials have been used for the construction of nanobombs including gold and silica nanoparticles as well as carbon nanotubes. Nanobombs are effective anticancer agents as the shock waves that are generated after local explosion inside the tumor kill cancer cells can also disrupt cancer pathways so that the effect spreads beyond the area of explosion.

Temperature change can be used to trigger explosion. Nanogels fabricated by light cross-linking exhibit abrupt volume expansion upon exposure to sudden temperature change, causing cell death.[36] In another approach, Nanoclusters (gold nanobombs) can be activated in cancer cells only by confining NIR laser pulse energy within the critical mass of the nanoparticles in the nanocluster.[37] Once the nanobombs are exploded and kill cancer cells, macrophages can effectively clear the cell debris and the exploded nanotube along with it.

Blending of supramolecular chemistry and mechanostereochemistry with mesoporous silica nanoparticles has led to a new class of materials that are biological nanoscale bombs with the potential to infiltrate cells and explode upon the pulling of a chemical trigger.[38] The triggers are initiated by changes in pH, light, and redox potentials, in addition to enzymatic catalysis. This approach has been tried in "in vitro" experiments where loaded mechanized silica nanoparticles are endocytosed selectively by cancer cells and an intracellular trigger causes release of a cytotoxin, effectively leading to apoptosis.

G. Nanoparticles, MRI, and Thermal Ablation of Tumors

Nanostructures with surface-bound ligands can be used for the targeted delivery and ablation of colorectal cancer (CRC), the third most common malignancy and the second most common cause of cancer-related mortality in the United States. Normal colonic epithelial cells as well as primary CRC and metastatic tumors all express a unique surface-bound guanylyl cyclase C (GCC), which binds the bacterial heat-stable enterotoxin (ST), a peptide. This makes GCC a potential target for metastatic tumor ablation using ST-bound nanoparticles in combination with thermal ablation with NIR or radiofrequency energy absorption.[39] Furthermore, the incorporation of iron or iron oxide nanoparticles into such structures would provide advantages for MRI.

Gold nanoshell-based, targeted, multimodal contrast agents in the NIR range are fabricated and utilized as a diagnostic and therapeutic probe for MRI, fluorescence optical imaging, and photothermal cancer therapy of breast carcinoma cells in vitro.[40] This may enable diagnosis as well as treatment of cancer during one hospital visit.

In the future, it may be possible for a patient to be screened for breast cancer using MRI techniques with engineered enhanced ferrites as the MRI contrast agent. Enhanced ferrites are a class of ferrites that are specially engineered to have enhanced magnetic or electrical properties and are created through the use of core-shell morphology. Magnetic nanoparticles are coupled to the radio frequency of the MRI, which converts the radio frequency into heat. If a tumor is detected, the physician could increase the power to the MRI coils and localized heating would destroy the tumor without damage to the surrounding healthy cells.

H. Gold Nanoparticle Plus Bombesin for Imaging and Therapy of Cancer

Bombesin (BBN) peptides have demonstrated high affinity toward gastrin-releasing peptide (GRP) receptors in vivo that are overexpressed in prostate, breast, and small-cell lung carcinoma. In vivo studies using gold nanoparticles (AuNPs)–BBN and its radiolabeled surrogate 198AuNP–BBN constructs are GRP-receptor-specific, showing accumulation with high selectivity in GRP-receptor-rich prostate tumors implanted in severe combined immunodeficient mice.[41] The intraperitoneal mode of delivery was found to be efficient as AuNP–BBN conjugates showed reduced RES organ uptake with concomitant increase in uptake at tumor targets. The selective uptake of this new generation of GRP-receptor-specific AuNP–BBN peptide analogs has clinical potential in molecular imaging using CT techniques as the contrast numbers in prostate tumor sites are severalfold higher as compared to the pretreatment group. They also provide synergistic advantages by combining molecular imaging with therapy of cancer.

I. Role of Nanoparticle-Based Imaging in Oncology Clinical Trials

Currently computed tomography (CT) scans are used as surrogate end points in cancer clinical trials. The size of the tumor gives only limited information about the effectiveness of therapy. New imaging agents could speed the clinical trials process as better imaging data could help oncologists better select the most effective therapy for a particular patient, and to provide real-time information about whether a therapy is working. Currently, oncologists and their patients must wait months to determine if a given therapy is working. Shorter clinical trials would mean that effective new drugs would reach patients quicker and ineffective drugs would be dropped from clinical trials sooner, allowing drug discoverers to focus their efforts on more promising therapies.

J. Concluding Remarks

Nanoparticles have shown promise for incorporating multiple functions including diagnosis and therapy of cancer. Imaging plays an important role in personalizing cancer therapy. Most of the work done in this area is still experimental and some challenges need to be resolved before clinical applications. These include the following:

- Preventing capture/removal of nanoparticles by the reticuloendothelial system
- Difficulties in selective targeting as well as penetration of tumor by systemic administration of anticancer nanostructures, which requires identification of receptors unique to a particular cancer
- Investigation of long-term fate and toxicity concerns of nanoparticles

Efforts are being made to use nanostructures to develop anticancer treatment strategies based on various mitochondrial targets that play vital roles in cancer development and progression. Cancer mitochondria-targeted multifunctional compounds have been identified that could provide an alternative strategy for the development of novel solutions for cancer diagnosis and therapy.[42]

VII. Personalized Nanoneurology

Nanoneurology is defined as the application of nanobiotechnology in neurology. Diseases of the nervous system are an important part of medicine. In spite of all the advances in neurology, particularly in the last decade of the twentieth century (Decade of the Brain), there are serious deficiencies in the understanding of the pathomechanism of several neurological disorders as well as the ability to diagnose and treat these disorders. Nanobiotechnology will have an impact on improving the understanding of the nervous system and developing new treatments for disorders of the nervous system.[43,44]

A. Nanoparticles for Neuroprotection

Nanoparticles can improve drug delivery to the central nervous system (CNS) and facilitate crossing of the blood–brain barrier (BBB) and more precisely target a CNS injury site. These technologies were described in detail in a special report on this topic.[45] Some nanoparticles have a neuroprotective effect.

QD technology has been used to gather information about how the CNS environment becomes inhospitable to neuronal regeneration following injury or degenerative events by studying the process of reactive gliosis. Other research is looking at how QDs might spur growth of neurites by adding bioactive molecules to the QDs, in a way to provide a medium that will encourage this growth in a directed way. Poly(lactic-*co*-glycolic acid) (PLGA) nanoparticles loaded with superoxide dismutase have a neuroprotective effect seen up to 6 h after H_2O_2-induced oxidative stress, which appears to be due to the stability of the encapsulated enzyme and its better neuronal uptake after encapsulation.[46]

Gold salts, known to have an immunosuppressive effect, have been considered for treatment of traumatic brain injury (TBI), which results in loss of neurons caused not only by the initial injury but also by the resulting neuroinflammation as a secondary effect. The systemic use of gold salts is limited by nephrotoxicity. However, implants of pure metallic gold release gold ions do not spread in the body, but are taken up by cells near the implant. This is a safer method of using gold ions to reduce local neuroinflammation. Release or dissolucytosis of gold ions from metallic gold surfaces requires the presence of disolycytes, that is, macrophages, and the process is limited by their number and activity. In one study, the investigators injected 20–45 μ gold particles into the neocortex of mice before generating a cryo-injury.[47] Comparison of gold-treated and untreated cryolesions showed that the release of gold reduced microgliosis and neuronal apoptosis accompanied by a transient astrogliosis and an increased neural stem cell response, indicating an antiinflammatory and neuroprotective effect. Intracerebral application of metallic gold as a pharmaceutical source of gold ions bypasses the BBB and enables direct drug delivery to inflamed brain tissue. The method of delivery is invasive and a gold implant could produce foreign body reaction, leading to an epileptic focus. This can be refined by the use of gold nanoparticles.

B. Peptide Nanostructures for Repair of the CNS

Peptide nanostructures containing bioactive signals offer novel therapies with potential impact on regenerative medicine. These nanostructures can be designed through self-assembly strategies and supramolecular chemistry, and can combine bioactivity for multiple targets with biocompatibility. It is also possible to multiplex their functions by using them to deliver proteins, nucleic acids, drugs, and cells. Self-assembling peptide nanostructures can facilitate regeneration of the CNS. Other self-assembling oligopeptide technologies and the progress made with these materials toward the development of potential therapies have been reviewed elsewhere.[48]

C. Nanobiotechnology for Brain Tumor Management

Several molecular biomarkers have been identified in brain tumors that carry diagnostic and prognostic information. In addition, some of these and other biomarkers predict the response of these tumors to particular chemotherapeutic approaches and are useful for personalized management. Nanobiotechnology has helped in the refinement of molecular diagnostics as well as the detection of biomarkers of brain tumors.

Nanoparticles may be especially helpful for the treatment of the disseminated and very aggressive brain tumors. Doxorubicin, an anticancer drug, bound to polysorbate-coated nanoparticles crosses the intact BBB, and reaches therapeutic concentrations in the brain.

Nanoparticles made of poly(butyl cyanoacrylate) (PBCA) or PLGA coated with polysorbate 80 or poloxamer 188 enable the transport of cytostatics such as doxorubicin across the BBB. Following intravenous injection to rats bearing intracranial glioblastoma, these particles loaded with doxorubicin significantly increased the survival times and led to a complete tumor remission in 20–40% of the animals.[49] Moreover, these particles considerably reduced the dose-limiting cardiotoxicity and also the testicular toxicity of this drug. The drug transport across the BBB by nanoparticles appears to be due to a receptor-mediated interaction with the brain capillary endothelial cells, which is facilitated by certain plasma apolipoproteins adsorbed by nanoparticles in the blood.

Another promising strategy is the use of biodegradable polymeric MAb nanobioconjugate that crosses the blood–brain–tumor barrier, targets tumor cells, and inhibits tumor angiogenesis by specifically blocking the synthesis of a tumor neovascular trimer protein, laminin-411.[50] Other examples of nanobiotechnology application for brain tumors are as follow:

- Ferumoxtran-10 (Combidex®), a dextran-coated iron oxide nanoparticle, provides image-guided brain surgery.
- Fullerenes with a fluorescent metal atom terbium improve detection of stray cells during brain tumor surgery and addition of lutetium enables delivery of a lethal dose of radiation.

VIII. Personalized Nanocardiology

Application of the principles of personalized medicine to management of cardiovascular diseases is referred to as "personalized cardiology."[51] The future of cardiovascular diagnosis is already being impacted by nanosystems that can both diagnose pathology and treat it with targeted delivery systems.[52]

A. Gene–Environment Interactions in Personalized Cardiovascular Medicine

The constantly growing volume of available data will require an organized interpretation of variations in DNA and mRNA as well as proteins both on the individual and population level. Nanobiotechnology will facilitate the study of gene–environment interactions, which are important in many human diseases. Genetic analysis of thousands of transcript abundance traits in human primary endothelial cell lines in response to proinflammatory oxidized phospholipids implicated in cardiovascular disease revealed that approximately one-third of most regulated transcripts showed evidence of gene–environment interactions.[53] The interactions resulted primarily from effects of distal-, trans-acting loci, but a striking example of a local gene–environment interaction was also observed for FGD6. Some of the distal interactions were validated by siRNA knockdown experiments. These findings are consistent with the possibility that gene–environment interactions are responsible, in part, for the failure of association studies to more fully explain common disease variation.

B. Use of Magnetic Nanoparticles as MRI Contrast Agents for Cardiac Disorders

Magnetic nanoparticles have been used as contrast agents for MRI and have refined molecular imaging. Targeted imaging of vascular inflammation or thrombosis may enable improved risk assessment of atherosclerosis by detecting plaques at high risk of acute complications.[54] Cell death in the heart can be imaged in vivo by using annexin-labeled magnetic nanoparticles, particularly AnxCLIO–Cy5.5.[55] Experimental studies have shown the feasibility of combination of diagnosis and therapy using magnetic nanoparticles. Magnetic nanoparticles, conjugated with plasmid DNA expressing enhanced green fluorescent protein and coated with chitosan, were injected into mice through the tail vein and directed to the heart by means of an external magnet without the need to functionalize the nanoparticles, and their location was confirmed by fluorescent imaging.[56] This approach requires further investigations before clinical applications can be considered.

C. Nanoparticles for Cardiovascular Imaging and Targeted Drug Delivery

The potential dual use of nanoparticles for both imaging and site-targeted delivery of therapeutic agents to cardiovascular disease offers great promise for individualizing therapeutics. Image-based therapeutics with site-selective agents should enable verification that the drug is reaching the intended target and a molecular effect is occurring. Experimental studies have shown that binding of paclitaxel to smooth muscle cells in culture has no effect in altering the growth characteristics of the cells. If paclitaxel-loaded nanoparticles are applied to the cells, however, specific binding elicits a substantial reduction in smooth muscle cell proliferation, indicating that selective targeting may be a requirement for effective drug delivery in this situation. Similar behavior has been demonstrated for doxorubicin-containing particles. Intravenous delivery of fumagillin (an antiangiogenic agent)-loaded nanoparticles targeted to $\alpha v\beta 3$-integrin epitopes on vasa vasorum in growing plaques results in marked inhibition of plaque angiogenesis in cholesterol-fed rabbits. The unique mechanism of drug delivery for highly lipophilic agents such as paclitaxel contained within emulsions depends on close apposition between the nanoparticle carrier and the targeted cell membrane and has been described as "contact-facilitated drug delivery." In contrast to liposomal drug delivery (generally requiring endocytosis), the mechanism of drug transport in this case involves lipid exchange or lipid mixing between the emulsion vesicle and the targeted cell membrane, which depends on the extent and frequency of contact between two lipidic surfaces. The rate of lipid exchange and drug delivery can be greatly increased by the application of clinically safe levels of ultrasound energy that increase the propensity for fusion or enhanced contact between the nanoparticles and the targeted cell membrane.

The combination of nanoparticle-based targeted drug delivery and molecular imaging with MRI has the potential to enable serial characterization of the molecular epitope expression based on imaging readouts. Monitoring and confirmation of therapeutic efficacy of the therapeutic agents at the targeted site would facilitate personalized medical regimens.

IX. Future Prospects of Personalized Nanomedicine

Personalized medicine is already being practiced in selected therapeutic areas at some medical centers in the United States and the use is expected to come into the mainstream during the next 5 years. The interest in personalized medicine is worldwide although the implementation may be delayed due to socio-economic factors in some developing Asian countries. Japan, with an advanced health-care system and a prominent position of research activity in genomic medicine, has good prospects for introduction of personalized medicine.

Technologies that are driving the development of personalized medicine are molecular diagnostics and sequencing. Another important development is the combination of diagnostics with therapeutics. Nanobiotechnology is playing an important role in all of these areas. By improving the understanding of disease at molecular level, nanobiotechnology will facilitate the development of rational therapies. Advances in nanomedicine will parallel that of personalized medicine and the interaction of both will justify the term personalized nanomedicine. Technical advances in miniaturization have made construction of nano-sized robots, nanobots, feasible. Nanobots may navigate the human body to diagnose disease and deliver treatment. Most of the advances have taken place in nanooncology, where the combination of diagnosis with delivery of therapeutic is a distinct advantage.

References

1. Jain KK. *Textbook of personalized medicine.* New York: Springer; 2009.
2. Jain KK. Role of nanobiotechnology in the development of personalized medicine. *Nanomedicine* 2009;**4**:249–52.
3. Jain KK. *Handbook of nanomedicine.* New York: Springer; 2008.
4. Jain KK. Nanodiagnostics: application of nanotechnology in molecular diagnostics. *Expert Rev Mol Diagn* 2003;**3**:153–61.
5. Jain KK. Applications of nanobiotechnology in clinical diagnostics. *Clin Chem* 2007;**53**:2002–9.
6. Zhang J, Lang HP, Huber F, et al. Rapid and label-free nanomechanical detection of biomarker transcripts in human RNA. *Nat Nanotechnol* 2006;**1**:214–20.
7. Ashley EA, Butte AJ, Wheeler MT, et al. Clinical assessment incorporating a personal genome. *Lancet* 2010;**375**:1525–35.
8. Jain KK. *DNA sequencing.* Basel: Jain PharmaBiotech Publications; 2011.
9. Blazej RG, Kumaresan P, Mathies RA. Microfabricated bioprocessor for integrated nanoliter-scale Sanger DNA sequencing. *PNAS* 2006;**103**:7240–5.
10. McCaughan F, Dear PH. Single-molecule genomics. *J Pathol* 2010;**220**:297–306.
11. Liu H, He J, Tang J, et al. Translocation of single-stranded DNA through single-walled carbon nanotubes. *Science* 2010;**327**:64–7.
12. Clarke J, Wu HC, Jayasinghe L, et al. Continuous base identification for single-molecule nanopore DNA sequencing. *Nat Nanotechnol* 2009;**4**:265–70.
13. Wanunu M, Morrison W, Rabin Y, et al. Electrostatic focusing of unlabelled DNA into nanoscale pores using a salt gradient. *Nat Nanotechnol* 2010;**5**:160–5.
14. Derrington IM, Butler TZ, Collins MD, et al. Nanopore DNA sequencing with MspA. *PNAS* 2010;**107**:16060–5.
15. Jain KK. *Handbook of biomarkers.* New York: Humana/Springer; 2010.
16. Lesniak W, Bielinska AU, Sun K, et al. Silver/dendrimer nanocomposites as biomarkers: fabrication, characterization, in vitro toxicity, and intracellular detection. *Nano Lett* 2005;**5**:2123–30.
17. Xing Y, Chaudry Q, Shen C, et al. Bioconjugated quantum dots for multiplexed and quantitative immunohistochemistry. *Nat Protoc* 2007;**2**:1152–65.
18. Jain KK. *Proteomics: technologies, markets and companies.* Basel: Jain PharmaBiotech Publications; 2011.
19. Charalambous A, Andreou M, Skourides PA. Intein-mediated site-specific conjugation of Quantum Dots to proteins in vivo. *J Nanobiotechnol.* 2009;**7**(1):9. doi:10.1186/1477-3155-7-9.
20. Jain KK. Nanotechnology in drug discovery. In: Seethala R, editor. *Handbook of drug screening.* 2nd ed. New York: Informa; 2009. pp. 464–75.
21. Hild W, Pollinger K, Caporale A, et al. G protein-coupled receptors function as logic gates for nanoparticle binding and cell uptake. *PNAS* 2010;**107**:10667–72.

22. Fichter KM, Flajolet M, Greengard P, Vu TQ. Kinetics of G-protein–coupled receptor endosomal trafficking pathways revealed by single quantum dots. *PNAS* 2010;**107**:18658–63.
23. Wang X, Inapagolla R, Kannan S, et al. Synthesis, characterization, and in vitro activity of dendrimer-streptokinase conjugates. *Bioconjug Chem* 2007;**18**:791–9.
24. Paleos CM, Tziveleka LA, Sideratou Z, Tsiourvas D. Gene delivery using functional dendritic polymers. *Expert Opin Drug Deliv* 2009;**6**:27–38.
25. Kojima C. Design of stimuli-responsive dendrimers. *Expert Opin Drug Deliv* 2010;**7**:307–19.
26. Winter PM, Cai K, Caruthers SD, et al. Emerging nanomedicine opportunities with perfluorocarbon nanoparticles. *Expert Rev Med Devices* 2007;**4**:137–45.
27. Bryson JM, Fichter KM, Chu WJ, et al. Polymer beacons for luminescence and magnetic resonance imaging of DNA delivery. *PNAS* 2009;**106**:16913–8.
28. Jain KK. Recent advances in nanooncology. *Technol Cancer Res Treat* 2008;**7**:1–13.
29. Thaxton CS, Elghanian R, Thomas AD, et al. Nanoparticle-based bio-barcode assay redefines "undetectable" PSA and biochemical recurrence after radical prostatectomy. *Proc Natl Acad Sci USA* 2009;**106**:18437–42.
30. Baker Jr. JR. Dendrimer-based nanoparticles for cancer therapy. *Hematology Am Soc Hematol Educ Program* 2009;708–19.
31. Santra S, Kaittanis C, Grimm J, Perez JM. Drug/dye-loaded, multifunctional iron oxide nanoparticles for combined targeted cancer therapy and dual optical/magnetic resonance imaging. *Small* 2009;**5**:1862–8.
32. Nair BG, Nagaoka Y, Morimoto H, et al. Aptamer conjugated magnetic nanoparticles as nanosurgeons. *Nanotechnology* 2010;**21**:455102.
33. McDevitt MR, Chattopadhyay D, Kappel BJ, et al. Tumor targeting with antibody-functionalized, radiolabeled carbon nanotubes. *J Nucl Med* 2007;**48**:1180–9.
34. Bagalkot V, Zhang L, Levy-Nissenbaum E, et al. Quantum dot-aptamer conjugates for synchronous cancer imaging, therapy, and sensing of drug delivery based on bi-fluorescence resonance energy transfer. *Nano Lett* 2007;**7**:3065–70.
35. Loo C, Lowery A, Halas N, West J, Drezek R. Immunotargeted nanoshells for integrated cancer imaging and therapy. *Nano Lett* 2005;**5**:709–11.
36. Lee Y, Park SY, Kim C, Park TG. Thermally triggered intracellular explosion of volume transition nanogels for necrotic cell death. *J Control Release* 2009;**135**:89–95.
37. Zharov VP, Galitovskaya EN, Johnson C, Kelly T. Synergistic enhancement of selective nanophotothermolysis with gold nanoclusters: potential for cancer therapy. *Lasers Surg Med* 2005;**37**:219–26.
38. Cotí KK, Belowich ME, Liong M, et al. Mechanised nanoparticles for drug delivery. *Nanoscale* 2009;**1**:16–39. doi:10.1039/B9NR00162J.
39. Fortina P, Kricka LJ, Surrey S, Grodzinski P. Nanobiotechnology: the promise and reality of new approaches to molecular recognition. *Trends Biotechnol* 2005;**23**:168–73.
40. Bardhan R, Chen W, Perez-Torres C, et al. Nanoshells with targeted simultaneous enhancement of magnetic and optical imaging and photothermal therapeutic response. *Adv. Funct. Mater.* 2009;**19**:3901–9.
41. Chanda N, Kattumuri V, Shukla R, et al. Bombesin functionalized gold nanoparticles show in vitro and in vivo cancer receptor specificity. *Proc Natl Acad Sci USA* 2010;**107**:8760–5.
42. Zhang E, Zhang C, Su Y, et al. Newly developed strategies for multifunctional mitochondria-targeted agents in cancer therapy. *Drug Discov Today* 2011;**16**:140–6.
43. Jain KK. Role of nanotechnology in developing new therapies for diseases of the nervous system (editorial). *Nanomedicine* 2006;**1**:9–12.
44. Jain KK. Current status and future prospects of nanoneurology. *J Nanoneurosci.* 2009;**1**:56–64.
45. Jain KK. *Drug delivery in CNS disorders.* Basel: Jain PharmaBiotech Publications; 2011.

46. Reddy MK, Wu L, Kou W, et al. Superoxide dismutase-loaded PLGA nanoparticles protect cultured human neurons under oxidative stress. *Appl Biochem Biotechnol* 2008;**151**:565–77.
47. Larsen A, Kolind K, Pedersen DS, et al. Gold ions bio-released from metallic gold particles reduce inflammation and apoptosis and increase the regenerative responses in focal brain injury. *Histochem Cell Biol* 2008;**130**:681–92.
48. Webber MJ, Kessler JA, Stupp SI. Emerging peptide nanomedicine to regenerate tissues and organs. *J Intern Med* 2010;**267**:71–88.
49. Kreuter J, Gelperina S. Use of nanoparticles for cerebral cancer. *Tumori* 2008;**94**:271–7.
50. Ding H, Inoue S, Ljubimov AV, et al. Inhibition of brain tumor growth by intravenous poly(β-Lmalic acid) nanobioconjugate with pH-dependent drug release. *PNAS* 2010;**107**:18143–8.
51. Jain KK. *Applications of Biotechnology in Cardiovascular Therapeutics.* New York: Springer; 2011.
52. Wickline SA, Neubauer AM, Winter P, et al. Applications of nanotechnology to atherosclerosis, thrombosis, and vascular biology. *Arterioscler Thromb Vasc Biol* 2006;**26**:435–41.
53. Romanoski CE, Lee S, Kim MJ, et al. Systems genetics analysis of gene-by-environment interactions in human cells. *Am. J Hum Genet* 2010;**86**:399–410.
54. Saraste A, Nekolla SG, Schwaiger M. Cardiovascular molecular imaging: an overview. *Cardiovasc Res* 2009;**83**:643–52.
55. Chen HH, Josephson L, Sosnovik DE. Imaging of apoptosis in the heart with nanoparticle technology. *Wiley Interdiscip Rev Nanomed Nanobiotechnol* 2011;**3**:86–99.
56. Kumar A, Jena PK, Behera S, et al. Multifunctional magnetic nanoparticles for targeted delivery. *Nanomedicine* 2010;**6**:64–9.

Tissue Engineering Using Magnetite Nanoparticles

Akira Ito and Masamichi Kamihira

Department of Chemical Engineering, Faculty of Engineering, Kyushu University, Nishi-ku, Fukuoka, Japan

I. Introduction 356
A. Functional Magnetite Nanoparticles 356
B. Tissue Engineering Using Functional Magnetite Nanoparticles 357
II. Magnetofection 359
A. DNA Transfection Using Functional Magnetite Nanoparticles 359
B. Viral Transduction Using Functional Magnetite Nanoparticles 360
III. Magnetic Patterning of Cell 362
A. Magnetic Patterning of Cells Using MCLs 363
B. Magnetic Patterning of Cells Using RGD-MCLs 365
C. Magnetic Patterning of Cells Using PEG-Mags 366
IV. Construction of 3D Tissue-Like Structures 368
A. Skin Tissue Engineering 368
B. Skeletal Muscle Tissue Engineering 373
C. Liver Tissue Engineering 380
D. Construction of Complex 3D Tissues 385
V. Conclusion 388
References 389

The major advantage of magnetic manipulation is "remote control." Magnetic labeling of cells with magnetic nanoparticles enables the manipulation of cells and also the control of cell functions by applying an external magnetic field. "Functional" magnetite nanoparticles were developed for cell manipulation using magnetic force, and the magnetite nanoparticles were applied to tissue-engineering processes, which are designated as magnetic force-based tissue engineering (Mag-TE). This chapter reviews recent progress in Mag-TE techniques, and the principles and utilities of the applications are discussed. This review covers three topics of magnetic cell manipulation using magnetite nanoparticles, including a magnetic force-based gene transfer technique (magnetofection), magnetic cell patterning using functional magnetite nanoparticles and micro-patterned magnetic field gradient concentrators, and finally applications for fabrication of tissue-like constructs in skin, liver, and muscle tissue engineering.

Progress in Molecular Biology
and Translational Science, Vol. 104
DOI: 10.1016/B978-0-12-416020-0.00009-7

I. Introduction

A. Functional Magnetite Nanoparticles

Since magnetic particles have "magnetic" properties that are not seen in other materials, they have been applied to various medical techniques such as cell separation,[1] drug or cell targeting,[2,3] magnetic resonance imaging (MRI),[4] and hyperthermia.[5] The magnetic particles most frequently used for cell separation are ferrites with a general composition of MFe_2O_3 (where M represents a divalent metal cation, such as Ni, Co, Mg, or Zn, and includes magnetite Fe_3O_4) and maghemite Fe_2O_3. For medical applications, the most important feature is nontoxicity of magnetic particles. Based on this criterion, magnetite nanoparticles have been mainly and extensively studied and are being used in an increasing number of biological and medical applications.[6,7]

In order to add an affinity and targeting ability for cells, the concepts involved in drug delivery systems were applied to magnetite nanoparticles and functionalized magnetite nanoparticles were developed. Three types of functionalized magnetite nanoparticles are illustrated in Fig. 1. Magnetite cationic liposomes (MCLs), in which 10 nm magnetite nanoparticles are encapsulated into 200 nm cationic liposomes, were developed to improve the accumulation of magnetite nanoparticles in target cells through electrostatic interactions between MCLs and the cell membrane.[8] Additionally, among cell-manipulating techniques, control of cell adhesion is one of the most important issues. To promote cell attachment, MCLs were modified with an RGD (Arg-Gly-Asp) peptide, an integrin recognition motif found in fibronectin,[9,10] and a well-studied cell adhesion peptide, designated RGD-MCLs.[11] The average particle size of RGD-MCLs was 240 nm, and this size was similar to that of the MCLs. As an opposite concept, development of functionalized magnetite nanoparticles possessing the ability to resist cell attachment enables spatial control of cell adhesion onto cultural substrates. One of the most useful

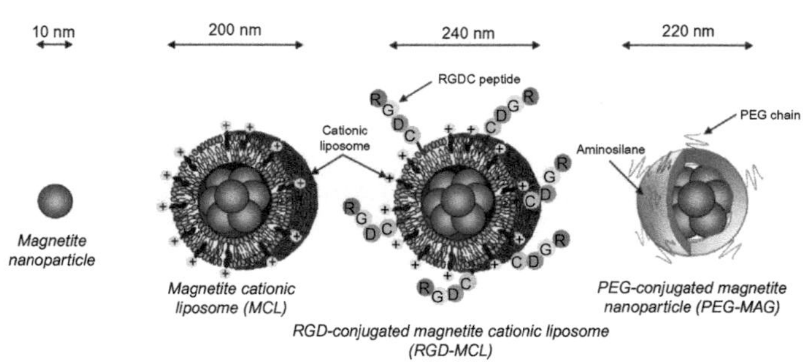

FIG. 1. Functional magnetite nanoparticles.

polymers to repel proteins is poly(ethylene glycol) (PEG). Surface modification with PEG leads to a significant reduction in the nonspecific interaction of biological molecules with the surface due to its high degree of hydrophilicity and chain flexibility.[12,13] Thus, 220 nm PEG-conjugated magnetite nanoparticles (PEG-Mags) were developed for spatial control of cell adhesion.[14]

B. Tissue Engineering Using Functional Magnetite Nanoparticles

Tissue engineering applies the principles of biology and engineering to the development of functional substitutes for damaged tissue.[15] There has been growing enthusiasm for tissue engineering, and this new technology has been a promising approach for overcoming the organ transplantation crisis resulting from donor organ shortage. Tissue engineering comprises the following processes (Fig. 2): (1) autologous cells isolated from healthy tissues or stem cells including embryonic stem (ES) cells and induced pluripotent stem (iPS) cells[16] are expanded to the required cell number; (2) genes of interest may be transferred into cells to enhance or modify cellular functions; (3) three-dimensional (3D) tissue-like structures are constructed, allowing 3D cell culture; in this step, if necessary, cells are cocultured with various cell types and/or patterned to mimic natural tissue structures; and (4) the cultured 3D constructs are transplanted into patients. Although overall technology of these processes in tissue engineering has been established, there is still plenty of room for improvement in each process.

Procedures to manipulate and remotely control cellular behavior can provide a powerful tool for tissue engineering. Magnetic manipulation offers such a tool, and the major advantage of magnetic manipulation is that it allows action from a distance. Dobson *et al.*[17,18] reported magnetic actuation for the mechanical conditioning of mesenchymal stem cells (MSCs) for tissue engineering and regenerative medicine. They used a range of magnetic particle sizes from 130 nm up to 4 μm and showed that the technique was effective for stimulation of intracellular calcium storage, membrane potential change, and upregulation of genes related to bone and cartilage formation in MSCs. In 2006, Ingber *et al.*[19] developed a magnetic force-based scaffold construction procedure. They used magnetic fields to position thrombin-coated magnetic nanoparticles in two-dimensional (2D) hexagonal arrays. The particles acted as nucleation sites for the ordered growth of fibrin, creating an ordered fibrin gel scaffold for endothelial cells. Moreover, magnetic manipulation presents distinct advantages for *in vivo* applications. In 2007, Wilhelm *et al.*[20] demonstrated that endothelial progenitor cells, which may facilitate angiogenesis and revascularization in ischemic sites, can be remotely guided both *in vitro* and *in vivo* by applying a magnetic force.

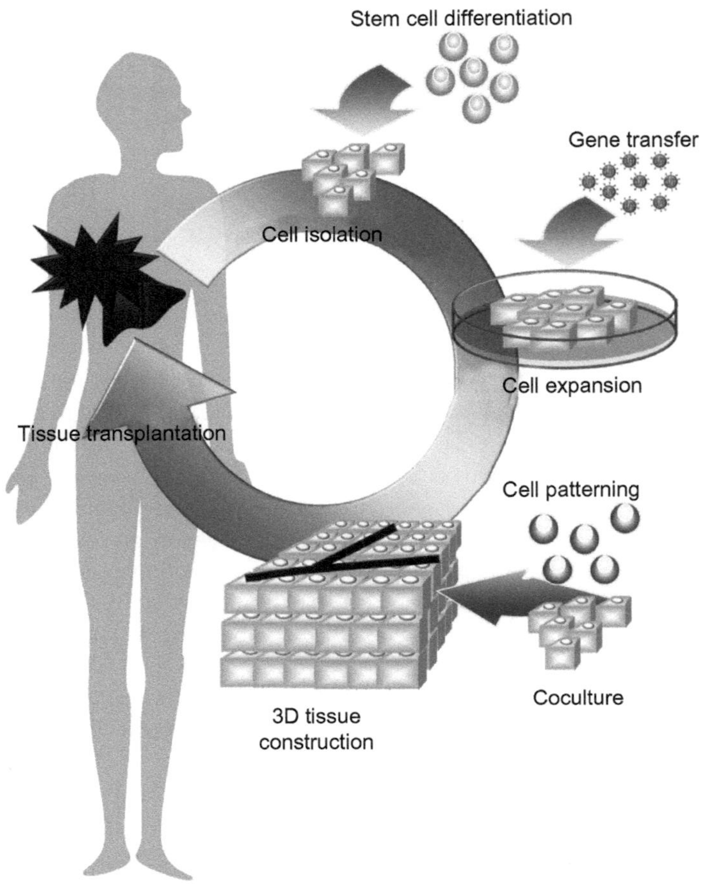

FIG. 2. Processes in tissue engineering.

From the viewpoint of bioprocess engineering, development of a methodology for physical manipulation of target cells is essential for tissue engineering in the next generation. A magnetic force was selected as a tool for physical manipulation, and target cells were manipulated using the functionalized magnetite nanoparticles. Thus, a novel cell-manipulating technology was developed using functionalized magnetite nanoparticles and magnetic force, designated as magnetic force-based tissue engineering (Mag-TE). This chapter focuses on Mag-TE techniques that have been applied to tissue-engineering processes: (1) gene transfer (magnetofection); (2) cell patterning; and (3) fabrication of tissue-like constructs.

II. Magnetofection

Growth factors stimulate cells for proliferation, differentiation, survival, and/ or extracellular matrix (ECM) synthesis, and they are therefore a key element of tissue engineering. However, many problems arise with the use of growth factors, including the transient effect of these proteins due to their relatively short biological half-lives and the difficulty in delivery to a specific injured site. Thus, gene delivery technology has become a crucial issue in recent years for establishing genetically manipulated cells including iPS cells,[16] and modification of cellular functions by overexpression of genes is being increasingly used in practice for tissue engineering and regenerative medicine.[21]

A. DNA Transfection Using Functional Magnetite Nanoparticles

The methods for gene delivery are generally classified into two categories: viral and nonviral. Nonviral methods using physicochemical properties, represented by electroporation[22] and lipofection,[23] have the advantages of simplicity and absence of a specific immune response, but the efficiency of gene introduction is limited due to a low transfection rate (see the chapter of Liu and Zhang in this volume for additional details). Therefore, further improvements with respect to the efficiency of DNA delivery are required. Magnetofection, in which gene transfection was magnetically achieved using magnetic particles, was developed as a new method for gene delivery.[24,25] Vector contact with target cells is the primary event in a successful transfection process. For magnetofection using plasmid DNA, complexes of DNA with cationic lipids or polymers were interacted with magnetic beads and attracted onto target cells by magnetic force to accumulate on the surface. Several research groups have independently developed magnetofection methods.[26] For example, in combination with polyethyleneimine (PEI), lipofectamine, or dioleoyl trimethylammonium propane (DOTAP)–cholesterol, magnetofection increased transgene expression levels.[24] Moreover, magnetofection showed high expression levels in target cells such as endothelial cells[27] and ES cells,[28] which are resistant to conventional transfection methods. Alternatively, since DNA (negatively charged) interacted with MCLs electrostatically due to their positive charge, investigations were carried out to see whether transfection efficiency could be enhanced by magnetofection that involves the use of plasmid DNA/ MCLs complexes (pDNA/MCL) and magnetic force.[29,30] The scheme of magnetofection using MCLs is illustrated in Fig. 3. The transfection efficiencies of the magnetofection technique by pDNA/MCL in fibroblasts and keratinocytes using reporter genes were 36- and 10-fold higher, respectively, than those of a lipofection technique by cationic liposomes.[29]

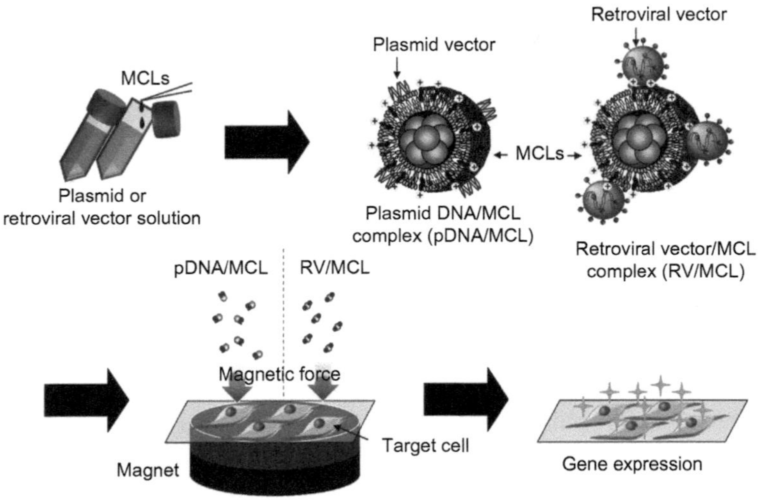

FIG. 3. Magnetofection using MCLs.

B. Viral Transduction Using Functional Magnetite Nanoparticles

Viruses are obvious candidates as gene transfer vectors since their ability to efficiently transfer viral nucleic acid into host cell is an important part of their life cycles. Consequently, several types of viruses have been used for introduction of genes into cells. The most commonly used viral vectors are based on retrovirus, adenovirus, and adeno-associated virus. Among transduction methods using viral vectors, retroviral vectors derived from RNA viruses can integrate into a gene of interest in the genome of host cells and have been widely used when stable and constant gene expression is required *in vitro* and *in vivo* for medical applications including tissue engineering. However, many problems still remain in gene delivery systems using retroviral vectors, including difficulty in preparations of viral vectors with high titer. Hughes *et al.*[31] proposed three strategies to concentrate infectious retroviral vectors from the supernatants of packaging cells, which was the first report on magnetic gene delivery, published in a peer-reviewed journal. Streptavidin-conjugated magnetic particles in conjunction with (1) antibodies directed against mouse fibronectin, (2) biotinylated lectins, or (3) biotin-modified packaging cell-surface proteins allow affinity-mediated magnetic concentration of retroviral vectors. On the other hand, because retroviral vectors spouted from packaging cells carry components of the cell membrane on their surface, electrostatic interactions between cationic liposomes or polymers and the retroviral vector are expected. To generate retroviral vector/magnetic particle complexes, Scherer *et al.*[24] developed transMAG^{PEI} composed of PEI-conjugated magnetic nanoparticles (average size of 200 nm). The magnetofection techniques based on electrostatic interaction using PEI-conjugated magnetic particles are now widely recognized and commercially available. To develop a new magnetic particle for gene delivery, MCLs were used to capture retroviral vectors and investigated whether retroviral vectors were concentrated by magnetic separation using MCLs.[32] MCLs were added to a solution containing a retroviral vector pseudotyped with vesicular stomatitis virus glycoprotein. The magnetic particles which captured the viral vectors were collected by a magnetic force and seeded into mouse neuroblastoma Neuro2a cells. As a result, the viral titer increased up to 55-fold, indicating that MCLs can capture and concentrate retroviral vectors. For magnetofection using retroviral vectors, retroviral vectors were magnetically labeled with MCLs, and the retroviral vector/MCL complexes were allowed to be attracted onto a monolayer of mouse myoblast C2C12 cells by placing a magnet under the culture plate.[33] The transduction efficiency was dramatically enhanced by increasing magnetite amount (0–150 ng) and magnetic field intensity (0–1010 G), indicating that the cellular uptake of magnetite nanoparticles is enhanced by physical interaction due to magnetic force. In general, cationic polymers, including polybrene, have been widely used for the enhancement of retroviral infection by increasing the flux of active viruses to the cells.[34] To evaluate the potential of magnetofection using MCLs, the transduction efficiencies were compared with a conventional method using polybrene.[33] As a result, the transduction efficiency was 6.7-fold higher for the magnetofection using MCLs than the conventional method using polybrene. These results indicate that magnetofection of retroviral vector using MCLs is applicable to gene therapy requiring stable and high expression of a target gene.

The transduction efficiency of adenoviral vectors is highly dependent on the coxsackievirus and adenovirus receptor (CAR) status of target cells. Unfortunately, many important target cells express little or no CAR.[35] Scherer *et al.*[24] demonstrated the efficacy of magnetofection on cells producing little or no CAR, such as NIH3T3 cells and primary human peripheral blood lymphocytes. By means of magnetofection using transMAG^{PEI}, they achieved a 500-fold enhancement of reporter gene expression as compared with standard infection with adenoviral vector in NIH3T3 cells.

Additionally, the magnetically labeled vectors can be directed to the desired regions for transduction by applying magnetic fields. Micro-patterns of gene-transduced cell regions were successfully created on a cellular monolayer using micro-patterned magnetic field gradient concentrators and retroviral vector/MCL complexes.[32] These results suggest that magnetofection provides a promising approach to capture viral vectors, thus achieving high transduction efficiency and the ability to deliver genes to a specifically injured site by applying a magnetic field. For *in vivo* gene therapies, magnetofection may become a promising choice where local treatment is required. Magnetic targeting of drugs has been used with some success in the treatment of cancer patients.[36] As compared with 2D magnetic targeting, efficient *in vivo* targeting has a 3D problem and would require more intense magnetic fields. However, gene delivery to surgically accessible sites such as vasculature can be improved by magnetic targeting.

III. Magnetic Patterning of Cell

Tissue engineering aims to create functional tissues using cells, growth factors, and biomaterials. In addition, if tissue-engineered architectures are completely similar to organs *in vivo*, tissue-engineered equivalents can be used for studies in cell biology or for evaluating the effects of drugs and toxins, which can lead to a reduction in the use of research animals. However, it is difficult to construct functional organs because tissue-engineered architectures are not completely similar to organs *in vivo*, in which cells are allocated with preciseness and complexity. Since tissues and organs *in vivo* are often composed of several types of cell layers, cell–cell interactions are important to maintain the normal physiology of organ systems. Therefore, technologies for fabricating functional tissue architectures by patterning several types of cells with complexity and preciseness are highly desired for tissue engineering. Micro-patterning of cells is a possible approach for this purpose.[37] In order to control specific cell adhesion on designed patterns, parameters of cultural substrates are varied in relative charge, hydrophilicity, and kind and density of immobilized adhesive proteins.[38] These methods can allocate cells precisely with high resolution. Although recent progress in surface chemistry enabled spatial control of cell adhesion onto substrates, these methods usually require specialized devices and time-consuming processes to fabricate the substrate. In these conventional methods, furthermore, since culture surfaces have to be chemically modified, they are highly restricted and may lead to limitation in applications for tissue engineering. Therefore, methodologies for fabrication of cell patterns on nonspecialized surfaces are required. On the other hand, physical cell-patterning methods such as inkjet printing[39,40] may not limit culture surface. However, these methods still require expensive apparatus and cause some other shortcomings; inkjet printing may cause cell damage due to high temperature and/or pressure. In this regard, a novel technique was developed based on a physical and robust method using magnetic force, in order to fabricate cell patterns on a nonspecialized surface. The schemes of magnetic cell patterning using functional magnetite nanoparticles are illustrated in Fig. 4.

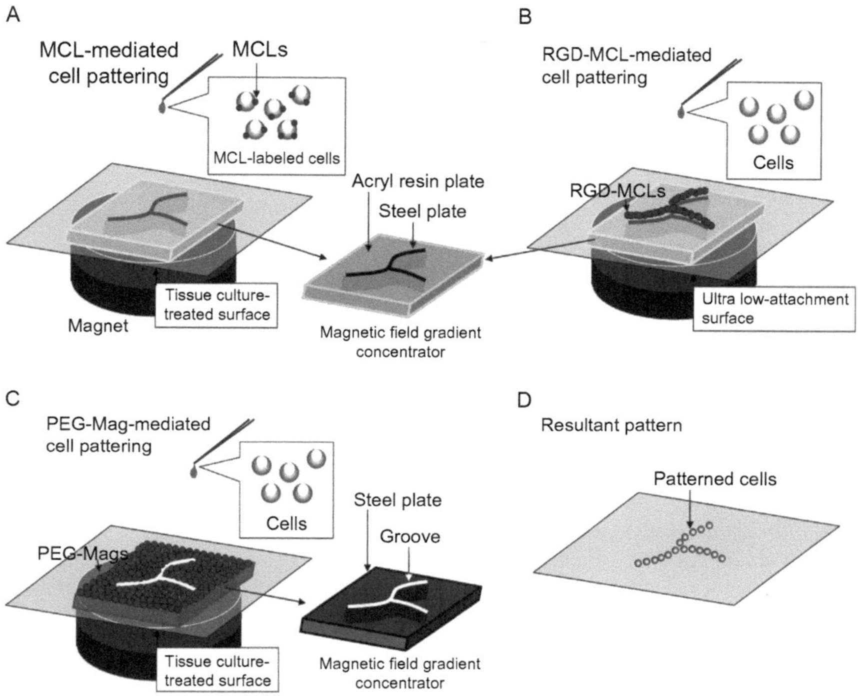

FIG. 4. Magnetic cell patterning using (A) MCLs, (B) RGD-MCLs, and (C) PEG-Mags. (D) The resultant pattern of cells from A, B, and C.

A. Magnetic Patterning of Cells Using MCLs

MCLs have been used to label a wide variety of mammalian cells magnetically. Cell types that have been labeled with MCLs, including different species (human, mouse, rat, canine), different cell types (primary cells, progenitor or stem cells, commonly used tumor cells), and their uptake amounts of magnetite are listed in Table I. On the contrary, Wilhelm *et al.*[51] proposed a cell-labeling method using "anionic" magnetic nanoparticles. Although this labeling method is very simple because of no modification of nanoparticle surface and no addition of transfection agent, cell-binding capacity is much lower than that of cationic liposomes. On the other hand, MCLs showed high cell-binding capacity due to the encapsulation of magnetite nanoparticles into cationic liposomes (Table I). To examine the pathway of MCLs into mammalian cells, target cell to MCL interactions were investigated both at 37 °C and at 4 °C. Membrane trafficking and the internalization process are known to be inhibited at 4 °C, so that only adhesion on cell membrane can occur. In contrast, incubation at 37 °C permits the endocytosis pathway. As a result, accumulation

TABLE I
CELLULAR UPTAKE OF MCLs

Cell types	Maximum MCL uptake (pg-magnetite/cell)	Reference
Primary cells		
Human umbilical vein endothelial cells	34.4	Ino *et al.*[41]
Human aortic endothelial cells	31.7	Ito *et al.*[42]
Human smooth muscle cells	23.3	Ito *et al.*[43]
Human dermal fibroblasts	13.7	Ino *et al.*[41]
Human mesenchymal stem cells	20.6	Shimizu *et al.*[44]
Rat bone marrow stromal cells	7.8	Shimizu *et al.*[45]
Rat cardiomyocytes	19.8	Shimizu *et al.*[46]
Canine urothelial cells	8.3	Ito *et al.*[43]
Cell lines		
Mouse NIH3T3 fibroblasts	18.8	Ito *et al.*[43]
Mouse C2C12 myoblasts	9.4	Yamamoto *et al.*[47]
Tumor cells		
Human HepG2 hepatoblastoma	48.9	Ito *et al.*[48]
Human U251-SP glioma	58.9	Le *et al.*[49]
Mouse RCC renal cell carcinoma	42.2	Shinkai *et al.*[50]
Rat T-9 glioma	54.4	Shinkai *et al.*[8]

of a large amount of the MCLs into NIH3T3 cells was observed at 37 °C as compared with that at 4 °C (authors' unpublished data), indicating that the major pathway of MCL uptake by cells was endocytosis. As shown in Table I, the amount of MCL uptake differed among cell types. Generally, tumor cell lines (e.g., human U251-SP glioma,[49] mouse RCC renal cell carcinoma[50]) showed a higher uptake of MCLs than primary cells. Because some tumor cells possess high endocytotic activity, the MCL uptake may depend on the endocytotic activity of the target cells. The maximum uptake amount of MCLs varied among cell types. Rat bone marrow stromal cells showed the lowest MCL uptake.[45] However, all cell types, including rat bone marrow stromal cells,[45] could be attracted to the magnetic field (4000 G), suggesting that MCL is a superior tool for universal magnetic labeling of cells. Thus, there is a great advantage in applying the technique, because cell-surface antigens and specific antibodies are not always available for each cell type.

The authors developed a simple and rapid cell-patterning technique using MCLs and magnetic force which enables the allocation of cells on an arbitrary surface, including biological gels.[52] The scheme of cell patterning using MCLs is illustrated in Fig. 4A. When thin steel plates (200 μm width), as a magnetic field gradient concentrator, placed on a magnet were laid under a cell culture surface, magnetically labeled cells were aligned on the surface where the magnetic field gradient concentrator was positioned. The line width of patterned cells could be adjusted by cell concentration, and patterned lines made of single cells were achieved by reducing the number of seeding cells. Furthermore, various cell patterns (curved, parallel, or crossing patterns) were successfully fabricated according to the line patterns of magnetic field gradient concentrators. Because cell patterning using magnetic force may not be limited by the property of culture surfaces, human umbilical vein endothelial cells (HUVECs) were patterned on Matrigel™, which can induce angiogenesis and subsequently result in the formation of patterned capillaries.[52] When HUVECs were seeded at low concentration, cells scattered in the culture area and did not connect with one another, while cells connected to each other and formed capillary-like structures when the same number of HUVECs were patterned in a line. To fabricate practical organs, it is necessary to allocate vascular endothelial cells in an arbitrary position, to allow the cells to connect with each other and to create vascularized organs having designed capillaries. These results indicate that magnetic pattering of HUVECs onto Matrigel™ is a possible approach in tissue engineering.

B. Magnetic Patterning of Cells Using RGD-MCLs

The RGD (Arg-Gly-Asp) peptide is an integrin recognition motif found in fibronectin[9,10] and one of the most extensively studied cell adhesion peptides. To promote cell attachment, an RGD motif-containing peptide was coupled to the phospholipid of MCLs (RGD-MCLs), and the RGD-MCLs were evaluated in terms of adhesion, spreading, cytoskeletal organization, and expression of fibronectin.[53] The scheme of cell patterning using RGD-MCLs is illustrated in Fig. 4B. A human keratinocyte cell line, HaCaT cells, which has a high anchorage dependency, was used as a model. The RGD-MCLs were added to an ultralow-attachment plate, whose culture surface is modified with a covalently bound hydrogel layer that is hydrophilic and neutrally charged, and then HaCaT cells were seeded to the plates. When RGD-MCLs were added at 20–25 μg/well, RGD-MCLs facilitated cell adhesion and proliferation. Several researchers have reported that cell adhesion mediated by RGD peptides depends on the concentration of immobilized RGD peptides.[54,55] Thus, higher concentrations of RGD-MCLs resulted in suitable conditions for cell adhesion.[53] Without RGD-MCLs, the cell aggregates floated in the media and few cells were observed to be attached to the plate. In the presence of MCLs (without RGD peptides), although cells were attached to the surface of the plates, most cells formed aggregates and very few spreading cells were observed, suggesting that positive charge derived from cationic liposomes facilitated cell attachment onto the culture surface but MCLs did not facilitate cell spreading without RGD peptides. On the other hand, when the cells were cultured on the surface in the presence of RGD-MCLs, HaCaT cells started to adhere on the surface within a day, and then proliferated. These results suggest that RGD-MCLs have two possible active effects on cell adherence: one is electrostatic interaction by positive charge of the cationic liposomes and the other is receptor-mediated interaction by the RGD peptides on the surface of RGD-MCLs. Actin stress fibers and pseudopodia in HaCaT cells that play an essential role in cell adhesion and migration were observed when HaCaT cells were seeded on the surface in the presence of RGD-MCLs. HaCaT cells grew to form cell–cell interactions and showed a well-developed cytoskeletal structure within the cells. Epithelial cells including HaCaT cells are critically dependent on ECM interactions for growth and survival. Fibronectin production by HaCaT cells cultured on the surface in the presence of RGD-MCLs was then investigated, and fibronectin production was apparently observed in the vicinity of the adhered cells. In addition, as cells grew and formed cell–cell interactions, the areas around and under the cells were fully covered with deposited fibronectin. Thus, the RGD-MCLs were shown to induce cell adhesion, spreading, cytoskeletal organization, and expression of fibronectin. When the lined magnetic field gradient concentrators with 200 μm width placed on a magnet were laid under a culture surface, HaCaT cells magnetically labeled with RGD-MCLs were aligned on the surface where the steel plate was positioned, resulting in magnetic cell patterning. Furthermore, magnetic field gradient concentrators were prepared to fabricate various patterns of cells. Acryl resin plates were cut by a laser beam using computer-aided design (CAD), and "M," "A," or "G" was engraved on the acryl resin plates (each character size: 10 mm × 10 mm), then steel plates were embedded into grooves in the devices. When these magnetic force gradient concentrators were placed on a magnet and HaCaT cells in the presence of RGD-MCLs were seeded onto nonattachment culture surface, the "M," "A," or "G" character patterned by the cells was successfully fabricated. These results suggest that CAD is a powerful tool for magnetic cell patterning because of the easiness to design complicated patterns. In this study, relatively large patterns of "M," "A," and "G" (10 mm × 10 mm) were fabricated. In order to fabricate more complicated patterns, a novel device using a laser beam with higher resolution and iron powder (not a steel plate) possessing a higher magnetic induction may be innovated.

C. Magnetic Patterning of Cells Using PEG-Mags

Recent progress in surface chemistry has enabled spatial control of cell adhesion onto cultural substrates by varying hydrophilicity, for example, by using PEG. The authors developed a novel cell-patterning procedure using PEG-Mags, in which magnetite nanoparticles (colloidal magnetite) coated with aminosilane (aminosilane-Mag) were modified with PEG and magnetic force.[14] The average particle size and zeta potential of PEG-Mags were 220 nm and −24.5 mV, respectively. The zeta potential of aminosilane-Mags (−2.5 mV) was higher than that of colloidal magnetite (−49.1 mV), because of the amino groups introduced on the surfaces of the aminosilane-Mags. The decreased zeta potential of PEG-Mags (−24.5 mV) indicated that the amino groups on aminosilane-Mags were coupled with PEG chains. Theoretically, more pronounced zeta potential values, whether negative or positive, tend to stabilize particle suspension.[56] Since the electrostatic repulsion between particles with the same electric charge prevents their aggregation, it is reasonable that the average particle size and the size distribution of PEG-Mags (zeta potential, −24.5 mV; particle size, 220 nm) were smaller than those of aminosilane-Mags (zeta potential, −2.5 mV; particle size, 915 nm).

The scheme of cell patterning using PEG-Mags is illustrated in Figs. 4C and 5. Using an array-patterned magnet, PEG-Mags were magnetically patterned on the surface of a tissue culture dish. The resultant substrate surface consisted of two regions: the PEG-Mag surface that acts as a cell-resistant region and the native substrate surface that promotes cell adhesion. When HaCaT cells were seeded onto the PEG-Mag-patterned surface, cells adhered only to the native substrate surface, resulting in cell patterning on the tissue culture dish. The patterned PEG-Mags were then washed away to expose the native substrate surface, and thereafter, when mouse myoblast C2C12 cells were seeded into the dish, cells adhered to the exposed substrate surface, resulting in a patterned coculture of heterotypic cells.[14] Moreover, it is worth noting that cell patterning of mouse fibroblast NIH3T3 cells on a monolayer of HaCaT cells (a layered coculture) was successfully achieved using PEG-Mags and magnetic force,[14] because the magnetic force-based cell-patterning procedure is not limited by the property of cultural substrate surfaces. This procedure provides a novel concept for cell patterning and may be useful for tissue engineering and cell biology.

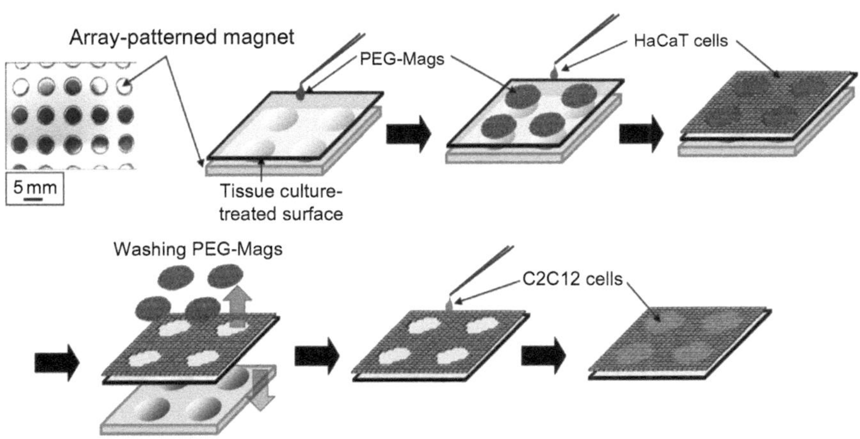

FIG. 5. PEG-Mag-mediated patterning of cells using an array-patterned magnet.

Furthermore, for medical applications, it is important to examine whether functional magnetite nanoparticles induce damage to target cells during the cellular uptake. The amount of PEG-Mags taken up by macrophage-like J774-1 cells at 2 h after addition of PEG-Mags at the concentration of 100 pg-magnetite/cell was 12 pg-magnetite/cell, while that of aminosilane-Mags taken up by the same cells was 88 pg-magnetite/cell.[14] These results indicated that the surface modification of magnetite nanoparticles with PEG caused inhibition of cellular uptake. Moreover, PEG-Mags showed almost no cytotoxic effect in NIH3T3 cells in the range of concentrations tested (0–10,000 pg/cell),[14] suggesting that the PEG modification provided the magnetite nanoparticles with biocompatibility and low toxicity.

IV. Construction of 3D Tissue-Like Structures

Conventionally, tissue engineering has been based on the seeding of cells onto 3D biodegradable scaffolds to reconstruct their native structure. Therefore, most efforts in tissue engineering may have been focusing on the scaffold design. The use of biodegradable scaffolds, however, poses problems such as insufficient cell migration into the scaffolds and inflammatory reaction due to the biodegradation of the scaffolds. Especially in muscle tissue engineering, because cell–cell interactions are essential for muscle differentiation, a 3D cell construct without artificial scaffolds may be more suitable. However, it is difficult to fabricate 3D tissue constructs without using 3D scaffolds, due to the lack of cell adherence via cell–cell junctions, particularly in the vertical direction. This nonadherence may be caused by the lack of ECM. As a scaffold-free method to construct tissue substitutes, Okano's group developed a cell sheet-based procedure.[57] They grafted a thermoresponsive polymer, poly (*N*-isopropylacrylamide), onto a cultural substrate surface. The cell layers grown on ECM deposited on the polymer were easily harvested as contiguous cell sheets by a change in temperature. Thus, this method of cell sheet engineering may be a promising approach to tissue engineering. Alternatively, the authors used magnetic force as a physical approach for enhancing layered cell–cell interactions. This section describes the magnetic force-based construction of 3D tissue-like structures.

A. Skin Tissue Engineering

Skin is a versatile organ functioning at the interface between humans and the external environment. This organ can be repaired naturally when damaged less than 40%, otherwise death may occur. The epidermis is one of only a few tissues for which it is possible to culture its principal cell (keratinocyte) and use these cultured cells to reconstitute stratified and differentiated human tissue. The pioneering work of Rheinwald and Green[58] allowed keratinocytes to be successfully cultured and subcultured on a feeder layer of mouse fibroblasts. Numerous efforts by various research groups have led to the development of commercially available defined media formulations, which enable keratinocytes to be cultured without a feeder layer. By these efforts, it has become possible to amplify keratinocytes to a clinically relevant number. The development of cultured skin substrates can be characterized by two different constructions: multilayered epithelial transplants (keratinocyte sheets)[59] and composite dermal–epidermal analogs.[60] These skin equivalents have been used clinically to repair burns and wounds.[61,62] Among them, the greatest advantage of keratinocyte sheets is that it enables the grafting of epidermal keratinocytes preserving sufficient proliferative capacity, because patients with epidermal damage such as whole body burns ultimately require an epidermis composed of keratinocytes.

1. Keratinocyte Sheets

After cultured epidermis was first reported by Rheinwald and Green in 1975,[58] O'Connor *et al.* achieved its clinical use.[61] They produced stratified keratinocyte sheets by allowing keratinocytes to undergo terminal differentiation during culture. Alternatively, a physical approach using magnetic attraction was taken to construct stratified keratinocytes.[63] When keratinocytes were seeded onto monolayer keratinocytes cultured on tissue culture plates, they did not form a multilayer sheetlike construct. Because proteases including trypsin are used for preparation of keratinocyte suspension, the ECMs may be digested. The authors investigated whether magnetically labeled keratinocytes could be accumulated using a magnet, and whether stratification is promoted by magnetic force to form a sheetlike 3D construct.[63] The scheme for construction of cell sheets using MCLs is illustrated in Fig. 6A. When keratinocyte sheets (cultured epidermis) fabricated by the method of Rheinwald and Green[58] consist of five or more cellular layers, they are sufficiently strong for recovery and transplantation. Therefore, to construct five-layered keratinocytes, magnetically labeled keratinocytes of fivefold confluency against the culture area were seeded into a well of the ultralow-attachment plates whose surface comprised a covalently bound hydrogel layer that is hydrophilic and neutrally charged, and a neodymium magnet was placed under the plate. In the absence of a magnet, keratinocytes with or without MCLs did not adhere onto the culture surface. In the presence of a magnet, in contrast, keratinocytes labeled with MCLs accumulated onto the well of the ultralow-attachment plates. The keratinocytes accumulated evenly throughout the wells and formed five-layered keratinocyte sheets. Histological studies revealed that the keratinocyte sheets fabricated by Mag-TE consisted of undifferentiated keratinocytes, which apparently differ from the epidermal sheets fabricated by the method of Rheinwald and Green.[58] It is speculated that the cell–cell adhesion

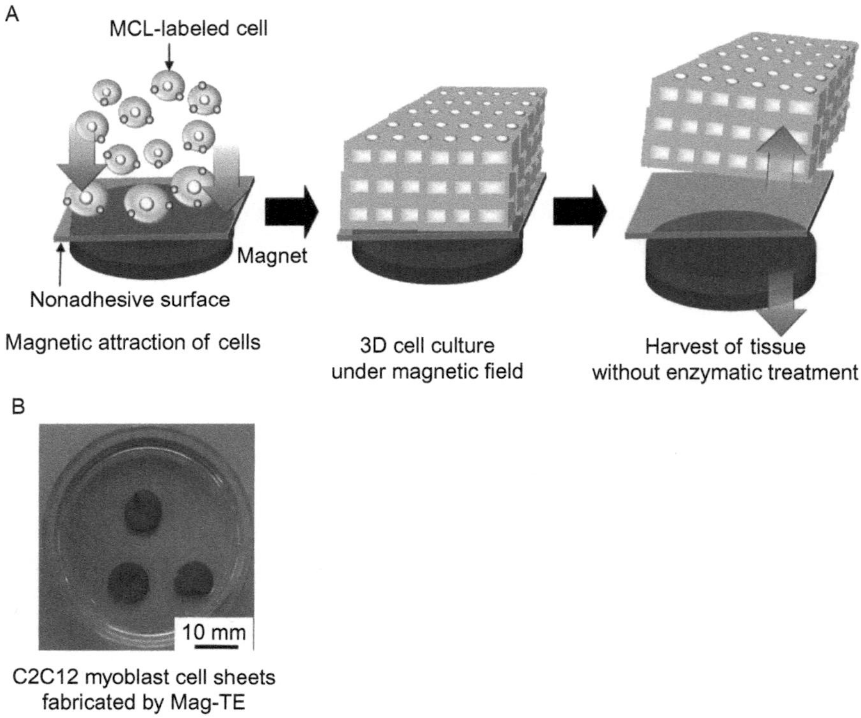

FIG. 6. Construction of cell sheets using MCLs. (A) Schematic illustration. (B) Photograph of representative cell sheets. Three C2C12 cell sheets were collected in a 35-mm tissue culture dish.

was caused by the very close placement of cells by the magnetic force, allowing 3D culture to produce ECMs. To the best of the authors' knowledge, this is the first time that multilayered undifferentiated keratinocyte sheets have been constructed. Although the authors have not investigated wound-healing effects of keratinocyte sheets constructed by Mag-TE *in vivo*, they speculate that undifferentiated keratinocytes in keratinocyte sheets produced by Mag-TE have greater effects on wound healing than cornified and anucleate keratinocytes fabricated by inducing terminal differentiation. Moreover, the culture medium was subsequently changed to the high-calcium medium, which was adjusted to a calcium concentration of 1.0 mM in order to induce stratification and terminal differentiation to construct thicker keratinocyte sheets. When the five-layered keratinocytes were cultured in high-calcium medium, they further stratified, producing 10-layer epidermal sheets.[63] These Mag tissue-engineered keratinocyte sheets fabricated using either low- or high-calcium media could be manipulated by tweezers, but keratinocyte sheets formed using high-calcium medium had significantly greater strength.

Dispase, a neutral protease from *Bacillus polymyxa*, is widely used to harvest multilayered keratinocyte sheets (cultured epidermis) from tissue culture dishes. In clinical use, extensive washing to remove dispase from keratinocyte sheets is required before they can be applied to wounds, because residual dispase is harmful to the wounded site. In industrial production of keratinocyte sheets, this washing is laborious and a technological barrier to automation of the process. Because ultralow-attachment plates were used, the keratinocyte sheets were harvested without enzymatic treatment. Moreover, magnetic force was used to make the recovery step easier, which could aid industrial production of keratinocyte sheets. Due to the magnetic force, the keratinocyte sheets labeled with MCLs floated up to the surface of the culture medium and stuck to a hydrophilically treated poly(vinylidene fluoride) membrane which was placed on top of a cylindrical magnet.[63] For this recovery step, the magnet used to harvest keratinocyte sheets can be substituted with an electromagnet. The authors have fabricated an electromagnet for transplantation of Mag-TE grafts.[44] The electromagnet consists of three parts: a foot switch, a generator, and a probe. A magnetic field was generated at the tip of the probe. Generation of the magnetic field was controlled by a foot switch; when the foot switch was pressed, the magnetic field which which formed by conducting electric current was generated and the magnetic field intensity was maintained while pressing the switch. The maximal magnetic field intensity at the tip of the probe was 450 G. When the foot switch was turned off, the magnetic field disappeared due to electric current conducted in the opposite direction, which allowed the cell sheets to be released from the tip of the probe. Using this device, Mag-TE MSC sheets were successfully harvested and transplanted onto an injured site.[44] Taken together, Mag-TE allowed the fabrication and harvest of keratinocyte sheets, and could be applied to tissue engineering for 3D tissues, including skin tissue engineering.

2. Gene-Engineered Keratinocyte Sheets

Gene therapy provides the potential for continuous production and delivery of therapeutic proteins. However, systemic or local administration of some viral vectors was found to elicit immune response and risk insertional mutagenesis. An alternative approach is the use of genetically modified cells and tissue-engineered grafts consisting of genetically modified cells. Skin is the biggest organ in the body, and the epidermis is a self-renewing tissue that is easily accessible and can provide large numbers of autologous cells to generate genetically modified skin substitutes. Based on these concepts, Lei *et al.*[64] used a retroviral vector to modify human keratinocytes with a gene encoding for human proinsulin and demonstrated that both keratinocytes and 3D skin equivalents were able to process proinsulin and secrete active insulin that promoted glucose uptake. These results suggest that gene-modified bioengineered skin can provide an alternative way of insulin delivery for treatment of diabetes.

Cytokines that are expressed during the wound-healing process have been used for genetic modification of keratinocytes. For example, platelet-derived growth factor (PDGF) is expressed in normal and wounded skin. Both isoforms of PGDF, PDGF-A and PDGF-B, are expressed in the epidermis, and their receptors are expressed in the dermis. Eming *et al.*[65] used retroviral transduction to overexpress the PDGF-A gene in human keratinocytes, and PDGF-A-modified keratinocytes were implanted as epithelial cell sheets in athymic mice. They showed that the connective tissue subadjacent to the graft was thicker and had more blood vessels at 1 week after grafting than control grafts. Alternatively, overexpression of vascular endothelial growth factor (VEGF) may be a good candidate for enhancing the functional performance of grafted keratinocytes. Dickens *et al.*[66] used cationic liposomes to overexpress VEGF in autologous keratinocytes cultured from the porcine donor. They observed upregulated levels and enhanced fibronectin deposition and found more endothelial cell tubular formations and higher rates of reepithelialization than in control grafts. These *ex vivo* gene transfer models may serve as a platform for vascular induction in full-thickness tissue repair.

In addition to hormones and cytokines, novel peptide therapeutics are increasingly making their way into clinical application.[67] One of these peptide drugs is antimicrobial peptides. Because skin is always exposed to invading microorganisms, it provides a protective barrier against infection. Antimicrobial peptides produced in skin tissue (e.g., defensins and cathelicidins[68]) were discovered and found to be highly effective for killing microorganisms directly. Although there is little sequence conservation between them, many of these peptides are short, cationic and able to form amphipathic structures.[68] They are thought to act by disrupting negatively charged bacterial cell membranes to which they are electrostatically attached. Upon binding, the hydrophobic face of the amphipathic structure may disrupt the lipid bilayer. The human beta defensin-3 (HBD-3), which possesses a broad spectrum of potent antimicrobial activity against both gram-negative and gram-positive bacteria, was identified by García *et al.* in 2001.[69] The advantage of antimicrobial peptides over other antimicrobial agents includes limited resistance. Defensins are naturally produced antimicrobial agents with low susceptibility to resistance, and thus gene therapy using defensins is a promising approach for treating infectious diseases.[70,71] Microbial infection represents a major problem in severely burned patients, causing skin graft failure and increase in the risk of mortality. Therefore, the HBD-3 gene was introduced into human keratinocytes, and a multilayered keratinocyte sheet was prepared overexpressing HBD-3 by the Mag-TE technique, to investigate the feasibility of the HBD-3-engineered cell sheet as a novel antibacterial therapy.[72]

Human keratinocyte cell line, HaCaT cells, was transduced with a retroviral vector encoding HBD-3 gene accompanied with enhanced green fluorescent protein (EGFP) gene as a reporter gene, and stable HBD-3-expressing clones were established using the limiting dilution method (designated as HaCaT/HBD-3 cells).[72] HBD-3 was strongly expressed in the HaCaT/HBD-3 cells, while no obvious expression of HBD-3 was detected in original HaCaT cells. Antimicrobial activity of HBD-3 expressed by HaCaT/HBD-3 cells was analyzed by adding *Escherichia coli* cells into the culture medium of HaCaT/HBD-3 cells. A rapid reduction in bacterial growth was observed. Subsequently, a slight, but not significant, decrease of bacterial growth was observed, and the percentage of bacterial growth at 24 h was 44.7%. Next, HBD-3-engineered keratinocyte sheets using HaCaT/HBD-3 cells were constructed by the Mag-TE technique (Fig. 6A).[72] Magnetically labeled HaCaT/HBD-3 cells were seeded into the wells of 24-well ultralow-attachment plates, and a magnet was placed under the plate. When the magnet under the 24-well ultralow-attachment plate was removed, the cells detached from the bottom of the well. The cells containing MCLs exhibited a black-brown color, which is the color of magnetite. Phase-contrast microscopy of the cross sections revealed that HaCaT/HBD-3 cells labeled with MCLs formed sheetlike structures with thickness ranging from 30 to 50 μm. Moreover, the cells within the multilayered sheets expressed EGFP, indicating that a gene-engineered cell sheet was successfully constructed by the Mag-TE technique. In this system, *E. coli* viability was significantly reduced and the bacterial growth was inhibited to 63.0% when *E. coli* cells were inoculated to the culture of HaCaT/HBD-3 cell sheets and cultured for 1 h. These results suggest that the magnetic force-based tissue-engineered keratinocyte sheet over-expressing HBD-3 can provide a protective shield against bacterial invasion.

B. Skeletal Muscle Tissue Engineering

Skeletal muscle tissue engineering is a promising approach for replacement of muscle tissues following traumatic injury, tumor ablation, or functional loss caused by muscle diseases such as muscular dystrophy.[73] In recent years, skeletal muscle tissues have also attracted much attention for bioactuator application. Herr and Dennis[74] designed a swimming robot using frog semi-tendinosus muscle and proposed a muscle-powered actuator. Through the exploitation of cell micro-pattering techniques, Feinberg *et al.*[75] fabricated muscle thin films using poly-dimethylsiloxane and cardiomyocytes for the construction of actuators and power devices. Tanaka *et al.*[76] created an on-chip cellular micropump, using cardiomyocyte sheets constructed by the cell sheet-engineering technique, as a prototype of applicative bio-microactuators.

To construct artificial skeletal muscles that are physiologically equivalent to native muscles, mimicking of the natural skeletal muscle is believed to be crucial. For this purpose, tissue-engineered skeletal muscle should have the following two structural features: (1) a high cell density that may lead to cell fusion, resulting in multinucleated myotube formation and (2) a highly unidirectional orientation that facilitates large muscular forces. Natural skeletal muscle tissue is composed of striated myotubes arranged in parallel alignment. A connective tissue covering, predominantly composed of collagen, tethers adjacent myotubes to form individual muscle fibers. In addition, satellite cells located beneath the basal lamina, at the periphery of muscle fibers, are believed to play an important role in the repair and replacement of damaged skeletal muscle cells.[77] Thus, it has been proposed that satellite cells are a key to the successful engineering of skeletal muscle tissue, and skeletal muscle tissue-like constructs have been produced from mammalian cells using established skeletal muscle myoblast cell lines, including C2C12 cells[78] (which are also known as satellite cells), or primary satellite cells isolated from neonatal rats.[79] Additionally, artificial scaffolds such as biodegradable sponges[80] and hydrogels composed of collagen[79] and/or Matrigel™[81] have been employed to fabricate these tissue-engineered constructs. The most common approach is a hydrogel-based procedure in which spontaneous 3D tissue formation can be induced from a mixture of myoblast cells and ECM precursors such as collagen and Matrigel™. ECM components play essential roles in the development and signaling of skeletal tissues and contribute to the enhancement of mechanical strengths with maintenance of tissue flexibility. Nevertheless, since the native skeletal muscle tissue is constructed with a high cell density, the use of synthetic scaffolds may interfere with cell–cell interactions, thereby resulting in the inhibition of multinucleated myotube formation. In addition, myoblast cells in skeletal tissues engineered by the hydrogel-based procedure were mainly distributed at the tissue periphery and were less compact than native skeletal tissues, which can limit the further development of skeletal muscle functionality. Consequently, successful skeletal muscle tissue engineering is associated with how a higher density of skeletal muscle cells can be achieved within artificial skeletal muscle tissue constructs. With the Mag-TE technique, target cells labeled with MCLs were accumulated using a magnet. Subsequently, stratification was promoted by the magnetic force, leading to the formation of multilayered sheetlike constructs without using any artificial scaffolds. This section reviews a procedure for the fabrication of highly dense and oriented muscle tissue constructs based on the Mag-TE technique.

1. Myoblast Cell Sheets

Myoblast cell sheets were constructed by a Mag-TE technique (Fig. 6B).[47] Mouse myoblast C2C12 cells labeled with MCLs were seeded into the wells of 24-well ultralow-attachment culture plates, and a magnet was placed on the reverse side of each plate. The magnetically labeled C2C12 cells were rapidly attracted to the magnet, and the cells accumulated uniformly at the bottom of each well and formed highly dense and multilayered sheetlike constructs composed of C2C12 cells without any scaffolds. Similar to previous results using keratinocytes (Section IV.A.1),[63] C2C12 cells in the absence of a magnet did not form evenly contiguous cell sheets, and instead preferentially formed small cell aggregates. Interestingly, unlike keratinocyte sheets, C2C12 cell sheets drastically shrank during the culture, and the resultant C2C12 cell sheets had a substantial strength. Histological observation revealed that the thickness of the cell sheets formed from 4×10^6 cells was 270 μm.[47] When a smaller number of cells were used, the cells formed aggregates rather than uniform cell sheets. Conversely, when a larger number of cells (6×10^6 cells) were used, the cell condition was not good owing to the depletion of nutrients after 24 h. Further development of the myoblast cell sheets will be discussed in Sections IV.B.4 and IV.D.

2. Myoblast Cell Strings

In native muscle tissues, skeletal muscle cells form a fibrous structure. Furthermore, tissue-engineered constructs need to be sufficiently thin to allow the supply of oxygen and nutrients. Therefore, the method of magnetic cell patterning (Section III.A) was modified for use in fabrication of string-shaped 3D tissue constructs,[47] to mimic skeletal muscle fiber bundles. The magnetic field gradient concentrator of 200 μm thickness was used to construct thin cellular strings. As already mentioned in Section III.A, the cell patterning using magnetic force may not limit the property of culture surfaces. Therefore, a patterning of C2C12 cells on nonadherent surface was conducted. When magnetically labeled cells with an excessive number against the magnetically restricted area were seeded onto the nonadherent surface and cultured for 1 day on the magnetic field gradient concentrator with a magnet, the cells aggregated in a line. After removal of the magnet, the string-like 3D tissue construct detached from the surface without enzymatic digestion and floated in the culture medium. The width and thickness of the cellular strings were successfully controlled within 200 μm by the magnetic field gradient concentrator, and they had sufficient strength to allow their manipulation with tweezers, suggesting that 3D cell–cell interactions were formed among the cells during 1-day cell cultivation in 3D manner by using magnetic force.[47] It was also demonstrated that both fibroblast and vascular endothelial string-like structures were fabricated and harvested without enzymatic digestion when the cells were patterned on nonadhesive surfaces,[52] suggesting that this novel methodology is a possible approach for fabricating tissue-engineered muscle fibers, tendons, and capillaries.

3. Myoblast Cell Rings

Although C2C12 cellular sheets and strings were successfully fabricated, they shrank considerably during a longer culture period. Since shrinkage of keratinocyte sheets or NIH3T3 cellular sheets was much slower than that of C2C12 cellular sheets, it is supposed that the drastic shrinking is a feature of tissues composed of myoblast cells. Shrinkage is a major problem for the differentiation of myoblast cells into skeletal muscle tissues *in vitro*, because the configurations of the artificial tissue constructs cannot be maintained during the culture period required for myogenic differentiation. In native muscles, the ends of the muscle tissues are attached to bone via tendons, and the tension generated between the tendons may facilitate the oriented differentiation of muscle fibers. Consequently, the fabricated C2C12 cell constructs should be fixed to tendon-like anchors to prevent shrinking and induce oriented differentiation during myogenic differentiation *in vitro*. Dennis and Kosnik[82] developed laminin-coated silk-suture anchors as artificial tendons. Alternatively, the formation of ring-shaped tissue constructs was induced for subsequent hooking around two pins, to maintain the shape and create the oriented and differentiated muscle structures.[47]

The scheme for construction of myoblast cellular rings using MCLs is illustrated in Fig. 7.[47] Ring-shaped tissue constructs were fabricated by utilizing the shrinkage feature of C2C12 cellular sheets. When a cylinder, as an anchor, was positioned at the center of a well, the cellular sheet drastically shrank to form a ringlike structure around the cylinder. In the absence of the magnetic force, the cells did not form a tissue construct, suggesting that the magnetic accumulation of the cells induced self-organization to form a cellular

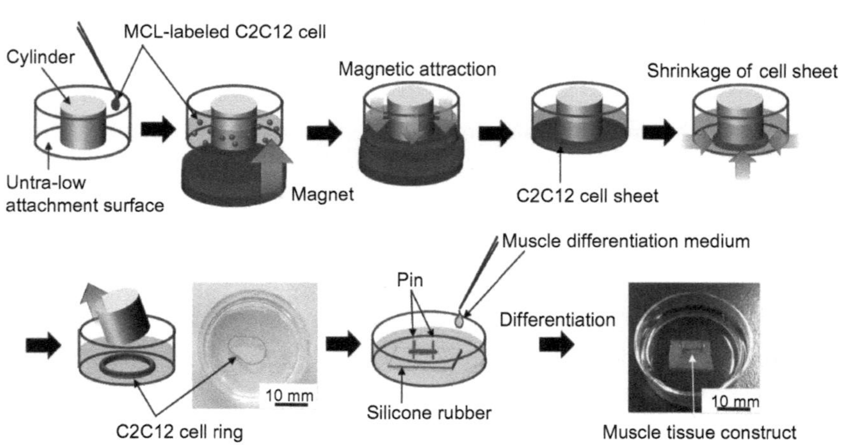

Fig. 7. Construction of myoblast cellular rings using MCLs.

sheet, eventually resulting in the formation of a cellular ring. The diameter and thickness of the cultured cellular rings were 12 mm and 120 μm, respectively. Interestingly, histological examination of cross sections of individual rings revealed that the cells were oriented in the direction of the circumference by the tension generated within the structure, which may mimic *in vivo* native myofibers that receive a tensile force from tendons. Thus, dense and oriented myoblast tissue constructs can be fabricated by this procedure. Subsequently, to ensure sufficient supply of oxygen and nutrients, each cellular ring was detached from the cylinder and transferred into a larger dish (a 35-mm tissue culture dish). The cellular ring was then hooked around two anchor pins separated by 8 mm and cultured in differentiation medium. Since muscle differentiation has been shown to improve in the presence of ECM proteins, especially laminin and collagen,[83] a gel mixture composed of type I collagen and Matrigel™ was used to coat each cellular ring. The cellular ring shrank to form a bundle between the anchors and retained the shape for over 3 weeks. After the 1-week culture in differentiation medium (day 7), the cells were oriented, fused, and multinucleated within the tissue constructs.[47] Semiquantitative RT-PCR and immunohistochemical analyses revealed that the myogenic differentiation marker myogenin was expressed in the tissue constructs, and the myogenin expression was located in the nuclei of the cells.[47]

To further evaluate differentiation of the cellular ring, the cells in the tissue were stained with specific antibodies against α-actinin and actin filament.[84] As a result, sarcomere structures were observed in the tissue construct, indicating that C2C12 cells successfully differentiated into skeletal muscle cells within the ring-shaped tissue construct. Western blot analysis revealed that the expressions of the later-stage muscle-specific proteins, myosin heavy chain and tropomyosin, increased from day 2 and then were maintained at high levels throughout the culture period. Moreover, activity of creatine kinase, which is involved in energy metabolism of muscle tissues during muscle contraction, increased from day 2 and continued to increase till day 17. It has been reported that proliferative myoblasts differentiate into myocytes, thereby inducing myogenin expression by cell cycle withdrawal, which subsequently leads to the expression of muscle-specific proteins, such as myosin heavy chain, tropomyosin, and creatine kinase.[85] Thus, the expression pattern of muscle-specific proteins in the artificial tissue construct was consistent with the expression in the normal myogenesis pathway. To evaluate contractile properties, the cellular rings were stimulated with electrical pulses.[84] The cellular rings did not generate obvious contractile forces on day 2. As muscle differentiation progressed within the tissue construct, the contractile force generated by electrical stimulation became progressively stronger. This response coincided with the expression patterns of muscle-specific proteins. The tissue constructs generated a maximum twitch force of 33.2 μN (1.06 mN/mm^2) on day 17, and the chronaxie which indicates their excitability was 0.72 ms. The force, however, corresponded to ~0.5% of adult mammalian skeletal muscle. Since many researchers have reported that continuous stimulation by electrical pulses[86] or mechanical stretching[87] enhances myogenesis of myoblast cells, the biological milieu and physical stimuli may be effective for mimicking natural skeletal muscle development and maturation. Interestingly, the Mag-TE cellular rings contracted rhythmically in response to relatively low-frequency electrical pulses (0.2, 0.5, and 1 Hz).[84] Therefore, it is worth noting that the artificial tissue constructs are applicable to electrically controlled bioactuators.

Taken together, these results indicate that the artificial skeletal muscle tissue constructs fabricated by the Mag-TE technique were physiologically functional, and these tissue-engineered skeletal muscles are applicable to regenerative medicine, drug screening, and bioactuator development.

4. Gene-Engineered Muscle Cell Sheets

In addition to keratinocytes (Section IV.A.2), myoblast cells may be a suitable model for cell-mediated gene delivery. Myoblast cells can be isolated from muscle tissues and expanded to a large number. The cells can be transfected or transduced with genes of interest, and injected back into muscles for transplantation. In addition, the unique biological feature of muscle cells is that transplanted myoblast cells undergo terminal differentiation and become part of myofibers by fusing with each other or fusing into preexisting myofibers, which serves as a stable source of transgene expression for long periods. For example, Hamamori *et al.*[88] established human erythropoietin (EPO)-secreting C2 myoblast cells. They demonstrated that myoblast cells could be transplanted in uremic mice and myoblast gene transfer could achieve sufficient and sustained delivery of functionally active EPO to correct anemia associated with renal failure in mice.

Successful transplantation relies on the vascularization of the grafts within the host. In muscle tissue engineering, the thickness of grafts has been limited by an inadequate infiltration of vessels upon implantation. Thus, the development of tissue-engineered muscle modified with proangiogenic factors has been a focus of the current study. To date, several angiogenesis-related growth factors, such as VEGF, basic fibroblast growth factor, PDGF, and transforming growth factor-β (TGF-β), have been found to be effective.[89] The best-studied molecule for therapeutic angiogenesis is VEGF. As angiogenesis in response to VEGF stimulation occurred in a dosage-dependent manner,[90] successful treatment requires the application of highly efficient gene transfer methods. In the Section II, it has been described that magnetofection, a gene delivery technique using magnetic particles, enhances transduction efficiency. In this section, the strategy to fabricate VEGF gene-engineered myoblast cell sheets possessing angiogenic potential will be exhibited, by using MCL-mediated retroviral gene transduction and the tissue fabrication technique described in Section IV.B.1.

VEGF gene-engineered C2C12 (C2C12/VEGF) cell sheets were fabricated by using MCLs and magnetic force.[33] After C2C12 cells were transduced with the VEGF gene using the MCL-mediated magnetofection technique described in Section II.B and Fig. 3, MCL-labeled C2C12/VEGF cells were accumulated onto a nonattachment culture surface using a magnetic force, and C2C12/VEGF cell sheets were formed. ELISA revealed that a high exogenous VEGF expression (240 ng/day) was observed in the C2C12/VEGF cell sheets,[33] indicating that the VEGF gene-engineered cell sheets were successfully fabricated by combining the magnetofection and Mag-TE techniques. To evaluate the angiogenic potential of VEGF gene-engineered cell sheets *in vivo*, C2C12/VEGF cell sheets were transplanted into subcutaneous spaces of athymic mice.[33] At 2 weeks posttransplantation, capillary vessels with a high density were observed in C2C12/VEGF cell sheet-derived tissues. The percentage of microvessel area within the tissues of C2C12/VEGF cell sheets was significantly higher than that of nontransduced C2C12 cell sheets. Histological study of transplants revealed that the graft formed cell-dense tissues containing magnetite nanoparticles. C2C12/VEGF cell sheet-derived tissues were composed of a core region containing magnetite-labeled cells and a surrounding thick peripheral region formed by cell proliferation where multinucleated myotubes were observed, indicating that this approach is applicable for the repair of skeletal muscle tissue defect. The cross-sectional area of C2C12/VEGF tissues was twofold larger than that of control C2C12 tissues, and it was confirmed that the newly formed vessels in the C2C12/VEGF cell sheet-derived tissue contained blood cells, indicating that they inosculated with the host's vasculature and had the functionality capable of metabolic exchange. On the other hand, when C2C12/VEGF cells without cell sheet formation were injected subcutaneously into athymic mice, cell assemblies at the transplanted sites were observed and the enhanced angiogenesis was confirmed in the site where the C2C12/VEGF cells were injected. However, their sizes were much smaller than that of the C2C12/VEGF cell sheet due to the dislocation of the injected cells. This indicated that the cell sheet formation allowed effective transplantation and *in vivo* tissue formation than direct cell injection. These results indicated that C2C12/VEGF cell sheet-derived tissue maintained a high cell density by promoting vascular network formation and produced thick tissues compared with the control grafts, suggesting that VEGF gene introduction was an effective strategy for the induction of angiogenesis into cell-dense constructs.

This section has described a novel magnetic biomanipulation technique combining gene transfer and tissue fabrication processes whose technological developments are considered important for the next generation of regenerative medicine. Before the clinical application, some problems should be cleared. As to the therapeutic efficiency of VEGF, VEGF alone is probably not sufficient to create mature and stable vasculature. Therefore, the magnetic VEGF gene transfer combined with a supportive angiogenic gene such as PDGF, which enhances the vessel stabilization,[91] would be necessary for the improvement of vessel functionality. Concerning the safety issue of retroviral vectors which may trigger oncogenesis by upregulating cellular proto-oncogenes, magnetofection using plasmid vectors may be a first candidate for clinical applications.

C. Liver Tissue Engineering

Today, many patients suffer from acute liver failure and hepatoma. Because this is an area of high unmet clinical need, there is an urgent need to develop techniques that will enable liver tissue engineering or generate a bioartificial liver, which will maintain liver function or offer the possibility of liver replacement. Liver tissue engineering is an innovative way of constructing an implantable liver and has the potential to overcome the shortage of organ donors for liver transplantation. In liver function, heterotypic interactions play a fundamental role. Since the liver is formed from the endodermal foregut and mesenchymal vascular structures, it may be functionally mediated by heterotypic interactions. Without nonparenchymal cells, freshly isolated hepatocytes lose their liver-specific functions shortly under *in vitro* culture conditions. In order to manipulate cell–cell interactions, various 2D coculture systems of hepatocytes and nonparenchymal cells have been developed.[92] As mentioned in Section III, recent progress in surface chemistry has enabled the spatial control of cell adhesion onto substrates, which has realized various 2D coculture systems. As a 2D coculture system, the authors took a genetic engineering approach.[93] The cell–cell interactions are mainly mediated by cytokines, ECMs, and cell–cell adhesions. Cell–cell adhesion mediated by various molecules is an important factor to regulate differentiation and proliferation. Especially, E-cadherin is a member of the classic cadherin family and is expressed mainly in epithelial cells and also in hepatocytes.[94] The extracellular domain on E-cadherin interacts in a homotypic calcium-dependent manner with E-cadherin molecules on neighboring cells, thereby facilitating cell–cell contact such as epithelial islands formed by epithelial cells. The E-cadherin gene was introduced into mesenchymal mouse fibroblast NIH3T3 cells to engineer heterotypic interactions between hepatocytes and mesenchymal cells, and the effects of coculture of E-cadherin-expressing NIH3T3 cells (designated as 3T3/E-cad) with hepatocytes[93] were investigated. In coculture with hepatocytes, 3T3/E-cad cells were incorporated into the cell islands formed by hepatocytes, and the frequency of cell–cell contacts between 3T3/E-cad cells and hepatocytes was enhanced. Furthermore, the frequency of cell–cell interactions between hepatocytes and fibroblasts promoted the expression of liver-specific functions in cultured hepatocytes, and the liver function was enhanced by the forced adhesion between 3T3/E-cad cells and hepatocytes.[93]

As an alternative method to enhance cell–cell interactions, a physical approach using magnetic attraction was applied. Moreover, the assembly of 3D tissues containing various cell types remains a challenge, and novel technologies are required to reconstruct the liver to function as it does *in vivo*. This would require a 3D construct containing various types of cells that could thrive beyond the cell type limitations of coculture. Two major difficulties obstruct the fabrication of 3D cocultures of heterotypic cells. One is nonadherence to heterotypic cells caused by the cell type limitation of coculture. To overcome this difficulty, Harimoto *et al.*[95] developed double-layered cocultures by cell sheet engineering. They reported that while trypsinized single endothelial cells do not attach to hepatocytes, endothelial cell sheets fabricated by cell sheet engineering could attach to hepatocyte monolayers. Another difficulty is to spatially control the positioning of target cells. Mironov *et al.*[40,96] developed a computer-aided jet-based cell printer which could place cells at specific sites on thermoresponsive gels and termed this "organ printing" to overcome the difficulties with spatial control. Alternatively, multilayered cocultures using a Mag-TE technique were developed. This section describes coculture methods using magnetic force to place magnetically labeled cells onto target cells and to promote heterotypic cell–cell adhesion to form a 3D construct.

1. Heterotypic Layers of Cocultured Hepatocytes and Endothelial Cells

Coculture systems of freshly isolated hepatocytes with nonparenchymal cells maintain hepatocyte functions for long periods.[92] The authors cocultured hepatocytes with endothelial cells by a layered coculture system using Mag-TE.[42] For this purpose, human aortic endothelial cells (HAECs) were magnetically labeled using MCLs, and then the labeled HAECs were placed onto a rat hepatocyte layer using magnetic force in order to investigate whether magnetic force promotes the adhesion of heterotypic cells.[42] HAECs accumulated on the hepatocyte layer in the center of the well in the same culture area where the smaller magnets were placed. When the excessive number of cells compared with the size of the magnets was applied, a multilayer of HAECs was formed on the hepatocyte monolayers where the magnet was located. HAECs labeled with or without MCLs did not attach to the hepatocytes in the absence of the magnet. These results suggest that the magnetic force can regulate the culture space and that multilayered constructs can be generated. In the presence of the magnet, HAECs remained attached to the hepatocyte layer when the magnet was removed on day 2 after coculture. Next, HAECs were seeded onto hepatocyte monolayers and a sufficiently large magnet compared with the culture area placed to form double cell layers throughout the wells.[42] In the presence of the magnet, HAECs evenly attached to the hepatocyte layer throughout the wells. Then, albumin expression of hepatocytes was measured to determine cellular function in the layered coculture system.[42] Albumin secretion in the single hepatocyte culture was undetectable beyond 5 days of culture. Albumin secretion was slightly enhanced in cocultures of HAECs, even in the absence of the magnet. On the other hand, layered cocultures in the presence of a magnetic force maintained a high level of albumin secretion at least for 8 days.

Although the precise mechanisms by which nonparenchymal cells modulate the hepatocyte phenotype remain to be elucidated, some new insights on the modes of cell signaling, the extent of cell–cell interaction, and the ratio of cell populations have been reported. For example, Bhandari *et al.*[97] reported that 3T3 fibroblast cells persist in cocultures with hepatocytes, but 3T3 cell-conditioned medium did not substitute for viable cocultured 3T3 cells in preserving hepatocyte function, suggesting that cell–cell interaction is essential for modulating hepatocyte functions. Potential mediators of cell–cell interactions include soluble factors such as cytokines and insoluble cell-associated factors such as ECMs. Chia *et al.*[98] have reported that TGF-β1 regulation in a coculture of hepatocytes with NIH3T3 cells within a 3D microenvironment is important for enhanced hepatocyte function. Moreover, recent advance in microfabrication has allowed for more precise control over cell–cell interactions. In 2007, Hui and Bhatia[99] demonstrated that maintenance of hepatocellular phenotypes by NIH3T3 cells required direct contact for a limited time (~ 18 h) followed by a sustained signal with an effective range of < 400 μm. Continued advances in microfabrication will allow further study of the role of cell communication in physiological processes. It is speculated that the tight and close interaction of cells using magnetic force caused deposition of ECMs and cytokines between the cell layers, thus enhancing liver function. Taken together, Mag-TE allows the spatial control of target cells and the adherence of heterotypic cells using magnetic force.

2. Construction of Heterotypic Cell Sheets of HepG2 and NIH3T3 Cells

As mentioned in Section IV.C.1, cell–cell interaction should be formed to promote the expression of liver functions. Therefore, although tissue engineering has been based on the seeding of cells onto 3D biodegradable scaffolds to reconstruct their native structure, a 3D cell construct without artificial scaffolds may be more suitable for liver tissue engineering. Therefore, magnetic attraction was used as a physical approach for enhancing layered cell–cell interactions. This section describes the heterotypic cell sheets of HepG2 (hepatocyte model) and NIH3T3 cells (stromal fibroblast model) constructed by a Mag-TE technique.[48] Magnetically labeled HepG2 and NIH3T3 cells using MCLs were both used to construct a heterotypic cell sheet. Fibroblasts are an integral component of all tissues; they contribute to tissue architecture by producing ECM that serves as scaffolding for various organ structures. In addition, fibroblasts are a rich source of growth factors for the self-stimulation and activation of other cell types in the microenvironment, and mouse 3T3 cells have been shown to induce a high level of albumin secretion by hepatocytes.[97] The authors then investigated whether MCL-labeled NIH3T3 cells could form a multilayered cell sheet containing sufficient ECMs to act as a stroma for HspG2 cells, and whether the 3D coculture of HepG2 cells with NIH3T3 cells enhanced albumin secretion, in order to assess the feasibility of this coculture system.

When MCL-labeled HepG2 or NIH3T3 cells were seeded onto the wells of ultralow-attachment plates and a magnet was placed under the plate, cells were rapidly attracted to the magnet and formed a sheetlike structure. In contrast, both HepG2 and NIH3T3 cells in the absence of a magnet did not form evenly contiguous cell sheets; nor did they attach to the surface of ultralow-attachment plates, but rather formed small cell aggregates. When the magnet was removed from the bottom of the plates, NIH3T3 cell sheets were detached from the bottom of the well of ultralow-attachment plates. Similar to keratinocyte sheets (Section IV.A.1) and myoblast cell sheets (Section IV.B.1), the NIH3T3 cell sheets had a sufficient strength. When a bar magnet was positioned on the surface of the culture medium, the NIH3T3 cell sheets floated up to the surface of the culture medium without disruption and stuck to the top of the magnet. In contrast, HepG2 cell sheets (as well as primary hepatocyte sheets in the authors' unpublished result) constructed by Mag-TE were broken, suggesting that HepG2 cell sheets (and primary hepatocyte sheets) are not sufficiently strong for recovery. Because ECMs produced by cells can provide a mechanical strength, ECM deposition within the NIH3T3 cell sheets was examined. Histological study revealed that NIH3T3 cells formed 7- to 8-layered cell sheets with a thickness of approximately 60 μm containing ECM components such as fibronectin and type I collagen.[48] Next, in order to construct a double-layered HepG2–NIH3T3 structure, HepG2 cells magnetically labeled with MCLs were seeded onto the NIH3T3 cell sheets (Fig. 8A).[48] Due to NIH3T3 cells, the resultant cell sheets had a sufficient strength to be recovered, and HepG2 cell layers were found to be attached to the NIH3T3 cell sheets, resulting in a double-layered cell sheet of HepG2 and NIH3T3 cells (designated as layered cell sheets). Alternatively, in order to construct "mixed cell sheets," magnetically labeled HepG2 and NIH3T3 cells were admixed and seeded into an ultralow-attachment plate (Fig. 8B).[48] Similar to layered cell sheets, mixed cell sheets were formed and showed a sufficient strength after 24 h of culture in the presence of a magnet. Cross sections of the cell sheets revealed that NIH3T3 cells and HepG2 cells were scattered within the cell sheets, and they formed cell sheets whose thickness was approximately equal to that of the layered cell sheet mentioned earlier.

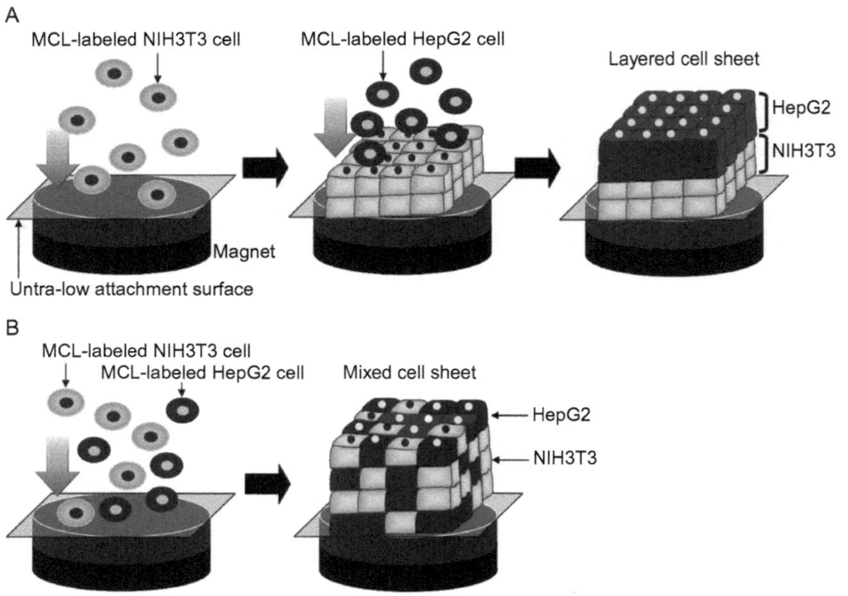

FIG. 8. Construction of heterotypic cell sheets of HepG2 and NIH3T3 cells. (A) Schematic illustration of the construction of "layered cell sheets." (B) Schematic illustration to construct "mixed cell sheets."

The hepatic albumin expression level was then measured to determine cellular function in the 3D coculture system constructed by Mag-TE.[48] Albumin secretion was enhanced in HepG2 cell sheets as compared with that in a HepG2 monolayer cells. It is known that spheroid culture is a possible approach to construct a 3D structure of hepatocytes and that albumin secretion is enhanced in the 3D spheroid culture as compared with that in the 2D monolayer culture. It is possible to say that cell sheets constructed by Mag-TE are a designed cell construct induced by magnetic force, rather than a spontaneously formed spheroid. Since designed structures can be constructed by Mag-TE, a structure with substantial thickness, which is impossible for spontaneously induced spheroids, could be constructed. As compared with HepG2 cell sheets, a slight but not significant enhancement of albumin secretion was observed in the layered cell sheets, whereas a high level of albumin secretion was observed in the mixed cell sheet.[48] For the layered cell sheets, heterotypic cell interactions could be achieved only on the boundary area between HepG2 and NIH3T3 cell sheets; on the other hand, for the mixed cell sheets, heterotypic cell interactions between HepG2 and NIH3T3 cells could be achieved throughout the cell sheets. Thus, it was speculated that the difference of albumin secretion between the layered cell sheets and the mixed cell sheets was caused by the frequency of cell–cell contacts and interactions between HepG2 and NIH3T3 cells.

As mentioned in Section IV.C.1, a magnetic force was applied to construct a layered coculture system of rat hepatocytes and HAECs. When rat hepatocytes and HAECs were cocultured on the surface of ultralow-attachment plates, the cells formed sheetlike structures but the cell sheets were not sufficiently strong for recovery. On the other hand, the cell sheets acquired sufficient strength using NIH3T3 fibroblasts producing ECM proteins, which may strengthen cell–cell interactions. The strength of the cell sheets is an important factor for tissue-engineered grafts. Generally, tissue-engineered grafts of scaffoldless cellular sheets including Mag-TE grafts, cell sheet-engineered grafts by Okano's method,[57] or cultured epidermis by Green's method[59] described in Section IV.A.1 are easily broken by physiological loads and it has been difficult to handle the grafts in transplantation. Therefore, the use of stromal cells producing ECM may be a promising approach for preparing scaffoldless cellular sheets. This section described a 3D heterotypic coculture system by using the Mag-TE technique, which may be applied to tissue engineering and cell biology to study cell–cell interactions.

D. Construction of Complex 3D Tissues

As described earlier, it is important to mimic the natural microenvironment for the construction of functional tissue substitutes. Because cells in normal tissues and organs are orderly arranged with surrounding homotypic and/or heterotypic cells to exert native functions, novel techniques to engineer more complex 3D tissues should be developed. The creation of vascularized tissues must be a first step to the engineering of more complex and thick tissue architectures. Since tissue-engineered grafts have a thickness limit of 100–200 μm, the limit must be overcome by creating functional blood vessels to supply cells with oxygen and nutrients and to remove waste products. One possible solution is vascularization of tissue-engineered constructs before transplantation. In this regard, Levenberg *et al.*[80] engineered 3D vascularized skeletal muscle constructs using myoblast cells, embryonic fibroblast cells, and vascular endothelial cells. It is worth noting that multicell cultures are difficult to handle but are very effective to create complex 3D tissues.

One of the most promising approaches to create vascularized tissues is the patterning of vascular endothelial cells. As mentioned in Section III, in principle, using the magnetic cell manipulation technique, cell patterns can be created irrespective of surface conditions. This section describes applications of the magnetic cell-patterning technique to form cell patterns on a monolayer of cells and multilayered cell sheets. Based on these techniques, the authors further attempted to fabricate complex 3D tissue constructs by combining the magnetic force-based cell sheet formation and cell-patterning procedures.

1. Cell Patterning onto Monolayer of Cells and Tissues

Magnetic field gradient concentrators with various thicknesses of magnetized steel plates sandwiched between acrylic resin plates were used to attract magnetically labeled target cells to the desired position for cell patterning. After C2C12 cells attained confluence, the culture dishes were placed on the cell-patterning devices (magnetic field gradient concentrators with a magnet), and then MCL-labeled HUVECs were seeded to the dishes. When a steel plate with thicknesses of 10 or 30 μm was used, the formation of a line of single cells was achieved,[100] indicating that this method facilitates single cell manipulation to construct tissue architectures. The use of steel plates with thicknesses of 100 and 200 μm gave rise to line widths of cell patterns that were consistent with the thickness of the steel plates.[100] These results indicated that line widths of cell patterns could be controlled by the thickness of the magnetized steel plates.

Alignment of cells in the direction of force production is important in skeletal muscle tissue engineering. As described in Section IV.B.2, when MCL-labeled C2C12 cells were linearly patterned on ultralow cell attachment surface using linear magnetic field gradient concentrators with a magnet, cells formed a linear construct but shrunk afterward. As an alternative method to support tissue retention during culture, fibroblast NIH3T3 cells were used as a stromal layer.[101] A small piece (2 × 2 mm) of collagen film was placed in two places in a 35-mm glass-bottom dish using fibrin as glue, and NIH3T3 cells were seeded and cultured until confluence. When MCL-labeled C2C12 cells were seeded onto a confluent NIH3T3 cell layer in the presence of the linear magnetic field gradient concentrator with a magnet, C2C12 cells were successfully patterned linearly on the NIH3T3 cell monolayer. Most of the C2C12 cells were restricted at the position where the magnetic field was present and attached onto the NIH3T3 cell monolayer. After 5–10 days from induction of differentiation, the NIH3T3 layer started to detach from the bottom of the dish in a sheetlike manner maintaining the cell–cell attachment. Detachment of the NIH3T3 cell layer rapidly progressed, but the place where the collagen film was present remained attached. After 3 days from the initiation of detachment of the NIH3T3 cell layer, a cylindrical construct was formed. The histological studies revealed the presence of multinucleated cells within the construct, indicating that myotubes were formed by cell fusion.[101] Myotubes oriented parallel to each other due to the linear patterning of C2C12 cells and probably internal force within the construct. In addition, immunofluorescent observation revealed the presence of sarcomere striation.[101] Western blot analysis showed that muscle proteins such as myogenin, myosin heavy chain, myosin light chain 2, and tropomyosin were expressed in the construct.[101] When the construct was stimulated by electric pulse, the construct exhibited active tension of $\sim 1\ \mu N$.[101] These results demonstrate that functional skeletal muscle tissue was formed by patterning of C2C12 cells onto NIH3T3 cells and allowing the NIH3T3 cell sheet to detach from the culture surface.

Because the magnetic force-based cell patterning is not limited to applications involving a monolayer of cells, the authors attempted to fabricate the skins on which HUVECs were patterned.[100] When MCL-labeled HUVECs were seeded onto skins dissected from neonatal rats or mice and the magnetic field gradient concentrator (thickness 200 μm) was placed, cell patterns with the line width of about 200 μm were formed on the skins. In this case, when steel plates with thicknesses less than 200 μm were used, cell patterns were not formed, which may be due to insufficient magnetic induction caused by the thick skin tissues. The thicknesses of dissected skins used in this study were 343 μm and 238 μm for rats and mice, respectively. In the authors' experience, cell patterns were not formed on thick skins dissected from adult rat (skin thickness 577 μm), suggesting that it is necessary to use stronger magnetic fields (>4000 G) for cell patterning on thick tissues such as human skins.

2. Incorporation of Patterned Vascular Endothelial Cells into 3D Tissue Constructs

Combining the magnetic force-based tissue fabrication technique and the magnetic patterning of cells, the authors fabricated 3D tissue constructs in which cellular organization is controlled by magnetic force.[100] First, the fabrication of cell sheets possessing patterned HUVECs was attempted. As described in Section IV.B.1 and Fig. 6, when MCL-labeled C2C12 cells were seeded into a culture dish in the presence of a magnetic field, the cells were rapidly attracted to the bottom of the dish and accumulated uniformly within the culture surface, and sediments formed cell sheets after 24 h of incubation. To fabricate arbitrary cell patterns on the cell sheets, magnetic field gradient concentrators with the letters "M," "A," or "G" were used as cell-patterning devices. The dish in which a cell sheet would form was placed on the magnetic force gradient concentrator with a magnet, and magnetically labeled HUVECs were seeded onto the cell sheets, resulting in successful patterning on the letter "M," "A," or "G" on C2C12 cell sheets.

In addition, the authors attempted to fabricate patterns of HUVECs within 3D tissue constructs.[100] The procedure is shown in Fig. 9A. C2C12 cells labeled with MCLs were rapidly attracted to the magnet and accumulated within the culture area to form a uniform multilayer of cells. Subsequently, MCL-labeled HUVECs were patterned on the C2C12 cell layer using the lined magnetic field gradient concentrator. After the magnetic seeding of C2C12 cells onto the first patterning of HUVECs, HUVECs were patterned again

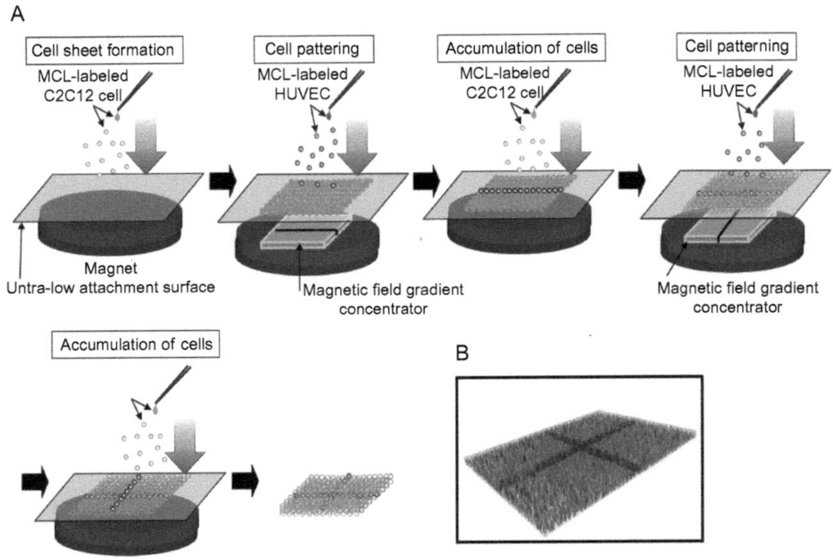

FIG. 9. Incorporation of patterned vascular endothelial cells into 3D tissue constructs. (A) Schematic illustration. (B) HUVECs and C2C12 cells were prestained with orange and green fluorescent probe, respectively, and the micrograph of resulting 3D tissue constructs were obtained by fluorescent microscopy. A merged and omnifocal image of a 3D view of the tissue constructs is shown.

using the magnetic field gradient concentrator. Then, C2C12 cells were magnetically layered onto the second pattern of HUVECs. Histological examinations of the tissue construct revealed that the first and second patterns of HUVECs were successfully created on the cell layers.[100] The 3D analysis of the tissue construct revealed that HUVECs were patterned to form a cross line in a C2C12 cell layer, and the cross pattern of HUVECs was embedded into the 3D tissue construct (Fig. 9B). These results suggest that the combination of magnetic force-based tissue fabrication and cell patterning is a promising approach to construct complex 3D tissue substitutes required for tissue engineering.

V. Conclusion

This chapter highlighted magnetofection, magnetic patterning of cells, and construction of 3D tissue-like structures. Among them, Mag-TE for constructing 3D structures has been extensively studied, and various kinds of other tissues such as retinal pigment epithelial cell sheets,[102] MSC sheets,[44] and cardiomyocyte sheets,[46] have been already generated. Tubular structures consisting of heterotypic layers of endothelial cells, smooth muscle cells, and fibroblasts have also been created.[43] In this approach, magnetically labeled cells formed a cell sheet onto which a cylindrical magnet was rolled, which was removed after a tubular structure was formed. If these processes can be scaled up, there is great potential for these techniques in the treatment of a variety of diseases and defects.

In the translational research, toxicology of functional magnetite nanoparticles is an important issue. The main requisite for a cell-labeling technique is to preserve the normal cell behavior. As for biocompatibility of MCLs, no toxic effects against proliferation of several cell types were observed within the range of magnetite concentrations tested (e.g., human keratinocytes,[63] < 50 pg-magnetite/cell; HUVECs,[41] HAECs,[42] human dermal fibroblasts,[41] human smooth muscle cells,[43] mouse fibroblast cells,[43] canine urothelial cells,[43] human MSCs,[44] and rat MSCs[45] < 100 pg/cell). Moreover, MCLs did not compromise MSC differentiation[44,45] or electrical connections of cardiomyocytes.[46] In addition, an *in vivo* toxicity of magnetite nanoparticles has been extensively studied. As an MRI contrast agent, Resovist^R^ was first applied clinically for detecting liver cancer, since Resovist^R^ is taken up rapidly by the reticuloendothelial system such as Kupffer cells of the liver compared with the uptake by cancer cells of the liver. In a preliminary study,[103] the authors investigated the toxicity of systemically administered MCLs (90 mg, i.p.) in mice; none of the 10 mice injected with MCLs died during the study. Transient accumulation of magnetite was observed in the liver and spleen of the mice, but the magnetite nanoparticles had been cleared from circulation by hepatic Kupffer cells in the spleen by the 10th day after administration.[103]

In conclusion, magnetic nanoparticles have been developed into "functional" magnetite nanoparticles which are highly promising tools for a wide spectrum of applications in tissue engineering. The proven lack of toxicity of the functional magnetite nanoparticles is expected to provide exciting tools in the near future for clinical tissue engineering and regenerative medicine.

References

1. Norell H, Zhang Y, McCracken J, Martins da Palma T, Lesher A, Liu Y, et al. CD34-based enrichment of genetically engineered human T cells for clinical use results in dramatically enhanced tumor targeting. *Cancer Immunol Immunother* 2010;**59**:851–62.
2. Hua MY, Yang HW, Chuang CK, Tsai RY, Chen WJ, Chuang KL, et al. Magnetic-nanoparticle-modified paclitaxel for targeted therapy for prostate cancer. *Biomaterials* 2010;**31**:7355–63.
3. Cheng K, Li TS, Malliaras K, Davis DR, Zhang Y, Marbán E. Magnetic targeting enhances engraftment and functional benefit of iron-labeled cardiosphere-derived cells in myocardial infarction. *Circ Res* 2010;**106**:1570–81.

4. Gao GH, Im GH, Kim MS, Lee JW, Yang J, Jeon H, et al. Magnetite-nanoparticle-encapsulated pH-responsive polymeric micelle as an MRI probe for detecting acidic pathologic areas. *Small* 2010;**6**:1201–4.
5. Sato M, Yamashita T, Ohkura M, Osai Y, Sato A, Takada T, et al. N-propionyl-cysteaminyl-phenol-magnetite conjugate (NPrCAP/M) is a nanoparticle for the targeted growth suppression of melanoma cells. *J Invest Dermatol* 2009;**129**:2233–41.
6. Ito A, Shinkai M, Honda H, Kobayashi T. Medical application of functionalized magnetic nanoparticles. *J Biosci Bioeng* 2005;**100**:1–11.
7. Corchero JL, Villaverde A. Biomedical applications of distally controlled magnetic nanoparticles. *Trends Biotechnol* 2009;**27**:468–76.
8. Shinkai M, Yanase M, Honda H, Wakabayashi T, Yoshida J, Kobayashi T. Intracellular hyperthermia for cancer using magnetite cationic liposomes: in vitro study. *Jpn J Cancer Res* 1996;**87**:1179–83.
9. Pierschbacher MD, Ruoslahti E. Cell attachment activity of fibronectin can be duplicated by small synthetic fragments of the molecule. *Nature* 1984;**309**:30–3.
10. Ruoslahti E, Pierschbacher MD. New perspectives in cell adhesion: RGD and integrins. *Science* 1987;**238**:491–7.
11. Ito A, Ino K, Kobayashi T, Honda H. The effect of RGD peptide-conjugated magnetite cationic liposomes on cell growth and cell sheet harvesting. *Biomaterials* 2005;**26**:6185–93.
12. Satomi T, Nagasaki Y, Kobayashi H, Otsuka H, Kataoka K. Density control of poly(ethylene glycol) layer to regulate cellular attachment. *Langmuir* 2007;**23**:6698–703.
13. Satomi T, Nagasaki Y, Kobayashi H, Tateishi T, Kataoka K, Otsuka H. Physicochemical characterization of densely packed poly(ethylene glycol) layer for minimizing nonspecific protein adsorption. *J Nanosci Nanotechnol* 2007;**7**:2394–9.
14. Akiyama H, Ito A, Kawabe Y, Kamihira M. Cell-patterning using poly (ethylene glycol)-modified magnetite nanoparticles. *J Biomed Mater Res A* 2010;**92**:1123–30.
15. Langer R, Vacanti JP. Tissue engineering. *Science* 1993;**260**:920–6.
16. Takahashi K, Yamanaka S. Induction of pluripotent stem cells from mouse embryonic and adult fibroblast cultures by defined factors. *Cell* 2006;**126**:663–76.
17. Cartmell SH, Dobson J, Verschueren SB, El Haj AJ. Development of magnetic particle techniques for long-term culture of bone cells with intermittent mechanical activation. *IEEE Trans Nanobioscience* 2002;**1**:92–7.
18. Dobson J, Cartmell SH, Keramane A, El Haj AJ. Principles and design of a novel magnetic force mechanical conditioning bioreactor for tissue engineering, stem cell conditioning, and dynamic in vitro screening. *IEEE Trans Nanobioscience* 2006;**5**:173–7.
19. Alsberg E, Feinstein E, Joy MP, Prentiss M, Ingber DE. Magnetically-guided self-assembly of fibrin matrices with ordered nano-scale structure for tissue engineering. *Tissue Eng* 2006;**12**:3247–56.
20. Wilhelm C, Bal L, Smirnov P, Galy-Fauroux I, Clément O, Gazeau F, et al. Magnetic control of vascular network formation with magnetically labeled endothelial progenitor cells. *Biomaterials* 2007;**28**:3797–806.
21. Sheyn D, Mizrahi O, Benjamin S, Gazit Z, Pelled G, Gazit D. Genetically modified cells in regenerative medicine and tissue engineering. *Adv Drug Deliv Rev* 2010;**62**:683–98.
22. Hojman P, Gissel H, Gehl J. Sensitive and precise regulation of haemoglobin after gene transfer of erythropoietin to muscle tissue using electroporation. *Gene Ther* 2007;**14**:950–9.
23. Felgner PL, Gadek TR, Holm M, Roman R, Chan HW, Wenz M, et al. Lipofection: a highly efficient, lipid-mediated DNA-transfection procedure. *Proc Natl Acad Sci USA* 1987;**84**:7413–7.
24. Scherer F, Anton M, Schillinger U, Henke J, Bergemann C, Krüger A, et al. Magnetofection: enhancing and targeting gene delivery by magnetic force in vitro and in vivo. *Gene Ther* 2002;**9**:102–9.

25. Mykhaylyk O, Antequera YS, Vlaskou D, Plank C. Generation of magnetic nonviral gene transfer agents and magnetofection in vitro. *Nat Protoc* 2007;**2**:2391–411.
26. Dobson J. Gene therapy progress and prospects: magnetic nanoparticle-based gene delivery. *Gene Ther* 2006;**13**:283–7.
27. Krötz F, Sohn HY, Gloe T, Plank C, Pohl U. Magnetofection potentiates gene delivery to cultured endothelial cells. *J Vasc Res* 2003;**40**:425–34.
28. Lee CH, Kim EY, Jeon K, Tae JC, Lee KS, Kim YO, et al. Simple, efficient, and reproducible gene transfection of mouse embryonic stem cells by magnetofection. *Stem Cells Dev* 2008;**17**:133–41.
29. Ino K, Kawasumi T, Ito A, Honda H. Plasmid DNA transfection using magnetite cationic liposomes for construction of multilayered gene-engineered cell sheet. *Biotechnol Bioeng* 2008;**100**:168–76.
30. Fujii Y, Kachi S, Ito A, Kawasumi T, Honda H, Terasaki H. Transfer of gene to human retinal pigment epithelial cells using magnetite cationic liposomes. *Br J Ophthalmol* 2010;**94**:1074–7.
31. Hughes C, Galea-Lauri J, Farzaneh F, Darling D. Streptavidin paramagnetic particles provide a choice of three affinity-based capture and magnetic concentration strategies for retroviral vectors. *Mol Ther* 2001;**3**:623–30.
32. Ito A, Takahashi T, Kameyama Y, Kawabe Y, Kamihira M. Magnetic concentration of a retroviral vector using magnetite cationic liposomes. *Tissue Eng Part C Methods* 2009;**15**:57–64.
33. Akiyama H, Ito A, Kawabe Y, Kamihira M. Genetically engineered angiogenic cell sheets using magnetic force-based gene delivery and tissue fabrication techniques. *Biomaterials* 2010;**31**:1251–9.
34. Davis HE, Morgan JR, Yarmush ML. Polybrene increases retrovirus gene transfer efficiency by enhancing receptor-independent virus adsorption on target cell membranes. *Biophys Chem* 2002;**97**:159–72.
35. Pickles RJ, Fahrner JA, Petrella JM, Boucher RC, Bergelson JM. Retargeting the coxsackievirus and adenovirus receptor to the apical surface of polarized epithelial cells reveals the glycocalyx as a barrier to adenovirus-mediated gene transfer. *J Virol* 2000;**74**:6050–7.
36. Lübbe AS, Bergemann C, Riess H, Schriever F, Reichardt P, Possinger K, et al. Clinical experiences with magnetic drug targeting: a phase I study with 4′-epidoxorubicin in 14 patients with advanced solid tumors. *Cancer Res* 1996;**56**:4686–93.
37. Liu WF, Chen CS. Cellular and multicellular form and function. *Adv Drug Deliv Rev* 2007;**59**:1319–28.
38. Falconnet D, Csucs G, Grandin HM, Textor M. Surface engineering approaches to micropattern surfaces for cell-based assays. *Biomaterials* 2006;**27**:3044–63.
39. Nakamura M, Kobayashi A, Takagi F, Watanabe A, Hiruma Y, Ohuchi K, et al. Biocompatible inkjet printing technique for designed seeding of individual living cells. *Tissue Eng* 2005;**11**:1658–66.
40. Mironov V, Visconti RP, Kasyanov V, Forgacs G, Drake CJ, Markwald RR. Organ printing: tissue spheroids as building blocks. *Biomaterials* 2009;**30**:2164–74.
41. Ino K, Ito A, Kumazawa H, Kagami H, Ueda M, Honda H. Incorporation of capillary-like structures into dermal cell sheets constructed by magnetic force-based tissue engineering. *J Chem Eng Jpn* 2007;**40**:51–8.
42. Ito A, Takizawa Y, Honda H, Hata K, Kagami H, Ueda M, et al. Tissue engineering using magnetite nanoparticles and magnetic force: heterotypic layers of cocultured hepatocytes and endothelial cells. *Tissue Eng* 2004;**10**:833–40.
43. Ito A, Ino K, Hayashida M, Kobayashi T, Matsunuma H, Kagami H, et al. Novel methodology for fabrication of tissue-engineered tubular constructs using magnetite nanoparticles and magnetic force. *Tissue Eng* 2005;**11**:1553–61.

44. Shimizu K, Ito A, Yoshida T, Yamada Y, Ueda M, Honda H. Bone tissue engineering with human mesenchymal stem cell sheets constructed using magnetite nanoparticles and magnetic force. *J Biomed Mater Res B Appl Biomater* 2007;**82**:471–80.
45. Shimizu K, Ito A, Honda H. Mag-seeding of rat bone marrow stromal cells into porous hydroxyapatite scaffolds for bone tissue engineering. *J Biosci Bioeng* 2007;**104**:171–7.
46. Shimizu K, Ito A, Lee JK, Yoshida T, Miwa K, Ishiguro H, et al. Construction of multi-layered cardiomyocyte sheets using magnetite nanoparticles and magnetic force. *Biotechnol Bioeng* 2007;**96**:803–9.
47. Yamamoto Y, Ito A, Kato M, Kawabe Y, Shimizu K, Fujita H, et al. Preparation of artificial skeletal muscle tissues by a magnetic force-based tissue engineering technique. *J Biosci Bioeng* 2009;**108**:538–43.
48. Ito A, Jitsunobu H, Kawabe Y, Kamihira M. Construction of heterotypic cell sheets by magnetic force-based 3-D coculture of HepG2 and NIH3T3 cells. *J Biosci Bioeng* 2007;**104**:371–8.
49. Le B, Shinkai M, Kitade T, Honda H, Yoshida J, Wakabayashi T, et al. Preparation of tumor-specific magnetoliposomes and their application for hyperthermia. *J Chem Eng Jpn* 2001;**34**:66–72.
50. Shinkai M, Le B, Honda H, Yoshikawa K, Shimizu K, Saga S, et al. Targeting hyperthermia for renal cell carcinoma using human MN antigen-specific magnetoliposomes. *Jpn J Cancer Res* 2001;**92**:1138–45.
51. Wilhelm C, Gazeau F. Universal cell labelling with anionic magnetic nanoparticles. *Biomaterials* 2008;**29**:3161–74.
52. Ino K, Ito A, Honda H. Cell patterning using magnetite nanoparticles and magnetic force. *Biotechnol Bioeng* 2007;**97**:1309–17.
53. Ito A, Akiyama H, Kawabe Y, Kamihira M. Magnetic force-based cell patterning using Arg-Gly-Asp (RGD) peptide-conjugated magnetite cationic liposomes. *J Biosci Bioeng* 2007;**104**:288–93.
54. Shu XZ, Ghosh K, Liu Y, Palumbo FS, Luo Y, Clark RA, et al. Attachment and spreading of fibroblasts on an RGD peptide-modified injectable hyaluronan hydrogel. *J Biomed Mater Res* 2004;**68**:365–75.
55. Burdick JA, Khademhosseini A, Langer R. Fabrication of gradient hydrogels using a microfluidics/photopolymerization process. *Langmuir* 2004;**20**:5153–6.
56. Avgoustakis K. Pegylated poly(lactide) and poly(lactide-co-glycolide) nanoparticles: preparation, properties and possible applications in drug delivery. *Curr Drug Deliv* 2004;**1**:321–33.
57. Yang J, Yamato M, Shimizu T, Sekine H, Ohashi K, Kanzaki M, et al. Reconstruction of functional tissues with cell sheet engineering. *Biomaterials* 2007;**28**:5033–43.
58. Rheinwald JG, Green H. Serial cultivation of strains of human epidermal keratinocytes: the formation of keratinizing colonies from single cells. *Cell* 1975;**6**:331–43.
59. Green H, Kehinde O, Thomas J. Growth of cultured human epidermal cells into multiple epithelia suitable for grafting. *Proc Natl Acad Sci USA* 1979;**76**:5665–8.
60. Bell E, Ehrlich HP, Buttle DJ, Nakatsuji T. Living tissue formed in vitro and accepted as skin-equivalent tissue of full thickness. *Science* 1981;**211**:1052–4.
61. O'Connor NE, Mulliken JB, Banks-Schlegel S, Kehinde O, Green H. Grafting of burns with cultured epithelium prepared from autologous epidermal cells. *Lancet* 1981;**1**:75–8.
62. Hull BE, Finley RK, Miller SF. Coverage of full-thickness burns with bilayered skin equivalents: a preliminary clinical trial. *Surgery* 1990;**107**:496–502.
63. Ito A, Hayashida M, Honda H, Hata K, Kagami H, Ueda M, et al. Construction and harvest of multilayered keratinocyte sheets using magnetite nanoparticles and magnetic force. *Tissue Eng* 2004;**10**:873–80.

64. Lei P, Ogunade A, Kirkwood KL, Laychock SG, Andreadis ST. Efficient production of bioactive insulin from human epidermal keratinocytes and tissue-engineered skin substitutes: implications for treatment of diabetes. *Tissue Eng* 2007;**13**:2119–31.
65. Eming SA, Yarmush ML, Morgan JR. Enhanced function of cultured epithelium by genetic modification: cell-based synthesis and delivery of growth factors. *Biotechnol Bioeng* 1996;**52**:15–23.
66. Dickens S, Vermeulen P, Hendrickx B, Van den Berge S, Vranckx JJ. Regulable vascular endothelial growth factor165 overexpression by ex vivo expanded keratinocyte cultures promotes matrix formation, angiogenesis, and healing in porcine full-thickness wounds. *Tissue Eng Part A* 2008;**14**:19–27.
67. Lien S, Lowman HB. Therapeutic peptides. *Trends Biotechnol* 2003;**21**:556–62.
68. Lehrer RI, Ganz T. Defensins of vertebrate animals. *Curr Opin Immunol* 2002;**14**:96–102.
69. García JR, Jaumann F, Schulz S, Krause A, Rodríguez-Jiménez J, Forssmann U, et al. Identification of a novel, multifunctional beta-defensin (human beta-defensin 3) with specific antimicrobial activity. Its interaction with plasma membranes of Xenopus oocytes and the induction of macrophage chemoattraction. *Cell Tissue Res* 2001;**306**:257–64.
70. Carretero M, Del Río M, García M, Escámez MJ, Mirones I, Rivas L, et al. A cutaneous gene therapy approach to treat infection through keratinocyte-targeted overexpression of antimicrobial peptides. *FASEB J* 2004;**18**:1931–3.
71. Sawamura D, Goto M, Shibaki A, Akiyama M, McMillan JR, Abiko Y, et al. Beta defensin-3 engineered epidermis shows highly protective effect for bacterial infection. *Gene Ther* 2005;**12**:857–61.
72. Ito A, Takahashi T, Kawabe Y, Kamihira M. Human beta defensin-3 engineered keratinocyte sheets constructed by a magnetic force-based tissue engineering technique. *J Biosci Bioeng* 2009;**108**:244–7.
73. Law PK, Goodwin TG, Fang Q, Hall TL, Quinley T, Vastagh G, et al. First human myoblast transfer therapy continues to show dystrophin after 6 years. *Cell Transplant* 1997;**6**:95–100.
74. Herr H, Dennis RG. A swimming robot actuated by living muscle tissue. *J Neuroeng Rehabil* 2004;**1**:6.
75. Feinberg AW, Feigel A, Shevkoplyas SS, Sheehy S, Whitesides GM, Parker KK. Muscular thin films for building actuators and powering devices. *Science* 2007;**317**:1366–70.
76. Tanaka Y, Morishima K, Shimizu T, Kikuchi A, Yamato M, Okano T, et al. An actuated pump on-chip powered by cultured cardiomyocytes. *Lab Chip* 2006;**6**:362–8.
77. Allen RE, Temm-Grove CJ, Sheehan SM, Rice G. Skeletal muscle satellite cell cultures. *Methods Cell Biol* 1997;**52**:155–76.
78. Gawlitta D, Boonen KJ, Oomens CW, Baaijens FP, Bouten CV. The influence of serum-free culture conditions on skeletal muscle differentiation in a tissue-engineered model. *Tissue Eng Part A* 2008;**14**:161–71.
79. Yan W, George S, Fotadar U, Tyhovych N, Kamer A, Yost MJ, et al. Tissue engineering of skeletal muscle. *Tissue Eng* 2007;**13**:2781–90.
80. Levenberg S, Rouwkema J, Macdonald M, Garfein ES, Kohane DS, Darland DC, et al. Engineering vascularized skeletal muscle tissue. *Nat Biotechnol* 2005;**23**:879–84.
81. Giraud MN, Ayuni E, Cook S, Siepe M, Carrel TP, Tevaearai HT. Hydrogel-based engineered skeletal muscle grafts normalize heart function early after myocardial infarction. *Artif Organs* 2008;**32**:692–700.
82. Dennis RG, Kosnik 2nd PE. Excitability and isometric contractile properties of mammalian skeletal muscle constructs engineered in vitro. *In Vitro Cell Dev Biol Anim* 2000;**36**:327–35.
83. Lawson MA, Purslow PP. Differentiation of myoblasts in serum-free media: effects of modified media are cell line-specific. *Cells Tissues Organs* 2000;**167**:130–7.

84. Yamamoto Y, Ito A, Fujita H, Nagamori E, Kawabe Y, Kamihira M. Functional evaluation of artificial skeletal muscle tissue constructs fabricated by a magnetic force-based tissue engineering technique. *Tissue Eng Part A* 2011;**17**(1–2):107–14.
85. Andres V, Walsh K. Myogenin expression, cell cycle withdrawal, and phenotypic differentiation are temporally separable events that precede cell fusion upon myogenesis. *J Cell Biol* 1996;**132**:657–66.
86. Fujita H, Nedachi T, Kanzaki M. Accelerated de novo sarcomere assembly by electric pulse stimulation in C2C12 myotubes. *Exp Cell Res* 2007;**313**:1853–65.
87. Powell CA, Smiley BL, Mills J, Vandenburgh HH. Mechanical stimulation improves tissue-engineered human skeletal muscle. *Am J Physiol Cell Physiol* 2002;**283**:C1557–65.
88. Hamamori Y, Samal B, Tian J, Kedes L. Myoblast transfer of human erythropoietin gene in a mouse model of renal failure. *J Clin Invest* 1995;**95**:1808–13.
89. Yancopoulos GD, Davis S, Gale NW, Rudge JS, Wiegand SJ, Holash J. Vascular-specific growth factors and blood vessel formation. *Nature* 2000;**407**:242–8.
90. Davies N, Dobner S, Bezuidenhout D, Schmidt C, Beck M, Zisch AH, et al. The dosage dependence of VEGF stimulation on scaffold neovascularisation. *Biomaterials* 2008;**29**:3531–8.
91. Korpisalo P, Karvinen H, Rissanen TT, Kilpijoki J, Marjomäki V, Baluk P, et al. Vascular endothelial growth factor-A and platelet-derived growth factor-B combination gene therapy prolongs angiogenic effects via recruitment of interstitial mononuclear cells and paracrine effects rather than improved pericyte coverage of angiogenic vessels. *Circ Res* 2008;**103**:1092–9.
92. Bhatia SN, Balis UJ, Yarmush ML, Toner M. Effect of cell-cell interactions in preservation of cellular phenotype: cocultivation of hepatocytes and nonparenchymal cells. *FASEB J* 1999;**13**:1883–900.
93. Ito A, Kiyohara T, Kawabe Y, Ijima H, Kamihira M. Enhancement of cell function through heterotypic cell-cell interactions using E-cadherin-expressing NIH3T3 cells. *J Biosci Bioeng* 2008;**105**:679–82.
94. Takeichi M. Cadherin cell adhesion receptors as a morphogenetic regulator. *Science* 1991;**251**:1451–5.
95. Harimoto M, Yamato M, Hirose M, Takahashi C, Isoi Y, Kikuchi A, et al. Novel approach for achieving double-layered cell sheets co-culture: overlaying endothelial cell sheets onto monolayer hepatocytes utilizing temperature-responsive culture dishes. *J Biomed Mater Res* 2002;**62**:464–70.
96. Mironov V, Boland T, Trusk T, Forgacs G, Markwald RR. Organ printing: computer-aided jet-based 3D tissue engineering. *Trends Biotechnol* 2003;**21**:157–61.
97. Bhandari RN, Riccalton LA, Lewis AL, Fry JR, Hammond AH, Tendler SJ, et al. Liver tissue engineering: a role for co-culture systems in modifying hepatocyte function and viability. *Tissue Eng* 2001;**7**:345–57.
98. Chia SM, Lin PC, Yu H. TGF-beta1 regulation in hepatocyte-NIH3T3 co-culture is important for the enhanced hepatocyte function in 3D microenvironment. *Biotechnol Bioeng* 2005;**89**:565–73.
99. Hui EE, Bhatia SN. Micromechanical control of cell-cell interactions. *Proc Natl Acad Sci USA* 2007;**104**:5722–6.
100. Akiyama H, Ito A, Kawabe Y, Kamihira M. Fabrication of complex three-dimensional tissue architectures using a magnetic force-based cell patterning technique. *Biomed Microdevices* 2009;**11**:713–21.
101. Fujita H, Shimizu K, Yamamoto Y, Ito A, Kamihira M, Nagamori E. Fabrication of scaffold-free contractile skeletal muscle tissue using magnetite-incorporated myogenic C2C12 cells. *J Tissue Eng Regen Med* 2010;**4**:437–43.

102. Ito A, Hibino E, Kobayashi C, Terasaki H, Kagami H, Ueda M, et al. Construction and delivery of tissue-engineered human retinal pigment epithelial cell sheets, using magnetite nanoparticles and magnetic force. *Tissue Eng* 2005;**11**:489–96.
103. Ito A, Nakahara Y, Tanaka K, Kuga Y, Honda H, Kobayashi T. Time course of biodistribution and heat generation of magnetite cationic liposomes in mouse model. *Jpn J Hyperthermic Oncol* 2003;**19**:151–9.

Peptide Nanoparticles for Oligonucleotide Delivery

Taavi Lehto,[*] Kariem Ezzat,[†] and Ülo Langel[*,†]

[*]*Laboratory of Molecular Biotechnology, Institute of Technology, University of Tartu, Tartu, Estonia*

[†]*Department of Neurochemistry, The Arrhenius Laboratories for Natural Sciences, Stockholm University, Stockholm, Sweden*

I. Introduction 397
II. Gene Therapy 399
A. Gene Delivery 400
B. Antisense 400
C. RNAi 400
D. Splice-Switching Therapeutics 401
III. Uptake Mechanisms of CPPs 401
IV. CPPs for Noncovalent Nanoparticle Formation 404
V. CPP-Based Nanoparticles and Their Applications 407
A. MPG, Pep, and CADY Peptides 407
B. Polyarginines 411
C. Transportans and PepFects 414
VI. Future Directions and Perspective with CPP-Based Nanoparticles 420
References 422

In the past two decades, different methods have emerged for intervention with gene expression, which can be generally referred to as gene therapy. Oligonucleotides (ONs) and their analogs form the basis of the molecules that can be used to modulate gene expression. Unfortunately, due to their physicochemical properties, these molecules require assistance in their intracellular delivery.

Cell-penetrating peptides (CPPs) are one class of nonviral delivery vectors that, because of their remarkable translocation properties, have been intensely utilized for the delivery of ON-based molecules, both *in vitro* and *in vivo*. This chapter concentrates on the applications of CPPs that directly form nanoparticles with different ONs and facilitate their intracellular delivery.

I. Introduction

Gene therapy and gene expression modulation seem today more accessible than ever and, consequently, many of these applications are slowly but firmly moving toward clinical applications. Most molecules used in these settings are

Progress in Molecular Biology
and Translational Science, Vol. 104
DOI: 10.1016/B978-0-12-416020-0.00010-3

1877-1173/11 $35.00

based on nucleic acids/oligonucleotides (ONs) and their analogs. These molecules have large molecular weight and high charge distribution, which generally make them impermeable over cellular membranes. Therefore, ONs require the assistance of delivery vectors in order to become bioavailable. Viral vectors are very efficient for the introduction of genes; however, their utilization is problematic because of their negative side effects, for example, immunogenicity and tumorigenicity.[1] Moreover, their cargo-carrying capacity is often limited with size, and they are not compatible with the transient delivery of shorter ONs and their analogs. This has raised significant interest toward nonviral delivery.[2,3]

One class of nonviral delivery vehicles that has recently been in the spotlight is based on short peptides, called cell-penetrating peptides (CPPs). CPPs are usually less than 30 amino acid long peptides, which are cationic and/or amphipathic and have a net positive charge.[4] These short peptides have been shown to vectorize a wide variety of cargos, including plasmids (plasmid DNA, pDNA), antisense ONs (asONs), splice-switching ONs (SSOs, also termed splice-correcting ONs (SCOs)), small-interfering RNAs (siRNA), decoy ONs, peptides, proteins, and nanoparticles, both *in vitro* and *in vivo* (as recently reviewed in Refs. 5–9). Even though the delivery properties of CPPs are undoubted, the uptake mechanism of CPPs is still controversial. Today it is accepted that most CPPs, especially when associated with cargo, utilize endocytic pathways to gain access to the cells.[10–12] Consequently, this pathway also determines that endosomal sequestration is the main rate-limiting step in the bioavailability of CPPs. Therefore, in order to create efficient CPP-based vectors, their endosomal entrapment has to be overcome to some extent.[13,14] Nevertheless, in some cases, direct nonvesicular penetration pathways might be responsible for the internalization of certain CPPs.[15]

There are two different approaches of how CPPs can be coupled to the cargo; covalent conjugation and noncovalent nanoparticle formation strategy. Covalent conjugation of ONs to the CPPs results in a defined chemical entity with known stoichiometry, which is preferred from the clinical point of view. Nevertheless, production of such conjugates on a large scale is complicated, as they are hard to synthesize, conjugate, and purify, apart from being very laborious and expensive. Moreover, this strategy is not compatible with plasmid and siRNA delivery. Another appealing strategy is to use the noncovalent complexes/nanoparticle formation approach.[15,16] In this approach, the CPP and ONs are simply mixed together to allow spontaneous nanoparticle formation by electrostatic/hydrophobic interactions. The formed particles as such facilitate the intracellular delivery. While the noncovalent strategy had generally gained less attention in the past, there have been significant advances in this field lately. In this overview, the authors will briefly introduce various gene therapy approaches while concentrating on the CPPs. We describe CPP chemical modifications that directly enable the formation of nanoparticles with different ON-based cargos and facilitate their intracellular delivery, both *in vitro* and *in vivo*.

II. Gene Therapy

Gene therapy approaches can be roughly divided into three types according to their therapeutic effect (Fig. 1):

- restoration of lost gene function by gene delivery via viral vectors or plasmids;
- silencing of disease causing genes by antisense, antigene, or RNAi approaches; and
- modification of gene function by interfering with the splicing machinery via SSOs.

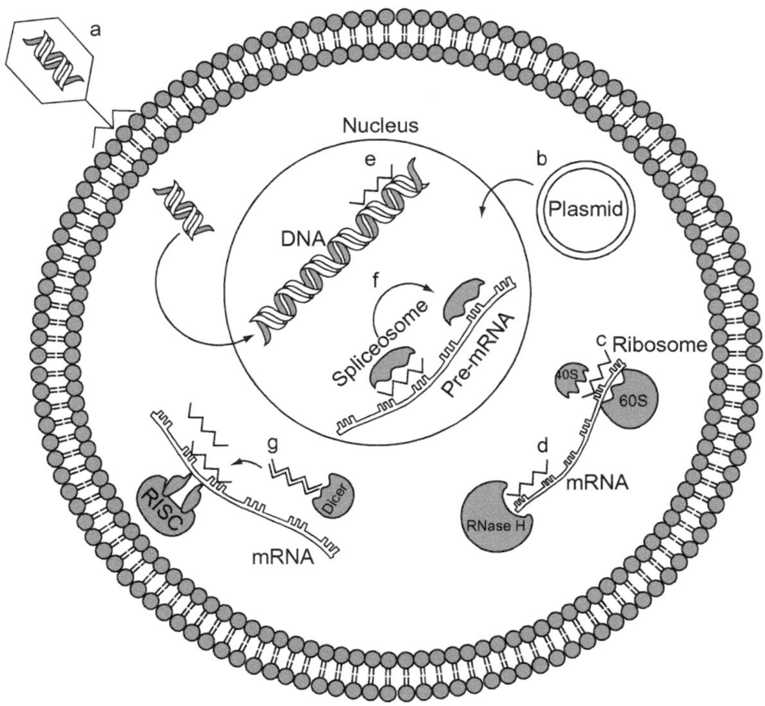

FIG. 1. Different gene therapy approaches. (a) Viral delivery and genome integration. (b) Plasmid delivery. (c) Antisense steric block of translation. (d) Antisense DNA/RNA hybrid and RNase H degradation. (e) Antigene. (f) Splice-switching ONs. (g) siRNA.

A. Gene Delivery

Loss of functional genes is the cause of not only several heritable diseases such as hemophilia, muscular dystrophy, and cystic fibrosis, but also more frequent disorders such as cardiovascular diseases and cancer. Thus, the delivery of functional genes that could restore normal phenotype has been the Holy Grail of gene therapy for decades. Viral vectors have been utilized for gene delivery; however, promising results have been tempered by the potential for insertional mutagenesis that might lead to severe leukemogenic side effects as reported recently.[17–19] Also, humoral immunity directed against the viral vector particle is generally observed.[1,20] Among the other safer alternatives for introducing genes to cells is the delivery of the desired gene containing pDNA via nonviral vectors, for instance, different peptides, cationic lipids, or polymers. When the pDNA enters the cell, it is transcribed and translated by the cellular machinery without the need of genome integration. While many of these vectors are very efficient for pDNA delivery, they are also often associated with toxic side effects. For example, cationic liposomes have been routinely used as delivery vectors for pDNA, but they have been shown to induce toxicity, both *in vitro* and *in vivo*.[21]

B. Antisense

There are two possible mechanisms for an antisense effect using asONs. When the target is a nuclear double-stranded DNA or genes, the approach is called the antigene strategy,[22] while the method that relies on targeting of the mRNA is called the antisense strategy. Antisense activity can be achieved either by blocking the binding sites for the 40S ribosomal subunit and for other translation initiation signals, or by the formation of a double-stranded DNA/RNA complex that renders the RNA susceptible to RNase H digestion[22] (Fig. 1). Natural DNA and RNA have been used for antisense approaches together with several chemically modified analogs that offer better annealing with RNA targets and possess enhanced serum stability. Examples of chemically modified ONs include phosphothiate DNA, 2'-*O*-methyl RNA(2′-OMe RNA), locked nucleic acid (LNA), peptide nucleic acid (PNA), and phosphorodiamidate morpholino oligo (PMO).[23]

C. RNAi

RNAi is a fundamental gene-silencing pathway in eukaryotic cells, where long pieces of double-stranded RNA are cleaved by an enzyme called dicer into shorter fragments called siRNAs that can cleave complementary mRNA sequences with the help of the RISC complex and argonaute 2 (see Fig. 1).[3] The proof-of-principle RNAi study was reported in 2001, which demonstrated that synthetic siRNA could achieve sequence-specific gene knockdown in mammalian cells, and this marked the birth of siRNA therapeutics.[24] What makes the RNAi approach more appealing compared to the antisense strategy is that siRNA cleaves the target mRNA in a catalytic manner, so lower doses are required to achieve gene knockdown. That is why intensive research has been carried out in the last decade to develop delivery vectors for siRNA therapeutics.[3]

D. Splice-Switching Therapeutics

Modification of gene function can be achieved by interfering with the splicing machinery; an approach termed splice switching.[25] Recent studies using high-throughput sequencing indicate that 95–100% of human pre-mRNAs have alternative splice forms.[26] Mutations that affect alternative pre-mRNA splicing have been linked to a variety of diseases, including genetic disorders and cancers, and SSOs can be used to silence mutations that cause aberrant splicing, thereby restoring correct splicing and function of the defective gene (Fig. 1).[25,27] SSOs are asONs, ranging from 15 to 25 bases in length, which do not activate RNase H and thereby prevent the cleavage of the pre-mRNA target before it can be spliced.[25,27] One example of a genetic disease amenable for SSO is Duchenne's muscular dystrophy (DMD). DMD is a neuromuscular genetic disorder that affects 1 in 3500 young boys worldwide[28] and is caused mainly by nonsense or frame-shift mutations in the dystrophin gene. SSOs are used to induce targeted "exon skipping" and to correct the reading frame of mutated dystrophin pre-mRNA such that shorter, partially functional dystrophin forms are produced.[29] SSOs targeting exon 51 are currently in human clinical trials in various parts of Europe to treat DMD.[30,31]

III. Uptake Mechanisms of CPPs

Although CPPs were discovered more than 15 years ago, the uptake mechanisms of CPPs are still controversial. Since receptor-independent cell penetration is the main character of CPPs, much research effort has gone into the identification of such a mechanism. In the early days of CPPs, the uptake of was mostly studied using high concentrations of fluorophore-labeled CPPs and cell fixation methods led to the conclusion that CPPs directly penetrate through cellular membranes in an energy-independent manner to gain access to the cytoplasm. Several models were suggested for such interaction, including the carpet model, the pore formation model, and the inverted micelle-mediated model, among others.[32,33]

However, the idea of direct translocation mechanism was challenged by several studies in the late 1990s and early 2000s. One of the most prominent studies was a paper published by Bernard Lebleue's laboratory in 2003 where

Richard *et al.* demonstrated that cell fixation, even at mild conditions, leads to the artifactual uptake of Tat and R9.[34] Furthermore, they showed that peptide uptake was inhibited by incubation at low temperature and cellular ATP pool depletion, which strongly suggested the involvement of endocytosis in the cellular internalization. Additionally, using fluorescence microscopy on living cells, they demonstrated a punctuated distribution pattern of the internalized CPPs, which was consistent with the distribution of endocytic vesicle and CPPs colocalized with common endocytic markers. Finally, it was shown that endocytic uptake was also responsible for the translocation of CPP/PNA conjugates.[34]

Based on these findings, as well as many others, endocytosis has been shown to be the main mechanism of uptake for various CPPs, especially when they are coupled to a large molecular weight cargo. For CPP-based nanoparticles, endocytosis has been implicated for most of the available platforms such as stearyl-Arg8, Tat-DRBD, and PepFects.[16,35–39] Several endocytic pathways have been suggested for the uptake of CPPs and their cargo, including classical clathrin-mediated endocytosis, caveolae-mediated endocytosis, and macropinocytosis[40] (Fig. 2). However, this has not ruled out the fact that some CPP-based nanoparticles are still able to gain access to the cell via direct translocation mechanism, as has been shown for Pep-1, MPG, and CADY which enter cells independently of the endosomal pathway.[41]

In our opinion, the discrepancies between different uptake mechanisms should not be taken as controversies, but rather as diversity of the ways in which a CPP and its cargo could gain entry into the cells. There are many biological and experimental factors that can contribute to this diversity. On the biological level, the cellular endocytic pathways and membrane rearrangement mechanisms are still not very well understood. Which factors direct certain cargos to which pathways and how different pathways might interact and overlap are far from settled. Additionally, different labs generally do not use exactly the same experimental setups when studying CPP uptake mechanisms. Depending on the nuances of their research, often divergent peptide concentrations, cell lines, incubation times, readout assays, cargo sizes, and cargo incorporation strategies (covalent or noncovalent) are used. These dissimilarities might inevitably lead to different conclusions about the details of the uptake mechanism because apparently each of these factors may have an influence of how the CPP and its cargo would interact with the cell surface and consequently which uptake pathways they will utilize. Also, a very important factor that affects the uptake mechanism is the interaction of CPP-based nanoparticles with different types of media *in vitro* and serum proteins *in vivo*. Nanoparticles being endowed with a very high surface area of interaction, their surface can be significantly altered by adsorption of ions from the medium and proteins from serum either *in vitro* or *in vivo*, forming a new surface of

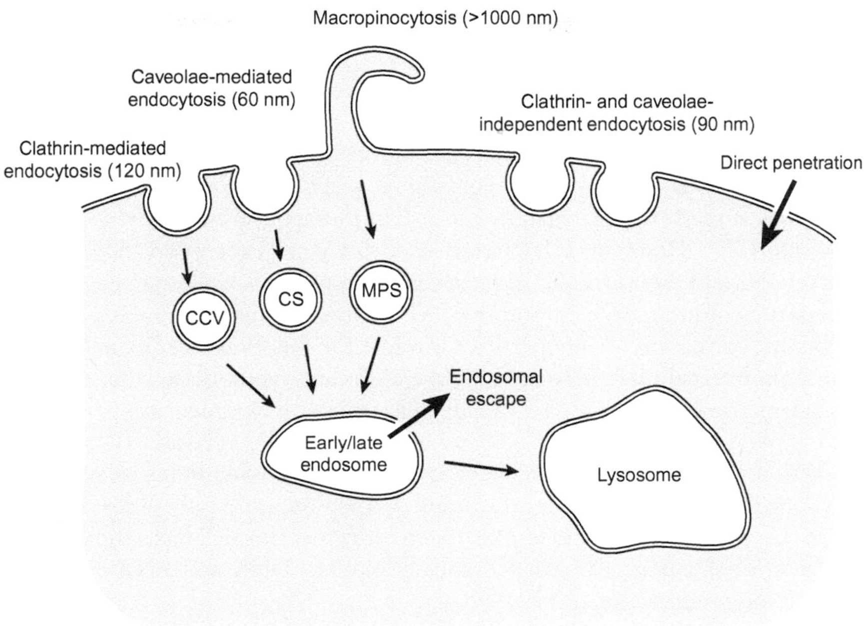

FIG. 2. Uptake mechanism of CPPs. Overview of the different uptake pathways of CPPs. CPPs are known to be internalized into cells by two pathways—either by different endocytic pathways or by direct translocation mechanism. Endocytosis comprises different pathways, including clathrin-mediated endocytosis, caveolae-mediated endocytosis, macropinocytosis, and clathrin- and caveolae-independent endocytosis. If CPP and the associated cargo are internalized by endocytosis, they end up in the endosomes and for the cargo to reach their active sites they need to escape endocytic compartments or they will proceed to the lysosomal degradation pathway.

interaction with biological membranes, which is different from that of the initial CPP nanoparticles that are biophysically studied in much simpler conditions. Protein adsorption may result in the exposure of amino acid residues, which are recognized by the cells as "cryptic epitopes" that have special modes of interaction at the cell surface.[42] Therefore, we believe that the uptake mechanism is not a straightforward, simple process that could be easily described by a single mechanism. It rather seems to be a much more complex process that is sensitive to different variables, which have to be thoroughly studied and specified for their implications on the uptake mechanism. Conclusively, we emphasize that there should be a new perspective when looking at the conflicting data coming out regarding the specific uptake mechanism; this has to be done from the perspective of diversity and complexity rather than rivalry and controversy.

IV. CPPs for Noncovalent Nanoparticle Formation

First nanoparticle therapeutics can be tracked back to 1950s when Jatzkewitz and colleagues synthesized a polyvinylpyrrolidone–mescalin conjugate which had a significantly extended half-life in the blood circulation.[43] Since then, different drugs have been conjugated to polymers and many of these nanoparticle platforms have been accepted as therapeutic agents by respective authorities.[44,45] However, compounds used for gene expression modulation, that is, ONs and their analogs, in most cases require noncovalent strategies for nanoparticle formation/incorporation. Very broadly, these materials should enable the formation of nanoparticles, protect the ON-based cargo, and facilitate their intracellular delivery.[2,3] While different synthetic agents, such as diethylaminoethyl-dextran and calcium phosphate, have been used for gene delivery since the 1960s, their delivery efficiency has been relatively low. Two most used platforms for ON delivery are based on cationic lipids or synthetic polymers and are known under the name of lipoplexes and polyplexes, respectively. These cationic moieties have been shown to form nanoparticles with ONs and their analogs by electrostatically binding DNA and RNA and condensing them into nanoparticles.[46,47]

CPPs are one group of nonviral vectors that have received intense interest in the past decades and have been readily used for the delivery of ONs and their analogs. In the field of CPPs, noncovalent nanoparticle formation strategy was first introduced by the group of Heitz and Divita in 1997. They showed that the amphipathic MPG peptide formed nanoparticles with DNA and these particles were efficiently delivered into cells.[48] Although this approach for CPPs was first reported for DNA, it has now been extended to a variety of other molecules, including pDNA, RNA, RNA analogs, peptides, proteins, and others.[14,15,49]

CPPs are known to facilitate nanoparticle formation with ONs predominantly by two integral mechanisms—electrostatic and hydrophobic interactions.[15] For example, upon the charge neutralization of negative charges of the pDNA backbone by electrostatic interactions, cationic moieties condense the ONs into compact, ordered particles containing only one or a few pDNA molecules.[50] Therefore, in theory, highly cationic moieties should be the most efficient for particle formation. However, too avid association between the vector and cargo is not usually preferred, as the dissociation of the cargo from the nanoparticle is equally crucial for the cargo to be available at its active site.[51,52] Importantly, in case of CPPs, it has also been shown that the positive charges of the cationic residues are responsible for the first association step with cellular membranes, so they play an integral role in the internalization mechanism of CPPs.[12,53,54] Lately, it has been recognized that hydrophobicity and hydrophobic interactions, including the adoption of secondary α-helical structures, also play important roles in CPP-mediated nanoparticle formation, stability, cellular internalization, and endosomal escape.[14,41,55] This is supported by the fact that all the successful reports showing CPP-mediated nanoparticle formation/delivery are based on at least partially amphipathic CPPs. Consequently, in order to create an efficient CPP-based vector, delicate balance between the electrostatic and hydrophobic properties is necessary.

In addition to these prerequisite properties of positive charges for charge neutralization and amphipathicity, it is known that CPPs compatible with nanoparticle formation require additional modifications in order to be biologically active.[48,56] For example, in case of MPG and Pip-1 peptides, it has been shown that C-terminal cysteamidation and N-terminal acetylation are crucial for their delivery efficiency. Cysteamidation is believed to render these CPPs more stable toward the serum proteases, and acetylation affects particle stability/formation, both enhancing the delivery properties. Also, in case of many other vehicles, it has been shown that the addition of hydrophobic residues to the CPPs, for example, stearic acid or cholesterol, renders the properties of CPPs more suited for nanoparticle formation and delivery. For instance, stearylation has been shown to significantly enhance the delivery of pDNA,[35,39,57] siRNAs,[37,58] and SCOs[38,57,59] upon noncovalent nanoparticle formation. Moreover, it is believed that CPPs designed to adopt secondary structures upon interaction with lipid membranes, such as α-helices, can be important in mediating the delivery of CPP-based nanoparticles and have been shown to be responsible for both direct translocation mechanism[60] and endosomal escape.[13] For example, in case of CADY peptide, it has been shown that α-helical nature of this peptide is crucial for its cellular uptake and delivery efficiency.[61] For an overview of the CPPs and modification that have been shown to allow nanoparticle formation with ONs, see Table I.

Finally, intense work on the nanoparticle-based ON delivery has led to the understanding of some general properties that are required from the ideal delivery vehicle. These requirements include (1) physicochemical properties, such as binding and condensing of the nucleic acids with suitable size, shape, surface charge, and stability in solution; (2) *in vitro* characteristics such as stability in the transfection media in the presence of serum, association with membrane and cellular internalization, endosomal escape, cytoplasmic trafficking, nuclear internalization (if necessary), and dissociation of the nanoparticles before or at the site of action of the cargo, either in the cytoplasm or in the nucleus; and (3) for *in vivo* barriers, stability and survival in the blood stream, penetration of the blood vessel wall and surrounding tissue, and, ideally, binding to the specific cells together with all of the further steps described for *in vitro*.

TABLE I

EXAMPLES OF CPPs AND MODIFICATIONS COMPATIBLE WITH NONCOVALENT DELIVERY OF ONs

CPPs	Sequence	Modification	Reference
MPG peptides			
MPG	Ac-GALFLGWLGAAGSTMGAPKKKRKV-Cya	Acetylation, cysteamidation	48
MPGΔ^{NLS}	Ac-GALFLAFLAAALSLMGLWSQPKKKRKV-Cya	Acetylation, cysteamidation	62
MPG-8	β-AFLGWLGAWGTMGWSPKKKRK-Cya	Cysteamidation	63
Chol-MPG-8	Chol-β-AFLGWLGAWGTMGWSPKKKRK-Cya	Cholesterol, cysteamidation	63
Pep peptides			
Pep-1	Ac-KETWWETWWTEWSQPKKKRKV-Cya	Acetylation, cysteamidation	56
Pep-2	Ac-KETWFETWFTEWSQPKKKRKV-Cya	Acetylation, cysteamidation	64
Pep-3	Ac-KWFETWFTEWPKKRK-Cya	Acetylation, cysteamidation	65
PEG-Pep-3	PEG-KWFETWFTEWPKKRK-Cya	PEGylation, cysteamidation	65
CADY peptides			
PPTG1	GLFRALLRLLRSLWRLLLRA	–	66
CADY	Ac-GLWRALWRLLRSLWRLLWRA-Cya	Acetylation, cysteamidation	61
Polyarginine peptides			
Stearyl-Arg8	Stearyl-RRRRRRRR-NH_2	Stearylation	67
Stearyl-Arg9	Stearyl-RRRRRRRRR-NH_2	Stearylation	57
Chol-Arg9	Chol-RRRRRRRRR-NH_2	Cholesterol	68
Stearyl-$(RxR)_4$	Stearyl-RXRRXRRXRRXR-NH_2	Stearylation	57
TP10 and PepFect peptides			
TP10	AGYLLGKINLKALAALAKKIL-NH_2	–	69
PepFect3, PF3	Stearyl-AGYLLGKINLKALAALAKKIL-NH_2	Stearylation	16
PepFect6, PF6	Stearyl-AGYLLGK[a]INLKALAALAKKIL-NH_2	Stearylation, trifluoromethylquinoline	37
PepFect14, PF14	Stearyl-AGYLLGKLLOOLAAAALOOLL-NH_2	Stearylation	38

Ac, N-terminal acetylation; Cya, C-terminal cysteamidation; PEG, polyethyleneglycol; stearyl, stearic acid; chol, cholesterol.
[a]Orthogonal trifluoromethylquinoline moiety.

V. CPP-Based Nanoparticles and Their Applications

CPPs have been extensively used for the delivery of ON-based molecules. Lately, significant improvements in the CPP-based vectors have been made and there is variety of different vectors that are compatible with ON delivery upon nanoparticle formation. Moreover, many of these vectors have also been shown to be efficient in different *in vivo* and disease models, raising hopes for further preclinical and clinical studies in the future. These advances will be discussed in the next section, starting from classical CPPs such as the MPG peptide to the new platform created by Langel and co-workers—PepFect peptides.

A. MPG, Pep, and CADY Peptides

These peptides are a group of amphipathic CPPs that have been created by the group of Divita for the delivery of various bioactive molecules, including ONs and their analogs, by the noncovalent nanoparticle formation approach (Table II).

1. MPG Peptides

MPG peptide was the first CPP reported to be able to vectorize cargo by forming noncovalent nanoparticles and was introduced by the group of Heitz and Divita in 1997.[48] This amphipathic MPG peptide, containing a hydrophobic domain derived from the fusion sequence of HIV gp41 and a hydrophilic domain derived from the nuclear localization sequence (NLS) of SV40 T-antigen, was capable of interacting with single- and double-stranded ONs and facilitating their intracellular delivery. They proposed that the charged NLS part was responsible for the interaction with ONs, while the hydrophobic part was supposed to facilitate the transport through the cell membrane. Also, an MPG excess over ONs indicates that multiple CPP molecules interacted with a single ON molecule and enabled the condensation and the subsequent nanoparticle formation. It was found that the particles were taken up by direct penetration (or non-endosomal pathway), and this theory still stands for MPG peptide, while in almost all the other cases, it is now widely believed that CPP-based nanoparticles are taken up through endocytic pathways.[15]

Not long after this early observation, the same group extended this strategy to deliver entire genes by vectorizing pDNA.[70] They showed that MPG efficiently interacted with pDNA by forming stable nanoparticles, and these particles facilitated the delivery of the luciferase-encoding plasmid into various cell lines. However, in this work, they also targeted a more relevant model. They showed that, by delivering a plasmid carrying the full-length antisense cDNA-encoding human cdc25C, they were able to arrest the cell cycle before the cells entered mitosis in a large population of cells. These biological effects were corroborated at the mRNA and protein levels, where the expression of

TABLE II
OVERVIEW OF AN ON DELIVERY WITH MPG, PEP, AND CADY PEPTIDES

CPP	Cargo	Gene/target	Effect	Reference
MPG peptides				
MPG	DNA	Labeled DNA	Intracellular delivery	48
MPG	pDNA	Luc/cdc25C	Luc expression and cell cycle arrest	70
MPGΔ^{NLS}	siRNA	Luc/GAPDH	Luc and GAPDH downregulation	62
MPGΔ^{NLS}	siRNA	Luc	Luc downregulation	
MPG8	siRNA	Cyclin B1	Inhibition of cell proliferation and tumor reduction upon intratumoral delivery *in vivo*	63
MPG8-Chol	siRNA	Cyclin B1	Tumor reduction upon systemic delivery and increase survival of the animals	63
Pep peptides				
Pep-1	Peptide protein, antibody		Intracellular delivery	56
Pep-2	asON	Cyclin B1	Cyclin B1 downregulation	64
Pep-3	asON	Cyclin B1	Cyclin B1 downregulation, tumor reduction after intratumoral delivery *in vivo*	65
PEG-Pep-3	asON	Cyclin B1	Tumor reduction after systemical delivery *in vivo*	65
CADY peptides				
ppTG1 ppTG20	pDNA	Luc	Luciferase expression *in vitro* and on systemic delivery *in vivo*	66
CADY	siRNA	GAPDH	GAPDH downregulation	61

luc, luciferase; GAPDH, glyceraldehyde 3-phosphate dehydrogenase; chol, cholesterol; NLS, nuclear localization signal.

cdc25C was substantially suppressed.[70] Strikingly, MPG/pDNA nanoparticles were also shown to mediate the delivery in the presence of serum, something that is still an obstacle to be overcome for most of the CPP-based platforms. It is possible that additional modifications comprised in the MPG peptide are responsible for this remarkable stability, namely, the aforementioned cysteamidation and acetylation.

In 2003, Simeone *et al.* studied the function of NLS sequence in the MPG structure further, and noticed that site mutation in the NLS sequence lowered the nucleolar uptake and decreased the delivery efficiency of pDNA.[62] They hypothesized that this phenomenon could enhance the cytoplasmic delivery of MPG and tried the novel MPG derivate, MPGΔ^{NLS}, in comparison with MPG for the delivery of siRNAs to induce RNAi-mediated gene silencing. They demonstrated that both MPG and MPGΔ^{NLS} facilitated siRNA delivery, but MPGΔ^{NLS} was much more effective for achieving higher luciferase gene knockdown at the lower siRNA concentrations. These results were also confirmed by targeting endogenous glyceraldehyde 3-phosphate dehydrogenase (GAPDH) mRNA, which also resulted in significant gene silencing. Although MPG also facilitated RNAi to some extent, it was shown that this modified MPGΔ^{NLS} was at least twofold more effective in siRNA delivery than MPG.[62]

More recently, Crombez *et al.* reported an MPG analog, MPG-8 that was designed to be more preferential for the siRNA delivery.[63] The MPG/siRNA nanoparticles targeting the cell cycle regulator cyclin B1(Cyc B1) mediated significant mRNA and protein knockdown in a wide variety of regular and refractory cell lines. Furthermore, these MPG-8/Cyc B1 siRNA particles were shown to efficiently inhibit tumor growth in the xenograft model upon intratumoral injection but not after systemic administration by intravenous injections. Interestingly, it was also shown that the aforementioned MPGΔ^{NLS} complexed with same siRNA did not affect the tumor growth to any great extent, indicating the need to fine-tune delivery vectors to suit certain needs. Thus in order to use MPG-8 in systemic delivery settings, it was further modified with cholesterol (Chol-MPG-8). Pharmacokinetic analysis of the fluorescently labeled siRNA complexed with MPG8 or Chol-MPG-8 compared to naked siRNA showed that both peptides increased the half-life of the siRNA and enabled rapid and wide tissue distribution. However, these effects were more pronounced in case of Chol-MPG-8. More importantly, Chol-MPG-8 was shown to enable tumor reduction after systemic administration in two different xenograft models by effectively inhibiting tumor growth and enabling substantially longer survival.[63]

In general, MPG-based nanoparticles seem to be able to internalize into cells by direct penetration. However, it has been reported that these particles could also utilize endocytic pathways to gain access to the cells. Recently, Veldhoen *et al.* studied MPGΔ^{NLS} for the delivery of siRNAs.[71] While reporting that MPGΔ^{NLS} mediated efficient siRNA-induced gene silencing, they additionally conducted studies to assign the uptake pathway of these particles. Interestingly, they found that endocytosis inhibitors had a significant negative effect on the gene-silencing efficiency. Moreover, confocal microscopy studies revealed the rather vesicular cytoplasmic pattern of the MPG/labeled-siRNA nanoparticles instead of diffuse distribution. Their data provided evidence that, depending on the conditions, endocytotic pathway can contribute to the uptake of MPGΔ^{NLS}/siRNA nanoparticles.[71]

2. Pep Peptides

In a search for CPP-based vectors that could also be used for the noncovalent delivery of other types of cargo such as peptides and proteins, Morris and colleagues reported a novel amphipathic CPP, Pep-1.[56] Pep-1 design is similar to that of the MPG peptides. However, the difference between MPG family and Pep-1 is in the hydrophobic domain. In Pep-1, this domain corresponds to tryptophan-rich cluster (KETWWETWWTEW). Pep-1 was shown to readily associate with both peptides and proteins and to form stable nanoparticles. Starting with peptide delivery, it was shown that 32- and 51-mer peptides were efficiently delivered into the cells. Interestingly, the same strategy was successful with full-length proteins. Pep-1 efficiently transported the green fluorescent protein and β-galactosidase, while the transfections were not affected by the presence of serum and these nanoparticles did not recruit the endocytic pathway for cellular internalization. Interestingly, Pep-1 did not mediate the delivery of cargo at concentrations lower than 0.5 μM and higher than 50 μM, showing that particle size is very important for successful delivery. Finally, the authors also demonstrated that Pep-1 facilitated delivery of full-length antibodies, by showing intracellular staining of different targets within the cell.[56]

In 2004, group of Divita also reported a second-generation Pep peptide, named Pep-2, for the delivery of PNA and the charged PNA analog HypNA-*p*PNA.[64] They showed how antisense HypNA-*p*PNA chimera targeting cyclic B1 noncovalently complexed with Pep-2 facilitated robust downregulation of cyclin B1 at both the mRNA and the protein levels and efficiently blocked the cell cycle in multiple cell lines, including the breast cancer cell line, and inhibited cell proliferation.

Later, Pep-2 peptide was further modified to get more efficient peptide vectors for PNA delivery both *in vitro* and *in vivo*, resulting in the Pep-3 peptide.[65] This optimized vector was used for the delivery of HypNA-*p*PNA, targeting the above-mentioned cyclin B1, and resulted in the substantial decrease of targeted protein in regular HeLa cell lines and in more difficult-to-transfect primary and suspension cells. Moreover, this setup was also extended to *in vivo* delivery. In a tumor xenograft model of prostate cancer, Pep-3/HypNA-*p*PNA antisense effectively inhibited the tumor growth after intratumoral injections. Moreover, they modified Pep-3 with polyethyleneglycols (PEGs) and this PEG–Pep-3 enabled tumor reduction after systemic administration of these particles.[65]

3. CADY Peptides

In the search for efficient peptide-based vectors for the delivery of plasmids, Rittner *et al.* designed a series of new basic amphipathic peptides, namely, ppTG1 and its derivates, which efficiently condensed pDNA into nanoparticles.[66] These nanoparticles were first shown to destabilize liposomal membranes, and structural studies showed that the reason for this could be the α-helical structure of the peptides imparted by the hydrophobic amino acid residues. This ppTG1 peptide and its derivate ppTG20 mediated efficient delivery of pDNA *in vitro* in multiple cell lines and with transfection efficiencies comparable with commercially used vectors. Most importantly, these nanoparticles were shown to facilitate gene delivery *in vivo* upon systemic administration, as reporter gene expression was increased in the lungs of the treated animals. To our knowledge, this was the first report that demonstrated the *in vivo* vectorization of pDNA by a single-component peptide (CPP) vector. However, these effects were also associated with acute toxicity, as many of the treated animals died in the course of the study.[66]

Based on this above-mentioned PPTG1 peptide, the group of Divita proposed a new generation of secondary amphipathic peptides—a CADY platform—for the delivery of siRNAs.[61] In this peptide, the PPTG1 peptide backbone was enriched and many amino acids were substituted to allow better interaction with siRNA and more preferred association with the lipid phase of the membrane. It seems that the secondary amphipathic structure is responsible for the delivery efficiency of CADY and it was shown to adopt and maintain its helical structure also in the membranes. CADY/siRNA nanoparticles targeting endogenous GAPDH mRNA efficiently mediated the knockdown of GAPDH in subnanomolar concentrations. Moreover, these effects were also obtained in very refractory cell lines, such as primary and suspension cell lines.[61] Nevertheless, no *in vivo* results have yet been published with this platform.

Conclusively, all these peptides—MPG, Pep, and CADY peptides—developed by the group of Divita have been for a long time the most successful CPP-based vectors that were shown to be applicable for the nanoparticle formation and delivery. While there is a great deal of controversy about how these peptides gain access to the cells, there is no doubt that many of these vectors are very efficiently transporting a variety of bioactive molecules into the cells, both *in vitro* and *in vivo*. These significant findings are compiled in Table II.

B. Polyarginines

Since the discovery of the delivery properties of argnine-rich peptides, they have been readily utilized for the delivery of ONs and their analogs. Here, polyarginine-based CPPs will be described, which have been shown to be applicable by the noncovalent nanoparticle formation strategy (Table III).

Futaki *et al.* reported in 2001 that stearic acid-modified octaarginine (Arg8) mediated an efficient vectorization of pDNA upon nanoparticle formation and facilitated expression of the reporter genes in variety of different cell lines.[67] More importantly, the transfection efficiency of stearyl-Arg8 was shown to be in a similar range as with the most widely used lipid reagent Lipofectamine™ 2000 (LF2000). It was found that, even though the parent peptide, Arg8, enabled the uptake of fluorescently labeled pDNA and mediated the expression of luciferase upon the transfection with luciferase-encoding plasmid to some extent, addition of stearic acid significantly increased both the uptake and

TABLE III
EXAMPLES OF POLYARGININE-FACILITATED DELIVERY

CPP	Cargo	Gene/target	Effect	Reference
Stearyl-Arg8	pDNA	Luc	Luc expression	67
Stearyl-Arg9	pDNA	Luc	Luc expression	57
Stearyl-(RxR)4	pDNA	Luc, EGFP	Luc and EGFP expression	57
Stearyl-Arg8 (in MEND)	siRNA	Luc	Luc downregulation	58
Stearyl-Arg8 (in MEND)	siRNA	Luc	Luc downregulation *in vivo* after intratumoral injection	72
Chol-Arg9	siRNA	VEGF	VEGF downregulation and tumor reduction in mice	68
RVG-Arg9	siRNA	GFP, SOD-1, siFvE^J	GFP and SOD-1 downregulation in brain. siFvE^J downregulation-mediated protection against JEV-induced encephalitis in mice	73
Tat-DRBD	siRNA	dGFP, dRFP, Luc, GAPDH, Oct-4	Efficient gene silencing in difficult-to-transfect primary and suspension cells and in mice luciferase model	74
Tat-DRBD	siRNA	EGFR, Akt2	EGFR and Akt2 downregulation induced tumor reduction and increased survival in mice	75

Luc, luciferase; dRFP, destabilized red fluorescent protein; (d)EGFP, (destabilized) enhanced green fluorescent protein; MEND, multifunctional envelope-type nanodevice; VEGF, vascular endothelial growth factor; SOD-1, superoxide dismutase 1; siFvE^J, envelope gene of Japanese encephalitis virus; JEV, Japanese encephalitis virus; EGFR, epidermal growth factor receptor; Akt2, RAC-beta serine/threonine-protein kinase; GAPDH, glyceraldehyde 3-phosphate dehydrogenase; chol, cholesterol.

biological effect. Physicochemical characterization of the complexes showed that both the Arg8 and the stearyl-Arg8 peptide formed nanoparticles with pDNA with a size of around 100 nm and a zeta potential of about 20 mV or more. Interestingly, atomic force microscopy analysis revealed that pDNA packaging/condensation efficiency differed, being much more pronounced in case of stearyl-Arg8.[35,67]

Other groups have also studied polyarginines and their possible application for direct nanoparticle formation and intracellular delivery of nucleic acids and their analogs. Our experience is that plain polyarginines, such as Arg9 without any modification, are not suited for the noncovalent strategy.[16,39,57] The addition of the aforementioned stearic acid modification to Arg9 also had an impact on the delivery properties of the vector, as it mediated pDNA delivery.[39,57] However, when the SSOs were used as cargos, stearyl-Arg9 did not have any effect on the splice switching efficiency.[16]

Interestingly, in case of polyarginines, it has been reported that modifying the polyarginine backbone with α-aminohexanoic acid had a significant impact on the nucleic acid delivery upon covalent conjugation.[76–79] Our group has lately reported that stearylation of the preclincally used $(RxR)_4$ peptide mediates both delivery of plasmids and SSOs[57] upon noncovalent complexation. We found that stearyl-$(RxR)_4$ forms stable nanoparticles with nucleic acids, and these particles facilitate efficient gene delivery in various cell lines. Additionally, these transfections were not associated with toxic side effects on the cell viability. Moreover, stearyl-$(RxR)_4$ efficiently vectorized SSOs, which mediated splice switching at a relatively low SSO concentrations. These effects were obtained at five times lower SCO concentrations when compared to the most efficient covalent conjugate $(RxR)_4$-PMO. As in the case with other vectors, the presence of serum substantially decreased the transfection efficacy of both plasmids and SSOs.[57] Hassane *et al.* recently worked also with stearyl-$(RxR)_4$/SSO nanoparticles and obtained very similar results. Additionally, looking at the mRNA level, they demonstrated that these particles enabled significant correction of the luciferase mRNA, while $(RxR)_4$-PMO at these concentrations was absolutely inactive. Moreover, in the liposome leakage assay they showed that stearyl-$(RxR)_4$ causes membrane destabilization in a pH-independent manner and this slightly increased the endosomal escape (Said Hassane Fatouma *et al.*, unpublished data).

Also utilizing the polyarginine platform, Nakamura *et al.* searched for the possibilities to make multifunctional envelope-type nano device (MEND) compatible with siRNA delivery.[58] They noticed that poly-L-lysine and protamine, which were successfully utilized for the MEND core formation earlier, did not enable the delivery of siRNAs. As a result, they opted to try stearyl-Arg8, which was initially used for the surface functionalization, for siRNA condensation. These nanoparticles achieved more than 80% silencing of the luciferase gene in stable luc-HeLa cells, without exerting any toxic side effects.[58] More recently, Hatakeyama *et al.* reported a slightly more complicated MEND system, where similarly stearyl-Arg8 was used to condense the siRNA; however, the liposome surface was additionally modified with PEG and GALA derivates. This system was shown to enable gene silencing after intratumoral injection *in vivo.*[72]

Lately, Kim *et al.* showed that cholesterol-modified Arg9 enabled efficient vectorization of siRNA by noncovalent nanoparticle formation.[68] Their aim was to target the vascular endothelial growth factor (VEGF), which is a primary stimulant of the development and maintenance of a vascular network in the vascularization of solid tumors, and designed an siRNA against VEGF. They demonstrated that this peptide interacts with siRNA and these particles mediated efficient knockdown of VEGF mRNA *in vitro*. Based on this, they sought to extend this platform to *in vivo* conditions by using a subcutaneous tumor xenocraft model. Intratumoral injections of Chol-Arg9/VEGF-siRNA nanoparticles mediated efficient VEGF knockdown and subsequent tumor reduction.[68]

In 2007, Kumar *et al.* published a study in *Nature*, where they reported a targeted siRNA delivery across the blood-brain barrier. In this report, rabies virus glycoprotein (RVG) was used as a targeting motif.[73] However, RVG peptide was additionally modified with Arg9 to facilitate the siRNA condensation and intracellular delivery. This RVG-Arg9 peptide facilitated efficient nanoparticle formation with siRNA and these particles mediated efficient gene knockdown *in vitro*. Strikingly, after systemic delivery in mice, these particles enabled efficient gene silencing in brain. Moreover, when using specific antiviral siRNA, these treatments allowed efficient protection against the fatal viral encephalitis in mice.[73]

More recently laboratory of Dowdy introduced a CPP-based platform for the delivery of siRNA applicable by noncovalent approach.[74] In this system, CPP motifs of the Tat peptide were expressed as a fusion protein with the double-stranded RNA-binding domain (Tat-DRBD). Tat-DRBD efficiently interacts with siRNA and enables efficient nanoparticle formation and intracellular delivery. This system was shown to enable efficient gene knockdown in various refractory cell lines, such as primary or suspension cells, without exerting any cytotoxic or immunogenic side effects. Moreover, they showed that this platform enabled silencing of a luciferase reporter gene *in vivo* after local intranasal delivery. Tat-DRBD system allows also tumor reduction in a tumor xenocraft model upon intratumoral delivery in mice.[75] However, *in vitro*, this system is only efficient if the GAGs have been removed from the treatment media, thereby indicating that the Tat-DRBD is very sensitive to the presence of serum proteins, and, possibly for this reason, no information about the efficiency after systemic administration has been reported.[74]

C. Transportans and PepFects

Transportans (TPs) are a class of chimeric CPPs created by our group that have been successfully used for ON delivery[80] both *in vitro* and *in vivo*.[81] Among the TP analogs, transportan 10 (TP10) has been shown to have most suitable delivery properties and has been most extensively utilized.[69]

EL Andaloussi *et al.* published the first successful report showing that TP10 is capable of vectorizing cargo in a noncovalent manner. They showed that TP10 complexed with decoy DNA ONs targeting Myc protein-mediated efficient cellular internalization and enhanced the biological effect by inhibiting cell proliferation.[82] Later, Kilk *et al.* showed that TP10 could be used as an uptake enhancer for polyethyleneimine (PEI)-mediated plasmid transfections and significantly enabled the lowering of the PEI concentrations required for efficient transfection. However, it was also shown that TP10 itself could not mediate plasmid delivery.[83] Accordingly, our later experience has shown that unmodified TP10, in most cases, is not capable of mediating the delivery of ONs, including plasmids,[39,83] SSOs,[16] and siRNA.[84] Interestingly, it seems that the inefficiency is not due to the lack of cellular uptake but rather to the fact that the complexes are sequestered in the endosomal compartments hindering the cargo from reaching their active site.[16,39] Moreover, we have lately seen that presence of serum completely abolished the uptake of TP10/plasmid particles.[39] Based on this, we believe that TP10 without any additional chemical modifications is not capable of mediating ON delivery by the noncovalent nanoparticle strategy. Therefore, we have recently studied different modifications that could render the delivery properties of TP10 more suited to the noncovalent approach (Table IV).

PepFects are a series of peptides based on the TP10 peptide, where TP10 has been either chemically modified with stearic acid and/or trifluoromethylquinoline-based moieties at different positions or the TP10 backbone has been modified. PepFect peptides and their utilization for the delivery of SSO, plasmids, and siRNAs are described below (also see Fig. 3).

1. PepFect3

PepFect3 (PF3) (also referred to as stearyl-TP10) is a TP10 peptide N-terminally modified with stearic acid. First of all, PF3 was shown to efficiently enable the vectorization of SSOs.[16] The stearic acid moiety increased the uptake of these particles and had a substantial effect on the splice switching

TABLE IV
Samples of PepFect Peptide Series-Facilitated Delivery

CPP	Cargo	Gene/target	Effect	Reference
PF3	SSO	Luc	Splicing correction in HeLa pLuc705	16
PF3	pDNA	Luc, EGFP	Luciferase and GFP expression *in vitro* and luciferase expression *in vivo* upon local delivery	39
PF4	SSO	Luc	Splicing correction in HeLa pLuc705	59
PF6	siRNA	Luc, EGFP, HPRT1, Oct-4, GAPDH	Efficient gene silencing in difficult-to-transfect primary and suspension cells and in mice upon systemic administration in two *in vivo* models	37
PF14	SSO	Luc, dystrophin	Splicing correction in HeLa pLuc705 and in a DMD cell model	38

Luc, luciferase; pDNA, plasmid DNA, SSO, splice-switching oligonucleotide; HeLa pLuc705, a model for splicing correction; EGFP, enhanced green fluorescent protein; siRNA, small-interfering RNA; HPRT1, hypoxanthine phosphoribosyltransferase 1; Oct-4, octamer-binding transcription factor 4; GAPDH, glyceraldehyde 3-phosphate dehydrogenase; DMD, Duchenne muscular dystrophy.

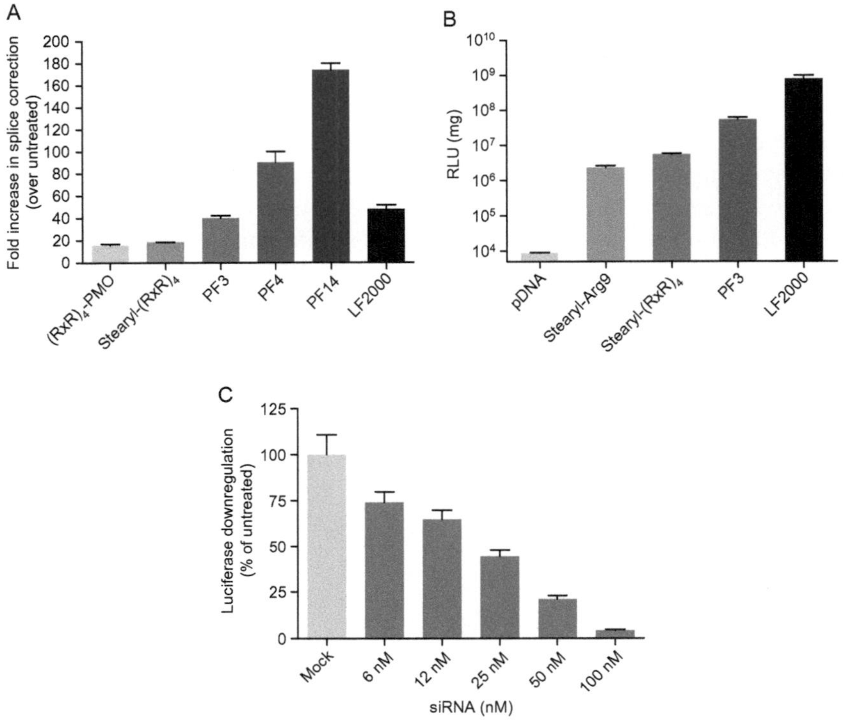

FIG. 3. Examples of the delivery of plasmids, SSOs, and siRNAs by PepFect peptides. (A) Splice switching efficiency in a HeLa pLuc705 model cell line, mediated by PepFect peptides, and comparison with stearyl-$(RxR)_4$, preclinically used $(RxR)_4$-PMO conjugate, and the cationic lipid LF2000. Concentration of SSOs was 200 nM in every experiment, except for $(RxR)_4$-PMO, which was used at 4 μM. (B) Plasmid delivery efficiency of stearyl-Arg9, stearyl-$(RxR)_4$, stearyl-TP10 (PF3), and comparison with LF2000 in CHO cells. (C) PF6-mediated siRNA delivery and dose-dependent gene silencing in luciferase-stable BHK21 cell line.

efficiency. Also, it enabled the maintenance of delivery efficiency in the presence of serum proteins. Transfection efficiency levels were comparable to the effects achieved via the commercially available agent LF2000. As compared to other peptide-based vehicles, these results were very impressive, as most of the conjugates that are used for the delivery of SSOs are active in micromolar concentrations, while PF3 was capable of mediating splice correction at the SSO levels of 200 nM. Compared to the leading CPP/SCO conjugates, PF3 was around to be 10-fold more efficient. Recently, Hassane *et al.* also utilized the PF3 peptide for vectorizing SSOs and obtained similar results. They also showed that stearic acid modification on the TP10 peptide enhances pH-independent destabilization of the artificial endosomes and, therefore, in addition to particle formation, PF3 also increases endosomal escape (Said Hassane Fatouma *et al.* unpublished data). We also investigated how the position of the stearic acid modification would affect the delivery efficiency. When the TP10 peptide was stearylated orthogonally on Lys^7, generating a PepFect4 (PF4) peptide, the SSO delivery efficiency was increased even further.[59]

Recently, we studied whether we could extend the delivery properties of PF3 to the much larger pDNA molecules.[39] PF3 efficiently formed nanoparticles with pDNA with a size of around 125 nm and slightly negative surface charge. These particles facilitated ubiquitous and uniform gene expression in the whole cell populations of a wide variety of adherent cell lines and these effects were also seen in the primary mouse embryonal fibroblasts (MEFs). These transfections were not dependent on the cell confluence and were not associated with any toxicity or immunogenicity *in vitro*. Interestingly, and in line with the results with SSOs, transfection efficiency was maintained in the presence of serum, which, together with the absence of toxicity and immunogenicity, is very important for implementation *in vivo*. In the next step, we wanted to test whether these PF3/plasmid nanoparticles could facilitate efficient gene delivery *in vivo* by administrating the particles locally by intramuscular or intradermal injections. Interestingly, these particles enhanced the gene expression in the muscle and skin of the mice by an order of magnitude as compared to naked plasmid, results which are in line with the best results achieved by nonviral delivery. Also, these effects were shown to be exerted in a dose-dependent manner. Importantly, these effects were not associated with any toxic or immunogenic side effects and did not affect the common hematological and clinical parameters.[39]

2. PepFect6

This peptide was originally designed for the reason that some of our stearylated CPPs were efficient in delivering of cargos that were destined to work in the nucleus, for example, pDNA and SSOs, but without exerting any potency for the delivery of siRNAs. As the RNAi machinery and the site of action of siRNAs reside in the cytoplasm, we hypothesized that the lack of biological effect has to derive from the endo-lysosomal sequestering of the particles, which does not enable efficient cytoplasmic delivery. Thus, a chloroquine analog (pH-titratable trifuoromethylquinoline moiety) with endosomolytic properties was introduced into the TP10 peptide on Lys^7, generating the PepFect5 (PF5) peptide.[37] This peptide was very efficient for siRNA delivery; however, its efficiency was significantly impaired by the presence of serum. For this reason, based on our knowledge that stearylation renders the TP10 peptide resistant to the presence of serum, we also introduced the stearyl group to the N-terminus of the PF5 peptide, generating the PepFect6 (PF6) peptide.[37]

The PF6 peptide enabled efficient condensation of siRNAs into nanoparticles with an average size of around 70–100 nm with a slight negative surface charge (− 7 to − 11 mV). First of all, delivery efficiency of these PF6/siRNA nanoparticles was screened in a battery of reporter cell lines encoding either luciferase or enhanced green fluorescent protein (EGFP) genes. Strikingly, these particles mediated extremely efficient gene knockdown in these cell lines, with inhibitory concentrations of 50% of the protein activity (IC_{50}) below 10 nM concentration of siRNA. Importantly, PF6/siRNA nanoparticles were significantly more efficient in gene silencing than different cationic liposomes, for example, LF2000 and LF RNAiMAX, which are known to be one of the most efficient measures to vectorize siRNAs. Moreover, PF6 fulfilled all the criteria for an extremely efficient transfection agent: it was not significantly hampered by the presence of serum; its transfection efficiency was relatively independent of cell confluence; it transfected the whole cell population in a uniform manner; and it mediated rapid and persistant RNAi responses over several days and enabled complete mRNA and protein knockdown in multiple treatments.

While there are relatively efficient vehicles to induce RNAi in regular adherent cell lines, including the above-mentioned cationic liposomes, efficient measures are lacking for primary and suspension cell lines. Thus, an endogenous gene, namely, hypoxanthine phosphoribosyltransferase 1 (HPRT1), was used as the target and the RNAi was studied in a battery of primary and suspension cell lines that are known to be very resistant to transfection by chemical means. PF6/HPRT1 siRNA nanoparticles induced significant gene silencing in the primary mouse fibroblasts (MEFs), human umbilical vein endothelial cells, T-lymphocytes (JURKATs), mouse embryonic stem cells, and others. As cationic liposomes were very inactive in these cell lines, a new promising siRNA delivery agent was used as a comparison: namely, the above-mentioned Tat-DRBD (Transductin reagent). In all these very refractory cell lines, PF6 was significantly more active than all the other reagents that were used for comparison. To see whether these treatments with PF6/siRNA nanoparticles cause any side effects, assessment of the cytotoxicity *in vitro* was carried out. We found that PF6 treatments were not associated with *in vitro* toxicity, as opposed to lipofection. These effects were corroborated by genome-wide microarray analysis and mass-spectroscopy-based proteomics, where PF6 affected the transcriptome and proteome to a significantly lesser extent. Moreover, in an *in vitro* immunogenicity assay, it was shown that PF6 did not cause the induction of innate immunity.

The fact that PF6/siRNA nanoparticles induced significant gene silencing in the cell cultures without any toxic or immunogenic side effects encouraged the investigators to evaluate the full potential of this platform for systemic delivery of siRNA *in vivo*. For this, two different *in vivo* models were used. By using HPRT1 as target, treatments with PF6/siRNA nanoparticles resulted in efficient silencing of HPRT1 as measured by PCR in the variety of organs, including liver, kidney, and lung. While RNAi-mediated silencing with other methods is mostly achieved in liver, targeting other organs is much more difficult. Interestingly, we observed $>60\%$ of silencing in kidney and lungs. Significantly, we were able to rule out that these treatments would be associated with any general toxic and immunogenic side effects.

Based on the strong RNAi responses in the liver, we opted to use another *in vivo* model for liver-specific silencing. In this model, mice were treated to stably express luciferase in the liver and then treated these animals via systemic injections of PF6/luc-siRNA nanoparticles at various doses. At the dose of 1 mg/kg siRNA, substantial knockdown of luciferase expression was achieved, and these effects lasted for more than a week. Moreover, substantial effects were seen at the dose of 0.2 mg/kg. Interestingly, these results were at least in line with the effects of the "gold standard" technique for liver transduction—a hydrodynamic injection technique.

Consequently, PF6 is a new potent vector for the siRNA transfections *in vitro*, as it efficiently transfects primary and suspension cell lines, in which most of the other vectors are not efficient enough. Importantly, PF6 is very efficient for the systemic delivery of siRNAs *in vivo*, with the delivery efficiency in line with the most potent delivery vectors available. Hence, it would be extremely interesting to test this system in disease-relevant models and develop optimized formulations for preclinical and clinical settings.[37]

Hassane *et al.* recently extended PF6-mediated delivery also to the SSOs (Said Hassane Fatouma *et al.*, unpublished data). It was shown that these particles were taken up in an energy-dependent manner and that clathrin-mediated endocytosis was the main route for internalization. Also, they showed that the increased biological effect was not dependent on the increased uptake of these particles and the uptake did not involve direct translocation. As the endosomal entrapment is the main limiting step in the bioavailability of CPPs, it was confirmed that PF6 substantially improves the endosomal release by showing that it causes high pH-dependent membrane destabilization of liposomes resembling the late endosomes and that addition of chloroquine did not enhance the biological effects as supposed to when particles are entrapped in the endosomes. Most importantly, PF6/SSO nanoparticles facilitated significant splice switching in the widely used HeLa pLuc705 reporter cell lines where it was extremely potent. Moreover, it was shown that it enabled the correction of almost all the transcripts of the mRNA, which is very promising.

3. PepFect14

In this peptide, in addition to stearic acid, we utilized ornithines as the main source of positive charges instead of lysines with slight modifications in the sequence of TP10.[38] Earlier reports had shown that poly-L-ornithine demonstrated superior transfection efficiency (up to 10-fold) compared to equivalent poly-L-lysine-based systems.[85] The superior efficiency was related to the higher affinity for DNA and the ability to make more stable complexes at lower charge ratios.[85] Furthermore, being a nonstandard amino acid, ornithine would be less prone to serum proteases, and thus would retain the activity in serum conditions. PepFect14 (PF14) has been tested only with SSOs so far, and it has proved very active in delivering SSOs to different cell lines including mdx mouse myotubes, a cell culture model of DMD. Interestingly, PF14 outperformed LF2000. Additionally, we tested the feasilbility of incoporation of PF14/SSO nanoparticles in solid formulation and whether such nanoparticles will remain active in the dry form. Solid dispersion was adopted to obtain uniform distribution of the nanoparticles over water-soluble excipients by solvent evaporation. Surprisingly, although we used relatively harsh drying conditions at elevated temperatures (50–60 °C) under vacuum, the resulting solid formulations were still active in the induction of splice switching even when stored at elevated temperatures for several months. To our knowledge, this was the first proof-of-concept that CPP-based nanoparticles are stable to the extent that they can be dried and incorporated in solid formulations that could be suitable for several therapeutic applications.[38]

VI. Future Directions and Perspective with CPP-Based Nanoparticles

Efficient measures are still lacking for the delivery of ON-based cargos, and CPP-based nanoparticle platform is a very promising addition to the repertoire of approaches to be used in these contexts. While the shortage of CPPs that could efficiently form nanoparticles and enable the intracellular delivery of ONs has hindered the progress of this approach, significant improvements in this field have been achieved in recent years.

Significant efforts have been made in the characterization of the CPP-based nanoparticles with ON-based cargos. While in most of the reports some of the main properties of these nanoparticles have been described, for instance, the particle size, shape, surface charge, etc., systematic information is still, for large part, missing. The reason for this is that often the studies are empirical and just look at whether one or the other vector would confer the biological activity. Consequently, there is much less information available of how exactly different vehicles, cargo, and their properties interact and how they would form particles with suitable properties for different purposes. Moreover, it is not clear how these results could be extrapolated from *in vitro* to *in vivo* conditions and what exactly is required in order to achieve the therapeutic goals and not to induce side effects. Therefore, there is a variety of questions that need to be answered in the future: (1) what are the requirements for CPPs that would be suitable for nanoparticle formation; (2) how the particle formation techniques and procedures exactly influence the formation of the nanoparticles and their properties; (3) which of these properties are crucial for the particles to be efficiently delivered into cells; (4) what kind of association and dissociation affinity would be most suited in general and for a particular cargo; (5) how exactly do the different modifications influence the nanoparticle formation; and (6) what kind of properties of the nanoparticles are pivotal for efficient delivery *in vivo*.

The main problem with CPP-based delivery has been that, while reasonable delivery efficiency has been achieved in regular adherent cell lines, the efficient targeting of the difficult-to-transfect cell lines, such as primary and suspension cells, has been very seldom achieved. Sensitivity to the presence of serum has been another aspect that has curbed the full potential of the CPP-based delivery vehicles. However, lately, many CPP-based platforms have been created that enable the transduction of very refractory cell lines in the presence of serum. Nevertheless, there is much room for improvements in this area and there is a need for vectors that would be more potent.

There have been significant advances in the CPP-mediated ON delivery *in vivo*. However, the number of successful reports is still a small and very potent vectors for systemic delivery are still lacking. Moreover, to a large part, these *in vivo* results have been achieved using less relevant reporter targets, which are well suited for screening but do not show the full potential of these delivery vehicles. Therefore, significant work is needed in order to screen and utilize disease-relevant systems in order to give a more therapeutic and translational background to these *in vivo* studies. Also, similar to many platforms of CPP/ON conjugates that are in preclinical trials, successful findings with CPP-based nanoparticles targeting relevant disease models need to be pushed into preclinical studies. This requires CPP-based vectors with improved delivery properties and very good safety profile, being nontoxic and nonimmunogenic, and well tolerated. Moreover, significant work has to be done in order to functionalize CPP-based vehicles with targeting moieties and stability enhancers *in vivo* (e.g., PEGylation) and on how to obtain these particles with exactly defined properties. Additionally, translating the promising results of the CPP platform into medicines at the bedside of patients requires the development of methods to formulate these nanoparticles into usable drug products that can be manufactured, transported, stored, and conveniently administered by patients. Tailored formulations have to be designed for different routes of administration and for different therapeutic needs.

Conclusively, although CPP-based nanoparticles have achieved substantial progress in recent years, much is still to be known about the perfect criteria and uptake mechanism, and more still to be done regarding the enhancement of efficiency, safety, and formulation. However, we believe that research in the coming years will hopefully solve most of the remaining issues and that this strategy could evolve into an important tool for delivering the dream of gene therapy.

Acknowledgments

The work presented in this chapter was supported by the Swedish Research Council (VR-NT); by the Center for Biomembrane Research, Stockholm; by the Knut and Alice Wallenberg's Foundation; by the EU through the European Regional Development Fund through the Center of Excellence in Chemical Biology, Estonia; by the targeted financing SF0180027s08 from the Estonian Government; by the DoRa Program of The European Social Fund; and by the Archimedes Foundation. We would like to thank Imre Mäger for carefully reading the chapter and for Fig. 2.

References

1. Thomas CE, Ehrhardt A, Kay MA. Progress and problems with the use of viral vectors for gene therapy. *Nat Rev Genet* 2003;**4**:346–58.
2. Pack DW, Hoffman AS, Pun S, Stayton PS. Design and development of polymers for gene delivery. *Nat Rev Drug Discov* 2005;**4**:581–93.
3. Whitehead KA, Langer R, Anderson DG. Knocking down barriers: advances in siRNA delivery. *Nat Rev Drug Discov* 2009;**8**:129–38.
4. Lindgren M, Langel Ü. Classes and prediction of cell-penetrating peptides. *Methods Mol Biol* 2011;**683**:3–19.
5. El-Andaloussi S, Holm T, Langel Ü. Cell-penetrating peptides: mechanisms and applications. *Curr Pharm Des* 2005;**11**:3597–611.
6. Mäe M, Langel Ü. Cell-penetrating peptides as vectors for peptide, protein and oligonucleotide delivery. *Curr Opin Pharmacol* 2006;**6**:509–14.
7. Brasseur R, Divita G. Happy birthday cell penetrating peptides: already 20 years. *Biochim Biophys Acta* 2010;**1798**:2177–81.
8. Fonseca SB, Pereira MP, Kelley SO. Recent advances in the use of cell-penetrating peptides for medical and biological applications. *Adv Drug Deliv Rev* 2009;**61**:953–64.
9. Hassane FS, Saleh AF, Abes R, Gait MJ, Lebleu B. Cell penetrating peptides: overview and applications to the delivery of oligonucleotides. *Cell Mol Life Sci* 2010;**67**:715–26.
10. Lundin P, Johansson H, Guterstam P, Holm T, Hansen M, Langel Ü, et al. Distinct uptake routes of cell-penetrating peptide conjugates. *Bioconjug Chem* 2008;**19**:2535–42.
11. Zorko M, Langel Ü. Cell-penetrating peptides: mechanism and kinetics of cargo delivery. *Adv Drug Deliv Rev* 2005;**57**:529–45.
12. Futaki S, Nakase I, Tadokoro A, Takeuchi T, Jones AT. Arginine-rich peptides and their internalization mechanisms. *Biochem Soc Trans* 2007;**35**:784–7.
13. El-Sayed A, Futaki S, Harashima H. Delivery of macromolecules using arginine-rich cell-penetrating peptides: ways to overcome endosomal entrapment. *AAPS J* 2009;**11**:13–22.
14. Mäe M, Andaloussi SE, Lehto T, Langel Ü. Chemically modified cell-penetrating peptides for the delivery of nucleic acids. *Expert Opin Drug Deliv* 2009;**6**:1195–205.

15. Crombez L, Morris MC, Deshayes S, Heitz F, Divita G. Peptide-based nanoparticle for ex vivo and in vivo dug delivery. *Curr Pharm Des* 2008;**14**:3656–65.
16. Mäe M, El Andaloussi S, Lundin P, Oskolkov N, Johansson HJ, Guterstam P, et al. A stearylated CPP for delivery of splice correcting oligonucleotides using a non-covalent co-incubation strategy. *J Control Release* 2009;**134**:221–7.
17. Hacein-Bey-Abina S, Garrigue A, Wang GP, Soulier J, Lim A, Morillon E, et al. Insertional oncogenesis in 4 patients after retrovirus-mediated gene therapy of SCID-X1. *J Clin Invest* 2008;**118**:3132–42.
18. Cavazzana-Calvo M. Basic research tries to decrease the risks of translational medicine. *Gene Ther* 2009;**16**:309–10.
19. Howe SJ, Mansour MR, Schwarzwaelder K, Bartholomae C, Hubank M, Kempski H, et al. Insertional mutagenesis combined with acquired somatic mutations causes leukemogenesis following gene therapy of SCID-X1 patients. *J Clin Invest* 2008;**118**:3143–50.
20. Kay MA, Glorioso JC, Naldini L. Viral vectors for gene therapy: the art of turning infectious agents into vehicles of therapeutics. *Nat Med* 2001;**7**:33–40.
21. Dass CR. Lipoplex-mediated delivery of nucleic acids: factors affecting in vivo transfection. *J Mol Med* 2004;**82**:579–91.
22. Rakoczy PE. Antisense DNA technology. *Methods Mol Med* 2001;**47**:89–104.
23. Bell NM, Micklefield J. Chemical modification of oligonucleotides for therapeutic, bioanalytical and other applications. *Chembiochem* 2009;**10**:2691–703.
24. Elbashir SM, Harborth J, Lendeckel W, Yalcin A, Weber K, Tuschl T. Duplexes of 21-nucleotide RNAs mediate RNA interference in cultured mammalian cells. *Nature* 2001;**411**:494–8.
25. Bauman J, Jearawiriyapaisarn N, Kole R. Therapeutic potential of splice-switching oligonucleotides. *Oligonucleotides* 2009;**19**:1–13.
26. Nilsen TW, Graveley BR. Expansion of the eukaryotic proteome by alternative splicing. *Nature* 2010;**463**:457–63.
27. Sazani P, Kole R. Therapeutic potential of antisense oligonucleotides as modulators of alternative splicing. *J Clin Invest* 2003;**112**:481–6.
28. Wood MJ, Gait MJ, Yin H. RNA-targeted splice-correction therapy for neuromuscular disease. *Brain* 2010;**133**:957–72.
29. Ivanova GD, Arzumanov A, Abes R, Yin H, Wood MJ, Lebleu B, et al. Improved cell-penetrating peptide-PNA conjugates for splicing redirection in HeLa cells and exon skipping in mdx mouse muscle. *Nucleic Acids Res* 2008;**36**:6418–28.
30. van Deutekom JC, Janson AA, Ginjaar IB, Frankhuizen WS, Aartsma-Rus A, Bremmer-Bout M, et al. Local dystrophin restoration with antisense oligonucleotide PRO051. *N Engl J Med* 2007;**357**:2677–86.
31. Kinali M, Arechavala-Gomeza V, Feng L, Cirak S, Hunt D, Adkin C, et al. Local restoration of dystrophin expression with the morpholino oligomer AVI-4658 in Duchenne muscular dystrophy: a single-blind, placebo-controlled, dose-escalation, proof-of-concept study. *Lancet Neurol* 2009;**8**:918–28.
32. Oren Z, Shai Y. Mode of action of linear amphipathic alpha-helical antimicrobial peptides. *Biopolymers* 1998;**47**:451–63.
33. Derossi D, Calvet S, Trembleau A, Brunissen A, Chassaing G, Prochiantz A. Cell internalization of the third helix of the Antennapedia homeodomain is receptor-independent. *J Biol Chem* 1996;**271**:18188–93.
34. Richard JP, Melikov K, Vives E, Ramos C, Verbeure B, Gait MJ, et al. Cell-penetrating peptides—a reevaluation of the mechanism of cellular uptake. *J Biol Chem* 2003;**278**:585–90.
35. Khalil IA, Futaki S, Niwa M, Baba Y, Kaji N, Kamiya H, et al. Mechanism of improved gene transfer by the N-terminal stearylation of octaarginine: enhanced cellular association by hydrophobic core formation. *Gene Ther* 2004;**11**:636–44.

36. Eguchi A, Dowdy SF. Efficient siRNA delivery by novel PTD-DRBD fusion proteins. *Cell Cycle* 2010;**9**:424–5.
37. EL Andaloussi S, Lehto T, Mager I, Rosenthal-Aizman K, Oprea II, Simonson OE, et al. Design of a peptide-based vector, PepFect6, for efficient delivery of siRNA in cell culture and systemically in vivo. *Nucleic Acids Res* 2011;**39**:3972–87.
38. Ezzat K, El Andaloussi S, Zaghloul EM, Lehto T, Lindberg S, Moreno PM, et al. PepFect 14, a novel cell-penetrating peptide for oligonucleotide delivery in solution and as solid formulation. *Nucleic Acids Res* 2011;**39**:5284–98.
39. Lehto T, Simonson OE, Mager I, Ezzat K, Sork H, Copolovici DM, et al. A peptide-based vector for efficient gene transfer in vitro and in vivo. *Mol Ther* 2011;**19**:1457–67.
40. Ezzat K, El Andaloussi S, Abdo R, Langel U. Peptide-based matrices as drug delivery vehicles. *Curr Pharm Des* 2010;**16**:1167–78.
41. Deshayes S, Konate K, Aldrian G, Crombez L, Heitz F, Divita G. Structural polymorphism of non-covalent peptide-based delivery systems: highway to cellular uptake. *Biochim Biophys Acta* 2010;**1798**:2304–14.
42. Lynch I, Dawson KA, Linse S. Detecting cryptic epitopes created by nanoparticles. *Sci STKE* 2006;**2006**:pe14.
43. Jatzkewitz H. Incorporation of physiologically-active substances into a colloidal blood plasma substitute. I. Incorporation of mescaline peptide into polyvinylpyrrolidone. *Hoppe Seylers Z Physiol Chem* 1954;**297**:149–56.
44. Davis ME, Chen ZG, Shin DM. Nanoparticle therapeutics: an emerging treatment modality for cancer. *Nat Rev Drug Discov* 2008;**7**:771–82.
45. Zhang L, Gu FX, Chan JM, Wang AZ, Langer RS, Farokhzad OC. Nanoparticles in medicine: therapeutic applications and developments. *Clin Pharmacol Ther* 2008;**83**:761–9.
46. Viola JR, El-Andaloussi S, Oprea II, Smith CIE. Non-viral nanovectors for gene delivery: factors that govern successful therapeutics. *Expert Opin Drug Deliv* 2010;**7**:721–35.
47. Glover DJ, Lipps HJ, Jans DA. Towards safe, non-viral therapeutic gene expression in humans. *Nat Rev Genet* 2005;**6**:299–310.
48. Morris MC, Vidal P, Chaloin L, Heitz F, Divita G. A new peptide vector for efficient delivery of oligonucleotides into mammalian cells. *Nucleic Acids Res* 1997;**25**:2730–6.
49. Meade BR, Dowdy SF. Enhancing the cellular uptake of siRNA duplexes following noncovalent packaging with protein transduction domain peptides. *Adv Drug Deliv Rev* 2008;**60**:530–6.
50. Bloomfield VA. DNA condensation by multivalent cations. *Biopolymers* 1997;**44**:269–82.
51. Schaffer DV, Fidelman NA, Dan N, Lauffenburger DA. Vector unpacking as a potential barrier for receptor-mediated polyplex gene delivery. *Biotechnol Bioeng* 2000;**67**:598–606.
52. Zelphati O, Szoka Jr. FC. Mechanism of oligonucleotide release from cationic liposomes. *Proc Natl Acad Sci USA* 1996;**93**:11493–8.
53. Nakase I, Takeuchi T, Tanaka G, Futaki S. Methodological and cellular aspects that govern the internalization mechanisms of arginine-rich cell-penetrating peptides. *Adv Drug Deliv Rev* 2008;**60**:598–607.
54. Nakase I, Tadokoro A, Kawabata N, Takeuchi T, Katoh H, Hiramoto K, et al. Interaction of arginine-rich peptides with membrane-associated proteoglycans is crucial for induction of actin organization and macropinocytosis. *Biochemistry* 2007;**46**:492–501.
55. Patel MM, Anchordoquy TJ. Contribution of hydrophobicity to thermodynamics of ligand-DNA binding and DNA collapse. *Biophys J* 2005;**88**:2089–103.
56. Morris MC, Depollier J, Mery J, Heitz F, Divita G. A peptide carrier for the delivery of biologically active proteins into mammalian cells. *Nat Biotechnol* 2001;**19**:1173–6.
57. Lehto T, Abes R, Oskolkov N, Suhorutšenko J, Copolovici DM, Mäger I, et al. Delivery of nucleic acids with a stearylated (RxR)(4) peptide using a non-covalent co-incubation strategy. *J Control Release* 2010;**141**:42–51.

58. Nakamura Y, Kogure K, Futaki S, Harashima H. Octaarginine-modified multifunctional envelope-type nano device for siRNA. *J Control Release* 2007;**119**:360–7.
59. EL Andaloussi S, Lehto T, Lundin P, Langel Ü. Application of PepFect peptides for the delivery of splice-correcting oligonucleotides. *Methods Mol Biol* 2011;**683**:361–73.
60. Konate K, Crombez L, Deshayes S, Decaffmeyer M, Thomas A, Brasseur R, et al. Insight into the cellular uptake mechanism of a secondary amphipathic cell-penetrating peptide for siRNA delivery. *Biochemistry* 2010;**49**:3393–402.
61. Crombez L, Aldrian-Herrada G, Konate K, Nguyen QN, McMaster GK, Brasseur R, et al. A new potent secondary amphipathic cell-penetrating peptide for siRNA delivery into mammalian cells. *Mol Ther* 2009;**17**:95–103.
62. Simeoni F, Morris MC, Heitz F, Divita G. Insight into the mechanism of the peptide-based gene delivery system MPG: implications for delivery of siRNA into mammalian cells. *Nucleic Acids Res* 2003;**31**:2717–24.
63. Crombez L, Morris MC, Dufort S, Aldrian-Herrada G, Nguyen Q, Mc Master G, et al. Targeting cyclin B1 through peptide-based delivery of siRNA prevents tumour growth. *Nucleic Acids Res* 2009;**37**:4559–69.
64. Morris MC, Chaloin L, Choob M, Archdeacon J, Heitz F, Divita G. Combination of a new generation of PNAs with a peptide-based carrier enables efficient targeting of cell cycle progression. *Gene Ther* 2004;**11**:757–64.
65. Morris MC, Gros E, Aldrian-Herrada G, Choob M, Archdeacon J, Heitz F, et al. A non-covalent peptide-based carrier for in vivo delivery of DNA mimics. *Nucleic Acids Res* 2007;**35**: e49.
66. Rittner K, Benavente A, Bompard-Sorlet A, Heitz F, Divita G, Brasseur R, et al. New basic membrane-destabilizing peptides for plasmid-based gene delivery in vitro and in vivo. *Mol Ther* 2002;**5**:104–14.
67. Futaki S, Ohashi W, Suzuki T, Niwa M, Tanaka S, Ueda K, et al. Stearylated arginine-rich peptides: a new class of transfection systems. *Bioconjug Chem* 2001;**12**:1005–11.
68. Kim WJ, Christensen LV, Jo S, Yockman JW, Jeong JH, Kim YH, et al. Cholesteryl oligoarginine delivering vascular endothelial growth factor siRNA effectively inhibits tumor growth in colon adenocarcinoma. *Mol Ther* 2006;**14**:343–50.
69. Soomets U, Lindgren M, Gallet X, Hällbrink M, Elmquist A, Balaspiri L, et al. Deletion analogues of transportan. *Biochim Biophys Acta* 2000;**1467**:165–76.
70. Morris MC, Chaloin L, Mery J, Heitz F, Divita G. A novel potent strategy for gene delivery using a single peptide vector as a carrier. *Nucleic Acids Res* 1999;**27**:3510–7.
71. Veldhoen S, Laufer SD, Trampe A, Restle T. Cellular delivery of small interfering RNA by a non-covalently attached cell-penetrating peptide: quantitative analysis of uptake and biological effect. *Nucleic Acids Res* 2006;**34**:6561–73.
72. Hatakeyama H, Ito E, Akita H, Oishi M, Nagasaki Y, Futaki S, et al. A pH-sensitive fusogenic peptide facilitates endosomal escape and greatly enhances the gene silencing of siRNA-containing nanoparticles in vitro and in vivo. *J Control Release* 2009;**139**:127–32.
73. Kumar P, Wu H, McBride JL, Jung KE, Kim MH, Davidson BL, et al. Transvascular delivery of small interfering RNA to the central nervous system. *Nature* 2007;**448**:39–43.
74. Eguchi A, Meade BR, Chang YC, Fredrickson CT, Willert K, Puri N, et al. Efficient siRNA delivery into primary cells by a peptide transduction domain-dsRNA binding domain fusion protein. *Nat Biotechnol* 2009;**27**:567–71.
75. Michiue H, Eguchi A, Scadeng M, Dowdy SF. Induction of in vivo synthetic lethal RNAi responses to treat glioblastoma. *Cancer Biol Ther* 2009;**8**:2306–13.
76. Abes R, Moulton HM, Clair P, Yang ST, Abes S, Melikov K, et al. Delivery of steric block morpholino oligomers by (R-X-R)4 peptides: structure-activity studies. *Nucleic Acids Res* 2008;**36**:6343–54.

77. Abes S, Moulton HM, Clair P, Prevot P, Youngblood DS, Wu RP, et al. Vectorization of morpholino oligomers by the (R-Ahx-R)4 peptide allows efficient splicing correction in the absence of endosomolytic agents. *J Control Release* 2006;**116**:304–13.
78. Jearawiriyapaisarn N, Moulton HM, Buckley B, Roberts J, Sazani P, Fucharoen S, et al. Sustained dystrophin expression induced by peptide-conjugated morpholino oligomers in the muscles of mdx mice. *Mol Ther* 2008;**16**:1624–9.
79. Moulton HM, Fletcher S, Neuman BW, McClorey G, Stein DA, Abes S, et al. Cell-penetrating peptide-morpholino conjugates alter pre-mRNA splicing of DMD (Duchenne muscular dystrophy) and inhibit murine coronavirus replication in vivo. *Biochem Soc Trans* 2007;**35**:826–8.
80. Pooga M, Hällbrink M, Zorko M, Langel Ü. Cell penetration by transportan. *FASEB J* 1998;**12**:67–77.
81. Pooga M, Soomets U, Hallbrink M, Valkna A, Saar K, Rezaei K, et al. Cell penetrating PNA constructs regulate galanin receptor levels and modify pain transmission in vivo. *Nat Biotechnol* 1998;**16**:857–61.
82. EL Andaloussi S, Johansson H, Magnusdottir A, Järver P, Lundberg P, Langel Ü. TP10, a delivery vector for decoy oligonucleotides targeting the Myc protein. *J Control Release* 2005;**110**:189–201.
83. Kilk K, El Andaloussi S, Järver P, Meikas A, Valkna A, Bartfai T, et al. Evaluation of transportan 10 in PEI mediated plasmid delivery assay. *J Control Release* 2005;**103**:511–23.
84. Lundberg P, El Andaloussi S, Sütlü T, Johansson H, Langel Ü. Delivery of short interfering RNA using endosomolytic cell-penetrating peptides. *FASEB J* 2007;**21**:2664–71.
85. Ramsay E, Gumbleton M. Polylysine and polyornithine gene transfer complexes: a study of complex stability and cellular uptake as a basis for their differential in-vitro transfection efficiency. *J Drug Target* 2002;**10**:1–9.

Nanoparticles in Molecular Diagnostics

Pedro V. Baptista,[*] Gonçalo Doria,[*] Pedro Quaresma,[*,†] Miguel Cavadas,[‡] Cristina S. Neves,[†] Inês Gomes,[‡] Peter Eaton,[†] Eulália Pereira,[†] and Ricardo Franco[‡]

[*]*Departamento de Ciências da Vida, Faculdade de Ciências e Tecnologia, Centro de Investigação em Genética Molecular Humana (CIGMH), Universidade Nova de Lisboa, Caparica, Portugal*

[†]*REQUIMTE, Departamento de Química e Bioquímica, Faculdade de Ciências da Universidade do Porto, R. Campo Alegre, Porto, Portugal*

[‡]*REQUIMTE, Departamento de Química, Faculdade de Ciências e Tecnologia, Universidade Nova de Lisboa, Caparica, Portugal*

I. Unique Properties of Nanoparticles for Molecular Diagnostics 428
A. Tailored Synthesis and Functionalization of Nanoparticles 429
B. AuNPs for Molecular Diagnostics 436
C. QDs and Fluorescent Silica Nanoparticles for Molecular Diagnostics 437
D. Magnetic Nanoparticles for Molecular Diagnostics 441
II. Molecular Diagnostics by Nucleic Acid Detection 445
A. Genetic Basis of Human Disorders 445
B. Current Nucleic Acid Diagnostics Techniques 446
C. Methods Based on the Optical Properties of AuNPs 449
D. Electrochemical Methods Based on AuNPs 455
E. Nucleic Acid Detection Methods Based on Silver/AuNPs 455
F. Methods Based on Fluorescent Nanoparticles 456
G. High-throughput Nucleic Acid Detection Assays 460
III. Molecular Diagnostics by Immunoassays 461
A. AuNPs-Based Immunoassays 461
B. QDs in Immunoassays 469
IV. Conclusions and Future Perspectives 471
References 473

The aim of this chapter is to provide an overview of the available and emerging molecular diagnostic methods that take advantage of the unique nanoscale properties of nanoparticles (NPs) to increase the sensitivity,

Progress in Molecular Biology
and Translational Science, Vol. 104
DOI: 10.1016/B978-0-12-416020-0.00011-5

detection capabilities, ease of operation, and portability of the biodetection assemblies. The focus will be on noble metal NPs, especially gold NPs, fluorescent NPs, especially quantum dots, and magnetic NPs, the three main players in the development of probes for biological sensing. The chapter is divided into four sections: a first section covering the unique physicochemical properties of NPs of relevance for their utilization in molecular diagnostics; the second section dedicated to applications of NPs in molecular diagnostics by nucleic acid detection; and the third section with major applications of NPs in the area of immunoassays. Finally, a concluding section highlights the most promising advances in the area and presents future perspectives.

Abbreviations: AFM, atomic force microscopy; Au nanoprobes, gold nanoparticles functionalized with thiol-modified oligonucleotides; AuNPs, gold nanoparticles; BSA, bovine serum albumin; CCD, charge-coupled device; CL, chemiluminescence; CTAB, cetyltrimethylammonium bromide; dsDNA, double-stranded DNA; EDC, 1-ethyl-3-(3-dimethylaminopropyl)-carbodiimide hydrochloride; ELISA, enzyme-linked immunosorbent assay; ET, energy transfer; Fab, fragment antigen-binding immunoglobulin region; FITC, fluorescein isothiocyanate; FRET, Förster resonance energy transfer; FS NP, fluorophore-loaded silica nanoparticles; GFP, green fluorescent protein; HCV, hepatitis C virus; IgG, immunoglobulin class G; IR, infrared; LOD, limit of detection; LSPR, localized surface plasmon resonance; MUA, 11-mercaptoundecanoic acid; NHS, *N*-hydroxysuccinimide; NIR, near-infrared; NP, nanoparticle; PCR, polymerase chain reaction; PEG, polyethylene glycol; PL, photoluminescence; POC, point-of-care; PSA, prostate-specific antigen; QD, quantum dot; SERS, surface-enhanced Raman scattering; SNP, single nucleotide polymorphism; SPR, surface plasmon resonance; ssDNA, single-stranded DNA; UV, ultraviolet.

I. Unique Properties of Nanoparticles for Molecular Diagnostics

Molecular diagnostics identify disease information at the genetic, protein, or metabolite level or assess the presence of a given pathogen, such as bacteria or viruses. The aim of this chapter is to provide an overview of the available and emerging molecular diagnostic methods that take advantage of the unique nanoscale properties of nanoparticles (NPs) to increase the sensitivity, detection capabilities, ease of operation, and portability of the biodetection assemblies.

The focus will be on noble metal NPs, especially gold NPs (AuNPs), fluorescent NPs, especially quantum dots (QDs), and magnetic NPs, the three most widely used categories of nanoparticle studied in the development of probes for biological sensing. Depending on their constitutive material, NPs can possess a number of distinctive properties such as high electron density and strong optical absorption (e.g., AuNPs), intense photoluminescence (PL) (semiconductor QD, e.g., CdSe or CdTe, and fluorescent silica NPs), or magnetic moment (e.g., magnetite NPs).[1] Applications to diagnostics of other types of nanomaterials, such as nanocarbon (e.g., carbon nanotubes and fullerenes), as well of other morphologies for the nanostructures, namely, nanotubes, nanowires, and nanocantilevers, has been dealt with in several recent books[2–4] and review articles.[5,6] In terms of the molecular analytical targets, special emphasis will be given to nucleic acids and antigens, the major detection targets in clinical assays.

Using the functional definition of Algar, Tavares, and Krull,[7] the role of NPs in biosensing can be distinguished between "integrated" and "nonintegrated" based on the design of the bioassay or biosensor. A *nonintegrated NP* is selectively introduced into bioanalysis as a consequence of biorecognition. Examples include the use of QDs as labels in microarrays,[8,9] or electroactive labels in assays based on anodic stripping voltammetry[10–14]; or the use of AuNPs as signal enhancement labels in detection by surface-enhanced Raman scattering (SERS).[15] In contrast, an *integrated NP* is present in a system throughout bioanalysis, simultaneously having a role in signal transduction and being a scaffold for biorecognition. In many cases, this requires the direct conjugation of affinity probes or biorecognition elements to the NP[7,16–18] or their co-assembly at an interface.[19] A prerequisite for these applications is suitable engineering of the NPs' interface, which will determine their interaction with the surrounding molecules. Nanoparticle functionalization is thus an extremely relevant feature for the specific application examples that will be described.

A. Tailored Synthesis and Functionalization of Nanoparticles

A large variety of methods to synthesize NPs of many different materials are currently available and have been recently reviewed.[20–23] In addition, commercially available NPs are increasingly common, and although the price is still high for most industrial applications, it is now possible to buy NPs with a wide range of sizes and good size dispersion for research applications. Many companies also offer a good variety of different surface functionalities, designed for different applications. As such, the need to synthesize NPs for bioassays has become less important, although the vast majority of the work published in this area utilizes NPs specifically synthesized for the particular application.

One of the major goals in the preparation of NPs is to obtain a material with a good homogeneity in physical properties, which ultimately means a tight control over size, shape, and surface properties. Depending on the particular synthetic method used and on the size range intended for the NPs, size dispersion can be as low as 5% for very small NPs (<5 nm) or 20% for larger NPs. Methods to prepare NPs may be either physical (chemical vapor deposition, laser ablation, grinding, etc.) or chemical (chemical reduction, coprecipitation, thermal decomposition, hydrolysis, etc.). The latter are far more common, due to high versatility and facile operation, whereas physical methods usually require expensive instrumentation. Solution methods commonly rely on the use of chemical species that adsorb onto the surface of NPs, usually known as capping agents. The capping agent performs a dual role of preventing uncontrolled growth and imparting surface properties that stabilize the NPs toward irreversible aggregation. The choice of capping agent is thus critical not only for the properties of the final NPs but also for downstream manipulation, including purification, size separation, and functionalization.

The choice of capping agent depends mainly on the NP material and the solvent used in the synthesis. The capping molecules should have a high affinity for the NPs and also impart either hydrophilic (for aqueous synthesis) or hydrophobic (for synthesis in organic solvents) properties to the NPs. In biological applications, aqueous synthesis is obviously favored, but in some cases, it may be advantageous to synthesize the NPs in organic media in order to achieve a better control over the physical and chemical properties and homogeneity of the NP colloidal solution. In this case, several methods are known to transfer NPs to an aqueous solution with the help of an appropriate transfer agent.[1]

Although in principle a strongly adsorbing capping agent will provide better colloidal stabilization to NPs, in some cases, it is necessary to use a weaker capping agent in the synthesis. For instance, many synthetic procedures for anisotropic NPs use the weak adsorbing surfactant cetyltrimethylammonium bromide (CTAB) as the capping agent. In this case, colloidal stability may be quite poor, and several experimental procedures have been developed to exchange the capping agent postsynthesis.

The adsorption of the capping agent at the surface of the NPs can be by chemisorption, electrostatic interaction, or hydrophobic interaction. Binding of these molecules to the NP surface is usually a dynamic and reversible process, and thus removal or exchange of the capping agent may occur either by excessive washing of the NPs or by addition of another capping molecule. Most frequently, insufficient capping of the NPs leads to irreversible aggregation, and it is thus to be avoided in all postsynthesis manipulation steps. In aqueous solution, colloidal stabilization is commonly due to a charged capping agent, and it is thus quite sensitive to ionic strength. In addition, frequently the charged group has acid–base properties, and so the NPs may be soluble only in a limited pH range, which will depend on the isoelectric point of the capped NPs.

Preferably, the capping agent should have a high affinity for the NP surface and form a dense layer around the NP. For example, bifunctional molecules containing a thiol headgroup, an aliphatic chain, and a charged end group are widely used capping agents for AuNPs. The thiol group is known to have a strong affinity toward noble metals, especially gold, but also for QDs, and thus adsorbs onto the surface of the NP. Dense packing of the aliphatic chains through hydrophobic interactions aids stabilization, and the terminal charged group improves colloidal stabilization in aqueous solution. The most common example of this type of bifunctional molecules is 11-mercaptoundecanoic acid, containing a carboxylate terminal group. Nevertheless, other charged groups such as sulfonate and trimethylammonium can also impart a good colloidal stability to either metal NPs or QDs.

Another common strategy for colloidal stabilization is the use of polymers containing strong adsorbing groups. Polymer capping is usually found to be very effective due to the entropic effect associated with multiple binding, but also to the steric stabilization provided by the macromolecule. Commonly used polymers are polyvinylpyrrolidone, poly(acrylic acid) (either modified or not with additional binding agents, such as amines), copolymers based on poly (maleic acid), polyethyleneimine, etc. Table I presents examples of common capping agents for several different materials of NPs.

Silica nanoparticles and core–shell nanoparticles with a silica shell are usually modified with an organosilane coupling agent, which will form a covalent bond to the silica via a condensation reaction with the hydrolyzed form of the methoxy or ethoxy moiety of the organosilane, although other absorption methods have been tried.[35] A large variety of organosilanes are easily available, allowing almost any functional group to be present on the particle surface after silanization, which can be further used for biomolecule attachment.

Further functionalization of the NPs with a sensing molecule is necessary in order to use the unique physical properties of NPs for detection. The sensing molecule may be a small molecule with a large affinity for the analyte, or a biomacromolecule, such as DNA, antibodies, or enzymes. Functionalization of the NPs with sensing molecules has been achieved by (i) electrostatic interaction (ii) chemisorption (e.g., by thiol or amine groups), (iii) covalent binding, and (iv) affinity-based systems.

Electrostatic interaction is the simplest of these methodologies and has been successfully used in many applications. The major drawbacks are that it is restricted to charged sensing molecules and that the adsorption is usually quite sensitive to environmental properties, such as ionic strength, pH, and the presence of other charged molecules in the solution.

Chemisorption of the biosensing species may be a better option for more robust NPs. For example, thiol-modified oligonucleotides have been used to functionalize either AuNPs or QDs, to specifically detect DNA and RNA, as extensively described in the following section of this chapter. Chemisorption of proteins and antibodies is also a known phenomenon, although in this case the process can be much more complex, since there are usually many chemical groups available in the protein for adsorption, giving rise to a statistical distribution of protein binding to NPs. It is also possible that direct chemisorption of a protein to the surface of the inorganic NP material may affect the biological activity of the protein, either by changing the tertiary structure of the protein or

TABLE I
Common Capping Agents for Nanoparticles of Different Materials

Capping agents	NP material	Refs.
11-Mercaptoundecanoic acid (MUA) 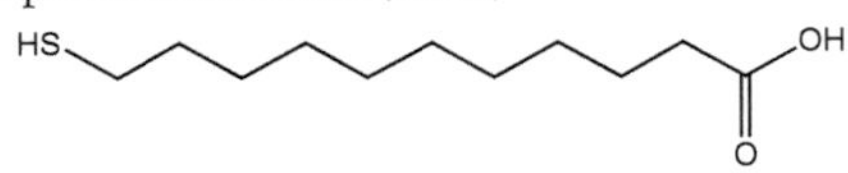 	Metals (Au, Ag, Pd, Pt, etc.), QDs	24
Dithiol 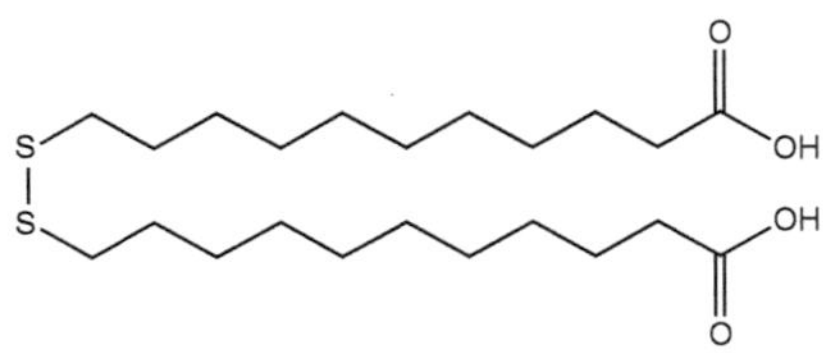 		25
Citric acid 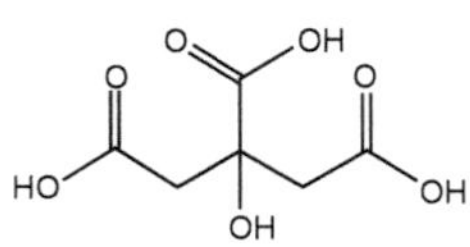 	Metals (Au, Ag), iron oxides	26,27
Polyethyeleneglycol (PEG) and functionalized derivatives 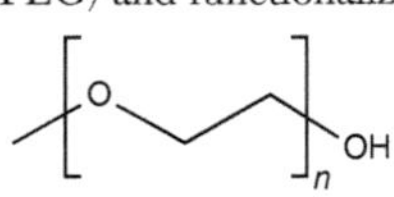 	Metals (Au, Ag, etc.), QDs, metal oxides, silica, etc.	28
Polyacrylic acid and amino-modified polyacrylic acid 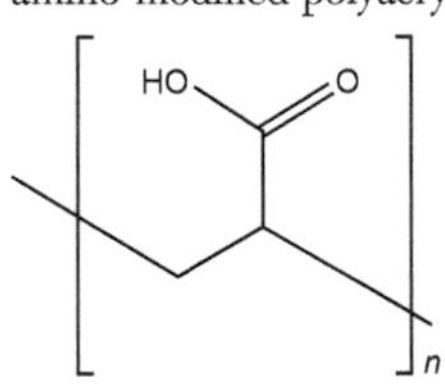 		29

(Continues)

TABLE I (*Continued*)

Capping agents	NP material	Refs.
Ascorbic acid 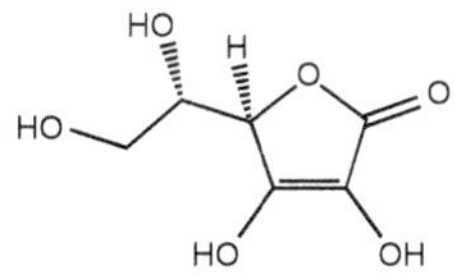	Iron oxides (magnetite, maghemite)	30
Catechol derivatives 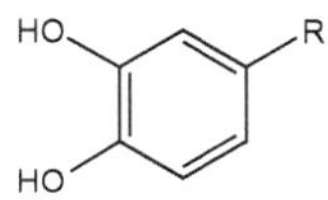		31,32
Oleic acid 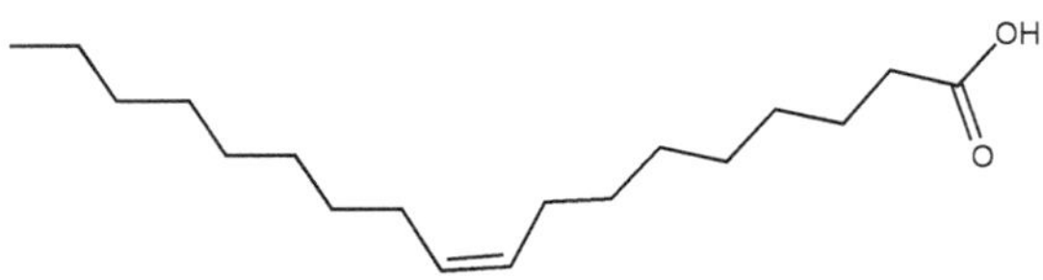		33
Trioctylphosphine oxide (TOPO) and trioctylphosphine (TOP) 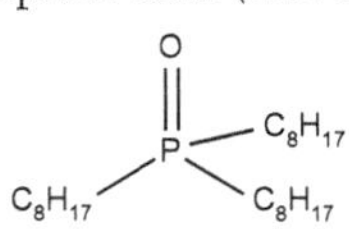 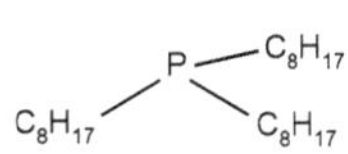	QDs	34

by blocking access to its reaction pocket. In addition, the protein can be desorbed in protein-rich sera.[36] It is thus usually preferred to use a linker between the NPs and the protein molecule, this linker being commonly the capping agent itself that may be coupled to a protein using well-known bioconjugation techniques. It should be mentioned that bioconjugation techniques are often preferred for covalent binding of a sensing molecule to NPs, even when the sensing molecule is not a biomacromolecule, since NP colloidal solutions can be unstable in the harsh conditions necessary for other chemical conjugation methods. Bioconjugation techniques are well known and have been thoroughly discussed elsewhere.[37] For example, the formation of an amide bond between NPs with carboxylate surface groups and available amine groups in proteins, using 1-ethyl-3-(3-dimethylaminopropyl)-

carbodiimide hydrochloride (EDC)/*N*-hydroxysuccinimide (NHS), is one of the most popular methodologies for covalent binding of proteins to NPs. Bioconjugation of thiol and amine groups is also easily accomplished using maleimide-derived coupling reagents and has been applied, for example, in the biofunctionalization of magnetite and fluorescent silica NPs.[38] Schemes for the chemical reactions of these functionalizations are depicted in Fig. 1.

Functionalization of NPs through affinity binding using well-known binding partners (e.g., avidin–biotin) is also a convenient way to bind biological molecules to NPs. This strategy requires the functionalization of the NPs with one of the binding partners and the functionalization of the biomolecule with the other binding partner. Since these binding partners are widely used in bioconjugation, it is simple to develop similar conjugation procedures for NPs. For example, biotinylated polyethylene glycol (PEG) was used as a capping agent for NPs that show a high affinity for avidin or avidin-functionalized species[39] (Fig. 2). The main disadvantage of this approach is the size of avidin (or one of their recombinant analogs, streptavidin, neutravidin, etc.) that adds to the size of the NPs and may lead to colloidal destabilization. In addition, since each avidin molecule has four binding sites for biotin, and the functionalized NPs also exhibit multiple binding sites for avidin, there is a high probability of interparticle cross-linking, which will decrease the binding ability

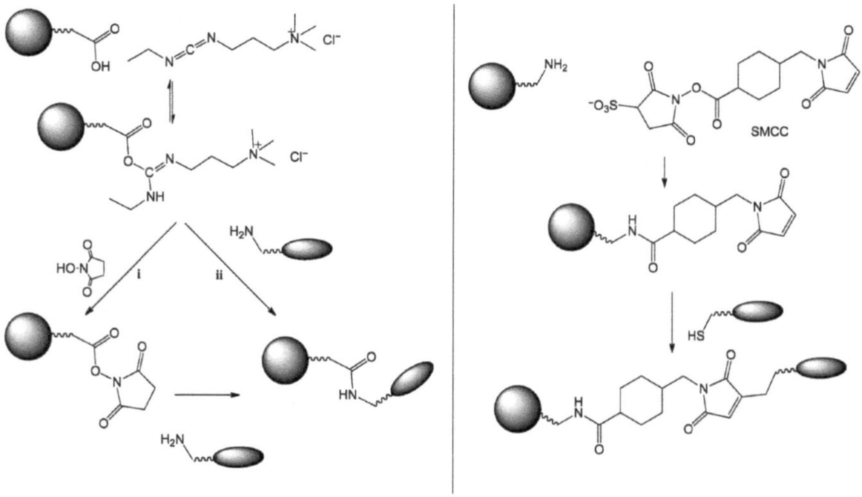

FIG. 1. Reactions schemes for the covalent binding of a biological macromolecule containing one amine group and a NP containing a carboxylate surface group using EDC/NHS coupling (left); and of a biological molecule containing a thiol group and a NP containing a surface amine group by maleimide coupling (right). Spheres represent nanoparticles, and oblates represent the biological molecule.

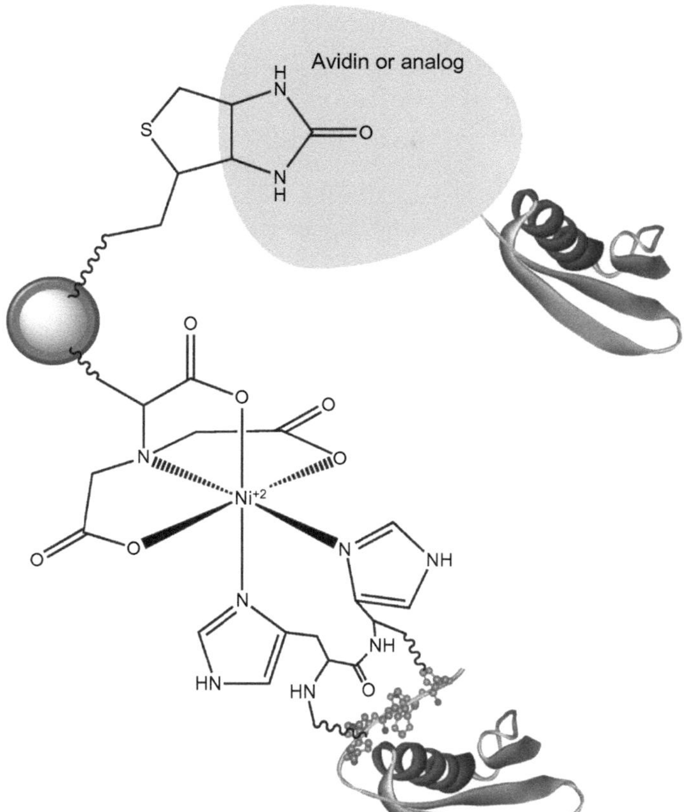

FIG. 2. Functionalization strategies for NP bioconjugation through affinity binding. *Top*: Biotinylated nanoparticle linked to a protein conjugated with avidin or an avidin analog. *Bottom*: Nanoparticle functionalized with Ni–NTA linked to the His-tag of a recombinant protein. (For color version of this figure, the reader is referred to the Web version of this chapter.)

toward the target biomolecule and reduce colloidal stability. Another example of using affinity binding in the modification of NPs is the functionalization with the Ni–NTA (nickel(II)–nitrilotriacetate complex, Fig. 2) which has a high affinity for His-tagged proteins. For example, Ni–NTA-modified magnetic particles were used for the magnetic-assisted purification of proteins directly from the cell lysate,[32] and attachment of recombinant horseradish peroxidase and ferredoxin-$NADP^+$ reductase to AuNPs did not affect enzyme activity.[40] Nevertheless, this method is restricted to recombinant proteins in which a His-tag is introduced.

B. AuNPs for Molecular Diagnostics

AuNPs have been used as radioactive labels *in vivo* since the 1950s, and since the 1980s, immunoconjugated AuNPs have found use as density probes in electron microscopy of biological samples. From the point of view of *in vivo* applications, AuNPs are practically nontoxic[41] and easy to functionalize with appropriate ligands. These ligands afford additional robustness to a delivery/imaging system in which the gold core imparts stability to the assembly while the functionalization monolayer allows tuning of surface properties, such as charge and hydrophobicity.[42] From the point of view of diagnostics applications, AuNPs surface-enhanced and distance- and refractive index (RI)-dependent spectroscopic properties make them unique players in the biodetection arena.

Three main approaches can be defined for biodiagnostics using AuNPs, based on the type of nanoscale properties involved[16]: (i) colorimetric sensing depending on inter-AuNP distance, representing the most developed approach, especially for specific nucleic acid sequence detection; (ii) surface-functionalized AuNPs providing highly selective probes that are most generally used in fluorescence quenching-based assays, although surface-enhanced based assays are also gaining importance; and (iii) electrochemical detection methods based on electrical signal enhancement provided by AuNPs.

1. AuNPs for Colorimetric Sensing

The use of AuNPs for sensing based on colorimetric assays is highly developed and has been used to detect specific DNA/RNA sequences and single nucleotide polymorphisms (SNPs). The basis for these assays is simply the color change from red to blue as spherical AuNPs with ca. 15 nm diameter change from a dispersed to an aggregated state. Recognition of the target analyte by base pairing can occur either through complementary oligomeric sequences that are bound by a thiol group to the AuNPs[16,43–47] or that are simply present in the same solution as unmodified AuNPs.[48–51] Section II contains a detailed overview of the applications of this method to specific DNA/RNA sequences and SNPs as a diagnostic tool. This type of simple and inexpensive colorimetric assay based on color change induced by AuNP aggregation has been also extensively used to quantitate proteins/antigens[52–54] and sugars[55] in solution.

2. AuNPs as Fluorescent Quenchers

The interesting PL properties of functionalized AuNPs can be explored for molecular diagnostic applications. In fact, AuNPs can cause fluorescence enhancement or fluorescence quenching of chromophores in its vicinity. Chromophores within ca. 5 nm of the surface of the AuNP have their fluorescence quenched, while chromophores at distances of ca. 10 nm or greater have their fluorescence enhanced up to 100-fold.[56] Chromophores within 5 nm of the surface interact electronically with the surface to donate the excited electrons to the metal, thus quenching the fluorescence by nonradiative pathways. Nevertheless, AuNPs can be highly efficient fluorescence quenchers over longer distances than conventional dye molecules for a variety of reasons.[57–59]

This quenching effect presented by AuNPs has been elegantly used by the Rotello and Bunz groups to develop array-based sensor platforms. This type of assays has been developed to identify several different proteins[60] and even for rapid and efficient whole-bacteria identification.[61] When used to detect proteins present in human serum, detection attained an identification accuracy of 100% in buffer and 97% in human serum.[62] These detection systems are based on specifically functionalized AuNP–fluorophore constructs. The constructs are made of AuNPs functionalized with positively charged ligands bearing different terminal functionalities (alkyl, cycloalkyl, aromatic, polar) working as the recognition element, and the anionic green fluorescent protein[62,63] or a synthetic polymer (PPE)[61] as the transducer element. The discerning capabilities of the method toward a protein or a cell rely on the judicious choice of terminal functionalizations of the AuNP responsible for direct interaction with the analyte (Fig. 3). The proposed approach is well suited for the detection of overall levels of serum proteins, whereas specific recognition of biomarkers, that is, proteins present in small quantities, is usually achieved by specific monoclonal antibodies.[62] In Section III, examples are presented for the use of the fluorescence quenching properties of AuNPs in highly sensitive immunological assays for the detection of clinically relevant analytes.

3. AuNPs in Electrochemical Applications

AuNPs are useful in electrochemical bioassays, in particular, to connect enzymes to electrode surfaces, mediate electrochemical reactions as redox catalysts, and amplify recognition signals for biological processes.[64,65] Examples of applications in DNA detection include direct detection of AuNPs anchored onto the surface of the genosensor, conductometric detection, and AuNPs as carriers of other AuNPs or of other electroactive labels.[66] Several application examples of different electrochemical AuNP-based detection methods for DNA and immunodetection can be found in the respective sections of this chapter.

C. QDs and Fluorescent Silica Nanoparticles for Molecular Diagnostics

QDs are one-dimensional semiconductor nanocrystals, usually with core diameters between 1 and 10 nm (see the chapter by Ishikawa and Biju in this volume). The first QDs were described around 30 years ago,[67] and early

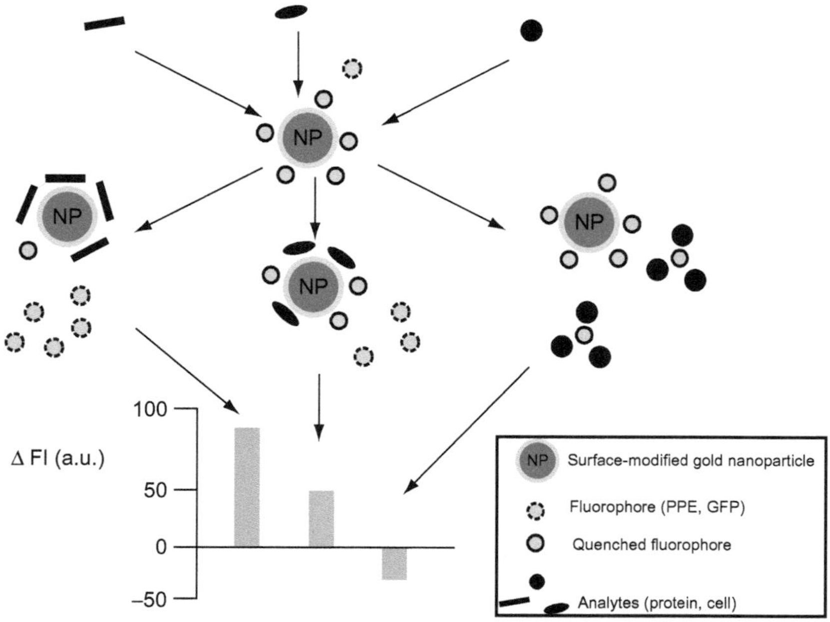

FIG. 3. Principle of operation of the fluorescence displacement protein and cell sensors, based on specifically functionalized AuNPs (recognition element). The level of displacement of the fluorophore (transducer element) from the surface of the AuNPs controls the level of fluorescence quenching by the AuNPs and, consequently, the fluorescence output. Higher affinity of the proteins or cells toward the recognition element yields higher fluorescence signal recovery as more transducer element is displaced from the surface. Higher affinity of the analyte toward the transducer elements can result in further fluorescence quenching by aggregation or by FRET.[17,60–63]

examples were usually nanocrystals of compounds formed from groups II–VI or II–V, (e.g., CdSe, InP). The unique properties of QDs come about from the nanoscopic confinement and semiconducting nature of the materials. This means that, unlike traditional fluorophores, it is the size of the QD, and to a less extent its chemical nature, that controls the absorption and emission wavelengths. Consequently, it is very simple to create QDs with different colors, even within the same synthetic process, for example, separating fractions with different sizes.[68] Moreover, their optical properties have several important differences compared to organic fluorophores. QDs can be extremely bright, having effectively 20–30 times the brightness of the best organic dyes,[69] and reach quantum yields approaching 100% for CdS/ZnS materials.[70] QDs are also often considerably more resistant to photobleaching than organic fluorophores.[71] QDs also combine broad absorption ranges with narrow emission ranges, which means that by using just one excitation source, emissions can be obtained at various wavelengths. The chief advantage of this property for molecular diagnosis is that it makes development of multiplexed assays considerably cheaper and simpler. The sum of these unique optical properties makes QDs highly suitable for sensing and diagnostic applications.[72] Figure 4 illustrates the unique optical properties of QDs.

Unfortunately, many QDs contain highly toxic metals, such as cadmium, lead, or mercury. Furthermore, the initial syntheses of QDs required a lot of energy and potentially dangerous starting materials and conditions, such as airless conditions at around 270 °C in trioctylphosphine oxide.[75] Unfortunately, use of this solvent, while yielding highly pure and monodisperse nanocrystals, leads to non-water-soluble QDs which must then have a replacement capping agent added to allow transfer to an aqueous solution.[76] Another issue with QDs is that they are particularly prone to a phenomenon called "blinking," in which the fluorophore switches between fluorescent and nonfluorescent states over time.[77] Although the rates of blinking can be decreased by careful synthesis,[78] this is a problem that still reduces the accuracy of QD-based assays.

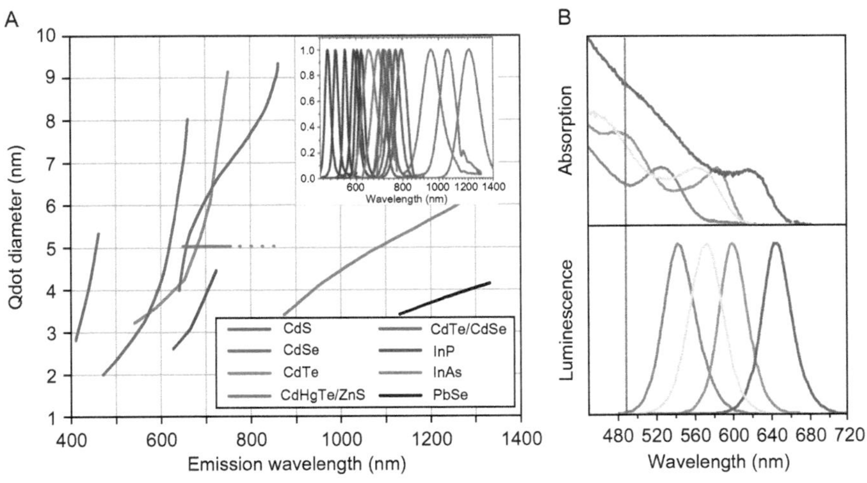

FIG. 4. (A) Emission maxima and sizes of QDs of different composition. The curves represent experimental data from the literature on the dependence of peak emission wavelength on the QD diameter. The range of emission wavelength is 400–1350 nm, with size varying from 2 to 9.5 nm (organic passivation/solubilization layer not included). All bands are typically around 30–50 nm (full width at half maximum). Inset: representative emission spectra for some materials. (B) Absorption (upper curves) and emission (lower curves) spectra of four CdSe/ZnS QDs samples. The blue vertical line indicates the 488 nm line of an argon ion laser, which can be used to efficiently excite all four types of QDs simultaneously. Figure from Ref. 73. Reprinted with permission from AAAS. Reprinted with permission from Ref. 74. Copyright 2001 American Chemical Society. (See Color Insert.)

Thus, the simplest synthesis methods that give the best QDs are rather inappropriate for *in vivo* applications, and therefore improvements of the synthesis to make more biologically compatible QDs by safer, less intensive routes which result in water-soluble and easily modified particles have been studied widely. These include replacement of the highly toxic precursors to CdSe-based QDs,[79] coating of the QDs to prevent leaching of toxic metal cations with polymers[29,80] or silica[74] or enclosing even within micelles,[81] as well as use of alternative semiconductor materials. Less toxic materials that have been used include zinc oxide[82,83] and silicon.[84,85]

In order to avoid chemical instability and degradation of QD fluorescence, chemical strategies have been developed, including the creation of a protecting silica shell around the QD. This transparent surface silanization enhances fluorescence of QDs, facilitates functionalization, and lowers leakage of heavy metal ions from the QDs.[74,86,87] Despite the huge amount of interest in QDs over the past 20 years, there has been a resurgence of interesting fluorescent labeling based on traditional fluorophores recently, but with a twist. Instead of being directly bonded to the molecule of interest, the fluorophores are encapsulated in an inert, optically transparent, nanosized container. Typically, silica nanoparticles are used to encapsulate the fluorescent molecules.[88] Silica is optically clear, simple to synthesize in nanodimensions, and highly biocompatible. The advantages of these particles over simple fluorescent labeling are mostly based on enhanced PL properties, including greater sensitivity and photostability. The sensitivity comes about because even a small nanoparticle (50 nm) might contain thousands of fluorophores.[89] The increased photostability compared to organic fluorophores is because the silica shell prevents oxidative species from reaching the fluorophores.[89,90] Another advantage of silica nanoparticles for molecular diagnostics applications is the simplified functionalization protocols, since silica surfaces are readily modified by silane grafting, which leads to simple modification by a wide range of biomolecules.[35,91,92] The primary advantages compared to QDs are their simple synthesis and biocompatibility. In particular, they may be synthesized in aqueous solutions, at low or room temperature,[93] and are widely considered to be markedly less toxic nanoparticles than many QDs.[94]

For these reasons, there is considerable amount of research describing the possibilities of using fluorophore-encapsulating silica NPs for molecular diagnostics applications. Moreover, the use of silica is not the only route to encapsulation of fluorophores in biocompatible nanoparticles, and various routes to organic polymer-based NPs have been explored.[95] Uses of these classes of photoluminescent NPs are described in Section II.F in Section II of this chapter.

D. Magnetic Nanoparticles for Molecular Diagnostics

Magnetic NPs can be used in sensors through (a) the direct application of tagged supports to the sensor and (b) the dispersion of the magnetic particles in the sample followed by their attraction by an external magnetic field onto the active detection surface (sensor).[96]

Magnetic particles have been evaluated for use in glucose biosensors in which glucose oxidase is immobilized on magnetite and applied to an electrochemical surface. The use of magnetic solids increases the biosensor's lifespan,[97] and the immobilization of enzymes on the magnetic NPs also increases the stability of the enzyme,[98] although at the expense of the progressive loss of the magnetic particles. Miniaturized configurations based on chip detection and magnetic beads manipulated via magnetoelectronics are used as a biosensor platform mainly for determining proteins and other biomolecules. One of the objectives of these devices is to decrease the limits of detection (LOD) so that they are useful for analyzing clinical samples.[99,100] The use of modified magnetic particles in sensing is targeted at developing more specific, stable, and reproducible devices.[101] In addition, magnetic particles (generally magnetite) provide a good microenvironment for retaining bioactivity, increase protein loading, and are a renewable alternative.[102] In DNA biosensing, techniques based on magnetic NPs achieve highly selective extraction and show low LODs, making them useful for hybridization and sequencing.[103–105]

1. Nanoparticles for Magnetic Separation

The use of magnetic beads for the separation of specific targets has been developed in the 1980s and since then numerous applications have been established, either using the large variety of magnetic beads commercially available or using magnetic micro- or nanoparticles specially prepared for a particular purpose. The technique is based on the surface modification of magnetic micro/nanoparticles with a bio/chemical moiety that specifically binds a biological target. These micro/nanoparticles are used to capture a particular analyte from a mixture that may then be easily separated by the application of a magnetic field and removal of the remaining solution. The pure analyte may afterwards be recovered by elution, provided that the binding between the magnetic particle and the ligand is reversible.

One of the major advantages of magnetic separation relative to other conventional separation techniques is its simplicity. The whole separation procedure may be performed in a test tube without the need of expensive apparatus. In addition, it is amenable to hyphenation with other separation/analytical techniques, such as microfluidics and mass spectrometry. In comparison with other separation techniques, for example, centrifugation, chromatography, etc., magnetic separations apply a low mechanical stress to the target molecule, enabling its use even with highly sensitive materials. In addition, magnetic NPs and magnetic manipulations show minimal or no interference with the chemical/biological processes, thus being compatible with a wide range of applications. Most commonly, magnetic separation has been applied to cell sorting using antibody modification of the magnetic particles, but it has also been used to capture proteins, nucleic acids, and small molecules from complex mixtures. In addition to magnetic separation, magnetic NPs may also be used to transport the molecule of interest in automated systems and also as probes for magnetic detection.

Guesdon and Avrameas[106] first described the preparation of magnetic hydrophilic agarose–polyacrylamide particles, (Magnogel-44®). These particles were relatively large and heterogenous in size (50–160 nm) and contained free amino groups, enabling antibodies to be attached by simple glutaraldehyde coupling.[107] Subsequently, Molday *et al.*[108] reported the preparation of iron-containing magnetic microparticles for the separation of red blood cells in 1977, with biological recognition being achieved by surface modification with antibodies or lectins. Subsequent to these early works, several improvements have been made, not only in the magnetic properties of the particles but also in their properties in recognition. There are several key aspects important for an effective magnetic separation: (a) strong response to a magnetic field gradient; (b) low size dispersion and thus a high uniformity of physical properties; (c) high chemical stability; (d) good dispersibility in a wide range of solvents; and (e) high specificity for the target molecule.

In order to be useful for magnetic separation, particles must have a strong magnetic dipole, and thus be strongly attracted to a magnetic field. Both ferromagnetic and ferrimagnetic materials fulfill this requirement, but in both cases, the particles have a permanent magnetic dipole and therefore attract each other, which can highly impair the dispersibility of the particles in the solvent. Superparamagnetic NPs have a better performance in magnetic separation since this type of particle does not have a permanent magnetic dipole but become magnetized in the presence of a magnetic field. Moreover, they have almost no remanence, meaning that, as soon as the magnetic field is removed, magnetization of the NPs returns to zero. Superparamagnetic behavior is observed in both ferromagnetic and ferrimagnetic particles below a certain critical size. Below this critical size, the thermal energy becomes comparable to the energy necessary to change magnetic states, leading to the randomization of magnetic dipoles in the absence of a magnetic field. Critical sizes have been calculated for several materials and are usually in the range 2–30 nm.

In addition to the superparamagnetic behavior, the particles should optimally have a strong magnetization. Several materials fulfill these magnetization requirements and thus may be used in magnetic separation, namely, metals, for example, Fe, Ni, Co; metallic alloys, for example, FePt, CoPt; and metal oxides, for example, magnetite (Fe_3O_4), maghemite (γ-Fe_2O_3), cobalt(II) oxide (CoO), etc., and methods are known to prepare superparamagnetic NPs of all these materials. Nevertheless, due to the low resistance to oxidation of metal NPs, most commonly metal oxides are used, predominantly magnetite NPs, although the maximum magnetization achievable is lower for magnetite than for metals or metal alloys. However, magnetite has other advantages: namely, several methods are known to synthesize magnetite colloids with good control over average size and size dispersion, and the critical size for magnetite is higher than for most other magnetic materials. Good size dispersion is important for magnetic separation, since at this scale magnetic properties are strongly dependent on size, and a large dispersion of sizes leads to a gradient of magnetic properties that decreases the efficacy of separation. On the other hand, Brownian motion opposes magnetic separation, and thus very small NPs may prove difficult to be separated by a magnetic field. Another major inconvenience of very small NPs is the predisposition to agglomeration due to the high surface area energy. The synthesis of composite NPs containing a magnetic material and a nonmagnetic material is one convenient way to overcome agglomeration problems. Most commonly, the nonmagnetic material is polymeric or silica, and two different strategies have been used: deposition of magnetic NPs at the surface of preprepared particles and preparation of magnetic core–nonmagnetic shell NPs[22,109–111]. This latter approach is advantageous since the final product provides a uniform surface for attachment of the binding moiety and avoids chemical interferences. The shell may also provide protection of the magnetic core toward chemical degradation as well as the surface properties necessary for a good dispersion in the solvent.

In addition to imparting solubility/stability properties to the NPs, the outer surface of the NPs must commonly be modified for attachment of the binding molecule. Many different procedures may be used to attach the capture molecule to the surface of magnetic particles, for instance, using amines, hydroxyls, thiols, or carboxyls groups available at the surface and conventional conjugation chemistry.[22]

2. Magnetic Separation and Concentration Strategies

As mentioned previously, magnetic separation is easily performed in a test tube, without any need for expensive apparatus, except for a permanent magnet. Nevertheless, separation may be improved by using a high-gradient magnetic field applied through magnetizable macrobeads or wires packed in a flow column[112,113] with a quadrupole magnetic sorter[114] or other more sophisticated devices.[115] In addition to these columnar flow devices, planar flow devices for use of magnetic separation in microfluidics platforms have been developed.[116–118] These devices have magnetic material or electromagnets patterned inside the channels (or in close contact with the channels), thus providing higher magnetic field gradients and hence a high efficiency of separation.

3. Molecular Diagnostics Applications

Magnetic separation/concentration is proving to be extremely useful to achieve the early detection of pathogenic organisms or biomarkers. In the vast majority of cases, these are present in trace amounts, and thus preconcentration and removal of possible interferences is critical to achieve the required detection limits.[119]

Several types of bioassays have been used with magnetic NPs: cell sorting and identification, nucleic acid processing and detection, immunoassays, and catalysis. Cell sorting and identification are well developed, and several systems are already commercially available. Commonly, NPs conjugated to antibodies that recognize a specific membrane surface antigen are used. For example, magnetic NPs for the separation of all major human leukocyte populations from peripheral blood are commercially available (e.g., EasySep® Stem Cell Technologies, Vancouver, Canada, www.stemcell.com; Dynal®, Invitrogen, Carlsbad, USA, www.invitrogen.com; IMAG® BD Biosciences, San Jose, USA, www.bdbiosciences.com; MACS® Miltenyi Biotec: Berfisch Gladbach, Germany, www.miltenyibiotec.com; MagCellect® R&D Systems, Minneapolis, USA, www.rndsystems.com). Similar strategies may be used to isolate stem cells,[120] cancer cells,[121] and bacterial cells.[122] Usually, the recovered cells are viable, but a negative selection procedure may be used in cases where modification of the desired cells with NPs is to be avoided.[123]

Antibody-functionalized magnetic particles are also widely used in the capture of proteins and immunoassays. Several methods may be used for the conjugation of antibodies or antibody fragments to magnetic NPs, based either on physical adsorption or on formation of chemical bonds. The latter are usually preferred since the resulting particles are more resistant to competitive displacement of the adsorbed antibody. In addition, covalent binding through the Fc region, leaving the antigen-binding site oriented to the solution, is also desirable, and several methods have been proposed to achieve this, such as the use of protein A as a linker, binding through the histidine- or lysine-rich regions located in the Fc region, etc. Use of streptavidin–biotin interactions is also a common binding strategy since many biotinylated antibodies are now commercially available.[72,124]

Superparamagnetic NPs for nucleic acid fishing are commercially available (e.g., geneMAG-RNA/DNA, Chemicell, Berlin, Germany, www.chemicell.com; Dynabeads®, Invitrogen, Carlsbad, USA, www.invitrogen.com; μMACS™, Miltenyi Biotec: Bergisch Gladbach, Germany, www.miltenyibiotec.com). The most common procedure is to use magnetic NPs functionalized with streptavidin to which a specific biotinylated oligonucleotide is linked. By an appropriate choice of the biotinylated oligonucleotide, it is possible to use magnetic separation to recover a specific type of nucleic acid, for example, mRNA or cDNA, from biological samples (cell cultures, biologic fluids, tissues, organs, etc.), to capture nucleic acids with a specific base sequence,[125,126] or to capture nucleic acid-binding proteins.[127]

Superparamagnetic NPs are also used in magnetic detection, and although this application is not as well developed as the separation techniques, recent developments are very promising in biosensing.[128] Magnetic sensing has several advantages compared to the more common optical sensing, because biological samples have no magnetic background. Therefore, magnetic sensing may be used even in turbid samples, since it is not affected by scattering, absorption, and autofluorescence inherent to many biological samples that usually affect optical sensing.

II. Molecular Diagnostics by Nucleic Acid Detection

Molecular diagnostics for genetic-based diseases or predisposition to a given condition or phenotype are rather intricate and so will be considered as the laboratory identification of specific nucleotide sequences that can be associated or linked to a given condition. Conversely, detection of a given pathogen's DNA/RNA signature, leading to unequivocal identification is rather straightforward and can be used as a laboratory confirmation of infection. Therefore, before focusing on the use of nanoparticles for diagnostics based on nucleic acids detection, we will present a short overview of the genetic basis of human disorders and the current molecular assays for its characterization.

A. Genetic Basis of Human Disorders

The unraveling of the human genome has revolutionized the understanding of many aspects of human disease at the molecular level and provided a host of opportunities to identify new disease markers.[129] Since then, several large-scale biological projects have emerged, such as the encyclopedia of DNA elements, which aims to identify all of the functional elements in the human genome,[130] or the International HapMap Project, which aims to determine the haplotype structure of human chromosomes in various ethnic groups.[131]

Several types of DNA variations can be found in the human genome (e.g., insertions and/or deletions of one or more bases, duplications, polymorphisms, etc.), but the most common are SNPs, accounting for around 68% of all detected changes.[132] All this individual genetic variability has been associated with individual susceptibility to several multifactorial diseases, such as cancer and diabetes, and to the individual response to therapeutics.[133,134] The use of molecular tools for genome characterization is thus becoming increasingly more important as the understanding of disease susceptibility and progression markers improves.

B. Current Nucleic Acid Diagnostics Techniques

Several technological platforms represent current nucleic acid diagnostics techniques, such as restriction enzyme analysis, hybridization assays, chromatographic and melting analysis, nucleotide sequencing, and enzymatic amplification (e.g., polymerase chain reaction, PCR). PCR has been considered the "gold standard" in nucleic acid detection and has become the most predominant technology in molecular diagnostics.[135–139] More recently, the development of real-time PCR has allowed target DNA quantification in real time through fluorescence.[136,140] Besides real-time PCR, several other variants of the PCR method have been developed to achieve a specific goal (e.g., multiplex PCR, asymmetric PCR, nested PCR, reverse transcription PCR, etc.).[141–149]

Hybridization assays are a fundamental tool in molecular genetics, taking advantage of the ability of individual single-stranded nucleic acid molecules to form double-stranded molecules, that is, to hybridize. Hybridization takes place when the interacting single-stranded molecules have a sufficiently high degree of base complementarity, according to Watson–Crick base pairing.[150] Standard hybridization assays use a labeled (e.g., fluorescent, radioactive) nucleic acid probe to identify a related DNA or RNA molecule, that is, with a significant high degree of sequence similarity within a sample of unlabeled nucleic acid molecules. Well-known hybridization assays that rely on such probes include Southern and Northern blots,[151,152] fluorescent *in situ* hybridization assays,[153] and more recently DNA-microarrays/chips technologies.[154] In the latter, a high-throughput DNA analysis can be performed due to highly density arrays of DNA oligonucleotides that are functionalized in chips made of glass, silicon, or plastic.

Along with the optical properties of AuNPs, other physicochemical properties such as electrochemical activity have been explored to develop highly sensitive methods for nucleic acid detection, some of which have already been successfully applied to biological samples. A list of these methods can be found in Table II, sorted by the detection technique, detection limit, and type of target (RNA, ssDNA, dsDNA, etc.). Among detection methods based on AuNPs, colorimetric approaches have been the most explored and, due to their simplicity and portability, are among the most promising for future diagnostics methods at point-of-care (POC). Some of these methods are based on the color change generated by the plasmon coupling between AuNPs upon aggregation, while others rely on the AuNP's exceptional scattering and/or absorbance properties.[46,47,49,156,177]

TABLE II

GOLD NANOPARTICLE-BASED METHODS FOR NUCLEIC ACIDS DETECTION

Detection technique	Detection limit	Target(s)	Application to biological samples	Refs.
Chemiluminescence	0.1 nM	ssDNA	n/a	155
Naked eye (dry-reagent dipstick)	2 fM, 0.16 nM	dsDNA, cDNA, SNP	Prostate-specific antigen, hepatitis C virus, GMO (35S promoter and nopaline synthase (NOS) terminator), mannose-binding lectin gene (MBL2)	156–160
Naked eye (electrostatic interactions with unmodified AuNPs)	< 100 fM, 10 nM	RNA, ssDNA, dsDNA, SNP	Unamplified HCV, RNA from clinical specimens, SNPs associated with a fatal arrhythmia known as long QT syndrome (KCNE1 gene), UV-induced mutagenic DNA dimers	49–51,161–166
Naked eye or CCD camera (sandwich hybridization)	0.4 fM	miRNA	miRNA from leaf and root of rice (*Oryzasativa L. ssp. indica*) seedlings	166
Naked eye or optical monitoring system (cross-linking aggregation)	50 pM–10 nM, 33 fM (using scattered light)	ssDNA, dsDNA, SNP	Methicillin-resistant *Staphylococcus aureus mecA* gene	167–172
UV–vis spectroscopy (aggregation induced by polymerase/nicking machine)	1 pM	ssDNA	M13 phage	173
Naked eye or UV–vis spectroscopy (non-cross-linking aggregation)	1–100 nM	ssDNA, dsDNA, RNA, SNP	K-ras oncogene; cytochrome p450 CYP2D6*4 haplotype; beta-thalassemia mutations, *M. tuberculosis*, FSY1 gene expression (*S. bayanus*)	43,44,174–176
Light scattering imaging (sandwich hybridization)	50–20 fM	ssDNA, dsDNA, RNA, SNP	High-, medium-, low- and nonexpressed genes in human brain tissue, methicillin-resistant *Staphylococcus aureus mecA* gene, coagulation genes associated with thrombotic (factor V, factor II, and MTHFR)	177–180

(*Continues*)

TABLE II (*Continued*)

Detection technique	Detection limit	Target(s)	Application to biological samples	Refs.
Electrochemical impedance spectroscopy	0.1 pM–0.24 nM	ssDNA, SNP	Cystic fibrosis	181,182
Electrogenerated chemiluminescence	5 pM	ssDNA	n/a	183
Microchip gel electrophoresis with microelectrode detector	5.7 amol	dsDNA	n/a	184
Conducting atomic force microscopy (AFM)	–	ssDNA	n/a	185
Differential pulse voltammetry	0.78 fM	ssDNA, dsDNA, SNP	Factor V Leiden mutation (MTHFR gene)	186
Potentiometric stripping analysis	32 pM–1.5 nM	ssDNA, SNP	n/a	187,188
Conductivity	10 pM	ssDNA	n/a	189
Cyclic voltammetry and chronocoulometry	10 fM	ssDNA	n/a	190
Magnetically induced direct electrochemical detection	660 nM	ssDNA, SNP	n/a	191
Atomic force microscopy	2.5 pM	ssDNA	n/a	192
Hyper-Rayleigh scattering (electrostatic interactions with unmodified AuNPs)	10 nM	ssDNA	n/a	193
Surface-enhanced Raman scattering	20 fM	DNA, RNA, SNP	n/a	194
Microarray SPR imaging (sandwich hybridization)	1.38 fM–10 pM	ssDNA, dsDNA, cDNA, SNP	Familiar breast and ovarian cancer (BRCA1 gene), p53 gene expression	195–197
Surface plasmon resonance (SPR) imaging (non-cross-linking aggregation)	32 nM	ssDNA	n/a	198
Fluorescence spectroscopy (quenching by AuNPs)	2 nM	ssDNA, mRNA	Melanoma cell lines	199
Infrared thermography	10–100 pM	ssDNA	n/a	200

C. Methods Based on the Optical Properties of AuNPs

1. Nonfunctionalized AuNPs

In the method developed by Li *et al.*, the differential propensity of single-stranded DNA (ssDNA) and double-stranded DNA (dsDNA) adsorption to AuNPs is exploited.[49] Due to the electrostatic interaction between the free bases of ssDNA and the negatively charged surface of AuNPs with a citrate capping, ssDNA confers increased stability to the AuNPs upon increasing the ionic strength. Based on this observation, Li and coworkers combined AuNPs with citrate capping with a PCR procedure, using ssDNA probes complementary to the amplicon. When PCR amplification is successful, the ssDNA probes will have a complementary target to hybridize and form a dsDNA that will not adsorb to the surface of AuNPs. Hence, the AuNPs will aggregate upon increasing ionic strength, changing color from red to blue. On the other hand, when PCR amplification fails, ssDNA probes will remain available to adsorb to the surface of AuNPs and prevent aggregation upon addition of an electrolyte (Fig. 5). Although this is a very simple and elegant approach,

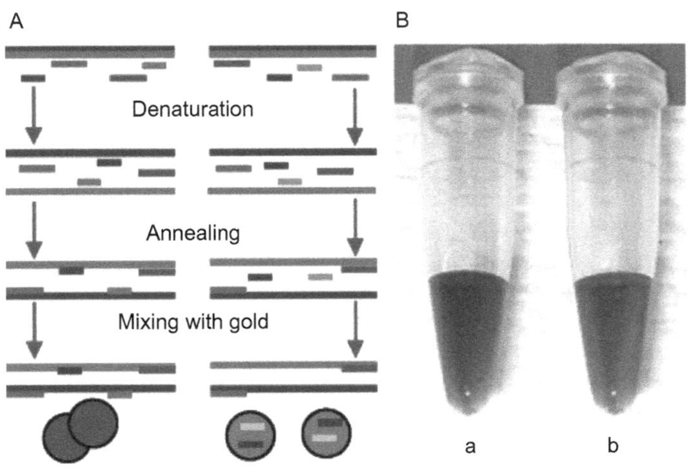

Fig. 5. Identification of PCR-amplified DNA target sequences by nonfunctionalized AuNPs. (A) Schematic of the detection protocol. The mixture of PCR product and probes is denatured and annealed below the melting temperature of the complementary probes followed by addition of AuNPs. The long blue and green lines represent the PCR-amplified DNA fragments, and the pink and light blue medium bars represent the excess PCR primers. The short blue and green bars are complementary probes that bind, resulting in AuNP aggregation (purple color). The short purple and orange bars are noncomplementary probes that do not bind and adsorb to the AuNP, preventing AuNP aggregation and leaving the solution pink. (B) Colorimetric results for complementary (a) and noncomplementary targets (b). (Reprinted with permission from Ref. 49, Copyright 2004 American Chemical Society.) (See Color Insert.)

it requires a fine-tuning of the PCR protocol in order for all primers (also in ssDNA form) to be consumed, such as to avoid a false negative. Moreover, a false negative can also arise from an amplification reaction that does not occur due to factors other than the lack of a complementary target DNA in solution (e.g., inactive polymerase, lack of dNTPs, etc.). A careful primer and probe design should also be taken into consideration to avoid hairpin or primer dimer formation that may lead to a false positive result. More recently, a similar approach using thiolated primers have also been described by Jung *et al.* which could help circumvent some of these limitations.[48] In this case, successful PCR amplification leads to the production of thiolated amplicons that are able to provide a greater stability to the bare AuNPs in solution, while excess thiolated primers only provide limited stability to salt-induced aggregation.

Other similar approaches have been developed taking advantage of the differential propensity of ssDNA and dsDNA adsorption to AuNPs. Namely, the differential ssDNA oligonucleotide probe adsorption to bare AuNPs upon target hybridization followed by salt-induced aggregation has been directly explored for the detection of SNPs[161,162] as well as of ultraviolet (UV)-induced mutagenic or carcinogenic DNA dimers.[163] Ma *et al.* have used unmodified gold nanorods in a CTAB solution and were able to achieve a target detection limit of around 0.1 pM.[164] Kanjanawarut *et al.* have used peptide nucleic acids instead of ssDNA oligonucleotide probes to achieve a more drastic differential aggregation between complementary and noncomplementary targets.[201] The Azzazy group has recently used unmodified AuNPs to develop an assay for direct detection of unamplified hepatitis C virus RNA isolated from clinical specimens.[51] The isolated RNA was added to a solution containing the complementary oligonucleotide probe, and after a denaturing and annealing cycle, unmodified 15 nm AuNPs were added. This extremely simple and inexpensive assay, which does not include an RT-PCR step, presented a detection limit of 50 copies/reaction and exhibited a sensitivity of 92% and a specificity of 89%.[51]

The chemiluminescence (CL) of ssDNA probes has also been used in conjunction with unmodified AuNPs, taking advantage of the capabilities of AuNPs to quench the CL of acridinium ester.[202] In the presence of a noncomplementary target, the chemiluminescent ssDNA probes are available to adsorb to the AuNPs surface and are therefore quenched by the AuNPs. Upon complementary target hybridization to the chemiluminescent ssDNA probes, their adsorption to the AuNPs surface is inhibited and CL detected.

2. Functionalized AuNPs: Cross-linking Approach

Functionalization of AuNPs with thiol-modified oligonucleotides led to the first application of AuNPs in nucleic acid detection.[46] In their approach, Mirkin and coworkers functionalized AuNPs with oligonucleotides modified with a thiol group at their 3'- and 5'-ends, whose sequences were contiguous and complementary to a target in a tail-to-tail (or head-to-tail) conformation, and used these probes to identify the target DNA sequence. When hybridization occurs between the sequences of the Au nanoprobes and the target DNA, the Au nanoprobes are drawn together and form a cross-linked network, leading to the aggregation of the AuNPs and subsequent color change of the solution (Fig. 6).

This change of color of the solution from red to blue can be observed by the naked eye in solution or spotted on reverse-phase silica plate or can be followed by visible spectroscopy. This method allows detection of single base mismatches by controlling the denaturation temperature of the cross-linked nanostructures, as they present a sharp melting transition.[204] This cross-linking approach has also been combined with a rolling circle amplification technique allowing the detection of single point mutations with 1 fM sensitivity.[205] Alternatively, Qin *et al.* have used an electrophoretic approach, where the screening is made by the differential electrophoretic migration of the different cross-linked conjugates (i.e., complementary targets generate a cross-linked conjugate which migrates less than the Au nanoprobes alone).[206]

3. Functionalized AuNPs: Non-cross-linking Approach

Since the development of the Au nanoprobes by Mirkin *et al.*, other methods for nucleic acid detection have been described using just one Au nanoprobe, instead of the two required for cross-linking. One of these methods has been developed by Sato *et al.*, which described a differential aggregation of the Au nanoprobe in presence of complementary, mismatched, or noncomplementary targets, induced by an increasing ionic strength of the solution.[47] This non-cross-linking method was first developed with synthetic oligonucleotides with sequences of the same length as the Au nanoprobe. So, when the complementary target sequences fully hybridize with the Au nanoprobe,

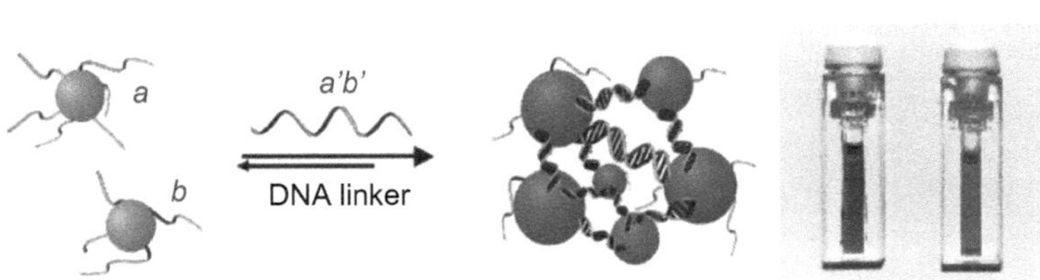

FIG. 6. Au nanoprobe-based cross-linking method. A mixture of AuNPs functionalized with noncomplementary DNA sequences (a, b) appears red in color. When a complementary DNA sequence (a'b') is added to the solution, the particles are reversibly aggregated causing a red shift in the surface plasmon absorbance, thus, appearing purple in color. (Reprinted from Ref. 203, Copyright (2006), with permission from Elsevier.) (See Color Insert.)

heteroduplexes with a blunt end are formed and the Au nanoprobe becomes less stable to increasing ionic strength and aggregates, changing color from red to blue. In the presence of noncomplementary targets or a target with a mismatch at the end of the Au nanoprobe, these blunt-end heteroduplexes are not formed and the Au nanoprobe remains dispersed and red, at the same ionic strength that led complementary/Au nanoprobe complexes to aggregate. Unfortunately, this approach has a limitation of application in real biological samples, since real biological targets are usually longer than the sequence of the Au nanoprobe, and hence do not confer a blunt-end heteroduplex hybridization that is essential for its application. To circumvent this limitation, Sato *et al.* used a PCR amplification followed by a single base primer extension approach to test biological samples and detect SNPs at room temperature.[174]

Following a simpler approach for nucleic acid detection in biological samples, Baptista *et al.* have developed a non-cross-linking method that can be used to detect specific DNA and/or RNA with a longer sequence than the Au nanoprobe sequence.[43,51,207–211] In this approach, the detection is achieved by color comparison upon salt addition between solutions containing the Au nanoprobe and either a complementary or a noncomplementary target sequence. While the presence of a complementary target prevents aggregation upon salt addition and the solution remains red, the noncomplementary targets do not prevent Au nanoprobe aggregation, resulting in a change of color from red to blue. A particularly successful clinical application of this strategy was in the rapid and sensitive detection of *Mycobacterium tuberculosis* (the etiologic agent of human tuberculosis) in clinical samples[16,43] (Fig. 7). This method has been successfully adapted to suit POC through the development of a optoelectronic platform[212–215] and to detect SNP/single point mutations.[44,210]

4. Multifunctional AuNPs

DNA and Raman-active dyes have been attached onto AuNPs, which assemble on a sensor surface only in the presence of a complementary target. If a patterned sensor surface of multiple DNA strands is used, the technique can detect millions of different DNA sequences simultaneously. The current nonoptimized detection limit of this method is 20 fM. This detection system has also been developed using chips dotted with DNA. If the targeted disease marker is present in the sample, its DNA will bind to the complementary strands of DNA on the chip. The chip is then treated with a silver-based solution, coating the AuNPs, and when exposed to a light scanner, the coating enhances the signal enough to detect minute amounts of DNA. Since the Raman band is narrower than the fluorescent band, it allows more dyes to detect more targets quickly. If the sequence of interest is present in the sample, it will bind to the DNA and cause the solution to change color. Labeling oligonucleotide targets with the AuNPs rather than the fluorophore probes

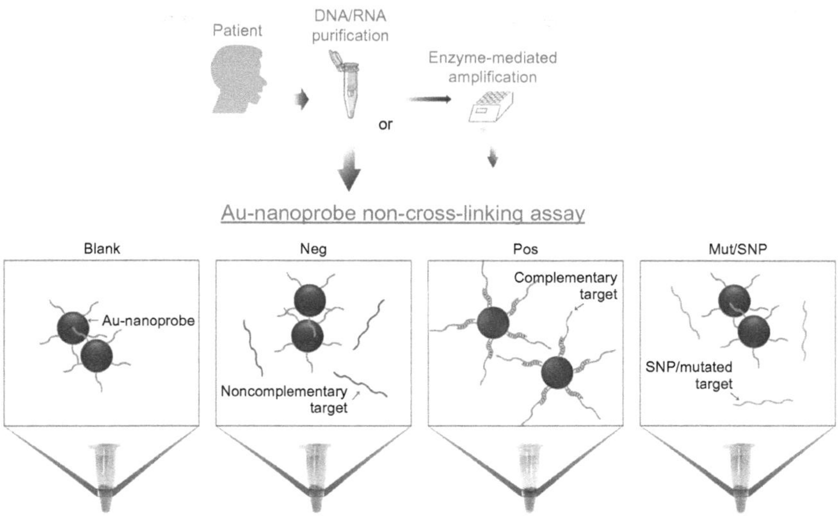

FIG. 7. Colorimetric sensing of rifampicin resistant *Mycobacterium tuberculosis* DNA based on a non-cross-linking aggregation method.[43] The method relies on visual comparison of test solutions before and after salt-induced Au nanoprobe aggregation. *Blank*: Au nanoprobe alone; *Neg*: negative sample in the presence of noncomplementary DNA; *Pos*: positive sample in the presence of complementary DNA; *Mut/SNP*: in the presence of a target harboring a SNP or a single point mutation. (See Color Insert.)

substantially alters the melting profiles of the targets from an array substrate. A AuNP-based DNA detection system has been constructed with 10 times more sensitivity and 100,000 times more specificity than current genomic detection systems. It is possible to coax DNA to aggregate and separate AuNPs on demand.[216,217]

The favorable optical properties of AuNPs have been further explored by different spectroscopic techniques. For instance, Au nanoprobes can be encoded with Raman reporter molecules, generating enhanced Raman signals that allow the detection of specific nucleic acids by SERS.[194,218,219] Additionally, the long-range plasmon coupling of AuNPs through a cross-linking hybridization approach generates an increase in SERS signal intensities by 40- to 200-fold.[220]

Au nanoprobes have also been used as color reporters. Glynou *et al.* have developed a method for detecting nucleic acid sequences through hybridization with Au nanoprobes in a chromatographic stripe (also known as dipstick).[156] In this approach, similar to the current pregnancy tests, biotinylated PCR products are hybridized to a specific oligo(dA)-tailed probe and loaded on the chromatographic stripe. As the buffer migrates through the stripe, the biotinylated PCR products are immobilized by streptavidin spotted in a specific location of the stripe. Finally, to detect the presence of the DNA target in the streptavidin spot, poly-dT Au nanoprobes are used to hybridize with the poly-dA probe. Other variants of this technique have also been described by Kalogianni *et al.* to fit a specific application (e.g., SNP detection and multiplex analysis).[157,158,221] However, although the detection procedure on this method is simple and easily translated into a POC, the method also requires a previous target DNA preparation protocol which is laborious and requires specialized human resources.

The stem–loop intramolecular base-pairing pattern that can occur in ssDNA, also known as hairpin structures, has been successfully used in molecular beacons for real-time PCR applications. These structures have also been used in conjugation with AuNPs to design new methods for specific DNA/RNA targets detection. Typically, the molecular beacons are constituted by an oligonucleotide that forms a hairpin loop and is functionalized with a fluorophore and a quencher dye, each functionalized in opposite ends. In the absence of a complementary target, the molecular beacon retains its original hairpin structure and the fluorophore is quenched by the quencher dye. Conversely, in the presence of a complementary target, the molecular beacon hybridizes with its complementary sequence and the hairpin structure is inhibited, as well as the quenching of the fluorophore. Following a similar approach, Harry *et al.* have used the quenching capabilities of AuNPs by functionalizing them with an ssDNA oligonucleotide that forms a hairpin structure and harbors a fluorophore dye at the opposite end to successfully detect gene expression in melanoma cells.[199] This approach allows the specific quantification of mRNA *in vivo*.

Other authors have replaced the fluorophore dye by other moieties that are precursors of a certain reaction, such that when these AuNPs conjugates are in a hairpin conformation (i.e., absence of a complementary target), the reporting reaction is blocked. For example, Zhang *et al.* developed a hairpin ssDNA probe dually labeled with a thiol at its 5′-end and a biotin at its 3′-end, respectively. This probe was immobilized on the AuNPs anchored by a gamma-globulin protein on a 96-well microplate. In the absence of the complementary DNA target, the immobilized probe with the stem–loop structure shields the biotin from being approached by a bulky horseradish peroxidase-linked streptavidin (streptavidin–HRP) conjugate due to steric hindrance. However, in the presence of target DNA, the hybridization between the hairpin DNA probe and the target DNA causes significant conformational change of the probe, which forces biotin away from the surface of AuNPs. As a result, the biotin becomes accessible by the streptavidin–HRP, and the target hybridization event can be sensitively detected via the HRP-catalyzed substrate 3,3′,5,5′-tetramethylbenzidine using a spectrophotometric method.[222] He *et al.* have combined a similar approach to a dipstick membrane, where the disruption of the hairpin structure by fully complementary target unlocks a blocker moiety (e.g., 11-mercaptoundecanol or dATP), allowing the AuNP conjugate to bind to the test zone of a dipstick membrane. This approach allowed the discrimination of perfect-matched DNA from single base-mismatched DNA targets, with 10 pM sensitivity.[159]

D. Electrochemical Methods Based on AuNPs

The electrochemical properties of Au nanoprobes can also be used to report hybridization in a microarray platform following different electrochemical strategies.[66] For instance, the hybridization of an Au nanoprobe with a sequence anchored onto the surface of a conventional electrochemical sensor can be directly assessed by stripping or differential pulse voltammetry, or through a conductometric technique. The electrochemical signal can be further enhanced by silver/gold reduction, or by using other AuNPs or electroactive labels attached to the Au nanoprobe. An extensive list of the use of different electrochemical methods based on AuNPs for nucleic acid detection can be found in Table II.

E. Nucleic Acid Detection Methods Based on Silver/AuNPs

Most nanoparticle-based methods for nucleic acids detection have so far relied only on AuNPs, mainly due to their ease of functionalization with thiol-modified oligonucleotides. Nonetheless, more recently, Mirkin *et al.* have also used Ag/Au core–shell NPs functionalized with thiol-modified oligonucleotides to detect nucleic acid targets following the same cross-linking approach that was developed with Au nanoprobes.[223] In their approach, the Ag/Au core–shell and Au nanoprobes presented identical results, but with different optical properties (i.e., different colors). This allowed establishing a two-color-change-based method for parallel SNP analysis, providing a convenient cross-check and assay control that offer a more accurate readout when compared to the single-color-change methods.

Apart from the Au/Ag core–shell NPs, silver NPs functionalized with thiol-modified oligonucleotides have also been used for the molecular detection of nucleic acids, following the cross-linking approach.[224]

The Baptista group extended their non-cross-linking approach to a dual-color non-cross-linking method, combining Au nanoprobes with gold/silver alloy nanoprobes.[45] Because the alloy- and the Au nanoprobes exhibit surface plasmon resonance (SPR) bands at different wavelengths and different aggregation profiles, this constitutes a dual-color detection system capable of characterizing two different targets in a single vial.[45]

F. Methods Based on Fluorescent Nanoparticles

Currently, a wide range of fluorescent nanoparticles have been prepared and are undergoing development or use in medicinal applications. The majority of these can be categorized into one of three classes: namely, QDs, fluorophore-loaded silica NPs, and fluorophore-loaded polymer-based NPs. Of these, QDs are certainly the mostly widely applied in both bench and clinical settings.

1. Quantum Dots

Through a diverse array of coatings and bioconjugation strategies, it is possible to use QDs as a scaffold for biorecognition events. The modulation of QD luminescence provides the transduction of these events by utilizing powerful spectroscopic techniques, such as Förster resonance energy transfer (FRET). In fact, FRET is the most widely used method to transduce biorecogition in bioassays with QDs.[225–230] In general, these function by exploiting energy transfer (ET) between the QD and an organic fluorophore by monitoring the FRET-stimulated emission caused when hybridization occurs, as schematically illustrated in Fig. 8.

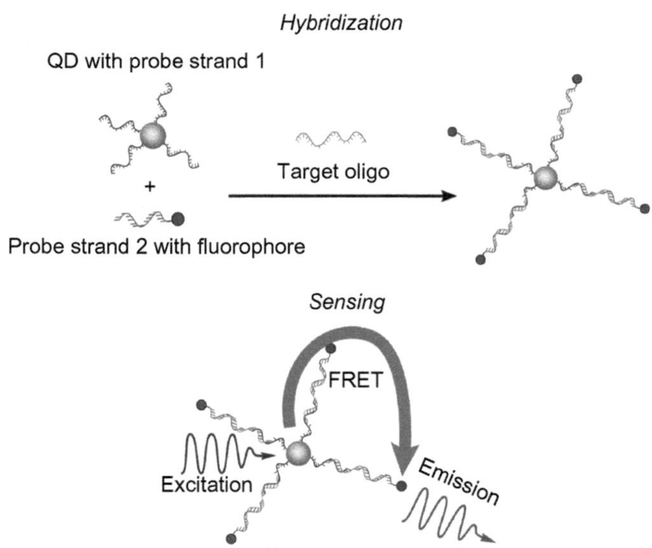

Fig. 8. Principle of FRET-based bioassay based on QDs. *Top*: The assay usually involves the QD bound to oligonucleotide 1, and an organic fluorophore bound to oligonucleotide 2. Sequences from 1 and 2 are complementary to either end of the target sequence. Successful hybridization binds the QD to one end of the target, and the fluorophore to the other. *Bottom*: The QD is excited by a laser and FRET occurs to the acceptor fluorophore. This energy transfer occurs only when hybridization is present. (For color version of this figure, the reader is referred to the Web version of this chapter.)

In the most common configuration, hybridization is carried out in a "sandwich" configuration, with two probe oligomers, one bound to a QD and the other bearing a fluorescent molecule, which then bind to either ends of the target sequence. An advantage of this type of assay is that the target molecule does not need to be modified.[229,231] Using strongly emitting QDs in this configuration, and a single-molecule fluorescence detector, detection of very small numbers of specific unlabeled oligonucleotide sequences is possible. However, such a sandwich technique is likely fundamentally limited to short target sequences, since the FRET donor and acceptor must be able to approach each other closely for ET to occur. This technique has been shown to be extremely sensitive, detecting target DNA at concentrations as low as 4.8 fM,[229] albeit with the drawback of the need for a rather complex and expensive fluorescence detection setup. Extension of this type of assay to a multiplexing model has been achieved.[231] In this way, it was possible to simultaneously detect the presence of two DNA target sequences, by monitoring the emission wavelengths of two fluorophores and the QD. Unfortunately, the assay is very complicated, in terms of both the chemistry involved (requiring modification of four probe oligonucleotides plus the QD) and the custom-built optical flow cell incorporating microfluidics. However, the assay has been successful in detecting either or both sequences from the HIV-1 and HIV-2 viruses. FRET assays for DNA detection based on molecular beacons[232] and QDs have also been developed. These rely on the ET between the QD and an acceptor located at the end of a stem-loop-like DNA structure that is linked to the QD at the opposite end.[233] Although FRET is the most extensively used transduction method for bioassays based on QD luminescence detection, other transduction methods exist.[7] For example, QD binding to oligomers can also be detected by a colocalization assay.[234] In this method, a sandwich configuration is again used, with two probe sequences in this case, each bound to QD of a different size, and thus different emission wavelengths. By probing colocalization of the two colors, hybridization could be detected. The authors were able to produce QDs with only one DNA probe molecule bound, with each interaction measured indicating a single probe-target hybridization reaction. Thus, they were able to use the technique to directly quantify the number of target sequences within a flow channel, and the lower LOD was around 30 fM. This technique was also extended to protein recognition.[234]

An alternative transduction approach is to use a quencher such as an AuNP to indicate hybridization. This approach was shown to be sensitive, but rather slow and involved NP-tagging of both probe and target sequences.[226,235] In a simpler approach of the same basic principle, Kuang *et al.* have also used the cross-linking approach by combining QD nanoprobes with Au nanoprobes, taking advantage of the fluorescence quenching of QDs by AuNPs upon the cross-linked hybridization induced by a complementary target to both types of nanoprobe.[236]

2. Fluorophore-Loaded Silica Nanoparticle

Fluorophore-loaded silica nanoparticles (FS NPs) follow a very different approach compared to that of QDs. Whereas QDs have intrinsic PL caused by their semiconducting nature, silica is in itself nonfluorescent. However, these fluorescent NPs typically encapsulate traditional organic fluorophore molecules. Just about any fluorophore (even inorganic ones, such as lanthanides)[237] can be used, so a wide range of fluorescence wavelengths, either visible or near-infrared,[238,239] can be obtained by simply changing the fluorophore. Encapsulation of the fluorescent molecules within the silica has a number of advantages over use of the bare molecule. The first benefit is in an increase in assay sensitivity. This is simply due to the fact the fluorescence from one FS NP will be many times that of a single fluorophore. Authors report that FS NPs contain hundreds,[240] thousands,[89] or tens of thousands[241] of fluorophores within each NP. There may be some loss of intensity associated with self-quenching from the dye and/or light absorption by the silica, resulting in some cases in a less than stoichiometric increase in brightness,[35,88] but even so, orders of magnitude increases in PL intensity are commonly found.[88,89,238] On the other hand, some authors have actually reported increased fluorescence emission per fluorophore after encapsulation.[238,242] Notably, this result was found with core–shell particles—with a fluorescent silica core, surrounded by a plain silica shell—so it seems likely that protection of the fluorescent silica from the medium is important for optimal optical properties. Importantly, photobleaching can be greatly reduced for a number of different dyes by encapsulation in silica, even without a core–shell architecture,[90] making the use of FS NPs for molecular diagnosis methods very attractive. This alone makes the particles extremely useful since photobleaching is one of the main sources of error in assays using PL. This property is thought to arise from the low oxygen penetration through the silica.[243] Owing to the ability to incorporate a wide range of dyes (either organic or inorganic),[241] the optical properties of FS NPs can be finely tuned, making them suitable for multiplexing,[244,245] although without the unique properties of QDs in this regard (i.e., exceptionally narrow emission profile, broad absorption). FS NPs are also immune to the blinking which can complicate the use of QDs.

Owing to these useful properties, and also because it is simple to replace fluorophores with fluorescent particles encapsulating the same fluorophores, the use of FS NPs in molecular diagnostic assays has been widely studied.[35,241,246,247] They have been used to create assays for oligonucleotides,[89,92] proteins,[248] and antibodies.[245,249]

FS NPs can be used to improve the sensitivity of DNA microarrays.[89,92,240] In this application, the particles are generally used to replace standard fluorescent dyes, in order to reveal the presence of bound DNA molecules on the microarray capture dots. However, because the FS NPs can have brightnesses far in excess of the organic dyes, they can greatly enhance sensitivity of the assays, in particular, when detecting low-abundance targets. Since the use of organic fluorophores in such systems is already well characterized and commercialized, replacing the dyes with FS NPs may be considerably simpler than switching to the use of QDs. For example, work in the Tan group[89] compared the use of silica NPs containing the dye $[Ru(bpy)_3]^{2+}$ with the use of a cyanine dye alone in microarray assays for a gene sequence from *Pseudomonas aeruginosa*. Using a standard commercially available microarray kit, they found that replacing the dye with the FS NPs led to an approximately 10-fold improvement in signal, suggesting that an order of magnitude increase in sensitivity could be obtained compared with cyanine staining. Similarly, a 10-fold improvement in signal intensity was found for replacement of an organic dye with core–shell NPs having a fluorescent silica layer. In this work,[89] improvement of protein microarrays was also studied, and the fluorescent nanoparticles were shown to give signals 80 times higher than the dye fluorescein isothiocyanate (FITC).

Because of the extremely high brightness of FS NPs, detection of DNA binding via staining with these advanced fluorophores can be extremely sensitive. Work from the Tan group also demonstrated how subfemtomolar concentrations of oligomers could be detected by a sandwich-type hybridization on a solid surface, the final probe strand being labeled with rhodamine-containing FS NPs.[92] This detection level is very low; however, the authors did use a large sample size, making it rather unpractical. In addition, in the same work, SNPs could be detected. Although direct detection of FS NP PL emission assays has been the most popular approach, other methods have been investigated in order to detect target hybridization. Core–shell FS NPs were used in an assay for alpha fetoprotein (AFP) based on fluorescence anisotropy.[249] Since a near-IR dye was used and the PL was rather immune to leaching and bleaching problems, due to the core–shell structure of the particles, it was possible to detect the protein in whole-blood samples. Linear response of the fluorescence anisotropy was found for AFP concentrations between 1.9 and 51.9 ng mL^{-1}. In the case of silica NPs doped with an electroactive dye such as $[Co(bpy)_3]^{3+}$, it is possible to detect the presence of the nanoparticles electrochemically, which has been used to develop an assay based on hybridization on a glassy carbon electrode.[250]

3. Fluorescent Polymer Nanoparticles

A third class of fluorescent nanoparticles is constituted by fluorescent polymeric nanoparticles (FP NPs). FP NPs can be made from a vast range of polymeric materials, including, but not limited to, polystyrene,[72] poly(methyl methacrylate),[95] and derivations thereof, natural biopolymers,[251] and many others, including conjugated polymers with intrinsic fluorescence.[252]

Nonfluorescent polymers can be used by covalent binding, or simple encapsulation of fluorophores. FP NPs can be prepared exhibiting a range of optical properties, in a wide range of sizes and surface functionalities. Depending on the polymeric material, it is often a simple matter to link probe biomolecules to the surface of the nanoparticles.[251,253] Despite these advantages, currently polymeric nanoparticles have seen little application to molecular diagnostic assays. Probably the major reason for this is the tendency they have to swell, and more importantly to change their swelling based on the nature of the solution. This is likely to change the fluorescence properties, since it can lead to changes in solvent access to encapsulated fluorophores, making it difficult to develop reliable quantitative assays with polymeric nanoparticles. Despite these drawbacks, FP NPs have been used for sensing protein–DNA binding,[254] development of a highly sensitive immunoassay for anthrax toxin,[255] a surface-based assay for adenovirus,[256] and others.[72]

G. High-throughput Nucleic Acid Detection Assays

As for the more traditional methods, high-throughput systems using AuNPs have relied on miniaturized platforms, such as "lab-on-chip." Strictly speaking "lab-on-chip" refers to the scaling down of several laboratory processes on a chip format, while micro total analysis systems (μTAS) is a broader term referring to integration of all laboratory processes required for an analysis on a single chip. However, in reality, many microfluidics researchers use the terms microfluidics, BioMEMS, BioChips, lab-on-chip, and μTAS interchangeably.[257] Lab-on-chip devices offer the advantages of a sophisticated analysis system, a small device size, faster sample analysis, higher throughput, portability, reduced reagent use, automation of complex bioanalytical processes, and ease of use, all of which are associated with decreased cost. Besides, closed architectures will reduce the potential for contamination, minimizing false positives. It is expected that the lab-on-chip will make DNA analysis widely available and provide opportunities for better screening of at-risk patients, tighter surveillance of disease recurrence, and better monitoring of treatment, saving lives and costs.

Nevertheless, integration of biochemical processes with sensors, or preferably with arrays of sensors into lab-on-chip devices, brings enormous challenges with respect to microfluidics, electronic interconnection, and interfacing.[258] Detection issues will arise when sensing systems are miniaturized, as the reduced analysis volumes mean a reduction in detection volumes, decreasing the number of analytes available for detection.[259] Moreover, many interfacial and hydrodynamic phenomena do not scale linearly from large to small implementations. When choosing lab-on-chip systems, it should be taken into account that some physicochemical processes, such as capillary forces, surface roughness, and chemical interactions of construction materials on reaction processes, become dominant at the microscale.[260]

Using a high-throughput approach, Mirkin *et al.* used Au nanoprobes to substitute the conventional fluorescent-labeled probes as target hybridization reporters, in a microarray platform.[177] In their method, light scattered by the AuNPs of Au nanoprobes is registered by a CCD camera and used as a signal for complementary target hybridization. Moreover, if an additional step of silver reduction is performed on the Au nanoprobe surface, the scattering intensity increases and, consequently, the sensitivity of the method also increases, allowing down to 200 fM of genomic DNA to be detected.[180] The detection of single base mismatches is also possible at controlled temperatures, and increased selectivity is observed when Au nanoprobes are used instead of standard fluorescent probes. This method is already being commercialized by Nanosphere, Inc. (known as Verigene® system) and was one of the first nanotechnology-based diagnostics platforms to be approved by the U.S. Food and Drug Administration.

III. Molecular Diagnostics by Immunoassays

A. AuNPs-Based Immunoassays

AuNP-based immunoassays described in this section rely on some degree of conjugation between the AuNPs and the antibody used for detection. The traditional and most widely used method of obtaining such bionanoconjugates is simply incubating citrate-capped AuNPs with the antibody of interest in basic medium.[261] Other methods of obtaining bioconjugation between the AuNPs and the antibodies are nevertheless available, most of them relying in electrochemical deposition of the NPs.

1. Electrochemical Methods

Electrochemical detection is by far the most widely used technique in AuNP-based immunoassays. One of the most interesting nanoscale properties of AuNPs is their extremely high surface area, imparting excellent catalytic properties to the nanoparticles and its conjugates and ensuring high functionalization rates. In an attempt to increase the available gold surface area for immune electrochemical detection, Lai *et al.*[262] obtained a sevenfold increase of the electrode surface area by first immobilizing in the electrode AuNPs with small diameters (5 nm) as tags to anti-immunoglobulin class G (IgG) (linked through an immobilized IgG to the electrode surface) and then promoting auto-catalyzed deposition of Au^{3+} on the surface.

The Limoges group was the first to report on the involvement of an electrochemical technique for the determination of a colloidal gold label in an immunoassay format.[10] A heterogeneous noncompetitive immunoassay format was developed in which the analyte (goat IgG) was indirectly measured by formation of a sandwich-type immunocomplex. Namely, the AuNP label in the secondary antibody was dissolved in an oxidative acidic solution, and the released Au (III) ions were quantitatively measured by anodic stripping voltammetry. This electrochemical technique is a very sensitive method for the determination of trace amounts of metal ions, and it is based upon the metal (Au^0) cathodic deposition on the surface of the electrode during a preconcentration period and subsequent metal stripping from the surface of the electrode by anodic oxidation. The dynamic range of the assay extended from 0.5 to 100 ng mL^{-1}.[10]

Regarding multiplex analysis, Leng *et al.*[12] recently developed a disposable chip for electrochemical immunoassay with capture antibodies, rabbit anti-goat IgG (RaH), and mouse anti-human IgG (MaH), immobilized by passive adsorption onto carbon electrodes. Subsequent addition to the chip of the model analytes HIgG and GIgG and capture AuNP-labeled RaH and MaH allowed sandwich-type immunocomplex formation. A linear correlation was obtained between the amount of analyte and the differential pulse voltammetric response, as can be seen in Fig. 9. This system seems particularly well suited for multiplex detection of several analytes in a complex sample, for example, in plasma samples, because of the ease of introduction of additional working electrodes by screenprinting on the disposable chip and the good sensitivity obtained, namely, a detection limit of 1.1 and 1.6 ng mL^{-1} for HIgG and GIgG, respectively.[12]

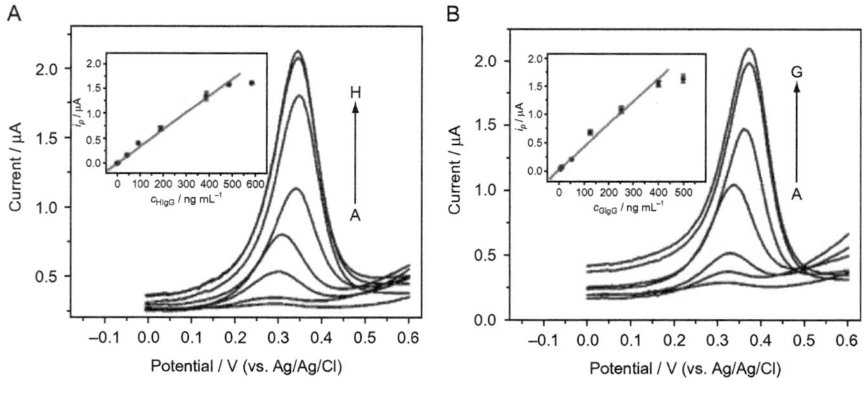

FIG. 9. DPV responses for the multiplexed detection of (A) HIgG and (B) GIgG using AuNP as a label. Insets: corresponding calibration curves. (Reprinted from Ref. 12, Copyright (2010), with permission from Elsevier.)

2. CL METHODS

CL, the emission of light by a chemical reaction, is an attractive detection method allowing high sensitivity and a wide linear response range, with low consumption of expensive reagents and requiring simple instrumentation.[263,264] CL detection methods using AuNP labels have been initially developed as "stripping" procedures, where the gold metal dissolution allows Au (III)-catalyzed luminol CL.[183,263] On a typical CL immunoassay, the human IgG analyte is first captured by the primary antibody and then sandwiched by a secondary antibody labeled with AuNPs. Gold metal is dissolved in an oxidative HCl–NaCl–Br_2 solution, and the Au(III) thus released in solution is quantitatively determined by the Au(III)-catalyzed luminol CL reaction.[265] Li *et al.* reported a similar CL immunoassay based on the sandwich-type immunoreaction coupled with the CL reaction of the Au(III)–luminol–H_2O_2 system. The Au(III), dissolved from rabbit–anti-goat IgG AuNP labels, served as an analyte in the CL reaction for the indirect measurement of goat–anti-human IgG, and 2% HNO_3–3.4% HCl was used for AuNP stripping.[264] This type of immunoassay was successfully applied in pig serum samples for the detection of antibodies against ApxIV, a virulence factor present in the blood of *Actinobacillus pleuropneumoniae* infected pigs, in good agreement (92%) with enzyme-linked immunosorbent assay (ELISA) results, and allowing the identification of infected animals.[266]

In a method using AuNP labeling coupled to nonstripping procedures, irregular shaped AuNP enhanced the CL of the luminol–H_2O_2 system with a detection limit of 17 pM for human IgG in plasma samples.[183] This method had the merit of avoiding the high CL background, but the synthesis of the so-called specially shaped irregular AuNPs proved to be rather cumbersome and lack of reproducibility among different batches was a concern. A report by Duan *et al.* on a nonstripping procedure involved the well-known citrate reduction method for spherical AuNP synthesis, without further NP modification. In this report, the catalytic reduction of $AgNO_3$ at the AuNP surface with concomitant luminol oxidation to its radical form and subsequent reaction with dissolved oxygen gave rise to the CL, which was used to quantify human IgG in plasma.[267] Although this method is simple and applicable to clinical diagnosis, it is not as sensitive as other *in situ* (based on plasma samples) AuNP-based CL immunoassays for the detection of human IgG.

Table III summarizes the main characteristics of several AuNP-based immunoassays described in the literature, comparing them with more standard immunoassays such as ELISA.

TABLE III
GOLD NANOPARTICLE-BASED IMMUNOASSAYS WITH DETECTION BY ELECTROCHEMICAL AND CHEMILUMINESCENT METHODS

Detection principle	Biological target	Detection limit/linear detection range	Refs.
AuNPs electrochemical stripping	Human IgM, rabbit IgG	1.5 ng mL^{-1}, 1.0 ng mL^{-1}/ 2.5–250 ng mL^{-1}	13
	Human IgG, goat IgG	0.5 ng mL^{-1}/5–5000 and 5–400 ng mL^{-1}	12
	Goat IgG	0.5 ng L^{-1}/0.5–100 ng mL^{-1}	10
AuNP label affords electrochemical signal amplification	Mouse IgG	1 fg mL^{-1}/1 fg mL^{-1}–10 μg mL^{-1}	268
Au (III) from AuNP stripping catalyses luminol CL	Goat–anti-human IgG	1.5 ng mL^{-1}/5–10000 ng mL^{-1}, good correlation with ELISA	264
	Human IgG	0.5 ng mL^{-1}/0.1–100 ng mL^{-1}	265
	ApxIV antibody in pig serum	1:40 000 (serum dilution), good agreement with ELISA (92%)	266
Catalytic activity of AuNPs on luminol CL	IgG in human plasma	17 pM (~ 2.7 ng mL^{-1})/ 0.05–10 nM, good correlation with RST[a]	183
		12.8 ng.mL^{-1}/25–5000 ng mL^{-1}	267

[a]RST: rate-scattering turbidimetry, a standard clinical method for IgG quantification.

3. AuNPs as Colored Probes and Scattering Enhancers

The optical response of noble metal NPs is typically characterized by strong absorption and scattering peaks not present in bulk materials. Thus, the simplest way of taking advantage of AuNP in immunoassays is by using their strong absorption in the visible range as an optical signal generator label. A traditional ELISA immunoassay to detect a breast cancer biomarker present in blood (the CA15-3 antigen) has recently been improved by using AuNPs as carriers of the signaling antibody anti-CA15-3-HRP (horseradish peroxidase) in order to achieve an amplification of the optical signal.[269] Higher sensitivity and a shorter assay time, as compared to classical ELISA procedures, were obtained in the 0–60 U mL^{-1} range.[269]

Also relying on the excellent optical response of AuNPs, a human serum albumin (HSA) immunoassay was developed in which a linear correlation was obtained between the logarithm of HSA concentration and the absorbance intensity, as detected with a common UV–vis spectrophotometer. The immunosensor was a chitosan-modified glass slide, where the abundant polymer amino groups served as linking sites (using glutaraldehyde) for anti-HAS antibody. The antigen (HSA) and AuNP-labeled anti-HSA were then added, to complete the sandwich-type immunocomplex.[270]

Gold nanoshells (AuNS) consist of a silica core surrounded by a thin shell of gold. By altering the relative dimensions of the core and shell, extinction maxima can be systematically varied from 700 to 1300 nm, and advantageously tuned to a wavelength that minimizes interference from absorbing molecules in clinical fluids.[271] Hirsch *et al.* have reported a rapid whole-blood immunoassay based on antibody–AuNS conjugates. The bionanoconjugates aggregate in the presence of subnanogram-per-milliliter quantities of various analytes, making it a less technically demanding assay than conventional ELISA.[272] The same type of AuNS was used to develop a solid-phase dot immunoassay with possible applications at POC.[273] The detection sensitivity for IgG target molecules was up to 0.25 ng in the case of 180 nm silica core/15 nm AuNS and up to 0.5–1 ng for 100/15 and 140/15 AuNS. These values are one order of magnitude higher than for the standard dot-assay protocol based on 15 nm AuNP conjugated with probing molecules.[273]

The bio-barcode assay[274] has been shown to be extraordinarily sensitive, with high sensitivity for both nucleic acid and protein targets, this in spite of its seemingly complex experimental setup.[275] In this method, initially developed by the Mirkin group, two types of nanoparticles are used: (i) a magnetic microparticle with recognition elements (antibodies or complementary DNA) for the target of interest and (ii) a AuNP with a second recognition agent (sandwich format) and hundreds of thiolated single-strand oligonucleotide barcodes. Figure 10 schematically illustrates the assay in the format it was used to detect prostate-specific antigen (PSA) in clinical samples, with 300 times more sensitivity than commercial PSA immunoassays.[54] Development of the assay relies on light scattering promoted by silver enhancement. The development method was recently improved by gold deposition, allowing the detection of as little as 3 fM of PSA in 10% serum.[276]

The scattering properties of silver NPs were also advantageously used for SERS detection of 2–5 nm AuNPs labeled with FITC and conjugated with anti-human IgG antibody. This system was used to detect human IgG *in situ*, immobilized in nitrocellulose strips, with comparable sensitivity to other AuNP-based immunoassays.[277] Another immunoassay based on SERS detection used 2-mercaptobenzothiazole as a tag.[278] The tag was used to aggregate AuNPs that were added to a goat–anti-mice IgG antibody solution. The tagged AuNP–antibody conjugates thus formed were utilized in a sandwich format assay with Raman detection of the tagged antibody with high selectivity.

The increasing interest of SERS and Raman-based immunoassays is due to the possibility of multiplexing, as each label has a unique vibrational fingerprint and thus many different proteins can be labeled for simultaneous determination.[15,279]

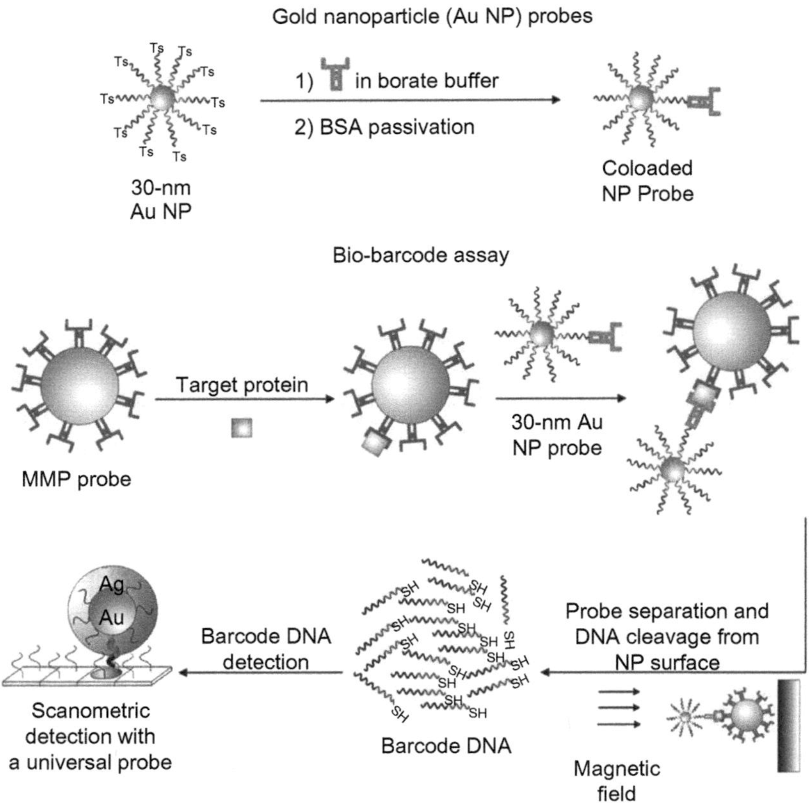

FIG. 10. Bio-barcode assay for the detection of PSA antigen in clinical samples. In this assay, the antigen is sandwiched between two antibodies, one conjugated to AuNPs and the other to magnetic microparticles. The AuNPs are previously derivatized with barcode DNA. The sandwich complex is separated magnetically and the barcode DNA is chemically cleaved. The released barcode DNA is captured by immobilized oligonucleotides that are complementary to part of the barcode DNA, and finally AuNP-labeled oligonucleotides complementary to another part of barcode DNA are added. Silver staining of the detection complex allows for scanometric detection (Ref. 54, reprinted with permission from the National Academy of Sciences USA). (For color version of this figure, the reader is referred to the Web version of this chapter.)

Other immunoassays based on less common light-scattering techniques have also produced promising results in terms of elevated sensitivity. For example, resonance Rayleigh scattering of AuNP bionanoprobes has allowed picomolar detection of transferrin. The combination of signal amplification bionanoprobes with the high sensitivity of this technique allowed a linear detection range from 85 pM to 3.4 nM.[280] Cancer biomarkers have also been detected using antibody-conjugated AuNP (both spherical and rod-like) coupled with dynamic light scattering.[281]

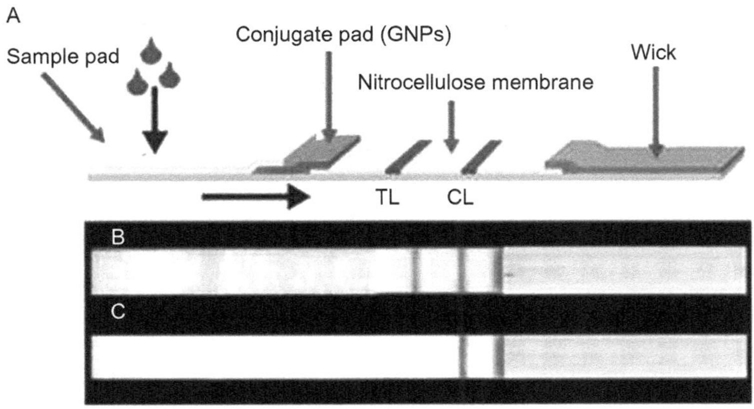

FIG. 11. (A) Design of a lateral-flow device showing how the four main components (sample pad, conjugate pad, nitrocellulose membrane, and wick) are interfaced by lamination. TL, test line and CL, control line. Result of a negative (B) and positive (C) competitive assay. Color development at the control line (CL) confirms correct operation of the device. GNP, gold nanoparticle (Ref. 271, reproduced by permission of The Royal Society of Chemistry. (For color version of this figure, the reader is referred to the Web version of this chapter.)

Lateral-flow devices are portable versions of immunoassays that can be used at POC, due to their low cost and simplicity of operation (Fig. 11). In their most common form, they consist of a porous white membrane striped with a line of antibodies or antigens, and interfaced with antibodies conjugated to an AuNP label that can be seen with the unaided eye. These types of platforms can support competitive or noncompetitive immunoassays. Generally, competitive immunoassays are used to detect low-molecular weight target molecules such as pesticides, hormones, and drugs, whereas a competitive assay format is used to detect high-molecular weight target molecules that can accommodate at least two antibody binding sites. Examples of the latter that have been detected with lateral-flow devices are PSA, ricin, and *Staphylococcus aureus*.[271] The excellent review by Wilson[271] contains the state of the art and discussion of this topic.

4. Plasmonics-Based Methods

Nanoparticle plasmonics deals with noble metal nanoparticles of various sizes, shapes, structures, and tunable plasmon resonances over the visible and near-infrared (NIR) spectral range.[282] Spherical AuNPs have a strong plasmon resonance absorption band around 520 nm, while anisotropic AuNPs present segregated localized surface plasmon resonance (LSPR) peaks, which in the case of nanorods and nanowires have longitudinal and transverse components.

These are known as transverse plasmon band and longitudinal plasmon band (LPB), respectively.[283] These bands are highly sensitive to the dielectric properties, that is, RI, of the surrounding media. Particularly, the LPB has been proved both theoretically and experimentally to be highly sensitive to changes in the dielectric properties in the vicinity of nanorods. Using a simple procedure with nanorods conjugated to goat IgG immobilized at mercaptosilane-modified glass surfaces, an optical immunoassay was developed in which the binding of anti-goat IgG could be quantified by the changes in the LPB induced by the immunocomplex formation. A linear correlation was observed between the concentration of rabbit anti-goat IgG and the intensity of LPB, with a detection limit as low as 0.1 pM.[284]

Due to their sensitivity as transducers of target binding events, AuNPs have been applied in real-time immunoassays, by monitoring the LSPR wavelength shifts in the optical extinction spectra upon target exposure and subsequent rinse. Kinetic binding rates and equilibrium constants of antibody–antigen systems could thus be obtained. In a report by Lee *et al.*, gold bipyramids were fixed on glass microscope slides, these were then conjugated to rabbit IgG, and finally, goat–anti-rabbit IgG was flowed and then rinsed with phosphate buffered saline. This experimental setup allowed the determination of the equilibrium constant for the system, with good agreement with standard antibody–antigen values. Additionally, the minimum analyte concentration necessary for obtaining a significant response from the biosensor was 10 nM, where the rate of extinction peak shift increased to 5.5×10^{-4} nm s^{-1}, a rate that is one order of magnitude higher than for an analyte concentration of 1 nM.[285]

Recently, a single-molecule immunoassay has been reported taking advantage of the extreme sensitivity of the bipyramidal nanoparticles' bright and sharp NIR scattering resonances to changes in the environment RI at their sharp edges. The method relies on an imaging spectrograph attached to a microscope making measurements on a flow cell. Blue shifts of an average 0.34 nm could be detected for goat–anti-rabbit IgG dissociation, when the rising buffer flowed, and the unbinding rate was found to be $7.9 \pm 1.3 \times 10^{-5}$ s^{-1}.[286]

5. Fluorescence Quenching by AuNPs

AuNPs can cause fluorescence enhancement or fluorescence quenching of molecules in their vicinity, depending on the distance between both systems.[287] The application of these interesting and diverse optical properties of AuNPs to immunoassays is still at its first steps, but interesting developments are expected as more basic studies unveil the complicated mechanisms underlying these processes. Recently, fluorescence quenching by AuNPs has been advantageously used in immunosensors in a variety of formats, including a typical sandwich assay,[288] and involving the utilization of magnetic beads.[289] The Feldmann group presented a competitive immunoassay for a bovine serum albumin-coupled drug (digoxigenin) with an LOD on the nanomolar range[290]; and a cardiac troponin T (cTnT) noncompetitive sandwich immunoassay that reached an LOD on the picomolar range,[291] the lowest value reported for this type of antigen detection assay. The principle of operation of the latter non-competitive method relies on two fragments of monoclonal antibodies for two different epitopes on the cTnT antigen. One of the antibodies is conjugated with AuNP and the other is labeled with a fluorescent dye, chosen so that its emission spectrum overlaps the AuNP SPR band. When the antigen is present in this homogeneous assay, each of the two monoclonal antibodies simultaneously binds to different epitopes, leading to AuNP quenching of the fluorescent emission of the dye (Fig. 12).

B. QDs in Immunoassays

Chan *et al.* reported on the synthesis of ZnS-capped CdSe QDs functionalized with mercaptoacetic acid for solubilization and protein conjugation. IgG molecules were covalently coupled to the QDs by the cross-linking agents EDC/NHS. When these nanoconjugates were incubated with a polyclonal antibody specific for the Fab region of the IgG, aggregation was clearly visible under an inverted microscope.[292] The observed aggregation of QDs was similar to latex immunoagglutination, a technique that is widely used in clinical diagnostics. This simple experiment showed that the QD-attached immunomolecules can recognize specific antibodies or antigens and was the first reported application of QDs in immunosensing.[292]

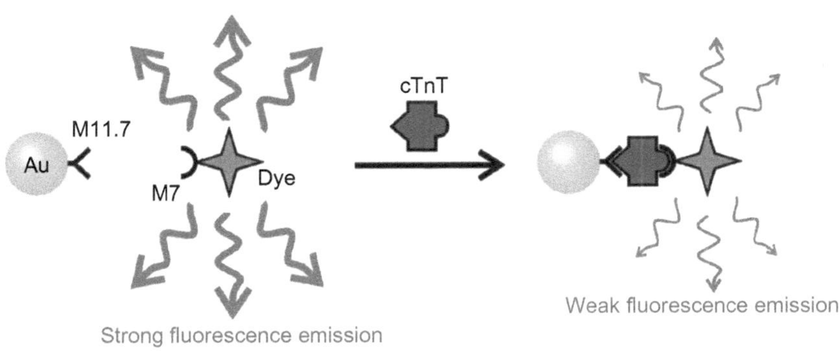

FIG. 12. Principle of operation of the homogenous sandwich immunoassay for cardiac troponin T (cTnT). AuNPs are functionalized with anti-cTnT M11.7 antibody fragments, and anti-cTnT M7 antibody fragments are labeled with fluorescent dye molecules. The assay relies on the fluorescence quenching of the dye by the AuNPs. (Reprinted with permission from Ref. 291. Copyright 2009 American Chemical Society.) (For color version of this figure, the reader is referred to the Web version of this chapter.)

An innovative conjugation approach between the QD and the antibody was used for the development of a sandwich-type immunoassay for staphylococcal enterotoxin B (SEB). The assay was run using polyclonal sheep anti-SEB antibody attached to the surface of a QD conjugated with a construct of the IgG-binding B domain of streptococcal protein G with a basic leucine zipper, and microtiter plates coated with monoclonal anti-SEB antibody. The lowest SEB concentration to provide a useful sign over background was ~ 2 ng mL^{-1}, and the signal increased until saturation was reached at ~ 30 ng mL^{-1}.[293]

QD-based immunoassays are becoming a feasible approach to rapid, sensitive, specific, and simple diagnostic of medically relevant samples. For example, Sun *et al.* reported on a 17β-estradiol indirect competitive immunoassay. The assay was performed using mostly commercially available components, including QD–streptavidin conjugates and biotynilated anti-rabbit IgG. The assay showed a good response for 17β-estradiol in the concentration range 0.01–10 000 ng mL^{-1}, and the LOD was 0.00542 ng mL^{-1}. Wastewater, tap water, and lake water were then analyzed (after a sample enrichment and purification step) and the results were compared with those from liquid chromatography–mass spectrometry, showing a good agreement between the two quantification methods.[294]

Multiplexed immunoassays are probably the most promising applications for QDs.[295,296] As a relatively new approach for accurate multiplexing analysis, the use of QD labels is still in need of further technical development to achieve superior analytical qualities.[297] To start with, the QD conjugation step is not perfect yet. Pathak *et al.* showed that the number of functional antibodies (able to bind the antigen) conjugated with QDs is very scarce when direct bioconjugation (surface attachment of the antibody thiol groups after disulfide reduction) with the QDs is used. The bioconjugation of streptavidin-coated QDs to biotinylated antibodies exhibited 10–20 times higher numbers of functional antibodies.[298] In addition, multiplexing can be impaired by the presence of FRET between different colored QDs. For example, Ghazani *et al.* reported on a protein microarray in which cancer tissue samples were labeled with three different colored QDs that could specifically detect three different cancer biomarkers. Under multiplexed conditions, the decrease of two of the QDs fluorescence signals was observed, with concomitant increase in one of them, implying crosstalk between the different QDs, which led to changes in their relative fluorescence intensities, probably caused by FRET.[299] It is also worth mentioning that while FRET between different QDs may be a problem in multiplexing, it has been employed for the detection of immunocomplex formation between two molecules, one labeled with QDs and the other with an organic dye, as previously described in this chapter for nucleic acid detection methods (Fig. 8).[297] In addition, because QDs emission spectra are much narrower and symmetric than from their organic counterparts, facilitating distinction between the emission of the QD donor and that of the acceptor, their combination should enable FRET-based immunoassays that are applicable to protein microarrays.[297]

IV. Conclusions and Future Perspectives

This chapter presented specialized applications of NPs in molecular diagnostics, with focus on applications that can be used at POC. Bringing NPs such as AuNPs, QDs, and magnetic nanoparticles to molecular diagnostic applications is an ongoing effort in the areas of chemistry, materials science, and molecular biology to provide better, faster, and more cost-effective and reliable assays. In spite of these efforts, many of the routine techniques are still time consuming and require expensive equipment (e.g., real-time PCR, DNA sequencing), expensive reagents (e.g., fluorescent dyes, DNA polymerases), and highly specialized laboratories and human resources. These factors contribute to restrict the use of molecular methods at POC or in a larger number of laboratories, especially in countries where fewer resources are available to healthcare.

Especially promising in lowering costs and increasing sensitivity in a simple format are AuNP-based methods. AuNPs are easy to obtain, functionalize, and further bioconjugate, generating detection probes. Within AuNP-based methods, those allowing detection of specific nucleic acid sequences in complex samples,[16] as well as immunoassays on a lateral-flow format,[271] are finding immediate applications for the analysis of clinical samples. The sensitivity of both types of assays has been much improved by combination with gold/silver nanoparticles[271,300] or by silver/gold deposition[54,276] to enhance light-scattering properties and further decrease detection limits. As pointed out by Wilson[271] when referring to lateral-flow assays, the minimum number of AuNPs that can be detected with the unaided eye on a white background is around 1×10^{10}, ideally allowing the detection of 10 fM of a target molecule. As the amount likely to be present in a typical biopsy sample is one million times less than this, simple techniques for signal enhancement such as PCR (target enhancement) and silver staining (AuNP signal enhancement) are likely to be extensively developed in the future.

A current trend in NP design is the preparation of composite nanoparticles, which combine different materials within the same NPs to confer multiple properties. For example, greatly increased Raman scattering profiles can be obtained via a combination of resonance Raman and surface-enhanced Raman, enabling ultrasensitive detection of gene sequences.[301] Such mixtures can also allow multiple modes of detection, that is, absorbance via the plasmon band of the AuNP or fluorescence emission of the semiconductor.[302] In another example, QDs have been utilized instead of dye molecules in an AuNP quenching assay. This assay allowed glucose detection in human serum or cell cultures, with high sensitivity (proportional to the concentration of glucose in the range 0.10–50 μM) and high selectivity relative to other sugars and most biological species present in serum.[303] The use of QD PL enhancement by AuNPs[304] has also paved the way for single-molecule detection. In this context, it is worth mentioning the pioneering work on single-molecule LSPR detection by monitoring antibody–antigen unbinding events through the scattering spectra of individual gold bipyramids.[286]

Another promising area of NP application is QD-encoded microbeads. These are typically silica microspheres embed with a combination of QDs. By using different combinations of only a few QDs of different size (and hence color), a library of microparticles with thousands of different spectral signatures can be created.[305] The aim is to enable broad multiplexing, and create a library of particles which can screen for a wide range of biomolecules simultaneously. For example, an assay to screen for 10 different SNPs in genes for cytochrome P450 has been developed with such a system.[306]

In order to use multifunctional NPs as contrast agents in both optical and magnetic techniques, hybrid fluorescent–magnetic nanoparticles may be created[307–310] that may be used simultaneously for the separation and detection of analytes. The idea behind multifunctional particles can be to enable not just multimodal imaging of the particles but also the ability for simultaneous diagnosis and therapy ("theranostics").[311,312]

The applicability of NP-based diagnostic methods at POC has been much improved by advances in electronics and engineering. Such advances have a positive impact in the field utilization of detection techniques that were only available in a laboratory setting, such as UV–vis and fluorimetry measurements.[212–215] Also, microfluidic systems allow miniaturization and integration of complex functions, which could move sophisticated diagnostic tools out of the laboratory.[313] Sample treatment and separation is also another area of research that will allow massive improvements in diagnostics applications. Namely, the use of magnetic separation conjugated to microfluidics will probably have a great impact in the development of POC diagnostic tools.[117]

Molecular diagnosis has special relevance in the management of cancer. Cancer outcomes are closely associated with early detection, and the increase in sensitivity provided by nanodiagnostics enhances the diagnosis limits, which, in turn, enables early cancer detection and thus more effective therapy.[297,314] In this context, multilabel detection coupled with high capture molecule density in immunosensors and arrays seems to be capable of detecting a wide range of protein concentrations with sensitivity ranging into the subpicogram-per-milliliter level. However, only a few of the newer ultrasensitive methods have been evaluated with real patient samples, which is key to establishing clinical sensitivity and selectivity. It is therefore urgent to define serum levels of proteins as biomarkers that can be used as indicators of the onset, existence, or progression of diseases, including cancer.[315,316]

Acknowledgments

The authors wish to thank Fundação para a Ciência e Tecnologia (FCT/MCTES), especially project PTDC/CTM-NAN/109877/2009, Portugal and the Luso-American Foundation (FLAD), Portugal, for continuing support to their research.

References

1. Sperling RA, Parak WJ. Surface modification, functionalization and bioconjugation of colloidal inorganic nanoparticles. *Philos Trans R Soc A Math Phys Eng Sci* 2010;**368**:1333–83.
2. Dai L, editor. *Carbon nanotechnology*. Amsterdam: Elsevier; 2006.
3. Kumar CSSR, editor. *Nanomaterials for Biosensors*. Nanotechnologies for Life Sciences, 10 vols. Weinheim: Wiley-VCH; 2007; 8.
4. Lhadi M, editor. *Hybrid nanocomposites for nanotechnology*. New York: Springer; 2009.
5. Datar R, Kim S, Jeon S, Hesketh P, Manalis S, Boisen A, et al. Cantilever sensors: nanomechanical tools for diagnostics. *Mrs Bulletin* 2009;**34**:449–54.
6. Jacobs CB, Peairs MJ, Venton BJ. Review: carbon nanotube based electrochemical sensors for biomolecules. *Anal Chim Acta* 2010;**662**:105–27.
7. Algar WR, Tavares AJ, Krull UJ. Beyond labels: a review of the application of quantum dots as integrated components of assays, bioprobes, and biosensors utilizing optical transduction. *Anal Chim Acta* 2010;**673**:1–25.
8. Geho D, Lahar N, Gurnani P, Huebschman M, Herrmann P, Espina V, et al. Pegylated, streptavidin-conjugated quantum dots are effective detection elements for reverse-phase protein microarrays. *Bioconjug Chem* 2005;**16**:559–66.
9. Gerion D, Chen FQ, Kannan B, Fu AH, Parak WJ, Chen DJ, et al. Room-temperature single-nucleotide polymorphism and multiallele DNA detection using fluorescent nanocrystals and microarrays. *Anal Chem* 2003;**75**:4766–72.
10. Dequaire M, Degrand C, Limoges B. An electrochemical metalloimmunoassay based on a colloidal gold label. *Anal Chem* 2000;**72**:5521–8.
11. Huang HP, Li JJ, Tan YL, Zhou JJ, Zhu JJ. Quantum dot-based DNA hybridization by electrochemiluminescence and anodic stripping voltammetry. *Analyst* 2010;**135**:1773–8.
12. Leng C, Lai G, Yan F, Ju H. Gold nanoparticle as an electrochemical label for inherently crosstalk-free multiplexed immunoassay on a disposable chip. *Anal Chim Acta* 2010;**666**:97–101.
13. Mao X, Baloda M, Gurung AS, Lin Y, Liu G. Multiplex electrochemical immunoassay using gold nanoparticle probes and immunochromatographic strips. *Electrochem Commun* 2008;**10**:1636–40.
14. Marin S, Merkoci A. Direct electrochemical stripping detection of cystic-fibrosis-related DNA linked through cadmium sulfide quantum dots. *Nanotechnology* 2009;**20**:055101.
15. Alvarez-Puebla RA, Liz-Marzan LM. SERS-based diagnosis and biodetection. *Small* 2010;**6**:604–10.
16. Baptista P, Pereira E, Eaton P, Doria G, Miranda A, Gomes I, et al. Gold nanoparticles for the development of clinical diagnosis methods. *Anal Bioanal Chem* 2008;**391**:943–50.

17. De M, Ghosh PS, Rotello VM. Applications of nanoparticles in biology. *Adv Mater* 2008;**20**:4225–41.
18. Laurent S, Forge D, Port M, Roch A, Robic C, Elst LV, et al. Magnetic iron oxide nanoparticles: synthesis, stabilization, vectorization, physicochemical characterizations, and biological applications. *Chem Rev* 2008;**108**:2064–110.
19. Kinge S, Crego-Calama M, Reinhoudt DN. Self-assembling nanoparticles at surfaces and interfaces. *Chemphyschem* 2008;**9**:20–42.
20. Cushing BL, Kolesnichenko VL, O'Connor CJ. Recent advances in the liquid-phase syntheses of inorganic nanoparticles. *Chem Rev* 2004;**104**:3893–946.
21. Goesmann H, Feldmann C. Nanoparticulate functional materials. *Angew Chem Int Ed* 2010;**49**:1362–95.
22. Lu AH, Salabas EL, Schuth F. Magnetic nanoparticles: synthesis, protection, functionalization, and application. *Angew Chem Int Ed* 2007;**46**:1222–44.
23. Park J, Joo J, Kwon SG, Jang Y, Hyeon T. Synthesis of monodisperse spherical nanocrystals. *Angew Chem Int Ed* 2007;**46**:4630–60.
24. Chen S, Liu L, Zhou J, Jiang S. Controlling antibody orientation on charged self-assembled monolayers. *Langmuir* 2003;**19**:2859–64.
25. Letsinger RL, Elghanian R, Viswanadham G, Mirkin CA. Use of a steroid cyclic disulfide anchor in constructing gold nanoparticle–oligonucleotide conjugates. *Bioconjug Chem* 2000;**11**:289–91.
26. Kimling J, Maier M, Okenve B, Kotaidis V, Ballot H, Plech A. Turkevich method for gold nanoparticle synthesis revisited. *J Phys Chem B* 2006;**110**:15700–7.
27. Turkevich J, Stevenson PC, Hillier J. A study of the nucleation and growth processes in the synthesis of colloidal gold. *Discuss Faraday Soc* 1951;**11**:55–75.
28. Prencipe G, Tabakman SM, Welsher K, Liu Z, Goodwin AP, Zhang L, et al. PEG branched polymer for functionalization of nanomaterials with ultralong blood circulation. *J Am Chem Soc* 2009;**131**:4783–7.
29. Wu XY, Liu HJ, Liu JQ, Haley KN, Treadway JA, Larson JP, et al. Immunofluorescent labeling of cancer marker Her2 and other cellular targets with semiconductor quantum dots. *Nat Biotechnol* 2003;**21**:41–6.
30. Chen LX, Liu T, Thurnauer MC, Csencsits R, Rajh T. Fe2O3 nanoparticle structures investigated by X-ray absorption near-edge structure, surface modifications, and model calculations. *J Phys Chem B* 2002;**106**:8539–46.
31. Peng S, Wang C, Xie J, Sun S. Synthesis and stabilization of monodisperse Fe nanoparticles. *J Am Chem Soc* 2006;**128**:10676–7.
32. Xu C, Xu K, Gu H, Zheng R, Liu H, Zhang X, et al. Dopamine as a robust anchor to immobilize functional molecules on the iron oxide shell of magnetic nanoparticles. *J Am Chem Soc* 2004;**126**:9938–9.
33. Sun S, Zeng H. Size-controlled synthesis of magnetite nanoparticles. *J Am Chem Soc* 2002;**124**:8204–5.
34. Basiruddin SK, Saha A, Pradhan N, Jana NR. Advances in coating chemistry in deriving soluble functional nanoparticle. *J Phys Chem C* 2010;**114**:11009–17.
35. Knopp D, Tang DP, Niessner R. Bioanalytical applications of biomolecule-functionalized nanometer-sized doped silica particles. *Anal Chim Acta* 2009;**647**:14–30.
36. Nitin N, Javier DJ, Richards-Kortum R. Oligonucleotide-coated metallic nanoparticles as a flexible platform for molecular imaging agents. *Bioconjug Chem* 2007;**18**:2090–6.
37. Hermanson GT. *Bioconjugate techniques.* 2nd ed. Oxford: Academic Press; 2008.
38. Lee J-H, Jun Y-w, Yeon S-I, Shin J-S, Cheon J. Dual-mode nanoparticle probes for high-performance magnetic resonance and fluorescence imaging of neuroblastoma. *Angew Chem Int Ed* 2006;**45**:8160–2.

39. Housni A, Cai H, Liu S, Pun SH, Narain R. Facile preparation of glyconanoparticles and their bioconjugation to streptavidin. *Langmuir* 2007;**23**:5056–61.
40. Abad JM, Mertens SFL, Pita M, FernÃ¡ndez VM, Schiffrin DJ. Functionalization of thioctic acid-capped gold nanoparticles for specific immobilization of histidine-tagged proteins. *J Am Chem Soc* 2005;**127**:5689–94.
41. Murphy CJ, Gole AM, Hunyadi SE, Stone JW, Sisco PN, Alkilany A, et al. Chemical sensing and imaging with metallic nanorods. *Chem Commun* 2008;544–57.
42. Ghosh P, Han G, De M, Kim CK, Rotello VM. Gold nanoparticles in delivery applications. *Adv Drug Deliv Rev* 2008;**60**:1307–15.
43. Baptista PV, Koziol-Montewka M, Paluch-Oles J, Doria G, Franco R. Gold-nanoparticle-probe-based assay for rapid and direct detection of Mycobacterium tuberculosis DNA in clinical samples. *Clin Chem* 2006;**52**:1433–4.
44. Doria G, Franco R, Baptista P. Nanodiagnostics: fast colorimetric method for single nucleotide polymorphism/mutation detection. *IET Nanobiotechnol* 2007;**1**:53–7.
45. Doria G, Larguinho M, Dias JT, Pereira E, Franco R, Baptista PV. Gold-silver-alloy nanoprobes for one-pot multiplex DNA detection. *Nanotechnology* 2010;**21**:255101.
46. Mirkin CA, Letsinger RL, Mucic RC, Storhoff JJ. A DNA-based method for rationally assembling nanoparticles into macroscopic materials. *Nature* 1996;**382**:607–9.
47. Sato K, Hosokawa K, Maeda M. Rapid aggregation of gold nanoparticles induced by non-cross-linking DNA hybridization. *J Am Chem Soc* 2003;**125**:8102–3.
48. Jung YL, Jung C, Parab H, Li T, Park HG. Direct colorimetric diagnosis of pathogen infections by utilizing thiol-labeled PCR primers and unmodified gold nanoparticles. *Biosens Bioelectron* 2010;**25**:1941–6.
49. Li H, Rothberg LJ. Label-free colorimetric detection of specific sequences in genomic DNA amplified by the polymerase chain reaction. *J Am Chem Soc* 2004;**126**:10958–61.
50. Li HX, Rothberg L. Colorimetric detection of DNA sequences based on electrostatic interactions with unmodified gold nanoparticles. *Proc Natl Acad Sci USA* 2004;**101**:14036–9.
51. Shawky SM, Bald D, Azzazy HME. Direct detection of unamplified hepatitis C virus RNA using unmodified gold nanoparticles. *Clin Biochem* 2010;**43**:1163–8.
52. Aili D, Selegard R, Baltzer L, Enander K, Liedberg B. Colorimetric protein sensing by controlled assembly of gold nanoparticles functionalized with synthetic receptors. *Small* 2009;**5**:2445–52.
53. Gupta S, Andresen H, Ghadiali JE, Stevens MM. Kinase-actuated immunoaggregation of peptide-conjugated gold nanoparticles. *Small* 2010;**6**:1509–13.
54. Thaxton CS, Elghanian R, Thomas AD, Stoeva SI, Lee JS, Smith ND, et al. Nanoparticle-based bio-barcode assay redefines "undetectable" PSA and biochemical recurrence after radical prostatectomy. *Proc Natl Acad Sci USA* 2009;**106**:18437–42.
55. Aslan K, Lakowicz JR, Geddes CD. Nanogold-plasmon-resonance-based glucose sensing. *Anal Biochem* 2004;**330**:145–55.
56. Eustis S, El-Sayed MA. Why gold nanoparticles are more precious than pretty gold: noble metal surface plasmon resonance and its enhancement of the radiative and nonradiative properties of nanocrystals of different shapes. *Chem Soc Rev* 2006;**35**:209–17.
57. Anger P, Bharadwaj P, Novotny L. Enhancement and quenching of single-molecule fluorescence. *Phys Rev Lett* 2006;**96**:113002.
58. Gersten JI. Theory of fluorophore-metallic surface interactions. In: Lakowicz Ga, editor. *Radiative decay engineering*, Vol. 8. New York: Plenum Press; 2005. pp. 197–221.
59. Kuhn S, Hakanson U, Rogobete L, Sandoghdar V. Enhancement of single-molecule fluorescence using a gold nanoparticle as an optical nanoantenna. *Phys Rev Lett* 2006;**97**:017402.
60. Bunz UHF, Rotello VM. Gold nanoparticle-fluorophore complexes: sensitive and discerning "noses" for biosystems sensing. *Angew Chem Int Ed* 2010;**49**:3268–79.

61. Phillips RL, Miranda OR, You CC, Rotello VM, Bunz UHF. Rapid and efficient identification of bacteria using gold-nanoparticle—poly(para-phenyleneethynylene) constructs. *Angew. Chem. Int. Ed.* 2008;**47**:2590–4.
62. De M, Rana S, Akpinar H, Miranda OR, Arvizo RR, Bunz UHF, et al. Sensing of proteins in human serum using conjugates of nanoparticles and green fluorescent protein. *Nat Chem* 2009;**1**:461–5.
63. You CC, Miranda OR, Gider B, Ghosh PS, Kim IB, Erdogan B, et al. Detection and identification of proteins using nanoparticle-fluorescent polymer 'chemical nose' sensors. *Nat Nanotechnol* 2007;**2**:318–23.
64. Astruc D. *Electron transfer and radical processes in transition metal chemistry.* Weinheim: Wiley-VCH; 1995.
65. Katz E, Willner I. Integrated nanoparticle-biomolecule hybrid systems: synthesis, properties, and applications. *Angew. Chem. Int. Ed.* 2004;**43**:6042–108.
66. Castaneda MT, Alegret S, Merkoci A. Electrochemical sensing of DNA using gold nanoparticles. *Electroanalysis* 2007;**19**:743–53.
67. Rossetti R, Brus L. Electron-hole recombination emission as a probe of surface-chemistry in aqueous Cds colloids. *J Phys Chem* 1982;**86**:4470–2.
68. Vossmeyer T, Katsikas L, Giersig M, Popovic IG, Diesner K, Chemseddine A, et al. Cds nanoclusters—synthesis, characterization, size-dependent oscillator strength, temperature shift of the excitonic-transition energy, and reversible absorbency shift. *J Phys Chem* 1994;**98**:7665–73.
69. Gao XH, Yang LL, Petros JA, Marshal FF, Simons JW, Nie SM. In vivo molecular and cellular imaging with quantum dots. *Curr Opin Biotechnol* 2005;**16**:63–72.
70. McBride J, Treadway J, Feldman LC, Pennycook SJ, Rosenthal SJ. Structural basis for near unity quantum yield core/shell nanostructures. *Nano Lett* 2006;**6**:1496–501.
71. Chan WCW, Maxwell DJ, Gao XH, Bailey RE, Han MY, Nie SM. Luminescent quantum dots for multiplexed biological detection and imaging. *Curr Opin Biotechnol* 2002;**13**:40–6.
72. Tallury P, Malhotra A, Byrne LM, Santra S. Nanobioimaging and sensing of infectious diseases. *Adv Drug Deliv Rev* 2010;**62**:424–37.
73. Michalet X, Pinaud FF, Bentolila LA, Tsay JM, Doose S, Li JJ, et al. Quantum dots for live cells, in vivo imaging, and diagnostics. *Science* 2005;**307**:538–44.
74. Gerion D, Pinaud F, Williams SC, Parak WJ, Zanchet D, Weiss S, et al. Synthesis and properties of biocompatible water-soluble silica-coated CdSe/ZnS semiconductor quantum dots. *J Phys Chem B* 2001;**105**:8861–71.
75. Murray CB, Norris DJ, Bawendi MG. Synthesis and characterization of nearly monodisperse Cde (E = S, Se, Te) semiconductor nanocrystallites. *J Am Chem Soc* 1993;**115**:8706–15.
76. Jamieson T, Bakhshi R, Petrova D, Pocock R, Imani M, Seifalian AM. Biological applications of quantum dots. *Biomaterials* 2007;**28**:4717–32.
77. Kuno M, Fromm DP, Hamann HF, Gallagher A, Nesbitt DJ. "On"/"off" fluorescence intermittency of single semiconductor quantum dots. *J Chem Phys* 2001;**115**:1028–40.
78. Mahler B, Spinicelli P, Buil S, Quelin X, Hermier JP, Dubertret B. Towards non-blinking colloidal quantum dots. *Nat Mater* 2008;**7**:659–64.
79. Rosenthal SJ, McBride J, Pennycook SJ, Feldman LC. Synthesis, surface studies, composition and structural characterization of CdSe, core/shell and biologically active nanocrystals. *Surf Sci Rep* 2007;**62**:111–57.
80. Kirchner C, Liedl T, Kudera S, Pellegrino T, Javier AM, Gaub HE, et al. Cytotoxicity of colloidal CdSe and CdSe/ZnS nanoparticles. *Nano Lett* 2005;**5**:331–8.
81. Dubertret B, Skourides P, Norris DJ, Noireaux V, Brivanlou AH, Libchaber A. In vivo imaging of quantum dots encapsulated in phospholipid micelles. *Science* 2002;**298**:1759–62.

82. Fonoberov VA, Alim KA, Balandin AA, Xiu FX, Liu JL. Photoluminescence investigation of the carrier recombination processes in ZnO quantum dots and nanocrystals. *Phys Rev B* 2006;**73**:165317.
83. Spanhel L. Colloidal ZnO nanostructures and functional coatings: a survey. *J Sol-Gel Sci Technol* 2006;**39**:7–24.
84. Erogbogbo F, Yong KT, Roy I, Xu GX, Prasad PN, Swihart MT. Biocompatible luminescent silicon quantum dots for imaging of cancer cells. *ACS Nano* 2008;**2**:873–8.
85. Warner JH, Hoshino A, Yamamoto K, Tilley RD. Water-soluble photoluminescent silicon quantum dots. *Angew Chem Int Ed* 2005;**44**:4550–4.
86. Parak WJ, Gerion D, Zanchet D, Woerz AS, Pellegrino T, Micheel C, et al. Conjugation of DNA to silanized colloidal semiconductor nanocrystalline quantum dots. *Chem Mater* 2002;**14**:2113–9.
87. Selvan ST, Tan TT, Ying JY. Robust, non-cytotoxic, silica-coated CdSe quantum dots with efficient photoluminescence. *Adv Mater* 2005;**17**:1620–5.
88. Tan WH, Wang KM, He XX, Zhao XJ, Drake T, Wang L, et al. Bionanotechnology based on silica nanoparticles. *Med Res Rev* 2004;**24**:621–38.
89. Lian W, Litherland SA, Badrane H, Tan WH, Wu DH, Baker HV, et al. Ultrasensitive detection of biomolecules with fluorescent dye-doped nanoparticles. *Anal Biochem* 2004;**334**:135–44.
90. Santra S, Zhang P, Wang KM, Tapec R, Tan WH. Conjugation of biomolecules with luminophore-doped silica nanoparticles for photostable biomarkers. *Anal Chem* 2001;**73**:4988–93.
91. Rossi LM, Shi LF, Quina FH, Rosenzweig Z. Stober synthesis of monodispersed luminescent silica nanoparticles for bioanalytical assays. *Langmuir* 2005;**21**:4277–80.
92. Zhao X, Tapec-Dytioco R, Tan W. Ultrasensitive DNA detection using highly fluorescent bioconjugated nanoparticles. *J Am Chem Soc* 2003;**125**:11474–5.
93. Nyffenegger R, Quellet C, Ricka J. Synthesis of fluorescent, monodisperse, colloidal silica particles. *J Colloid Interface Sci* 1993;**159**:150–7.
94. Zhang T, Stilwell JL, Gerion D, Ding L, Elboudwarej O, Cooke PA, et al. Cellular effect of high doses of silica-coated quantum dot profiled with high throughput gene expression analysis and high content cellomics measurements. *Nano Lett* 2006;**6**:800–8.
95. Dullens RPA, Claesson M, Derks D, van Blaaderen A, Kegel WK. Monodisperse core-shell poly(methyl methacrylate) latex colloids. *Langmuir* 2003;**19**:5963–6.
96. Aguilar-Arteaga K, Rodriguez JA, Barrado E. Magnetic solids in analytical chemistry: a review. *Anal Chim Acta* 2010;**674**:157–65.
97. Li J, Song S, Li D, Su Y, Huang Q, Zhao Y, et al. Multi-functional crosslinked Au nanoaggregates for the amplified optical DNA detection. *Biosens Bioelectron* 2009;**24**:3311–5.
98. Rossi LM, Quach AD, Rosenzweig Z. Glucose oxidase-magnetite nanoparticle bioconjugate for glucose sensing. *Anal Bioanal Chem* 2004;**380**:606–13.
99. De Palma R, Reekmans G, Liu CX, Wirix-Speetjens R, Laureyn W, Nilsson O, et al. Magnetic bead sensing platform for the detection of proteins. *Anal Chem* 2007;**79**:8669–77.
100. Loaiza OA, Campuzano S, Pedrero M, Pividori MI, Garcia P, Pingarron JM. Disposable magnetic DNA sensors for the determination at the attomolar level of a specific enterobacteriaceae family gene. *Anal Chem* 2008;**80**:8239–45.
101. Hu P, Huang CZ, Li YF, Ling J, Liu YL, Fei LR, et al. Magnetic particle-based sandwich sensor with DNA-modified carbon nanotubes as recognition elements for detection of DNA hybridization. *Anal Chem* 2008;**80**:1819–23.
102. Li JP, Gao HD. A renewable potentiometric immunosensor based on Fe3O4 nanoparticles immobilized anti-IgG. *Electroanalysis* 2008;**20**:881–7.

103. Erdem A, Karadeniz H, Mayer G, Famulok M, Caliskan A. Electrochemical sensing of aptamer-protein interactions using a magnetic particle assay and single-use sensor technology. *Electroanalysis* 2009;**21**:1278–84.
104. Huska D, Hubalek J, Adam V, Vajtr D, Horna A, Trnkova L, et al. Automated nucleic acids isolation using paramagnetic microparticles coupled with electrochemical detection. *Talanta* 2009;**79**:402–11.
105. Karadeniz H, Erdem A, Kuralay F, Jelen F. Indicator-based and indicator-free magnetic assays connected with disposable electrochemical nucleic acid sensor system. *Talanta* 2009;**78**:187–92.
106. Guesdon J-L, Avrameas S. Magnetic solid phase enzyme-immunoassay. *Immunochemistry* 1977;**14**:443–7.
107. Antoine JC, Ternynck TRS, Rodrigot M, Avrameas S. Lymphoid cell fractionation on magnetic polyacrylamide-agarose beads. *Immunochemistry* 1978;**15**:443–52.
108. Molday RS, Yen SPS, Rembaum A. Application of magnetic microspheres in labelling and separation of cells. *Nature* 1977;**268**:437–8.
109. Hao R, Xing R, Xu Z, Hou Y, Gao S, Sun S. Synthesis, functionalization, and biomedical applications of multifunctional magnetic nanoparticles. *Adv Mater* 2010;**22**:2729–42.
110. Jeong U, Teng X, Wang Y, Yang H, Xia Y. Superparamagnetic colloids: controlled synthesis and niche applications. *Adv Mater* 2007;**19**:33–60.
111. Sandhu A, Handa H, Abe M. Synthesis and applications of magnetic nanoparticles for biorecognition and point of care medical diagnostics. *Nanotechnology* 2010;**21**:442001.
112. Hatch GP, Stelter RE. Magnetic design considerations for devices and particles used for biological high-gradient magnetic separation (HGMS) systems. *J Magn Magn Mater* 2001;**225**:262–76.
113. Rheinlander T, Kotitz R, Weitschies W, Semmler W. Magnetic fractionation of magnetic fluids. *J Magn Magn Mater* 2000;**219**:219–28.
114. Moore LR, Rodriguez AR, Williams PS, McCloskey K, Bolwell BJ, Nakamura M, et al. Progenitor cell isolation with a high-capacity quadrupole magnetic flow sorter. *J Magn Magn Mater* 2001;**225**:277–84.
115. Earhart CM, Wilson RJ, White RL, Pourmand N, Wang SX. Microfabricated magnetic sifter for high-throughput and high-gradient magnetic separation. *J Magn Magn Mater* 2009;**321**:1436–9.
116. Gijs MAM. Magnetic bead handling on-chip: new opportunities for analytical applications. *Microfluid Nanofluid* 2004;**1**:22–40.
117. Gijs MAM, Lacharme F, Lehmann U. Microfluidic applications of magnetic particles for biological analysis and catalysis. *Chem Rev* 2010;**110**:1518–63.
118. Pamme N. Magnetism and microfluidics. *Lab Chip* 2006;**6**:24–38.
119. Peter JF, Otto AM. Magnetic particles as powerful purification tool for high sensitive mass spectrometric screening procedures. *Proteomics* 2010;**10**:628–33.
120. McCloskey KE, Moore LR, Hoyos M, Rodriguez A, Chalmers JJ, Zborowski M. Magnetophoretic cell sorting is a function of antibody binding capacity. *Biotechnol Prog* 2003;**19**:899–907.
121. Lara O, Tong XD, Zborowski M, Chalmers JJ. Enrichment of rare cancer cells through depletion of normal cells using density and flow-through, immunomagnetic cell separation. *Exp Hematol* 2004;**32**:891–904.
122. Xia N, Hunt TP, Mayers BT, Alsberg E, Whitesides GM, Westervelt RM, et al. Combined microfluidic-micromagnetic separation of living cells in continuous flow. *Biomed Microdevices* 2006;**8**:299–308.
123. Kim JS, Ligler FS. Utilization of microparticles in next-generation assays for microflow cytometers. *Anal Bioanal Chem* 2010;**398**:2373–82.

124. Arruebo M, Valladares M, Gonzalez-Fernandez A. Antibody-conjugated nanoparticles for biomedical applications. *J Nanomater* 2009. Article ID 439389, 24 pages.
125. Wang J, Kawde A-N, Erdem A, Salazar M. Magnetic bead-based label-free electrochemical detection of DNA hybridization. *Analyst* 2001;**126**:2020–4.
126. Wang J, Xu D, Erdem A, Polsky R, Salazar MA. Genomagnetic electrochemical assays of DNA hybridization. *Talanta* 2002;**56**:931–8.
127. Chowdhury EH, Akaike T. Bio-functional inorganic materials: an attractive branch of gene-based nano-medicine delivery for 21st century. *Curr Gene Ther* 2005;**5**:669–76.
128. Haun JB, Yoon TJ, Lee H, Weissleder R. Magnetic nanoparticle biosensors. *Wiley Interdiscip Rev Nanomed Nanobiotechnol* 2010;**2**:291–304.
129. Venter JC, Adams MD, Myers EW, Li PW, Mural RJ, Sutton GG, et al. The sequence of the human genome. *Science* 2001;**291**:1304–51.
130. Feingold EA, Good PJ, Guyer MS, Kamholz S, Liefer L, Wetterstrand K, et al. The ENCODE (ENCyclopedia of DNA elements) Project. *Science* 2004;**306**:636–40.
131. Altshuler D, Brooks LD, Chakravarti A, Collins FS, Daly MJ, Donnelly P, et al. A haplotype map of the human genome. *Nature* 2005;**437**:1299–320.
132. Weber JL, David D, Heil J, Fan Y, Zhao CF, Marth G. Human diallelic insertion/deletion polymorphisms. *Am J Hum Genet* 2002;**71**:854–62.
133. Bell J. Predicting disease using genomics. *Nature* 2004;**429**:453–6.
134. Dykes CW. Genes, disease and medicine. *Br J Clin Pharmacol* 1996;**42**:683–95.
135. Abramson RD, Myers TW. Nucleic acid amplification technologies. *Curr Opin Biotechnol* 1993;**4**:41–7.
136. Arya M, Shergill IS, Williamson M, Gommersall L, Arya N, Patel HRH. Basic principles of real-time quantitative PCR. *Expert Rev Mol Diagn* 2005;**5**:209–19.
137. Erlich HA, Gelfand D, Sninsky JJ. Recent advances in the polymerase chain-reaction. *Science* 1991;**252**:1643–51.
138. Mullis K, Faloona F, Scharf S, Saiki R, Horn G, Erlich H. Specific enzymatic amplification of DNA in vitro—the polymerase chain-reaction. *Cold Spring Harb Symp Quant Biol* 1986;**51**:263–73.
139. Newton CR, Graham A, Heptinstall LE, Powell SJ, Summers C, Kalsheker N, et al. Analysis of any point mutation in DNA—the amplification refractory mutation system (arms). *Nucleic Acids Res* 1989;**17**:2503–16.
140. Mackay J, Landt O. Real-time PCR fluorescent chemistries. *Methods Mol Biol* 2007;**353**:237–61.
141. Enomoto Y, Yoshikawa T, Ihira M, Akimoto S, Miyake F, Usui C, et al. Rapid diagnosis of herpes simplex virus infection by a loop-mediated isothermal amplification method. *J Clin Microbiol* 2005;**43**:951–5.
142. Gill P, Ghaemi A. Nucleic acid isothermal amplification technologies—a review. *Nucleosides Nucleotides Nucleic Acids* 2008;**27**:224–43.
143. Gyllensten UB, Erlich HA. Generation of single-stranded DNA by the polymerase chain reaction and its application to direct sequencing of the HLA-DQA locus. *Proc Natl Acad Sci USA* 1988;**85**:7652–6.
144. Kaneko H, Iida T, Aoki K, Ohno S, Suzutani T. Sensitive and rapid detection of herpes simplex virus and varicella-zoster virus DNA by loop-mediated isothermal amplification. *J Clin Microbiol* 2005;**43**:3290–6.
145. Nagamine K, Watanabe K, Ohtsuka K, Hase T, Notomi T. Loop-mediated isothermal amplification reaction using a nondenatured template. *Clin Chem* 2001;**47**:1742–3.
146. Notomi T, Okayama H, Masubuchi H, Yonekawa T, Watanabe K, Amino N, et al. Loop-mediated isothermal amplification of DNA. *Nucleic Acids Res* 2000;**28**:e63.

147. Parida M, Sannarangaiah S, Dash PK, Rao PVL, Morita K. Loop mediated isothermal amplification (LAMP): a new generation of innovative gene amplification technique; perspectives in clinical diagnosis of infectious diseases. *Rev Med Virol* 2008;**18**:407–21.
148. Sanchez JA, Pierce KE, Rice JE, Wangh LJ. Linear-after-the-exponential (LATE)-PCR: an advanced method of asymmetric PCR and its uses in quantitative real-time analysis. *Proc Natl Acad Sci USA* 2004;**101**:1933–8.
149. SantaLucia J. A unified view of polymer, dumbbell, and oligonucleotide DNA nearest-neighbor thermodynamics. *Proc Natl Acad Sci USA* 1998;**95**:1460–5.
150. Turner DH. Thermodynamics of base pairing. *Curr Opin Struct Biol* 1996;**6**:299–304.
151. Alwine JC, Kemp DJ, Stark GR. Method for detection of specific RNAs in agarose gels by transfer to diazobenzyloxymethyl-paper and hybridization with DNA probes. *Proc Natl Acad Sci USA* 1977;**74**:5350–4.
152. Southern EM. Detection of specific sequences among DNA fragments separated by gel-electrophoresis. *J Mol Biol* 1975;**98**(3):503–17.
153. Volpi EV, Bridger JM. FISH glossary: an overview of the fluorescence in situ hybridization technique. *Biotechniques* 2008;**45**:385–6.
154. Dufva M. Introduction to microarray technology. *Methods Mol Biol* 2009;**529**:1–22.
155. Niazov T, Pavlov V, Xiao Y, Gill R, Willner I. DNAzyme-functionalized Au nanoparticles for the amplified detection of DNA or telomerase activity. *Nano Lett* 2004;**4**:1683–7.
156. Glynou K, Ioannou PC, Christopoulos TK, Syriopoulou V. Oligonucleotide-functionalized gold nanoparticles as probes in a dry-reagent strip biosensor for DNA analysis by hybridization. *Anal Chem* 2003;**75**:4155–60.
157. Kalogianni DP, Bravou V, Christopoulos TK, Ioannou PC, Zoumbos NC. Dry-reagent disposable dipstick test for visual screening of seven leukemia-related chromosomal translocations. *Nucleic Acids Res* 2007;**35**:e23.
158. Litos IK, Ioannou PC, Christopoulos TK, Traeger-Synodinos J, Kanavakis E. Genotyping of single-nucleotide polymorphisms by primer extension reaction in a dry-reagent dipstick format. *Anal Chem* 2007;**79**:395–402.
159. He YQ, Zeng K, Gurung AS, Baloda M, Xu H, Zhang XB, et al. Visual detection of single-nucleotide polymorphism with hairpin oligonucleotide-functionalized gold nanoparticles. *Anal Chem* 2010;**82**:7169–77.
160. Kalogianni DP, Koraki T, Christopoulos TK, Ioannou PC. Nanoparticle-based DNA biosensor for visual detection of genetically modified organisms. *Biosens Bioelectron* 2006;**21**:1069–76.
161. Lee H, Joo SW, Lee SY, Lee CH, Yoon KA, Lee K. Colorimetric genotyping of single nucleotide polymorphism based on selective aggregation of unmodified gold nanoparticles. *Biosens Bioelectron* 2010;**26**:730–5.
162. Lee H, Kang T, Yoon KA, Lee SY, Joo SW, Lee K. Colorimetric detection of mutations in epidermal growth factor receptor using gold nanoparticle aggregation. *Biosens Bioelectron* 2010;**25**:1669–74.
163. Kim JH, Chung BH. Naked eye detection of mutagenic DNA photodimers using gold nanoparticles. *Biosens Bioelectron* 2011;**26**:2805–9.
164. Ma ZF, Tian L, Wang TT, Wang CG. Optical DNA detection based on gold nanorods aggregation. *Anal Chim Acta* 2010;**673**:179–84.
165. Li HX, Rothberg LJ. DNA sequence detection using selective fluorescence quenching of tagged oligonucleotide probes by gold nanoparticles. *Anal Chem* 2004;**76**:5414–7.
166. Liang RQ, Li W, Li Y, Tan CY, Li JX, Jin YX, et al. An oligonucleotide microarray for microRNA expression analysis based on labeling RNA with quantum dot and nanogold probe. *Nucleic Acids Res* 2005;**33**:e17.

167. Charrier A, Candoni N, Liachenko N, Thibaudau F. 2D aggregation and selective desorption of nanoparticle probes: a new method to probe DNA mismatches and damages. *Biosens Bioelectron* 2007;**22**:1881–6.
168. Charrier A, Candoni N, Thibaudau F. DNA detection method based on the two-dimensional aggregation and selective desorption of nanoparticle probes. *J Phys Chem B* 2006;**110**:12896–900.
169. Elghanian R, Storhoff JJ, Mucic RC, Letsinger RL, Mirkin CA. Selective colorimetric detection of polynucleotides based on the distance-dependent optical properties of gold nanoparticles. *Science* 1997;**277**:1078–81.
170. Murphy D, O'Brien P, Redmond G. Sub-picomole colorimetric single nucleotide polymorphism discrimination using oligonucleotide-nanoparticle conjugates. *Analyst* 2004;**129**:970–4.
171. Reynolds RA, Mirkin CA, Letsinger RL. Homogeneous, nanoparticle-based quantitative colorimetric detection of oligonucleotides. *J Am Chem Soc* 2000;**122**:3795–6.
172. Storhoff JJ, Lucas AD, Garimella V, Bao YP, Muller UR. Homogeneous detection of unamplified genomic DNA sequences based on colorimetric scatter of gold nanoparticle probes. *Nat Biotechnol* 2004;**22**:883–7.
173. Beissenhirtz MK, Elnathan R, Weizmann Y, Willner I. The aggregation of Au nanoparticles by an autonomous DNA machine detects viruses. *Small* 2007;**3**:375–9.
174. Sato K, Hosokawa K, Maeda M. Non-cross-linking gold nanoparticle aggregation as a detection method for single-base substitutions. *Nucleic Acids Res* 2005;**33**:e4.
175. Baptista PV. Principles in genetic risk assessment. *J Ther Clin Risk Manag* 2005;**1**:15–20.
176. Sato K, Onoguchi M, Sato Y, Hosokawa K, Maeda M. Non-cross-linking gold nanoparticle aggregation for sensitive detection of single-nucleotide polymorphisms: optimization of the particle diameter. *Anal Biochem* 2006;**350**:162–4.
177. Taton TA, Mirkin CA, Letsinger RL. Scanometric DNA array detection with nanoparticle probes. *Science* 2000;**289**:1757–60.
178. Bao YP, Huber M, Wei TF, Marla SS, Storhoff JJ, Muller UR. SNP identification in unamplified human genomic DNA with gold nanoparticle probes. *Nucleic Acids Res* 2005;**33**:e15.
179. Huber M, Wei TF, Muller UR, Lefebvre PA, Marla SS, Bao YP. Gold nanoparticle probe-based gene expression analysis with unamplified total human RNA. *Nucleic Acids Res* 2004;**32**:e137.
180. Storhoff JJ, Marla SS, Bao P, Hagenow S, Mehta H, Lucas A, et al. Gold nanoparticle-based detection of genomic DNA targets on microarrays using a novel optical detection system. *Biosens Bioelectron* 2004;**19**:875–83.
181. Bonanni A, Esplandiu MJ, del Valle M. Impedimetric genosensing of DNA polymorphism correlated to cystic fibrosis: a comparison among different protocols and electrode surfaces. *Biosens Bioelectron* 2010;**26**:1245–51.
182. Li XM, Fu PY, Liu JM, Zhang SS. Biosensor for multiplex detection of two DNA target sequences using enzyme-functionalized Au nanoparticles as signal amplification. *Anal Chim Acta* 2010;**673**:133–8.
183. Wang ZP, Hu JQ, Jin Y, Yao X, Li JH. In situ amplified chemiluminescent detection of DNA and immunoassay of IgG using special-shaped gold nanoparticles as label. *Clin Chem* 2006;**52**:1958–61.
184. Shiddiky MJA, Shim YB. Trace analysis of DNA: preconcentration, separation, and electrochemical detection in microchip electrophoresis using Au nanoparticles. *Anal Chem* 2007;**79**:3724–33.
185. Nogues C, Cohen SR, Daube S, Apter N, Naaman R. Sequence dependence of charge transport properties of DNA. *J Phys Chem B* 2006;**110**:8910–3.

186. Ozsoz M, Erdem A, Kerman K, Ozkan D, Tugrul B, Topcuoglu N, et al. Electrochemical genosensor based on colloidal gold nanoparticles for the detection of Factor V Leiden mutation using disposable pencil graphite electrodes. *Anal Chem* 2003;**75**:2181–7.
187. Wang J, Polsky R, Xu DK. Silver-enhanced colloidal gold electrochemical stripping detection of DNA hybridization. *Langmuir* 2001;**17**:5739–41.
188. Wang J, Xu DK, Kawde AN, Polsky R. Metal nanoparticle-based electrochemical stripping potentiometric detection of DNA hybridization. *Anal Chem* 2001;**73**:5576–81.
189. Ah CS, Yun YJ, Park HJ, Jung SK, Kim W-J, Ha DH, et al. Electric detection of DNA hybridization by nanoparticle nanoswitch. *Curr Appl Phys* 2006;**6**:e157–60.
190. Zhang J, Song SP, Zhang LY, Wang LH, Wu HP, Pan D, et al. Sequence-specific detection of femtomolar DNA via a chronocoulometric DNA sensor (CDS): effects of nanoparticle-mediated amplification and nanoscale control of DNA assembly at electrodes. *J Am Chem Soc* 2006;**128**:8575–80.
191. Castaneda MT, Merkoci A, Pumera M, Alegret S. Electrochemical genosensors for biomedical applications based on gold nanoparticles. *Biosens Bioelectron* 2007;**22**:1961–7.
192. Bui MPN, Baek TJ, Seong GH. Gold nanoparticle aggregation-based highly sensitive DNA detection using atomic force microscopy. *Anal Bioanal Chem* 2007;**388**:1185–90.
193. Ray PC, Darbha GK, Ray A, Walker J, Hardy W. Gold nanoparticle based FRET for DNA detection. *Plasmonics* 2007;**2**:173–83.
194. Cao YWC, Jin RC, Mirkin CA. Nanoparticles with Raman spectroscopic fingerprints for DNA and RNA detection. *Science* 2002;**297**:1536–40.
195. He L, Musick MD, Nicewarner SR, Salinas FG, Benkovic SJ, Natan MJ, et al. Colloidal Au-enhanced surface plasmon resonance for ultrasensitive detection of DNA hybridization. *J Am Chem Soc* 2000;**122**:9071–7.
196. Li YA, Wark AW, Lee HJ, Corn RM. Single-nucleotide polymorphism genotyping by nanoparticle-enhanced surface plasmon resonance imaging measurements of surface ligation reactions. *Anal Chem* 2006;**78**:3158–64.
197. Yao X, Li X, Toledo F, Zurita-Lopez C, Gutova M, Momand J, et al. Sub-attomole oligonucleotide and p53 cDNA determinations via a high-resolution surface plasmon resonance combined with oligonucleotide-capped gold nanoparticle signal amplification. *Anal Biochem* 2006;**354**:220–8.
198. Sato Y, Sato K, Hosokawa K, Maeda M. Surface plasmon resonance imaging on a microchip for detection of DNA-modified gold nanoparticles deposited onto the surface in a non-cross-linking configuration. *Anal Biochem* 2006;**355**:125–31.
199. Harry SR, Hicks DJ, Amiri KI, Wright DW. Hairpin DNA coated gold nanoparticles as intracellular mRNA probes for the detection of tyrosinase gene expression in melanoma cells. *Chem Commun* 2010;**46**:5557–9.
200. Cerruti MG, Sauthier M, Leonard D, Liu D, Duscher G, Feldheim DL, et al. Gold and silica-coated gold nanoparticles as thermographic labels for DNA detection. *Anal Chem* 2006;**78**:3282–8.
201. Kanjanawarut R, Su XD. Colorimetric detection of DNA using unmodified metallic nanoparticles and peptide nucleic acid probes. *Anal Chem* 2009;**81**:6122–9.
202. Xu QF, Liu J, He ZK, Yang S. Superquenching acridinium ester chemiluminescence by gold nanoparticles for DNA detection. *Chem Commun* 2010;**46**:8800–2.
203. Thaxton CS, Georganopoulou DG, Mirkin CA. Gold nanoparticle probes for the detection of nucleic acid targets. *Clin Chim Acta* 2006;**363**:120–6.
204. Storhoff JJ, Elghanian R, Mucic RC, Mirkin CA, Letsinger RL. One-pot colorimetric differentiation of polynucleotides with single base imperfections using gold nanoparticle probes. *J Am Chem Soc* 1998;**120**:1959–64.

205. Li JS, Deng T, Chu X, Yang RH, Jiang JH, Shen GL, et al. Rolling circle amplification combined with gold nanoparticle aggregates for highly sensitive identification of single-nucleotide polymorphisms. *Anal Chem* 2010;**82**:2811–6.
206. Qin WJ, Yim OS, Lai PS, Yung LYL. Dimeric gold nanoparticle assembly for detection and discrimination of single nucleotide mutation in Duchenne muscular dystrophy. *Biosens Bioelectron* 2010;**25**:2021–5.
207. Baptista P, Doria G, Henriques D, Pereira E, Franco R. Colorimetric detection of eukaryotic gene expression with DNA-derivatized gold nanoparticles. *J Biotechnol* 2005;**119**:111–7.
208. Conde J, de la Fuente J, Baptista P. RNA quantification using gold nanoprobes—application to cancer diagnostics. *J Nanobiotechnol* 2010;**8**:5.
209. Costa P, Amaro A, Botelho A, Inácio J, Baptista PV. Gold nanoprobe assay for the identification of mycobacteria of the Mycobacterium tuberculosis complex. *Clin Microbiol Infect* 2010;**16**:1464–9.
210. Doria G, Baumgartner BG, Franco R, Baptista PV. Optimizing Au-nanoprobes for specific sequence discrimination. *Colloids Surf B Biointerfaces* 2010;**77**:122–4.
211. Veigas B, Machado D, Perdigao J, Portugal I, Couto I, Viveiros M, et al. Au-nanoprobes for detection of SNPs associated with antibiotic resistance in Mycobacterium tuberculosis. *Nanotechnology* 2010;**21**:415101.
212. Bernacka-Wojcik I, Senadeera R, Wojcik PJ, Silva LB, Doria G, Baptista P, et al. Inkjet printed and "doctor blade" TiO2 photodetectors for DNA biosensors. *Biosens Bioelectron* 2010;**25**:1229–34.
213. Martins R, Baptista P, Raniero L, Doria G, Silva L, Franco R, et al. Amorphous/nanocrystalline silicon biosensor for the specific identification of unamplified nucleic acid sequences using gold nanoparticle probes. *Appl Phys Lett* 2007;**90**:023903.
214. Silva LB, Baptista P, Raniero L, Doria G, Martins R, Fortunato E. Characterization of optoelectronic platform using an amorphous/nanocrystalline silicon biosensor for the specific identification of nucleic acid sequences based on gold nanoparticle probes. *Sensor Actuat B-Chem* 2008;**132**:508–11.
215. Silva LB, Veigas B, Doria G, Costa P, Inácio J, Martins R, et al. Portable optoelectronic biosensing platform for identification of mycobacteria from the Mycobacterium tuberculosis complex. *Biosens Bioelectron* 2011;**26**:2012–7.
216. Hazarika P, Ceyhan B, Niemeyer CM. Reversible switching of DNA-gold nanoparticle aggregation. *Angew Chem Int Ed* 2004;**43**:6469–71.
217. Park SJ, Taton TA, Mirkin CA. Array-based electrical detection of DNA with nanoparticle probes. *Science* 2002;**295**:1503–6.
218. Kang T, Yoo SM, Yoon I, Lee SY, Kim B. Patterned multiplex pathogen DNA detection by Au particle-on-wire SERS sensor. *Nano Lett* 2010;**10**:1189–93.
219. Zhang H, Harpster MH, Park HJ, Johnson PA, Wilson WC. Surface-enhanced Raman scattering detection of DNA derived from the West Nile virus genome using magnetic capture of Raman-active gold nanoparticles. *Anal Chem* 2010;**83**:254–60.
220. Qian X, Zhou X, Nie S. Surface-enhanced Raman nanoparticle beacons based on bioconjugated gold nanocrystals and long range plasmonic coupling. *J Am Chem Soc* 2008;**130**:14934–5.
221. Litos IK, Ioannou PC, Christopoulos TK, Traeger-Synodinos J, Kanavakis E. Multianalyte, dipstick-type, nanoparticle-based DNA biosensor for visual genotyping of single-nucleotide polymorphisms. *Biosens Bioelectron* 2009;**24**:3135–9.
222. Zhang YY, Tang ZW, Wang J, Wu H, Maham AH, Lin YH. Hairpin DNA switch for ultrasensitive spectrophotometric detection of DNA hybridization based on gold nanoparticles and enzyme signal amplification. *Anal Chem* 2010;**82**:6440–6.

223. Cao YC, Jin RC, Thaxton S, Mirkin CA. A two-color-change, nanoparticle-based method for DNA detection. *Talanta* 2005;**67**:449–55.
224. Lee JS, Lytton-Jean AKR, Hurst SJ, Mirkin CA. Silver nanoparticle-oligonucleotide conjugates based on DNA with triple cyclic disulfide moieties. *Nano Lett* 2007;**7**:2112–5.
225. Jiang GX, Susha AS, Lutich AA, Stefani FD, Feldmann J, Rogach AL. Cascaded FRET in conjugated polymer/quantum dot/dye-labeled DNA complexes for DNA hybridization detection. *ACS Nano* 2009;**3**:4127–31.
226. Lee J, Choi Y, Kim J, Park E, Song R. Positively charged compact quantum dot-DNA complexes for detection of nucleic acids. *Chemphyschem* 2009;**10**:806–11.
227. Medintz IL, Mattoussi H. Quantum dot-based resonance energy transfer and its growing application in biology. *Phys Chem Chem Phys* 2009;**11**:17–45.
228. Patolsky F, Gill R, Weizmann Y, Mokari T, Banin U, Willner I. Lighting-up the dynamics of telomerization and DNA replication by CdSe-ZnS quantum dots. *J Am Chem Soc* 2003;**125**:13918–9.
229. Zhang CY, Yeh HC, Kuroki MT, Wang TH. Single-quantum-dot-based DNA nanosensor. *Nat Mater* 2005;**4**:826–31.
230. Zhou D, Ying L, Hong X, Hall EA, Abell C, Klenerman D. A compact functional quantum dot-DNA conjugate: preparation, hybridization, and specific label-free DNA detection. *Langmuir* 2008;**24**:1659–64.
231. Zhang CY, Hu J. Single quantum dot-based nanosensor for multiple DNA detection. *Anal Chem* 2010;**82**:1921–7.
232. Heyduk T, Heyduk E. Molecular beacons for detecting DNA binding proteins. *Nat Biotechnol* 2002;**20**:171–6.
233. Medintz IL, Berti L, Pons T, Grimes AF, English DS, Alessandrini A, et al. A reactive peptidic linker for self-assembling hybrid quantum dot-DNA bioconjugates. *Nano Lett* 2007;**7**:1741–8.
234. Agrawal A, Zhang CY, Byassee T, Tripp RA, Nie SM. Counting single native biomolecules and intact viruses with color-coded nanoparticles. *Anal Chem* 2006;**78**:1061–70.
235. Dyadyusha L, Yin H, Jaiswal S, Brown T, Baumberg JJ, Booy FP, et al. Quenching of CdSe quantum dot emission, a new approach for biosensing. *Chem Commun* 2005;3201–3.
236. Kuang H, Zhao S, Chen W, Ma W, Yong Q, Xu L, et al. Rapid DNA detection by interface PCR on nanoparticles. *Biosens Bioelectron* 2011;**26**:2495–9.
237. Xu Y, Li QG. Multiple fluorescent labeling of silica nanoparticles with lanthanide chelates for highly sensitive time-resolved Immunofluorometric assays. *Clin Chem* 2007;**53**:1503–10.
238. Herz E, Burns A, Bonner D, Wiesner U. Large stokes-shift fluorescent silica nanoparticles with enhanced emission over free dye for single excitation multiplexing. *Macromol Rapid Commun* 2009;**30**:1907–10.
239. Xu J, Liang J, Li J, Yang W. Multicolor dye-doped silica nanoparticles independent of FRET. *Langmuir* 2010;**26**:15722–5.
240. Zhou XC, Zhou JZ. Improving the signal sensitivity and photostability of DNA hybridizations on microarrays by using dye-doped core-shell silica nanoparticles. *Anal Chem* 2004;**76**:5302–12.
241. Yan JL, Estevez MC, Smith JE, Wang KM, He XX, Wang L, et al. Dye-doped nanoparticles for bioanalysis. *Nano Today* 2007;**2**:44–50.
242. Herz E, Marchincin T, Connelly L, Bonner D, Burns A, Switalski S, et al. Relative quantum yield measurements of coumarin encapsulated in core-shell silica nanoparticles. *J Fluoresc* 2010;**20**:67–72.
243. Santra S, Liesenfeld B, Bertolino C, Dutta D, Cao ZH, Tan WH, et al. Fluorescence lifetime measurements to determine the core-shell nanostructure of FITC-doped silica nanoparticles: an optical approach to evaluate nanoparticle photostability. *J Lumin* 2006;**117**:75–82.

244. Chen X, Esteĺvez MC, Zhu Z, Huang Y-F, Chen Y, Wang L, et al. Using aptamer-conjugated fluorescence resonance energy transfer nanoparticles for multiplexed cancer cell monitoring. *Anal Chem* 2009;**81**:7009–14.
245. Wang L, Yang C, Tan W. Dual-luminophore-doped silica nanoparticles for multiplexed signaling. *Nano Lett* 2004;**5**:37–43.
246. Sharma P, Brown S, Walter G, Santra S, Moudgil B. Nanoparticles for bioimaging. *Adv Colloid Interface Sci* 2006;**123–126**:471–85.
247. Smith JE, Wang L, Tan WT. Bioconjugated silica-coated nanoparticles for bioseparation and bioanalysis. *Trac-Trend Anal Chem* 2006;**25**:848–55.
248. Tapec R, Zhao XJJ, Tan WH. Development of organic dye-doped silica nanoparticles for bioanalysis and biosensors. *J Nanosci Nanotechnol* 2002;**2**:405–9.
249. Deng T, Li JS, Jiang JH, Shen GL, Yu RQ. Preparation of near-IR fluorescent nanoparticles for fluorescence-anisotropy-based immunoagglutination assay in whole blood. *Adv Funct Mater* 2006;**16**:2147–55.
250. Zhu NN, Cai H, He PG, Fang YZ. Tris(2,2′-bipyridyl)cobalt(III)-doped silica nanoparticle DNA probe for the electrochemical detection of DNA hybridization. *Anal Chim Acta* 2003;**481**:181–9.
251. Tallury P, Kar S, Bamrungsap S, Huang YF, Tand WH, Santra S. Ultra-small water-dispersible fluorescent chitosan nanoparticles: synthesis, characterization and specific targeting. *Chem Commun* 2009;2347–9.
252. Wu C, Szymanski C, McNeill J. Preparation and encapsulation of highly fluorescent conjugated polymer nanoparticles. *Langmuir* 2006;**22**:2956–60.
253. Wu C, Schneider T, Zeigler M, Yu J, Schiro PG, Burnham DR, et al. Bioconjugation of ultrabright semiconducting polymer dots for specific cellular targeting. *J Am Chem Soc* 2010;**132**:15410–7.
254. Taylor JR, Fang MM, Nie SM. Probing specific sequences on single DNA molecules with bioconjugated fluorescent nanoparticles. *Anal Chem* 2000;**72**:1979–86.
255. Tang S, Moayeri M, Chen Z, Harma H, Zhao J, Hu H, et al. Detection of anthrax toxin by an ultrasensitive immunoassay using europium nanoparticles. *Clin Vaccine Immunol* 2009;**16**:408–13.
256. Valanne A, Huopalahti S, Soukka T, Vainionpaa R, Lovgren T, Harma H. A sensitive adenovirus immunoassay as a model for using nanoparticle label technology in virus diagnostics. *J Clin Virol* 2005;**33**:217–23.
257. Weigl BH, Bardell RL, Cabrera C. Introduction to microfluidic techniques. In: Marks RS, Cullen DC, Karube I, Lowe CR, Weetall HH, editors. *Handbook of biosensors and biochips*. Cambridge: John Wiley & Sons, Ltd. 2007.
258. Laureyn W, Stakenborg T, Jacobs P. Genetic and other DNA-based biosensor applications. In: Marks RS, Cullen DC, Karube I, Lowe CR, Weetall HH, editors. *Handbook of biosensors and biochips*. Cambridge: John Wiley & Sons, Ltd. 2007. pp. 1–19.
259. Kuswandi B, Nuriman, Huskens J, Verboom W. Optical sensing systems for microfluidic devices: a review. *Anal Chim Acta* 2007;**601**:141–55.
260. Whitesides GM. The origins and the future of microfluidics. *Nature* 2006;**442**:368–73.
261. Chu X, Xiang ZF, Fu X, Wang SP, Shen GL, Yu RQ. Silver-enhanced colloidal gold metalloimmunoassay for Schistosoma japonicum antibody detection. *J Immunol Methods* 2005;**301**:77–88.
262. Lai LJ, Yang YW, Lin YK, Huang LL, Hsieh YH. Surface characterization of immunosensor conjugated with gold nanoparticles based on cyclic voltammetry and X-ray photoelectron spectroscopy. *Colloids Surf B Biointerfaces* 2009;**68**:130–5.
263. Fan AP, Cao ZJ, Li HA, Kai M, Lu JZ. Chemiluminescence platforms in immunoassay and DNA analyses. *Anal Sci* 2009;**25**:587–97.

264. Li Z-P, Wang Y-C, Liu C-H, Li Y-K. Development of chemiluminescence detection of gold nanoparticles in biological conjugates for immunoassay. *Anal Chim Acta* 2005;**551**:85–91.
265. Fan AP, Lau CW, Lu JZ. Magnetic bead-based chemiluminescent metal immunoassay with a colloidal gold label. *Anal Chem* 2005;**77**:3238–42.
266. Hu DH, Han HY, Zhou R, Dong F, Bei WC, Jia F, et al. Gold(III) enhanced chemiluminescence immunoassay for detection of antibody against ApxIV of Actinobacillus pleuropneumoniae. *Analyst* 2008;**133**:768–73.
267. Duan CF, Yu YQ, Cui H. Gold nanoparticle-based immunoassay by using non-stripping chemiluminescence detection. *Analyst* 2008;**133**:1250–5.
268. Das J, Aziz MA, Yang H. A nanocatalyst-based assay for proteins: DNA-free ultrasensitive electrochemical detection using catalytic reduction of p-nitrophenol by gold-nanoparticle labels. *J Am Chem Soc* 2006;**128**:16022–3.
269. Ambrosi A, Airó F, Merkoçi A. Enhanced gold nanoparticle based ELISA for a breast cancer biomarker. *Anal Chem* 2009;**82**:1151–6.
270. Zhang S-B, Wu Z-S, Guo M-M, Shen G-L, Yu R-Q. A novel immunoassay strategy based on combination of chitosan and a gold nanoparticle label. *Talanta* 2007;**71**:1530–5.
271. Wilson R. The use of gold nanoparticles in diagnostics and detection. *Chem Soc Rev* 2008;**37**:2028–45.
272. Hirsch LR, Jackson JB, Lee A, Halas NJ, West J. A whole blood immunoassay using gold nanoshells. *Anal Chem* 2003;**75**:2377–81.
273. Khlebtsov B, Khlebtsov N. Enhanced solid-phase immunoassay using gold nanoshells: effect of nanoparticle optical properties. *Nanotechnology* 2008;**19**:435703.
274. Nam JM, Thaxton CS, Mirkin CA. Nanoparticle-based bio-bar codes for the ultrasensitive detection of proteins. *Science* 2003;**301**:1884–6.
275. Hill HD, Mirkin CA. The bio-barcode assay for the detection of protein and nucleic acid targets using DTT-induced ligand exchange. *Nat Protoc* 2006;**1**:324–36.
276. Kim D, Daniel WL, Mirkin CA. Microarray-based multiplexed scanometric immunoassay for protein cancer markers using gold nanoparticle probes. *Anal Chem* 2009;**81**:9183–7.
277. Manimaran M, Jana NR. Detection of protein molecules by surface-enhanced Raman spectroscopy-based immunoassay using 2–5 nm gold nanoparticle lables. *J Raman Spectrosc* 2007;**38**:1326–31.
278. Song CY, Wang ZY, Yang J, Zhang RH, Cui YP. Preparation of 2-mercaptobenzothiazole-labeled immuno-Au aggregates for SERS-based immunoassay. *Colloids Surf B Biointerfaces* 2010;**81**:285–8.
279. Matschulat A, Drescher D, Kneipp J. Surface-enhanced Raman scattering hybrid nanoprobe multiplexing and imaging in biological systems. *ACS Nano* 2010;**4**:3259–69.
280. Cai HH, Yang PH, Feng J, Cai JY. Immunoassay detection using functionalized gold nanoparticle probes coupled with resonance Rayleigh scattering. *Sensors Actuat B Chem* 2009;**135**:603–9.
281. Liu X, Dai Q, Austin L, Coutts J, Knowles G, Zou J, et al. A one-step homogeneous immunoassay for cancer biomarker detection using gold nanoparticle probes coupled with dynamic light scattering. *J Am Chem Soc* 2008;**130**:2780–2.
282. Jain PK, Huang XH, El-Sayed IH, El-Sayed MA. Noble metals on the nanoscale: optical and photothermal properties and some applications in imaging, sensing, biology, and medicine. *Acc Chem Res* 2008;**41**:1578–86.
283. Khlebtsov NG. Anisotropic properties of plasmonic nanoparticles: depolarized light scattering, dichroism, and birefringence. *J Nanophotonics* 2010;**4**:041587.
284. Huang HW, Tang CR, Zeng YL, Yu XY, Liao B, Xia XD, et al. Label-free optical biosensor based on localized surface plasmon resonance of immobilized gold nanorods. *Colloids Surf B Biointerfaces* 2009;**71**:96–101.

285. Lee S, Mayer KM, Hafner JH. Improved localized surface plasmon resonance immunoassay with gold bipyramid substrates. *Anal Chem* 2009;**81**:4450–5.
286. Mayer KM, Hao F, Lee S, Nordlander P, Hafner JH. A single molecule immunoassay by localized surface plasmon resonance. *Nanotechnology* 2010;**21**:255503.
287. Dulkeith E, Ringler M, Klar TA, Feldmann J, Javier AM, Parak WJ. Gold nanoparticles quench fluorescence by phase induced radiative rate suppression. *Nano Lett* 2005;**5**:585–9.
288. Peng ZF, Chen ZP, Jiang JH, Zhang XB, Shen GL, Yu RQ. A novel immunoassay based on the dissociation of immunocomplex and fluorescence quenching by gold nanoparticles. *Anal Chim Acta* 2007;**583**:40–4.
289. Ao LM, Gao F, Pan BF, He R, Cui DX. Fluoroimmunoassay for antigen based on fluorescence quenching signal of gold nanoparticles. *Anal Chem* 2006;**78**:1104–6.
290. Mayilo S, Ehlers B, Wunderlich M, Klar TA, Josel HP, Heindl D, et al. Competitive homogeneous digoxigenin immunoassay based on fluorescence quenching by gold nanoparticles. *Anal Chim Acta* 2009;**646**:119–22.
291. Mayilo S, Kloster MA, Wunderlich M, Lutich A, Klar TA, Nichtl A, et al. Long-range fluorescence quenching by gold nanoparticles in a sandwich immunoassay for cardiac troponin T. *Nano Lett* 2009;**9**:4558–63.
292. Chan WCW, Nie SM. Quantum dot bioconjugates for ultrasensitive nonisotopic detection. *Science* 1998;**281**:2016–8.
293. Goldman ER, Anderson GP, Tran PT, Mattoussi H, Charles PT, Mauro JM. Conjugation of luminescent quantum dots with antibodies using an engineered adaptor protein to provide new reagents for fluoroimmunoassays. *Anal Chem* 2002;**74**:841–7.
294. Sun MM, Du LY, Gao SQ, Bao YH, Wang SH. Determination of 17 beta-oestradiol by fluorescence immunoassay with streptavidin-conjugated quantum dots as label. *Steroids* 2010;**75**:400–3.
295. Bruchez M, Moronne M, Gin P, Weiss S, Alivisatos AP. Semiconductor nanocrystals as fluorescent biological labels. *Science* 1998;**281**:2013–6.
296. Seydack M. Nanoparticle labels in immunosensing using optical detection methods. *Biosens Bioelectron* 2005;**20**:2454–69.
297. Rousserie G, Sukhanova A, Even-Desrumeaux K, Fleury F, Chames P, Baty D, et al. Semiconductor quantum dots for multiplexed bio-detection on solid-state microarrays. *Crit Rev Oncol Hematol* 2010;**74**:1–15.
298. Pathak S, Davidson MC, Silva GA. Characterization of the functional binding properties of antibody conjugated quantum dots. *Nano Lett* 2007;**7**:1839–45.
299. Ghazani AA, Lee JA, Klostranec J, Xiang Q, Dacosta RS, Wilson BC, et al. High throughput quantification of protein expression of cancer antigens in tissue microarray using quantum dot nanocrystals. *Nano Lett* 2006;**6**:2881–6.
300. Aveyard J, Nolan P, Wilson R. Improving the sensitivity of immunoassays by tuning gold nanoparticles to the tipping point. *Anal Chem* 2008;**80**:6001–5.
301. Liu Y, Zhong M, Shan G, Li Y, Huang B, Yang G. Biocompatible ZnO/Au nanocomposites for ultrasensitive DNA detection using resonance Raman scattering. *J Phys Chem B* 2008;**112**:6484–9.
302. Jin YD, Gao XH. Plasmonic fluorescent quantum dots. *Nat Nanotechnol* 2009;**4**:571–6.
303. Tang B, Cao L, Xu K, Zhuo L, Ge J, Li Q, et al. A new nanobiosensor for glucose with high sensitivity and selectivity in serum based on fluorescence resonance energy transfer (FRET) between CdTe quantum dots and Au nanoparticles. *Chem-Eur J* 2008;**14**:3637–44.
304. Maye MM, Gang O, Cotlet M. Photoluminescence enhancement in CdSe/ZnS-DNA linked-Au nanoparticle heterodimers probed by single molecule spectroscopy. *Chem Commun* 2010;**46**:6111–3.

305. Han MY, Gao XH, Su JZ, Nie S. Quantum-dot-tagged microbeads for multiplexed optical coding of biomolecules. *Nat Biotechnol* 2001;**19**:631–5.
306. Xu HX, Sha MY, Wong EY, Uphoff J, Xu YH, Treadway JA, et al. Multiplexed SNP genotyping using the Qbead (TM) system: a quantum dot-encoded microsphere-based assay. *Nucleic Acids Res* 2003;**31**:e43.
307. de Dios AS, Diaz-Garcia ME. Multifunctional nanoparticles: analytical prospects. *Anal Chim Acta* 2010;**666**:1–22.
308. Kim HN, Lee MH, Kim HJ, Kim JS, Yoon J. A new trend in rhodamine-based chemosensors: application of spirolactam ring-opening to sensing ions. *Chem Soc Rev* 2008;**37**:1465–72.
309. Pellegrino T, Kudera S, Liedl T, Javier AM, Manna L, Parak WJ. On the development of colloidal nanoparticles towards multifunctional structures and their possible use for biological applications. *Small* 2005;**1**:48–63.
310. Yiu HHP, Niu HJ, Biermans E, van Tendeloo G, Rosseinsky MJ. Designed multifunctional nanocomposites for biomedical applications. *Adv Funct Mater* 2010;**20**:1599–609.
311. Ho Y-P, Leong KW. Quantum dot-based theranostics. *Nanoscale* 2010;**2**:60–8.
312. Smith AM, Duan HW, Mohs AM, Nie SM. Bioconjugated quantum dots for in vivo molecular and cellular imaging. *Adv Drug Deliv Rev* 2008;**60**:1226–40.
313. Yager P, Edwards T, Fu E, Helton K, Nelson K, Tam MR, et al. Microfluidic diagnostic technologies for global public health. *Nature* 2006;**442**:412–8.
314. Farrell D, Alper J, Ptak K, Panaro NJ, Grodzinski P, Barker AD. Recent advances from the national cancer institute alliance for nanotechnology in cancer. *ACS Nano* 2010;**4**:589–94.
315. Gomez-Hens A, Fernandez-Romero JM, Aguilar-Caballos MP. Control of tumor markers using nanotechnology. *Mini Rev Med Chem* 2009;**9**:1064–74.
316. Rusling JF, Kumar CV, Gutkind JS, Patel V. Measurement of biomarker proteins for point-of-care early detection and monitoring of cancer. *Analyst* 2010;**135**:2496–511.

Nanoparticles in Cancer Chemotherapy

Deboshri Banerjee and
Shiladitya Sengupta

BWH-HST Center for Biomedical Engineering, Department of Medicine, Brigham and Women's Hospital, Harvard Medical School, Harvard-MIT Division of Health Science and Technology, Cambridge, Massachusetts, USA

I. Introduction 490
II. Physicochemical Properties of Nanoparticles and Impact on Tumor Targeting 490
A. Active Targeting of Nanoparticles to Tumor 493
B. Types of Targeting Moieties in Active Tumor Targeting 494
C. Targeting Cellular Signal Transduction Pathways 495
III. Engineering Nanoparticles for Therapeutic and Imaging Purposes 496
A. Polymeric Nanoconjugates 498
B. Lipid-Based Nanoparticles 499
C. Carbon Nanovectors 501
D. Inorganic Nanoparticles 501
E. Modified Natural Nanoparticles 502
IV. Shape–Size Effect of Nanoparticles 502
V. Current Status of Nanoparticles in Clinics 504
VI. Conclusion 504
References 504

Nanotechnology has evolved as an exciting platform in the field of anticancer research with promises to improve the pharmacology of current cancer therapeutics. Nanoparticles confer several advantages over that of free drugs, including their capability to carry high payloads of drugs, with prolonged half-life and reduced toxicity of the drugs, and increased targeting efficiency. The wide variety of nanovectors, coupled with the different methods available to conjugate or encapsulate therapeutic and/or imaging agents within, provide us with opportunities to fine-tune the pharmacological properties of these agents and open up new vistas in anticancer research. Here, we will discuss the physicochemical properties of different nanoparticles, their impact on tumor targeting, and their current status in the clinics with respect to cancer chemotherapy.

Progress in Molecular Biology
and Translational Science, Vol. 104
DOI: 10.1016/B978-0-12-416020-0.00012-7

I. Introduction

Cancer remains one of the deadliest diseases in today's world, with more than 1.5 million estimated new cases and over half a million deaths estimated in 2010.[1] Outstanding progress has been made in the field of cancer biology, which has helped us immensely to gain insight into the fundamentals of the disease.[2] However, challenge remains in our ability to translate our understanding of fundamental cancer biology to clinics. Despite the advances in the development of cancer therapeutics, it is increasingly becoming evident that cytotoxic chemotherapies will still remain the mainstay in cancer management. However, current chemotherapies are associated with nonspecific drug delivery resulting in off-target side effects with associated toxicity, often limiting their potential. This calls for a paradigm shift in the mode of treatment with the development of novel therapeutic strategies.

Over the past two decades, nanotechnology-based approaches have emerged as an exciting field with promises to remedy this limitation, due to their prolonged half-life and increased targeting efficiency.[3] These advantages are made possible by several parameters, including their size that allows them to intrinsically home in to tumors through enhanced permeability and retention (EPR) effect, their ability to evade the immune system and improve the drug's half-life, thereby significantly lowering its effective dose 50 (ED_{50}), and their potential for selective targeting due to their high surface density.[4–6] For these reasons, nanotherapeutics is emerging as the new paradigm in anticancer research.

In this chapter, we will focus on the physicochemical properties of nanoparticles and how they impact tumor targeting, followed by a discussion on the different types of nanoformulations studied to date. A short discussion on the important aspect of the shapes of nanoparticles and how they can affect their behavior will follow. We will conclude the chapter with a discussion on the current status of nanoparticles in cancer chemotherapy in clinics.

II. Physicochemical Properties of Nanoparticles and Impact on Tumor Targeting

Since nanoparticles were first proposed by Marty *et al.* in 1978 as novel drug delivery systems,[7] their use as anticancer agents has exploded during the 1980s. Nanotechnology in cancer therapy includes an arsenal of nanosized materials engineered using a top-down or bottom-up approach, generally ranging in dimensions from one to a few hundred nanometers in at least one dimension.[8] The resulting nanoparticles possess unique properties based on

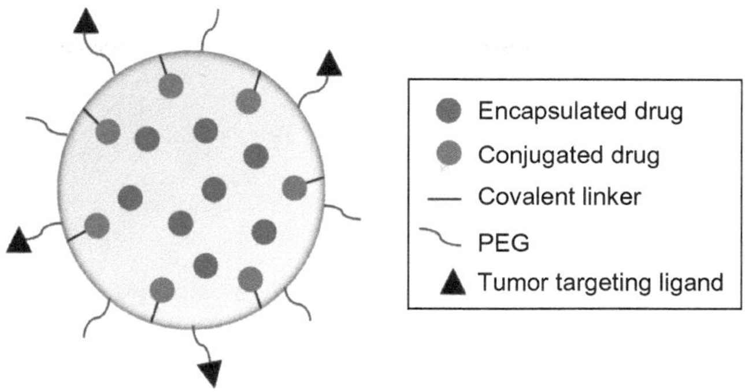

FIG. 1. Schematic diagram of a nanoparticle. (See Color Insert.)

their shape, size, and surface modifications. They are designed to carry therapeutic drugs and imaging agents, which are loaded on or within the nanocarriers by chemical conjugation or simply by encapsulation (Fig. 1). The nanoparticle-conjugated or encapsulated molecules (hydrophobic/hydrophilic), proteins, peptides, and nucleic acids possess the following advantages: (a) improved stability and solubility of encapsulated molecules as compared to the free entities, and (b) release in a more controlled manner over time to maintain drug concentration within a therapeutic window.[9,10] In addition, the surface of the nanoparticles can be modified to increase their blood-circulation half-life and improve their biodistribution profile. Designing "stealth" nanoparticles by surface functionalization with polyethylene glycol (PEG) prevents adherence of serum proteins to the nanoparticle surface, which is necessary to avoid clearance by the reticuloendothelial system (RES) and to have an increased circulation time.[11–13] Long circulation times are essential for an efficient delivery of the nanoparticles to the tumor site. Furthermore, attachment of tumor-targeting ligands to the nanoparticle surface can result in enhanced uptake at the targeted tissues.[14,15] The small size of the nanoparticles also plays a crucial role in delivery of drugs to tumor tissues.[16] With all these advantages, the net result is lowered systemic toxicity of the therapeutic agent and increased local concentration at the tissue of interest, resulting in a higher therapeutic index for the drug molecule.

Targeting of tumor tissues can be achieved by either "passive" or "active" targeting,[15] as shown in Fig. 2. The leaky vasculature associated with tumors contributes to the phenomenon of passive targeting by nanoparticles.[5] The tumor vessels have increased permeability due to aberrant angiogenesis, that is, the formation of new capillaries from existing vessels, thus allowing nanoparticles with diameters less than 200 nm to passively extravasate into the tumor sites through the EPR effect.[17] These nanoparticles are subsequently cleared

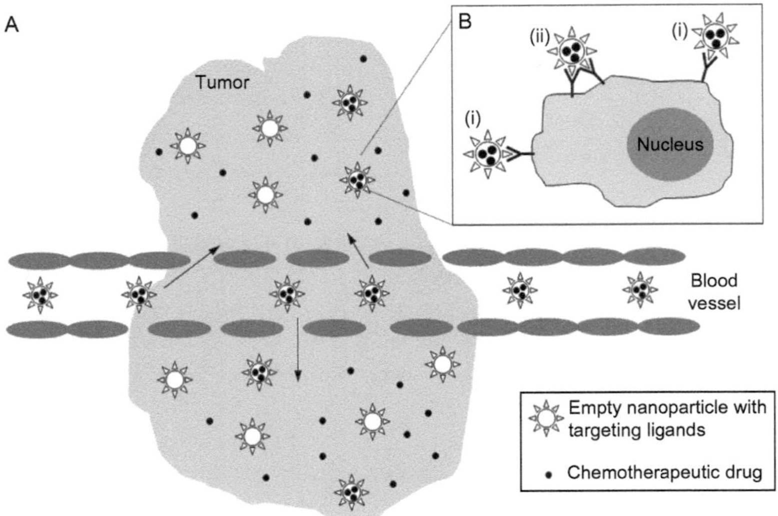

FIG. 2. Active versus passive targeting of nanoparticles to cancer tissues. (A) Nanoparticle extravasation through leaky vessels via the EPR effect in passive targeting. (B) (inset) Active targeting of receptors on an individual cell by ligands conjugated to the nanoparticles, showing (i) monovalent receptor–ligand interaction and (ii) multivalent receptor–ligand interaction.

by the liver. Although many factors, including surface area and chemical modifications, can affect the nanoparticle biodistribution, size remains the limiting factor in achieving passive targeting of the tumor sites. It has been observed that nanoparticles with sizes less than 10 nm are cleared by the kidney, whereas those larger than 200 nm often accumulate in the extracellular space and fail to reach the cancer environment.[4] Poor lymphatic drainage mechanisms in tumors allow the nanoparticles to be retained in the vicinity of the tumor cells and allow them to release their cargo in a sustained manner.[5] For example, the polymer-conjugated angiogenesis inhibitor TNP-470 (caplostatin) was found to accumulate selectively in the tumor vessels by the EPR effect and inhibit hyperpermeability of tumor blood vessels.[18,19] In studies published from our laboratory, we have shown that nanoparticle-conjugated chemotherapeutic agents such as doxorubicin (Dox)[20–22] and angiogenic small-molecule inhibitors[23] can preferentially home into tumors by the EPR effect, resulting in selective vascular shutdown and inhibition of tumor growth.

The passive targeting approach has formed the basis of clinical trials so far. However, the phenomenon suffers from several limitations. In certain tumors that do not exhibit the EPR effect, reaching the tumor tissue via passive targeting can be a problem.[2,5] Also, vessel permeability may not be uniform throughout a single tumor, making it difficult for therapeutics to diffuse uniformly to all cells.[24] Furthermore, lack of control of the process of drug delivery may induce multiple-drug resistance (MDR), a situation where chemotherapy fails in patients due to resistance of cancer cells to one or more drugs.[25]

Hence, it should be noted that passive targeting alone is not always sufficient in targeting the tumor sites and therefore is often used in conjunction with active targeting. This combination ensures that nanoparticles are retained in the tumor tissues following their extravasation from leaky vessels.[15] Active targeting of tumor tissues is achieved by chemically arraying ligands—molecules that bind to specific cell surface receptors—on the surface of nanoparticles. These ligands can recognize and selectively bind to receptors specifically expressed on tumor cells and vessels. The high surface area to volume ratio of the nanoparticles leads to high local density of ligands for targeting. Following ligand–receptor interactions, the receptor-bound nanocarriers are internalized by the cells, followed by release of their payloads (i.e., therapeutics or imaging agents) inside the cells.[26,27]

A. Active Targeting of Nanoparticles to Tumor

The most effective cancer therapeutic would be the one capable of reducing and eliminating tumors without causing any damage to the surrounding healthy tissues. This could be rendered possible by incorporating specific tools in the nanoparticle design that can selectively target the tumor via the phenomenon of "active" targeting. An increase in site specificity and selective internalization can improve treatment efficacy and decrease treatment-induced side effects in patients.

Peripherally conjugated targeting ligands specifically recognize a surface marker, for example, an antigen or receptor, overexpressed on target cancer cells relative to normal healthy cells. Receptor density on cell surface plays a crucial role for efficient delivery of the nanocarriers to the targeted cells. For example, in a study reported by Park *et al.*, the presence of 10^5 copies of ErbB2 receptors on a cell was found to be necessary for improved therapeutic efficacy of an ErbB2-targeted liposomal formulation of Dox in comparison to its non-targeted version in a breast cancer model.[28] Targeting ligands are important for the mechanism of cellular uptake. Binding of nanoparticle-bound ligands to cell surface receptors may cause receptor-mediated internalization, which has often shown an increased therapeutic effect of the released drugs.[26,27,29,30] On the contrary, if the ligands attach to a noninternalizing receptor, drugs released from their nanocarriers will have a high local concentration outside the target cells. But diffusion of the drugs to the interior of the target cells may not be homogeneous, resulting in a decreased therapeutic efficacy of the released drug.[31]

Higher binding affinity is generally known to increase targeting efficiency, although there are evidences where ligands bind to receptors so strongly that penetration of the nanoparticles into the tissue is prevented.[32] In addition, "multivalency" may provide an important way to enhance the binding affinity (or avidity) of ligands to receptors as compared to monovalent ligand–receptor binding affinity[33] (shown in Fig. 2). Multivalency, as the term suggests, is collective binding of multiple copies of a ligand on the surface of a nanoparticle to cellular receptors, resulting in a very high affinity for the nanoparticle to the target cell. For example, in an attempt to target folic acid receptors overexpressed on certain cancer cells, a virus-based nanoparticle decorated by about 60 folic acid moieties was found to increase the avidity of the nanoconjugates by approximately 300 times over free folic acid.[34]

B. Types of Targeting Moieties in Active Tumor Targeting

There are different types of targeting agents, such as proteins (antibodies and fragments thereof), ligands for receptors (peptides, carbohydrates, and other small molecules), and nucleic acid aptamers. Antibody-based targeting of tumor tissue has been a field of active research over the past three decades, the first study being reported in 1981,[35] owing to their high affinity for cell surface antigens. Since then, a number of clinical trials have demonstrated the feasibility of antibody-based targeting,[36,37] and 17 monoclonal antibodies (mAb) have been approved by the U.S. Food and Drug Administration (FDA) for use in patients. Trastuzumab (or Herceptin®), for example, is a monoclonal antibody that binds to the ErbB2 receptors. It has been approved for the treatment of breast cancer.[38] Recent studies in the field of antibody engineering have led to the development of recombinant mAbs from viruses and yeast, partially mouse and partially human chimeric antibodies, "fully" human mAbs using phage display and genetic engineering, and antibody fragments.

Although antibodies and antibody fragments have been found to be effective targeting agents, there are inherent problems associated with their immunogenicity (ability to generate an immune response), decreased receptor affinity due to inadequate conjugation methods, circulating free antigens or changes in cell surface antigens, insufficient tumor penetration, and nonspecific binding of antibodies to cellular receptors. New technologies are being explored to enhance the selectivity and efficacy of targeting agents while attempting to overcome the shortcomings associated with existing targeting moieties. Fusion proteins, created by the fusion of two or more genes to develop a new protein with desired properties, have been found to exhibit highly enhanced affinity for target receptors.[39] Antibody mimetics have been designed, such as affibodies, that can be engineered to bind specifically to target proteins in a conformational-sensitive manner, and avimers that bind to targeted receptors through multivalent interactions.

In addition to protein-based targeting agents, high-throughput approaches have been used to generate aptamers, which are short, three-dimensional synthetic RNA or DNA oligonucleotides (15–40 nucleotides long) or peptides (10–20 amino acids long) that bind their target with high affinity and specificity, hence earning their nickname of "chemical antibodies."[40–42] The main lacuna of aptamers is that they are easily degraded by cellular nucleases and proteinases. This lacuna can easily be remedied by chemical modifications of the aptamers, making them an ideal candidate for novel nanotherapeutics with improved targeting power and therapeutic index.[42] In a recent study, it has been found that aptamers conjugated to the surface of docetaxel-encapsulating nanoparticles can efficiently and specifically target antigens on the surface of prostate cancer cells *in vivo*.[43]

Certain cell proliferation markers, such as human endothelial receptors, transferring receptors, and folate receptors, are highly overexpressed on certain tumor cells and have been used as significant targets for cancer therapeutics. The epidermal growth factor (EGF) has been found to be a promising targeting agent of the EGF receptor (EGFR) overexpressed on different tumor cells, including breast, lung, colorectal, and brain cancers.[44,45] Similarly, the vitamin folic acid and the iron-transfer protein transferrin have been used to decorate the surfaces of nanoparticles for targeting folate receptors and transferrin receptors, respectively, overexpressed on various tumor cells.[46,47] One major drawback of targeting the transferrin and folate receptors, though, is that they are also expressed in large numbers in rapidly growing healthy epithelial and endothelial cells, which in turn may give rise to nonspecific targeting with decreased effectiveness in drug delivery and increased toxicity.

Peptides have recently emerged as targeting agents owing to their ease of synthesis and purification. The integrin family, particularly the $\alpha v\beta 3$ integrins, has been the most widely studied target for cancer chemotherapeutics. A synthetic peptide consisting of arginine-glycine-aspartic acid (RGD) residues has been used as a ligand conjugated to nanoparticles for targeting the cell adhesion $\alpha v\beta 3$ integrins on endothelial cells. Recent studies are further optimizing integrin targeting by engineering novel peptide moieties which bind with better affinity to integrins than current RGD tags.[48,49]

C. Targeting Cellular Signal Transduction Pathways

Signal transduction pathways regulate many important cellular functions such as growth, differentiation, metabolism, and survival. Many of these signaling pathways are altered in human cancer. Basic research in cancer biology has unveiled the interconnections between different oncogenic signaling pathways, and extensive studies in the fields of chemistry and biology have led to the development of various small molecule inhibitors of these pathways.

Receptor tyrosine kinases (RTKs) are high-affinity cell-surface receptors that play a major role in the signaling pathways governing the fundamental cellular processes of proliferation, differentiation, metabolism, survival, and intercellular communications.[50] They bind to their corresponding growth factors, triggering the downstream signaling cascade. Examples of RTKs that have been implicated in cancer include the EGFR, fibroblast growth factor receptor (FGFR), vascular endothelial growth factor receptor (VEGFR), and platelet-derived growth factor receptor (PDGFR). As a result, the RTKs and their growth factors have become a rational target for therapeutic interventions using specific antibodies and small-molecule inhibitory drugs. Trastuzumab (Herceptin®, Genentech Inc.) is a monoclonal antibody approved by the FDA that targets the human epidermal growth factor receptor-2 (HER2) for the treatment of HER2-overexpressing metastatic breast cancer. On the other hand, imatinib (Gleevec®, Novartis) is a small-molecule tyrosine kinase inhibitor of PDGFR approved by the FDA for treatment of chronic myelogenous leukemia. Among the other major signaling pathways implicated in cancer are the mitogen-activated protein kinase (MAPK) and the phosphatidylinositol 3-kinase (PI3K) signaling pathways.

Although there are few reports on the application of nanotechnology in targeting signal transduction pathways in cancer, it seems logical to combine inhibitors that target the signaling pathways with nanoparticle-based tumor targeting. Basu *et al.* provides a detailed review on the role of different signaling pathways in cancer and the role of nanotechnology in targeting these pathways.[51] In a recent study, Basu *et al.* have reported the use of poly (lactic-*co*-glycolic acid) (PLGA) copolymer for chemically conjugating PD98059, a selective MAPK inhibitor, for targeting the MAPK signaling pathway.[52] The resulting nanoparticles selectively resulted in melanoma regression in a mouse model. In a consecutive study, Harfouche *et al.* encapsulated LY294002, an inhibitor of the PI3K pathway, in a PLGA copolymer. They used novel zebrafish melanoma and breast adenocarcinoma tumor xenograft models to demonstrate the antiangiogenic effect of these nanoparticles.[23]

III. Engineering Nanoparticles for Therapeutic and Imaging Purposes

We have discussed so far how nanoparticles have impacted drug delivery by selectively targeting the tumor tissue. In this section, we will discuss the different types of nanoparticles that have been studied to date to formulate therapeutic and diagnostic agents for treatment and imaging purposes and their modifications, along with their advantages and limitations. Several nanovectors have been reported thus far in mediating anticancer therapy and imaging the tumor environment. These include an arsenal of synthetic and natural nanoparticles such as polymeric conjugates and polymeric nanoparticles; liposomes and micelles; synthetic organic nanoparticles such as dendrimers; carbon-based nanostructures such as carbon nanotubes (CNTs) and polyhydroxylated fullerenes; inorganic nanoparticles of gold, silver, and iron oxide; quantum dots (QDs); viral capsids and ferritin (Fig. 3). The plethora of nanovectors allows researchers to fine-tune the properties of the drugs depending on their target. Further fine-tuning is also possible depending on the method of drug–nanovector coupling, thus offering the potential to engineer revolutionary therapeutics in the field of cancer.

In addition to enhancing the therapeutic efficacy of chemotherapeutic agents, nanoparticles have also been extensively studied in the field of medical imaging. Medical imaging has undergone tremendous research and development over the past few decades, with the introduction of techniques such as magnetic resonance imaging (MRI), computed tomography (CT), ultrasonography, nuclear medicine scanning, and optical fluorescence imaging.[53,54]

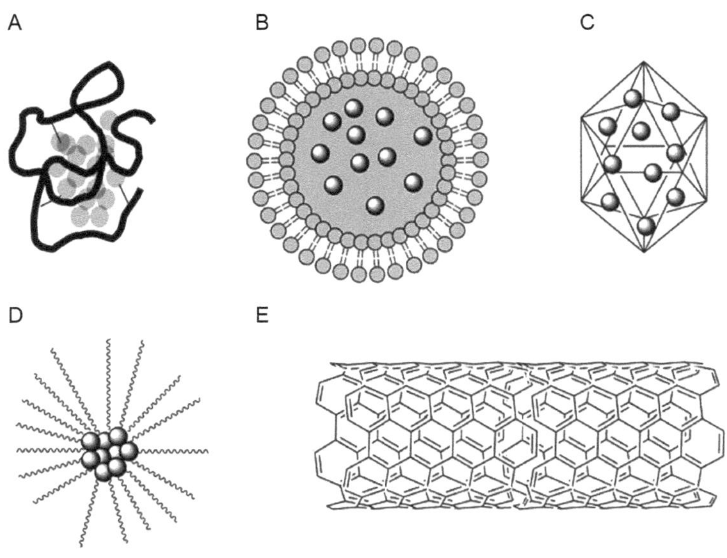

FIG. 3. Examples of nanovectors that are used to deliver cancer therapeutics and imaging agents. (A) Polymer-drug conjugates. (B) Liposomes with encapsulated drugs. (C) Viral-based nanoparticles with enclosed drugs. (D) Polymeric micelles with entrapped hydrophobic drugs. (E) A single-walled carbon nanotube.

The possibility of noninvasive and accurate monitoring of tumor response has led to a growing interest in the use of these techniques in cancer research. Use of nanoparticles offer several advantages in this area of research, including their capability of carrying high payloads of therapeutic and diagnostic agents, improved contrast, and longer circulation times in the body.

A. Polymeric Nanoconjugates

A diverse family of polymers has been studied for the engineering of nanoparticle-based drug delivery agents since the earliest reports in 1979 describing their use in cancer therapy.[55] Polymers chemically conjugated to drugs are regarded as new chemical entities owing to their distinctive pharmacokinetic profile as compared to the parent drugs. Polymeric nanoparticles can also be designed to encapsulate drugs without any chemical modification. Encapsulated drugs can be released in a controlled manner from the polymer matrix by diffusion or through surface or bulk erosion, while release of conjugated drugs requires cleavage of the covalent bonds under biological conditions. Some key examples of polymer-based nanoconjugates for anticancer therapy have been prepared from structures such as *N*-(2-hydroxypropyl) methacrylamide (HPMA), PLGA, polysaccharides (e.g., chitosan), and dendrimers, to name a few.

HPMA copolymers are hydrophilic substances that have been extensively studied. An HPMA copolymer conjugated to the angiogenesis inhibitor TNP-470 (Caplostatin) was found to selectively accumulate in the tumor microvasculature, resulting in decreased tumor growth rate in human melanoma and lung carcinoma mice models.[19] Interestingly, such formulation of TNP-470 prevented it from crossing the blood–brain barrier, thus overcoming the neurotoxicity often associated with chemotherapeutic drugs. Specific peptide sequences have been conjugated to HPMA copolymers for active targeting of the $\alpha v\beta 3$ integrin in tumor-associated vasculature. Radionuclide-labeled, cyclized, RGD peptide-tagged, HPMA copolymer-based nanoconjugates have been designed that provide the potential for targeted delivery of radionuclides and drugs to solid tumors for diagnostic and therapeutic applications.[56,57] The conjugates exhibited increased tumor retention times and rapid clearance from normal tissues, thus reducing systemic toxicity associated with standard therapeutics.

PLGA copolymers have been extensively used in the field of cancer research, owing to their biodegradability and biocompatibility, resulting in their FDA approval. PLGA is synthesized by the copolymerization of two different monomers, lactic acid and glycolic acid, and can be further modified chemically for conjugation or simple encapsulation of drugs in a nanoparticle formulation. PLGA-based nanoparticles have also been used to engineer complex nanosystems. In a novel strategy reported by Sengupta *et al.*, temporal targeting of tumor cells and the tumor vasculature was achieved using a nanoscale delivery system, described as a "nanocell," that comprised of a core PLGA nanoparticle encapsulated within a PEG-linked lipid envelop.[22] In this nanostructure, the chemotherapeutic drug Dox was covalently attached to the inner PLGA core, and the antiangiogenic agent combretastatin was trapped within the outside lipid envelope. After disruption of the outer envelope inside a tumor, release of combretastatin led to vascular collapse and intratumoral trapping of the nanoparticles, which subsequently released the chemotherapeutic drug in response to local hypoxia, resulting in significant regression of various tumors including melanoma. The nanocell is an example of nanoparticles being engineered based on understanding of the disease and can be fine-tuned to optimize chemotherapy, thus shifting the paradigm from conventional antiangiogenic treatments.

Polysaccharides (e.g., chitosan) and dendrimers can also be considered as polymeric systems and have been studied as carriers of anticancer agents for therapeutic applications. Chitosan is a commercially available cationic linear polysaccharide, which has found a variety of applications in pharmaceuticals and biomedicine. Dendrimers, on the other hand, possess a branched structure that endows it with certain unique properties such as ease of chemical conjugation, biocompatibility, high water solubility, and easy renal clearance due to their small size.[5,58] Different dendrimeric systems are now under investigation for novel cancer treatments.[58] For example, in a recent study, chitosan nanoparticles have shown significant inhibition of tumor growth and induction of tumor necrosis in a mouse hepatocellular carcinoma xenograft model.[59]

In a study reported recently from our laboratory, structure–activity relationship was exploited to design a novel cisplatin-containing polymeric nanoparticle system that showed significant tumor growth delay in breast and lung cancer models and tumor regression in a genetically engineered ovarian cancer model.[60] The polymer used in this study was glucosamine-functionalized poly (isobutylene maleic acid), which self-assembled into nanosized structures upon complexation with cisplatin.

B. Lipid-Based Nanoparticles

Lipid-based nanocarriers, such as liposomes and micelles, possess attractive biological properties, including biocompatibility, biodegradability, and the ability to entrap both hydrophobic and hydrophilic drugs.[5] However, they pose several challenges, such as burst release of the encapsulated drug, instability of the nanocarrier, and rapid clearance from the bloodstream by the RES defense mechanism, which make their translation into clinics a challenging task.

Liposomes are FDA-approved spherical structures consisting of phospholipid bilayers with an enclosed aqueous phase that can carry a range of chemotherapeutic drugs.[61] Liposomes are often coated with PEG to improve their circulation times. Doxil is an example of a clinically approved PEGylated liposome encapsulating Dox, which is currently on the market for the treatment of refractory Kaposi's sarcoma, recurrent breast cancer, and ovarian cancer. DaunoXome and Myocet, on the other hand, are examples of clinically approved non-PEGylated liposomes that encapsulate the chemotherapeutic drugs daunorubicin and Dox, respectively.

Liposomes have also been extensively used in imaging studies. Endothelial αvβ3-targeted, gadolinium ion-containing paramagnetic liposomes have been used for imaging tumor angiogenesis by MRI.[62] Intravenous administration of these nanoparticles provided detailed and enhanced imaging of the αvβ3-expressing tumor vasculature in rabbit carcinomas. A combination of two or more modalities (i.e., bimodal or higher), either in imaging or in therapeutic applications, can function synergistically to provide complementary information. In a novel approach, bimodal (MRI and fluorescence detectable) paramagnetic liposomal nanoparticle was functionalized with two angiogenesis-specific targeting ligands—an αvβ3-specific RGD peptide sequence and a galectin-1-specific designer peptide anginex.[63] This strategy of dual-ligand targeting provided synergistic targeting effects *in vitro*, significantly improving the uptake of these nanoparticles as compared to those modified for single-ligand targeting.

Apart from MRI and optical imaging, liposomes have also been used to encapsulate contrast agents for CT imaging. For example, Samei *et al.* have reported the use of a long-circulating liposomal system encapsulating traditional iodinated contrast agent for micro-CT imaging of rats implanted with R3230AC mammary carcinoma.[64]

Micelles, on the other hand, consist of lipid monolayers with a hydrophilic shell enclosing a hydrophobic core in a spherical structure. For example, Benny *et al.* reported that conjugation of TNP-470 to monomethoxy–PEG–polylactic acid copolymer resulted in the formation of nanopolymeric micelles, named Lodamin.[65] On oral administration, the conjugate was found to accumulate selectively in tumors, inhibiting tumor growth, angiogenesis, and proliferation, without causing any neurological impairment in tumor-bearing mice. In another study based on targeting the αvβ3 integrin, Nasongkla *et al.* designed a cyclic RGD pentapeptide conjugated, Dox-loaded, poly(ε-caprolactone) (PCL)–PEG nanopolymeric micelles, which showed high efficiency in targeting tumor cells derived from human Kaposi's sarcoma *in vitro*.[66]

C. Carbon Nanovectors

Carbon-based nanostructures, such as carbon nanofibers, nanotubes, and fullerenes, have received considerable attention for cancer research in the past. This is due to several advantages, including their (1) high mechanical strength and surface area, (2) numerous sites for chemical or physical conjugation, (3) lightweight properties, and (4) ease of scalability and manufacturing.[67] However, issues related to their biocompatibility, renal clearance, and toxicology have since limited their use in biomedical applications. Chaudhuri *et al.* have reported the use of Dox-conjugated, water-soluble fullerene analogs which showed significant antitumor efficacy in a mouse model.[21]

D. Inorganic Nanoparticles

Inorganic nanoparticles, such as the ones derived from gold, silver, and iron oxide, possess unique optical, electrical, magnetic, and photothermal properties, which have been harnessed in numerous biomedical applications. The paramagnetic properties of iron oxide nanoparticles have been harnessed for therapeutic and imaging applications.[68] For example, Maeng *et al.* reported the designing of Dox-loaded, folate-receptor-targeted, superparamagnetic iron oxide nanoparticles which significantly inhibited tumor growth, yet, surprisingly, did not increase systemic cytotoxicity often associated with heavy metals, most likely due to their selective localization in tumors.[69] Drevs *et al.* have reported the use of the dynamic enhanced MRI technique for studying the effects of PTK787/ZK 222584, a specific VEGF RTK inhibitor, on the anatomy and functional properties of tumor vessels.[70] Dextran-associated superparamagnetic iron oxide nanoparticles (Endorem) were used in this study to detect the partial tumor blood volume in a murine renal cell carcinoma model. Reichardt *et al.* have also used the MRI technique and superparamagnetic nanoparticles in imaging the antiangiogenic effects of a small-molecule VEGF RTK inhibitor in a drug-resistant human adenocarcinoma model.[71]

QDs, which are fluorescent nanocrystals made of inorganic semiconductor materials, have gained prominence in biomedical imaging because of their unique photostable and fluorescence properties. We have observed that QD complexes conjugated to $\alpha v\beta 3$ integrin-binding cyclic RGD peptide for endothelial cell targeting *in vitro* showed increased uptake into the cells as compared to QDs that were conjugated to a control RAD (arginine-alanine-aspartic acid) peptide (negative control). Paramagnetic QDs have been engineered as a bimodal (MRI and fluorescence) imaging probe. Such QDs functionalized by $\alpha v\beta 3$-specific RGD peptide have been used for successful targeting of human endothelial cells *in vitro*.[72]

E. Modified Natural Nanoparticles

Natural nanoparticles, such as viruses and ferritin, have several advantages over synthetic nanoparticles, including precise dimensions, possible evasion by the immune system, biocompatibility, and biodegradability.[53] A number of viral nanoparticles (VNPs) have been developed for targeted delivery and imaging purposes, for example, cowpea mosaic virus (CPMV) and bacteriophages such as MS2 and Qβ, and their modes of cellular uptake have been studied.[73] Multivalent display of fluorescent tags on CPMV facilitated intravital visualization of human fibrosarcoma-mediated tumor angiogenesis in a CAM model.[74] In a different study, covalent attachment of a VEGF receptor-1-specific peptide to fluorescently labeled CPMV enabled efficient targeting of the VNPs to VEGFR-1-expressing tumor *in vivo.*[75]

Apart from VNPs, protein cages such as the iron-containing ferritin have also been studied in medical imaging.[53] Crich *et al.* have reported the MRI visualization of tumor angiogenesis *in vivo* by targeting the neural cell adhesion molecules expressed on tumor endothelium with a highly sensitive gadolinium-containing apoferritin probe.[76]

IV. Shape–Size Effect of Nanoparticles

Physically diverse nanostructures are increasingly being studied for potential applications in cancer chemotherapy. Although a number of design parameters have been optimized in engineering nanoparticles, such as the choice of material, surface chemistry, and particle size, limited knowledge exists on the shape effects of these nanoparticles in tuning the biological outcomes when used in drug delivery. This is perhaps due to a lack of available techniques to produce nonspherical particles. New studies that have emerged in the recent years have indeed shown the importance of particle shape in governing the performance of drug-delivering nanocarriers.[77,78]

In a very interesting study published by Chaudhuri *et al.*, it was reported that the shape of carbon-based nanostructures markedly affects the chemotherapeutic potential of Dox.[20] The authors showed differential effects between Dox conjugated to single-walled CNTs versus spherical fullerenols (Full) on angiogenesis. Both empty and Dox-conjugated fullerenols exerted antiangiogenic effects in zebrafish and mouse melanoma models. In contrast, empty and Dox-conjugated CNTs exerted a proangiogenic effect both *in vitro* and *in vivo*. Mechanistic studies showed differential activation of $\alpha v\beta 3$ integrins and downstream PI3K signaling between CNT and fullerenol in endothelial cells, which implicates the role of nanovector shapes in mediating drug fate and underlies the importance of choosing the right vector to obtain optimal therapeutic index.

TABLE 1

NANOPARTICLE-BASED DRUGS CURRENTLY ON THE MARKET OR UNDERGOING CLINICAL EVALUATION

Active drug	Product name	Nanovector	Indications
Daunorubicin	DaunoXome	Non-PEGylated liposomes	Kaposi's sarcoma
Doxorubicin	Myocet	Non-PEGylated liposomes	Combination therapy of recurrent breast cancer
Doxorubicin	Doxil/Caelyx	PEGylated liposomes	Recurrent breast cancer, ovarian cancer, refractory Kaposi's sarcoma
Platinate	ProLindac	HPMA copolymer	Recurrent ovarian cancer (Phase II)
Paclitaxel	Xyotax	Polyglutamate	Non-small-cell lung cancer (Phase III)
Paclitaxel	Abraxane	Albumin-bound nanoparticles	Metastatic breast cancer
Anti-CD20 mAb conjugated to chelator tiuxetan and radioisotope (Yttrium 90 or Indium 111)	Zevalin	Immunoconjugate	B-cell non-Hodgkin's lymphoma
Anti-CD20 mAb conjugated to Iodine-131	Bexxar	Immunoconjugate	Follicular lymphoma
PEGylated L-asparginase	Oncaspar	Polymer–protein conjugate	Acute lymphoblastic leukemia
Paclitaxel	Genexol-PM	Polymeric micelle	Advanced pancreatic cancer (Phase II)
Paclitaxel	NK105	Micelle	Recurrent Gastric cancer (Phase II)
Doxorubicin	NK911	Micelle	Metastatic pancreatic cancer (Phase I)

V. Current Status of Nanoparticles in Clinics

Lipid-based nanoparticles are the forerunners of nanosized drug delivery systems, the first studies being reported as early as the 1960s.[79] Further research over the next few decades resulted in the designing of more sophisticated drug delivery nanostructures,[5,9] which culminated in the approval of Doxil, PEGylated liposomes encapsulating Dox, by the FDA for the treatment of various types of cancer.[80] Since then, other nanoparticle-based formulations have entered clinical trials and have received FDA approval for the delivery of cancer therapeutics or are currently undergoing clinical evaluation (Table 1). Currently, there are over 400 reported clinical trials evaluating the use of nanoparticle-based carriers in anticancer therapy (http://www.clinicaltrials.gov).

VI. Conclusion

Nanotechnology has emerged as an exciting field in the area of cancer research owing to multiple advantages, including their capacity to carry multiple therapeutic and imaging agents, to offer longer circulation time, to increase the therapeutic index of chemotherapeutics, and to reduce toxicity by targeted delivery of drugs to the tumor site. Moreover, with the various types of nanovectors available, many of which are approved by the FDA, along with the various conjugation methods available for coupling drugs and diagnostic agents to the nanoparticles, there is endless opportunity to fine-tune nanotherapeutics to meet our needs. Clearly, with the advent of nanothechnology, there is a huge potential for devising increasingly novel cancer therapeutics that can eventually be translated from the bench to the bedside.

References

1. *American Cancer Society: Cancer facts and figures.* 2011.
2. Ferrari M. Cancer nanotechnology: opportunities and challenges. *Nat Rev Cancer* 2005;**5**:161–71.
3. Alexis F, Pridgen EM, Langer R, Farokhzad OC. Scha ferKorting M, editor. *Drug delivery, handbook of experimental pharmacology.* Springer-Verlag: Berlin Heidelberg; 2010. pp. 55–86.
4. Davis ME, Chen Z, Shin DM. Nanoparticle therapeutics: an emerging treatment modality for cancer. *Nat Rev Drug Discov* 2008;**7**:771–82.
5. Peer D, et al. Nanocarriers as an emerging platform for cancer therapy. *Nat Nanotechnol* 2007;**2**:751.
6. Suri SS, Fenniri H, Singh B. Nanotechnology-based drug delivery systems. *J Occup Med Toxicol* 2007;**2**:16.
7. Marty JJ, Oppenheim RC, Speiser P. Nanoparticles—a new colloidal drug delivery system. *Pharm Acta Helv* 1978;**53**:17.

8. Farokhzad OC, Langer R. Impact of nanotechnology on drug delivery. *ACS Nano* 2009;**3**:16–20.
9. Langer R. Drug delivery and targeting. *Nat Nanotechnol* 1998;**392**:5–10.
10. Moghimi SM. Recent developments in polymeric nanoparticle engineering and their applications in experimental and clinical oncology. *Anticancer Agents Med Chem* 2006;**6**:553–61.
11. Bhadra D, Bhadra S, Jain S, Jain NK. A PEGylated dendritic nanoparticle carrier of fluorouracil. *Int J Pharm* 2003;**257**:111–24.
12. Gref R, et al. 'Stealth' corona-core nanoparticles surface modified by polyethylene glycol (PEG): influences of the corona (PEG chain length and surface density) and of the core composition on phagocytic uptake and plasma protein adsorption. *Colloids Surf B Biointerfaces* 2000;**18**:301–13.
13. Otsuka H, Nagasaki Y, Kataoka K. PEGylated nanoparticles for biological and pharmaceutical aplications. *Adv Drug Deliv Rev* 2003;**55**:403–19.
14. Brannon-Peppas L, Blanchette JO. Nanoparticle and targeted systems for cancer therapy. *Adv Drug Deliv Rev* 2004;**56**:1649.
15. Byrne JD, Betancourt T, Brannon-Peppas L. Active targeting schemes for nanoparticle systems in cancer therapeutics. *Adv Drug Deliv Rev* 2008;**60**:1615–26.
16. Brigger I, Dubernet C, Couvreur P. Nanoparticles in cancer therapy and diagnosis. *Adv Drug Deliv Rev* 2002;**54**:631–51.
17. Couvreur P, Vauthier C. Nanotechnology: intelligent design to treat complex disease. *Pharm Res* 2006;**23**:1417–50.
18. Satchi-Fainaro R, et al. Inhibition of vessel permeability by TNP-470 and its polymer conjugate, caplostatin. *Cancer Cell* 2005;**7**:251–61.
19. Satchi-Fainaro R, et al. Targeting Angiogenesis with a conjugate of HPMA copolymer and TNP-470. *Nat Med* 2004;**10**:255–61.
20. Chaudhuri P, Harfouche R, Soni S, Hentschel DM, Sengupta S. Shape effect of carbon nanovectors on angiogenesis. *ACS Nano* 2010;**4**:574–82.
21. Chaudhuri P, Paraskar A, Soni S, Mashelkar RA, Sengupta S. Fullerenol-cytotoxic conjugates for cancer chemotherapy. *ACS Nano* 2009;**3**:2505–14.
22. Sengupta S, et al. Tenporal targeting of tumour cells and neovasculature with a nanoscale delivery system. *Nature* 2005;**436**:568–72.
23. Harfouche R, et al. Nanoparticle-mediated targeting of phosphatidylinositol-3-kinase signaling inhibits angiogenesis. *Angiogenesis* 2009;**12**(4):325–38.
24. Jain RK. Barriers to drug delivery in sloid tumors. *Sci Am* 1994;**271**:58–65.
25. Gottesman MM, Fojo T, Bates SE. Multidrug resistance in cancer: role of ATP-dependent transporters. *Nat Rev Cancer* 2002;**2**:48–58.
26. Iinuma H, et al. Intracellular targeting therapy of cisplatin-encapsulated transferrin-polyethylene glycol liposome on peritoneal dissemination of gastric cancer. *Int J Cancer* 2002;**99**:130–7.
27. Kobayashi T, et al. Effect of transferrin receptor-targeted liposomal doxorubicin in P-glycoprotein-mediated drug resistant tumor cells. *Int J Pharm* 2007;**329**:94–102.
28. Park JW, et al. Anti-HER2 immunoliposomes: enhanced efficacy attributable to targeted delivery. *Clin Cancer Res* 2002;**8**:1172–81.
29. Drummond DC, Hong K, Park JW, Benz CC, Kirpotin DB. Liposome targeting to tumors using vitamin and growth factor receptors. *Vitam Horm* 2000;**60**:285–332.
30. Lopes de Menezes DE, Pilarski LM, Allen TM. In vitro and in vivo targeting of immunoliposomal doxorubicin to human B-cell lymphoma. *Cancer Res* 1998;**58**:3320–30.
31. Sapra P, Allen TM. Internalizing antibodies are necessary for improved therapeutic efficacy of antibody-targeted liposomal drugs. *Cancer Res* 2002;**62**:7190–4.
32. Adams GP, et al. High affinity restricts the localization and tumor penetration of single-chain Fv antibody molecules. *Cancer Res* 2001;**61**:4750–5.

33. Hong S, et al. The binding avidity of a nanoparticle-based multivalent targeted drug delivery platform. *Chem Biol* 2007;**14**:107–15.
34. Destito G, Yeh R, Rae CS, Finn MG, Manchester M. Folic acid-mediated targeting of cowpea mosaic virus particles to tumor cells. *Chem Biol* 2007;**14**:1152–62.
35. Warenius HM, Galfre G, Bleehen NM, Milstein C. Attempted targeting of a monoclonal antibody in a human tumour xenograft system. *Eur J Cancer Clin Oncol* 1981;**17**:1009–15.
36. VonMehren M, Adams GP, Weiner LM. Monoclonal antibody therapy for cancer. *Annu Rev Med* 2003;**54**:343–69.
37. Weiner LM, Adams GP. New approaches to antibody therapy. *Oncogene* 2000;**19**:6144–51.
38. Albanell J, Baselga J. Trastuzumab, a humanized anti-HER2 monoclonal antibody, for the treatment of breast cancer. *Drugs Today* 1999;**35**:931–46.
39. Peer D, Zhu P, Carman CV, Lieberman J, Shimaoka M. Selective gene silencing in activated leukocytes by targeting siRNAs to the integrin lymphocyte function-associated antigen-1. *Proc Natl Acad Sci USA* 2007;**104**:4095–100.
40. Crawford M, Woodman R, Ko FP. Peptide aptamers: tools for biology and drug discovery. *Brief Funct Genomic Proteomic* 2003;**2**:72.
41. Khati M. The future of aptamers in medicine. *J Clin Pathol* 2010;**63**(6):480–7.
42. Lee JH, Yigit MV, Mazumdar D, Lu Y. Molecular diagnostic and drug delivery agents based on aptamer-nanomaterial conjugates. *Adv Drug Deliv Rev* 2010;**62**:592–605.
43. Farokhzad OC, et al. Targeted nanoparticle-aptamer bioconjugates for cancer chemotherapy in vivo. *Proc Natl Acad Sci USA* 2006;**103**:6315–20.
44. Laskin JJ, Sandler AB. Epidermal growth factor receptor inhibitors in lung cancer therapy. *Semin Respir Crit Care Med* 2004;**25**(Suppl. 1):17–27.
45. Nicholson RI, Gee JM, Harper ME. EGFR and cancer prognosis. *Eur J Cancer* 2001;**37** (Suppl. 4):S9–S15.
46. Bellocq NC, Pun SH, Jensen GS, Davis ME. Transferrin-containing, cyclodextrin polymer-based particles for tumor-targeted gene delivery. *Bioconjug Chem* 2003;**14**:1122–32.
47. Kukowska-Latallo JF, et al. Nanoparticle targeting of anticancer drug improves therapeutic response in animal model of human epithelial cancer. *Cancer Res* 2005;**65**:5317–24.
48. Buehler A, et al. cNGR: a novel homing sequence for CD13/APN targeted molecular imaging of murine cardiac angiogenesis in vivo. *Arterioscler Thromb Vasc Biol* 2006;**26**:2681.
49. Xie J, Shen Z, Li KC, Danthi N. Tumor angiogenic endothelial cell targeting by a novel integrin-targeted nanoparticle. *Int J Nanomedicine* 2007;**2**:479.
50. Schlessinger J. Cell signaling by receptor tyrosine kinase. *Cell* 2000;**103**:211–25.
51. Basu S, Chaudhuri P, Sengupta S. Targeting oncogenic signaling pathways by exploiting nanotechnology. *Cell Cycle* 2009;**8**:3480–7.
52. Basu S, et al. Nanoparticle-mediated targeting of MAPK signaling predisposes tumor to chemotherapy. *Proc Natl Acad Sci USA* 2009;**106**:7957.
53. Cormode DP, Jarzyna PA, Mulder WJ, Fayad ZA. Modified natural nanoparticles as contrast agents for medical imaging. *Adv Drug Deliv Rev* 2010;**62**:329–38.
54. Cristofanilli M, Charnsangavej C, Hortobagyi GN. Angiogenesis modulation in cancer research: novel clinical approaches. *Nat Rev Drug Discov* 2002;**1**:415.
55. Couvreur P, Kante B, Roland M, Speiser P. Adsorption of antineoplastic drugs to polyalkylcyanoacrylate nanoparticles and their release in calf serum. *J Pharm Sci* 1979;**68**:1521.
56. Mitra A, et al. Polymeric conjugates of mono- and bi-cyclic alphaVbeta3 binding peptides for tumor targeting. *J Control Release* 2006;**114**:175.
57. Mitra A, et al. Targeting tumor angiogenic vasculature using polymer-RGD conjugates. *J Control Release* 2005;**102**:191.
58. Gillies ER, Frechet JM. Dendrimers and dendritic polymers in drug delivery. *Drug Discov Today* 2005;**10**:35.

59. Xu Y, Wen Z, Xu Z. Chitosan nanoparticles inhibit the growth of human hepatocellular carcinoma xenografts through an antiangiogenic mechanism. *Anticancer Res* 2009;**29**:5103.
60. Paraskar AS, et al. Harnessing structure-activity relationship to engineer a cisplatin nanoparticle for enhanced antitumor efficacy. *Proc Natl Acad Sci USA* 2010;**107**:12435–40.
61. Torchilin VP. Recent advances with liposomes as pharmaceutical carriers. *Nat Rev Drug Discov* 2005;**4**:145.
62. Sipkins DA, et al. Detection of tumor angiogenesis in vivo by alphaVbeta3-targeted magnetic resonance imaging. *Nat Med* 1998;**4**:623.
63. Kluza E, et al. Synergistic targeting of alphavbeta3 integrin and galectin-1 with heteromultivalent paramagnetic liposomes for combined MR imaging and treatment of angiogenesis. *Nano Lett* 2010;**10**:52–8.
64. Samei E, et al. Micro-CT imaging of breast tumors in rodents using a liposomal, nanoparticle contrast agent. *Int J Nanomedicine* 2009;**4**:277.
65. Benny O, et al. An orally delivered small-molecule formulation with antiangiogenic and anticancer activity. *Nat Biotechnol* 2008;**26**:799.
66. Nasongkla N, et al. Crgd functionalized polymer micelles for targeted doxorubicin delivery. *Angew Chem Int Ed Engl* 2004;**43**:6323–7.
67. Murugesan S, Mousa SA, O'Connor LJ, Lincoln DW, Linhardt RJ. Carbon inhibits vascular endothelial growth factor- and fibroblast growth factor-promoted angiogenesis. *FEBS Lett* 2007;**581**:1157.
68. Mccarthy JR, Kelly KA, Sun EY, Weissleder R. Targeted delivery of multifunctional magnetic nanoparticles. *Nanomedicine (Lond)* 2007;**2**:153.
69. Maeng JH, et al. Multifunctional doxorubicin loaded superparamagnetic iron oxide nanoparticles for chemotherapy and magnetic resonance imaging in liver cancer. *Biomaterials* 2010;**31**:4995–5006.
70. Drevs J, et al. PTK787/ZK 222584, a specific vascular endothelial growth factor-receptor tyrosine kinase inhibitor, affects the anatomy of the tumor vascular bed and the functional vascular properties as detected by dynamic enhanced magnetic resonance imaging. *Cancer Res* 2002;**62**:4015.
71. Reichardt W, Hu-Lowe D, Torres D, Weissleder R, Bogdanov Jr. A. Imaging of VEGF receptor kinase inhibitor-induced antiangiogenic effects in drug-resistant human adenocarcinoma model. *Neoplasia* 2005;**7**:847.
72. Mulder WJ, et al. Quantum dots with a paramagnetic coating as a bimodal molecular imaging probe. *Nano Lett* 2006;**6**:1.
73. Manchester M, Singh P. Virus-based nanoparticles (VNPs): platform technologies for diagnostic imaging. *Adv Drug Deliv Rev* 2006;**58**:1505.
74. Lewis JD, et al. Viral nanoparticles as tools for intravital vascular imaging. *Nat Med* 2006;**12**:354.
75. Brunel FM, et al. Hydrazone ligation strategy to assemble multifunctional viral nanoparticles for cell imaging and tumor targeting. *Nano Lett* 2010;**10**:1093–7.
76. Geninatti CS, et al. Magnetic resonance visualization of tumor angiogenesis by targeting neural cell adhesion molecules with the highly sensitive gadolinium-loaded apoferritin probe. *Cancer Res* 2006;**66**:9196.
77. Champion JA, Katare YK, Mitragotri S. Particle shape: a new design parameter for micro- and nanoscale drug delivery carriers. *J Control Release* 2007;**121**:3–9.
78. Geng Y, et al. Shape effects of filaments vs. spherical particles in flow and drug delivery. *Nat Nanotechnol* 2007;**2**:249–55.
79. Bangham AD, Standish MM, Watkins JC. Diffusion of univalent ions across the lamellae of swollen phospholipids. *J Mol Biol* 1965;**13**:238–52.
80. James JS. DOXIL approved by FDA. *AIDS Patient Care* 1995;**9**:306.

Nanoparticles in Gene Therapy: Principles, Prospects, and Challenges

Chunxi Liu and Na Zhang

Department of Pharmaceutics, School of Pharmaceutical Science, Shandong University, Shandong Province, People's Republic of China

I. Introduction 510
II. Barriers and Strategies to Gene Delivery 511
A. Extracellular Barriers and the Strategies to Improve Stability of Gene-Loaded NPs 512
B. Intracellular Barriers and the Corresponding Strategies 514
III. NPs Used in Gene Therapy 521
A. Lipid-Based NPs 521
B. Polymer-Based NPs 530
C. Polyion Complex (PIC) Micelle 539
D. Hybrid NPs 542
E. PLGA Nanoparticles 545
F. Peptide-Based NPs 547
G. Inorganic NPs 549
IV. Challenges and Prospects 551
References 552

With the development of genomic technologies, the prospect for gene therapy has progressed rapidly. The major challenge of gene delivery is to improve the transfection efficiencies of the nonviral carriers. Among various nonviral gene vectors, nanoparticles (NPs) offer an ideal platform for the incorporation of all the desirable characteristics into a single gene delivery system. In this chapter, some of the most popularly used lipid- and polymer-based gene delivery vectors and their components are discussed in terms of their characteristics, advantages, and limitations. Although substantial progress has been made, further development of sophisticated delivery systems is foreseen for the nonviral vectors' application in gene therapy. Therefore, we have described several effective strategies to enhance the transfection efficiency of nonviral gene vectors. It is hoped that this review will provide an impetus to the expansion of this promising field of nanomedicine.

Progress in Molecular Biology
and Translational Science, Vol. 104
DOI: 10.1016/B978-0-12-416020-0.00013-9

I. Introduction

The clinical management of various fatal diseases, including genetic diseases,[1] cancer,[2] cardiovascular diseases,[3] and viral infections such as acquired immune deficiency syndrome (AIDS),[4] with plasmid DNA (pDNA), oligonucleotides (ODNs), or small interfering RNA (siRNA) has emerged as a potential approach in medical sciences.[5] On 17 March 2011, it was reported in *The Lancet Neurology* that treating Parkinson's disease with gene therapy has been shown to be successful in clinical trials for the first time.[6] Gene therapy is regarded as one of the most promising therapeutic approaches, as it has the potential to treat disorders at the nucleic acid level by altering or manipulating genes. Gene therapy is now an important part of pharmaceutical development as evidenced by number of clinical trials undertaken globally.[7] However, gene therapy is currently available only in a research setting. The U.S. Food and Drug Administration (FDA) has not yet approved any gene therapy products for sale in the United States. Actually, the primary challenge for gene therapy is to develop a method that would efficiently deliver a therapeutic gene to selected cells where proper gene expression at therapeutic levels can be achieved.[8] Otherwise, it may lose its therapeutic value and even cause side effects through unexpected interactions.

An appropriate gene delivery system is required for their successful cellular uptake and efficient gene transfer. Viral vectors are able to mediate gene transfer with high efficiency and the possibility of long-term gene expression, depending on their unique abilities of protecting genetic materials, crossing cellular membranes, escaping from endosomes, and transporting genetic materials into the nucleus. However, serious safety concerns including acute immune response, immunogenicity, and insertion mutagenesis in gene therapy clinical trials have been raised. Additional practical challenges are the size limitations of the inserted DNA and difficult large-scale production of pharmaceutical-grade products. Nonviral approaches as alternative gene transfer vehicles to the popular viral vectors have received significant attention because of their favorable properties, including lack of immunogenicity, low toxicity, and potential for tissue specificity.[9]

Among various nonviral gene vectors, nanoparticle (NPs) offer an ideal platform for the incorporation of all desirable characteristics into a single gene delivery system.[10] Pharmaceutical NPs were first described in the 1970s. Since then, a considerable amount of research has been conducted in laboratories around the world on the use of NPs for drug and DNA delivery. Their nanometer size ranges offer the most attractive advantage for gene delivery, which allows more efficient delivery of therapeutic agents into target sites.

Besides, NPs can provide targeted (cellular/tissue) delivery of gene, sustainment of gene effect in target tissue, and improvement of the stability of therapeutic agents against enzymatic degradation (nucleases).

Ideally, a gene delivery system should be stable, biocompatible, nontoxic, and capable of highly efficient transport of exogenous genetic material into the tissue-specific site. Although significant progress has been made in the basic science and applications of various nonviral gene delivery systems, their transfection efficacy is far from satisfactory. Besides, their toxicity at high concentrations and their potential interaction with negatively charged molecules in serum and on cell surface are major obstacles to achieving efficient transfection. A variety of functional nonviral vectors have been designed to overcome these barriers.

By far, the most frequently studied strategy for nonviral gene delivery is the incorporation of DNA into condensed particles based on cationic lipids or cationic polymers.[11] Besides, other carriers such as peptides, dendrimers, and magnetic NPs have also been used as carriers in gene delivery. The details of these will be discussed in the following sections. In this chapter, some of the most popularly used gene delivery vectors and their components are discussed in terms of their characteristics, advantages, and limitations. We also describe strategies to enhance the transfection efficiency of nonviral gene vectors. Further, we anticipate that this discussion will kindle fresh thoughts and efforts towards expansion of this promising field of nanomedicine.

II. Barriers and Strategies to Gene Delivery

Although nonviral delivery systems have potentially considerable advantages over their viral counterparts, it is difficult to achieve the expected biological effects of nucleic acids by traditional delivery strategies. This inefficiency of gene delivery primarily results from the inability of these vectors to surpass the numerous obstacles encountered from the site of administration to localization in the cell nucleus.[12] These barriers include the physical and chemical stability of DNA and its carrying vehicle in the extracellular space. Besides instability, nucleic acids should go across extracellular, cellular, and intracellular biological membranes, undergo uptake by endocytosis, and escape from the endosomal and nuclear localization of nucleic acids for transcription (Fig. 1).

As particles encounter each of these barriers, they are subjected to a certain probability of success or failure in overcoming each. The cumulative probability of success for the entire journey is reflected in the transfection efficiency for a given system.[13] Generally, the barriers can be divided into extracellular and

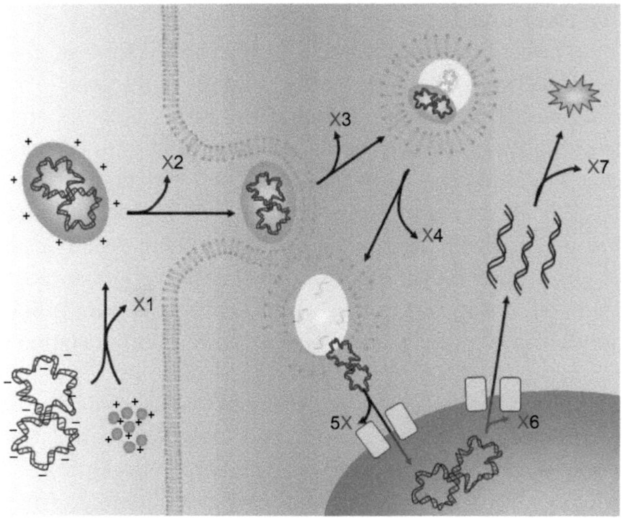

FIG. 1. Barriers to nonviral gene delivery. (1) Transgenic DNA can be lost due to incomplete complex formation with cationic polymer/lipid. (2) Complexes may be cleared from the circulation before they are able to bind to the cell surface. (3) Some of the complexes bound to the cell surface will not be internalized. (4) Following endocytosis, a portion of DNA may be degraded within the acidic late endosomes and lysosomes. (5) DNA successfully escaping the endosomal compartment may be further degraded by cytoplasmic DNAse. (6) A portion of the DNA reaching the nucleus may be unable to induce transcription. (7) Some of the exported mRNA may be incapable of translation into useful transgenic protein. (For color version of this figure, the reader is referred to the Web version of this chapter.)

intracellular barriers (Fig. 2). The aim of this section is to give an overview of the main barriers that have been identified as limiting gene transfer using NP-based nonviral vectors. Some novel strategies to overcome these barriers will be described simultaneously.

A. Extracellular Barriers and the Strategies to Improve Stability of Gene-Loaded NPs

Most of the extracellular barriers can influence the biodistribution of the administered vectors from the point of delivery to the intended target cells,[14] and they will vary profoundly depending on the route of delivery. For intravenous administration, at first the nucleic acids have to go sequentially across the vasculature wall and intercellular tissue junction, which are the major challenges that limit the delivery of particles to reach the target tissue. When particles reached the target cells, cell binding, cell entry/endocytosis, endosomolysis, and finally entry to the nuclear happened. If the target site is located in

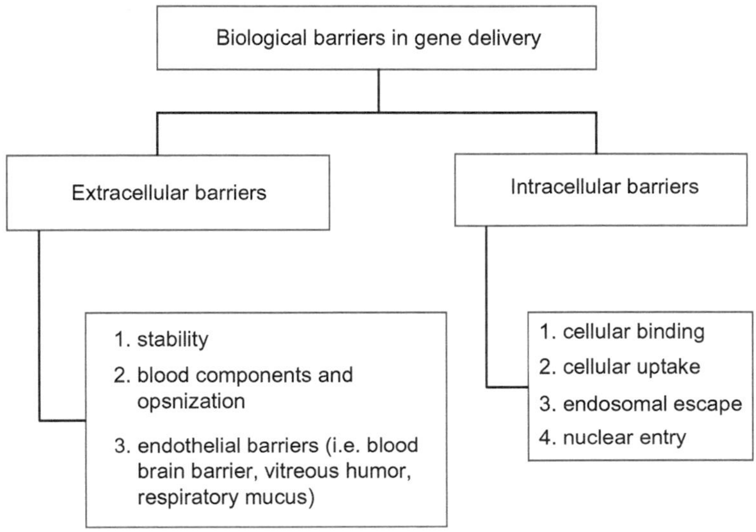

FIG. 2. Schematic representation of biological barriers to gene delivery.

the central nervous system (CNS), the tight junctions between endothelial cells of CNS vessels (the so-called blood–brain barrier, BBB) have to be overcome. Second, nucleic acids may face various enzymes which may degrade them immediately. During their delivery to the target cells, interaction with blood proteins is another physiological barrier encountered in circulation. Positively charged NPs tend to be adsorbed by the polyanionic proteins such as proteoglycans, which are proteins covalently cross-linked with carboxylic or sulfated glycosaminoglycans, serum albumin, and other extracellular proteins, thereby forming large aggregates and getting rapidly cleared by the mononuclear phagocyte system.[15] Likewise, after their local administration (e.g., in arterial walls, vitreous of the eye, joints, dermis, extracellular matrix in tumor), the gene-loaded NPs may also interact with the extracellular matrix components, such as proteoglycans which might bind the positively NPs, affect their mobility in tissue, and limit their access to target cells,[16] and, furthermore, modify the transfection ability of the complexes. These extracellular barriers are the main reasons for the lower efficiency of *in vivo* gene transfer than *in vitro* transfer.[17]

To overcome these barriers and achieve high gene tranfection efficiency, increasing the stability of nonviral delivery systems is very important. The stability is related to the chemical stability of the DNA as well as the physical stability of the delivery system. First of all, the delivery system must provide protection to the therapeutic gene in the cellular environment by encapsulating or electrostatically binding them, since serum proteases and nucleases in blood and intrastitial fluids and variations in pH can rapidly degrade DNA. Typically, these kinds of degradation has generally been at least partially overcome by condensing the DNA with a variety of cationic substrates, such as cationic liposomes and cationic polymers, thereby protecting the DNA from nuclease degradation and keeping the particles stable during systemic circulation.[18]

To solve the instability arising from interactions of positively charged NPs with endogenous negatively charged molecules, the concept of obtaining steric-stabilized complexes by enshrouding NPs with hydrophilic polymers has been recently accepted. The surface coating could prevent the nonspecific interactions of gene-loaded NPs with plasma components, macromolecules, and cells. This approach also shows the additional benefits of increasing the solubility of the gene-loaded NPs and decreasing their tendency to aggregate in the presence of salts. Incorporation of poly(ethylene glycol) (PEG) on the surface of NPs (PEGylation) confers them with stealth properties by creating a hydrophilic steric barrier that delays opsonization and rapid recognition by the reticuloendothelial system (RES).[19] The high flexibility and hydrophilicity of the PEG chain provide the NPs stability in the colloidal suspension through interparticle steric repulsion and prevent interaction of the NPs with macromolecules in the body.[20]

Other materials used as steric stabilizers are poly-*N*-(2-hydroxypropyl) methacrylamide (pHPMA)[21] and poloxamers.[22] Recently, another principle for shielding by ligand density was introduced. Previously, transferring-PEI (800) complexes were found to have a reduced zeta potential compared to ligand-free PEI(800).[23] When the ligand density was further increased, the zeta potential was reduced to near neutral and the nonspecific interactions with cells such as erythrocytes were inhibited. The toxicity of the complexes was also reduced when applied *in vivo*.[24] The combination strategies for shielding transfection-gene-loaded NPs from nonspecific interactions with strategies of active targeting will finally enable target-specific gene delivery and expression.

B. Intracellular Barriers and the Corresponding Strategies

For gene delivery in particular, intracellular barriers represent an extra concern when designing the formulation. A prerequisite for efficient gene transfer is the delivery of nucleic acids from outside the cell to the nucleus. However, this goal is challenging, as the synthetic vectors need to facilitate a number of distinct steps as depicted schematically in Fig. 3, including cell-specific binding, internalization (e.g., endocytosis), escape from endocytic vesicles, and nuclear entry.

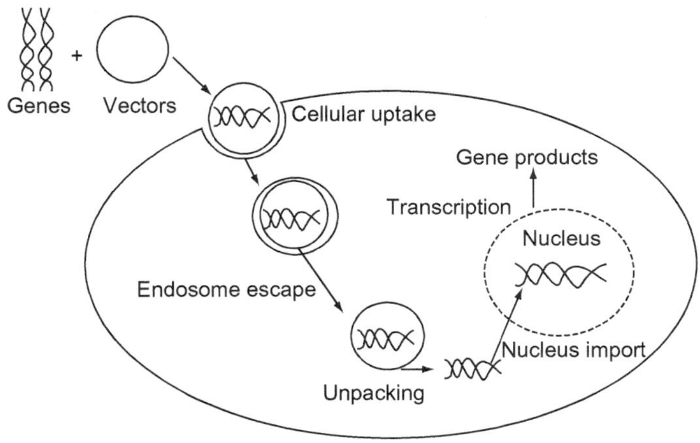

FIG. 3. Schematic of DNA delivery pathways after cell uptake.

1. Cellular Binding

Once the gene vectors reach the vicinity of the target cells, the first barrier encountered is the cell membrane, a lipid dynamic bilayer that effectively provides protection for the cells against the extracellular matrix and regulates the transportation of ions, nutrients, and wastes.[25] Crossing plasma membrane is considered to be one of the most critical limiting steps for efficient DNA transfection. Nonviral vectors can interact with the cell membrane through receptor-mediated and non-receptor-mediated binding.[26] Non-receptor-mediated binding is the electrophilic attraction between the cationic vectors and the highly anionic cell membrane surface proteoglycans, namely, heparin sulfate proteoglycans (HSPGs). These proteoglycans are involved in a variety of important cellular processes, including differentiation, adhesion, and migration. The composition of HSPGs includes a protein core with one or more glycosaminoglycans attached which can attract and bind to the positively charged vectors to trigger internalization. Cationic lipids and cationic polymers are usually applied in gene delivery for adsorptive endocytosis.[27]

Alternatively, delivery to a specific cell type can also be achieved via conjugating a specific ligand. The modified NPs with specific ligands can be efficiently taken up by target cells via receptor-mediated endocytosis. The most used ligands are listed in Table I. Intracellular transport and processing after the receptor mediates endocytosis vary markedly between different receptor–ligand systems and different cell types.[28] In past few years, receptor-mediated intracellular drug delivery has received major attention in modern gene delivery therapeutics, and, although the approach is just at the nascent stage, it shows good promise in developing gene-based pharmaceuticals for the effective management of gene-associated disorders in near future.[29]

TABLE I
TARGET RECEPTORS EMPLOYED FOR RECEPTOR MEDIATED ENDOCYTOSIS

Receptors	Ligands	Cells
Transferring receptor	Transferring	Various
Asialoglycoproteins receptor	Asialoglycoproteins	Hepatocytes
	Galactose	Liver-parenchymal cell
	Lactose	Hepatocytes
Mannose receptor	Mannose	Macrophages
Epidermal growth factor receptor (EGFR)	EGF	Various
Folate receptor (FR)	Folate	Various
Low density lipoprotein (LDL) receptor	LDL	Various
Integrins	Arg-Gly-Asp (RGD) peptides	Tumor endothelia
Aminopeptidase N (APN)	Asn-Gly-Arg (NGR) peptides	Tumor endothelia

2. Cellular Uptake

It is probable that binding of gene vectors to cells results in the two-dimensional aggregation of various cell surface proteins. This aggregation of membrane components is expected to be a strong stimulus for internalization by processes generally referred to as endocytosis.[12] Traditionally, endocytosis can be subdivided into two broad categories: phagocytosis (the uptake of large particles) and pinocytosis (the uptake of fluids and solutes). Phagocytosis is typically restricted to specialized professional phagocytes such as macrophages, neutrophils, monocytes, and dendritcells. Conversely, pinocytosis occurs in all types of cells and has multiple forms. Four classes of pinocytosis have been described to date (Fig. 4): (i) clathrin-mediated endocytosis (CME); (ii) caveolae-mediated endocytosis; (iii) macropinocytosis; and (iv) clathrin- and caveolae-independent endocytosis.[27]

All these mechanisms have been implicated as pathways for internalization of different types of cargo, and the differentiation in the composition of cargos leads to the heterogeneity of these four pathways in respect of the size of vesicles formed, the intracellular fate of the internalized particles, and the factors regulating vesicle formation. Although the relative contribution of each pathway in particle internalization has been poorly defined to date, more and more studies report that the different routes of entry strongly affect transfection efficiency of the gene vectors.[30] Usually, clathrin-dependent internalization follows a well-defined pathway where mature endosomes fuse with lysosomes which contain a variety of digestive enzymes, resulting in complete degradation of cargos.[31] On the other hand, caveolae-mediated pathway has attracted tremendous attention since it has the ability to avoid lysosomal degradation of the delivered genes.[32] As a matter of fact, caveolae-mediated endocytosis may serve as a promising alternative to CME.

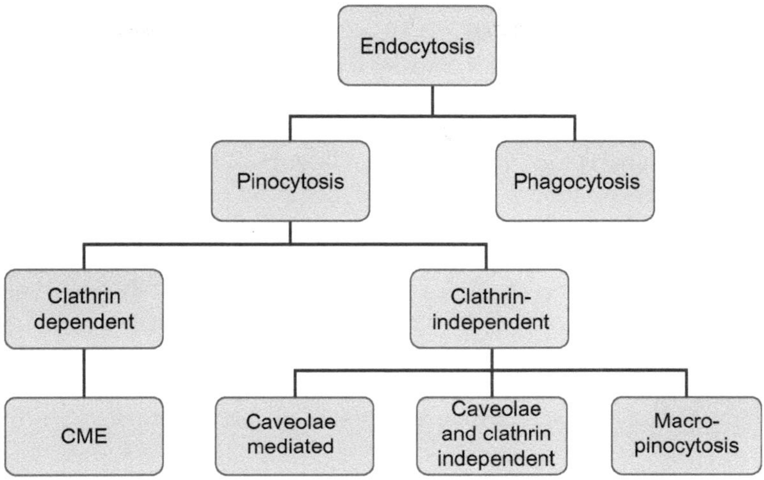

FIG. 4. Different uptake pathways in nonviral gene delivery.

As the significance of caveolae for gene delivery has become clear, the approach of introducing ligands onto the NP has become promising for the construction of nonviral gene vectors to target caveolae-mediated endocytosis.[33] It has been reported that cyclic RGD ligands could facilitate the caveolae-mediated endocytosis of thiolated c(RGDfK)-PEG-b-PLL micelles without high endosomal-disrupting properties and thus improve transfection efficiency.[34] Similar to RGD peptide, NGR peptide, another tripeptide identified by in vivo phage display screenings, was also reported to mediate gene-loaded NPs targeting the human umbilical vein endothelial cell (HUVEC) through caveolae-mediated endocytosis.[35]

It is evident that the processes of the NPs entering the cell often depend on the cell type and structure of the NP itself. This suggests that NPs may need to be tailored based on the targeted cell type.[36] Recent studies have further demonstrated that physicochemical properties of the NPs, such as particle size, shape, molecular weight (MW), surface charge, composition, and stability, affect their cellular internalization and trafficking.[36] These factors make it evident that greater consideration of the uptake and trafficking mechanisms is required in designing vectors that could maximize the transfection efficiency. Nevertheless, any single property of NPs may not be the only dominant factor influencing the uptake. The investigation of the uptake behavior through the variation of physicochemical properties may need to be pursued on a case-by-case basis for each type of nanocarriers.

3. Endosomal Escape

After being internalized, the particles tend to be trapped in the intracellular vesicle, named endosome, and then the endosome will subsequently traffic through the cell. In the late stages, endosomes fuse with lysosomes (the main degradative compartments in the cell) that have even lower pH environments with higher concentration of nucleases to digest the contents.[37] The particles face the major challenges of the acidic pH of endosomes/lysosomes, the digestive enzymes of lysosomes, and the endosomal membrane. Therefore, facilitating the endosomal escape and ensuring the release of the gene become very important limiting steps in achieving an effective gene therapy.[38]

Two routes to promote escape from endosomes have been proposed for lipid- and polymer-based gene carriers (Fig. 5)[5]. Liposomes formulated with fusogenic lipids/peptides facilitate endosome escape by interacting with the

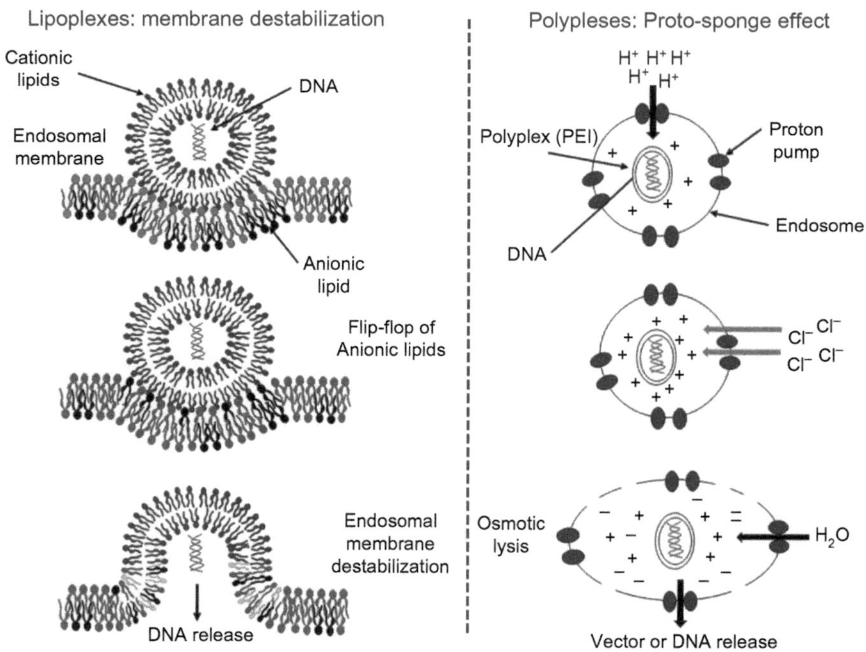

Fig. 5. Mechanisms of endosomal escape of lipoplexes and polyplexes. During this process, anionic lipids that can form a charge-neutral ion pair with the cationic lipids of the liposomes play an important role in the dissociation of the complex and the release of DNA into the cytoplasm. Alternative polymeric systems typically utilize tertiary amine functionalities to cause more and more protons to be pumped inside the endosomes, accompanied by an influx of Cl^- ions. As a result, endosomal rupture occurs by increasing the osmotic pressure during acidification, which is termed the "proton sponge" mechanism. (For color version of this figure, the reader is referred to the Web version of this chapter.)

endosomal lipid membrane followed by its disruption. For example, DOPE acts as a fusogenic agent that is converted from an anionic to a helical conformation at the acidic endosomal pH. This new α-helical structure leads to fusion of the gene vector with endosomes or lysosomes membrane, and then promotes endosomal escape.[39] Polymeric systems typically utilize tertiary amine functionalities to cause endosomal rupture by increasing the osmotic pressure during acidification. Among the various cationic polymers, polyethylenimine (PEI) based gene vector can gain more efficient transfection by relying on its intrinsic capacity of the so-called proton sponge effect. This effect is mediated by agents with high buffering capacity which causes more and more protons to be pumped inside the endosomes, accompanied by the influx of Cl^- to buffer the endosomal environment, and builds up osmotic pressure with the accumulation of water in the endosomes. Moreover, it has been shown that the synthetic pH-sensitive anionic polymer, poly(propylacrylic acid) (PPAA), is efficient in the endosomal escape of nonviral vectors and improves their transfection efficiency.[40] pH sensitivity combined with osmotic swelling finally leads to the liberation of the content. Poly-L-histidine also possesses a strong pH buffering effect, due to the presence of several aminoethyl groups in the structure. Therefore, this efficient endosome-disrupting property make poly-L-histidine a promising delivery system.[41] Poly(amido amine)s (PAAs) are a group of water-soluble synthetic polymers designed to be biodegradable and biocompatible. Most PAAs possess buffering capacities in the pH range 7.4–5.1 which are even higher than that of PEI, a property that may favorably contribute to endosomal escape.[38]

In addition, other peptides, such as GALA (glutamic acid-alanine-leucine-alanine), KALA (lysine-alanine-leucine-alanine), histidine, histidine-rich peptide, polylysine coupled with histidine, and so on, have been designed. Once inside the endosome, these peptides can either display the proton pump effect to cause lysis or fuse with the endosomal membrane leading to pore formation.[42] For instance, cationic lipids complexed with a negatively charged GALA–cargo conjugates could escape from the endosomes with the help of GALA.[43] An octaarginine-modified multifunctional envelope-type nano device (R8-MEND) was previously reported to be an efficient nonviral vector for the delivery of pDNA. KALA could enhance the endosomal escape of R8-MEND through a mechanism independent of fusion between the MEND coat and the endosomal membrane.[44]

4. Nuclear Entry

Ultimately, after escaping from the endocytic compartment, for successful DNA transfection the DNA must be efficiently trafficked to the nucleus. The proper DNA expression cannot be achieved unless the DNA crosses the nuclear envelope (NE) and enters the nucleus.[24] The mechanism of DNA nuclear translocation and whether the DNA is still associated with the delivery system are not fully understood but appear to depend on the type of delivery vehicle employed. At least the following three possible routes exist for DNA transport to the nucleus: (i) During the cell division stage, the nuclear membrane is temporarily ruptured, which allows the DNA to diffuse into the nuclear region; (ii) Molecules of < 9 nm or MW less than 60 kDa can passively diffuse through nuclear pores[45]; (iii) Particles less than 25 nm can be actively transported through nuclear pore complexes (NPCs).

A gene vector is generally much larger than 9 nm, thus the second route is not feasible and there are only two other pathways to be exploited. The first one is entry during the period of cell division. During this period, the NE is temporarily broken down, and therefore the rate of nuclear uptake and the functional size of the transport channels are greater. By comparing transfection efficiencies of both cationic lipid and polymer-mediated gene delivery at various stages in the cell cycle, it was indicated that reaching the nucleus could be accomplished more easily in proliferating cells than in quiescent cells.[46]

However, in quiescent cells the NE acts as the gate keeper around the nucleus that impedes nuclear import of DNA and nonviral gene delivery systems. The NPs entering the nucleus through the NPC (Fig. 6) in the NE is now recognized as an important mechanism for regulating gene expression, which has been implicated in numerous studies.

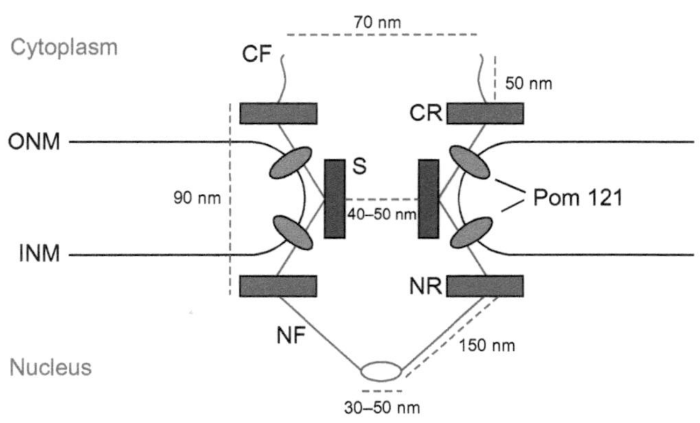

FIG. 6. The nuclear pore complex (NPC). The NPC is formed in the channel between the cytoplasm and nucleus where the inner (INM) and outer membrane (ONM) are fused. The NPC consists of eight spokes (S), connected to flanking rings (CR and NR), from which the cytoplasmic (CF) and nuclear filaments (NF) emerge. The nuclear filaments are joined at the distal end to form a basketlike structure. (For color version of this figure, the reader is referred to the Web version of this chapter.)

This active transport of targeting exogenous DNA toward the nucleus and enhancing nuclear uptake through the NPC is regulated by specific nuclear import and export systems such as nuclear localization signal (NLS) peptides. NLS peptides can be associated with their DNA cargo through ionic interactions, covalent attachment, or site-specific attachment using peptide nucleic acid (PNA) clamp.[42] Zhang *et al.* proved that the complex formation between pDNA and PEG-acridine/NLS-PEG-acridine mixture (9:1 w/w) taken in an amount equal to 5–6 NLS peptides per DNA molecule significantly enhanced both lipoplex and polyplex transfection efficiency and increased the number of transfected bEnd.3 endothelial cells in the presence of serum.[47] In addition, the NLS peptide can also be attached to the DNA-condensing agent of the nonviral gene delivery carrier. Jeon *et al.* conjugated a NLS peptide to poly (D,L-lactide-*co*-glycolide) (PLGA) copolymer and demonstrated that slowly degradable PLGA nanospheres could increase the nuclear localization and transfection efficiency of foreign DNA.[48]

III. NPs Used in Gene Therapy

A. Lipid-Based NPs

1. Cationic Lipids

Since their first use for gene therapy introduced by Felgner *et al.* in 1987,[49] a large number of cationic lipids have been synthesized and used for delivery of nucleic acids into cells and still constitute one of the major techniques for gene delivery.[50] Cationic lipids are positively charged amphiphilic molecules which have four basic constituents: (i) the cationic polar head group, which plays the important role in the self-assembly interaction with polyanionic DNA usually via the protonation of one (monovalent lipid) or several (multivalent lipid) amino groups, is a highly critical determinant for transfection and cytotoxic properties; (ii) a hydrophobic domain (comprising saturated or unsaturated alkyl chains or cholesterol) influences the physical properties of the lipid bilayer, such as flexibility and rate of lipid exchange. Therefore, the hydrophobicity of the lipid moiety also has a crucial effect on the gene transfer efficiency[51]; (iii) a linker functionality connecting the polar head group with the nonpolar tail (often one or more amino acids, ether or ester bond, amide, or carbamate) is undoubtedly an important component that determines the chemical stability and biodegradability and transfection efficiency of the lipid. The linker groups should be biodegradable yet strong enough to survive in a biological environment; (iv) a backbone domain separates the head group from the hydrophobic domain, and acts as a scaffold on which the cationic lipid is built (the most common backbone is glycerol-based)[52] (see Fig. 7).

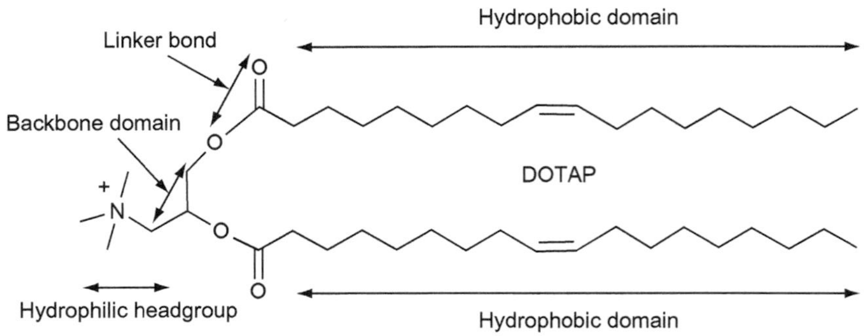

FIG. 7. Schematic representation of DOTAP, a commonly used cationic lipid for gene delivery.

The ester linkage between hydrophobic lipid anchors and the cationic head group is likely to provide biodegradability to cationic lipids. The use of cleavable linkers sensitive to stimuli such as decrease in pH (which occurs during endocytosis), changes in redox potential, and various enzymes (esterases and peptidases) have been investigated to trigger decomplexation of the lipoplex and facilitate the internalized DNA release. The linker can also provide sites for the introduction of novel side chains to enhance targeting, cell uptake, and trafficking.[51] In addition, a hydrophobic spacer between the cationic head group and hydrophobic moieties is a key component for improving pDNA delivery.[53]

Until now, a large number of cationic lipids have been reported, such as quaternary ammonium detergents, cationic derivatives of cholesterol and diacylglycerol, lipid derivatives of polyamines, and so on (Table II). Among them, dioleylpropyl trimethylammonium chloride (DOTMA) and dioleoyl trimethylammonium propane (DOTAP) are two of the most popular cationic lipids. Recently, extensive efforts have been devoted to synthesize novel cationic lipid agents that promote higher transfection efficiency. Mathieu *et al.* prepared a cationic liposomal vector incorporating a novel cationic lipid *N′*,*N′*-dioctadecyl-*N*-4, 8-diaza-10-aminodecanoylglycine amide (DODAG) and the neutral co-lipid 1,2-dioleyl-*sn*-glycerol-3-phosphoethanolamine (DOPE) as the lipid component. It was shown that this novel liposomal vector exhibited higher level of transfection efficiency in OVCAR-3, IGROV-1, and HeLa cells with minimum cellular toxicity both in the presence or absence of serum.[54] Usually, the metabolizable cationic lipids are more favorable and exhibit less cytotoxicity than the nonbiodegradable ones.[55] Based on this conclusion, a novel single-tailed cationic lipid, 6-lauroxyhexyl lysinate (LHLN), was designed and synthesized based on 6-lauroxyhexyl ornithinate (LHON), which showed lower cytotoxicity than that of DOTAP.[56] Both LHLN and LHON have two

TABLE II

A Selection of Some Structures (Cationic and Neutral Lipids) Used for Gene Delivery

Subgroups	Characteristic	Structure
Monovalent aliphatic lipids	Polar head groups contain a single amine function	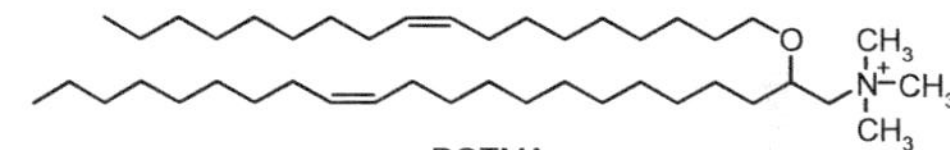 DOTMA 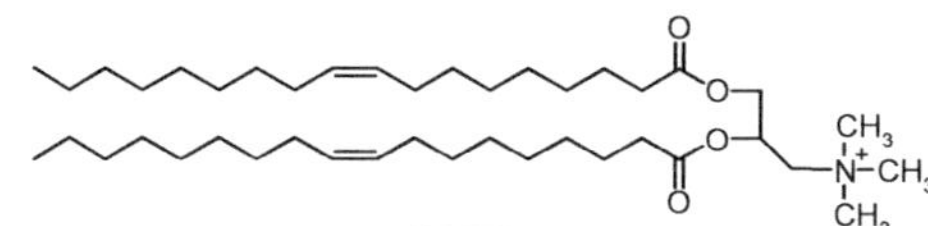 DOTPA 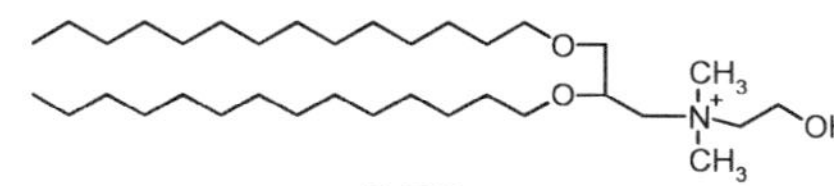 DMRIE 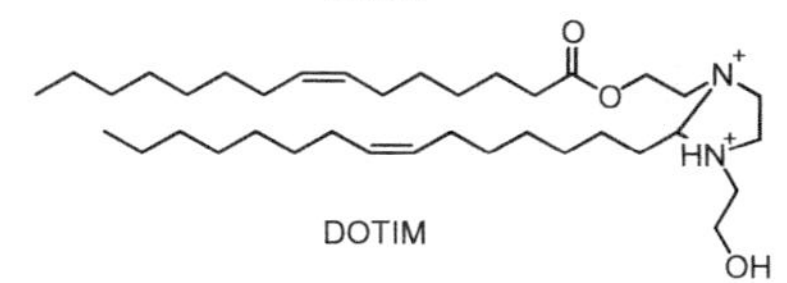 DOTIM

(*Continues*)

TABLE II (*Continued*)

Subgroups	Characteristic	Structure
	Have a single tail; biodegradable	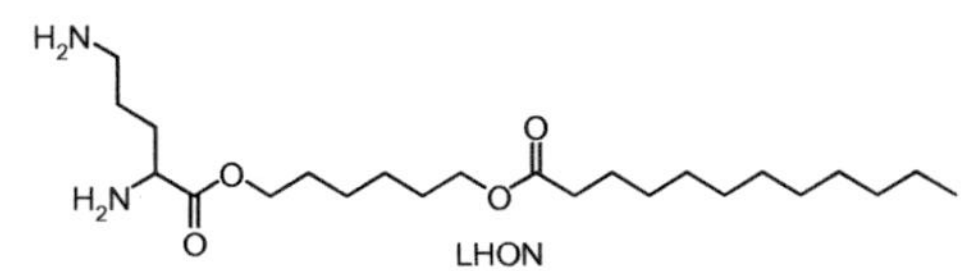 LHON
		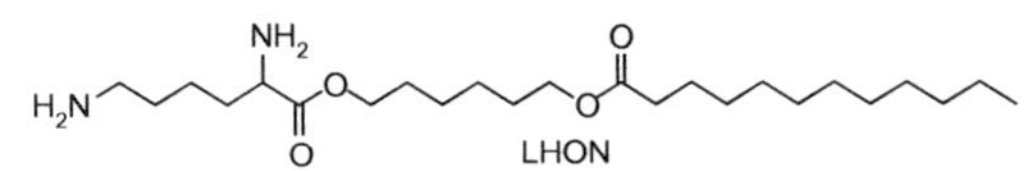 LHON
Multivalent aliphatic lipids	Polar head groups contain several amine functions such as the spermine group	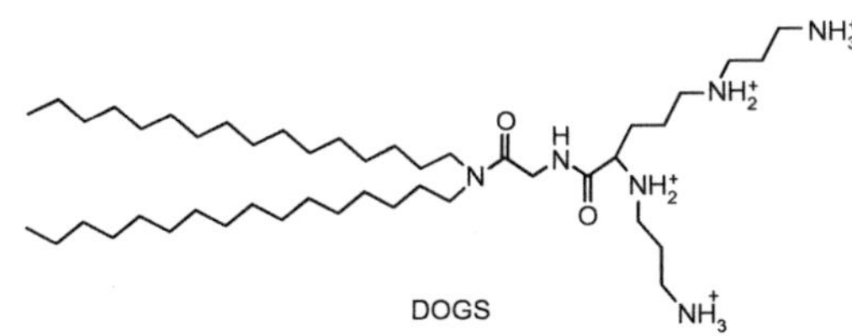 DOGS

Cationic cholesterol derivatives	Have a cholesterol backbone	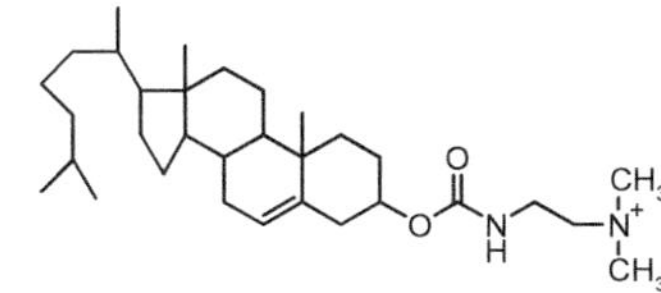 DC-Chol 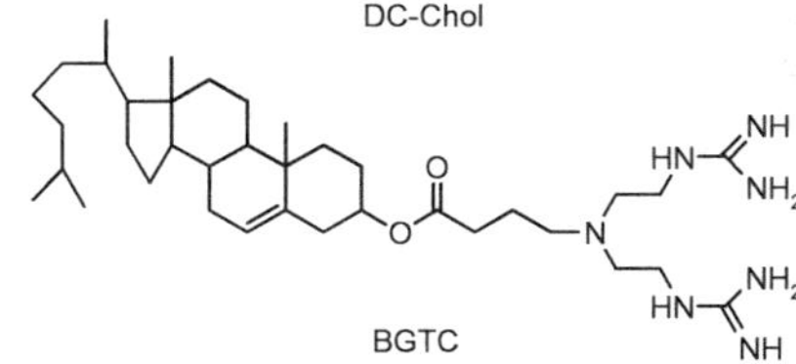 BGTC
Helper lipid	–	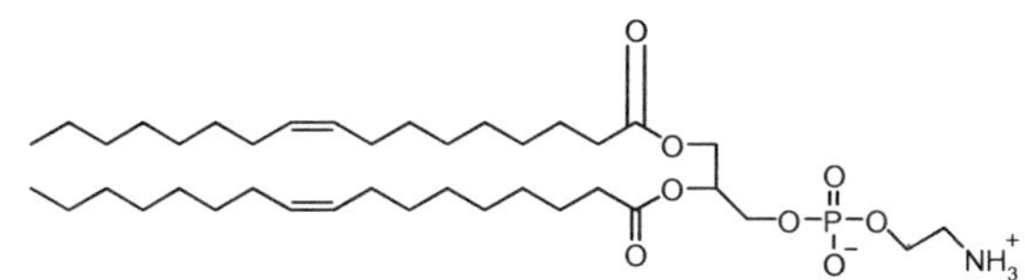 DOPE

biodegradable ester bonds in their structure. The difference between LHLN and its analog LHON is that the head group of LHLN is lysine whereas that of LHON is ornithinate. The cytotoxicity of LHLN as well as its application in the preparation of cationic SLNs for gene delivery was further investigated.[35,57–59]

For transfection application, the cationic lipids are usually employed to form positively charged liposomes by combination with a neutral helper lipid such as DOPE or cholesterol.[60] The important role of helper lipids can be summarized as follows: (1) they weaken the interaction between ions and cationic surface to facilitate the removal of counter ions from the lipid surface and the lipoplexes formation with DNA; (2) they provide good miscibility with the cationic lipid and a better surface charge distribution, resulting in a more complete and balanced packaging of DNA by the lipids; and (3) they enhance the stability of the lipoplex.[61] In addition, DOPE is useful to enhance endosome escape as a fusion lipid by promoting the transition from a lamellar to a hexagnonal phase.[50]

2. Cationic Liposomes

Cationic liposomes are generally composed of a positively charged lipidic component, such as DOTMA, DOTAP, or dimethylaminoethane carbamoyl cholesterol (DC-Chol), and a "helper" lipid, such as DOPE and cholesterol, which provides added fusogenicity and/or stability to the lipoplexes.[49] Cationic liposomes can protect DNA from enzymatic degradation, subsequently make the vector system more stable in the presence of serum, and deliver DNA into cells by interacting with the negatively charged cell membrane.

Recently, there has been a flurry of experimental work expecting to clarify the relationship between the supramolecular structures and the transfection efficiency of cationic lipoplexes. Using synchrotron X-ray diffraction, it has been demonstrated that most lipoplexes form a multilayered structure (L_{α}^{C}) comprising DNA monolayers sandwiched between lipid bilayers. Much more rarely, a different columnar inverted hexagonal structure (H_{II}^{C}) with single DNA strands encapsulated in lipid tubules is also observed (Fig. 8).[62,63] It was reported that H_{II}^{C} lipoplexes fuse and release DNA when in contact with anionic vesicles including the cellular membranes and endosomal vesicles, while L_{α}^{C} lipoplexes remain stable when in contact with these vesicle membranes.[64] Thus, it is hypothesized that lipoplexes containing those cationic lipids that are capable of promoting inverted hexagonal phase structures can fuse with the endosomal membrane and facilitate the following uptake of nucleic acid–cationic lipid lipoplexes into cells.[65] However, it is still not clear whether the hexagonal phase is beneficial for transfection. Koynova *et al.* have deduced a strategy for optimizing lipofection: Lamellar lipoplex formulations with compositions close to the lamellar–nonlamellar phase boundary, which could easily undergo phase transition upon mixing with cellular anionic lipids

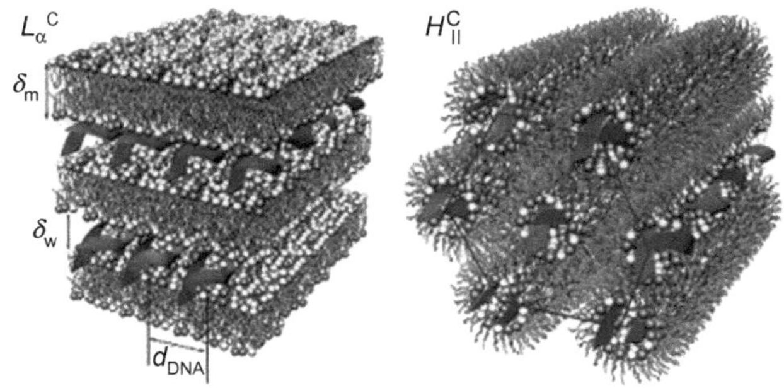

FIG. 8. Schematic depictions of two equilibrium structures of lipoplexes: lamellar (L_α^C) and the inverted hexagonal (H_{II}^C) phase. This picture is cited from Ref. 67.

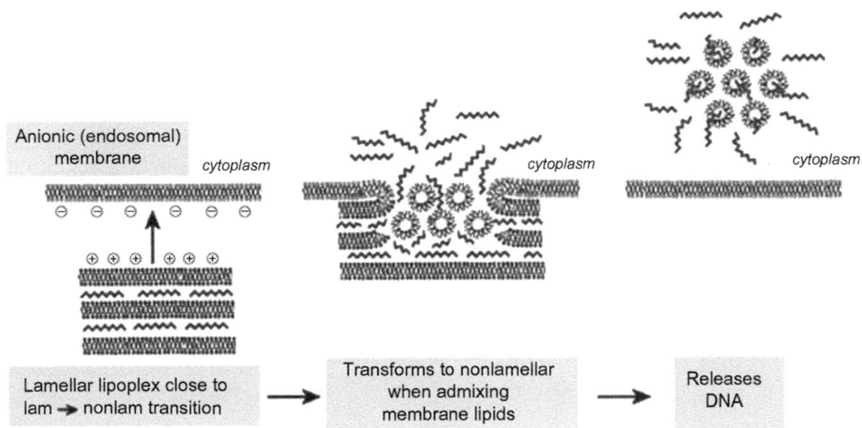

FIG. 9. Hypothetical scheme for efficiency of DNA delivery deduced by Koynova. This picture is cited from Ref. 66.

(Fig. 9), are predicted to be especially efficient because they would be equally successful in transporting DNA into cells. They also identified that one of the very effective and widely explored lipid transfection formulations, DOTAP/DOPE 1:1 (mol/mol), exhibits precisely this kind of behavior.[66] It was illustrated by Ewert *et al.* that an H_{II}^C phase of complexes of DNA with mixtures of a novel multivalent cationic lipid, named MVLBG2 and DOPC, respectively, efficiently transfected mouse and human cells in culture.[67]

3. Cationic Solid Lipid NPs

Recently, solid lipid NPs (SLNs) have been developed and proved to be a useful vehicle for gene delivery. SLNs are basically composed of high melting point lipids that act as a solid core, covered by surfactants. The main ingredients of SLNs (i.e., triglycerides, partial glycerides, fatty acids, steroids) are generally recognized as safe, are physiologically well tolerated, and have already been approved for pharmaceutical application in humans.[68] The use of these materials may also lead to an advantageous toxicity profile of SLN.[69] Same as with cationic liposomes, positively charged SLNs can bind with polyanionic DNA via electrostatic force, leading to SLN–DNA complexes. However, cationic SLNs formulated from the same cationic lipid are smaller in diameter than the corresponding cationic liposomes (88 vs. 148 nm) and are more dominant for *in vitro* transfection performance.[70] In addition, SLNs have good stability, which facilitates the industrial elaboration and the manipulation for different processes such as lyophilization.[71]

Cationic SLNs are promising nonviral gene delivery carriers suitable for systemic administration. However, the main limitations of cationic SLNs are their low and only transient expression levels owing to their inability to support the amplification, cell-to-cell transmission, and the toxicity at high doses especially when used repeatedly, possibly inducing inflammatory reactions *in vivo.*[72] Decreasing the cytotoxicity of cationic lipids or surfactants has become one of the hot topics in the research field of SLNs for gene delivery. For example, the cationic lipid cetyltrimethylammonium bromide (CTAB) is widely used in gene delivery systems and shows good performance in loading DNA and promoting gene transfection. However, it exhibits higher cytotoxicity than its double-tailed counterparts. To overcome this problem, divalent cations such as Ca^{2+} have been incorporated in the SLN formulations to reduce the amount of CTAB used, thereby decreasing the corresponding cytotoxicity.[73] Well-tolerated and highly efficient *in vitro* transfection can also be achieved by optimizing the cationic lipid and matrix lipid composition of SLNs. The DOTAP/DNA ratio is the crucial one since it influences the condensation of DNA with SLN.[74] Doroud *et al.* have prepared SLNs formulated by cetyl palmitate, cholesterol, DOTAP, and Tween 80 via a melt emulsification method followed by high-shear homogenization. It was indicated that efficiency/cytotoxicity ratio of SLN–DNA formulations was comparable to that of linear PEI–25 kDa DNA.[75]

Vighi *et al.* compared two types of cationic composition, namely, Esterquart 1 (EQ1) or Protamine, which were used to fabricate SLN–EQ1 and SLN–Protamine, respectively. The results demonstrated that compared with SLN–EQ1, SLN–Protamine was able to condense pDNA more efficaciously and showed less cytotoxicity and more efficiency in the transfection of the Na1300 cell line. These findings were attributed to the different surface composition of the two samples and, in particular, to the localization of the protamine on the surface of the particle as against EQ1 in the lipid core. In conclusion, the results here suggest that not only the zeta potential but also the surface composition might be affecting the pDNA condensation proprieties and thus the transfection efficiency of nonviral gene nanocarriers.[76]

4. Cationic Emulsions

In recent years, it has been demonstrated that emulsions, which are lipid-based delivery systems composed of oil, surfactants, and cationic lipids, have more favorable physical stability and biological activities than traditional cationic liposomes as the gene carrier *in vivo*.[77] Emulsions are composed of a hydrophobic oil phase surrounded by the cationic lipid, and have high potential as nonviral gene carriers. Such nanosized emulsion systems have the advantages of biocompatibility, biodegradability, physical stability, and low toxicity.[78] Additionally, high-pressure homogenization, easy processibility, and low production costs provide the possibility of large-scale production of emulsions required for *in vivo* clinical applications.[78,79] Physicochemical characteristics of the emulsions can be controlled by varying their constituents such as the cationic lipid, co-emulsifier, and core oil. Also, a sustained circulation or cell-specific targeted emulsions can provide prolonged circulation in blood and/or improve their therapeutic effect with reduced a side effects. Adding cationic surfactants to these dispersed systems makes them suitable for gene delivery. The presence of cationic surfactants results in the formation of positively charged droplets which promote strong electrostatic interactions between the emulsion and the anionic nucleic acid phosphate groups (Fig. 10).[80]

Since the first cationic emulsion formulation containing castor oil and DC-Chol was used to deliver a plasmid through a portal vein injection in mice,[81] different oils such as medium-chain triglycerides, soybean oil, squalene, and poppy seed oil (lipiodol) have been incorporated as the inner phase of the emulsions.[82] In addition, the cationic property of the emulsion is imposed by the cationic lipid or surfactant. Stearylamine, DOTAP, DOTMA, and DC-Chol

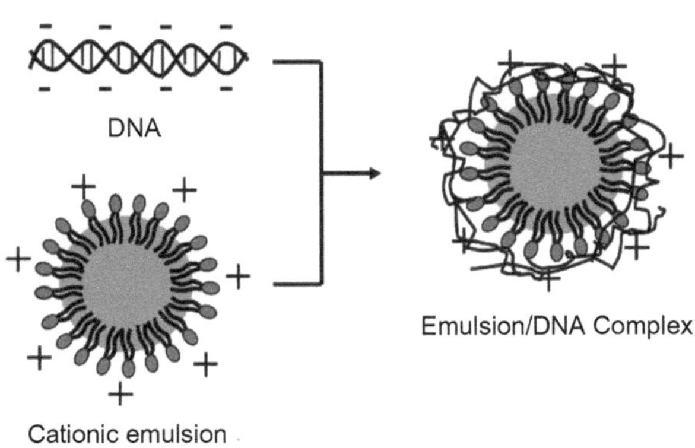

Fig. 10. Cationic emulsion system acting as a DNA nanocarrier. (For color version of this figure, the reader is referred to the Web version of this chapter.)

are the most commonly used cationic surfactants in the formulations.[83] It was also reported that a cationic lipid containing an aspartate or glutamate backbone, a lysine head group, and two alkyl tails could yield efficient gene expression with lower cytotoxicity.[84]

Besides, amino acid-based surfactants that mimic natural lipoamino acids, such as lauroyl-arginine methyl ester, were reported to be valuable cationic surfactants with rich self-aggregation properties, biodegradability, and low toxicity, which could be used in the preparation of cationic lipid-based emulsions.[85] Recently, stable nanoemulsions were prepared and modified by conjugating fatty acids with cationic amino acids including lysine, arginine, and histidine with the help of carbodiimide. The transfection efficiencies of amino acid-modified nanoemulsions were in the order lysine > arginine > histidine, and low cytotoxicities were observed for these three amino acid-modified nanoemulsions.[82]

However, emulsions are thermodynamically unstable dispersions, and much effort must be devoted for developing the optimized emulsion-based gene delivery systems, exhibiting safe and sustained gene expression both *in vitro* and *in vivo*.

B. Polymer-Based NPs

Cationic polymers represent the second group of cationic synthetic vectors commonly used in gene delivery. Same as with cationic lipids, upon mixing with DNA (at physiological pH), cationic polymers can "self-assemble," condense, and neutralize DNA molecules to form nanosized complexes, often called polyplexes. Polyplexes are emerging as potential gene delivery vectors because of their versatility and present a few advantages over lipid-based systems such as their relatively small size and narrow distribution, higher protection against nucleases, and easy control of the physical factors (e.g., hydrophilicity and charge).[86] Typically, polyplexes are more stable than lipoplexes.[87] In addition, the cationic residues within the polyplexes can enhance the binding with the cell and may also mediate the transfer of nucleic acid to cytoplasm by disruption of the vesicular membranes.

Cationic polymers were introduced in the same year as the cationic lipids by Wu and Wu[88] and further expanded by Behr and coworkers in 1995.[89] Till now, a variety of cationic polymers such as poly-L-lysine (PLL), PEI, poly(di-methlylaminoethyl methacrylate) (PDMAEMA), and poly(amidoamine) (PAMAM) have been utilized (Table III). Among them, PLL and PEI are the most widely studied polymers for gene delivery. In fact, PLLs, as first-generation cationic polymers, were quite inefficient in terms of endosomal escape and transfection efficiency. PEI and PAMAM can successfully mediate endosome disruption by acting as proton sponges, whereas their nonbiodegradability tends to result in higher cytotoxicity.[90]

TABLE III
STRUCTURES OF SOME WIDELY USED GENE-PACKAGING POLYMERS

Polycations	Structures
Poly(lysine) (PLL)	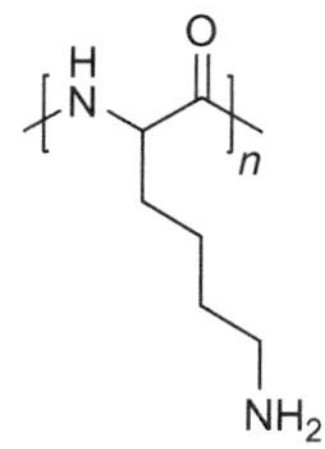
Branched poly (ethylenimine) (b-PEI)	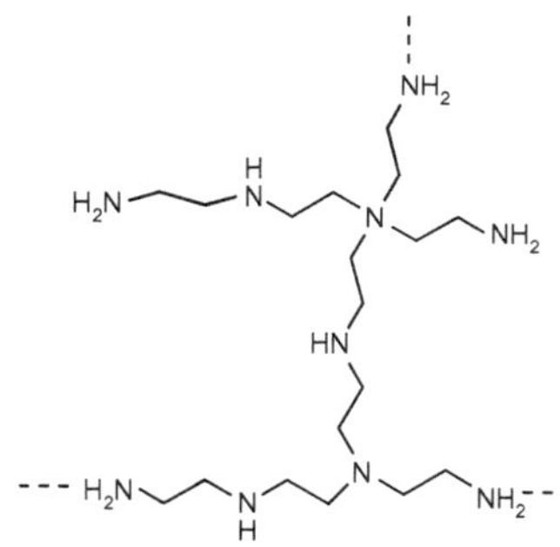
Linear poly (ethylenimine) (l-PEI)	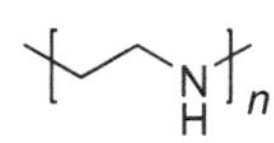
Poly(2-(dimethylamino) ethyl methacrylate) (PDMAEMA)	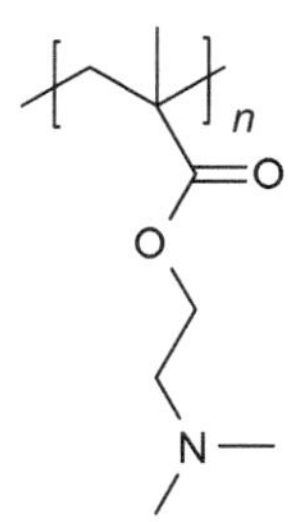

(*Continues*)

TABLE III (*Continued*)

Polycations	Structures
Poly(amidoamine) (PAMAM)	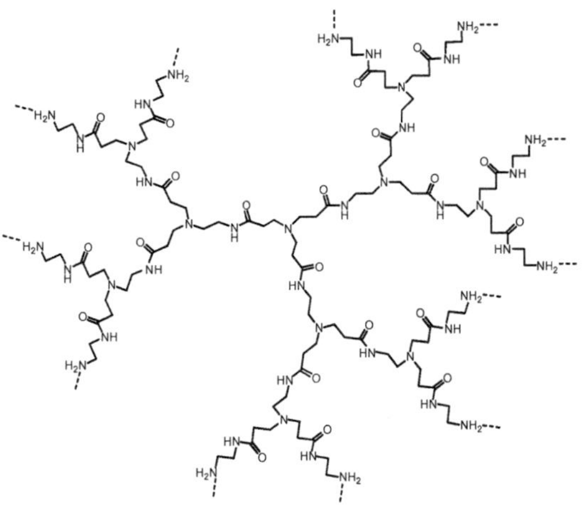
Chitosan	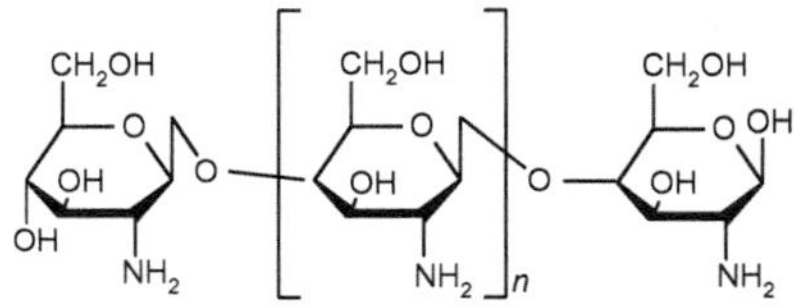

1. Poly-L-Lysine

PLL polymers are one of the first cationic polymers to be employed for gene transfer. The peptide structure contributes to the biodegradable nature of PLL, which is very useful for *in vivo* applications. The primary amine groups of lysine in the structure can spontaneously interact, through electrostatic interactions, with the negatively charged phosphate groups of DNA to form polyplexes. PLL with low MW (less than 3 kDa) cannot form stable complexes with DNA, while PLL with high MW is more suitable for gene delivery via systemic injection but exhibits relatively high cytotoxicity.[91] In addition, PLL has poor transfection ability when applied alone or without modifications because PLL/DNA polyplexes are rapidly bound to plasma proteins and cleared from the circulation. Thus, searching for more efficient cationic polymers or modified PLLs has been main the task in recent years.

2. Polyethylenimine

PEI is one of the most popularly used and efficient examples of cationic polymers capable of gene transfection because PEI-based polyplexes are the most efficient in intracellular delivery of nucleic acids in a variety of cells *in vitro* and tissues *in vivo* compared with other types of cationic polymers. Since the first successful PEI-mediated oligonucleotide transfer conducted in 1995,[89] PEI has been derivatized to improve the physicochemical and biological properties of polyplexes. Branched PEI (bPEI) and linear PEI (lPEI) can both be used effectively for gene delivery. bPEI contains a higher percentage of primary amines and is more amenable to modifications. lPEI is generally preferred for *in vivo* applications because of its low cytotoxicity. Several linear PEI transfection agents have been made commercially available, including ExGen500 and jetPEI™.[92] Moreover, jetPEI™ is selected for gene delivery applications and preclinical studies as part of "The Polyplus in vivo Consortium."[93]

In the structure of bPEI, there are high densities of primary, secondary, and tertiary amines in the ratio of 1:2:1 with p*Ka* values around the physiological pH, of which the majority are nonprotonated at the physiological pH providing remarkable buffering capacity. More importantly, these nonprotonated amines can exert the so-called proton sponge effect to enable the PEI to escape from the endosome. It was reported that the excess free PEI present in the colloidal suspension substantially contributed to efficient gene expression, as it was internalized with the polyplexes and aids in the proton sponge effect.[94] However, they also mediated toxic effects in a dose-dependent manner. Therefore, purified polyplexes without free PEI have to be applied at increased concentrations to achieve high transfection levels, but they exhibit a greatly improved toxicity profile.[95] The MW, degree of branching, ionic strength of the solution, zeta potential, particle size, configuration, and the charge ratio of polymer to DNA used can affect the transfection efficiency and toxicity of PEI polyplexes. The primary amines are mainly responsible for the high degree of DNA binding, but they also contribute to the toxicity during transfection, while the secondary and tertiary amino groups provide good buffering capacity to the system. The increased toxicity is caused by aggregation and adherence on the cell surface, which results in significant necrosis. The high charge density can effectively increase the transfection efficiency, but it simultaneously contributes to increased cytotoxicity. Actually, the efficacy of the transfection system results from a balance between the transfection efficiency and cytotoxicity.

PEI has been extensively modified in an effort to improve the transfection efficiency. Amongst these, PEGylation is the most utilized method which creates a hydrophilic exterior that reduces interactions of the polyplex with plasma proteins and erythrocytes. However, the gene expression is reduced, which can be attributed to the decreased surface charge of PEI-g-PEG copolymers and the subsequent reduced interaction with the cell membrane. Furthermore, the reduced transfection efficiency also results from the hindered gene transfer steps that follow after cellular uptake.[91]

PEG-conjugated PEI and low-MW PEI have lower toxicity, but also have lower transfection efficiency than high-MW PEI (22 or 25 kDa). To circumvent such a problem, low-MW PEI have been conjugated to each other by some degradable linkers for the development of higher efficiency and lower cytotoxicity nonviral vectors in recent years.[96] It has been reported that the disulfide bonds are stable in the oxidative extracellular environment but can be degraded rapidly in the reductive intracellular environment by glutathione. Thus, polymers linked with disulfide bonds can be useful for selectively releasing the loaded gene in the cytosol.[97] The introduction of disulfide bonds through DSP and DTBP should promote reversion of the high-MW transfection complexes back to their low-MW counterparts.[98] Deng *et al.* introduced the disulfide bond to link less toxic, short PEI chains (2 kD), resulting in two such linked PEI samples (MW $\sim$ 7 kD, PEI-7K-L and $\sim$ 400 kD, PEI-400K-L). The results revealed that PEI-7K-L was less cytotoxic and more effective in gene transfection than both PEI-25K and Lipofectamine 2000 in *in vitro* gene transfection. However, PEI-400K-L has no gene transfection efficiency even though it is nontoxic.[99]

3. Dendrimers

Compared to conventional polyamine transfection agents, a variety of positively charged dendrimers that have the advantages of low toxicity, high transfection, and ease of manufacturing have been investigated extensively for their ability as effective gene vectors. Dendrimers have a unique highly branched molecular architecture that is globular and possess repeating units emanating from a central core.[100] This defined structure, the inner cavities to encapsulate guest molecules, and controllable multivalent functionalities in their inner or outer part (the chapter by Imae and Tsai in this volume) make dendrimers attractive for gene and drug delivery.[101] Dendrimers are superior to other systems due to their unique characteristics such as uniformity, monodispersity, and the ability to functionalize their terminal groups with various targeting agents.[102] They can be synthesized by convergent or divergent methods and the resulting dendrimers grow in a geometrically progressive fashion (Fig. 11).

Since Tomalia and coworkers first proposed starburst polyamidoamine (PAMAM) dendrimers for gene delivery,[103] a rapid increase of interest in the chemistry of dendrimers has been observed. PAMAM bearing tertiary amino groups in their interior exhibit the proton sponge effect, fulfilling to a significant extent the criteria for gene transfection of cells.[104] PAMAM dendrimers

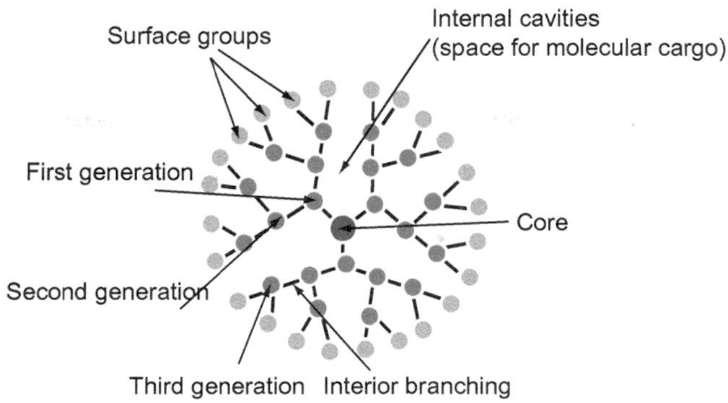

FIG. 11. General structure of dentrimers. (See Color Insert.)

with six generations proved to be more efficient in gene delivery than any others and have been commercialized.[105] Now, a transfection reagent called SuperFect™ consisting of activated dendrimers (fractured generation-6 PAMAM) is commercially available for *in vitro* applications. SuperFect can carry a larger amount of genes than viruses. In addition, SuperFect–DNA complexes provide higher stability and more efficient transport of DNA into the nucleus than liposomes. The high transfection efficiency of dendrimers may not only be attributed to their well-defined shape but also to the low p*K* of the amines. The low p*K* permits the dendrimers to buffer the pH change in the endosomal compartment.[106]

PAMAM dendrimers are the most commonly encountered due to their high transfection efficiency. However, they are not biodegradable, thus causing significant problems *in vivo*. Typical approaches to optimize dendritic gene delivery for *in vivo* use involve the surface modification of PAMAM backbone, either with arginine[107] or hydroxyl groups.[108] It was reported that PAMAM dendrimers functionalized with L-arginine (PAMAM-Arg) by an ester bond rather than amide bond showed equivalent transfection efficiency to PEI but lower toxicity, which could attributed to the faster degradation of the ester bond avoiding carrier accumulation in the tissue.[109] In an attempt to lower the cytotoxicity of PAMAM, PAMAM-OH dendrimers were prepared by hydroxylation of its primary amino groups. The absence of surface primary amino groups in PAMAM-OH renders this polymer nearly neutral, which might be advantageous in terms of cytotoxicity but unable to form complexes with DNA. Therefore, quaternized PAMAM-OH (QPAMAM-OH) has been designed to overcome this problem. Although the transfection efficiency of QPAMAM-OH/DNA polyplexes was lower by one order of magnitude than that of the parent PAMAM, they exhibited significantly reduced cytotoxicity.

Other strategies to reduce their toxicity are PEG or targeting ligand modification. The influence of surface modification on the cytotoxicity of PAMAM dendrimers was examined by conjugating PEG2000 to PAMAM, which showed a marked decrease in the cytotoxicity.[110] Alternatively, Kim and coworkers have reported that a novel PAMAM-PEG-PAMAM triblock copolymer could form highly water-soluble polyplexes with pDNA and finally achieve high transfection efficiency comparable to that of PEI in 293 cells.[111]

Despite the extensive interest in the pharmaceutical applications of dendrimers, the clinical applicability of a dendrimer-based gene transfection agent is yet to be established and many basic principles of applicability are still highly debated.[112] Therefore, dendrimers have a long way to go before they can enter clinical applications. The success of the applications of dentrimers is likely to depend on the continuing development of novel materials for dendrimer synthesis.[113] The ability to functionalize the terminal groups and structures offers endless possibilities to solve all the problems. Overall, the reports available to date certainly suggest that dendrimer-based delivery systems hold great promise and potential in gene delivery.

4. Chitosan

Among nonviral vectors, chitosan and its derivatives have been developed *in vitro* and *in vivo* for DNA and siRNA delivery systems. Chitosan is a linear biodegradable polysaccharide consisting of repeating D-glucosamine (deacetylated unit) and *N*-acetyl-D-glucosamine units (acetylated unit) linked via (1–4) glycosidic bonds (Table III).[114] Mumper *et al.* were the first to study the potential of chitosan for in vitro DNA delivery.[115] Now chitosan is widely applied in pharmaceutical industry and produced commercially by deacetylation of chitin which is the structural element in the exoskeleton of crustaceans (crabs, shrimp, etc.). Positively charged chitosan backbone and negatively charged DNA lead to the spontaneous formation of nanosized complexes (polyplexes) in the aqueous milieu. Besides, each deacetylated subunit of chitosan contains a primary amine group with a p*K*a value of about 6.5, which makes it soluble in acidic media and insoluble at neutral and alkaline pH values. It is biodegradable and can be readily digested either by lysozymes or by chitinases in the physiological environment. More importantly, chitosan is practically nontoxic to animals as well as humans. Recently, chitosan has been proposed as a safer alternative to other nonviral vectors for gene delivery.[116]

However, the application of gene delivery using chitosan remains limited owing to its significantly lower transfection efficiency compared with other cationic gene delivery agents such as PEI, dendrimers, or cationic lipids. This is attributed primarily to its minimal solubility and low buffering capacity at physiological pH, leading to poor endosomal escape of the gene carrier and inefficient cytoplasmic decoupling of the complexed nucleic acid. The binding affinity of chitosan to DNA, stability, and the transfection efficiency of the chitosan/DNA complexes depend on a series of formulation parameters, such as the MW of chitosan, degree of deacetylation (DD), stoichiometry of the chitosan/DNA complex (N/P ratio, charge ratio of amine (chitosan) to phosphate (DNA)), chitosan salt form, pH, serum, additives, preparation techniques of chitosan/nucleic acid particles, and routes of administration.[117]

It is generally accepted that high-MW chitosans are superior to those with low MW in enhancing the stability of complexes. The stable complexes can protect DNA in the cellular endosomal/lysosomal compartments but restrict the release of DNA inside the cells, resulting in low or delayed expression.[118] Consequently, an intermediate degree of stability needs to be achieved with chitosan of an appropriate MW for extracellular DNA protection (better with high MW) versus efficient intracellular DNA release (better with low MW) in order to obtain high levels of transfection.[119] The degree of DD signifies the percentage of deacetylated primary amine groups along the molecular chain,[114] and must exceed 65% in order to obtain stable complexes with DNA that transfect target cells in vitro.[118] Taking together the MW and DD for chitosan, maximum expression levels can be obtained by simultaneously lowering the MW and increasing DD, or lowering DD and increasing the MW. Ishii *et al.*[120] reported that the transfection efficiency of chitosan/DNA complexes increased at charge ratios of 3 and 5, and decreased at higher charge ratios in SOJ cells. Furthermore, to improve the poor water solubility of chitosan at physiological pH and its low transfection efficiency, several chitosan derivatives have been synthesized, such as PEG-chitosan, alkylated chitosan (ACS), *N*,*N*,*N*-trimethylated chitosan (TMC), and thiolated chitosan.[121] It was reported that chitosan grafted with PEG (MW 5 kDa, grafting degree 9.6%) could prevent aggregation of the polyplex even in the presence of serum and bile.[23]

Over the years, though numerous innovative chitosan-derived vectors have been generated and the exciting *in vitro* and *in vivo* data thereof have been published, no accredited chitosan vector usable in clinical practices has so far been recognized. Based on recent studies, it is still an uphill struggle to search for clinically applicable chitosan vectors with high transfectability in the near future. Once such vectors are found, there will be fascinating opportunities for the delivery of a wide range of exogenous genes and therapeutic agents for both medical and scientific purposes.[121]

5. Cyclodextrin

Cyclodextrins (CDs) are a series of natural, cyclic, water-soluble oligosaccharides composed of six (α-CD), seven (β-CD), or eight (γ-CD) D(+)-glucose units linked by α-1,4 linkages. They have a torus-like architecture with an apolar interior cavity which can induce an inclusion complex with a hydrophobic moiety.[122] Various molecules can be fitted into the cavities of CDs to form supramolecular inclusion complexes. Therefore, natural CDs and modified CDs have recently attracted considerable attention as gene delivery vectors due to their excellent biocompatibility and unique molecular architecture.[123]

As nucleic acids are negatively charged hydrophilic macromolecules, they cannot be trapped in native CDs cavities. Therefore, they have to be incorporated by cations or modified by other cationic polymers. A number of cationic polymers such as linear and branched PEI, and PAMAM dendrimer, have been extensively studied. The use of a linear β-CD-containing cationic polymer for gene delivery was first developed by Davis's group in 1999.[124] So far, they have developed a set of such CD-containing polymers and studied the structural effects of the polymers on gene delivery.[125–127] Generally, it was reported that the CD-containing cationic polymers had lower cytotoxicity and efficient gene transfection in cell cultures. In addition, the most important feature of the CD-containing cationic polymer-based gene delivery system is that the polyplexes formed between the polymers and DNA can be further modified by inclusion complex formation since there are many CDs moieties.[128] CDs functionalized with PAMAM dendrimers showed about 100 times higher luciferase gene expression than unfunctionalized PAMAM or noncovalent mixtures of PAMAM and CDs.[129] Recently, Burckbuchler *et al.* constructed novel and efficient nonviral vectors constituted by CD polymers (polyβ-CD) complexing with adamantyl cationic derivatives (Ada). This gene vector performed the same order of in vitro transfection levels as with DOTAP on two cell lines (HepG2 and HEK293).[130] Tang *et al.* have successfully linked β-CD to low-MW PEI (MW 600) to conjugate a PEI-CD (PC) polymer. The developed PC polymer was soluble in water and biodegradable, and was capable of mediating efficient gene transfection in cultured neurons and in the central nervous system.[131]

Most recently, another new class of CD-containing gene carriers was designed based on cationic polyrotaxanes. Different from the conventional polycations which have long sequences of covalently bonded repeating units, this new cationic polyrotaxanes composed of multiple oligoethylenimine (OEI) grafted β-CDs that were threaded on a poly(propylene oxide) (PPO) block and blocked on the two ends of a pluronic (PEO-PPO-PEO) triblock copolymer[132] (Fig. 12).

The cationic supramolecular gene delivery vector fabricated from the cationic polyrotaxanes with multiple OEI-grafted β-CD showed good DNA binding ability, low cytotoxicity, and high gene transfection efficacy similar to that of high-MW PEI (25 kDa). It was also verified that the cationic supramolecules grafted with longer OEI showed higher gene transfection efficiency in cell cultures.[132] Cationic polyrotaxanes could effectively condense pDNA to NPs and showed low cytotoxicity and high transfection efficiency in BHK-21 and MES-SA cells.[133] In addition, Shuai *et al.*[134] constructed supramolecular

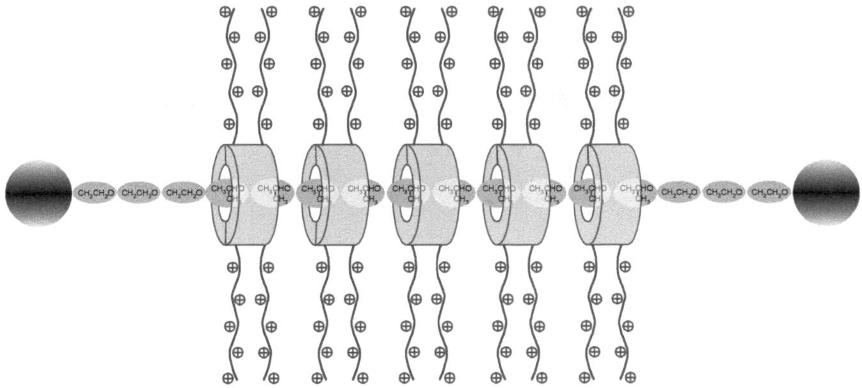

FIG. 12. Structures of cationic polyrotaxanes with multiple OEI-grafted β-CD. (For color version of this figure, the reader is referred to the Web version of this chapter.)

inclusion complexes by threading α-CD molecules over ternary block copolymers composed of PEG, poly(epsilon-caprolactone) (PCL), and PEI. It was indicated that transfection efficiencies of this inclusion complex were in the same order of magnitude as PEI. However, its toxicity was 100 times lower than PEI as observed by 3-(4,5)-dimethylthiahiazo (-z-y1)-3,5-di- phenytetrazoliumromide (MTT) assay.

The development of CD-based supramolecular biomaterials for gene delivery is an emerging area, which still faces many challenges. Although a number of proof-of-concept studies have been demonstrated, smarter material designs are expected for a better balance of higher functions and performances and lower toxicity, both in vitro and in vivo. Safety of the delivery systems is still a major concern. Intensive *in vitro* and *in vivo* biocompatibility and biodegradability studies are desired for supramolecular gene delivery systems.

C. Polyion Complex (PIC) Micelle

To increase both the transfection ability and the circulation half-life of polyplexes, coating with PEG has become one of the most popular modifications.[90] Kataoka *et al.* introduced polyion complex (PIC) micelles using PLL-PEG block copolymer, in which PLL segments and pDNA formed a hydrophobic core by electrostatic interactions and PEG played a role as a surrounding hydrophilic shell layer.[135] It is to be noted that PIC micelles will be benefit not only for use in a gene delivery system but also from the standpoint of the basic study on DNA condensation (Fig. 13).[136]

PIC micelles demonstrated such remarkable properties as excellent colloidal stability in protein aqueous media, high solubility in aqueous media, high tolerance toward nuclease degradation, minimal interaction with biological

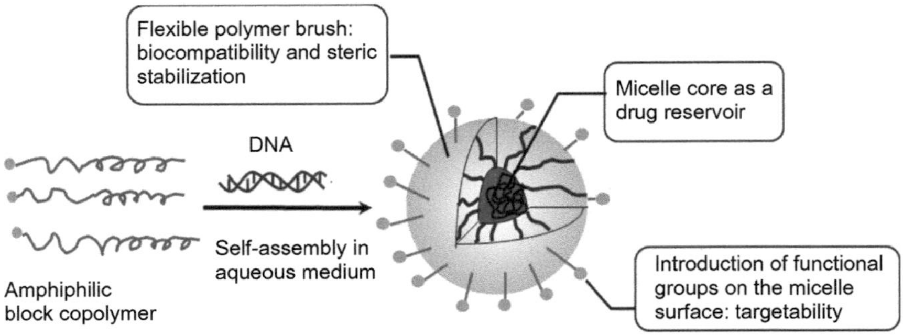

FIG. 13. Polyion complex (PIC) micelles as intelligent nanocarriers for gene delivery. As delivery vehicles for DNA, PIC micelles demonstrated remarkable properties such as excellent colloidal stability in protein aceous media, high solubility in aqueous media, high tolerability toward nuclease degradation, minimal interaction with biological components including proteins and cells, and prolonged blood circulation compared to the conventional polyplex and lipoplex systems. Furthermore, the termination-functionalized PEG also provides the possibility of targeting modification. (See Color Insert.)

components including proteins and cells, and prolonged blood circulation compared to conventional polyplex and lipoplex systems. Furthermore, the termination-functionalized PEG also provide the possibility of targeting modification. Other diblock copolymers are made up of PEG and cationic polymers or peptides such as PEI,[137] poly(phosphoaromidate) (PPA),[138] chitosan,[139] and PAMAM in both dendrimer or linear forms.[140]

The novel PIC micelles with smart functions such as stimuli sensitivity have emerged as promising carriers that enhance the efficacy of genes with minimal side effects. The use of the chemical or physical stimuli-sensitive vectors in conjunction with the genes expressed under the control of the specific promoters to organ or tissue type and chemical or physical signals will lead to a highly specific gene transfer to cells at the diseased site. Kataoka *et al.* have developed a novel PIC micelle with a disulfide cross-linked core. In this micelle, a certain fraction of the lysine residue of the PEG-PLL block copolymer was replaced by thiol groups to form a disulfide cross-linking network structure in the micelle core after DNA complex formation. As a result, the micelle could efficiently release the loaded DNA in response to the reductive condition mimicking the intracellular environment. More importantly, this type of micelle induced higher transfection efficiency in the cell.[141] Li *et al.* synthesized an acid-labile block copolymer consisting of PEG and poly(2-(dimethylamino)ethyl methacrylate) segments (PEG-*b*-PDMAEMA) connected through cyclic ortho ester linkage, which condensed with pDNA to form a PIC micelle with an acid-triggered reversible PEG shield. The acid sensitivity of the ortho ester linkage could be greatly enhanced by changing the structure of the cyclic ortho ester, which might enable extracellular deshielding of the polyplexes in solid tumor tissues and promote cellular uptake and transfection efficiency.[142]

For enhancing the transfection efficiency, the polymer/pDNA complexes must facilitate endosomal escape so that the delivered DNA can be released from the endosome into the cytosol. To create the desirable proton sponge effect, histidine residues were introduced to the PLL backbone. Histidylated PLL showed better transfection efficiency than the PLL/chloroquine mixture.[143] Kim *et al.*[144] have produced poly(ethylene glycol)-*co*-poly(L-lysine)-*g*-histidine (PEG-PLL-*g*-his) multiblock copolymers (MBCs); the p*K* of the conjugated imidazoles was found to be 4.75, which would facilitate buffering at low pH environments of the late endosome/lysosome. The ability of the imidazoles to protonate and destabilize membrane vesicles was confirmed by the use of bafilomycin A1. It was shown that MBCs produced about 5 times higher transfection efficiency in murine smooth muscle cells compared to the treated cells. A novel block copolymer with two amino groups with higher and lower p*K*a values in each of the monomer units was designed to solve the issues of stability and buffering capacity simultaneously (Fig. 14). The PEG-PAsp (DET)/PEG-PAsp(DPT) block copolymer possessed multiple cationic groups with distinctive p*K*a in the side chain, by the unique feature of the regulated location of primary and secondary amino groups in the side chain, and the block copolymer enabled both sufficient binding with DNA and the endosome buffering capacity (proton sponge effect).[145]

An A-B-C triblock copolymer, tandemly aligning two types of polycations with different pKa values in a single polymer strand, was designed and widely used. Kataoka *et al.* have developed a triblock copolymer PEG-*b*-PMPA-*b*-PLL

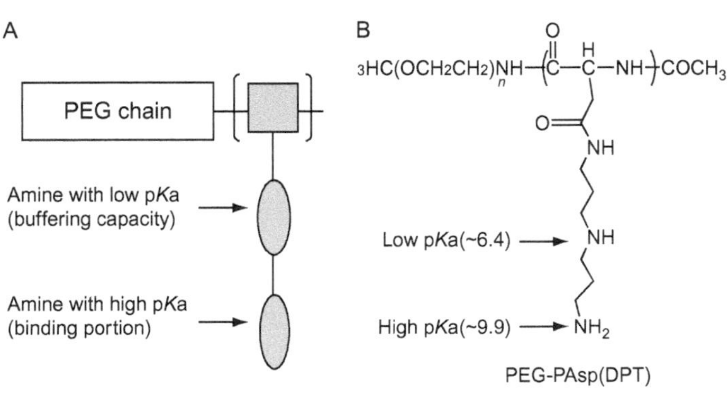

FIG. 14. Block copolymer with amino group with distinctive p*K*a values in the side chain (A) and PEG-PAsp(DPT) block copolymer (B).

in which PEG was used as the biocompatible A segment, poly[(3-morpholinopropyl) aspartamide] (PMPA), was used as the low-p*K*a B segment with buffering capacity, and PLL was used as the high-p*K*a C segment to condense the DNA to construct the biocompatible polyplexes (polyplex micelles) satisfying the stability, biocompatibility and transfection ability criteria. An ^{1}H-NMR study revealed that the PLL segment preferentially contributed to the DNA condensation and the uncomplexed PMPA segment remained in the micelle. Importantly, due to the buffering capacity of the uncomplexed PMPA segment in the polyplex micelle, the PEG-*b*-PMPA-*b*-PLL PIC micelle exhibited higher transfection efficiency against HuH-7 cells than the PIC micelle from PEG-*b*-PLL or the mixture of PEG-*b*-PLL and PEG-*b*-PMPA.[146]

Linking targeting ligands to the polymer chain is another effective pathway to obtain enhanced transfection efficiency. Wakebayashi *et al.* synthesized a block copolymer (lactose-PEG-PAMA) to construct a PIC micellar-type gene vector potentially useful for selective transfection of hepatic cells.[147] Results showed that the transfection of lactosylated PIC micelle is much higher than that of the control PIC micelle without lactose moieties against HepG2 cells possessing asialoglycoprotein (ASGP) receptors. Oishi *et al.* designed a novel A-B-C triblock copolymer for constructing a pH-responsive and targetable gene vector. The copolymer, lactosylated poly(ethyleneglycol)-*b*-poly(silamine)-*b*-poly [2-(*N*, *N*-dimethylamino)ethyl methacrylate] (Lac-PEG-PSAO-PAMA), consisted of lactosylated PEG (A segment), a pH-responsive polyamine segment (B segment), and a DNA-condensing polyamine segment (C segment). Lac-PEG-PSAO-PAMA spontaneously associated with pDNA to form three-layered polyplex micelles with a PAMA/pDNA core, an uncomplexed PSAO inner shell, and a lactosylated PEG outer shell. The micelles exhibited a specific cellular uptake into HuH-7 cells (hepatocytes) through ASGP receptor-mediated endocytosis and achieved a far more efficient transfection ability of a reporter gene compared to the Lac-PEG-PSAO/pDNA and Lac-PEG-PAMA/pDNA polyplex micelles.[148]

D. Hybrid NPs

1. Liposome–Polycation–DNA (LPD) NPs

Cationic liposomes had been widely used in gene delivery, but their tendency to form large DNA/liposome complexes creates difficulties for in vivo studies and clinical trials. The DNA molecules in the complexes are not well condensed, and often exist in an extended conformation covered by lipids.[149] In contrast to cationic liposomes, polycations such as PLL, histone, and protamine are known to condense DNA in highly compact structures of about 30–100 nm in diameter. Therefore, LPD NPs with small particle sizes were first developed by Leaf Huang's group for pDNA delivery.[150] DNA precondensed with polycations is mixed with cationic liposomes and the components spontaneously rearranged to form a virus-like structure, with the condensed DNA located inside the lipid membranes.[151] PLL, protamine sulfate, and PEI are often used as polycations to condense DNA into a compact core. The cationic liposomes used are composed of neutral lipid DOPE and cationic lipids, such as DOTAP, DC-Chol, DOTMA, and so on.[152] Apart from cationic LPDs, polycation-condensed DNA was also entrapped into anionic liposomes via charge interaction by Leaf Huang's group.[153] Due to its structural similarity to cationic LPD, this kind of vector was named LPDII (liposome-entrapped polycation-condensed DNA, Fig. 15). The surface of LPD could be modified by various ligands to increase cellular uptake via a specific receptor-mediated pathway.[154,155]

Recently, a novel multifunctional enveloped nanodevice (MEND, Fig. 16) was proposed which had a structure similar to that of LPD.[156–159] The first step of DNA condensation allowed protection of DNA from DNase, size control, and improvement in packaging efficiency. In the second step, complexes were incorporated into lipid envelopes so that the DNA core and lipid envelope existed as separate structures, rather than a disordered mixture, to control topology.[157] PLL, protamine, and stearyl octaarginine (STR-R8) were often used to condense DNA for MEND and showed higher transfection efficiency.[156] The lipid envelope was often prepared with neutral DOPE and anionic lipid CHEMS, as they both could enhance the internalization efficiency into the cytoplasmic space by their fusogenic properties and then facilitate the nuclear delivery of DNA.[159,160] Transferrin (Tf), sugar, peptide, and antibodies could be used as target ligands equipped on the lipid envelope.[158] Besides vectors assembled with one anionic lipid layer, a multilayered NP called tetra-lamellar multifunctional enveloped nano device (T-MEND) was constructed that achieved dramatic levels of transgene expression in nondividing cells.[161]

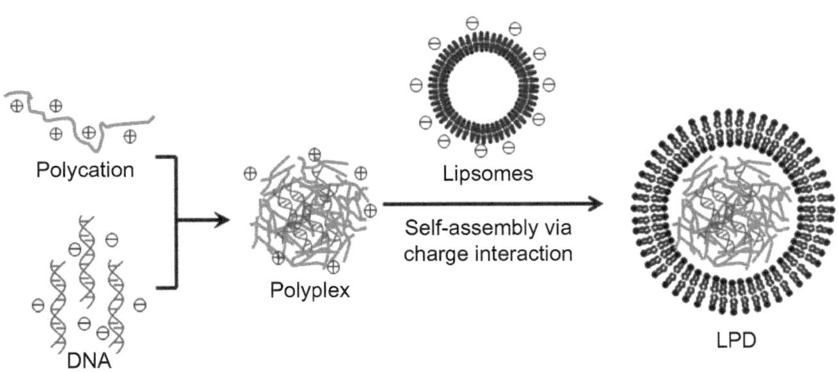

FIG. 15. The assembly process of LPDII. (See Color Insert.)

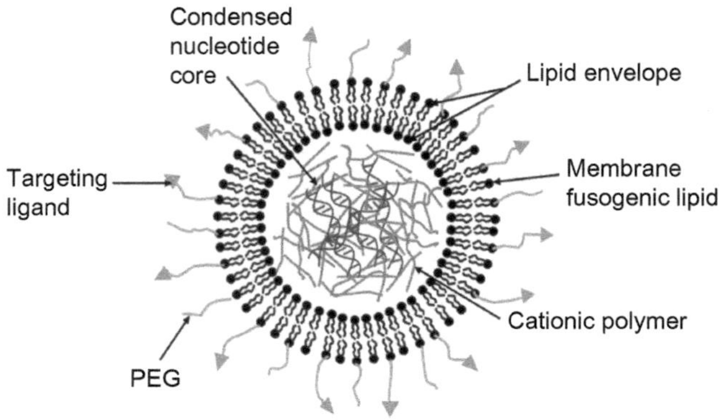

FIG. 16. The structure of MEND. The MEND consisted of a condensed DNA core and an anionic lipid envelope structure equipped with various functional devices such as PEG for long blood circulation, ligand for specific targeting, a protein transduction domain peptide to increase intracellular availability, and fusogenic lipids to enhance endosomal escape. (See Color Insert.)

The critical structural elements of the T-MEND were a DNA–polycation condensed core coated with two nuclear membrane fusogenic inner lipid layers and two endosome–fusogenic outer lipid layers.

2. Multilayered NPs

In conclusion, with its virus-like structure, LPD is of great value to design new alternatives to recombinant viruses for gene therapy applications in the future. Multilayered NPs can be assembled based on electrostatic or other intermolecular forces using a layer-by-layer (LbL) technique with alternate deposition of polycations and polyanions (including DNA or other nucleic acid). The assembly process of multilayered NPs via the LbL technique is illustrated in Fig. 17.

Multilayered gene vectors can be assembled with alternate layers of the required functional materials, including ligands for specific targeting, fusogenic lipids or pH-sensitive polymers for endosomal escape, and a NLS for enhanced nuclear delivery. To date, multilayered, multifunctional gene vectors have been widely used to increase the loading dose of DNA, control the release of DNA in different environments, target specific sites, and enhance cellular uptake.[162] Kakade *et al.*[163] reported that the loading dose of active pDNA was increased to 7.9 μg mg^{-1} of PLGA particles when they deposited three bilayers of PEI/DNA on the surface of PLGA microparticles and the transfection efficiency increased with increasing dose. When environment-sensitive polymers, such as hydrolytically degradable poly(β-amino ester), pH-sensitive PAA, histidine-modifying polycations, or reductively degradable disulfide-containing

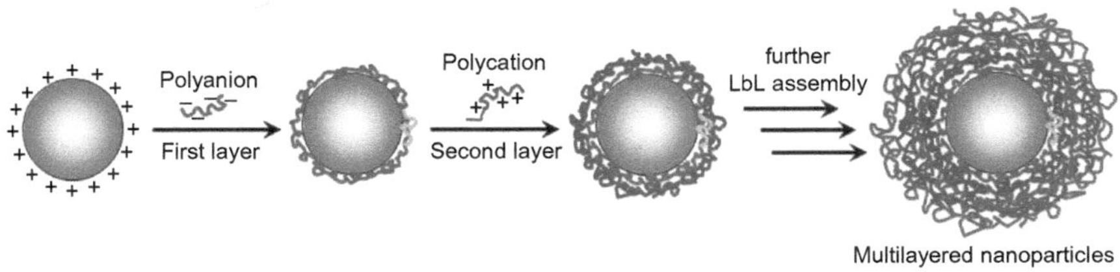

FIG. 17. The assembly process of multilayered nanoparticles via the LbL technique. (For color version of this figure, the reader is referred to the Web version of this chapter.)

polymers, were incorporated, the multilayered NPs could control the release rate of DNA in the cells. It was reported that a multilayered quaternary polyplex (PEI/DNA/PAA/poly(*N*-(8-aminooctyl)-acrylamide) (P8Am)) could significantly increase the gene transfection efficiency.[164] In addition, immobilization of targeting agents and PEG on the films of multilayered vectors were widely used to obtain targeting specificity and prolonged circulation time. Khopade and coworkers assembled PEG2000-DSPE (MW~2000) on the particles fabricated by poly(styrene sulfonate) (PSS) and a fourth-generation PAMAM dendrimer (4G PAMAM). The surface-modified NPs showed reduced adhesion to the human serum albumin.[165] The microcapsules composed of poly(vinyl galactose ester-*co*-methacryloxyethyl trimethylammonium chloride) (PGEDMC) and PSS modified with peanut agglutinin lectin could specifically recognize the hepatic cells.[166]

E. PLGA Nanoparticles

Although the above-mentioned vectors result in relatively high efficiency for gene delivery, maintaining therapeutic concentrations for a prolonged time becomes the main obstacle for their further application.[167] A promising polymer for providing sustained pDNA delivery is PLGA, which together with polylactic acid (PLA) has been approved for human use by the FDA. Biodegradable PLGA particles are biocompatible and have the capacity to protect pDNA from nuclease degradation and increase pDNA stability.[168] In addition, the polymeric matrix allows precise control over the release kinetics of the gene from NPs. Typically, for gene delivery, pDNA is usually encapsulated into the anionic PLGA NPs. The commonly used methods to encapsulate DNA into PLGA NPs always require harsh manufacturing conditions such as sonication or high-shear agitation, which tend to result in the inactivation of the pDNA.[169] Recently, it was reported that the most efficient transfection efficiencies generated by biodegradable polymeric microparticles could be achieved by first forming complexes of the pDNA with a cationic polymer such as PEI, PLL, dentrimer, and chitosan prior to entrapping the complex into the microparticles.[170]

Intra *et al.* firstly prepared PLGA microparticles encapsulating PAMAM/DNA dendriplexes which have low toxicity, high pDNA loading efficiencies, and high transfection efficiencies, which are even not reduced in the presence of serum.[171] The folate-linked multipolymeric NPs were feasible as efficient targeted carriers for gene delivery. For example, the polylysine-condensed DNA (PD) was encapsulated in folate-modified poly(ethylene glycol)-poly(D, L-lactic-*co*-glycolic acid) NPs (PELGA-NPs) to form polymer–polycationic peptide–DNA (PPD) NPs.[172] It was indicated that cellular uptake of PPD NPs was significantly enhanced by the folate modification in the in vitro transfection study compared to unmodified NPs.

Another approach to fabricate PLGA gene carriers makes use of surface adsorption. The surface adsorption of DNA on PLGA NPs is a relatively new approach. DNA is a polyanionic molecule which can be loaded on a polycationic surface via ionic interaction. Zou *et al.* developed new bioadhesive PLGA NPs which could efficiently bind DNA onto their surface by non-ionic interaction using carbopol as the bioadhesive agent. It was indicated that the bioadhesive PLGA NPs hold many advantages as gene delivery systems such as DNA-binding efficiency ($> 80\%$) at an optimal concentration, DNA protection from enzymatic degradation, better buffering capacity, and, most importantly, a higher transfection efficiency in A549 cells.[173]

Cationic surfactants such as CTAB and dimethyldidodecylammonium bromide (DMAB),[174] polymers such as PEI, or combination of both have been used for preparation of cationic NPs which can facilitate DNA loading and also enhance intracellular uptake of the particles through the effective interaction with negatively charged cell membrane. PEI has been widely used as a cationic polymer for the preparation of cationic PLGA particles. Son *et al.* developed two kinds of PLGA-based gene delivery systems, namely, DNA-encapsulated PLGA NPs (PLGA-E) and surface-adsorbed DNA on PLGA-BPEI NPs (PLGA-BPEI). It was demonstrated that PLGA-BPEI NPs could efficiently transfect HEK293 cells 6.7 times higher than PLGA-E NPs.[175] Kim *et al.* successfully designed and fabricated a nonviral vector system, that is, biodegradable PLGA NPs modified with a polyplexed PEI/DNA coating, that exhibited higher gene transfection efficiency in human mesenchymal stem cells (hMSCs).[176]

However, there are several limitations to PLGA particles for DNA delivery system development.[177] First, because of the large size and hydrophilic character of DNA, encapsulation of pDNA in hydrophobic PLGA microspheres is a challenge. Second, it is well known that the hydrolysis of PLGA may substantially decrease the pH in PLGA microspheres, potentially resulting in exposing pDNA to the harsh processes and solvents and DNA degradation. Third, the rate of DNA release is often very slow, and therefore also the subsequent antigen production in case of DNA vaccination, which may prevent an optimal immune response. Given the limitations of PLGA NPs for DNA delivery, there is a clear need for developing new types of biodegradable NPs. Until recently, the main focus was on the delivery of the DNA, but now sustained and controlled gene expression is becoming increasingly important.[178] As a consequence, the biodegradable PLGA NPs are also excellent potential candidates for the intracellular delivery of gene therapeutics, due to their low cytotoxicity.

F. Peptide-Based NPs

Short basic peptides containing positively charged amino acids such as histidine, lysine, and arginine can interact with the negatively charged phosphate backbone of DNA through electrostatic interactions and form peptide/DNA complexes.[179] The positive charge exhibited by peptide/DNA complexes facilitates the interaction with cell membranes and internalization into the cell, overcoming membrane barriers. In addition, the amphiphilicity of peptides favors their conformational changes in acidic environments to escape from endosome/lysosome, thereby allowing nuclear gene delivery and expression.[180] DNA-condensing peptides could prevent DNA from being degraded by cytosolic nucleases as well as prolong the half-life of the targeted nucleic acid. The typical example of cationic peptide is PLL, which has been discussed above. Here we will introduce other widely used cationic peptides.

Another example is protamine. Protamine sulfate has a small MW of 4000 and highly positively charged due to the presence of 21 arginine residues. The multivalent nature of the polyamines facilitates charge interactions with the phosphate backbone of DNA and provides complete protection of DNA from nuclease degradation.[83] Protamine sulfate was used to substitute PLL in further studies of LPD formulation, because protamine could drastically condense DNA into spherical, electrodense particles of 100 nm which could take up a high content of lipid.[181] In addition, the sequence of protamine was considered to mimic NLS, which could facilitate nuclear uptake through the nuclear pore complex.[182] In a previous study, three layers gene vectors were constructed. First, anionic DNA was condensed by cationic protamine to form protamine/DNA nanometric complexes. Then, a small-sized anionic SLN was absorbed on the cationic complex matrix. The obtained three-layer nanovectors could efficiently transfect DNA into A549 cells.[183]

The α-helical KALA peptide (derived from the influenza HA-2 subunit, which enables the virus to infuse into the cell membrane) is one of the cationic peptides used successfully in gene delivery.[42] KALA peptide is an α-helical amphiphilic peptide and can bind to DNA, destabilize membranes, and mediate DNA transfection.[21] It was reported that the efficiency of the peptide vector depends on the hydrophobic portion which plays a major role in aggregation and endosomal escape.[83] Peptides with a large hydrophobic region can form large and stable aggregates which can be internalized by endocytosis and remain resistant to digestion in lysosomal vesicles, as a result of high transfection efficiency. In a recent study, it was shown that amphiphilic oligopeptides, which were cholesterol-conjugated oligohistidine and oligoarginine peptides, could self-assemble into cationic micelles in aqueous solution and induce higher gene expression than the peptides in both HepG2 and HEK293 cell lines.[184] Min *et al.* synthesized a new peptide, KALA-Antennapedia (K-Antp). In this configuration, Antp is designed to provide the cell permeation capacity and NLS, while the KALA moiety promotes cellular entry of the peptide–DNA complex. It was indicated that an optimal K-Antp/DNA formula was nearly 400–600-fold more efficient than Antp or poly-lysine-Antp (L-Antp) in gene delivery, and comparable or superior to a commercial liposome. In addition, incorporation of PEI into the system or treatment with chloroquine further increased the efficiency of K-Antp-mediated gene delivery.[185] Choi *et al.* described a peptide/siRNA complex containing the cell-penetrating peptide derived from natural protamine, termed low-MW protamine (LMWP). In vivo studies demonstrated that the peptide could carry and localize siRNA inside tumors and inhibit the expression of VEGF through systemic application of the peptide complex, thereby suppressing tumor growth. Moreover LMWP/siRNA did not exert measurable immunestimulatory effect.[186]

Peptide gene carriers have been mainly explored in vitro; however, their in vivo behavior is still under investigation. The ideal peptide-mediated gene delivery system must satisfy a number of requirements. The bound, condensed peptide/DNA complex must be stable to the extracellular as well as the intracellular environment. Once internalized, the complex must be impervious to degradative enzymes. Finally, the peptides must promote the nuclear translocation of pDNA.[187] This, in turn, will allow the development of approaches able to overcome each identified limiting step. Recently, genetically engineered biomacromolecules, called "biomimetic vectors" (DBVs), were designed to mimic viral characteristics in order to overcome the cellular barriers associated with the targeted gene transfer. For example, a novel vector was genetically engineered to contain at precise locations (a) four tandem repeating units of N-terminal domain of histone H2A to condense DNA into stable nanosized particles suitable for cellular uptake, (b) a model targeting motif to target HER2 and enhance internalization of NPs, and (c) a pH-responsive synthetic fusogenic peptide to disrupt endosome membranes and promote escape of the NPs into the cytosol. The results demonstrate that the biopolymer was able to condense pDNA into nanosize particles, protect pDNA from serum endonucleases, target HER2 positive cancer cells but not HER2 negative ones, efficiently disrupt endosomes, and effectively reach the cell nucleus of target cells to mediate gene expression.[188]

G. Inorganic NPs

1. Quantum Dots

Luminescent semiconductor nanocrystals (QDs) stand out among the research tools in chemistry, physics, and biology as one of the most exciting developments.[189] QDs are typically synthesized from binary combinations of a variety of semiconductor materials (ZnS, CdS, CdSe, InP, CdTe, PbS, PbTe). CdSe/ZnS and CdTe/ZnS are the most studied and widely used QDs. When excited at different UV and near-IR wavelengths, the QDs will appear in different colors.[190] QDs have a long lifetime in vivo, and their fluorescence can be maintained for up to several months without light-induced bleaching. Therefore, QDs are emerging as a powerful class of visualization tools for biolabeling and bioimaging applications (see the chapter by Ishikawa and Biju in this volume).

Recently, surface modification on QDs has led to the development of a new generation of probes with integrated functionalities of labeling and drug/gene delivery.[191,192] QDs can be creatively combined with various NPs for gene delivery to achieve a synergistic effect so that the particles can be successfully delivered to the targeted sites with significantly reduced cytotoxicity, and simultaneously monitoring in real time the fluorescent intensity. This synergistic use of QDs with NPs holds tremendous potential for biochemical medicine.[193] CdSe QD–amphipol nanocomplex,[194] cysteamine–capped CdTe QD vectors,[195] 2-vinylpyridine-functionalized silicon QDs,[196] and QD–peptide conjugates[197] have been recently reported by researchers for DNA/siRNA delivery. However, the optimistic outlook for such an application should be tempered by the fact that QDs are toxic. The CdSe and CdSe/ZnS QD-based vectors pose a potential risk to biological systems and cannot be used in human experiments. To address the toxicity problem of CdSe QDs, the innate, less toxic nanodiamond was modified with 800 Da PEI (PEI800). The low cytotoxic PEI800-modified nanodiamonds could mediate the transfection of luciferase plasmid (pLuc) and GFP plasmid in HeLa cells though the efficiency was two- to threefold lower than that of PEI25K.[13] Alternately, ZnO QDs are noncytotoxic but their blue fluorescence is unfit for bioimaging.[198] Recently, Zhang *et al.* designed low cytotoxic ZnO QD-based nonviral vectors, poly(2-(dimethylamino)ethyl methacrylate) (PDMAEMA)-*co*-PMAA)-capped ZnO QDs to transfer pDNA and label live cells. It was indicated that these ZnO QDs could mediate an efficient transfer of pDNA into COS-7 cells with much lower cytotoxicity, at the same time allowing real-time imaging of gene transfection.[199]

2. Magnetic NPs

Magnetic NPs (MNPs) can also be used in gene transfection. The appearance of MNPs used as a delivery vector was in the late 1970s.[200] They typically comprise a magnetic core, an organic coating, and multiple functional

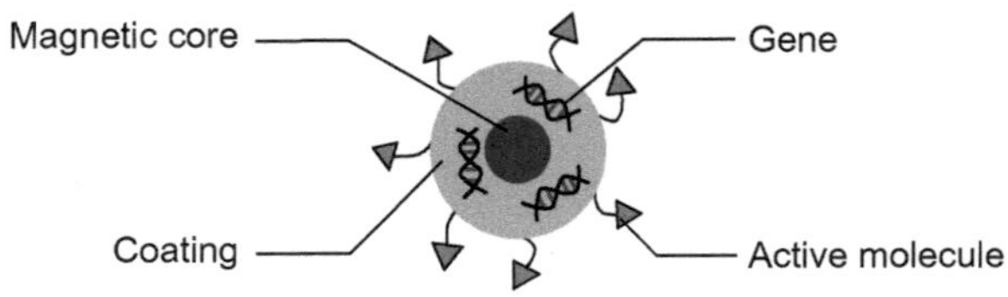

FIG. 18. The basic structure of MNPs.

molecules on its surface. The basic structure of MNPs is showed in Fig. 18. For biomedical applications, the magnetic particles should be completely covered and as dense as possible in order to protect the iron oxide core against contacting with blood proteins and phagocytosis-associated receptors, as well as leaching nucleic acid.

Magnetic cores can be formed by various magnetic materials including pure iron; chromium and cobalt metals; alloys such as $CoPt_3$, FePt, FeZn; and iron oxides such as magnetite (Fe_3O_4) and maghemite (γ-Fe_2O_3).[201] These magnetic core NPs need to be protected against corrosion by acid or base, agglomeration, and oxidition.[202] Coating with proper materials can isolate the core from the environment and protect it against destruction. These coating materials are mainly cationic polymers, such as PEI, PAMAM, and chitosan, where negatively charged nucleic acids can be adsorbed or condensed. Besides, these surface coatings can (1) protect against iron oxide core agglomeration; (2) provide chemical handles for the conjugation of genes, targeting ligands, and reporter moieties, and (3) limit nonspecific cell interactions. Additionally, polymeric coatings have been engineered to enhance superparamagnetic iron oxide NP (SPION) pharmacokinetics, endosomal release, and tailored gene-loading and release behaviors.[201] MNP coating can be achieved via a number of approaches, including in situ coating, post-synthesis adsorption and post-synthesis end grafting, while these cationic polymer coatings are mainly gained by surface absorption.[203] Except for condensation of DNA, surface coatings also provide handles for the modification of several functional molecules. These molecules include hydrophilic molecules to provide long blood circulation time in vivo such as PEG, and target molecules to increase the local accumulation and achieve active targeting such as small-molecule ligands, monoclonal antibodies, peptidomimetics, aptamers, and polysaccharides.

MNPs have attracted much interest in the study of gene delivery systems. They contain intrinsic magnetic properties, which enable tracking through magnetic resonance imaging (MRI). In the gene delivery field, MNPs can protect the nucleic acids against enzymatic degradation and facilitate cellular internalization and endosomal release. Moreover, MNP-based transfection (termed magnetofection) can result in very rapid transfection and improved transduction efficiency both in vitro and in vivo under static magnetic fields. Another major advantage of this method is that the magnetic fields may be configured to target specific sites in vivo and increase the residence time of the vector at the target.[204,205]

Ensenauer *et al.* compared the utility of magnetofection with lipofection procedures for siRNA delivery to primary and immortalized mammalian fibroblasts. Lipofectamine 2000 and magnetofection achieved high transfection efficiencies comparable to similar gene silencing effects of about 80%.[206] Pan *et al.* reported the cationic lipid-coated MNPs associated with transferrin for gene delivery. MNPs were surface-coated with cationic lipids and then combined with PEI-condensed pDNA, followed by transferring by electrical attraction. DNA was protected and the magnetic complexes exhibited up to 300-fold higher transfection activity compared to common cationic liposomes or cationic polymer complexes in the presence of a magnetic field.[207] Silvio Dutz *et al.* employed carboxymethyldextran (CMD) as the coating material for the iron oxide NPs. The dextran shell on these clusters can be clearly seen under transmission electron microscopy (TEM). The increased stability against agglomeration was achieved over a period of more than 7 days.[208] Bhattarai and coworkers developed *N*-hexanoyl chloride-modified chitosan-stabilized iron oxide NPs. Results showed that compared with unmodified iron oxide NPs, those with *N*-hexanoyl chitosan modification were much easier to disperse in an aqueous environment, and the cytotoxicity was still very low even at a concentration 50-fold higher than the concentration required for high-efficiency transfection.[209]

IV. Challenges and Prospects

The success of gene therapy depends greatly on the delivery systems. The future of gene therapy is expected to see a combination of both genetic and NP engineering. Cationic lipids and polymers are widely used as transfection reagents. Each of them has advantages and disadvantages. Despite the excitement of the new breakthroughs in gene therapy, this field is still in its infancy. Although in vitro studies and tests on animals have shown the usefulness of NPs as gene delivery vehicles, only a relatively small number of clinical trials of gene therapy have been completed or are under way.

PEGylation can significantly decrease nonspecific distribution and unexpected immune response. However, the surface aqueous phase formed by the PEG moiety inhibits the interaction of the gene carrier with the tumor cell surface. Furthermore, PEGylation improves the stability of NPs, which results in poor endosomal escape via membrane fusion as well as degradation of cargos in lysosomes and digestive compartments. These serious issues regarding the use of PEG in gene delivery for cancer are referred to as the "PEG dilemma."[158] Therefore, a cleavable linkage between the PEG moiety and the cationic lipid or polymer is preferred. Under certain intracellular environment, the stimulatory linkage can be cleaved and the positively charged surface exposed to trigger endocytosis and endosomal escape. Modification of gene carriers with ligands may be another efficient way to achieve cell-specific delivery.

In addition, the interaction between gene and cationic carriers should be strong enough to make sure that the complexes are stable before cellular internalization. Thereafter, the complexes commonly locate in endosomes. The endosomal escape is an important step, which can be enhanced by the proton sponge effect of polyplexes or lipid fusion by lipoplexes. Then the loaded nucleic acids are released from the complexes, diffuse into cytoplasm, and finally translocate into the nucleus. Any problems in each step may cause the whole delivery process to fail. In this area, a great deal of work needs to be done so that we can have a comprehensive understanding of the nucleic acid delivery process.

It is likely that no single material, modification, or method will be universally applicable. At present, there are no NPs that are completely safe and nontoxic. Further improvements are definitely needed, particularly in the areas of particle charge, shapes, sizes as well as particle degradation in order to achieve NPs that are nontoxic, non-immunogenic, and highly stable, with high transfection efficiency. The great flexibility inherent in the use of NP-mediated gene therapy allows the selection of the best possible combination of factors for maximum effectiveness. There is still a long way before us before we can reach a perfect design of an NP-based nucleic acid delivery system.

References

1. Arumugam P, Malik P. Genetic therapy for Beta-thalassemia: from the bench to the bedside. *Hematology Am Soc Hematol Educ Program* 2010;**2010**:445–50.
2. Cao S, Cripps A, Wei MQ. New strategies for cancer gene therapy: progress and opportunities. *Clin Exp Pharmacol Physiol* 2010;**37**:108–14.
3. Reynolds PN. Gene therapy for pulmonary hypertension: prospects and challenges. *Expert Opin Biol Ther* 2011;**11**:133–43.
4. Huelsmann P, Hofmann A, Knoepfel S, Popp J, Rauch P, Di Giallonardo F, et al. A suicide gene approach using the human pro-apoptotic protein tBid inhibits HIV-1 replication. *BMC Biotechnol* 2011;**11**:4.
5. Pathak A, Patnaik S, Gupta KC. Recent trends in non-viral vector-mediated gene delivery. *Biotechnol J* 2009;**4**:1559–72.
6. Lewitt PA, Rezai AR, Leehey MA, Ojemann SG, Flaherty AW, Eskandar EN, et al. AAV2-GAD gene therapy for advanced Parkinson's disease: a double-blind, sham-surgery controlled, randomised trial. *Lancet Neurol* 2011;**10**:309–19.

7. Edelstein ML, Abedi MR, Wixon J. Gene therapy clinical trials worldwide to 2007—an update. *J Gene Med* 2007;**9**:833–42.
8. Van Tendeloo VF, Van Broeckhoven C, Berneman ZN. Gene therapy: principles and applications to hematopoietic cells. *Leukemia* 2001;**15**:523–44.
9. Davis ME. Non-viral gene delivery systems. *Curr Opin Biotechnol* 2002;**13**:128–31.
10. Ditto AJ, Shah PN, Yun YH. Non-viral gene delivery using nanoparticles. *Expert Opin Drug Deliev* 2009;**6**:1149–60.
11. Gao X, Kim KS, Liu DX. Nonviral gene delivery: what we know and what is next. *AAPS J* 2007;**9**:E92–E104.
12. Wiethoff CM, Middaugh CR. Barriers to nonviral gene delivery. *J Pharm Sci* 2003;**92**:203–17.
13. Adler AF, Leong KW. Emerging links between surface nanotechnology and endocytosis: impact on nonviral gene delivery. *Nano Today* 2010;**5**:553–69.
14. Read ML, Logan A, Seymour LW. Barriers to gene delivery using synthetic vectors. *Adv Genet* 2005;**53**:19–46.
15. Alexis F, Pridgen E, Molnar LK, Farokhzad OC. Factors affecting the clearance and biodistribution of polymeric nanoparticles. *Mol Pharm* 2008;**5**:505–15.
16. Pitkanen L, Ruponen M, Nieminen J, Urtti A. Vitreous is a barrier in nonviral gene transfer by cationic lipids and polymers. *Pharm Res* 2003;**20**:576–83.
17. Ruponen M, Honkakoski P, Ronkko S, Pelkonen J, Tammi M, Urtti A. Extracellular and intracellular barriers in non-viral gene delivery. *J Control Release* 2003;**93**:213–7.
18. Belting M, Sandgren S, Wittrup A. Nuclear delivery of macromolecules: barriers and carriers. *Adv Drug Deliv Rev* 2005;**57**:505–27.
19. Betancourt T, Byrne JD, Sunaryo N, Crowder SW, Kadapakkam M, Patel S, et al. PEGylation strategies for active targeting of PLA/PLGA nanoparticles. *J Biomed Mater Res A* 2009;**91A**:263–76.
20. Li SD, Huang L. Gene therapy progress and prospects: non-viral gene therapy by systemic delivery. *Gene Ther* 2006;**13**:1313–9.
21. Dash PR, Read ML, Fisher KD, Howard KA, Wolfert M, Oupicky D, et al. Decreased binding to proteins and cells of polymeric gene delivery vectors surface modified with a multivalent hydrophilic polymer and retargeting through attachment of transferrin. *J Biol Chem* 2000;**275**:3793–802.
22. Lemieux P, Guerin N, Paradis G, Proulx R, Chistyakova L, Kabanov A, et al. A combination of poloxamers increases gene expression of plasmid DNA in skeletal muscle. *Gene Ther* 2000;**7**:986–91.
23. Jiang X, Dai H, Leong KW, Goh SH, Mao HQ, Yang YY. Chitosan-g-PEG/DNA complexes deliver gene to the rat liver via intrabiliary and intraportal infusions. *J Gene Med* 2006;**8**:477–87.
24. Huang M, Fong CW, Khor E, Lim LY. Transfection efficiency of chitosan vectors: effect of polymer molecular weight and degree of deacetylation. *J Control Release* 2005;**106**:391–406.
25. Kuo WT, Huang HY, Huang YY. Intracellular trafficking, metabolism and toxicity of current gene carriers. *Curr Drug Metab* 2009;**10**:885–94.
26. Lechardeur D, Verkman AS, Lukacs GL. Intracellular routing of plasmid DNA during non-viral gene transfer. *Adv Drug Deliv Rev* 2005;**57**:755–67.
27. Liu C, Zhang N. Pharmaceutical strategies enhancing cell penetration efficiencies of non-viral gene delivery systems. *Curr Gene Ther* 2009;**9**:267–90.
28. Rawat A, Vaidya B, Khatri K, Goyal AK, Gupta PN, Mahor S, et al. Targeted intracellular delivery of therapeutics: an overview. *Pharmazie* 2007;**62**:643–58.
29. Kircheis R, Wightman L, Wagner E. Design and gene delivery activity of modified polyethylenimines. *Adv Drug Deliv Rev* 2001;**53**:341–58.

30. Nam HY, Kwon SM, Chung H, Lee SY, Kwon SH, Jeon H, et al. Cellular uptake mechanism and intracellular fate of hydrophobically modified glycol chitosan nanoparticles. *J Control Release* 2009;**135**:259–67.
31. Bareford LM, Swaan PW. Endocytic mechanisms for targeted drug delivery. *Adv Drug Deliv Rev* 2007;**59**:748–58.
32. Douglas KL. Toward development of artificial viruses for gene therapy: a comparative evaluation of viral and non-viral transfection. *Biotechnol Prog* 2008;**24**:871–83.
33. Sahay G, Alakhova DY, Kabanov AV. Endocytosis of nanomedicines. *J Control Release* 2010;**145**:182–95.
34. Oba M, Aoyagi K, Miyata K, Matsumoto Y, Itaka K, Nishiyama N, et al. Polyplex micelles with cyclic RGD peptide ligands and disulfide cross-links directing to the enhanced transfection via controlled intracellular trafficking. *Mol Pharm* 2008;**5**:1080–92.
35. Liu C, Yu W, Chen Z, Zhang J, Zhang N. Enhanced gene transfection efficiency in CD13-positive vascular endothelial cells with targeted poly (lactic acid)-poly (ethylene glycol) nanoparticles through caveolae-mediated endocytosis. *J Control Release* 2011;**151**(2):162–75.
36. Perumal OP, Inapagolla R, Kannan S, Kannan RM. The effect of surface functionality on cellular trafficking of dendrimers. *Biomaterials* 2008;**29**:3469–76.
37. Little SR, Kohane DS. Polymers for intracellular delivery of nucleic acids. *J Mater Chem* 2008;**18**:832–41.
38. Varkouhi AK, Scholte M, Storm G, Haisma HJ. Endosomal escape pathways for delivery of biologicals. *J Control Release* 2011;**151**(3):220–8.
39. Hatakeyama H, Ito E, Akita H, Oishi M, Nagasaki Y, Futaki S, et al. A pH-sensitive fusogenic peptide facilitates endosomal escape and greatly enhances the gene silencing of siRNA-containing nanoparticles in vitro and in vivo. *J Control Release* 2009;**139**:127–32.
40. Cheung CY, Murthy N, Stayton PS, Hoffman AS. A pH-sensitive polymer that enhances cationic lipid-mediated gene transfer. *Bioconjug Chem* 2001;**12**:906–10.
41. Asayama S, Hamaya A, Sekine T, Kawakami H, Nagaoka S. Aminated poly(L-histidine) as new pH-sensitive DNA carrier. *Nucleic Acids Symp Ser (Oxf)* 2004;229–30.
42. Martin ME, Rice KG. Peptide-guided gene delivery. *AAPS J* 2007;**9**:E18–29.
43. Nakase I, Kogure K, Harashima H, Futaki S. Application of a fusiogenic peptide GALA for intracellular delivery. *Methods Mol Biol* 2011;**683**:525–33.
44. El-Sayed A, Masuda T, Khalil I, Akita H, Harashima H. Enhanced gene expression by a novel stearylated INF7 peptide derivative through fusion independent endosomal escape. *J Control Release* 2009;**138**:160–7.
45. Allen TD, Cronshaw JM, Bagley S, Kiseleva E, Goldberg MW. The nuclear pore complex: mediator of translocation between nucleus and cytoplasm. *J Cell Sci* 2000;**113**(Pt 10):1651–9.
46. Wiethoff CM, Middaugh CR. Barriers to nonviral gene delivery. *J Pharm Sci* 2003;**92**:203–17.
47. Zhang H, Mitin A, Vinogradov SV. Efficient transfection of blood-brain barrier endothelial cells by lipoplexes and polyplexes in the presence of nuclear targeting NLS-PEG-acridine conjugates. *Bioconjug Chem* 2009;**20**:120–8.
48. Jeon O, Lim HW, Lee M, Song SJ, Kim BS. Poly(L-lactide-co-glycolide) nanospheres conjugated with a nuclear localization signal for delivery of plasmid DNA. *J Drug Target* 2007;**15**:190–8.
49. Felgner PL, Gadek TR, Holm M, Roman R, Chan HW, Wenz M, et al. Lipofection: a highly efficient, lipid-mediated DNA-transfection procedure. *Proc Natl Acad Sci USA* 1987;**84**:7413–7.
50. Wasungu L, Hoekstra D. Cationic lipids, lipoplexes and intracellular delivery of genes. *J Control Release* 2006;**116**:255–64.

51. Montier T, Benvegnu T, Jaffres PA, Yaouanc JJ, Lehn P. Progress in cationic lipid-mediated gene transfection: a series of bio-inspired lipids as an example. *Curr Gene Ther* 2008;**8**:296–312.
52. Zhi D, Zhang S, Wang B, Zhao Y, Yang B, Yu S. Transfection efficiency of cationic lipids with different hydrophobic domains in gene delivery. *Bioconjug Chem* 2010;**21**:563–77.
53. Obata Y, Saito S, Takeda N, Takeoka S. Plasmid DNA-encapsulating liposomes: effect of a spacer between the cationic head group and hydrophobic moieties of the lipids on gene expression efficiency. *Biochim Biophys Acta* 2009;**1788**:1148–58.
54. Mevel M, Kamaly N, Carmona S, Oliver MH, Jorgensen MR, Crowther C, et al. DODAG; a versatile new cationic lipid that mediates efficient delivery of pDNA and siRNA. *J Control Release* 2010;**143**:222–32.
55. Lv H, Zhang S, Wang B, Cui S, Yan J. Toxicity of cationic lipids and cationic polymers in gene delivery. *J Control Release* 2006;**114**:100–9.
56. Tang F, Hughes JA. Synthesis of a single-tailed cationic lipid and investigation of its transfection. *J Control Release* 1999;**62**:345–58.
57. Liu C, Chen Z, Yu W, Zhang N. Novel cationic 6-lauroxyhexyl lysinate modified poly(lactic acid)-poly(ethylene glycol) nanoparticles enhance gene transfection. *J Colloid Interface Sci* 2011;**354**:528–35.
58. Yu W, Liu C, Liu Y, Zhang N, Xu W. Mannan-modified solid lipid nanoparticles for targeted gene delivery to alveolar macrophages. *Pharm Res* 2010;**27**:1584–96.
59. Yu W, Liu C, Ye J, Zou W, Zhang N, Xu W. Novel cationic SLN containing a synthesized single-tailed lipid as a modifier for gene delivery. *Nanotechnology* 2009;**20**:215102.
60. Martin B, Sainlos M, Aissaoui A, Oudrhiri N, Hauchecorne M, Vigneron JP, et al. The design of cationic lipids for gene delivery. *Curr Pharm Des* 2005;**11**:375–94.
61. Zuhorn IS, Hoekstra D. On the mechanism of cationic amphiphile-mediated transfection. To fuse or not to fuse: is that the question? *J Membr Biol* 2002;**189**:167–79.
62. Safinya CR. Structures of lipid-DNA complexes: supramolecular assembly and gene delivery. *Curr Opin Struct Biol* 2001;**11**:440–8.
63. Ewert KK, Ahmad A, Evans HM, Safinya CR. Cationic lipid-DNA complexes for non-viral gene therapy: relating supramolecular structures to cellular pathways. *Expert Opin Biol Ther* 2005;**5**:33–53.
64. Koltover I, Salditt T, Radler JO, Safinya CR. An inverted hexagonal phase of cationic liposome-DNA complexes related to DNA release and delivery. *Science* 1998;**281**:78–81.
65. Ma B, Zhang S, Jiang H, Zhao B, Lv H. Lipoplex morphologies and their influences on transfection efficiency in gene delivery. *J Control Release* 2007;**123**:184–94.
66. Koynova R, Wang L, Tarahovsky Y, MacDonald RC. Lipid phase control of DNA delivery. *Bioconjug Chem* 2005;**16**:1335–9.
67. Ewert KK, Evans HM, Zidovska A, Bouxsein NF, Ahmad A, Safinya CR. A columnar phase of dendritic lipid-based cationic liposome-DNA complexes for gene delivery: hexagonally ordered cylindrical micelles embedded in a DNA honeycomb lattice. *J Am Chem Soc* 2006;**128**:3998–4006.
68. Pedersen N, Hansen S, Heydenreich AV, Kristensen HG, Poulsen HS. Solid lipid nanoparticles can effectively bind DNA, streptavidin and biotinylated ligands. *Eur J Pharm Biopharm* 2006;**62**:155–62.
69. Muller RH, Mader K, Gohla S. Solid lipid nanoparticles (SLN) for controlled drug delivery—a review of the state of the art. *Eur J Pharm Biopharm* 2000;**50**:161–77.
70. Tabatt K, Kneuer C, Sameti M, Olbrich C, Muller RH, Lehr CM, et al. Transfection with different colloidal systems: comparison of solid lipid nanoparticles and liposomes. *J Control Release* 2004;**97**:321–32.

71. Choi SH, Jin SE, Lee MK, Lim SJ, Park JS, Kim BG, et al. Novel cationic solid lipid nanoparticles enhanced p53 gene transfer to lung cancer cells. *Eur J Pharm Biopharm* 2008;**68**:545–54.
72. Somia N, Verma IM. Gene therapy: trials and tribulations. *Nat Rev Genet* 2000;**1**:91–9.
73. Liu C, Ye J, Zhang N, Ma C, Luan F. Preliminary studies on cationic solid lipid nanoparticles/pDNA binary complex. *Chin Pharm J* 2008;**43**:1050–6.
74. del Pozo-Rodriguez A, Delgado D, Solinis MA, Gascon AR, Pedraz JL. Solid lipid nanoparticles: formulation factors affecting cell transfection capacity. *Int J Pharm* 2007;**339**:261–8.
75. Doroud D, Vatanara A, Zahedifard F, Gholami E, Vahabpour R, Rouholamini Najafabadi A, et al. Cationic solid lipid nanoparticles loaded by cysteine proteinase genes as a novel anti-leishmaniasis DNA vaccine delivery system: characterization and in vitro evaluations. *J Pharm Pharm Sci* 2010;**13**:320–35.
76. Vighi E, Ruozi B, Montanari M, Battini R, Leo E. pDNA condensation capacity and in vitro gene delivery properties of cationic solid lipid nanoparticles. *Int J Pharm* 2010;**389**:254–61.
77. Kwon SM, Nam HY, Nam T, Park K, Lee S, Kim K, et al. In vivo time-dependent gene expression of cationic lipid-based emulsion as a stable and biocompatible non-viral gene carrier. *J Control Release* 2008;**128**:89–97.
78. Nam HY, Park JH, Kim K, Kwon IC, Jeong SY. Lipid-based emulsion system as non-viral gene carriers. *Arch Pharm Res* 2009;**32**:639–46.
79. Choi WJ, Kim JK, Choi SH, Park JS, Ahn WS, Kim CK. Low toxicity of cationic lipid-based emulsion for gene transfer. *Biomaterials* 2004;**25**:5893–903.
80. Barut KD, Coskun Ari FF, Oner F. Development and characterization of a cationic emulsion formulation as a potential pDNA carrier system. *Turk J Chem* 2005;**29**:27–40.
81. Hara T, Tan Y, Huang L. In vivo gene delivery to the liver using reconstituted chylomicron remnants as a novel nonviral vector. *Proc Natl Acad Sci USA* 1997;**94**:14547–52.
82. Liu CH, Yu SY. Cationic nanoemulsions as non-viral vectors for plasmid DNA delivery. *Colloids Surf B Biointerfaces* 2010;**79**:509–15.
83. Zhang S, Xu Y, Wang B, Qiao W, Liu D, Li Z. Cationic compounds used in lipoplexes and polyplexes for gene delivery. *J Control Release* 2004;**100**:165–80.
84. Obata Y, Suzuki D, Takeoka S. Evaluation of cationic assemblies constructed with amino acid based lipids for plasmid DNA delivery. *Bioconjug Chem* 2008;**19**:1055–63.
85. Pinazo A, Infante MR, Izquierdo P, Solans C. Synthesis of arginine-based surfactants in highly concentrated water-in-oil emulsions. *J Chem Soc Perkin Trans* 2000;**2**:1535–9.
86. Fattal E, Bochot A. State of the art and perspectives for the delivery of antisense oligonucleotides and siRNA by polymeric nanocarriers. *Int J Pharm* 2008;**364**:237–48.
87. Al-Dosari MS, Gao X. Nonviral gene delivery: principle, limitations, and recent progress. *AAPS J* 2009;**11**:671–81.
88. Wu GY, Wu CH. Receptor-mediated in vitro gene transformation by a soluble DNA carrier system. *J Biol Chem* 1987;**262**:4429–32.
89. Boussif O, Lezoualc'h F, Zanta MA, Mergny MD, Scherman D, Demeneix B, et al. A versatile vector for gene and oligonucleotide transfer into cells in culture and in vivo: polyethylenimine. *Proc Natl Acad Sci USA* 1995;**92**:7297–301.
90. Tros de Ilarduya C, Sun Y, Duzgunes N. Gene delivery by lipoplexes and polyplexes. *Eur J Pharm Sci* 2010;**40**:159–70.
91. Morille M, Passirani C, Vonarbourg A, Clavreul A, Benoit JP. Progress in developing cationic vectors for non-viral systemic gene therapy against cancer. *Biomaterials* 2008;**29**:3477–96.
92. Mintzer MA, Simanek EE. Nonviral vectors for gene delivery. *Chem Rev* 2009;**109**:259–302.
93. Swami A, Kurupati RK, Pathak A, Singh Y, Kumar P, Gupta KC. A unique and highly efficient non-viral DNA/siRNA delivery system based on PEI-bisepoxide nanoparticles. *Biochem Biophys Res Commun* 2007;**362**:835–41.

94. Boeckle S, von Gersdorff K, van der Piepen S, Culmsee C, Wagner E, Ogris M. Purification of polyethylenimine polyplexes highlights the role of free polycations in gene transfer. *J Gene Med* 2004;**6**:1102–11.
95. Saul JM, Linnes MP, Ratner BD, Giachelli CM, Pun SH. Delivery of non-viral gene carriers from sphere-templated fibrin scaffolds for sustained transgene expression. *Biomaterials* 2007;**28**:4705–16.
96. Breunig M, Lungwitz U, Liebl R, Goepferich A. Breaking up the correlation between efficacy and toxicity for nonviral gene delivery. *Proc Natl Acad Sci USA* 2007;**104**:14454–9.
97. Oupicky D, Parker AL, Seymour LW. Laterally stabilized complexes of DNA with linear reducible polycations: strategy for triggered intracellular activation of DNA delivery vectors. *J Am Chem Soc* 2002;**124**:8–9.
98. Gosselin MA, Guo W, Lee RJ. Efficient gene transfer using reversibly cross-linked low molecular weight polyethylenimine. *Bioconjug Chem* 2001;**12**:989–94.
99. Deng R, Yue Y, Jin F, Chen Y, Kung HF, Lin MC, et al. Revisit the complexation of PEI and DNA - how to make low cytotoxic and highly efficient PEI gene transfection non-viral vectors with a controllable chain length and structure? *J Control Release* 2009;**140**:40–6.
100. Gillies ER, Frechet JM. Dendrimers and dendritic polymers in drug delivery. *Drug Discov Today* 2005;**10**:35–43.
101. Boas U, Heegaard PM. Dendrimers in drug research. *Chem Soc Rev* 2004;**33**:43–63.
102. Xu Q, Wang CH, Pack DW. Polymeric carriers for gene delivery: chitosan and poly (amidoamine) dendrimers. *Curr Pharm Des* 2010;**16**:2350–68.
103. Tomalia DA, Baker H, Dewald JR, Hall M, Kallos G, Martin S, et al. A new class of polymers: starburst-dendritic macromolecules. *Polym J* 1985;**17**:117–32.
104. Paleos CM, Tziveleka LA, Sideratou Z, Tsiourvas D. Gene delivery using functional dendritic polymers. *Expert Opin Drug Deliv* 2009;**6**:27–38.
105. Eliyahu H, Barenholz Y, Domb AJ. Polymers for DNA delivery. *Molecules* 2005;**10**:34–64.
106. Klajnert B, Bryszewska M. Dendrimers: properties and applications. *Acta Biochim Pol* 2001;**48**:199–208.
107. Choi JS, Nam K, Park JY, Kim JB, Lee JK, Park JS. Enhanced transfection efficiency of PAMAM dendrimer by surface modification with L-arginine. *J Control Release* 2004;**99**:445–56.
108. Lee JH, Lim YB, Choi JS, Lee Y, Kim TI, Kim HJ, et al. Polyplexes assembled with internally quaternized PAMAM-OH dendrimer and plasmid DNA have a neutral surface and gene delivery potency. *Bioconjug Chem* 2003;**14**:1214–21.
109. Nam HY, Hahn HJ, Nam K, Choi WH, Jeong Y, Kim DE, et al. Evaluation of generations 2, 3 and 4 arginine modified PAMAM dendrimers for gene delivery. *Int J Pharm* 2008;**363**:199–205.
110. Jevprasesphant R, Penny J, Jalal R, Attwood D, McKeown NB, D'Emanuele A. The influence of surface modification on the cytotoxicity of PAMAM dendrimers. *Int J Pharm* 2003;**252**:263–6.
111. Kim TI, Seo HJ, Choi JS, Jang HS, Baek JU, Kim K, et al. PAMAM-PEG-PAMAM: novel triblock copolymer as a biocompatible and efficient gene delivery carrier. *Biomacromolecules* 2004;**5**:2487–92.
112. Majoros IJ, Williams CR, Baker Jr. JR. Current dendrimer applications in cancer diagnosis and therapy. *Curr Top Med Chem* 2008;**8**:1165–79.
113. Hussain M, Shchepinov M, Sohail M, Benter IF, Hollins AJ, Southern EM, et al. A novel anionic dendrimer for improved cellular delivery of antisense oligonucleotides. *J Control Release* 2004;**99**:139–55.
114. Mao S, Sun W, Kissel T. Chitosan-based formulations for delivery of DNA and siRNA. *Adv Drug Deliv Rev* 2010;**62**:12–27.

115. Mumper RJ, Wang J, Claspell JM, Rolland AP. Novel polymeric condensing carriers for gene delivery. *Proc Intl Symp Control Rel Bioact Mater* 1995;**22**:178–9.
116. Guang Liu W, De Yao K. Chitosan and its derivatives—a promising non-viral vector for gene transfection. *J Control Release* 2002;**83**:1–11.
117. Lavertu M, Methot S, Tran-Khanh N, Buschmann MD. High efficiency gene transfer using chitosan/DNA nanoparticles with specific combinations of molecular weight and degree of deacetylation. *Biomaterials* 2006;**27**:4815–24.
118. Koping-Hoggard M, Tubulekas I, Guan H, Edwards K, Nilsson M, Varum KM, et al. Chitosan as a nonviral gene delivery system. Structure-property relationships and characteristics compared with polyethylenimine in vitro and after lung administration in vivo. *Gene Ther* 2001;**8**:1108–21.
119. Strand SP, Lelu S, Reitan NK, de Lange Davies C, Artursson P, Varum KM. Molecular design of chitosan gene delivery systems with an optimized balance between polyplex stability and polyplex unpacking. *Biomaterials* 2010;**31**:975–87.
120. Ishii T, Okahata Y, Sato T. Mechanism of cell transfection with plasmid/chitosan complexes. *Biochim Biophys Acta* 2001;**1514**:51–64.
121. Lai WF, Lin MC. Nucleic acid delivery with chitosan and its derivatives. *J Control Release* 2009;**134**:158–68.
122. Uekama K, Hirayama F, Irie T. Cyclodextrin drug carrier systems. *Chem Rev* 1998;**98**:2045–76.
123. Li J, Xiao H, Zhong Y. Drug carrier systems based on water-soluble cationic beta-cyclodextrin polymers. *Int J Pharm* 2004;**278**:329–42.
124. Gonzalez H, Hwang SJ, Davis ME. New class of polymers for the delivery of macromolecular therapeutics. *Bioconjug Chem* 1999;**10**:1068–74.
125. Reineke TM, Davis ME. Structural effects of carbohydrate-containing polycations on gene delivery. 1. Carbohydrate size and its distance from charge centers. *Bioconjug Chem* 2003;**14**:247–54.
126. Reineke TM, Davis ME. Structural effects of carbohydrate-containing polycations on gene delivery. 2. Charge center type. *Bioconjug Chem* 2003;**14**:255–61.
127. Popielarski SR, Mishra S, Davis ME. Structural effects of carbohydrate-containing polycations on gene delivery. 3. Cyclodextrin type and functionalization. *Bioconjug Chem* 2003;**14**:672–8.
128. Zhu L, Mahato RI. Lipid and polymeric carrier-mediated nucleic acid delivery. *Expert Opin Drug Deliv* 2010;**7**:1209–26.
129. Svenson S, Tomalia DA. Dendrimers in biomedical applications—reflections on the field. *Adv Drug Deliv Rev* 2005;**57**:2106–29.
130. Burckbuchler V, Wintgens V, Leborgne C, Lecomte S, Leygue N, Scherman D, et al. Development and characterization of new cyclodextrin polymer-based DNA delivery systems. *Bioconjug Chem* 2008;**19**:2311–20.
131. Tang GP, Guo HY, Alexis F, Wang X, Zeng S, Lim TM, et al. Low molecular weight polyethylenimines linked by beta-cyclodextrin for gene transfer into the nervous system. *J Gene Med* 2006;**8**:736–44.
132. Li J, Loh XJ. Cyclodextrin-based supramolecular architectures: syntheses, structures, and applications for drug and gene delivery. *Adv Drug Deliv Rev* 2008;**60**:1000–17.
133. Yang C, Li H, Wang X, Li J. Cationic supramolecules consisting of oligoethylenimine-grafted alpha-cyclodextrins threaded on poly(ethylene oxide) for gene delivery. *J Biomed Mater Res A* 2009;**89**:13–23.
134. Shuai X, Merdan T, Unger F, Kissel T. Supramolecular gene delivery vectors showing enhanced transgene expression and good biocompatibility. *Bioconjug Chem* 2005;**16**:322–9.

135. Kabanov AV, Kabanov VA. DNA complexes with polycations for the delivery of genetic material into cells. *Bioconjug Chem* 1995;**6**:7–20.
136. Sun X, Zhang N. Cationic polymer optimization for efficient gene delivery. *Mini Rev Med Chem* 2010;**10**:108–25.
137. Oishi M, Sasaki S, Nagasaki Y, Kataoka K. pH-responsive oligodeoxynucleotide (ODN)-poly (ethylene glycol) conjugate through acid-labile beta-thiopropionate linkage: preparation and polyion complex micelle formation. *Biomacromolecules* 2003;**4**:1426–32.
138. Jiang X, Dai H, Ke CY, Mo X, Torbenson MS, Li Z, et al. PEG-b-PPA/DNA micelles improve transgene expression in rat liver through intrabiliary infusion. *J Control Release* 2007;**122**:297–304.
139. Chan P, Kurisawa M, Chung JE, Yang YY. Synthesis and characterization of chitosan-g-poly (ethylene glycol)-folate as a non-viral carrier for tumor-targeted gene delivery. *Biomaterials* 2007;**28**:540–9.
140. Lin C, Zhong Z, Lok MC, Jiang X, Hennink WE, Feijen J, et al. Linear poly(amido amine) s with secondary and tertiary amino groups and variable amounts of disulfide linkages: synthesis and in vitro gene transfer properties. *J Control Release* 2006;**116**:130–7.
141. Miyata K, Kakizawa Y, Nishiyama N, Harada A, Yamasaki Y, Koyama H, et al. Block catiomer polyplexes with regulated densities of charge and disulfide cross-linking directed to enhance gene expression. *J Am Chem Soc* 2004;**126**:2355–61.
142. Lin S, Du FS, Wang Y, Ji SP, Liang DH, Yu L, et al. An acid-labile block copolymer of PDMAEMA and PEG as potential carrier for intelligent gene delivery systems. *Biomacromolecules* 2008;**9**:109–15.
143. Benns JM, Choi JS, Mahato RI, Park JS, Kim SW. pH-sensitive cationic polymer gene delivery vehicle: N-Ac-poly(L-histidine)-graft-poly(L-lysine) comb shaped polymer. *Bioconjug Chem* 2000;**11**:637–45.
144. Bikram M, Ahn CH, Chae SY, Lee MY, Yockman JW, Kim SW. Biodegradable poly(ethylene glycol)-co-poly(L-lysine)-g-histidine multiblock copolymers for nonviral gene delivery. *Macromolecules* 2004;**37**:1903–16.
145. Miyata K, Oba M, Nakanishi M, Fukushima S, Yamasaki Y, Koyama H, et al. Polyplexes from poly(aspartamide) bearing 1,2-diaminoethane side chains induce pH-selective, endosomal membrane destabilization with amplified transfection and negligible cytotoxicity. *J Am Chem Soc* 2008;**130**:16287–94.
146. Fukushima S, Miyata K, Nishiyama N, Kanayama N, Yamasaki Y, Kataoka K. PEGylated polyplex micelles from triblock catiomers with spatially ordered layering of condensed pDNA and buffering units for enhanced intracellular gene delivery. *J Am Chem Soc* 2005;**127**:2810–1.
147. Wakebayashi D, Nishiyama N, Yamasaki Y, Itaka K, Kanayama N, Harada A, et al. Lactose-conjugated polyion complex micelles incorporating plasmid DNA as a targetable gene vector system: their preparation and gene transfecting efficiency against cultured HepG2 cells. *J Control Release* 2004;**95**:653–64.
148. Oishi M, Kataoka K, Nagasaki Y. pH-responsive three-layered PEGylated polyplex micelle based on a lactosylated ABC triblock copolymer as a targetable and endosome-disruptive nonviral gene vector. *Bioconjug Chem* 2006;**17**:677–88.
149. Sternberg B, Sorgi FL, Huang L. New structures in complex formation between DNA and cationic liposomes visualized by freeze-fracture electron microscopy. *FEBS Lett* 1994;**356**:361–6.
150. Gao X, Huang L. Potentiation of cationic liposome-mediated gene delivery by polycations. *Biochemistry* 1996;**35**:1027–36.
151. Li S, Huang L. In vivo gene transfer via intravenous administration of cationic lipid-protamine-DNA (LPD) complexes. *Gene Ther* 1997;**4**:891–900.

152. Tan Y, Whitmore M, Li S, Frederik P, Huang L. LPD nanoparticles—novel nonviral vector for efficient gene delivery. *Methods Mol Med* 2002;**69**:73–81.
153. Lee RJ, Huang L. Folate-targeted, anionic liposome-entrapped polylysine-condensed DNA for tumor cell-specific gene transfer. *J Biol Chem* 1996;**271**:8481–7.
154. Chen Y, Sen J, Bathula SR, Yang Q, Fittipaldi R, Huang L. Novel cationic lipid that delivers siRNA and enhances therapeutic effect in lung cancer cells. *Mol Pharm* 2009;**6**:696–705.
155. Li SD, Huang L. Surface-modified LPD nanoparticles for tumor targeting. *Ann N Y Acad Sci* 2006;**1082**:1–8.
156. Kogure K, Moriguchi R, Sasaki K, Ueno M, Futaki S, Harashima H. Development of a non-viral multifunctional envelope-type nano device by a novel lipid film hydration method. *J Control Release* 2004;**98**:317–23.
157. Kogure K, Akita H, Yamada Y, Harashima H. Multifunctional envelope-type nano device (MEND) as a non-viral gene delivery system. *Adv Drug Deliv Rev* 2008;**60**:559–71.
158. Hatakeyama H, Akita H, Harashima H. A multifunctional envelope type nano device (MEND) for gene delivery to tumours based on the EPR effect: A strategy for overcoming the PEG dilemma. *Adv Drug Deliv Rev* 2011;**63**:152–60.
159. Kogure K, Akita H, Harashima H. Multifunctional envelope-type nano device for non-viral gene delivery: concept and application of Programmed Packaging. *J Control Release* 2007;**122**:246–51.
160. Moriguchi R, Kogure K, Akita H, Futaki S, Miyagishi M, Taira K, et al. A multifunctional envelope-type nano device for novel gene delivery of siRNA plasmids. *Int J Pharm* 2005;**301**:277–85.
161. Akita H, Kudo A, Minoura A, Yamaguti M, Khalil IA, Moriguchi R, et al. Multi-layered nanoparticles for penetrating the endosome and nuclear membrane via a step-wise membrane fusion process. *Biomaterials* 2009;**30**:2940–9.
162. Li P, Zhang N. Layer-by-layer self-assembly vectors for gene delivery. *Curr Gene Ther* 2011;**11**:58–73.
163. Kakade S, Manickam DS, Handa H, Mao G, Oupicky D. Transfection activity of layer-by-layer plasmid DNA/poly(ethylenimine) films deposited on PLGA microparticles. *Int J Pharm* 2009;**365**:44–52.
164. Ke JH, Young TH. Multilayered polyplexes with the endosomal buffering polycation in the core and the cell uptake-favorable polycation in the outer layer for enhanced gene delivery. *Biomaterials* 2010;**31**:9366–72.
165. Khopade AJ, Caruso F. Surface-modification of polyelectrolyte multilayer-coated particles for biological applications. *Langmuir* 2003;**19**:6219–25.
166. Zhang F, Wu Q, Chen ZC, Zhang M, Lin XF. Hepatic-targeting microcapsules construction by self-assembly of bioactive galactose-branched polyelectrolyte for controlled drug release system. *J Colloid Interface Sci* 2008;**317**:477–84.
167. Prabha S, Labhasetwar V. Nanoparticle-mediated wild-type p53 gene delivery results in sustained antiproliferative activity in breast cancer cells. *Mol Pharm* 2004;**1**:211–9.
168. Wang D, Robinson DR, Kwon GS, Samuel J. Encapsulation of plasmid DNA in biodegradable poly(D, L-lactic-co-glycolic acid) microspheres as a novel approach for immunogene delivery. *J Control Release* 1999;**57**:9–18.
169. Oster CG, Kissel T. Comparative study of DNA encapsulation into PLGA microparticles using modified double emulsion methods and spray drying techniques. *J Microencapsul* 2005;**22**:235–44.
170. Zhang XQ, Intra J, Salem AK. Comparative study of poly (lactic-co-glycolic acid)-poly ethyleneimine-plasmid DNA microparticles prepared using double emulsion methods. *J Microencapsul* 2008;**25**:1–12.

171. Intra J, Salem AK. Fabrication, characterization and in vitro evaluation of poly(D, L-lactide-co-glycolide) microparticles loaded with polyamidoamine-plasmid DNA dendriplexes for applications in nonviral gene delivery. *J Pharm Sci* 2010;**99**:368–84.
172. Nie Y, Zhang Z, Li L, Luo K, Ding H, Gu Z. Synthesis, characterization and transfection of a novel folate-targeted multipolymeric nanoparticles for gene delivery. *J Mater Sci Mater Med* 2009;**20**:1849–57.
173. Zou W, Liu C, Chen Z, Zhang N. Studies on bioadhesive PLGA nanoparticles: a promising gene delivery system for efficient gene therapy to lung cancer. *Int J Pharm* 2009;**370**:187–95.
174. Fay F, Quinn DJ, Gilmore BF, McCarron PA, Scott CJ. Gene delivery using dimethyldidodecylammonium bromide-coated PLGA nanoparticles. *Biomaterials* 2010;**31**:4214–22.
175. Son S, Kim WJ. Biodegradable nanoparticles modified by branched polyethylenimine for plasmid DNA delivery. *Biomaterials* 2010;**31**:133–43.
176. Kim JH, Park JS, Yang HN, Woo DG, Jeon SY, Do HJ, et al. The use of biodegradable PLGA nanoparticles to mediate SOX9 gene delivery in human mesenchymal stem cells (hMSCs) and induce chondrogenesis. *Biomaterials* 2011;**32**:268–78.
177. Abbas AO, Donovan MD, Salem AK. Formulating poly(lactide-co-glycolide) particles for plasmid DNA delivery. *J Pharm Sci* 2008;**97**:2448–61.
178. van Gaal EV, Hennink WE, Crommelin DJ, Mastrobattista E. Plasmid engineering for controlled and sustained gene expression for nonviral gene therapy. *Pharm Res* 2006;**23**:1053–74.
179. Bloomfield VA. DNA condensation. *Curr Opin Struct Biol* 1996;**6**:334–41.
180. Henke E, Perk J, Vider J, de Candia P, Chin Y, Solit DB, et al. Peptide-conjugated antisense oligonucleotides for targeted inhibition of a transcriptional regulator in vivo. *Nat Biotechnol* 2008;**26**:91–100.
181. Sorgi FL, Bhattacharya S, Huang L. Protamine sulfate enhances lipid-mediated gene transfer. *Gene Ther* 1997;**4**:961–8.
182. Ziemienowicz A, Gorlich D, Lanka E, Hohn B, Rossi L. Import of DNA into mammalian nuclei by proteins originating from a plant pathogenic bacterium. *Proc Natl Acad Sci USA* 1999;**96**:3729–33.
183. Ye J, Wang A, Liu C, Chen Z, Zhang N. Anionic solid lipid nanoparticles supported on protamine/DNA complexes. *Nanotechnology* 2008;**19**:285708.
184. Guo XD, Tandiono F, Wiradharma N, Khor D, Tan CG, Khan M, et al. Cationic micelles self-assembled from cholesterol-conjugated oligopeptides as an efficient gene delivery vector. *Biomaterials* 2008;**29**:4838–46.
185. Min SH, Kim DM, Kim MN, Ge J, Lee DC, Park IY, et al. Gene delivery using a derivative of the protein transduction domain peptide, K-Antp. *Biomaterials* 2010;**31**:1858–64.
186. Choi YS, Lee JY, Suh JS, Kwon YM, Lee SJ, Chung JK, et al. The systemic delivery of siRNAs by a cell penetrating peptide, low molecular weight protamine. *Biomaterials* 2010;**31**:1429–43.
187. Kim HH, Choi HS, Yang JM, Shin S. Characterization of gene delivery in vitro and in vivo by the arginine peptide system. *Int J Pharm* 2007;**335**:70–8.
188. Wang Y, Mangipudi SS, Canine BF, Hatefi A. A designer biomimetic vector with a chimeric architecture for targeted gene transfer. *J Control Release* 2009;**137**:46–53.
189. Ho YP, Leong KW. Quantum dot-based theranostics. *Nanoscale* 2010;**2**:60–8.
190. Yong KT, Roy I, Ding H, Bergey EJ, Prasad PN. Biocompatible near-infrared quantum dots as ultrasensitive probes for long-term in vivo imaging applications. *Small* 2009;**5**:1997–2004.
191. Qi L, Gao X. Emerging application of quantum dots for drug delivery and therapy. *Expert Opin Drug Deliv* 2008;**5**:263–7.
192. Zrazhevskiy P, Sena M, Gao X. Designing multifunctional quantum dots for bioimaging, detection, and drug delivery. *Chem Soc Rev* 2010;**39**:4326–54.

193. Michalet X, Pinaud FF, Bentolila LA, Tsay JM, Doose S, Li JJ, et al. Quantum dots for live cells, in vivo imaging, and diagnostics. *Science* 2005;**307**:538–44.
194. Qi L, Gao X. Quantum dot-amphipol nanocomplex for intracellular delivery and real-time imaging of siRNA. *ACS Nano* 2008;**2**:1403–10.
195. Li D, Li G, Guo W, Li P, Wang E, Wang J. Glutathione-mediated release of functional plasmid DNA from positively charged quantum dots. *Biomaterials* 2008;**29**:2776–82.
196. Klein S, Zolk O, Fromm MF, Schrodl F, Neuhuber W, Kryschi C. Functionalized silicon quantum dots tailored for targeted siRNA delivery. *Biochem Biophys Res Commun* 2009;**387**:164–8.
197. Walther C, Meyer K, Rennert R, Neundorf I. Quantum dot-carrier peptide conjugates suitable for imaging and delivery applications. *Bioconjug Chem* 2008;**19**:2346–56.
198. Viswanatha R, Chakraborty S, Basu S, Sarma DD. Blue-emitting copper-doped zinc oxide nanocrystals. *J Phys Chem B* 2006;**110**:22310–2.
199. Zhang P, Liu W. ZnO QD@PMAA-co-PDMAEMA nonviral vector for plasmid DNA delivery and bioimaging. *Biomaterials* 2010;**31**:3087–94.
200. Widder KJ, Senyel AE, Scarpelli GD. Magnetic microspheres: a model system of site specific drug delivery in vivo. *Proc Soc Exp Biol Med* 1978;**158**:141–6.
201. Veiseh O, Gunn JW, Zhang M. Design and fabrication of magnetic nanoparticles for targeted drug delivery and imaging. *Adv Drug Deliv Rev* 2010;**62**:284–304.
202. Zhang Y, Kohler N, Zhang M. Surface modification of superparamagnetic magnetite nanoparticles and their intracellular uptake. *Biomaterials* 2002;**23**:1553–61.
203. Laurent S, Forge D, Port M, Roch A, Robic C, Vander Elst L, et al. Magnetic iron oxide nanoparticles: synthesis, stabilization, vectorization, physicochemical characterizations, and biological applications. *Chem Rev* 2008;**108**:2064–110.
204. McBain SC, Yiu HH, Dobson J. Magnetic nanoparticles for gene and drug delivery. *Int J Nanomedicine* 2008;**3**:169–80.
205. Dobson J. Gene therapy progress and prospects: magnetic nanoparticle-based gene delivery. *Gene Ther* 2006;**13**:283–7.
206. Ensenauer R, Hartl D, Vockley J, Roscher A, Fuchs U. Efficient and gentle siRNA delivery by magnetofection. *Biotech Histochem* 2011;**86**(4):226–31.
207. Pan X, Guan J, Yoo JW, Epstein AJ, Lee LJ, Lee RJ. Cationic lipid-coated magnetic nanoparticles associated with transferrin for gene delivery. *Int J Pharm* 2008;**358**:263–70.
208. Dutza S, Andräa W, Hergta R, Müllera R, Oestreichc C, Schmidta C, et al. Influence of dextran coating on the magnetic behaviour of iron oxide nanoparticles. *J Magn Magn Mater* 2007;**311**:51–4.
209. Bhattarai SR, Kim SY, Jang KY, Lee KC, Yi HK, Lee DY, et al. N-hexanoyl chitosan-stabilized magnetic nanoparticles: enhancement of adenoviral-mediated gene expression both in vitro and in vivo. *Nanomedicine (UK)* 2008;**4**:146–54.

Cell Delivery of Therapeutic Nanoparticles

JoEllyn McMillan,* Elena Batrakova,† and Howard E. Gendelman*

*Department of Pharmacology and Experimental Neuroscience, Nebraska Medical Center, Omaha, Nebraska, USA

†Department of Pharmaceutical Sciences, Nebraska Medical Center, Omaha, Nebraska, USA

I. Translational Pathways for Cell-Based Nanomedicines 564
II. Nanomedicines for the Diagnosis and Treatment of Cancer, Infectious and Degenerative Diseases, and Tissue Repair 567
A. Cancer 567
B. Infectious Diseases 570
C. Degenerative Diseases 570
D. Immunomodulation 571
E. Regenerative Medicine and Facilitation of Wound Repair 571
F. Ocular Drug Delivery 571
G. Imaging and Diagnostics 572
III. Chemical Composition, Structure, Function, and Manufacture 572
A. Functionalized Fullerenes and Carbon Nanotubes 572
B. Liposomes 573
C. Polymeric Micelles 573
D. Polymeric Nanospheres 574
E. Dendrimers 574
F. Polymer-Coated Nanocrystals 575
G. Nanoshells 575
H. SPIO Nanoparticles 575
I. Quantum Dots 576
IV. Targeted Drug Delivery 576
V. Nanomedicine and Vaccines 577
VI. Nanodevices and Cell Reprogramming 577
VII. Nanomedicine and Stem Cells 578
VIII. Potential of Nanotherapeutics 579
A. Improved Pharmacokinetics 579
B. Reduced Toxicity 579
C. Targeted Tissue Delivery 580
D. Subcellular Localization 580
IX. Nanotoxicology: Immunogenicity, Cytotoxicity, and Generation of ROS 581
X. Unique Challenges for Translational Nanomedicine 583
A. Assessing Efficacy 583
B. Unique Toxicological Issues 584

Progress in Molecular Biology and Translational Science, Vol. 104
DOI: 10.1016/B978-0-12-416020-0.00014-0

1877-1173/11 $35.00

XI. Development of Cell-Based Nanomedicine: A Perspective 585
A. Cell-Specific Targeting 585
B. Particle Uptake.. 585
C. Subcellular Localization, Drug Stability, and Drug Release 586
D. Cells as Trojan Horses for Drug Delivery.................................... 587
E. Future Perspectives.. 588
References.. 588

Nanomedicine seeks to manufacture drugs and other biologically relevant molecules that are packaged into nanoscale systems for improved delivery. This includes known drugs, proteins, enzymes, and antibodies that have limited clinical efficacy based on delivery, circulating half-lives, or toxicity profiles. The < 100 nm nanoscale physical properties afford them a unique biologic potential for biomedical applications. Hence they are attractive systems for treatment of cancer, heart and lung, blood, inflammatory, and infectious diseases. Proposed clinical applications include tissue regeneration, cochlear and retinal implants, cartilage and joint repair, skin regeneration, antimicrobial therapy, correction of metabolic disorders, and targeted drug delivery to diseased sites including the central nervous system. The potential for cell and immune side effects has necessitated new methods for determining formulation toxicities. To realize the potential of nanomedicine from the bench to the patient bedside, our laboratories have embarked on developing cell-based carriage of drug nanoparticles to improve clinical outcomes in infectious and degenerative diseases. The past half decade has seen the development and use of cells of mononuclear phagocyte lineage, including dendritic cells, monocytes, and macrophages, as Trojan horses for carriage of anti-inflammatory and anti-infective medicines. The promise of this new technology and the perils in translating it for clinical use are developed and discussed in this chapter.

I. Translational Pathways for Cell-Based Nanomedicines

Evasion of mononuclear phagocytes' (MP; monocyte, macrophage, and dendritic cell) antimicrobial immune responses, including complement, interferon, and phagolysomal fusion among others, represent effective strategies for any pathogen to circumvent immune defenses. Although microbes commonly secrete metabolites that can effect MP phagocytosis, in practice, this does little to impair this critical cell function.[1–4] Indeed, pathogens are readily phagocytosed and enter phagosomes (Table I). This vesicle then undergoes a series of fission and fusion events, associated with alterations of the surrounding membrane and vacuolar content.[1–3] The microbicidal microenvironment is associated with pH reduction, hydrolytic enzymes, defensins, and generation of toxic oxidative compounds. Many microbes have developed strategies to survive in MP and replicate intracellularly, including effecting the production of

TABLE I

THERAPIES FOR PATHOGEN SEQUESTRATION IN MONONUCLEAR PHAGOCYTES

Pathogen	Treatment
Bacteria	
Mycobacterium tuberculosis	Isoniazid, rifampin, pyrazinamide, ethambutol, streptomycin
Mycobacterium leprae	Dapsone, rifampicin, clofazimine
Legionella pneumophila	Azithromycin, moxifloxacin
Listeria monocytogenes	Penicillin, ampicillin, trimethoprim/sulfamethoxazole
Brucella spp.	Streptomycin, doxycycline, gentamicin, rifampin, (ciprofloxacin, cotrimoxazole)
Salmonella spp.	Ampicillin, gentamicin, trimethoprim/sulfamethoxazole, ceftriaxone, amoxicillin, or ciprofloxacin
Shigella spp.	Ampicillin, trimethoprim/sulfamethoxazole, ciprofloxacin
Coxiella burnetii	Tetracycline, doxycycline, erythromycin, rifampin, Q-vax vaccine (CSL)
Anaplasma phagocytophilum	Doxycycline, rifampin
Ehrlichia chaffeensis	Doxycycline, rifampicin
Bacillus anthracis	Penicillin, fluoroquinolones (ciprofloxacin), doxycycline
Francisella tularensis	Streptomycin, gentamicin, doxycycline, chloramphenicol, fluoroquinolones
Aeromonas hydrophila	Chloramphenicol, florfenicol, tetracyclines, sulfonamide, nitrofuran derivatives, pyrodinecarboxylic acids
Rhodococcus equi	Erythromycin, azithromycin, clarithromycin, ciprofloxacin, vancomycin, aminoglycosides, rifampin, imipenem, meropenem, linezolid, penicillin G, ampicillin, carbenicillin, cefazolin
Protozoa	
Leishmania spp.	Pentavalent antimonials (sodium stibogluconate, meglumine antimoniate), amphotericin B, paromomycin
Toxoplasma gondii	*Acute*: pyrimethamine, sulfadiazine, clindamycin, spiramycin
	Latent: atovaquone, clindamycin, Artemisia annua
Fungi	
Histoplasma capsulatum	Amphotericin B, itraconazole (1 year), ketoconazole (mild)
Cryptococcus neoformans	Fluconazole, amphotericin B, flucytosine, AmBisome 4®
Viruses	
Human immunodeficiency virus	Antiretroviral drugs (NRTIs, PIs, NNRTIs)[a]
Ross River virus	No specific antivirals, analgesics, anti-inflammatories, anti-pyretics
Dengue virus	No specific antivirals, acetaminophen

(Continues)

TABLE I (*Continued*)

Pathogen	Treatment
Human cytomegalovirus	Ganciclovir
Herpes Simplex virus type I/II	Acyclovir, valacyclovir, famciclovir, penciclovir
Highly pathogenic avian influenza virus	Oseltamivir
Human T cell leukemia virus type 1	Zidovudine, interferon alpha, CHOP (cyclophosphamide, hydroxydaunorubicin, vincristine, prednisone), arsenic trioxide
Monkeypox virus	Smallpox vaccination preventative
Vaccinia virus	VIG (vaccinia immune globulin intravenous), cidofovir (proposed)

[a]NRTI, nucleoside reverse transcriptase inhibitor; PI, protease inhibitor; NNRTI, non-nucleoside reverse transcriptase inhibitor.

reactive oxygen species (ROS). Microbes use MP as "Trojan horses" for their dissemination while remaining undetected by host immune surveillance and entering into tissues including the central nervous system (CNS). Fusion of phagosomes with early or late endosomes and/or lysosomes occurs where microbes are destined for elimination and clearance. However, during chronic infection, the system breaks down and the microbe can survive in these compartments for months to years. Thus, the search for a means to best deliver antimicrobial drugs directly to organelles that harbor infection is critically important.[5–8] We recently demonstrated that recycling endosomes harbor microbes such as the human immunodeficiency virus (HIV) capable of pathogen replication. Interestingly, it is these exact compartments where drug-laden nanoparticles are directed (Fig. 1). This, at least in theory, provides a boost to microbial clearance in that the drug can be delivered to the site of active pathogen growth. Independent works were performed recently and in parallel within our laboratories demonstrating that catalase nanozymes can be used effectively for treatment of an experimental model of Parkinson's disease (PD), leading to neuroprotective outcomes.[10–13] These results taken together provide a clear path for translational studies in cell-based nanomedicine for human investigations and ultimate use (Fig. 2). The subsequent sections of this review outline the pathways through which this may be achieved. Reviews of the advances made in nanomedicine overall, with a particular focus on how they may be redirected for cell-based therapies, are illustrated.

II. Nanomedicines for the Diagnosis and Treatment of Cancer, Infectious and Degenerative Diseases, and Tissue Repair

Nanomedicine research has focused, in large measure, on cancer and infectious diseases. Of the nanomedical products developed, reformulated pharmaceuticals remain in majority.[14] Several are in clinical trials for bioimaging and cancer[15] treatments. These include Doxil®, a polyethyleneglycol (PEG) liposomal version of the anticancer drug doxorubicin,[16] Abraxane®, an albumin conjugated nanoparticle version of the anticancer drug paclitaxel, and Rapamune®, a micellar nanoformulation of rapamycin.[17,18] These nanoformulated drugs have exhibited fewer unwanted side effects and improved therapeutic indices over their non-formulated drug counterparts.[19–22]

A. Cancer

The most highly studied area of nanoparticle drug delivery is in cancer treatment. A variety of nanoparticulate systems have been developed for cancer therapeutics including functionalized liposomes, albumin-based particles,

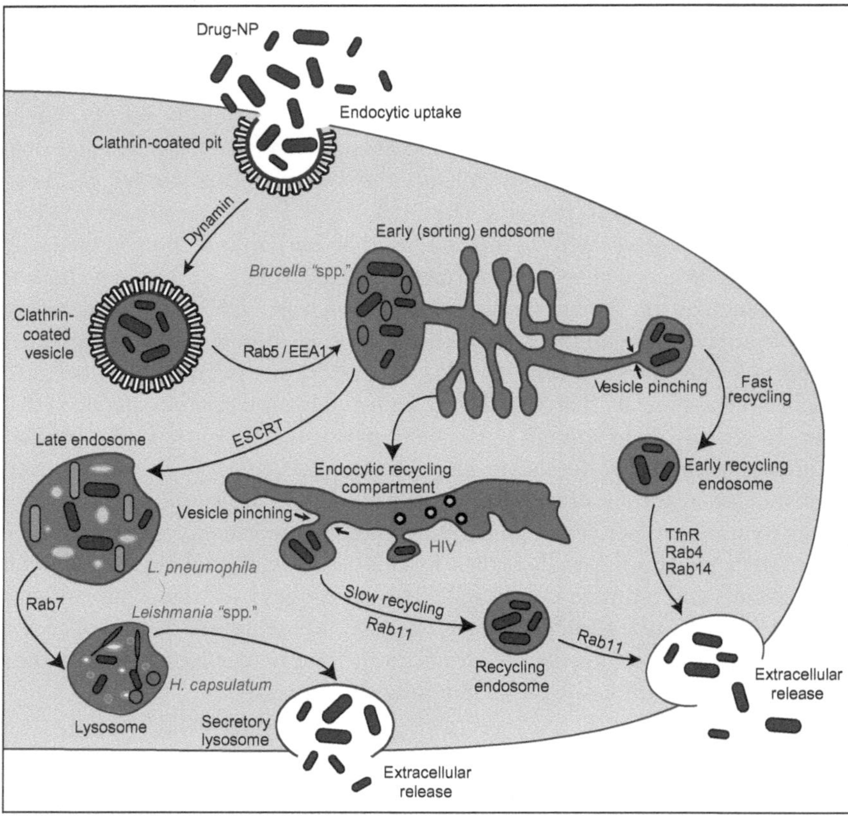

FIG. 1. Intracellular pathways of nanoparticles and pathogens in macrophages. Drug nanoparticles (shown in red) enter macrophages via clathrin-coated pits and are then transported to the early endosome compartment. From the early endosome compartment, the particles can have three different fates: (1) fast recycling via Rab4^{+} or 14^{+} endosomes; (2) trafficking to late endosomes, regulated in part by endosomal sorting complex required for transport (ESCRT) machinery for eventual release as a secretory lysosome; or (3) transport to the recycling endosome compartment where they will be stored for long periods and slowly recycled via Rab11^{+} endosomes. Pathogens are maintained in early endosomes (*Brucella* spp.), late endosomes (*L. pneumophila*), nonacidic lysosomes (*Leishmania* spp.; *H. capsulatum*), or recycling endosomes (HIV). (Figure adapted with permission from Ref. 9.) (See Color Insert.)

polymeric micelles, dendrimers, gold nanoparticles, and cell-based nanoparticle delivery systems.[23–25] Nanoformulations for delivery of the chemotherapeutic drugs doxorubicin and paclitaxel are Food and Drug Administration (FDA) approved. PEGylated liposomal formulations of doxorubicin (Doxil®, Caelyx®) can extend the half-life of the drug dramatically and decrease the cardiotoxicity.[23,25] The albumin-based paclitaxel nanoformulation Abraxane®

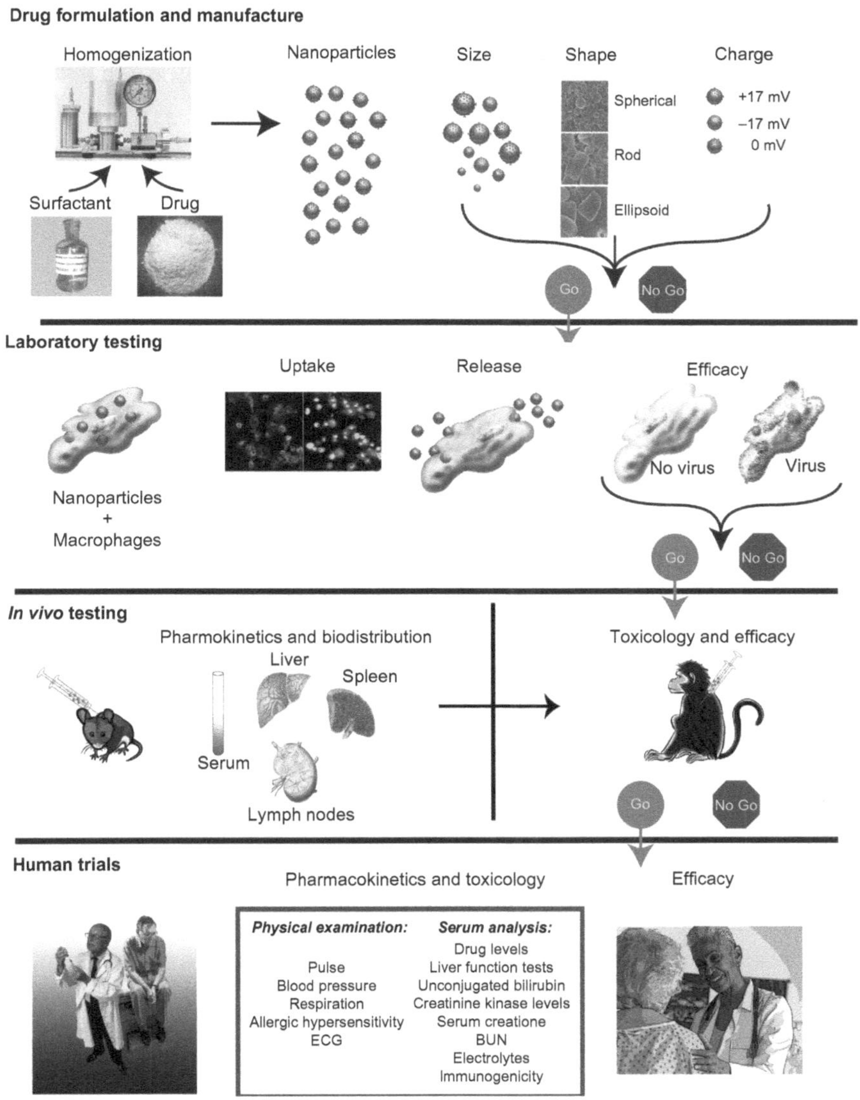

FIG. 2. Bench-to-bedside development of nanoformulated drugs. Nanoformulated drugs hold considerable promise for treatment of human disease and their development for cell-based carriage is illustrated herein. Independent of application, neuroprotective or anti-inflammatory, antitumor, or antimicrobial agents are packaged into particles with surfactant coats that target circulating mononuclear phagocytes (MP; dendritic cells, monocytes, and macrophages). The nanoparticles (NPs) are developed dependent on size, shape, charge, and surfactant coating to facilitate uptake by MP. MP scavenge the particles and traffic them into endocytic compartments and as such act as delivery vehicles. Laboratory tests measure uptake and release of drug-NP. In the case of HIV-1 infection, the drug-NP housed in macrophages would release antiretroviral drugs, leading to inhibition of viral replication. This may be measured by attenuation of cytotoxicity including giant cell formulation, reverse transcriptase activity, and HIV-1 p24 antigen expression. Select formulations are used for drug pharmacokinetics (PK) and toxicology to demonstrate sustained drug levels and extended efficacy. Final therapeutic use in humans will be dependent upon toxicity measures, PK responses, hypersensitivity, immunogenicity, and other untoward side effects. (See Color Insert.)

targets tumor cells by engaging the endothelial gp60 receptor and the albumin-binding protein. Secondary toxicities are reduced compared to Cremophor® or ethanol drug suspensions.[25] Recently, targeted multifunctional anticancer nanoparticles for tumor imaging and delivery were developed to track drug distribution and monitor therapeutic efficacy.[25–28]

B. Infectious Diseases

Nanosystems are being used or developed for treatment of infectious diseases. These include nanoemulsion, niosome (a nonionic surfactant-based liposome), polymeric nanoparticle, dendrimer, liposomal and poly(lactic-*co*-glycolic acid) (PLGA) nanobead delivery systems, RNA/siRNA nanoparticles, and DNA vaccines. In particular, antimicrobials are being placed into polymer-coated crystalline nanoparticles, homogenized particulate suspensions, cholesterol-conjugated amphiphilic peptide self-assembled particles, composite hydrogel/glass particles, liposomes, PLGA, cationic, and pDNA-coated gold nanoparticles, and developed for human treatment for a broad range of microbial infections including tuberculosis.[29–41] Nanoformulated drugs are being designed to specifically deliver therapeutics to sites of infection and in regions of the body that are often difficult to reach using traditional or available treatments.

C. Degenerative Diseases

Use of nanoparticulate systems to treat autoimmune and neurodegenerative disorders include approaches for treatment of rheumatoid arthritis. These include anti-TNF-α-Fab′ with a PEG block, soluble C60 fullerenes for their free radical scavenging abilities, targeted micelles of camptothecin, and *N*-(2-hydroxypropyl)methacrylamide copolymer–dexamethasone conjugate.[42–45] In regard to neurodegenerative diseases such as Alzheimer's disease (AD) and PD, amyotrophic lateral sclerosis (ALS), and Huntington's disease, delivery systems are being manufactured to overcome the blood–brain barrier (BBB) and deliver drugs to the diseased areas of the brain. Metal chelators conjugated to nanoparticles cross the BBB and chelate metal ions in brains of AD patients and are designed to reduce amyloid-beta aggregates.[46–48] Copolymeric *N*-isopropylacrylamide, *N*-*tert*-butylacrylamide nanoparticles, and *N*-acetyl-L-cysteine capped quantum dots also reduce amyloid-beta aggregation.[49,50] Solid lipid nanoparticles and nanostructured lipid carriers have been used to deliver bromocriptine and apomorphine to the brain and improve locomotor skills in animal models of PD.[51,52] A nanoparticle delivery system for the human glial cell line-derived neurotrophic factor gene improves locomotor activity, reduces dopaminergic neuronal loss, and enhances monoamine neurotransmitter levels in a rat model of PD.[53,54]

D. Immunomodulation

Nanoformulated drugs and nanoparticle treatments can improve the efficacy of traditional immunosuppressants and regulate the host immune responses to reduce graft versus host disease. The rapamycin nanocrystalline formulation Rapamune® has greater efficacy over traditional rapamycin treatment as an immunosuppressant in transplant patients.[17,55] Development of nanoformulations of rapamycin for use in arterial stents and for oral delivery has been described and tested in animal models of diseases.[56,57] Other immunomodulatory strategies include targeting nanoparticles to dendritic cells to develop specific immune responses against cancer cells, autoimmune disorders, and infectious diseases.[58,59]

E. Regenerative Medicine and Facilitation of Wound Repair

Recent studies have described the use of nanoparticles for enhancing wound repair and tissue regeneration. Wigglesworth *et al.*[60] reported that liposomes containing glycolipids with Gala1-3Galb1-4GlcNAc-R epitopes applied to skin wounds in mice induced local activation of complement and its chemotactic factors with resultant recruitment of macrophages. This resulted in a two fold increase in healing time. Kwan *et al.* described the use of silver nanoparticles to promote wound healing and improve the tensile properties of skin.[61] Use of nanocomposites in scaffolds to promote bone and cartilage regeneration is under clinical trials.[62,63] Lei *et al.* used caged nanoparticle encapsulation to load DNA/polyethyleneimine (PEI) polyplexes into hyaluronic acid and fibrin hydrogels for gene therapy.[64] Others used nanotubes and nanofibers to promote axonal growth.[65]

F. Ocular Drug Delivery

The protective barriers of the eye make drug delivery difficult without tissue damage. Poor drug absorption and penetration of drugs to intraocular tissues limit the delivery of drugs. Use of nanoparticles and nanosuspensions for drug delivery to the intraocular tissues is being developed.[66] One example is cross-linked polymer nanosuspensions of dexamethasone, which show enhanced anti-inflammatory activity in a model of rabbit eye irritation.[67] Chitosan nanoparticles increase bioavailability of 5-fluorouracil to the aqueous humor in rabbits.[68] In other model systems, solid lipid nanoparticles containing methazolamide yield prolonged therapeutic efficacy in reducing intraocular pressure in glaucoma compared to commercial eye drop preparations.[69] Magnetic hyperthermia was used with superparamagnetic $Mn_{0.5}Zn_{0.5}Fe_2O_4$ nanoparticles to induce heat shock proteins throughout the vitreous body as a neuroprotective strategy for glaucoma.[70] Delivery of macromolecules and plasmid DNA to the nuclei of retinal and corneal cells can be achieved using PEGylated nucleolin-binding peptide nanoparticles.[71]

G. Imaging and Diagnostics

A range of nanoscale systems has been developed for bioimaging. These include liposomes, quantum dots, magnetic nanoparticles, and dendrimers.[72] These nanoparticulate imaging systems are amenable to bioimaging by single-photon emission computed tomography, positron emission tomography, magnetic resonance imaging (MRI), fluorescence microscopy, computed tomography, and ultrasound.

Use of superparamagnetic iron oxide (SPIO) nanoparticles for MRI imaging to distinguish between normal and cancerous liver tissue was successfully developed.[73] SPIO particles are used, for example, for imaging prostate cancer metastases and atherosclerotic plaques.[74,75] Other nanoformulations in development for cancer detection include ferritin nanocages, dendrimers, liposomes, nanoshells, nanotubes, emulsions, and quantum dots.[76–78] By actively targeting these nanoparticles to specific cell surface receptors, the sensitivity and accuracy of imaging are improved in animal models of human disease.[18,79,80] Perfluorocarbon nanoparticle carriers were developed for targeting ligands for a_Vb_3-integrin to visualize aortic neovasculature during the development of atherosclerotic plaques in rabbits.[81] Nanoparticle systems also offer the possibility of multiple imaging modalities. For example, quantum dots in combination with SPIO particles have been used in animals to provide for both optical imaging and MRI.[82] To determine drug delivery to a target site, dendrimers can provide targeted delivery of both drug and optical dyes.[24]

III. Chemical Composition, Structure, Function, and Manufacture

As described above, a range of other nanoparticular systems, which include functionalized fullerenes and carbon nanotubes, liposomes, iron oxide nanoparticles, polymeric micelles, dendrimers, nanoshells, polymeric nanospheres, nanobins, quantum dots, and polymer-coated nanocrystals, among others, are being applied to improve human disease outcomes[72,81,83,84] These systems are discussed in relation to their potential for use in cancer and infectious disease diagnosis and treatment.

A. Functionalized Fullerenes and Carbon Nanotubes

Fullerenes and carbon nanotubes have been well studied since their discovery in 1985. However, their use in nanomedicine has been limited by their lack of water solubility. Functionalization of fullerenes and carbon nanotubes has increased their water solubility and attractiveness for use in biomedical applications as vehicles for nanodrug delivery.[83] To functionalize fullerenes and carbon nanotubes for drug delivery and ROS quenching and for use as MRI contrast agents, functional moieties were developed which include dendrimers, amino acids, peptides, proteins, liposomes, polyamines, polymers, carboxyl groups, and gadolinium.[76,83,85–88] Because of their ability to release substantial vibrational energy upon exposure to near-infrared radiation (NIR), multiwalled carbon nanotubes (MWCNTs) were studied for photothermal cancer ablation.[89] MWCNTs coated with Pluronic F127 when administered to mice bearing RENCA tumors localized to the tumor sites. Treatment of the tumors with an NIR laser resulted in a 10-fold reduction in tumor size and increased survival rate in mice treated with 100-μg MWCNT compared to saline-treated mice. The authors proposed this treatment as a potential method for reducing the amount of NIR radiation needed for ablation of embedded cancers and limiting damage to surrounding dermal tissue.

B. Liposomes

Liposomes are vesicles composed of a lipid bilayer surrounding a hollow core into which drugs or other molecules can be loaded for delivery to tumors or other disease sites.[88,90,91] They can be composed of natural phospholipids or other surfactants. Hydrophilic molecules can be carried in the aqueous interior of the liposome, while hydrophobic molecules can be dissolved in the lipid membrane.[92] Thus, liposomes can carry both hydrophobic and hydrophilic drugs and molecules to a target site. Delivery of the drug is accomplished when the liposome fuses with the lipid membrane of a cell, releasing its contents into the cell cytoplasm. Liposomes can be coated with a functionalized polymer, creating a nanobin, to improve targeted drug delivery.[93] Long-circulating liposomes (or "stealth liposomes") can be obtained by coating with polyethylenglycol chains (PEGylated liposomes).[94] The most important application of liposomal delivery is AmBisome®, an injectable liposomal formulation of amphotericin B that consists of the drug dissolved in the lipid bilayer of unilamellar liposomes composed of soy phosphatidylcholine, cholesterol, and distearoyl phosphatidylglycerol. In clinical trials, AmBisome® showed equal or improved efficacy and fewer side effects than amphotericin B in the treatment of febrile neutropenia, cryptococcal meningitis, and histoplasmosis.[95]

C. Polymeric Micelles

When amphiphiles are placed in water, they form micelles with their hydrophobic tails forming a core surrounded by a hydrophilic shell.[96] Polymeric micelles are made with amphiphilic polymers such as the block copolymers poly(ethyleneglycol)-*b*-poly(e-caprolactone)(PEG-*b*-PCL),[22] poly(styrene) or PLGA.[96] FDA-approved block copolymers of poly(ethylene) oxide-polypropylene oxide are the most commonly used for targeted drug delivery.[86] Hydrophobic drugs can be suspended in the core and protected from the surrounding environment by the hydrophilic shell. The hydrophilic shell serves to disperse the micelles in an aqueous environment and imparts the unique pharmaceutical behavior of the drug-micellar suspension. Polymeric micelles have been extensively studied as injectables to deliver poorly water-soluble drugs such as paclitaxel and amphotericin B.[97–99] Adding ligands to the ends of the polymer chains can target the micelles to specific cell types, such as T cells, macrophages, and dendritic cells.[86] Potential use of polymeric micelles composed of Pluronic block copolymers containing dextran for oral delivery of drugs, such as cyclosporine A, has also been described.[96]

D. Polymeric Nanospheres

Polymeric nanospheres are uniform spherical structures less than a micron in size made from nonbiodegradable or biodegradable polymers. The aqueous polymer is dispersed in an organic phase and then cross-linked to form spherical structures.[100] Drug molecules can be entrapped in the interior of a hollow nanosphere or incorporated into the matrix of a solid nanosphere. Nanospheres made from tyrosine-derived triblock copolymers are able to deliver lipophilic therapeutics[101–103] and, when incorporated into gels, are effective agents for transdermal drug delivery.[104] Their use as targeted imaging agents was described recently by Li *et al.*[105] The fluorescence activity of the polymers poly[9,9-bis(2-(2-(2-methoxyethoxy)ethoxy)ethyl) fluorenyldivinylene] (PFV) and the PFV derivative (PFVBT) containing 10 mol% 2,1,3-benzothiadiazole (BT) when conjugated and functionalized with the peptide arginine–glycine–aspartic acid was utilized to diagnose human epidermal growth factor receptor 2 (HER2)-positive and integrin-positive cancer cells *in vitro*.

E. Dendrimers

Dendrimers are large, complex molecules with a well-defined branched chemical structure. They are monodisperse, highly symmetric, highly branched, and generally spherical.[24] Their composition comprises a series of branched chains around a central core, with surface functional groups on the exterior. Void spaces between the chains allow carriage of drugs or molecules for imaging.[106] The most utilized dendrimers are clusters made of poly(amidoamine) (PAMAM) units[107] and are assembled in layers, or generations. Functionalization of the dendrimers can be accomplished by linking a targeting moiety to the surface structure. By attaching sugar moieties, such as mannose, to the surface of polypropyleneimine dendrimers, the antituberculosis drug rifampicin was delivered directly to macrophages and its hemolytic side effects were reduced.[108]

F. Polymer-Coated Nanocrystals

Coating of crystalline hydrophobic molecules with polymers and surfactants has been accomplished by wet-milling and high-pressure homogenization processes.[34,36,109–111] The polymeric coating of the nanocrystals prevents aggregation and helps in establishing a stable nanosupension. Stability of the nanoformulations is dependent on both the drug and surfactant composition. Polymer-coated nanocrystals of antiretroviral drugs such as indinavir, atazanavir, ritonavir, and efavirenz are being developed by our laboratory for macrophage-based delivery to sites of HIV infection and sequestration.[34–36,109,112] Crystalline drug coated with polymer was readily taken up by macrophages in culture and was effective in reducing HIV infection of the cells. Further, macrophages loaded with nanoformulated indinavir were able to deliver an effective antiviral dose to an area of HIV infection in a mouse model of HIV encephalitis.

G. Nanoshells

Nanoshells are spherical particles consisting of a dielectric core surrounded by a thin metallic shell, most commonly gold.[113,114] Because of their optical and chemical properties, these particles have been used for biomedical imaging and cancer treatment. PEG coating of the gold shell minimized nonspecific uptake by macrophages and improved *in vivo* bioavailability.[115] Other functional groups can be added, such as receptor antibodies, to target the nanoshells to specific cells.[115] Photothermal ablation can then eliminate those cells containing the particles.[116] Nanoshells have also been designed that carry antitumor drugs such as doxorubicin and combretastatin to allow concerted action of the two drugs at tumor sites.[117]

H. SPIO Nanoparticles

Iron oxide particles in the range of 1–100 nm possess superparamagnetic properties that make them attractive for biomedical imaging, diagnosis, and therapeutics in addition to their long-standing use in separation technologies.[118,119] SPIO particles consist of a core of magnetite or maghemite with a coating of polysaccharides, polymers, or monomers. Functional groups can be attached to the surface coating to achieve targeted delivery of the particles for imaging specific cell and tissue sites.[120–123] Beduneau *et al*. covalently linked IgG and Fab′2 fragments to SPIO particles to facilitate uptake by macrophages *in vitro*. Intravenous administration of the IgG-linked SPIO to mice demonstrated sustained distribution to lymphoid tissue over 24 h.[124]

I. Quantum Dots

Quantum dots are semiconductors with spatially confined excitons that afford them unique optical and electrical properties.[125] Their distinct fluorescence spectra make them valuable tools for biomedical imaging. Gao *et al.*[126] demonstrated the utility of quantum dots for simultaneous visualization of several target sites by injecting quantum dots with differing emission wavelengths into three different subcutaneous sites in a mouse. Using a single excitation wavelength, a different color emission for each injection site was observed. Bioactive targeting molecules can be conjugated to the surface of the quantum dots to image specific cells and structures, such as tumor cells, cell membrane receptors, and cellular organelles.[122,127–129] Because of the stability of their fluorescence emission, quantum dots are increasingly being studied to label intracellular compartments. Quantum dots were used to label endosomal compartments,[130,131] F-actin filaments,[132,133] mortalin, and P-glycoprotein.[134,135] The use of quantum dots for imaging in human disease, however, is limited by their potential heavy metal toxicity.[7,136] Methods to reduce their toxicity include PEGylation and encapsulation by micelles.[113]

IV. Targeted Drug Delivery

A major focus of nanomedicine research has been to target nanoparticle drug delivery to specific sites. Targeted delivery could simultaneously increase the efficacy of the therapeutic treatment and decrease unwanted side effects.[72] One hurdle in the use of nanoparticulate drug delivery has been the uptake of nanoparticles by the cells of the reticuloendothelial system (RES). To minimize removal of nanoparticles by the cells of the RES, coating the particles with PEG is beneficial.[14,72] While this may prevent nonspecific uptake by phagocytic cells, more direct targeting strategies are used to focus the nanoparticles to a specific site of cellular action. This is being accomplished by the addition of targeting ligands to the surface of nanoparticles, enabling their recognition by specific receptors on the surface of cancer cells or other target cells.[72,137] Ligand-functionalized nanocarriers allow large quantities of drug to be delivered to a cell upon interaction of the ligand with its receptor. By attaching the ligand to the nanocarrier rather than to the drug, the therapeutic activity of the drug is maintained. Targeted nanoparticle delivery also can provide a more efficient delivery into tumor sites. Specifically, targeting of drug nanoparticles to folate receptors overexpressed on cancer cells, human growth factor receptors, and integrins involved in angiogenesis and atherosclerosis have been moved from *in vitro* testing to *in vivo* testing.[76,138–142]

The addition of functional groups on nanoparticles can also serve to target the particle to specific intracellular regions in order to enhance drug function and reduce drug toxicity. Thus, by changing the surface charge of cerium nanoparticles, they could be localized to the cytoplasm or the lysosomes of cancer cells with a corresponding change in cytotoxicity.[143] Attachment of a nucleolin-binding peptide to PEGylated nanoparticles effectively delivered a green fluorescent protein to the nucleus of retinal and corneal cells.[71]

V. Nanomedicine and Vaccines

Nanoparticles are used as adjuvants for vaccines, especially peptide and DNA vaccines.[144] Because of their ability to release their entrapped cargo over extended time frames, they have shown potential as vaccine delivery vehicles. Coatings on the surface of nanoparticles or specific polymer composition can increase uptake across mucosal layers and provide oral and intranasal delivery of vaccines.[145–147] A recent study reported the successful intranasal delivery of an antitumor vaccine using amphiphilic poly(g-glutamic acid) nanoparticles.[148] Biodegradable PLGA particles can elicit a Th1 humoral response[144,149] and downregulate Th2 responses in a mouse model of Type 1 allergy.[144,150] Lipid A liposomes are used as adjuvants for malarial vaccines.[151] Liposomal vaccines can simultaneously activate the major histocompatibility complex class I and II pathways and induce antibody and cellular immune responses.[151] Specific targeting of nanoparticulate vaccines to dendritic cells is being explored as a way to enhance the immunogenicity of respiratory virus vaccines.[152] SPIO nanoparticles can facilitate delivery of a malaria DNA vaccine into eukaryotic cells.[153] Application of an external magnetic field served to enhance transfection efficiency.

Nonbiodegradable nanoparticles (latex, gold, silica, polystyrene) are also used as adjuvants. These particles remain at the site of injection for extended time periods, with potential for enhanced immunogenicity.[144] For example, gold nanoparticles were used in Phase I studies for the delivery of hepatitis B and malaria DNA vaccines.[154]

VI. Nanodevices and Cell Reprogramming

Cell-based therapies using adult, embryonic, or pluripotent stem cells as drug carriers are under development.[155,156] To induce cell differentiation for cancer therapeutics and tissue regeneration, adjuvant drugs and growth factors are coadministered.[155–158] The growth factors, in particular, need to be maintained at high systemic levels for clinical benefit.[157,158] Targeted induction of stem cell growth and differentiation for specific tissue engineering is possible by using scaffolds for drug and growth factor delivery.[159–163] Human mesenchymal stem cells (MSCs) were induced to differentiate by embedding them in PLGA nanosphere-encapsulated PLGA microspheres containing different types of growth factors and transplanting the microspheres into nude mice.[164] This system can deliver MSCs to any desired tissue site and facilitate their differentiation.

Genetic engineering of donor cells to produce their own support can be accomplished by nonviral vector loading of growth factors by nanoparticle delivery.[165] As proof of concept, PAMAM dendrimers functionalized with peptides that exhibit high affinity for MSCs were used for gene transfection of enhanced green fluorescent protein with no cytotoxicity.[165] Nanoparticles containing adjuvant drugs conjugated to the surface of hematopoietic stem cells resulted in increased *in vivo* repopulation and used lower doses of adjuvant than with systemic administration.[166] In a different vein, Zhou *et al.* used synthetic low-density lipoprotein nanoparticles for targeted delivery of the tyrosine kinase inhibitor imatinib to chronic myeloid leukemia (CML) stem cells as a potential treatment for eradication of previously resistant CML cells.[167]

Loading stem cells with magnetic nanoparticles has provided a means to track the migration of cells to sites of disease.[168,169] In addition, they can be used as magnetic nanoparticle-based vector systems for transfection of therapeutic biomolecules into stem cells while simultaneously allowing the *in vivo* migration of the cells to be tracked.[168,169] The carriage of the magnetic nanoparticles did not alter the viability and differentiation potential of Schwann, olfactory ensheathing, oligodendrocyte progenitor, and human neural stem cells (NSCs).[170–172]

VII. Nanomedicine and Stem Cells

Stem cells have been proposed as drug delivery vehicles for chemotherapeutic agents and gene therapy. NSCs have been studied for delivery of neurotrophic factors to the CNS.[173–175] In response to disease and injury in cases of AD, PD, cancer, stroke, and multiple sclerosis, NSCs readily migrate to sites of tissue damage. Thus they could be used to deliver neurotrophic factors to diseased and damaged areas in the CNS to promote neuron integrity and regeneration.[11] MSCs offer advantages for delivery of therapeutic agents in regenerative medicine and cancer treatment. They are relatively easy to isolate, can differentiate into a wide variety of functional cell types, can be expanded extensively in culture, are not immunogenic, possess immunosuppressant and anti-inflammatory properties upon transplantation, and can migrate to damaged tissues, tumors, and areas of metastases. MSCs have been engineered using nanoparticle delivery vectors to produce antitumor proteins such as TRAIL and were shown to successfully deliver the protein to an intracranial glioma in a mouse xenograft model.[176–181] In other tumor models, MSCs modified to produce interferon beta were introduced with a resultant reduction in tumor growth.[177,180,181] The potential of using MSCs as vehicles for the delivery of drug-loaded nanoparticles was shown using polylactic acid nanoparticles and lipid nanocapsules loaded with coumarin-6. Marrow-isolated adult multilineage inducible cells loaded with these nanoparticles retained their migratory and differentiation capabilities and migrated to the tumor site in a mouse model of glioma where they were detectable for at least 7 days.[182] Using another delivery strategy, NeutrAvidin-coated nanoparticles were attached to MSCs containing a biotinylated plasma membrane and remained attached to the surface for up to 2 days.[183] These nanoparticulate cellular patches may provide a novel means of delivering nanotherapeutics to tumors using stem cells.

VIII. Potential of Nanotherapeutics

Nanoparticles can improve pharmacokinetics and biodistribution profiles that lead to increased efficacy and reduced undesirable side effects.[21] This may be achieved through increased intestinal uptake, reduced liver metabolism, increased drug half-life, active targeting to a site of disease, or improved accessibility to sites of disease. The utility of nanoparticles in improving pharmacokinetics, reducing unwanted side effects, and improving delivery to disease sites has been demonstrated for a number of nanodrug delivery systems.[21]

A. Improved Pharmacokinetics

Reformulation of drugs into nanoformulations has been done to increase plasma half-life and enhance the oral bioavailability of several drugs, including those used to treat infectious diseases. As an example, amphotericin B, a poorly water-soluble drug, is used to treat fungal infections and leishmaniasis. Its oral absorption, however, is poor because of its insolubility, instability at acidic pH, molecular size, and P-glycoprotein export activity in intestinal epithelial cells. Several investigators have described the development of nanoparticlute forms of the drug that have increased oral bioavailability and decreased toxicity in animal models.[30,184]

B. Reduced Toxicity

Nanoformulations of drugs can serve to decrease toxic side effects that limit their therapeutic efficacy. In particular, reduction in toxic side effects is an important driver for the clinical use of nanoformulations of anticancer drugs and antifungal drugs. As an example, the anticancer drug doxorubicin can elicit undesirable side effects, which include cardiotoxicity and myelosuppression.[185] Cardiotoxicity of doxorubicin is reduced with use of the nanoformulated version Doxil®.[19] The liposomal formulation of amphotericin B, AmBisome®, exhibits not only improved efficacy but also decreased toxicity compared to unformulated drug.[186]

C. Targeted Tissue Delivery

The development of targeted drug delivery systems that elicit fewer toxic side effects and improved pharmacokinetics is an area of intense focus. Much of the focus on development of targeted nanoparticles has been for delivery of antineoplastic drugs; however, targeted delivery for other diseases is also being explored. Several studies have described the development of targeted nanoparticle delivery of vaccines to dendritic cells.[41,59,187–189]

To reduce uptake of nanoparticles by the cells of the RES, "stealth" carriers have been developed. These carrier systems contain PEG polymers on their outer surface. PEGylated nanoparticles are not easily recognized by the phagocytic cells of the RES system and thus can avoid accumulation in the RES organs such as the liver, spleen, and lymph nodes and deliver their cargo to the intended site of action more efficiently.[14]

As an extension of targeted therapy, the use of cells to carry nanoparticles across biologic barriers such as the BBB has been proposed. MSCs loaded with coumarin-6-polylactic acid nanoparticles and coumarin-6-LNCs migrate to a mouse tumor site and there release the nanoparticles and differentiate.[182] Similarly, several nanoparticle systems are being developed to take advantage of the phagocytic capability of macrophages and their ability to cross biological barriers such as the BBB. As an example, macrophages have been shown to carry "nanozymes" to sites of injury in an animal model of PD.[12] In addition, macrophages loaded with nanoparticles containing antiretroviral drugs have been shown to migrate to the site of HIV infection in the CNS in an animal model of HIV encephalitis and to reduce the rate of viral infection.[112] By targeting macrophages for drug uptake *in vivo*, delivery of drugs to areas of disease or viral infection that are protected from the action of free drug in circulation can be achieved.

D. Subcellular Localization

How nanoparticles are taken up by cells and their intracellular trafficking is determined in large part by their physicochemical properties. Studies using polystyrene beads demonstrated different uptake processes for fluorescent beads of 78 nm and carboxylated spheres of 50 nm. Uptake of the larger fluorescent beads was by a passive, nonphagocytic mechanism,[190] while the smaller carboxylated spheres were opsonized and recognized by scavenger receptors on macrophages.[191] In other studies, opsonization of particles decreased cellular uptake,[192] suggesting that the physicochemical properties of the nanoparticles and cell-specific characteristics influence how nanoparticles are taken up by cells. Endocytosis is a major process by which nanoparticles are taken up by cells and has been demonstrated to occur for gold nanoparticles, iron oxide particles, quantum dots, and carbon nanotubes.[130,131,193–195] Once the particles are internalized, intracellular trafficking is dependent on the physicochemical properties of the particles, such as surface charge, polymer coat, and surface modifications.[85] Konan and coworkers observed that meso-tetra-(4-hydroxyphenyl)porphyrin (*p*-THPP)-PLGA nanoparticles were accumulated intracellularly 1.3-fold better than *p*-THPP-poly (D,L-lactide) (PLA) nanoparticles,[196] suggesting that the presence of glycolic acid, imparting a hydrophilic character to the nanoparticle, enhanced cellular sequestration. Cerium nanoparticles were taken up by a clathrin-dependent mechanism and trafficked throughout the cell, supporting the use of these nanoparticles as intracellular scavengers of reactive oxygen and reactive nitrogen species.[197] Most nanoparticles taken up by endocytosis through either clathrin- or caveolae-dependent mechanisms remain in the endosomes; however, use of specific targeting ligands on the surface of gold nanoparticles was successful in bypassing endosomal uptake and targeting the nanoparticles to specific cellular compartments such as the nucleus.[8] By varying alkyl spacer lengths at the termini of dendrimers, Huang *et al.* could direct the targeting of the dendrimers to either the nucleus or cytosol,[198] suggesting that nanoparticle delivery could be targeted to specific intracellular compartments in addition to specific cell types. In additional studies, Bale *et al.*[199] used hydrophobic silica nanoparticles to deliver active proteins intracellularly to a variety of cell types without extended entrapment in the endocytic compartment, which maintained the protein's biologic activity.

IX. Nanotoxicology: Immunogenicity, Cytotoxicity, and Generation of ROS

Nanotoxicology was coined as a term in 2004–2005[137,200,201] although concerns about the adverse affects of nanomaterials on human and environmental health were voiced several years earlier.[202] The field came about to study the adverse effects of engineered nanomaterials on living organisms and the ecosystems, with the goal to prevent and eliminate the adverse responses and evaluate the risk/benefit ratio of using nanoparticles in medical settings. Much about the interaction of small particles with biological systems is gleaned from studies of ultrafine particles; however, the uniqueness of the physicochemical properties of nanoparticles may result in unpredictable biological interactions. The small size of nanoparticles puts them in the range of viruses, DNA, and proteins and thus their subcellular interactions can reflect the interactions of these molecules. Gold nanocluster compounds 1.4 nm in size were observed to induce cell death by intercalation into the major groove of DNA, while larger gold nanoparticles were nontoxic.[203] The subcellular localization of quantum dots appears to depend on size, as quantum dots in the range of 2.1 nm localized to the nucleus while larger quantum dots (3.4 nm) remained in the cytoplasm.[204,205] Single-walled carbon nanotubes have been shown to interfere with the microtubular system of cells and thus interfere with cell division.[206,207] The shape of nanoparticles in combination with their small size may play a role in how they are perceived by cells. Cationic PEG nanoparticles that are rod-shaped are better internalized than particles of other shapes, suggesting that their resemblance to rod-shaped bacteria may enhance their recognition by cells of the immune system.[208] The toxicity of nanoparticles is not only dependent on their size and shape, but is also correlated with the relative surface area of administered nanoparticles. Clearance of titanium dioxide nanoparticles administered by inhalation to rats was proportional to the surface area of the particles rather than to the mass of the particles.[209] Also, Monteiller and coworkers found that the proinflammatory activity of low-solubility, low-toxicity nanoparticles was proportional to surface area rather than mass.[210] Another component of nanoparticles that can contribute to the overall biologic response is the actual surface coating of the particles. Endotoxin adherence to nanoparticle surfaces,[211] surfactant composition of the nanoparticles,[212] and transition-metal contamination during synthesis[213] can induce adverse effects of administered nanoparticles. Furthermore, aggregation of the nanoparticles can mean that the toxic response to the nanoparticles is dependent on the size of aggregates rather than size of nanoparticles.[214]

Cell-based delivery can positively affect the biodistribution of the nanoparticles and their efficacy and toxicity profiles.[137] The immunogenicity of certain types of nanoparticles is being taken advantage of in the development of adjuvants for vaccines.[215] In contrast, potential deleterious effects of nanoparticles on phagocytic cells can interfere with innate and adaptive immune responses. Nanoparticles such as fullerenes and quantum dots have been observed to induce autophagy in phagocytic cells and could lead to insufficient innate and adaptive immune responses.[216–218]

Nanoparticles have been proposed as inducers of oxidative stress.[219–224] Two potential mechanisms for eliciting excessive cellular oxidative stress have been presented, ROS generation by metal contaminants in nanoparticles, and nanoparticles as sources of oxidizing equivalents. Metals are used during the manufacture of carbon nanotubes and are present as contaminants in the final product. Metal ions such as Fe^{2+} can participate in the one-electron oxidation of oxygen to produce ROS such as superoxide radicals, hydrogen peroxide, and hydroxyl radicals.[225] In animals treated with carbon nanotubes, oxidation of cellular proteins, DNA, and lipids has been observed[226,227] and could be exacerbated by maintaining animals on a vitamin E deficient diet.[228] Suh and colleagues[84] proposed several events in which nanoparticles could produce cell toxicity and ROS both outside and inside the cell. (1) Nanoparticles smaller than a cell produce ROS externally and destabilize the cell membrane. (2) Internalized nanoparticles create ROS internally. (3) Particle dissolution after internalization results in ROS production. (4) Mechanical damage of organelles such as lysozomes, endoplasmic reticulum, and the nucleus results in increased ROS production. (5) Different functional groups and surface chemistry on different types of nanoparticles can affect the interaction between the nanoparticles and their surroundings. (6) The size of the nanoparticle can determine the toxic event, with larger particles damaging the cell membrane and smaller internalized particles affecting cells intracellularly. (7) Nanoparticle shape may determine cell surface effects and internalization. (8) Dissolution products of the nanoparticles externally can affect their interaction with the cell. These interactions are very complex, and applying systems biology to understand them has been proposed.[84]

X. Unique Challenges for Translational Nanomedicine

While the potential for therapeutic benefits from nanomedicines is great, the translation of basic research into clinically used drugs can be complicated.[217] Specifically, how should the safety and efficacy of nanomaterials be defined? Regulatory guidelines currently in existence are generally suitable for nanomedicines; however, there can be confusion over guidelines related to animal study design, therapeutic efficacy and potency, dosimetrics across species, and drug delivery to the target site.[217] Many of the studies needed to develop nanoparticle formulations for drug delivery are similar to those for small molecules.[229,230] However, in some instances, the nanoparticles may interfere with common tests, and other means of evaluation of suitability will have to be considered. For example, dendrimers exhibit a high false-positive rate in a common endotoxin test.[229,230] Thus some tests designed for small molecules may not be suitable for evaluating nanoparticulates, and alternatives will need to be developed.

A. Assessing Efficacy

Different nanomaterials have different physicochemical properties that can impact their activity in biologic systems. The size of the nanoparticles can affect biodistribution and potential toxicity.[229,230] For example, smaller gold nanoparticles

(10 nm) exhibit a wider biodistribution profile than larger (50 and 250 nm) particles following intravenous injection,[231,232] with the smaller particles being detected in liver, spleen, kidney, testis, thymus, heart, lung, and brain and the larger particles being present in only spleen and liver. Others have shown that smaller polypropylene nanoparticles stabilized with Pluronic (poloxamer 188) could enter the lymphatic system more readily than larger particles following intradermal injection.[233] Pharmacokinetics and distribution can also be influenced by aggregation of the particles and their association with endogenous proteins.[229,230] The pharmacokinetics and biodistribution of dendrimers is also greatly influenced by parameters such as molecular weight and polymer chain length, with smaller sizes exhibiting more rapid renal clearance than larger sizes.[234]

For each nanoformulation, characteristics such as surface charge, shape, stability, particle density, and solubility, in addition to the unique properties of a particular type of formulation, may differ and may influence the pharmacokinetics, biodistribution, and efficacy of the nanoparticles. The oil/water partitioning coefficient affects the biodistribution of drugs in an emulsion, while size, charge, surface properties, and lipid type affect the biodistribution of liposomal formulations.[87,229] If specific mechanisms are required to release the active drug from its carrier, the bioavailability and biodistribution of the drug will be affected by the activity of the release mechanisms.[229,230]

The choice of animal species for preclinical testing is critical. Animal models of human disease are used to predict human efficacy of potential therapeutics and can be used for preclinical testing of nanoparticle drug delivery and efficacy. For certain degenerative and infectious diseases, appropriate animal models may be available, for example, the 1-methyl-4-phenyl-1,2,3,6-tetrahydropyridine (MPTP) mouse model for PD[12,235] and humanized mouse models for HIV infection and neurological disorders.[236,237] Testing of nanoformulations of enzymes and antiretroviral drugs in these animal systems will provide evidence for estimating clinical efficacy in the human diseases.

B. Unique Toxicological Issues

Because of the unique properties of nanomaterials, unexpected, adverse responses may arise that would not be predicted from what is known about the compounds on a larger scale.[238] In addition, the use of standard toxicological screening assays may not be appropriate for all nanoparticulate systems. A number of articles have reviewed the need for specific *in vitro* assays to test specific nanoparticle systems for determining oxidative stress and ROS production, proinflammatory activity, and genotoxicity.[223,239,240] Traditional and proteomics-based assays can be used to assess target cell function in response to nanoparticle treatment.[223,241–243] Nanoparticulate drug systems are generally composed of both carrier and drug. Thus, the pharmacokinetics and toxicological properties of both the carrier and drug need to be assessed.[229,230,244] Further, nanoparticles tend to accumulate in cells of the phagocytic lineage,[244] and thus may cause unwanted side effects in tissues with high numbers of monocytes and macrophages (liver, spleen, bone marrow, lymphatics).

XI. Development of Cell-Based Nanomedicine: A Perspective

A. Cell-Specific Targeting

As a means of targeting drug delivery, efforts to develop cell-based carriers for nanomedicines are underway. Use of immunocytes, MP (monocytes, macrophages, dendritic cells), lymphocytes, neutrophils, and stem cells has been proposed. The ability to target drugs to specific sites of disease, prolong drug half-life by sequestering drug away from hepatic metabolism, release drug slowly, and decrease drug toxicity are all advantages provided by cell-based drug carriage. In addition, immunocytes and stem cells can migrate to sites of injury and tumors and act as Trojan horses to deliver drug across biological barriers such as the BBB. Despite their potential, there are limitations that need to be overcome for cell-based nanomedicine delivery to be a viable clinical treatment paradigm. Drug loading into the cells can be low, thus sufficient numbers of cells must migrate to the site of release in order to deliver a therapeutically effective drug dose. Once inside the cells, the drug must be released from the carrier into the extracellular space to be effective. A slow, steady release of drug from the carrier cell is required rather than a quick, bolus release. The ability of the cells to migrate and function should not be compromised by the presence of the nanomaterial and drug.

B. Particle Uptake

Nanocarriers for drugs are commonly composed of an outer polymer shell and inner core for drug carriage. The outer core imparts stability to the nanosuspension, determines particle circulation time, and defines the interaction of the nanoparticle with the surrounding environment and cell surfaces. Charged carriers are generally taken up rapidly by mononuclear phagocytes, immunocytes, and stem cells through interaction with plasma membrane receptors.[245–251] These receptors include mannose receptor, complement and Fc receptors that recognize mannans and integrins, and complement and antibody opsonized particles. Positively charged particles are taken up by macrophages somewhat better than negatively charged particles,[36,167] perhaps through interaction with the negatively charged plasma membrane. To improve cell loading of nanocarriers with neutral and hydrophobic shells, such as PEG, attachment of targeting vectors has proven to be beneficial.[245,252–254]

C. Subcellular Localization, Drug Stability, and Drug Release

Once inside cell carriers, the stability, retention, and controlled release of the drug can depend on the intracellular trafficking of the nanocarrier. By avoiding entry of drug-loaded nanocarriers into lysosomes, disintegration of the drug is reduced.[9,110] Nanoparticles of ritonavir prepared by high-pressure homogenization and coated with poloxamer 188, *N*-(carbonyl-methoxypolyethyleneglycol 2000)-1,2 *sn*-glycero-3-phosphoethanolamine (mPEG-DSPE), and (1-oleoyl-2-[6-[(7-nitro-2-1,3-benzoxadiazol-4-yl) amino]hexanoyl]-3-trimethylammonium propane were taken up by human monocyte-derived macrophages (MDM) via a clathrin-mediated mechanism. They were trafficked to recycling endosomes where they remained intact and were released through Rab 11- and Rab14-dependent mechanisms. The nanoparticles that were released retained their ability to prevent HIV-1 infection of MDM cultures. It is noteworthy that these particles were positively charged, and that divergent fates for positively and negatively charged particles inside macrophages have been described.[254] In another example, positively charged block copolymers (PEI-PEG and polylysine (PL)-PEG) provided protection for "nanozymes" from lysosomal degradation following uptake by MDM, whereas a negatively charged block copolymer (poly-L-glutamic acid (PGLU)-PEG) did not.[167]

The release of uploaded nanoencapsulated drugs from the carrier cells at the target site is also an area of active study. A controlled release of the drug from the cell carrier is desired in order to provide a sustained dose and duration of exposure at the diseased site. Cell residence time and extracellular conditions at the site of disease can provide release of the drug from the cell carrier. Macrophages are well known to release compounds from intracellular vesicles in response to disease stimuli, and this mechanism could be applied to release of nanocarriers of drugs.[13] Other stimuli for nanocarrier release could include an increase in intracellular Ca^{++}[255] and mild hyperthermia in the case of anticancer treatment.[256]

The structure and composition of the nanocarrier itself will also affect the intracellular stability and toxicity and cellular release of the drug nanoparticle. Nanozymes containing catalase were taken up and released best by macrophages when the negatively charged block copolymer PGLU-PEG was used and the nanozymes were not cytotoxic; however, the enzyme was readily degraded in the lysosomal compartment of the cells.[167] When a positively charged block copolymer was used, cytotoxicity of the nanozymes was increased and cell loading and release were lower; however, enzymatic activity of the enzyme was retained.

D. Cells as Trojan Horses for Drug Delivery

For cells to be effective carriers of nanomedicines, they have to effectively target to the site of disease. For CNS disorders, they also need to be able to penetrate the BBB to elicit their therapeutic effect. Peripheral diseases, and, in particular, many neurological disorders, have an inflammatory component that can actively recruit macrophages. Through the processes of diapedesis, chemotaxis, margination, and extravasation, macrophages can cross the BBB. These cells, loaded with drug nanoparticles, can then deliver the drug to the site of disease in the CNS. In an experimental mouse model of neuroAIDS, mouse bone marrow macrophages (BMMs) loaded with indinavir nanoparticles and injected intravenously into SCID mice migrated to a site of induced HIV encephalitis in the brain and reduced the rate of infection compared to untreated animals.[112] Indinavir levels at the site of infection were elevated compared to levels in noninfected brain tissue 14 days after administration of indinavir nanoparticle-loaded BMMs. As another example, significant amounts of catalase (2.1% of the injected dose) were detected in the brain after injection of catalase nanozyme-loaded mouse BMM into mice treated with MPTP, an experimental model of PD.[13] Macrophages loaded with nanoparticles were also observed to migrate to areas of myocardial infarction, spinal cord injury, cerebral ischemia, and cancer.[257–262] Other cell types such as NSCs also have the potential to migrate to areas of disease in the CNS and thus could serve as potential drug nanoparticle delivery vehicles.[173–175]

While these methods require preloading of the cells with drug nanoparticles and then administration to the diseased host, an alternative approach is being explored for delivery of antiretroviral drug nanoparticles. Polymeric nanoparticles of the antiretroviral drugs indinavir, ritonavir, atazanavir, and efavirenz have been made by wet-milling and high-pressure homogenization.[34–36] These particles are rapidly taken up by MP in culture and are retained by the cells for up to 15 days. Antiretroviral efficacy in HIV-infected MP is observed through 15 days postinfection. Preliminary *in vivo* studies in mice demonstrated atazanavir levels in serum exceeding 100 ng/ml and in liver and spleen exceeding 500 ng/g up to 14 days after subcutaneous administration of poloxamer 188-coated atazanavir nanocrystals (unpublished observations). This proof of concept suggests that, when administered by parenteral injection, polymer-coated antiretroviral nanoparticles can be taken up by macrophages and delivered to reservoir sites of viral infection including the CNS.

E. Future Perspectives

Cell delivery of drug nanoparticles has been demonstrated for animal models of disease. However, these methods have not yet been tested in a clinical setting. For clinical use, cell carriers could be harvested from the patient's blood by apharesis, loaded with drug, and then readministered to the patient. Stem cells could be harvested from bone marrow, propagated, loaded with drug, and then adoptively transferred to the patient. In another setting, drug nanoparticles that were targeted for selective uptake by macrophages could be administered to patients. The patient's own phagocytic cells would take up the administered nanoparticles and deliver them to sites of injury and disease. By targeting the drug nanoparticles to specific cell carriers, drug uptake and delivery could be greatly improved.

Acknowledgments

This work was supported in part by the Carol Swarts MD Neuroscience Research Laboratory Fund, the Frances and Louie Blumkin Foundation, the Community Neuroscience Pride Research Initiative, the Alan Baer Charitable Trust, and the National Institutes of Health Grants P20 DA026146, R01 NS36126, P01 NS31492, 2R01 NS034239, P20 RR15635, P01 MH64570, and P01 NS43985 (to HEG) and 1RO1 NS057748 (to EVB). The authors declare no competing financial interests.

References

1. Ernst JD. Bacterial inhibition of phagocytosis. *Cell Microbiol* 2000;**2**:379–86.
2. Rosenberger CM, Finlay BB. Phagocyte sabotage: disruption of macrophage signalling by bacterial pathogens. *Nat Rev Mol Cell Biol* 2003;**4**:385–96.
3. Shin JS, Abraham SN. Co-option of endocytic functions of cellular caveolae by pathogens. *Immunology* 2001;**102**:2–7.
4. Skendros P, Pappas G, Boura P. Cell-mediated immunity in human brucellosis. *Microbes Infect* 2011;**13**:134–42.
5. Breunig M, Bauer S, Goepferich A. Polymers and nanoparticles: intelligent tools for intracellular targeting? *Eur J Pharm Biopharm* 2008;**68**:112–28.
6. Chen K, Conti PS. Target-specific delivery of peptide-based probes for PET imaging. *Adv Drug Deliv Rev* 2010;**62**:1005–22.
7. Hild WA, Breunig M, Goepferich A. Quantum dots—nano-sized probes for the exploration of cellular and intracellular targeting. *Eur J Pharm Biopharm* 2008;**68**:153–68.
8. Nativo P, Prior IA, Brust M. Uptake and intracellular fate of surface-modified gold nanoparticles. *ACS Nano* 2008;**2**:1639–44.
9. Kadiu I, Nowacek A, McMillan J, Gendelman HE. Macrophage endocytic trafficking of antiretroviral nanoparticles. *Nanomedicine* 2011;**6**:975–94.
10. Zhao Y, Haney MJ, Klyachko NL, Li S, Booth SL, Higginbotham SM, et al. Polyelectrolyte complex optimization for macrophage delivery of redox enzyme nanoparticles. *Nanomedicine* 2011;**6**:25–42.

11. Batrakova EV, Gendelman HE, Kabanov AV. Cell-mediated drug delivery. *Expert Opin Drug Deliv* 2011;**8**:415–33.
12. Brynskikh AM, Zhao Y, Mosley RL, Li S, Boska MD, Klyachko NL, et al. Macrophage delivery of therapeutic nanozymes in a murine model of Parkinson's disease. *Nanomedicine* 2010;**5**:379–96.
13. Batrakova EV, Li S, Reynolds AD, Mosley RL, Bronich TK, Kabanov AV, et al. A macrophage-nanozyme delivery system for Parkinson's disease. *Bioconjug Chem* 2007;**18**:1498–506.
14. Wagner V, Dullaart A, Bock AK, Zweck A. The emerging nanomedicine landscape. *Nat Biotechnol* 2006;**24**:1211–7.
15. Gradishar WJ. Albumin-bound paclitaxel: a next-generation taxane. *Expert Opin Pharmacother* 2006;**7**:1041–53.
16. Cattel L, Ceruti M, Dosio F. From conventional to stealth liposomes: a new Frontier in cancer chemotherapy. *J Chemother* 2004;**16**(Suppl. 4):94–7.
17. MacDonald A. Improving tolerability of immunosuppressive regimens. *Transplantation* 2001;**72**:S105–12.
18. Mendelsohn J. A national cancer clinical trials system for targeted therapies. *Sci Transl Med* 2011;**3**: 75cm78.
19. Allen TM, Cullis PR. Drug delivery systems: entering the mainstream. *Science* 2004;**303**:1818–22.
20. Emerich DF, Thanos CG. The pinpoint promise of nanoparticle-based drug delivery and molecular diagnosis. *Biomol Eng* 2006;**23**:171–84.
21. Igarashi E. Factors affecting toxicity and efficacy of polymeric nanomedicines. *Toxicol Appl Pharmacol* 2008;**229**:121–34.
22. Yanez JA, Forrest ML, Ohgami Y, Kwon GS, Davies NM. Pharmacometrics and delivery of novel nanoformulated PEG-b-poly(epsilon-caprolactone) micelles of rapamycin. *Cancer Chemother Pharmacol* 2008;**61**:133–44.
23. Misra R, Acharya S, Sahoo SK. Cancer nanotechnology: application of nanotechnology in cancer therapy. *Drug Discov Today* 2010;**15**:842–50.
24. Sampathkumar SG, Yarema KJ. Targeting cancer cells with dendrimers. *Chem Biol* 2005;**12**:5–6.
25. Seigneuric R, Markey L, Nuyten DS, Dubernet C, Evelo CT, Finot E, et al. From nanotechnology to nanomedicine: applications to cancer research. *Curr Mol Med* 2010;**10**:640–52.
26. Bothun GD, Lelis A, Chen Y, Scully K, Stoner MA. *Multicomponent folate-targeted magnetoliposomes: design, characterization, and cellular uptake. Nanomedicine* 2011 [Epub ahead of print].
27. Deshayes S, Maurizot V, Clochard MC, Baudin C, Berthelot T, Esnouf S, et al. "Click" conjugation of peptide on the surface of polymeric nanoparticles for targeting tumor angiogenesis. *Pharm Res* 2011;**28**:1631–42.
28. Taratula O, Garbuzenko O, Savla R, Wang YA, He H, Minko T. Multifunctional nanomedicine platform for cancer specific delivery of siRNA by superparamagnetic iron oxide nanoparticles-dendrimer complexes. *Curr Drug Deliv* 2011;**8**:59–69.
29. Griffiths G, Nystrom B, Sable SB, Khuller GK. Nanobead-based interventions for the treatment and prevention of tuberculosis. *Nat Rev Microbiol* 2010;**8**:827–34.
30. Italia JL, Yahya MM, Singh D, Ravi Kumar MN. Biodegradable nanoparticles improve oral bioavailability of amphotericin B and show reduced nephrotoxicity compared to intravenous Fungizone. *Pharm Res* 2009;**26**:1324–31.
31. Kim SS, Peer D, Kumar P, Subramanya S, Wu H, Asthana D, et al. RNAi-mediated CCR5 silencing by LFA-1-targeted nanoparticles prevents HIV infection in BLT mice. *Mol Ther* 2010;**18**:370–6.

32. Manandhar KD, Yadav TP, Prajapati VK, Kumar S, Rai M, Dube A, et al. Antileishmanial activity of nano-amphotericin B deoxycholate. *J Antimicrob Chemother* 2008;**62**:376–80.
33. Mihu MR, Sandkovsky U, Han G, Friedman JM, Nosanchuk JD, Martinez LR. The use of nitric oxide releasing nanoparticles as a treatment against Acinetobacter baumannii in wound infections. *Virulence* 2010;**1**:62–7.
34. Nowacek AS, Balkundi S, McMillan J, Roy U, Martinez-Skinner A, Mosley RL, et al. Analyses of nanoformulated antiretroviral drug charge, size, shape and content for uptake, drug release and antiviral activities in human monocyte-derived macrophages. *J Control Release* 2010;**150**:204–11.
35. Nowacek AS, McMillan J, Miller R, Anderson A, Rabinow B, Gendelman HE. Nanoformulated antiretroviral drug combinations extend drug release and antiretroviral responses in HIV-1-infected macrophages: implications for neuroAIDS therapeutics. *J Neuroimmune Pharmacol* 2010;**5**:592–601.
36. Nowacek AS, Miller RL, McMillan J, Kanmogne G, Kanmogne M, Mosley RL, et al. NanoART synthesis, characterization, uptake, release and toxicology for human monocyte-macrophage drug delivery. *Nanomedicine* 2009;**4**:903–17.
37. Rosa Borges A, Wieczorek L, Johnson B, Benesi AJ, Brown BK, Kensinger RD, et al. Multivalent dendrimeric compounds containing carbohydrates expressed on immune cells inhibit infection by primary isolates of HIV-1. *Virology* 2010;**408**:80–8.
38. Sosnik A, Carcaboso AM, Glisoni RJ, Moretton MA, Chiappetta DA. New old challenges in tuberculosis: potentially effective nanotechnologies in drug delivery. *Adv Drug Deliv Rev* 2010;**62**:547–59.
39. Wang H, Xu K, Liu L, Tan JP, Chen Y, Li Y, et al. The efficacy of self-assembled cationic antimicrobial peptide nanoparticles against Cryptococcus neoformans for the treatment of meningitis. *Biomaterials* 2010;**31**:2874–81.
40. Wang J, Feng SS, Wang S, Chen ZY. Evaluation of cationic nanoparticles of biodegradable copolymers as siRNA delivery system for hepatitis B treatment. *Int J Pharm* 2010;**400**:194–200.
41. Xiang SD, Selomulya C, Ho J, Apostolopoulos V, Plebanski M. Delivery of DNA vaccines: an overview on the use of biodegradable polymeric and magnetic nanoparticles. *Wiley Interdiscip Rev Nanomed Nanobiotechnol* 2010;**2**:205–18.
42. Barnes T, Moots R. Targeting nanomedicines in the treatment of rheumatoid arthritis: focus on certolizumab pegol. *Int J Nanomedicine* 2007;**2**:3–7.
43. Koo OM, Rubinstein I, Onyuksel H. Actively targeted low-dose camptothecin as a safe, long-acting, disease-modifying nanomedicine for rheumatoid arthritis. *Pharm Res* 2010;**28**:776–87.
44. Liu XM, Miller SC, Wang D. Beyond oncology–application of HPMA copolymers in non-cancerous diseases. *Adv Drug Deliv Rev* 2010;**62**:258–71.
45. Yudoh K, Karasawa R, Masuko K, Kato T. Water-soluble fullerene (C60) inhibits the development of arthritis in the rat model of arthritis. *Int J Nanomedicine* 2009;**4**:217–25.
46. Cui Z, Lockman PR, Atwood CS, Hsu CH, Gupte A, Allen DD, et al. Novel D-penicillamine carrying nanoparticles for metal chelation therapy in Alzheimer's and other CNS diseases. *Eur J Pharm Biopharm* 2005;**59**:263–72.
47. Liu G, Men P, Kudo W, Perry G, Smith MA. Nanoparticle-chelator conjugates as inhibitors of amyloid-beta aggregation and neurotoxicity: a novel therapeutic approach for Alzheimer disease. *Neurosci Lett* 2009;**455**:187–90.
48. Liu G, Men P, Perry G, Smith MA. Nanoparticle and iron chelators as a potential novel Alzheimer therapy. *Methods Mol Biol* 2010;**610**:123–44.
49. Cabaleiro-Lago C, Quinlan-Pluck F, Lynch I, Lindman S, Minogue AM, Thulin E, et al. Inhibition of amyloid beta protein fibrillation by polymeric nanoparticles. *J Am Chem Soc* 2008;**130**:15437–43.

50. Xiao L, Zhao D, Chan WH, Choi MM, Li HW. Inhibition of beta 1–40 amyloid fibrillation with N-acetyl-L-cysteine capped quantum dots. *Biomaterials* 2010;**31**:91–8.
51. Esposito E, Fantin M, Marti M, Drechsler M, Paccamiccio L, Mariani P, et al. Solid lipid nanoparticles as delivery systems for bromocriptine. *Pharm Res* 2008;**25**:1521–30.
52. Hsu SH, Wen CJ, Al-Suwayeh SA, Chang HW, Yen TC, Fang JY. Physicochemical characterization and in vivo bioluminescence imaging of nanostructured lipid carriers for targeting the brain: apomorphine as a model drug. *Nanotechnology* 2010;**21**:405101.
53. Huang R, Ke W, Liu Y, Jiang C, Pei Y. The use of lactoferrin as a ligand for targeting the polyamidoamine-based gene delivery system to the brain. *Biomaterials* 2008;**29**:238–46.
54. Xu S, Chen M, Yao Y, Zhang Z, Jin T, Huang Y, et al. Novel poly(ethylene imine) biscarbamate conjugate as an efficient and nontoxic gene delivery system. *J Control Release* 2008;**130**:64–8.
55. Warrington JS, Shaw LM. Pharmacogenetic differences and drug-drug interactions in immunosuppressive therapy. *Expert Opin Drug Metab Toxicol* 2005;**1**:487–503.
56. Bisht S, Feldmann G, Koorstra JB, Mullendore M, Alvarez H, Karikari C, et al. In vivo characterization of a polymeric nanoparticle platform with potential oral drug delivery capabilities. *Mol Cancer Ther* 2008;**7**:3878–88.
57. Liu K, Cao G, Zhang X, Liu R, Zou W, Wu S. Pretreatment with intraluminal rapamycin nanoparticle perfusion inhibits neointimal hyperplasia in a rabbit vein graft model. *Int J Nanomedicine* 2010;**5**:853–60.
58. Clawson C, Huang CT, Futalan D, Martin Seible D, Saenz R, Larsson M, et al. Delivery of a peptide via poly(D, L-lactic-co-glycolic) acid nanoparticles enhances its dendritic cell-stimulatory capacity. *Nanomedicine* 2010;**6**:651–61.
59. Klippstein R, Pozo D. Nanotechnology-based manipulation of dendritic cells for enhanced immunotherapy strategies. *Nanomedicine* 2010;**6**:523–9.
60. Wigglesworth KM, Racki WJ, Mishra R, Szomolanyi-Tsuda E, Greiner DL, Galili U. Rapid recruitment and activation of macrophages by anti-Gal/{alpha}-Gal Liposome interaction accelerates wound healing. *J Immunol* 2011;**186**:4422–32.
61. Kwan KH, Liu X, To MK, Yeung KW, Ho CM, Wong KK. Modulation of collagen alignment by silver nanoparticles results in better mechanical properties in wound healing. *Nanomedicine* 2011;**7**:497–504.
62. Cai L, Guinn AS, Wang S. Exposed hydroxyapatite particles on the surface of photo-crosslinked nanocomposites for promoting MC3T3 cell proliferation and differentiation. *Acta Biomater* 2011;**7**:2185–99.
63. Kon E, Delcogliano M, Filardo G, Busacca M, Di Martino A, Marcacci M. Novel nanocomposite multilayered biomaterial for osteochondral regeneration: a pilot clinical trial. *Am J Sports Med* 2011;**39**:1180–90.
64. Lei Y, Rahim M, Ng Q, Segura T. Hyaluronic acid and fibrin hydrogels with concentrated DNA/PEI polyplexes for local gene delivery. *J Control Release* 2011;**153**:255–61.
65. Olakowska E, Woszczycka-Korczynska I, Jedrzejowska-Szypulka H, Lewin-Kowalik J. Application of nanotubes and nanofibres in nerve repair A review. *Folia Neuropathol* 2010;**48**:231–7.
66. Das S, Suresh PK. Nanosuspension: a new vehicle for the improvement of the delivery of drugs to the ocular surface. Application to amphotericin B. *Nanomedicine* 2011;**7**:242–7.
67. Rafie F, Javadzadeh Y, Javadzadeh AR, Ghavidel LA, Jafari B, Moogooee M, et al. In vivo evaluation of novel nanoparticles containing dexamethasone for ocular drug delivery on rabbit eye. *Curr Eye Res* 2010;**35**:1081–9.
68. Nagarwal RC, Singh PN, Kant S, Maiti P, Pandit JK. Chitosan nanoparticles of 5-fluorouracil for ophthalmic delivery: characterization, in-vitro and in-vivo study. *Chem Pharm Bull (Tokyo)* 2011;**59**:272–8.

69. Li R, Jiang S, Liu D, Bi X, Wang F, Zhang Q, et al. A potential new therapeutic system for glaucoma: solid lipid nanoparticles containing methazolamide. *J Microencapsul* 2011;**28**:134–41.
70. Jeun M, Jeoung JW, Moon S, Kim YJ, Lee S, Paek SH, et al. Engineered superparamagnetic Mn0.5Zn0.5Fe2O4 nanoparticles as a heat shock protein induction agent for ocular neuroprotection in glaucoma. *Biomaterials* 2011;**32**:387–94.
71. Binder C, Cashman SM, Read SP, Kumar-Singh R. Nuclear targeted delivery of macromolecules to retina and cornea. *J Gene Med* 2011;**13**:158–70.
72. Koo OM, Rubinstein I, Onyuksel H. Role of nanotechnology in targeted drug delivery and imaging: a concise review. *Nanomedicine* 2005;**1**:193–212.
73. Saini S, Edelman RR, Sharma P, Li W, Mayo-Smith W, Slater GJ, et al. Blood-pool MR contrast material for detection and characterization of focal hepatic lesions: initial clinical experience with ultrasmall superparamagnetic iron oxide (AMI-227). *AJR Am J Roentgenol* 1995;**164**:1147–52.
74. Margolis DJ, Hoffman JM, Herfkens RJ, Jeffrey RB, Quon A, Gambhir SS. Molecular imaging techniques in body imaging. *Radiology* 2007;**245**:333–56.
75. Schmitz SA, Taupitz M, Wagner S, Wolf KJ, Beyersdorff D, Hamm B. Magnetic resonance imaging of atherosclerotic plaques using superparamagnetic iron oxide particles. *J Magn Reson Imaging* 2001;**14**:355–61.
76. Caruthers SD, Wickline SA, Lanza GM. Nanotechnological applications in medicine. *Curr Opin Biotechnol* 2007;**18**:26–30.
77. Vucic E, Sanders HM, Arena F, Terreno E, Aime S, Nicolay K, et al. Well-defined, multifunctional nanostructures of a paramagnetic lipid and a lipopeptide for macrophage imaging. *J Am Chem Soc* 2009;**131**:406–7.
78. Lin X, Xie J, Niu G, Zhang F, Gao H, Yang M, et al. Chimeric ferritin nanocages for multiple function loading and multimodal imaging. *Nano Lett* 2011;**11**:814–9.
79. Barreto JA, O'Malley W, Kubeil M, Graham B, Stephan H, Spiccia L. Cancer research: nanomaterials: applications in cancer imaging and therapy (adv. Mater. 12/2011). *Adv Mater* 2011;**23**:H2.
80. Barreto JA, O'Malley W, Kubeil M, Graham B, Stephan H, Spiccia L. Nanomaterials: applications in cancer imaging and therapy. *Adv Mater* 2011;**23**:H18–40.
81. Moghimi SM, Hunter AC, Murray JC. Nanomedicine: current status and future prospects. *FASEB J* 2005;**19**:311–30.
82. Mackay PS, Kremers GJ, Kobukai S, Cobb JG, Kuley A, Rosenthal SJ, et al. Multimodal imaging of dendritic cells using a novel hybrid magneto-optical nanoprobe. *Nanomedicine* 2011;**7**:489–96.
83. Partha R, Conyers JL. Biomedical applications of functionalized fullerene-based nanomaterials. *Int J Nanomedicine* 2009;**4**:261–75.
84. Suh WH, Suslick KS, Stucky GD, Suh YH. Nanotechnology, nanotoxicology, and neuroscience. *Prog Neurobiol* 2009;**87**:133–70.
85. Singh S. Nanomedicine-nanoscale drugs and delivery systems. *J Nanosci Nanotechnol* 2010;**10**:7906–18.
86. Mallipeddi R, Rohan LC. Progress in antiretroviral drug delivery using nanotechnology. *Int J Nanomedicine* 2010;**5**:533–47.
87. Hall JB, Dobrovolskaia MA, Patri AK, McNeil SE. Characterization of nanoparticles for therapeutics. *Nanomedicine* 2007;**2**:789–803.
88. Leucuta SE. Nanotechnology for delivery of drugs and biomedical applications. *Curr Clin Pharmacol* 2010;**5**:257–80.
89. Burke A, Ding X, Singh R, Kraft RA, Levi-Polyachenko N, Rylander MN, et al. Long-term survival following a single treatment of kidney tumors with multiwalled carbon nanotubes and near-infrared radiation. *Proc Natl Acad Sci USA* 2009;**106**:12897–902.

90. Buse J, El-Aneed A. Properties, engineering and applications of lipid-based nanoparticle drug-delivery systems: current research and advances. *Nanomedicine* 2010;**5**:1237–60.
91. Zhang L, Gu FX, Chan JM, Wang AZ, Langer RS, Farokhzad OC. Nanoparticles in medicine: therapeutic applications and developments. *Clin Pharmacol Ther* 2008;**83**:761–9.
92. Sharma A, Sharma US. Liposomes in drug delivery: progress and limitations. *Int J Pharm* 1997;**154**:123–40.
93. Lee SM, Chen H, O'Halloran TV, Nguyen ST. "Clickable" polymer-caged nanobins as a modular drug delivery platform. *J Am Chem Soc* 2009;**131**:9311–20.
94. Gabizon A, Catane R, Uziely B, Kaufman B, Safra T, Cohen R, et al. Prolonged circulation time and enhanced accumulation in malignant exudates of doxorubicin encapsulated in polyethylene-glycol coated liposomes. *Cancer Res* 1994;**54**:987–92.
95. Moen MD, Lyseng-Williamson KA, Scott LJ. Liposomal amphotericin B: a review of its use as empirical therapy in febrile neutropenia and in the treatment of invasive fungal infections. *Drugs* 2009;**69**:361–92.
96. Francis MF, Cristea M, Winnik FM. Polymeric micelles for oral drug delivery: why and how. *Pure Appl Chem* 2004;**76**:1321–5.
97. Ali I, Rahis U, Salim K, Rather MA, Wani WA, Haque A. Advances in nano drugs for cancer chemotherapy. *Curr Cancer Drug Targets* 2011;**11**:135–46.
98. Kwon GS. Polymeric micelles for delivery of poorly water-soluble compounds. *Crit Rev Ther Drug Carrier Syst* 2003;**20**:357–403.
99. Zhang W, Shi Y, Chen Y, Ye J, Sha X, Fang X. Multifunctional Pluronic P123/F127 mixed polymeric micelles loaded with paclitaxel for the treatment of multidrug resistant tumors. *Biomaterials* 2011;**32**:2894–906.
100. Gref R, Minamitake Y, Peracchia MT, Trubetskoy V, Torchilin V, Langer R. Biodegradable long-circulating polymeric nanospheres. *Science* 1994;**263**:1600–3.
101. Sheihet L, Chandra P, Batheja P, Devore D, Kohn J, Michniak B. Tyrosine-derived nanospheres for enhanced topical skin penetration. *Int J Pharm* 2008;**350**:312–9.
102. Sheihet L, Dubin RA, Devore D, Kohn J. Hydrophobic drug delivery by self-assembling triblock copolymer-derived nanospheres. *Biomacromolecules* 2005;**6**:2726–31.
103. Sheihet L, Piotrowska K, Dubin RA, Kohn J, Devore D. Effect of tyrosine-derived triblock copolymer compositions on nanosphere self-assembly and drug delivery. *Biomacromolecules* 2007;**8**:998–1003.
104. Batheja P, Sheihet L, Kohn J, Singer AJ, Michniak-Kohn B. Topical drug delivery by a polymeric nanosphere gel: formulation optimization and in vitro and in vivo skin distribution studies. *J Control Release* 2011;**149**:159–67.
105. Li K, Zhan R, Feng SS, Liu B. Conjugated polymer loaded nanospheres with surface functionalization for simultaneous discrimination of different live cancer cells under single wavelength excitation. *Anal Chem* 2011;**83**:2125–32.
106. Boas U, Heegaard PM. Dendrimers in drug research. *Chem Soc Rev* 2004;**33**:43–63.
107. Maiti PK, Li Y, Cagin T, Goddard 3rd WA. Structure of polyamidoamide dendrimers up to limiting generations: a mesoscale description. *J Chem Phys* 2009;**130**:144902.
108. Kumar PV, Asthana A, Dutta T, Jain NK. Intracellular macrophage uptake of rifampicin loaded mannosylated dendrimers. *J Drug Target* 2006;**14**:546–56.
109. Dou H, Destache CJ, Morehead JR, Mosley RL, Boska MD, Kingsley J, et al. Development of a macrophage-based nanoparticle platform for antiretroviral drug delivery. *Blood* 2006;**108**:2827–35.
110. Dou H, Morehead J, Destache CJ, Kingsley JD, Shlyakhtenko L, Zhou Y, et al. Laboratory investigations for the morphologic, pharmacokinetic, and anti-retroviral properties of indinavir nanoparticles in human monocyte-derived macrophages. *Virology* 2007;**358**:148–58.

111. Kingsley JD, Dou H, Morehead J, Rabinow B, Gendelman HE, Destache CJ. Nanotechnology: a focus on nanoparticles as a drug delivery system. *J Neuroimmune Pharmacol* 2006;**1**:340–50.
112. Dou H, Grotepas CB, McMillan JM, Destache CJ, Chaubal M, Werling J, et al. Macrophage delivery of nanoformulated antiretroviral drug to the brain in a murine model of neuroAIDS. *J Immunol* 2009;**183**:661–9.
113. Cuenca AG, Jiang H, Hochwald SN, Delano M, Cance WG, Grobmyer SR. Emerging implications of nanotechnology on cancer diagnostics and therapeutics. *Cancer* 2006;**107**:459–66.
114. Mody VV, Siwale R, Singh A, Mody HR. Introduction to metallic nanoparticles. *J Pharm Bioallied Sci* 2010;**2**:282–9.
115. Loo C, Lowery A, Halas N, West J, Drezek R. Immunotargeted nanoshells for integrated cancer imaging and therapy. *Nano Lett* 2005;**5**:709–11.
116. O'Neal DP, Hirsch LR, Halas NJ, Payne JD, West JL. Photo-thermal tumor ablation in mice using near infrared-absorbing nanoparticles. *Cancer Lett* 2004;**209**:171–6.
117. Sengupta S, Eavarone D, Capila I, Zhao G, Watson N, Kiziltepe T, et al. Temporal targeting of tumour cells and neovasculature with a nanoscale delivery system. *Nature* 2005;**436**:568–72.
118. Perez JM. Iron oxide nanoparticles: hidden talent. *Nat Nanotechnol* 2007;**2**:535–6.
119. Thorek DL, Chen AK, Czupryna J, Tsourkas A. Superparamagnetic iron oxide nanoparticle probes for molecular imaging. *Ann Biomed Eng* 2006;**34**:23–38.
120. Fan C, Gao W, Chen Z, Fan H, Li M, Deng F. Tumor selectivity of stealth multi-functionalized superparamagnetic iron oxide nanoparticles. *Int J Pharm* 2011;**404**:180–90.
121. Hong G, Yuan R, Liang B, Shen J, Yang X, Shuai X. Folate-functionalized polymeric micelle as hepatic carcinoma-targeted MRI-ultrasensitive delivery system of antitumor drugs. *Biomed Microdevices* 2008;**10**:693–700.
122. Jia H, Titmuss S. Polymer-functionalized nanoparticles: from stealth viruses to biocompatible quantum dots. *Nanomedicine* 2009;**4**:951–66.
123. Liao C, Sun Q, Liang B, Shen J, Shuai X. *Targeting EGFR-overexpressing tumor cells using Cetuximab-immunomicelles loaded with doxorubicin and superparamagnetic iron oxide. Eur J Radiol.* 2010 [Epub ahead of print].
124. Beduneau A, Ma Z, Grotepas CB, Kabanov A, Rabinow BE, Gong N, et al. Facilitated monocyte-macrophage uptake and tissue distribution of superparmagnetic iron-oxide nanoparticles. *PLoS One* 2009;**4**:e4343.
125. Hardman R. A toxicologic review of quantum dots: toxicity depends on physicochemical and environmental factors. *Environ Health Perspect* 2006;**114**:165–72.
126. Gao X, Cui Y, Levenson RM, Chung LW, Nie S. In vivo cancer targeting and imaging with semiconductor quantum dots. *Nat Biotechnol* 2004;**22**:969–76.
127. Geraldo DA, Duran-Lara EF, Aguayo D, Cachau RE, Tapia J, Esparza R, et al. Supramolecular complexes of quantum dots and a polyamidoamine (PAMAM)-folate derivative for molecular imaging of cancer cells. *Anal Bioanal Chem* 2011;**400**:483–92.
128. Ho YP, Leong KW. Quantum dot-based theranostics. *Nanoscale* 2010;**2**:60–8.
129. Prasuhn DE, Susumu K, Medintz IL. Multivalent conjugation of peptides, proteins, and DNA to semiconductor quantum dots. *Methods Mol Biol* 2011;**726**:95–110.
130. Hanaki K, Momo A, Oku T, Komoto A, Maenosono S, Yamaguchi Y, et al. Semiconductor quantum dot/albumin complex is a long-life and highly photostable endosome marker. *Biochem Biophys Res Commun* 2003;**302**:496–501.
131. Jaiswal JK, Mattoussi H, Mauro JM, Simon SM. Long-term multiple color imaging of live cells using quantum dot bioconjugates. *Nat Biotechnol* 2003;**21**:47–51.
132. Mansson A, Sundberg M, Balaz M, Bunk R, Nicholls IA, Omling P, et al. In vitro sliding of actin filaments labelled with single quantum dots. *Biochem Biophys Res Commun* 2004;**314**:529–34.

133. Wu X, Liu H, Liu J, Haley KN, Treadway JA, Larson JP, et al. Immunofluorescent labeling of cancer marker Her2 and other cellular targets with semiconductor quantum dots. *Nat Biotechnol* 2003;**21**:41–6.
134. Kaul Z, Yaguchi T, Kaul SC, Hirano T, Wadhwa R, Taira K. Mortalin imaging in normal and cancer cells with quantum dot immuno-conjugates. *Cell Res* 2003;**13**:503–7.
135. Sukhanova A, Devy J, Venteo L, Kaplan H, Artemyev M, Oleinikov V, et al. Biocompatible fluorescent nanocrystals for immunolabeling of membrane proteins and cells. *Anal Biochem* 2004;**324**:60–7.
136. Jamieson T, Bakhshi R, Petrova D, Pocock R, Imani M, Seifalian AM. Biological applications of quantum dots. *Biomaterials* 2007;**28**:4717–32.
137. Shvedova AA, Kagan VE, Fadeel B. Close encounters of the small kind: adverse effects of man-made materials interfacing with the nano-cosmos of biological systems. *Annu Rev Pharmacol Toxicol* 2010;**50**:63–88.
138. Quintana A, Raczka E, Piehler L, Lee I, Myc A, Majoros I, et al. Design and function of a dendrimer-based therapeutic nanodevice targeted to tumor cells through the folate receptor. *Pharm Res* 2002;**19**:1310–6.
139. Schmieder AH, Winter PM, Caruthers SD, Harris TD, Williams TA, Allen JS, et al. Molecular MR imaging of melanoma angiogenesis with alphanubeta3-targeted paramagnetic nanoparticles. *Magn Reson Med* 2005;**53**:621–7.
140. Shukla R, Thomas TP, Peters JL, Desai AM, Kukowska-Latallo J, Patri AK, et al. HER2 specific tumor targeting with dendrimer conjugated anti-HER2 mAb. *Bioconjug Chem* 2006;**17**:1109–15.
141. Winter PM, Caruthers SD, Kassner A, Harris TD, Chinen LK, Allen JS, et al. Molecular imaging of angiogenesis in nascent Vx-2 rabbit tumors using a novel alpha(nu)beta3-targeted nanoparticle and 1.5 tesla magnetic resonance imaging. *Cancer Res* 2003;**63**:5838–43.
142. Winter PM, Morawski AM, Caruthers SD, Fuhrhop RW, Zhang H, Williams TA, et al. Molecular imaging of angiogenesis in early-stage atherosclerosis with alpha(v)beta3-integrin-targeted nanoparticles. *Circulation* 2003;**108**:2270–4.
143. Asati A, Santra S, Kaittanis C, Perez JM. Surface-charge-dependent cell localization and cytotoxicity of cerium oxide nanoparticles. *ACS Nano* 2010;**4**:5321–31.
144. Brunner R, Jensen-Jarolim E, Pali-Scholl I. The ABC of clinical and experimental adjuvants—a brief overview. *Immunol Lett* 2010;**128**:29–35.
145. Krishnamachari Y, Geary SM, Lemke CD, Salem AK. Nanoparticle delivery systems in cancer vaccines. *Pharm Res* 2011;**28**:215–36.
146. Chadwick S, Kriegel C, Amiji M. Delivery strategies to enhance mucosal vaccination. *Expert Opin Biol Ther* 2009;**9**:427–40.
147. Chadwick S, Kriegel C, Amiji M. Nanotechnology solutions for mucosal immunization. *Adv Drug Deliv Rev* 2010;**62**:394–407.
148. Matsuo K, Koizumi H, Akashi M, Nakagawa S, Fujita T, Yamamoto A, et al. Intranasal immunization with poly(gamma-glutamic acid) nanoparticles entrapping antigenic proteins can induce potent tumor immunity. *J Control Release* 2011;**152**:310–6.
149. San Roman B, Irache JM, Gomez S, Tsapis N, Gamazo C, Espuelas MS. Co-encapsulation of an antigen and CpG oligonucleotides into PLGA microparticles by TROMS technology. *Eur J Pharm Biopharm* 2008;**70**:98–108.
150. Scholl I, Weissenbock A, Forster-Waldl E, Untersmayr E, Walter F, Willheim M, et al. Allergen-loaded biodegradable poly(D, L-lactic-co-glycolic) acid nanoparticles down-regulate an ongoing Th2 response in the BALB/c mouse model. *Clin Exp Allergy* 2004;**34**:315–21.
151. Santos-Magalhaes NS, Mosqueira VC. Nanotechnology applied to the treatment of malaria. *Adv Drug Deliv Rev* 2010;**62**:560–75.

152. Adair BM. Nanoparticle vaccines against respiratory viruses. *Wiley Interdiscip Rev Nanomed Nanobiotechnol* 2009;**1**:405–14.
153. Al-Deen FN, Ho J, Selomulya C, Ma C, Coppel R. Superparamagnetic nanoparticles for effective delivery of malaria DNA vaccine. *Langmuir* 2011;**27**:3703–12.
154. Fuller DH, Loudon P, Schmaljohn C. Preclinical and clinical progress of particle-mediated DNA vaccines for infectious diseases. *Methods* 2006;**40**:86–97.
155. Alper J. Geron gets green light for human trial of ES cell-derived product. *Nat Biotechnol* 2009;**27**:213–4.
156. Dimos JT, Rodolfa KT, Niakan KK, Weisenthal LM, Mitsumoto H, Chung W, et al. Induced pluripotent stem cells generated from patients with ALS can be differentiated into motor neurons. *Science* 2008;**321**:1218–21.
157. Berger C, Berger M, Hackman RC, Gough M, Elliott C, Jensen MC, et al. Safety and immunologic effects of IL-15 administration in nonhuman primates. *Blood* 2009;**114**:2417–26.
158. Thompson JA, Lee DJ, Cox WW, Lindgren CG, Collins C, Neraas KA, et al. Recombinant interleukin 2 toxicity, pharmacokinetics, and immunomodulatory effects in a phase I trial. *Cancer Res* 1987;**47**:4202–7.
159. Kim H, Suh H, Jo SA, Kim HW, Lee JM, Kim EH, et al. In vivo bone formation by human marrow stromal cells in biodegradable scaffolds that release dexamethasone and ascorbate-2-phosphate. *Biochem Biophys Res Commun* 2005;**332**:1053–60.
160. Kloxin AM, Kasko AM, Salinas CN, Anseth KS. Photodegradable hydrogels for dynamic tuning of physical and chemical properties. *Science* 2009;**324**:59–63.
161. Nisbet DR, Moses D, Gengenbach TR, Forsythe JS, Finkelstein DI, Horne MK. Enhancing neurite outgrowth from primary neurones and neural stem cells using thermoresponsive hydrogel scaffolds for the repair of spinal cord injury. *J Biomed Mater Res A* 2009;**89**:24–35.
162. Park H, Temenoff JS, Tabata Y, Caplan AI, Mikos AG. Injectable biodegradable hydrogel composites for rabbit marrow mesenchymal stem cell and growth factor delivery for cartilage tissue engineering. *Biomaterials* 2007;**28**:3217–27.
163. Yang Q, Peng J, Guo Q, Huang J, Zhang L, Yao J, et al. A cartilage ECM-derived 3-D porous acellular matrix scaffold for in vivo cartilage tissue engineering with PKH26-labeled chondrogenic bone marrow-derived mesenchymal stem cells. *Biomaterials* 2008;**29**:2378–87.
164. Park JS, Yang HN, Woo DG, Jeon SY, Park KH. The promotion of chondrogenesis, osteogenesis, and adipogenesis of human mesenchymal stem cells by multiple growth factors incorporated into nanosphere-coated microspheres. *Biomaterials* 2011;**32**:28–38.
165. Santos JL, Pandita D, Rodrigues J, Pego AP, Granja PL, Balian G, et al. Receptor-mediated gene delivery using PAMAM dendrimers conjugated with peptides recognized by mesenchymal stem cells. *Mol Pharm* 2010;**7**:763–74.
166. Stephan MT, Moon JJ, Um SH, Bershteyn A, Irvine DJ. Therapeutic cell engineering with surface-conjugated synthetic nanoparticles. *Nat Med* 2010;**16**:1035–41.
167. Zhou P, Hatziieremia S, Elliott MA, Scobie L, Crossan C, Michie AM, et al. Uptake of synthetic Low Density Lipoprotein by leukemic stem cells–a potential stem cell targeted drug delivery strategy. *J Control Release* 2010;**148**:380–7.
168. Pickard MR, Barraud P, Chari DM. The transfection of multipotent neural precursor/stem cell transplant populations with magnetic nanoparticles. *Biomaterials* 2011;**32**:2274–84.
169. Roger M, Clavreul A, Venier-Julienne MC, Passirani C, Montero-Menei C, Menei P. The potential of combinations of drug-loaded nanoparticle systems and adult stem cells for glioma therapy. *Biomaterials* 2011;**32**:2106–16.
170. Bulte JW, Zhang S, van Gelderen P, Herynek V, Jordan EK, Duncan ID, et al. Neurotransplantation of magnetically labeled oligodendrocyte progenitors: magnetic resonance tracking of cell migration and myelination. *Proc Natl Acad Sci USA* 1999;**96**:15256–61.

171. Dunning MD, Lakatos A, Loizou L, Kettunen M, ffrench-Constant C, Brindle KM, et al. Superparamagnetic iron oxide-labeled Schwann cells and olfactory ensheathing cells can be traced in vivo by magnetic resonance imaging and retain functional properties after transplantation into the CNS. *J Neurosci* 2004;**24**:9799–810.
172. Guzman R, Uchida N, Bliss TM, He D, Christopherson KK, Stellwagen D, et al. Long-term monitoring of transplanted human neural stem cells in developmental and pathological contexts with MRI. *Proc Natl Acad Sci USA* 2007;**104**:10211–6.
173. Martinez-Serrano A, Bjorklund A. Ex vivo nerve growth factor gene transfer to the basal forebrain in presymptomatic middle-aged rats prevents the development of cholinergic neuron atrophy and cognitive impairment during aging. *Proc Natl Acad Sci USA* 1998;**95**:1858–63.
174. Martinez-Serrano A, Hantzopoulos PA, Bjorklund A. Ex vivo gene transfer of brain-derived neurotrophic factor to the intact rat forebrain: neurotrophic effects on cholinergic neurons. *Eur J Neurosci* 1996;**8**:727–35.
175. Muller FJ, Snyder EY, Loring JF. Gene therapy: can neural stem cells deliver? *Nat Rev Neurosci* 2006;**7**:75–84.
176. Menon LG, Kelly K, Yang HW, Kim SK, Black PM, Carroll RS. Human bone marrow-derived mesenchymal stromal cells expressing S-TRAIL as a cellular delivery vehicle for human glioma therapy. *Stem Cells* 2009;**27**:2320–30.
177. Nakamizo A, Marini F, Amano T, Khan A, Studeny M, Gumin J, et al. Human bone marrow-derived mesenchymal stem cells in the treatment of gliomas. *Cancer Res* 2005;**65**:3307–18.
178. Sonabend AM, Ulasov IV, Tyler MA, Rivera AA, Mathis JM, Lesniak MS. Mesenchymal stem cells effectively deliver an oncolytic adenovirus to intracranial glioma. *Stem Cells* 2008;**26**:831–41.
179. Stoff-Khalili MA, Rivera AA, Mathis JM, Banerjee NS, Moon AS, Hess A, et al. Mesenchymal stem cells as a vehicle for targeted delivery of CRAds to lung metastases of breast carcinoma. *Breast Cancer Res Treat* 2007;**105**:157–67.
180. Studeny M, Marini FC, Champlin RE, Zompetta C, Fidler IJ, Andreeff M. Bone marrow-derived mesenchymal stem cells as vehicles for interferon-beta delivery into tumors. *Cancer Res* 2002;**62**:3603–8.
181. Studeny M, Marini FC, Dembinski JL, Zompetta C, Cabreira-Hansen M, Bekele BN, et al. Mesenchymal stem cells: potential precursors for tumor stroma and targeted-delivery vehicles for anticancer agents. *J Natl Cancer Inst* 2004;**96**:1593–603.
182. Roger M, Clavreul A, Venier-Julienne MC, Passirani C, Sindji L, Schiller P, et al. Mesenchymal stem cells as cellular vehicles for delivery of nanoparticles to brain tumors. *Biomaterials* 2010;**31**:8393–401.
183. Cheng H, Kastrup CJ, Ramanathan R, Siegwart DJ, Ma M, Bogatyrev SR, et al. Nanoparticulate cellular patches for cell-mediated tumoritropic delivery. *ACS Nano* 2010;**4**:625–31.
184. Kayser O, Lemke A, Hernandez-Trejo N. The impact of nanobiotechnology on the development of new drug delivery systems. *Curr Pharm Biotechnol* 2005;**6**:3–5.
185. Brannon-Peppas L, Blanchette JO. Nanoparticle and targeted systems for cancer therapy. *Adv Drug Deliv Rev* 2004;**56**:1649–59.
186. Pound MW, Townsend ML, Dimondi V, Wilson D, Drew RH. Overview of treatment options for invasive fungal infections. *Med Mycol* 2011;**49**:561–80.
187. Chen J, Li Z, Huang H, Yang Y, Ding Q, Mai J, et al. Improved antigen cross-presentation by polyethyleneimine-based nanoparticles. *Int J Nanomedicine* 2011;**6**:77–84.
188. Kasturi SP, Skountzou I, Albrecht RA, Koutsonanos D, Hua T, Nakaya HI, et al. Programming the magnitude and persistence of antibody responses with innate immunity. *Nature* 2011;**470**:543–7.

189. Reddy ST, Swartz MA, Hubbell JA. Targeting dendritic cells with biomaterials: developing the next generation of vaccines. *Trends Immunol* 2006;**27**:573–9.
190. Geiser M, Rothen-Rutishauser B, Kapp N, Schurch S, Kreyling W, Schulz H, et al. Ultrafine particles cross cellular membranes by nonphagocytic mechanisms in lungs and in cultured cells. *Environ Health Perspect* 2005;**113**:1555–60.
191. Nagayama S, Ogawara K, Minato K, Fukuoka Y, Takakura Y, Hashida M, et al. Fetuin mediates hepatic uptake of negatively charged nanoparticles via scavenger receptor. *Int J Pharm* 2007;**329**:192–8.
192. Xing X, He X, Peng J, Wang K, Tan W. Uptake of silica-coated nanoparticles by HeLa cells. *J Nanosci Nanotechnol* 2005;**5**:1688–93.
193. Chithrani BD, Ghazani AA, Chan WC. Determining the size and shape dependence of gold nanoparticle uptake into mammalian cells. *Nano Lett* 2006;**6**:662–8.
194. Parak WJ, Boudreau R, Le Gros M, Gerion D, Zanchet D, Micheel CM, et al. Cell motility and metastatic potential studies based on quantum dot imaging of phagocytic tracks. *Adv Mater* 2002;**14**:882–5.
195. Zhang S, Li J, Lykkotrafitis G, Bao G, Suresh S. Size-dependent endocytosis of nanoparticles. *Adv Mater* 2008;**21**:419–24.
196. Konan YN, Chevallier J, Gurny R, Allemann E. Encapsulation of p-THPP into nanoparticles: cellular uptake, subcellular localization and effect of serum on photodynamic activity. *Photochem Photobiol* 2003;**77**:638–44.
197. Singh S, Kumar A, Karakoti A, Seal S, Self WT. Unveiling the mechanism of uptake and subcellular distribution of cerium oxide nanoparticles. *Mol Biosyst* 2010;**6**:1813–20.
198. Huang K, Voss B, Kumar D, Hamm HE, Harth E. Dendritic molecular transporters provide control of delivery to intracellular compartments. *Bioconjug Chem* 2007;**18**:403–9.
199. Bale SS, Kwon SJ, Shah DA, Banerjee A, Dordick JS, Kane RS. Nanoparticle-mediated cytoplasmic delivery of proteins to target cellular machinery. *ACS Nano* 2010;**4**:1493–500.
200. Donaldson K, Stone V, Tran CL, Kreyling W, Borm PJ. Nanotoxicology. *Occup Environ Med* 2004;**61**:727–8.
201. Oberdorster G, Oberdorster E, Oberdorster J. Nanotoxicology: an emerging discipline evolving from studies of ultrafine particles. *Environ Health Perspect* 2005;**113**:823–39.
202. Service RF. Is nanotechnology dangerous? *Science* 2000;**290**:1526–7.
203. Pan Y, Neuss S, Leifert A, Fischler M, Wen F, Simon U, et al. Size-dependent cytotoxicity of gold nanoparticles. *Small* 2007;**3**:1941–9.
204. Nabiev I, Mitchell S, Davies A, Williams Y, Kelleher D, Moore R, et al. Nonfunctionalized nanocrystals can exploit a cell's active transport machinery delivering them to specific nuclear and cytoplasmic compartments. *Nano Lett* 2007;**7**:3452–61.
205. Maysinger D, Lovric J, Eisenberg A, Savic R. Fate of micelles and quantum dots in cells. *Eur J Pharm Biopharm* 2007;**65**:270–81.
206. Porter AE, Gass M, Muller K, Skepper JN, Midgley PA, Welland M. Direct imaging of single-walled carbon nanotubes in cells. *Nat Nanotechnol* 2007;**2**:713–7.
207. Sargent LM, Shvedova AA, Hubbs AF, Salisbury JL, Benkovic SA, Kashon ML, et al. Induction of aneuploidy by single-walled carbon nanotubes. *Environ Mol Mutagen* 2009;**50**:708–17.
208. Gratton SE, Ropp PA, Pohlhaus PD, Luft JC, Madden VJ, Napier ME, et al. The effect of particle design on cellular internalization pathways. *Proc Natl Acad Sci USA* 2008;**105**:11613–8.
209. Oberdorster G, Ferin J, Lehnert BE. Correlation between particle size, in vivo particle persistence, and lung injury. *Environ Health Perspect* 1994;**102**(Suppl. 5):173–9.

210. Monteiller C, Tran L, MacNee W, Faux S, Jones A, Miller B, et al. The pro-inflammatory effects of low-toxicity low-solubility particles, nanoparticles and fine particles, on epithelial cells in vitro: the role of surface area. *Occup Environ Med* 2007;**64**:609–15.
211. Vallhov H, Qin J, Johansson SM, Ahlborg N, Muhammed MA, Scheynius A, et al. The importance of an endotoxin-free environment during the production of nanoparticles used in medical applications. *Nano Lett* 2006;**6**:1682–6.
212. Connor EE, Mwamuka J, Gole A, Murphy CJ, Wyatt MD. Gold nanoparticles are taken up by human cells but do not cause acute cytotoxicity. *Small* 2005;**1**:325–7.
213. Kagan VE, Tyurina YY, Tyurin VA, Konduru NV, Potapovich AI, Osipov AN, et al. Direct and indirect effects of single walled carbon nanotubes on RAW 264.7 macrophages: role of iron. *Toxicol Lett* 2006;**165**:88–100.
214. Murdock RC, Braydich-Stolle L, Schrand AM, Schlager JJ, Hussain SM. Characterization of nanomaterial dispersion in solution prior to in vitro exposure using dynamic light scattering technique. *Toxicol Sci* 2008;**101**:239–53.
215. Sheng WY, Huang L. Cancer immunotherapy and nanomedicine. *Pharm Res* 2011;**28**:200–14.
216. Seleverstov O, Zabirnyk O, Zscharnack M, Bulavina L, Nowicki M, Heinrich JM, et al. Quantum dots for human mesenchymal stem cells labeling. A size-dependent autophagy activation. *Nano Lett* 2006;**6**:2826–32.
217. Stern ST, Zolnik BS, McLeland CB, Clogston J, Zheng J, McNeil SE. Induction of autophagy in porcine kidney cells by quantum dots: a common cellular response to nanomaterials? *Toxicol Sci* 2008;**106**:140–52.
218. Yamawaki H, Iwai N. Cytotoxicity of water-soluble fullerene in vascular endothelial cells. *Am J Physiol Cell Physiol* 2006;**290**:C1495–502.
219. Marano F, Hussain S, Rodrigues-Lima F, Baeza-Squiban A, Boland S. Nanoparticles: molecular targets and cell signalling. *Arch Toxicol* 2011;**85**:733–41.
220. Borm PJ, Muller-Schulte D. Nanoparticles in drug delivery and environmental exposure: same size, same risks? *Nanomedicine* 2006;**1**:235–49.
221. Kovacic P, Somanathan R. Biomechanisms of nanoparticles (toxicants, antioxidants and therapeutics): electron transfer and reactive oxygen species. *J Nanosci Nanotechnol* 2010;**10**:7919–30.
222. Mocan T, Clichici S, Agoston-Coldea L, Mocan L, Simon S, Ilie IR, et al. Implications of oxidative stress mechanisms in toxicity of nanoparticles (review). *Acta Physiol Hung* 2010;**97**:247–55.
223. Stone V, Johnston H, Schins RP. Development of in vitro systems for nanotoxicology: methodological considerations. *Crit Rev Toxicol* 2009;**39**:613–26.
224. Wessels A, Van Berlo D, Boots AW, Gerloff K, Scherbart AM, Cassee FR, et al. Oxidative stress and DNA damage responses in rat and mouse lung to inhaled carbon nanoparticles. *Nanotoxicology* 2011;**5**:66–78.
225. Jomova K, Valko M. Advances in metal-induced oxidative stress and human disease. *Toxicology* 2011;**283**:65–87.
226. Bergamaschi E, Bussolati O, Magrini A, Bottini M, Migliore L, Bellucci S, et al. Nanomaterials and lung toxicity: interactions with airways cells and relevance for occupational health risk assessment. *Int J Immunopathol Pharmacol* 2006;**19**:3–10.
227. Shvedova AA, Kisin ER, Porter D, Schulte P, Kagan VE, Fadeel B, et al. Mechanisms of pulmonary toxicity and medical applications of carbon nanotubes: two faces of Janus? *Pharmacol Ther* 2009;**121**:192–204.
228. Shvedova AA, Kisin ER, Murray AR, Gorelik O, Arepalli S, Castranova V, et al. Vitamin E deficiency enhances pulmonary inflammatory response and oxidative stress induced by single-walled carbon nanotubes in C57BL/6 mice. *Toxicol Appl Pharmacol* 2007;**221**:339–48.

229. Zolnik BS, Stern ST, Kaiser JM, Heakal Y, Clogston JD, Kester M, et al. Rapid distribution of liposomal short-chain ceramide in vitro and in vivo. *Drug Metab Dispos* 2008;**36**:1709–15.
230. Zolnik BS, Sadrieh N. Regulatory perspective on the importance of ADME assessment of nanoscale material containing drugs. *Adv Drug Deliv Rev* 2009;**61**:422–7.
231. De Jong WH, Hagens WI, Krystek P, Burger MC, Sips AJ, Geertsma RE. Particle size-dependent organ distribution of gold nanoparticles after intravenous administration. *Biomaterials* 2008;**29**:1912–9.
232. Sonavane G, Tomoda K, Makino K. Biodistribution of colloidal gold nanoparticles after intravenous administration: effect of particle size. *Colloids Surf B Biointerfaces* 2008;**66**:274–80.
233. Reddy ST, van der Vlies AJ, Simeoni E, Angeli V, Randolph GJ, O'Neil CP, et al. Exploiting lymphatic transport and complement activation in nanoparticle vaccines. *Nat Biotechnol* 2007;**25**:1159–64.
234. Kaminskas LM, Boyd BJ, Karellas P, Krippner GY, Lessene R, Kelly B, et al. The impact of molecular weight and PEG chain length on the systemic pharmacokinetics of PEGylated poly l-lysine dendrimers. *Mol Pharm* 2008;**5**:449–63.
235. Hutter-Saunders JA, Gendelman HE, Mosley RL. Murine motor and behavior functional evaluations for acute 1-methyl-4-phenyl-1,2,3,6-tetrahydropyridine (MPTP) intoxication. *J Neuroimmune Pharmacol.* 2011 [Epub ahead of print].
236. Dash PK, Gorantla S, Gendelman HE, Knibbe J, Casale GP, Makarov E, et al. Loss of neuronal integrity during progressive HIV-1 infection of humanized mice. *J Neurosci* 2011;**31**:3148–57.
237. Gorantla S, Sneller H, Walters L, Sharp JG, Pirruccello SJ, West JT, et al. Human immunodeficiency virus type 1 pathobiology studied in humanized BALB/c-Rag2−/−gammac−/− mice. *J Virol* 2007;**81**:2700–12.
238. Fischer HC, Chan WC. Nanotoxicity: the growing need for in vivo study. *Curr Opin Biotechnol* 2007;**18**:565–71.
239. Nel A, Xia T, Madler L, Li N. Toxic potential of materials at the nanolevel. *Science* 2006;**311**:622–7.
240. Maynard AD, Warheit DB, Philbert MA. The new toxicology of sophisticated materials: nanotoxicology and beyond. *Toxicol Sci* 2011;**120**(Suppl. 1):S109–29.
241. Selbach M, Schwanhausser B, Thierfelder N, Fang Z, Khanin R, Rajewsky N. Widespread changes in protein synthesis induced by microRNAs. *Nature* 2008;**455**:58–63.
242. Huang X, Stone DK, Yu F, Zeng Y, Gendelman HE. Functional proteomic analysis for regulatory T cell surveillance of the HIV-1-infected macrophage. *J Proteome Res* 2010;**9**:6759–73.
243. Schwanhausser B, Gossen M, Dittmar G, Selbach M. Global analysis of cellular protein translation by pulsed SILAC. *Proteomics* 2009;**9**:205–9.
244. Moghimi SM, Hunter AC, Murray JC. Long-circulating and target-specific nanoparticles: theory to practice. *Pharmacol Rev* 2001;**53**:283–318.
245. Fujiwara M, Baldeschwieler JD, Grubbs RH. Receptor-mediated endocytosis of poly(acrylic acid)-conjugated liposomes by macrophages. *Biochim Biophys Acta* 1996;**1278**:59–67.
246. Ishihara T, Izumo N, Higaki M, Shimada E, Hagi T, Mine L, et al. Role of zinc in formulation of PLGA/PLA nanoparticles encapsulating betamethasone phosphate and its release profile. *J Control Release* 2005;**105**:68–76.
247. Juliano RL, Stamp D. The effect of particle size and charge on the clearance rates of liposomes and liposome encapsulated drugs. *Biochem Biophys Res Commun* 1975;**63**:651–8.
248. Krieger M. Molecular flypaper and atherosclerosis: structure of the macrophage scavenger receptor. *Trends Biochem Sci* 1992;**17**:141–6.
249. Lee KD, Hong K, Papahadjopoulos D. Recognition of liposomes by cells: in vitro binding and endocytosis mediated by specific lipid headgroups and surface charge density. *Biochim Biophys Acta* 1992;**1103**:185–97.

250. Miller CR, Bondurant B, McLean SD, McGovern KA, O'Brien DF. Liposome-cell interactions in vitro: effect of liposome surface charge on the binding and endocytosis of conventional and sterically stabilized liposomes. *Biochemistry* 1998;**37**:12875–83.
251. Nishikawa K, Arai H, Inoue K. Scavenger receptor-mediated uptake and metabolism of lipid vesicles containing acidic phospholipids by mouse peritoneal macrophages. *J Biol Chem* 1990;**265**:5226–31.
252. Jain S, Mishra V, Singh P, Dubey PK, Saraf DK, Vyas SP. RGD-anchored magnetic liposomes for monocytes/neutrophils-mediated brain targeting. *Int J Pharm* 2003;**261**:43–55.
253. Tempone AG, Perez D, Rath S, Vilarinho AL, Mortara RA, de Andrade Jr. HF. Targeting Leishmania (L.) chagasi amastigotes through macrophage scavenger receptors: the use of drugs entrapped in liposomes containing phosphatidylserine. *J Antimicrob Chemother* 2004;**54**:60–8.
254. Thiele L, Merkle HP, Walter E. Phagocytosis and phagosomal fate of surface-modified microparticles in dendritic cells and macrophages. *Pharm Res* 2003;**20**:221–8.
255. Sollner T, Bennett MK, Whiteheart SW, Scheller RH, Rothman JE. A protein assembly-disassembly pathway in vitro that may correspond to sequential steps of synaptic vesicle docking, activation, and fusion. *Cell* 1993;**75**:409–18.
256. Ikehara Y, Niwa T, Biao L, Ikehara SK, Ohashi N, Kobayashi T, et al. A carbohydrate recognition-based drug delivery and controlled release system using intraperitoneal macrophages as a cellular vehicle. *Cancer Res* 2006;**66**:8740–8.
257. Hofstetter CP, Schwarz EJ, Hess D, Widenfalk J, El Manira A, Prockop DJ, et al. Marrow stromal cells form guiding strands in the injured spinal cord and promote recovery. *Proc Natl Acad Sci USA* 2002;**99**:2199–204.
258. Leek RD, Lewis CE, Whitehouse R, Greenall M, Clarke J, Harris AL. Association of macrophage infiltration with angiogenesis and prognosis in invasive breast carcinoma. *Cancer Res* 1996;**56**:4625–9.
259. Lewis JS, Landers RJ, Underwood JC, Harris AL, Lewis CE. Expression of vascular endothelial growth factor by macrophages is up-regulated in poorly vascularized areas of breast carcinomas. *J Pathol* 2000;**192**:150–8.
260. Mahmood A, Lu D, Lu M, Chopp M. Treatment of traumatic brain injury in adult rats with intravenous administration of human bone marrow stromal cells. *Neurosurgery* 2003;**53**:697–702. [discussion 702–693].
261. Orlic D, Kajstura J, Chimenti S, Jakoniuk I, Anderson SM, Li B, et al. Bone marrow cells regenerate infarcted myocardium. *Nature* 2001;**410**:701–5.
262. Orlic D, Kajstura J, Chimenti S, Limana F, Jakoniuk I, Quaini F, et al. Mobilized bone marrow cells repair the infarcted heart, improving function and survival. *Proc Natl Acad Sci USA* 2001;**98**:10344–9.

Index

A

Agarose gel electrophoresis separation method, SWCNT, 213–214
Albumin, 274–275
Alternating current dielectrophoresis separation method, 212–213
Antibody conjugated delivery systems, 117–118
Antisense strategy, 400
Arc discharge (AD) process. *See also* Multi-wall carbon nanotubes (MWCNTs); Single-wall carbon nanotubes (SWCNTs)
  apparatus, 195–196
  description, 195
  schematic representation, 195, 197
Arg-Gly-Asp peptides, 81–82
AuroShellT microparticles, 344–345

B

Biomedical applications, CNTs
  diagnostic imaging, 224–227
  drug delivery application, 231
  electrochemical sensors, 227–228
  gene delivery, 231–232
  photothermal treatments, 229–230
  tissue regenerations
    neural, 232
    scaffolds, bone regeneration, 232
    self-organized neuronal networks, 232, 233
Biomimetic vectors, 548
Biomolecule-liposome interaction strength measurement
  dilution assays, 23
  DSC, 21
  ITC, 21–24
  partial enthalpy assays, 23
  pressure perturbation calorimetry, 24–25
  quartz crystal microbalance, 25
  reaction assays, 23
Bombesin (BBN), 346
Boron-neuron capture therapy (BNCT), dendrimers, 119–120
Branched polyethylenimine (bPEI), 533

C

CADY peptides, 410–411
Cancer targeting delivery, dendrimers
  antibody, 117–118
  EPR effect, 116–117
  folic acid, 117
  glycodendrimer, 118
Carbon nanotubes (CNTs). *See also* Multi-wall carbon nanotubes (MWCNTs); Single-wall carbon nanotubes (SWCNTs)
  biomedical applications
    diagnostics, 224–228
    drug delivery application, 231
    electrochemical sensors, 227–228
    gene delivery, 231–232
    photothermal treatments, 229–230
    tissue regenerations, 232–233
  dispersion, 188–189
  dissolution, 186–187
  electrical and optical properties, 181–186
  functionalization
    covalent, 219–221
    non-covalent, 221–223
    uses, 217–218
  hemi-spherical cap structure, 179–180
  micelle formation, 187, 188
  purification methods, 208–211
  separation methods
    chromatography, 214–215
    DGU, 215–217
    electrophoresis, 212–214
  sp^2 hybridization, 177
  Stone–Wales defect, 180

Carbon nanotubes (CNTs). *See also* Multi-wall carbon nanotubes (MWCNTs); Single-wall carbon nanotubes (SWCNTs) (*cont.*)
  structural characterization, 176–194
  substitutional defects, 180
  surfactants, 187
  synthesis
    arc discharge (AD), 195–199
    carbon source transformation, 194
    chirality control, 205–207
    CVD, 202–204
    laser ablation (LA), 199–202
    relative yield assessment, 195
    selectivity, 194–195
  toxicity
    cell cultures, 190, 191
    functionalized CNTs, 190–194
    pristine, inhalation of, 189–190
Carbon nanovectors, 501
Cardiology, personalized. *See* Nanocardiology
Caveolae-dependent endocytosis, 67–68
Cell-based nanomedicine
  cell-specific targeting, 585
  particle uptake, 585–586
  subcellular localization, drug stability, and drug release, 586–587
  translational pathways
    intracellular, 564–567, 568
    nanoformulated drug development, 564–567, 569
    therapies for pathogen, mononuclear phagocytes, 564–567
  Trojan horses, drug delivery, 587
Cell-penetrating peptides (CPPs)
  CADY peptides, 410–411
  covalent conjugation, 398–399
  delivery properties, 421
  endosomal sequestration, 398
  gene therapy
    approaches, 399–400
    gene delivery, 400
    RNAi, 400–401
    splice-switching therapeutics, 401
  MPG peptide, 407–409
  noncovalent nanoparticle formation
    charge neutralization and amphipathicity, 405
    electrostatic and hydrophobic interactions, 404–405
    and modifications, oligonucleotides, 405, 406
    requirements, 405
    secondary α–helical structures, 404–405
  PepFects
    delivery of SSO, plasmids, and siRNAs, 415, 416
    PF6, 417–419
    PF10, 415–417
    PF14, 419–420
  Pep peptide, 409–410
  polyarginines, 411–414
  primary and suspension cells, 421
  transportans, 414–420
  uptake mechanisms
    endocytosis, 402, 403
    models, 401
    types of media and serum proteins, 402–403
Cell surface labeling, QD
  electrostatic nonspecific adsorption, 60–61
  nonspecific interaction, schematic illustration, 61
  specific
    antibodies and secondary antibodies, 63–65
    ligands, fusion proteins, 65–66
    peptides, membrane proteins, 66–67
Cellular signal transduction pathways
  growth, differentiation, metabolism, and survival, 495–496
  PD98059, 496
  receptor tyrosine kinases (RTKs), 496
Chemical vapor deposition (CVD)
  catalytic material used, 202
  hydrocarbon gas decomposition, 202
  nanoparticle size factor, 202–203
  nanotube synthesis
    base-growth mechanism, 203
    cobalt-molybdenum catalyst (CoMoCAT), 205–206
    floating catalyst, 203–204
    HiPCO, 204, 205
    plasma enhanced CVD, 204
    supported catalyst, 203
    tip-growth mechanism, 203
  precursor used, 202
Chitosan, 499
  definition, 536
  high molecular weight, 537

pharmaceutical industry, 536
properties, 536
Clathrin-mediated endocytosis, 67–68
ClearReadT, 330
CNTs. *See* Carbon nanotubes (CNTs)
Compressed fluid (CF) technology, vesicle preparation
advantages, 38
CF definition, 37–38
DELOS-susp procedure
cholestrol rich vesicles, 41
drawbacks, 41
schematic representation, 40–41
unilamellar antibiotic loaded liposomes, 41, 42
injection and decompression method, 39
supercritical liposome method, 39
supercritical reverse phase evaporation (scRPE) method, 39
super fluids phospholipid nanosome (SFS-CFN), 39
use, 38
Contact-facilitated drug delivery, 340
Covalent functionalization, CNTs, 219–221
CVD. *See* Chemical vapor deposition (CVD)
Cyclodextrin (CD), polymer-based NPs, 537–539

D

de Gennes's dense packing, 112
Dendrimers, 499
biomedical toxicity reduction approaches, 108–110
BNCT, 119–120
cancer targeting delivery
antibody, 117–118
EPR effect, 118
folic acid, 117
glycodendrimer, 118
definition, 102
drug delivery systems
chemical binding, 114–116
physical doping, 110–114
ethylene glycol dendrimer effect, 112–113
gene carriers, 122–123
gene delivery
photochemical internalization, 125–126
vectors used, 123
in vitro restrictions, 123
in vivo targeted hepatic, 125
gene therapy, 122–126
imaging agents
molecular probes, 126–128
MRI, 128–129
optical fluorescence imaging, 131–132
X-ray CT, 129–131
monodispersity, 103–105
multivalent behavior, 105
organic nanoparticles, 103
PAMAM, 103, 104
PDT, 120–122
photothermal therapy, 122
poly(propylenemine), 103, 104
polymer-based NPs, 534–536
vs. proteins, characteristic property, 103
second generation, two functional branch, 102
star dendrimer synthesis, 109
therapeutic nanoparticles, 574
in vitro cytotoxicity, 106
in vivo toxicity, 107–108
Density gradient ultracentrifugation (DGU), 215–217
DNA transfection, 359
Drug delivery systems, dendrimers
chemical binding, 114–116
physical doping
drug loading, 110–113
drug release mechanism, 113–114
Drug-loaded vesicle preparation
Bangham method, 35
compressed fluid (CF) technology
advantages, 38
definition, 37–38
DELOS-susp procedure, 40–41
injection and decompression method, 39
supercritical liposome method, 39
supercritical reverse phase evaporation (scRPE) method, 39
super fluids phospholipid nanosome (SFS-CFN), 39
use, 38
dehydration-rehydration procedure, 36
detegent depletion method, 36
microfluidics, 37
postformation processing, 36
REV method, 36
solvent injection method, 36
thin film hydration method, 35

E

Emulsification-solvent diffusion method, polymer nanoparticles, 302–303
Emulsions
  cationic, DNA nanocarrier, 529
  composition, 529
  physicochemical characteristics, 529
  stable nanoemulsion preparation, 530
  thermodynamically unstable dispersions, 530
Enhanced permeability retention (EPR) effect, 116–117
EPR effect. *See* Enhanced permeability retention (EPR) effect
Epstein–Barr virus nuclear antigen 1 (EBNA1), 255–256
Escherichia coli (E.Coli)
  concentration *vs.* fluorescence intensity, 148–149
  detection and separation, 148
  drug ampicillin resistance, 149
  shiga toxins, 149–151
  type 1 pili protein structures, 147–148

F

Ferumoxtran-10, 349
Fluorescent nanoparticles
  fluorophore-loaded silica, 458–459
  polymers, 459–460
  quantum dots (QDs), 456–457
Folate receptors, 81
Folic acid dendrimers, 117
Förster resonance energy transfer (FRET)
  QD-organic and inorganic dyepairs, 90–91
  QD-phthalocyanine conjugates, 88–89
  QD-porphine conjugates, 89–90
Functional fusion protein
  via PHA polymerase (PhaC)
    enzyme immobilization, 315–316
    fused with C terminal, 316
    immunoglobulin (IG) purification, 314
    streptavidin beads preparation, 315
  via phasins (PhaP), 316–318
Functionalized fullerenes and carbon nanotubes, therapeutics, 572–573
Functional magnetite nanoparticles
  complex 3D tissue construction, 385–388
  DNA transfection, 359
  magnetic patterning
    schematic representation, 362, 363
    using MCLs, 363–365
    using PEG-Mags, 366–368
    using RGD-MCLs, 365–366
  nontoxicity, 356
  tissue engineering
    cell-manipulating technology, 358
    liver, 380–385
    magnetic manipulation, 357
    processes, 357, 358
    skeletal muscle, 373–380
    skin, 368–373
  viral transfection, 360–362

G

Gene delivery
  cationic and neutral lipids, 522–526
  extracellular barrier and strategy, 512–514
  ideal system properties, 511
  inefficiency, 511
  intracellular barrier
    cellular binding, 515
    cellular uptake, 516–517
    endosomal escape mechanism, 518–519
    nuclear entry, 519–521
  nonviral
    barriers, 511, 512
    different uptake pathways, 516, 517
  target receptors, receptor mediated endocytosis, 515, 516
Gene therapy
  antisense, 400
  approaches, 399–400
  cell-penetrating peptides (*see* Cell-penetrating peptides (CPPs))
  dendrimers, 122–126
  gene delivery, 400
  nanoparticles (*see* Nanoparticles (NPs))
  RNAi, 400–401
  splice-switching therapeutics, 401
Glycodendrimer, 118
Glyconanoparticles (GNPs). *See* Metallic carbohydrate-based nanoparticles
Glycoside cluster effect, 105
Gold nanoparticles (AuNPs)
  affinity, Gb3 globotriose-functionalized, 151
  based immunoassays

chemiluminescent methods, 463
colored probes and scattering enhancers, 464–467
electrochemical methods, 461–462
fluorescence quenching, 468–469
plasmonics-based methods, 467–468
biomedicine application, 159
charge and hydrophobicity, 436
colorimetric sensing, 436
cross-linking approach, functionalized, 450–451
dextran-coated, 149
electrochemical applications, 437
electrochemical methods, 455
fluorescent quenchers, 436–437
multifunctional, 452–455
non-cross-linking approach, functionalized, 451–452
nonfunctionalized, 449–450
thiol-ended glycoconjugates, 143
viral fusion protein interaction, 161, 162

H

High pressure carbon monoxide (HiPCO) conversion process, 204, 205
Hybrid nanoparticles
liposome-polycation-DNA (LPD), 542–544
multilayered, 544–545
Hydrophobic phosphoinositide 3-kinase (PI3K) inhibitor, 309–310

I

Imatinib, 496
Inorganic nanoparticles, 501
MNPs, 549–551
QDs, 549

K

Kataura plot, 183, 184

L

Laser ablation (LA)
continuous and pulsed laser utilization, 200
description, 199
design, 199–200
eutectic diagram, 201
frequency variation, 200–201
inert gas pressure and composition, 201
scooter model, 201
lego^{TM} chemistry, 2–3
Linear polyethylenimine (lPEI), 533
Lipid-based nanoparticles
emulsions
composition, 529
DNA nanocarrier, 529
physicochemical characteristics, 529
stable nanoemulsion preparation, 530
thermodynamically unstable dispersions, 530
lipids
constituents, 521
dioleoyl trimethylammonium propane (DOTAP), 521, 522
transfection application, 526
liposomes, 526–527
SLNs (*see* Solid lipid nanoparticles (SLNs))
Liposome-polycation-DNA (LPD) hybrid nanoparticles, 542–544
Liposomes, 500. *See also* Vesicles
based drug products
AmBisome®, 28
amphotericin B antifungal agent, 28
Definity®, 28
DepoDur®, 28
Doxil®, 27
doxorubicin anticancer drug, 27
Octocog alfa®, 28
Visudyne® therapy, 28
drug
carrier features, 28–29
encapsulation, factors affecting, 30–32
physical and chemical instability, 6–7
therapeutic nanoparticles, 573
Liver tissue engineering
cell-cell interactions, 380–381
cocultured hepatocytes and endothelial cells, 381–382
E-cadherin, 380–381
HEPG2 and NIH3T3 cells, 382–385
Living cells, QD delivery
caveolae-dependent endocytosis, 67–68
clathrin-mediated endocytosis, 67–68
macropinocytosis, 67–68

Living cells, QD delivery (*cont.*)
nonspecific
carbohydrates, 77–78
lipids and liposomes, 68–69
peptides, 71–76
phagocytosis, 67–68
polymers, 69–71
proteins, 76–77
small molecules, 77
specific
antibodies, 78–80
Arg-Gly-Asp peptides, 81–82
ligands, 80–81
toxins, 82

M

Macropinocytosis mechanism, 67–68
Magnetic nanoparticles (MNPs)
basic structure, 549–550
gene delivery systems, 550–551
magnetic cores, 550
magnetofection, 550–551
MRI contrast agents, 350
surface coatings, 550
Magnetic patterning
schematic representation, 362, 363
using magnetite cationic liposomes (MCLs), 363–365
using PEG-Mags, 366–368
using RGD-magnetite cationic liposomes (MCLs), 365–366
Magnetic resonance imaging (MRI)
dendrimers, 128–129
metallic carbohydrate-based nanoparticles
glyo-ferrites, 154–155
molecular probes, clinical use, 151
neurological disorder diagnosis, 153
NPCP-Cy5.5-CTX nanoprobes, 153, 154
superparamagnetic GNPs, 154–155
Magnetite cationic liposomes (MCLs), 356–357
Magnetite nanoparticles. *See* Functional magnetite nanoparticles
Magnetofection
DNA transfection, 359
viral transfection, 360–362
Manno-glyconanoparticles (GNPs), 160–161
Metallic carbohydrate-based nanoparticles
biosensing field, 144–145
cell detection
hyaluronic acid (HA)-encapsulated iron oxide nanoparticles, 146
QDs, 146–147
scanometric approach, 145
MRI
glyo-ferrites, 154–155
molecular probes, clinical use, 151
neurological disorder diagnosis, 153
NPCP-Cy5.5-CTX nanoprobes, 153, 154
superparamagnetic GNPs, 154–155
nonmetallic polysaccharide-based nanoparticles, 164–165
optical imaging, 156–158
pathogen detection, 147–151
polysaccharide-encapsulated metallic nanoparticles, 143–144
therapeutic applications
anti-adhesion agents, 159, 164
DC-SIGN lectin, 160–161
galactosyl ceramide (GalCer), 160
gold NPs-viral fusion protein interaction, 161, 162
high-mannose glycans, gp120, 160–161
human immunodeficiency virus (HIV), 159–160
Streptococcus pneumoniae, vaccine preparation, 163
Vero cell toxin inhibition assay, 150
in vitro diagnostics, 151, 152
Micelles, 500, 539–542
Microbial polyhydroxyalkanoate
bioavailability, 310
biopolymer granules production, 318
chemical structures, 308, 309
functional fusion protein
via PHA polymerase (PHAC), 314–316
via phasins (PhaP), 316–318
granule-associated proteins, 313
molecular weight, 308
SEM images, RBITC-loaded, 309–310
targeted drug delivery carriers
hepatocellular carcinoma cells, 311–312
lipid-soluble colorant RBITC, 311–312
phagocytic clearance, 312–313
TGX221 model drug, 309–310
Microfluidics, vesicle production, 37
MNPs. *See* Magnetic nanoparticles (MNPs)
Modified natural nanoparticles, 502
Molecular probes, dendrimers

acetylcholinesterase (AChE) detection, 127
amperometric enzyme electrode, 127
biosensors, DNA hybridization, 126–127
Molecular self-assembly
characteristics, 3–5
definition, 3
MPG peptide
endocytotic pathway, 409
oligonucleotide interaction and intracellular delivery, 407
pharmacokinetic analysis, 409
Multifunctional enveloped nanodevice (MEND), 543–544
Multifunctional proteins
de novo synthesis, 258
design, 259–261
directed evolution, 257, 258
physicochemical characterization, 262–264
production
animal cell expression systems, 262
biological systems, 261–262
rational design, 258–259
Multilayerd nanoparticles, 544–545
Multivalency, 494
Multi-wall carbon nanotubes (MWCNTs)
arc discharge (AD) process, 197–198
cell sizes, 177
feature, 183
gas phase oxidation, purification, 208–209
inter-wall interactions, 183
MWCNTs. *See* Multi-wall carbon nanotubes (MWCNTs)
Myoblast cell, tissue engineering
rings, 376–378
sheets, 374–375
strings, 375

N

Nanobiotechnology. *See* Personalized medicine
Nanocardiology
gene-environment interactions, 350
magnetic nanoparticles uasge, MRI contrast agents, 350
nanoparticles, cardiovascular imaging and targeted drug delivery, 350–351
Nanodevices and cell reprogramming, 577–578
Nanomedicines
cancer, 567–570
cell-based, 564–567
degenerative diseases, 570
imaging and diagnostics, 572
immunomodulation, 571
infectious diseases, 570
and nanobiotechnology (*see* Personalized medicine)
ocular drug delivery, 571–572
regenerative medicine and wound repair facilitation, 571
and stem cells, 578–579
and vaccines, 577
Nanoneurology
brain tumor management, nanobiotechnology, 349
nanoparticles, neuroprotection, 347–348
peptide nanostructures, CNS repair, 348
Nanooncology
cancer diagnostics-therapeutics, nanobiotechnology, 342–343
gold nanoparticle-bombesin, 346
gold nanoshells, cancer thermal ablation, 343–344
nanobombs, cancer, 345
nanoparticles, MRI, and thermal ablation, tumors, 345–346
quantum dot conjugates-cancer imaging, therapy, and sensing, 343
radiolabeled carbon nanotubes, 343
role of nanoparticle-based imaging, 346
treatment development, cancer detection, 341–342
Nanoparticles (NPs)
carbon nanovectors, 501
chemotherapeutic agents, therapeutic efficacy, 497–498
clinical evaluation, 503, 504
CPP (*see* Cell-penetrating peptides (CPPs))
fluorescent silica and QDs, 437–440
gold
charge and hydrophobicity, 436
colorimetric sensing, 436
electrochemical applications, 437
fluorescent quenchers, 436–437
hybrid, 542–545
inorganic, 501
integrated, 429
lipid-based, 499–500
cationic emulsions, 529–530
cationic lipids, 521–526

Nanoparticles (NPs) (*cont.*)
cationic liposomes, 526–527
cationic solid lipid NPS, 527–528
liposome-polycation-DNA (LPD) hybrid, 542–544
magnetic, 441–445
basic structure, 549–550
gene delivery systems, 550–551
magnetic cores, 550
magnetofection, 550–551
surface coatings, 550
magnetite (*see* Functional magnetite nanoparticles)
modified natural, 502
molecular diagnostics
AuNPs-based immunoassays, 461–469
Au optical properties, 449–455
electrochemical methods, Au, 455
fluorescent, 427–489
genetic basis of human disorders, 445–446
high-throughput nucleic acid detection assays, 460–461
nucleic acid detection methods, silver/Au, 455
nucleic acid diagnostics techniques, 446–448
QDs in immunoassays, 469–471
multilayerd, 544–545
nanovectors, 496–497
nonintegrated, 429
peptide-based
amphiphilicity, 547
biomimetic vectors, 548
α-helical KALA, 547–548
protamine, 547
physicochemical properties
active targeting, 493–494
active *vs.* passive targeting, 491–492
schematic diagram, 490–491
targeting cellular signal transduction pathways, 495–496
types of targeting moieties, 494–495
PLGA
biodegradable, 545–546
DNA encapsulation methods, 545–546
gene carrier fabrication, 546
limitations, DNA delivery, 546–547
polymer (*see* Polymer nanoparticles)
polymer-based
chitosan, 536–537
cyclodextrin (CD), 537–539
dendrimers, 534–536
gene-packaging polymer, structures of, 530, 531
polyethylenimine (PEI), 533–534
poly-l-lysine (PLL), 532
polymeric nanoconjugates, 498–499
QDs, 549
shape-size effect, 502–503
solid lipid, 527–528
tailored synthesis and functionalization
adsorption, capping agent, 430
affinity binding, 434–435
chemical reactions, 431–434
chemisorption, 431–434
electrostatic interaction, 431
homogeneity, physical properties, 429–430
Nanoparticle tracking analysis (NTA), 12–13
Nanoproteomics, 336
Nanoshells, 575
Natural self-assembling protein nanoparticles
bioactives molecules, 270
characteristic features and limitations, 265–267
low efficiency, gene transduction, 270
nanocarriers/nanocontainers, 269
natural cellular tropism modification, 269–270
structure, 265
vaccines, 268
N-(2-hydroxypropyl) methacrylamide (HPMA) copolymers, 498
Niosomes, 7
NLS. *See* Nuclear localization signals (NLS)
Non-covalent functionalization, CNTs, 221–223
Nonmetallic polysaccharide-based nanoparticles, 164–165
Nonspecific endocytosis, QD
carbohydrates, 77–78
lipids and liposomes, 68–69
peptides
advantage, 71
arginine-rich, 71–72
insect neuropeptides, 73–76
NLS, 72–73
versatility, 76
phagocytosis, 67–68
polymers, 69–71
proteins, 76–77

proton sponge effect, 69–71
small molecules, 77
Nonviral self-assembling proteins
bacterial microcompartments (BMC), 271
bovine serum albumin (BSA), 273
enzyme encapsulation, 272
icosahedral enzyme complex, 272
inclusion bodies, 273
protein–protein interactions, 271
NTA. *See* Nanoparticle tracking analysis (NTA)
Nuclear localization signals (NLS), 72–73

P

PAMAM dendrimer. *See* Poly(amido amine) (PAMAM) dendrimer
PDT. *See* Photodynamic therapy (PDT)
Pep-1, 409–410
Pep-2, 410
PepFect6 (PF6), 417–419
PepFect10 (PF10), 415–417
PepFect14 (PF14), 419–420
Peptide-based nanoparticles
amphiphilicity, 547
biomimetic vectors, 548
α-helical KALA, 547–548
protamine, 547
Peptide nanoparticles. *See* Cell-penetrating peptides (CPPs)
Perfluorocarbon nanoparticles, 340
Personalized medicine
bioconjugated quantum dots, 333
biomarkers, 332
definitions, 326
diagnostics and therapeutics
Hercep test, 333–334
nanobiotechnology and pharmacogenetics, 334
intracellular delivery, nucleic acids, 340–341
molecular diagnostics and sequencing, 351–352
nanobiotechnology-based drug discovery, 336–337
-nanobiotechnology relationships, 326, 327
nanocardiology (*see* Nanocardiology)
nanoneurology (*see* Nanoneurology)
nanooncology (*see* Nanooncology)
nanoparticle-based drug delivery, 338–340
nanoparticle-based drug development, 337–338
nanotechnology-on-a-chip, 327
perfluorocarbon nanoparticles, imaging and targeted drug delivery, 340
pharmacogenomics and nanobiotechnology, 335
pharmacoproteomics and nanobiotechnology, 335–336
role of nanodiagnostics
cantilever arrays, 328
nanobiochips, 328–329
in sequencing, 330–332
single nucleotide polymorphism detection, 329–330
Phagocytosis, 67–68
Pharmacogenetics, 334
Pharmacogenomics, 335
Pharmacoproteomics, 335–336
Photochemical internalization (PCI) gene delivery, 125–126
Photodynamic therapy (PDT)
dendrimers, 120–122
QD
cancer, 84–85
vs. conventional photosensitizing (PS) drugs, 85–87
first and second generation PS drug, 84
fluorescence imaging, 82–83
-organic and inorganic dyepairs, 90–91
photophysical and photochemical process, 83–84
-phthalocyanine conjugates, 88–89
-porphine conjugates, 89–90
principle, 82–83
reactive oxygen species (ROS), 83–84
use, 82–83
PIC micelles. *See* Polyion complex (PIC) micelles
Pinocytosis, 516
PLGA copolymers, 498–499
PLGA nanoparticles
biodegradable, 545–546
DNA encapsulation methods, 545–546
gene carrier fabrication, 546
limitations, DNA delivery, 546–547
Poly(amido amine) (PAMAM) dendrimer
efficiency, gene transfer, 124
ibuprofen transport, 111–112
ketoprofen, drug carrier, 112

Poly(amido amine) (PAMAM) dendrimer (*cont.*)
  membrane damage mechanism, cationic, 106
  second generation, 103, 104
  in vivo toxicity, 107–108
Polyion complex (PIC) micelles
  disulfide cross-linked core, 540–541
  DNA condensation study, 539, 540
  properties, 539–540
  stimuli sensitivity, 540–541
  transfection of, lactosylated, 542
  triblock copolymer, 541–542
Polymer-based nanoparticles
  chitosan, 536–537
  cyclodextrin (CD), 537–539
  dendrimers, 534–536
  gene-packaging polymer, structures of, 530, 531
  polyethylenimine (PEI), 533–534
  poly-l-lysine (PLL), 532
Polymer-coated nanocrystals, 575
Polymeric micelles, 573–574
Polymeric nanospheres, 574
Polymer nanoparticles
  Abraxane and doxorubicin-Trans-drug, 301
  advantages, active substance delivery systems, 301
  definition of, 300
  drug loading and drug encapsulation efficiency, 306–307
  drug release, 307
  emulsification-solvent diffusion method, 302–303
  microbial polyhydroxyalkanoate (PHA)
    biopolymer granules production, 318
    as drug delivery carriers, 308–310
    functionalization, 313–318
    targeted drug delivery carriers, 311–313
  monomer polymerization, 304–305
  particle size, size distribution, and morphology, 305–306
  preparation, natural polymers involvement, 301
  solvent displacement method, 303–304
  solvent evaporation method, 302
  stuctures, 300
  in vitro stability performances, 307
Pressure perturbation calorimetry, biomolecule-liposome, 24–25
Protamine, 250
Protein nanoparticles
  antigen–antibody binding, 249
  blood–brain barrier (BBB), 257
  cell binding and internalization, 254
  de novo synthesis, 258
  design, 259–261
  directed evolution, 257, 258
  drug-nucleic acid interaction, 250–253
  endosomal escape, 254–255
  intracytosolic mobility, 256
  medical applications
    in clinical trials, 276–277
    in the market, 274–276
    preclinical models, human diseases, 277–281
  molecular delivery vehicles, 249
  new biological entities (NBE), 248–249
  new chemical entities (NCE), 248
  nonviral self-assembling proteins
    bacterial microcompartments (BMC), 271
    bovine serum albumin (BSA), 273
    enzyme encapsulation, 272
    icosahedral enzyme complex, 272
    inclusion bodies, 273
    protein–protein interactions, 271
  nuclear DNA delivery and expression, 256–257
  peptide motif selection, 250, 251
  physicochemical characterization, 262–264
  production
    animal cell expression systems, 262
    biological systems, 261–262
  protein stability, serum, 254
  rational design, 258–259
  schematic representation, drug delivery and gene therapy, 259, 260
  vector stability in the cytosol, 255–256
  virus-like particles (VLPs)
    bioactives molecules, 270
    characteristic features and limitations, 265–267
    low efficiency, gene transduction, 270
    nanocarriers/nanocontainers, 269
    natural cellular tropism modification, 269–270
    structure, 265
    vaccines, 268
Proton sponge, 255

Q

QDs. *See* Quantum dots (QDs)
Quantum dots (QDs)
cell surface labeling
nonspecific, 60–62
specific, 62–67
delivery, living cells
caveolae-dependent endocytosis, 67–68
clathrin-mediated endocytosis, 67–68
macropinocytosis, 67–68
nonspecific, 68–78
phagocytosis, 67–68
specific, 78–82
vs. fluorescent dyes, bioimaging advantages
CdSe/ZnS, absorption and luminescence spectra, 55–57
Goeppert–Mayer unit, 57
limitations, 55–57
and fluorescent silica nanoparticles, 437–440
in immunoassays, 469–471
nanobiotechnology, 54–55
PDT
cancer, 84–85
vs. conventional photosensitizing (PS) drugs, 85–87
photosensitizer hybrids, 87–91
principle, 82–83
reactive oxygen species (ROS), 83–84
use, 82–83
preparation, bioconjugated
core synthesis, 57–59
shell formation, 59
therapeutic nanoparticles, 576
Quartz crystal microbalance, biomolecule-liposome, 25

R

Receptor tyrosine kinases (RTKs), 496
Reverse phase evaporation (REV) method, 36
REV method. *See* Reverse phase evaporation (REV) method
RNAi, 400–401

S

Scooter model, 201
Self-assembled amphiphilic molecules, 5–6
Single-molecule genomics, 330–331
Single photon emission computed tomography (SPECT), dendrimers, 130–131
Single-wall carbon nanotubes (SWCNTs)
arc discharge (AD) process
inspection, 198
metal catalyst incorporation, 198
Raman spectroscopy, 199
yield, 198
armchair (ac), 177–178
cell sizes, 177
chirality, 177, 178
covalent functionalization
non-defective sites, 220–221
schematic representation, 219–220
density of states (DOS), 181, 182
electrophoretic separation
ac dielectrophoresis method, 212–213
agarose gel electrophoresis, 213–214
approach, 212–213
capillary electrophoresis, 213–214
field emission X-ray source technology, 227
fluoresce, 183
functionalization
covalent, 219–221
fluorine, 220–221
limitations, 220–221
non-covalent, 221–223
organic, 220–221
high aspect ratio, 182
(n,m) indexes, 178
Kataura plot, 183, 184
NIR fluorescence microscopy, 224–225
optical transition, 183
purification
before *vs.* after, Raman spectra, 211, 212
chirality specific, 217
CVD, 210
liquid phase oxidation, 208–209
microfiltration procedure, 209–210
size exclusion chromatography, 209–210
sonication process, 209–210
radial breathing mode (RBM), 184–185, 186
Raman spectroscopy, 225–226
sorting, DGU, 215, 216
structural defects, 179
tangential and disorder-induced modes, 184–185, 186
VHS, 182
zigzag (zz), 177–178

Skeletal muscle tissue engineering
gene-engineered muscle cell sheets, 378–380
myoblast cell rings, 376–378
myoblast cell sheets, 374–375
myoblast cell strings, 375
structural features, 373–374
Skin tissue engineering
gene therapy, 371–373
keratinocyte sheets, 369–371
Solid lipid nanoparticles (SLNs), 527–528
Solvent displacement method, 303–304
Solvent evaporation method, 302
SPIO nanoparticles, therapeutics, 575
Splice-switching therapeutics, 401
Stone–Wales defect, 180
Streptococcus pneumoniae, vaccine preparation, 163
Streptococcus suis (S. suis), 149
Supramolecular chemistry
categories, 2–3
lego™ chemistry, 2–3
SWCNTs. *See* Single-wall carbon nanotubes (SWCNTs)

T

Therapeutic nanoparticles
chemical composition, structure, function, and manufacture
dendrimers, 574
functionalized fullerenes and carbon nanotubes, 572–573
liposomes, 573
nanoshells, 575
polymer-coated nanocrystals, 575
polymeric micelles, 573–574
polymeric nanospheres, 574
quantum dots, 576
SPIO nanoparticles, 575
improved pharmacokinetics, 579
nanodevices and cell reprogramming, 577–578
nanomedicines
cancer, 567–570
cell-based, 564–567
degenerative diseases, 570
imaging and diagnostics, 572
immunomodulation, 571
infectious diseases, 570
ocular drug delivery, 571–572
regenerative medicine and wound repair facilitation, 571
and stem cells, 578–579
and vaccines, 577
nanotoxicoogy, 581–583
reduced toxicity, 579–580
subcellular localization, 580–581
targeted drug delivery, 576–577
targeted tissue delivery, 580
Tissue engineering, magnetite nanoparticles
cell-manipulating technology, 358
liver, 380–385
magnetic manipulation, 357
processes, 357, 358
skeletal muscle, 373–380
Trastuzumab, 496

V

Van Hove singularities (VHS), 182
Verigener F5/F2/MTHFR nucleic acid test, 334
Verigene Warfarin Metabolism nucleic acid test, 334
Vesicles
biomolecule-liposome interaction strength
differential scanning calorimetry (DSC), 21
dilution assays, 23
isothermal titration calorimetry (ITC), 21–24
partial enthalpy assays, 23
pressure perturbation calorimetry, 24–25
quartz crystal microbalance, 25
reaction assays, 23
catanionic, 7
characteristics, 32
description, 2
drug
carrier features, 28–29
content analysis, 25–27
encapsulation, factors affecting, 30–32
ionic, 32–33
loading, active trapping technique, 33
water-soluble, encapsulation of, 32
factors, 8
FDA regulations, 34
hydrophobic effect, 5
and lipid bilayer, 2
molecular self-assembly

characteristics, 3–5
definition, 3
morphological features, 15–18
niosomes, 7
optimal performance, 34
preparation
drug-loaded vesicles (*see* Drug-loaded vesicle preparation)
lipid molecular structures used, 6
production and postformation processing, 35
response, external stimulus, 8–9
size and distribution
dynamic light scattering (DLS), 10, 11
hydrodynamic radius, particle, 11
NTA, 12–13
pharmaceutical carrier, 11
Stokes radius, 11
vesicle stability, 11
structural homogeneity, 8
supramolecular structures
membrane supramolecular organization, 18–20
self-assembled amphiphilic molecules, 5–6
surfactants, 7
types, 7–8
Zeta potential, 13–15
VHS. *See* Van Hove singularities (VHS)
Viral transfection, 360–362

Z

Zeta potential, vesicles, 13–15

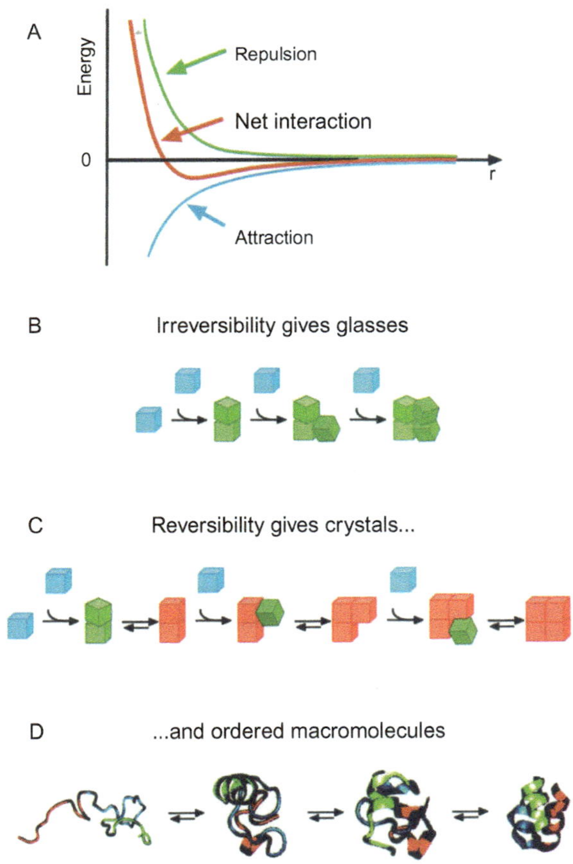

ELIZONDO *ET AL.*, FIG. 2.

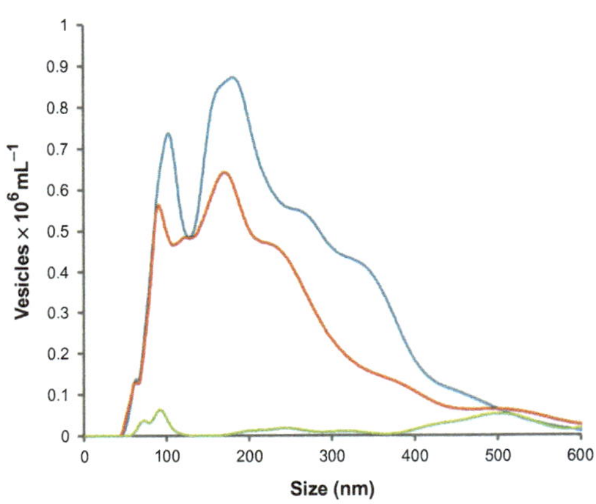

ELIZONDO *ET AL.*, FIG. 8.

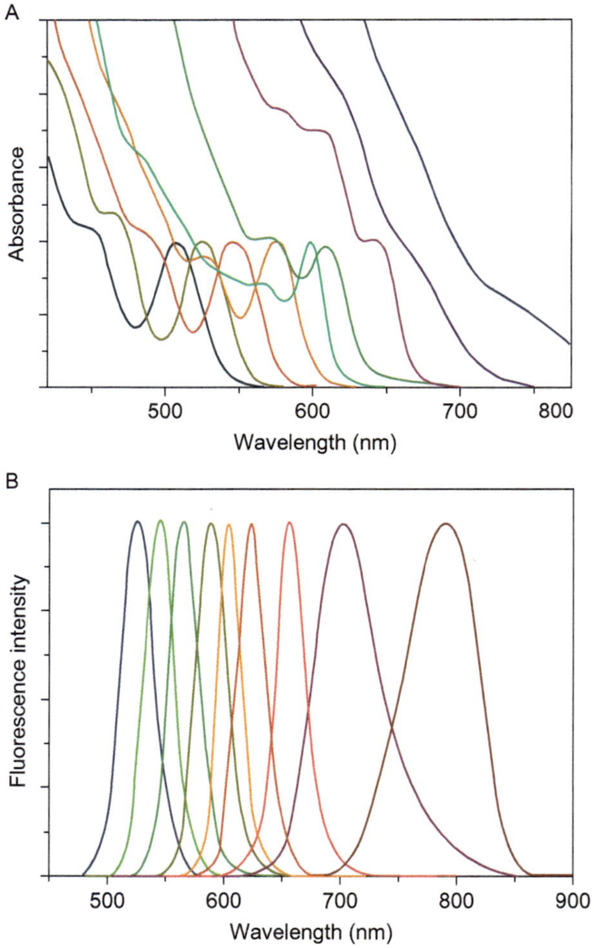

ISHIKAWA AND BIJU, FIG. 2.

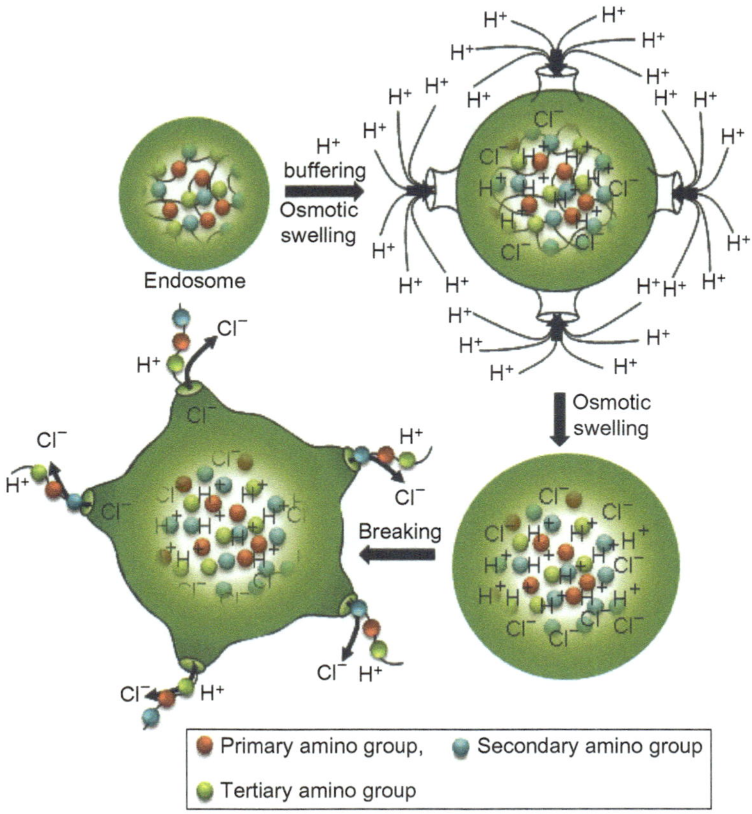

ISHIKAWA AND BIJU, FIG. 5.

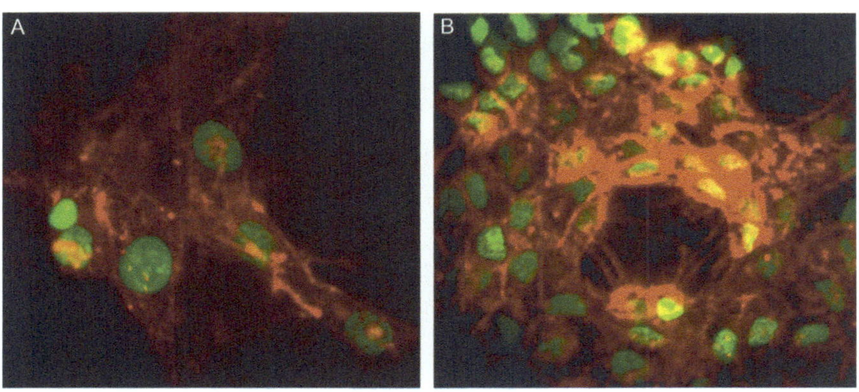

ISHIKAWA AND BIJU, FIG. 6.

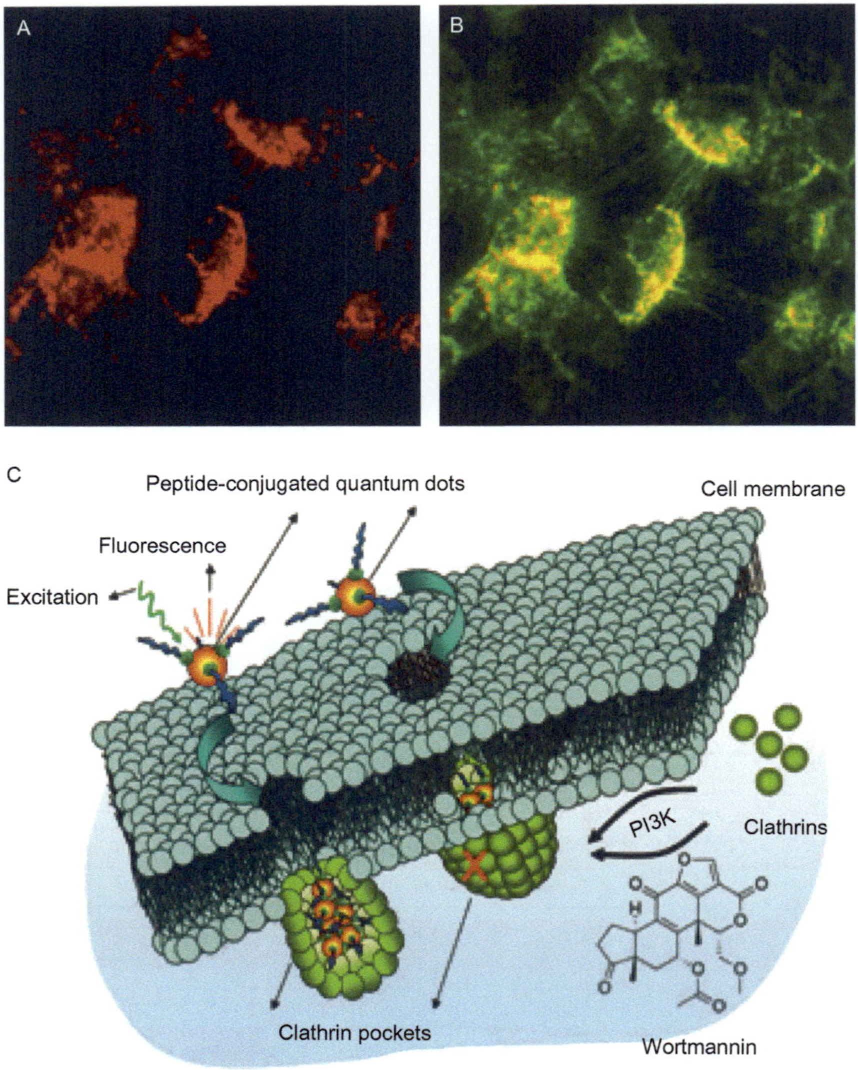

ISHIKAWA AND BIJU, FIG. 7.

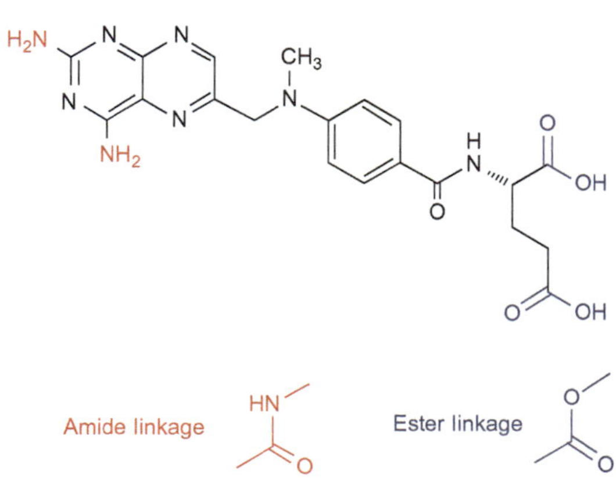

TSAI AND IMAE, FIG. 6.

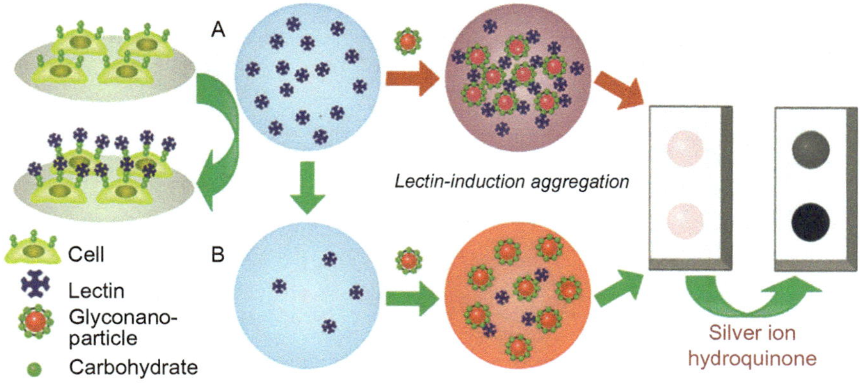

MARRADI *ET AL.*, FIG. 2.

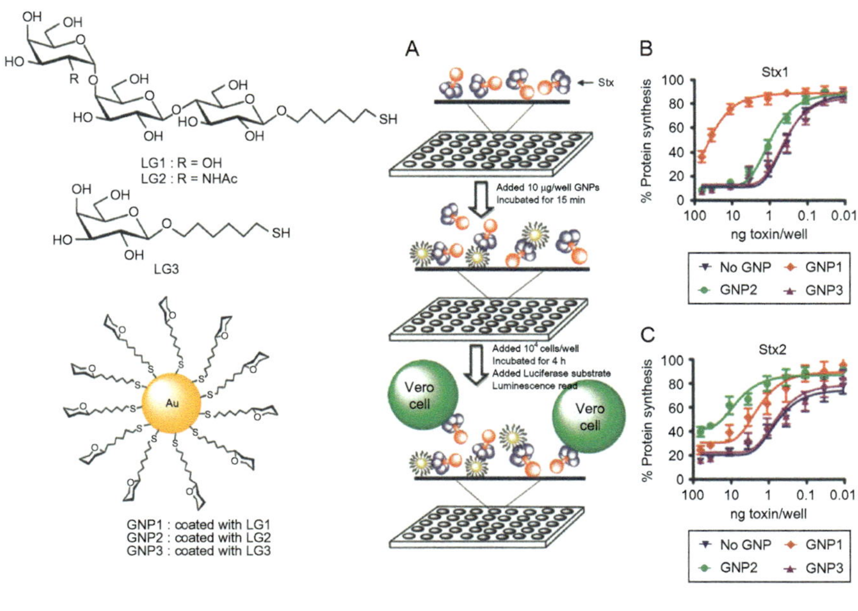

MARRADI *ET AL.*, FIG. 3.

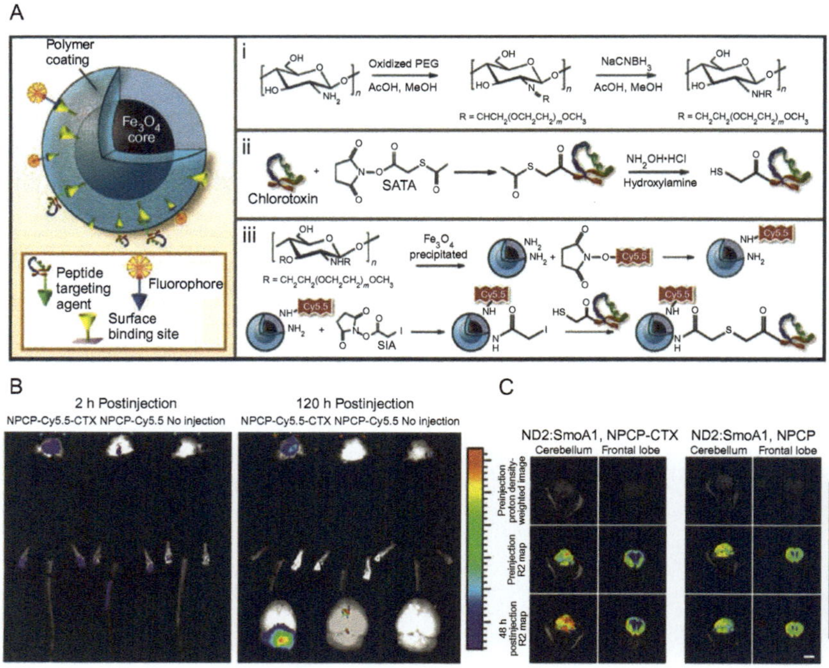

Marradi *et al.*, Fig. 4.

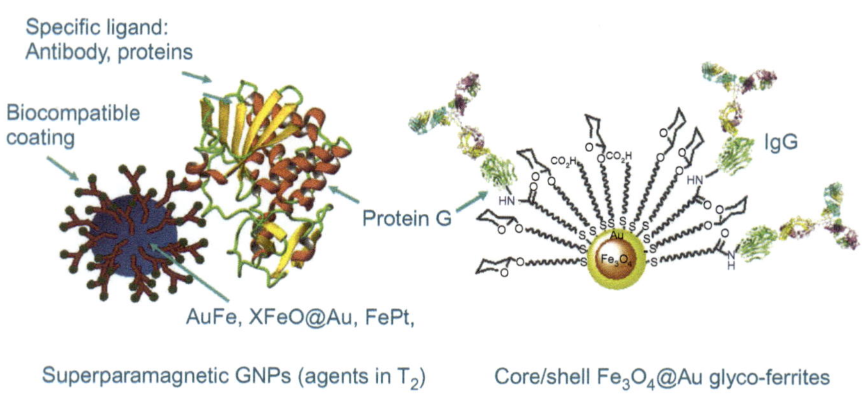

Marradi *et al.*, Fig. 5.

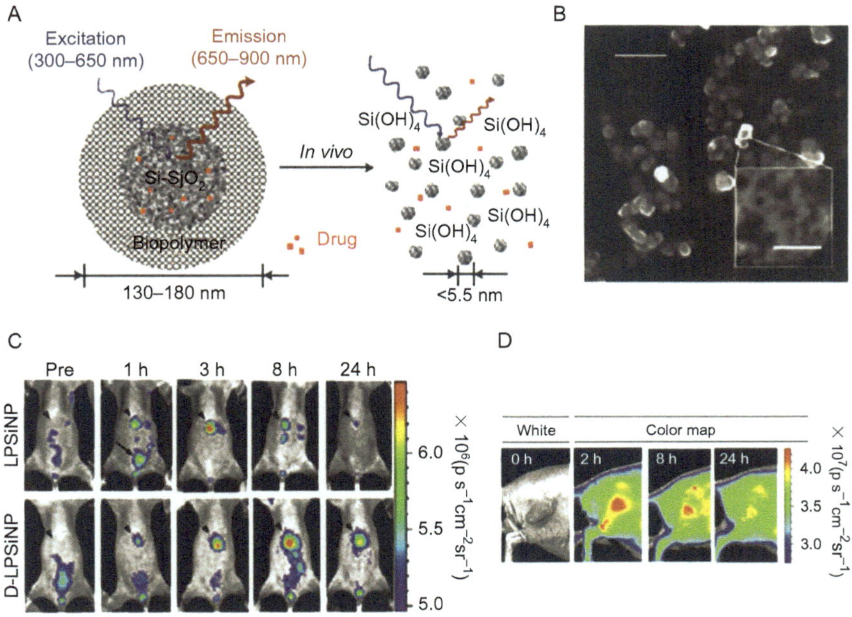

MARRADI *ET AL.*, FIG. 6.

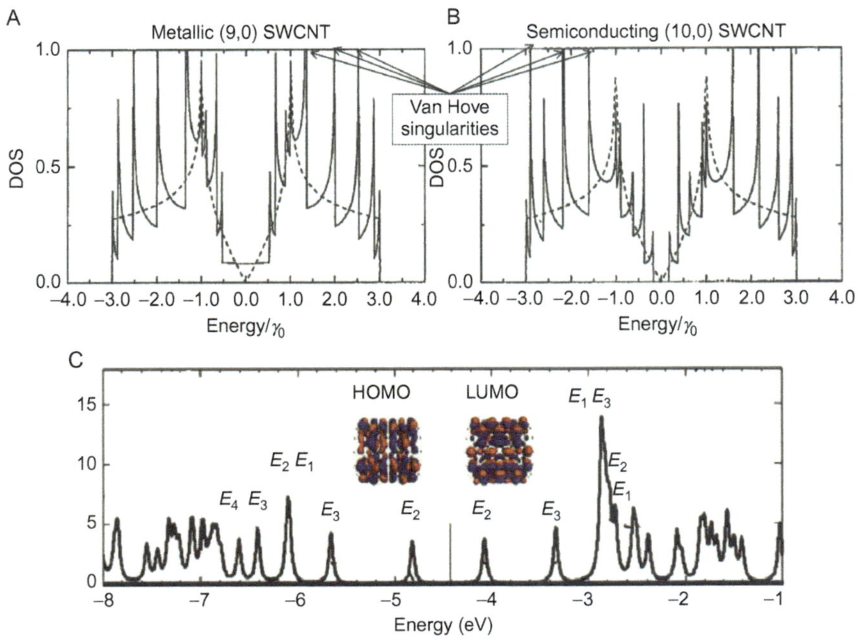

GOMEZ-GUALDRÓN *ET AL.*, FIG. 3.

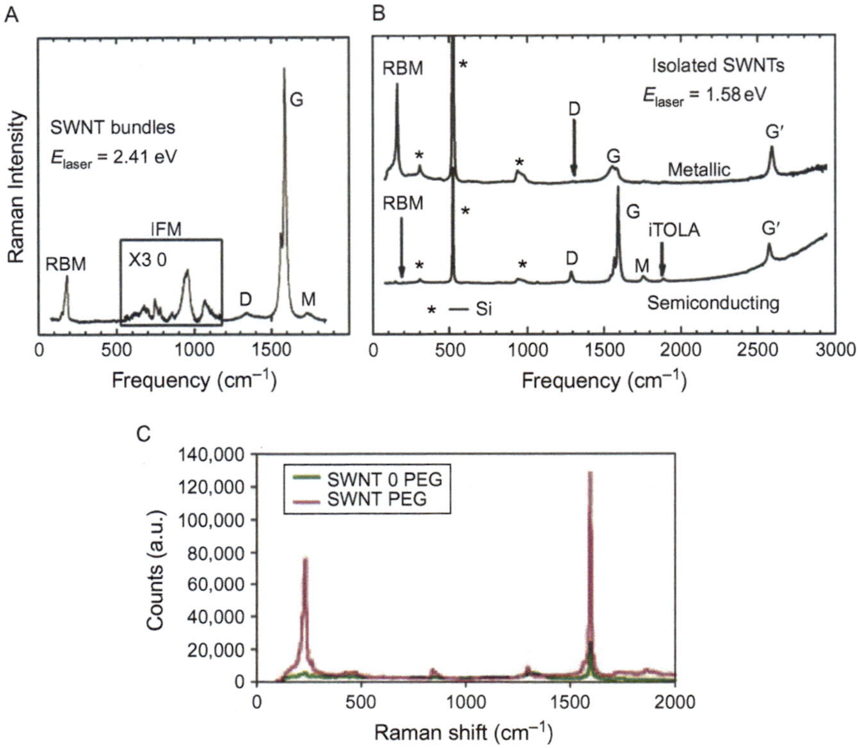

GOMEZ-GUALDRÓN *ET AL.*, FIG. 5.

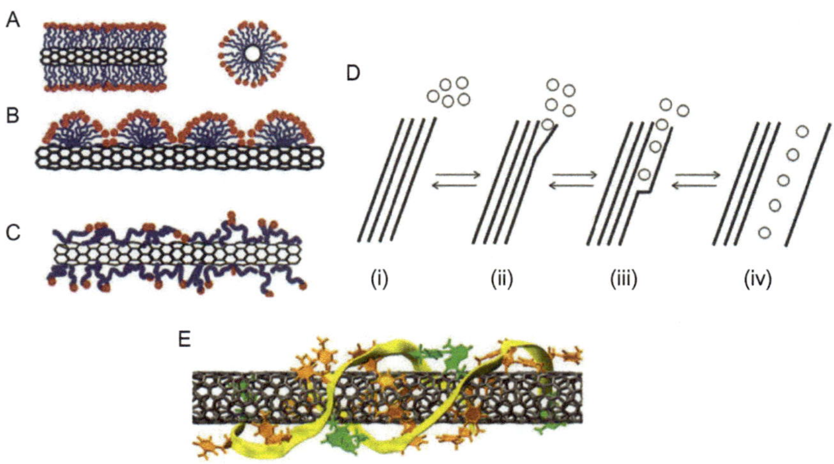

GOMEZ-GUALDRÓN *ET AL.*, FIG. 6.

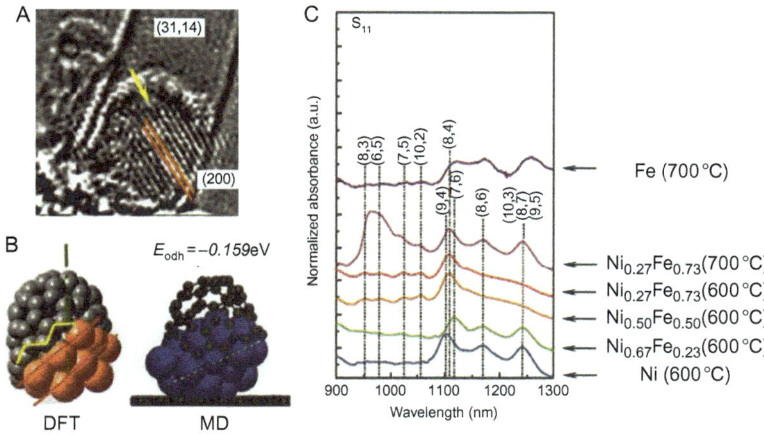

Gomez-Gualdrón *et al.*, Fig. 13.

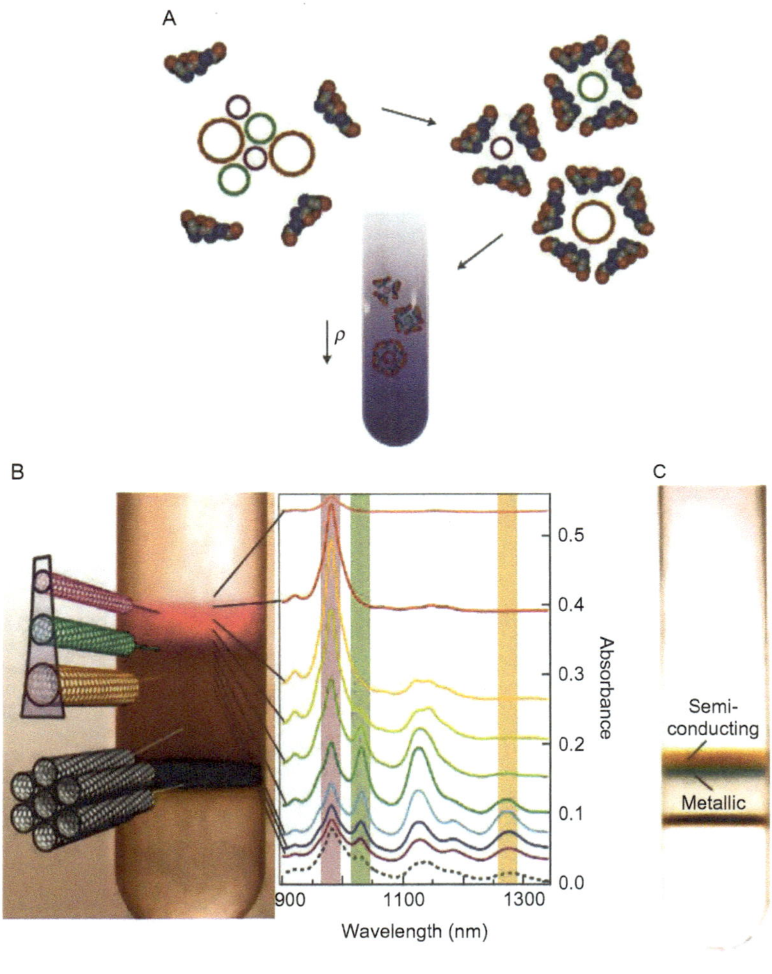

Gomez-Gualdrón *et al.*, Fig. 16.

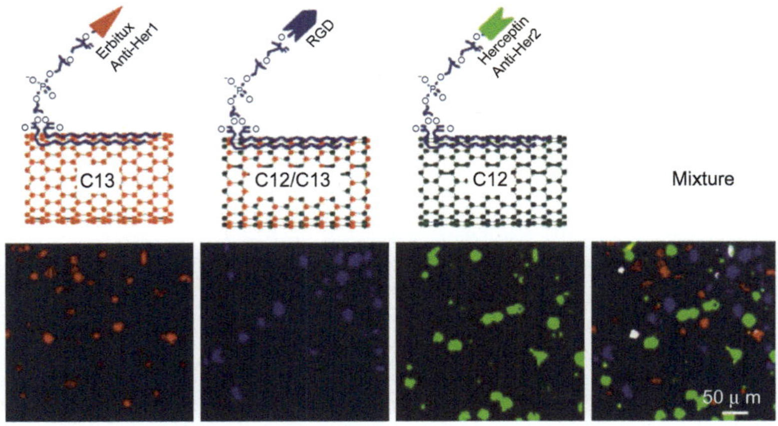

Gomez-Gualdrón *et al.*, Fig. 21.

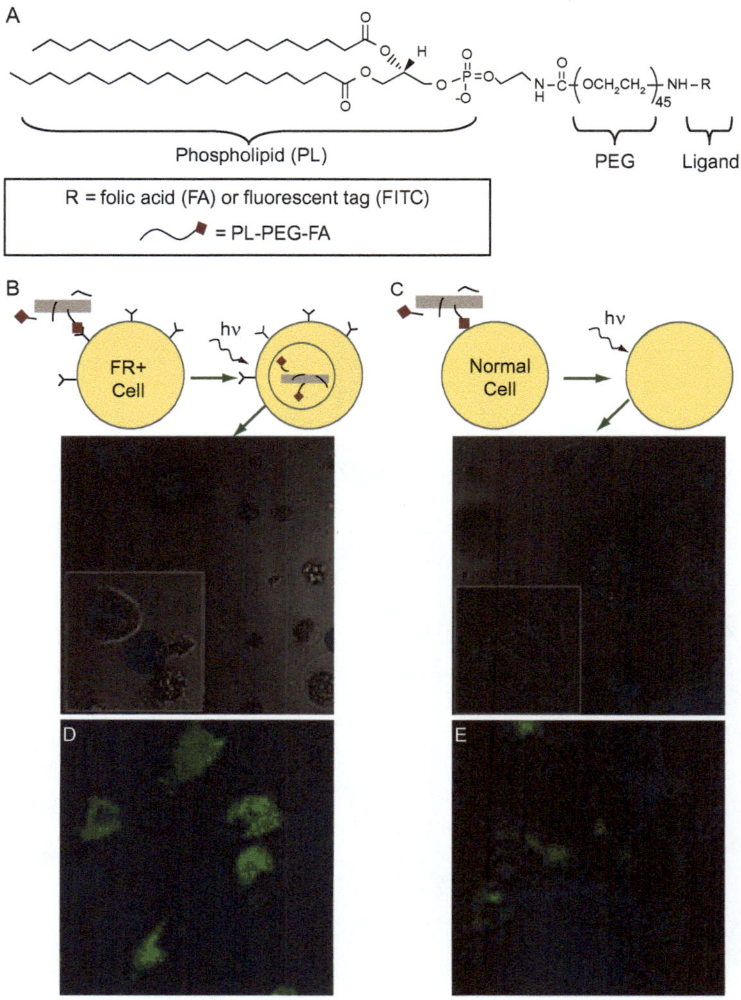

Gomez-Gualdrón *et al.*, Fig. 23.

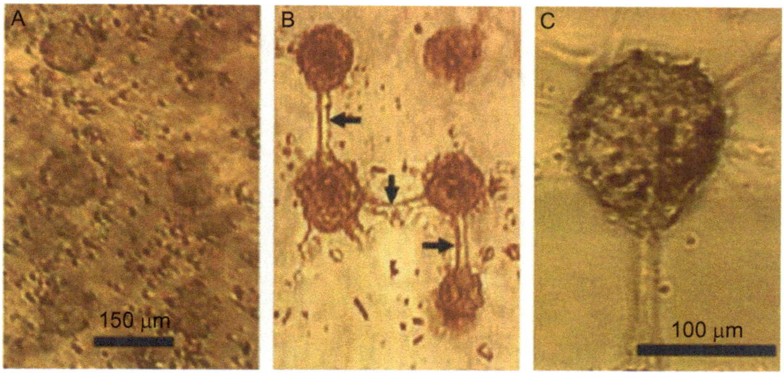

GOMEZ-GUALDRÓN *ET AL.*, FIG. 25.

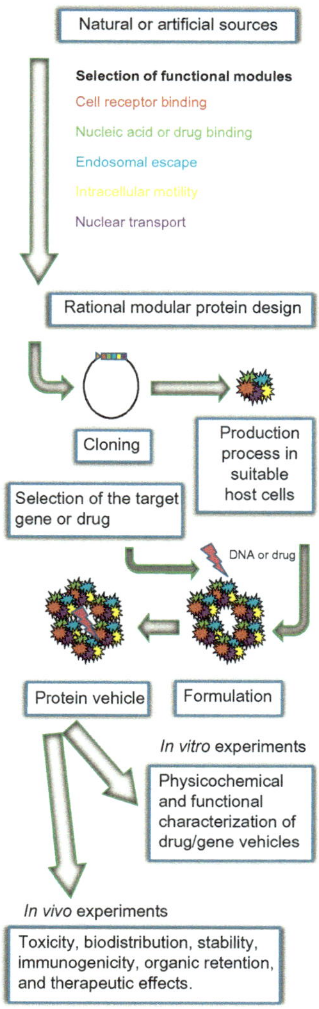

DOMINGO-ESPÍN *ET AL.*, FIG. 1.

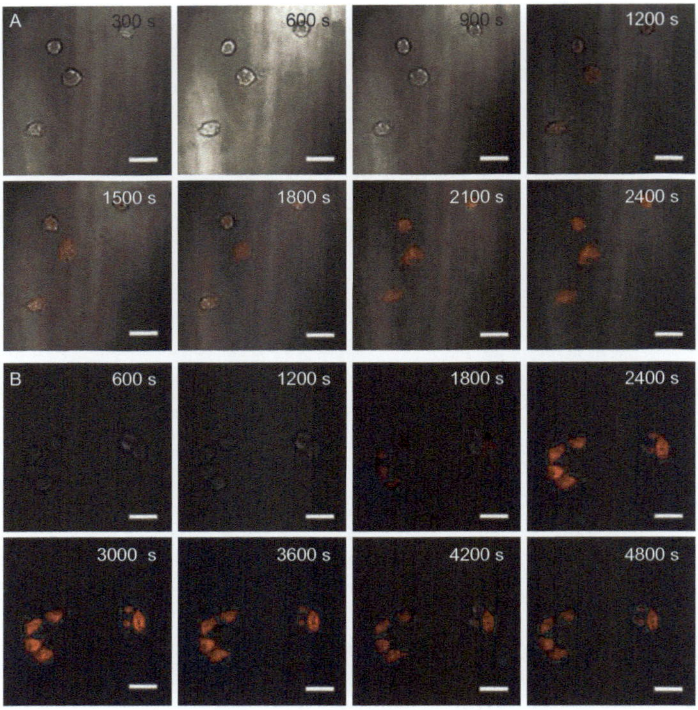

LU *ET AL.*, FIG. 9.

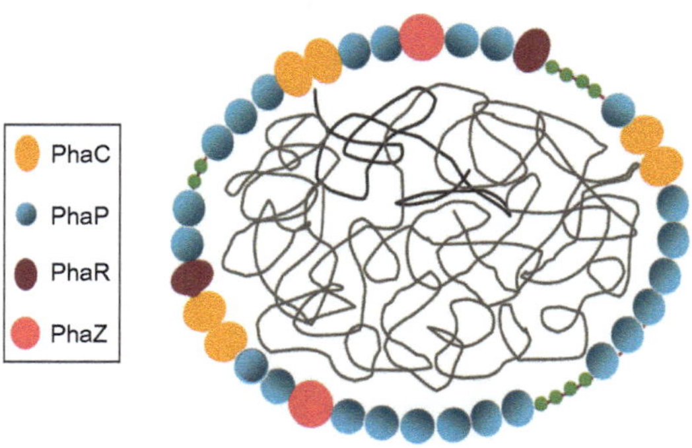

LU *ET AL.*, FIG. 10.

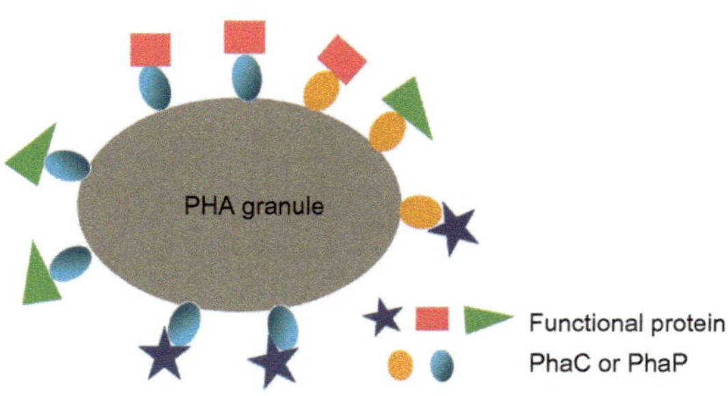

LU *ET AL.*, FIG. 11.

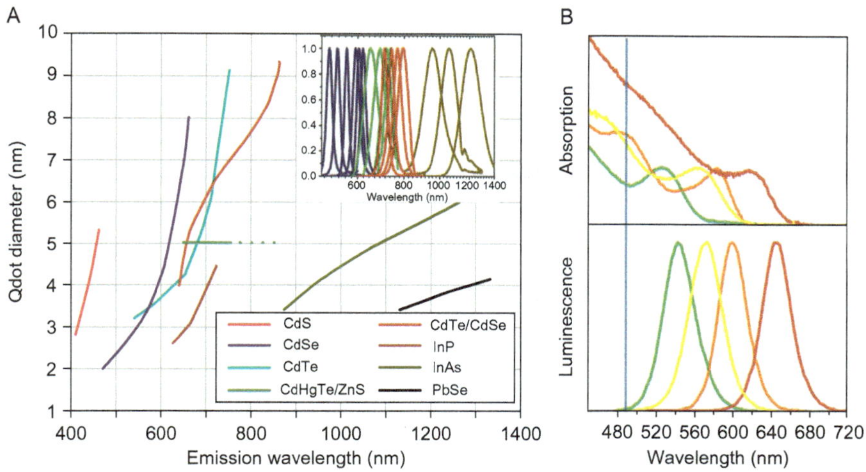

BAPTISTA *ET AL.*, FIG. 4.

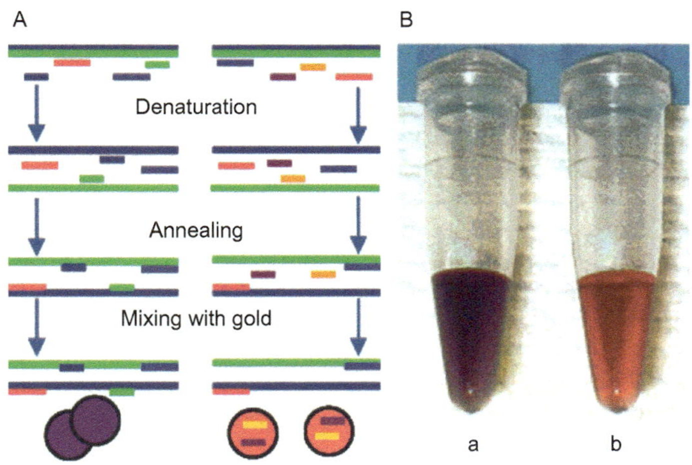

BAPTISTA *ET AL.*, FIG. 5.

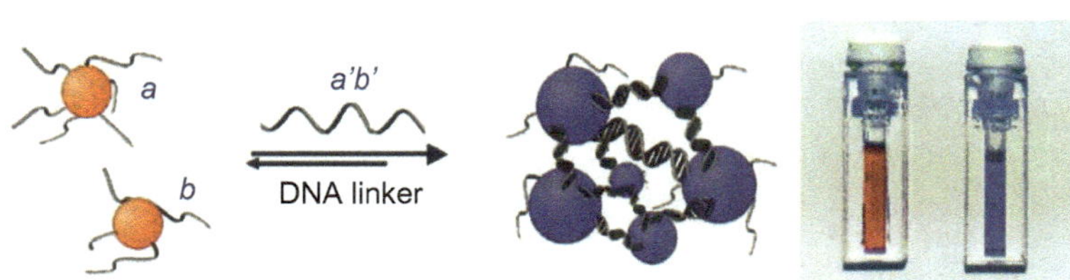

BAPTISTA *ET AL.*, FIG. 6.

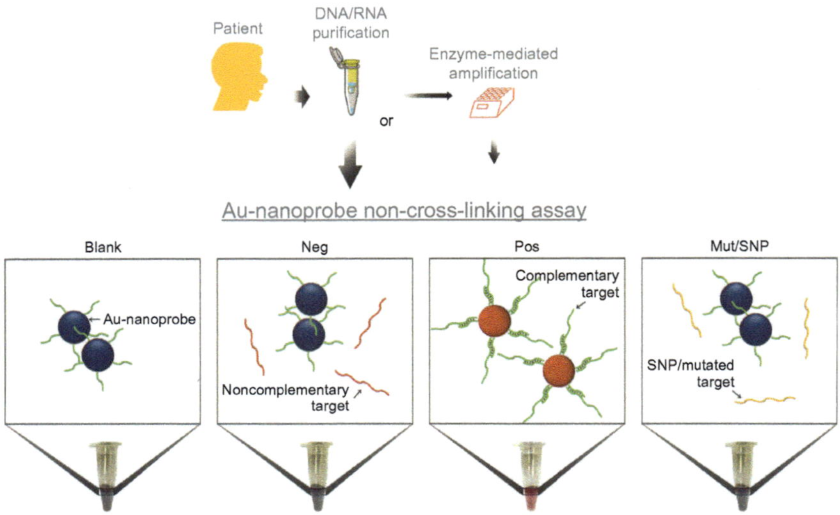

Baptista *et al.*, Fig. 7.

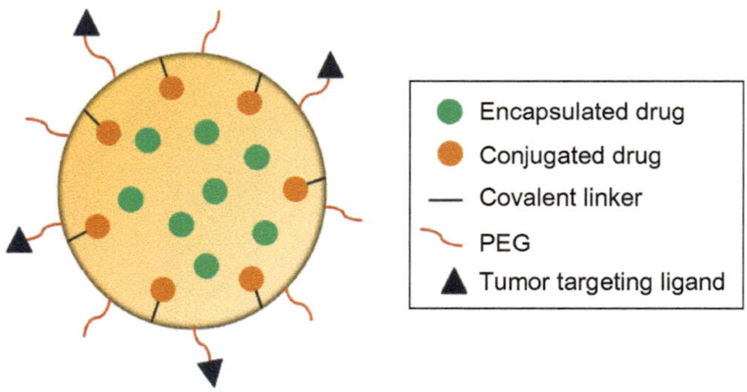

Banerjee and Sengupta, Fig. 1.

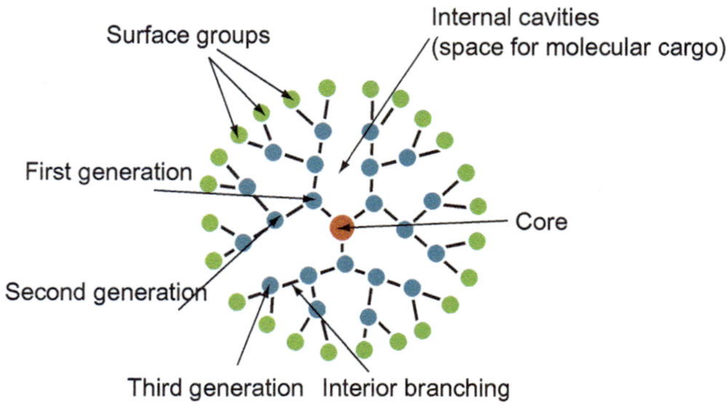

Liu and Zhang, Fig. 11.

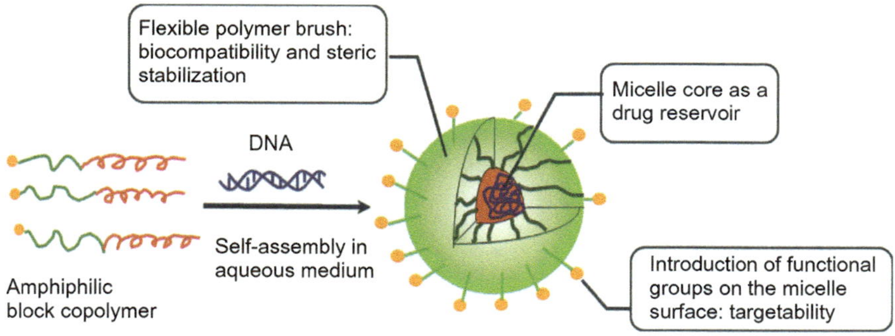

Liu and Zhang, Fig. 13.

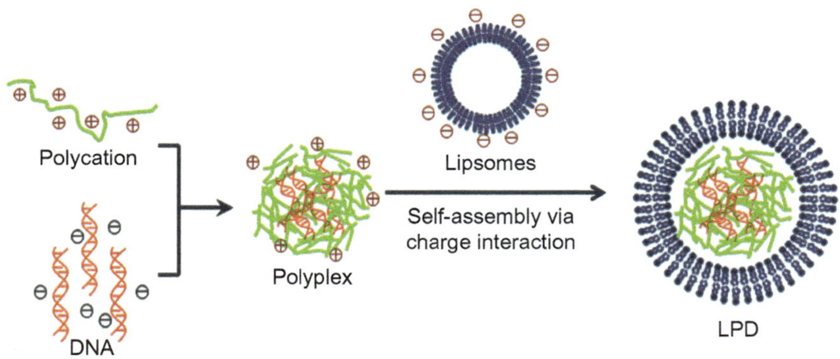

Liu and Zhang, Fig. 15.

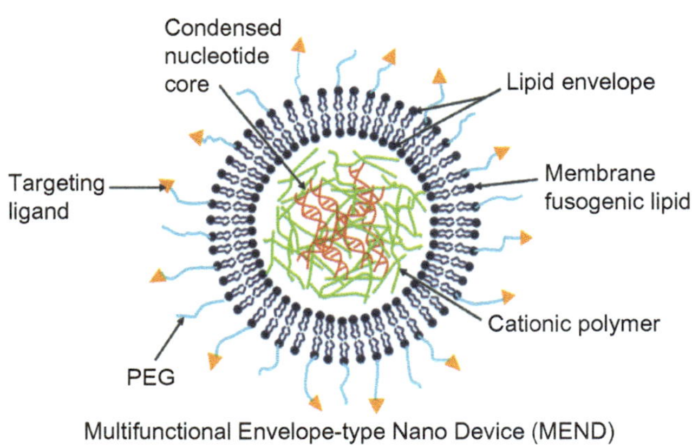

Liu and Zhang, Fig. 16.

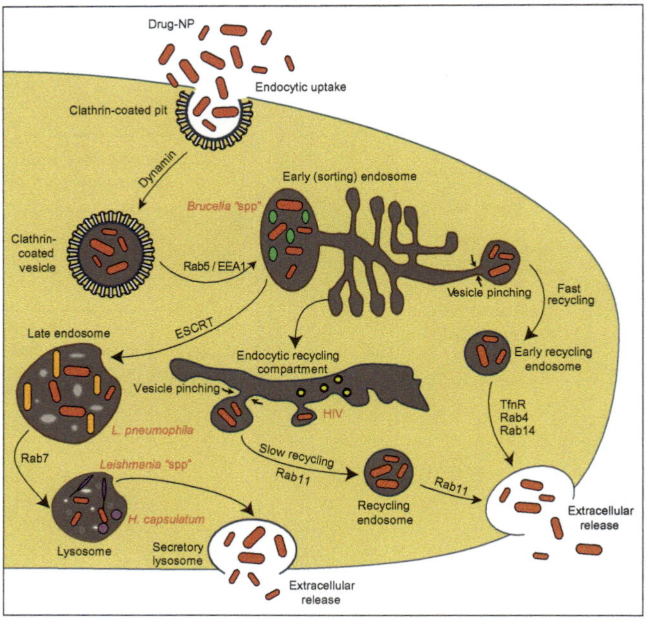

McMillan *et al.*, Fig. 1.

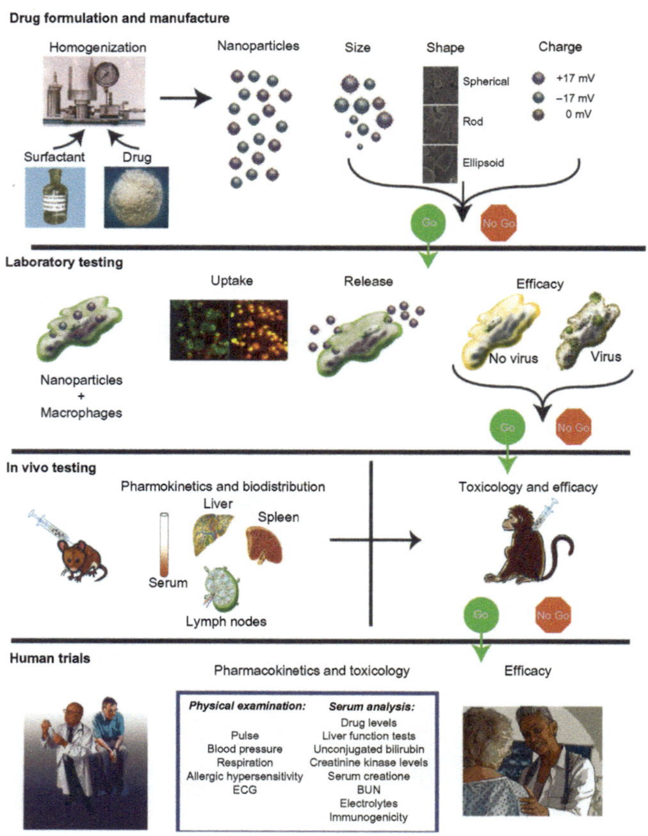

McMillan *et al.*, Fig. 2.

FSC
www.fsc.org
MIX
Papier | Fördert
gute Waldnutzung
FSC® C083411

Kolibri 360 GmbH
Ferdinand-Jühlke-Straße 7
99095 Erfurt, Deutschland
produktsicherheit@kolibri360.de